The Minerals, Metals & Materials Series

The TMS 2017 Annual Meeting Supplemental Proceedings is a collection of papers from the TMS 2017 Annual Meeting & Exhibition, held February 26–March 2 in San Diego, California, USA. The papers in this volume represent 21 symposia from the meeting. This volume, along with the other proceedings volumes published for the meeting, and archival journals, such as *Metallurgical and Materials Transactions* and *Journal of Electronic Materials*, represents the available written record of the 80 symposia held at TMS 2017.

The Minerals, Metals & Materials Society
Editor

TMS 2017 146th Annual Meeting & Exhibition Supplemental Proceedings

TMS

Springer

Editor
The Minerals, Metals & Materials Society
Pittsburgh, PA
USA

ISSN 2367-1181 ISSN 2367-1696 (electronic)
The Minerals, Metals & Materials Series
ISBN 978-3-319-51492-5 ISBN 978-3-319-51493-2 (eBook)
DOI 10.1007/978-3-319-51493-2
TMS owns copyright; Springer has full publishing rights

Library of Congress Control Number: 2016960778

Printed on acid-free paper

This Springer imprint is published by Springer Nature
The registered company is Springer International Publishing AG
The registered company address is: Gewerbestrasse 11, 6330 Cham, Switzerland

Contents

Part I
2017 Symposium on Functional Nanomaterials: Emerging Nanomaterials and Nanotechnology

High-Performance Supercapacitors Based on Hierarchical VOx Microspheres Forming from Hyperbranched Nanoribbons

Chuang Wei, Hong-Yi Li, Zhao Yang and Bing Xie

Abstract Novel VOx nanomaterials as high performance electrode materials of supercapacitors were successfully synthesized by solvothermal method using V_2O_5, H_2O_2, $(NH_4)_2SO_4$ and EG (Ethylene Glycol). These as-prepared nanomaterials are hierarchical microspheres with diameter of ~5 μm, which formed from hyperbranched growth of nanoribbons. According to XRD and TEM, these hierarchical VOx microspheres consist of V_6O_{13} with metallic conductivity and VO_2. These materials exhibited a tremendous specific capacitance of 581 F/g, with corresponding volumetric specific capacitance of 3.94 F/cm^3, at 0.6 A/g in the potential range of 0 to 1.2 V when used as supercapacitor electrodes in a solution of 1 M $LiNO_3$. The energy density is as high as 29 Wh/kg, which is much higher than those of many other symmetrical supercapacitors. In addition, the capacity retention of 65% was achieved even after 2000 cycles, demonstrating high performance of vanadium oxide nanomaterials used in supercapacitors.

Keywords Supercapacitor · VOx · Nanomaterial · Microsphere · Nanoribbons

Introduction

In recent years, supercapacitors have attracted considerable attention due to their energy density higher than conventional capacitors. The supercapacitors are also known for their higher power density and longer cycle life than that of batteries [1–3]. Vanadium oxides, such as V_2O_5, VO_2, and V_2O_3 etc., have been investigated widely as materials for energy storage [4]. The theoretical specific capacitance and potential window of vanadium oxides are relatively high owning to the multiple valance from V(II) to V(V). Furthermore, most of vanadium oxides (except VO_2) are lamellar

C. Wei (✉) · H.-Y. Li (✉) · Z. Yang · B. Xie
College of Materials Science and Engineering, Chongqing University,
Chongqing 400044, China
e-mail: hongyi.li@cqu.edu.cn

The Minerals, Metals & Materials Society, *TMS 2017 146th Annual Meeting & Exhibition Supplemental Proceedings*, The Minerals, Metals & Materials Series,
DOI 10.1007/978-3-319-51493-2_1

structure, which contributes to ions insertion/extraction during the charge/discharge process, improving specific capacitance of supercapacitors. Therefore, vanadium oxides are promising materials as electrode materials for supercapacitors.

Vanadium oxide is usually made into nanomaterials to improve electrochemical properties of supercapacitors by increasing the specific surface of electrode materials. For example, 0D nanomaterial of V_2O_5 nanoparticles [5], 1D nanomaterials of V_2O_5 nanowires [6], and 2D nanomaterials VO_2 nanobelts [7] have been synthesized and applied in supercapacitors. However, these low dimensional nanomaterials are more likely to agglomerate severely leading to inferior cycling stability of supercapacitors. So vanadium oxides have been prepared as 3D nanomaterials to avoid this phenomenon, such as starfruit-like VO_2 [8] and V_2O_5 nanosheets based 3D architecture [9]. But the specific capacitances of supercapacitors even based on 3D pure vanadium oxide nanomaterials (200–300 F/g) still cannot approach that of RuO_2 due to the lower electric conductivity of common vanadium oxides.

In this work, 3D mesoporous hierarchical microspheres, composed of VO_2 and uncommon vanadium oxide V_6O_{13} which is metallically conductive [10], have been synthesized. These microspheres also possess high specific surface and high density and thus exhibits a high specific capacitance (581 F/g) and high volumetric energy density (3.94 F/cm^3). The cycling stability of these microspheres is good due to the stable 3D nanostructure of hierarchical VOx microspheres. This work proves a strategy to improve the conductivity and stability of electrode materials, which has potential application in the future.

Experiment

Synthesis of Samples

These hierarchical VOx microspheres were prepared as follows: 4.5 mmol V_2O_5 powder was dissolved in 32.64 mL 10% H_2O_2 solution with vigorously magnetic stirring at 25 °C for 2 h in dark. 6 mL obtained reddish solution was mixed with 15 mL ethylene glycol (EG) and 15 mL de-ionized water (Millipore Aquelix) and 0.396 g $(NH_4)_2SO_4$, then stirred for about 10 min in order to obtain homogeneous solution. After that, the mixture was transferred into a 50 mL Teflon container and sealed in an autoclave, which was placed in a electrical oven and heated at 120 °C for 24 h, which was then cooled down to room temperature in the air. The black precipitates existed in the bottom of Teflon were collected by centrifugation, followed by washed with de-ionized water and anhydrous alcohol for 3 times, respectively. The precipitates were dried in vacuum at 80 °C for 12 h and then calcinated in tube furnace at 350 ° for 3 h in flowing N_2 by heating from room temperature to 350 °C at the rate of 10 °C/min. After cooling down to room temperature in N_2 atmosphere, hierarchical VOx microspheres were finally obtained. All reagents used in the experiments are of analytical grade and used without any further purification.

Characterization

The chemical composition of products was characterized by X-ray diffraction (XRD; Rigaku D/Max 2500 PC, Cu Kα, Japan). The morphology and microstructure of hierarchical VOx microspheres were characterized by field emission scanning electron microscopy (FESEM; FEI Nova 400, Netherland) and transmission electron microscopy (TEM; Zeiss LIBRA 200, Germany). The focused ion beam (FIB) cutting was conducted in Zeiss AURIGA FIB (Germany).

Electrochemical Measurements

CHI660E electrochemical workstation was used to measure the electrochemical performance. The working electrode was prepared by mixing the active materials, acetylene black and polyvinylidene fluoride (PVDF) with a mass ratio of 70:20:10. The mixture then was made into slurry by using 1-methyl-2-pyrrolidinone (NMP) as a solvent. Subsequently, the slurry of mixture was spread on a round foam nickel with a diameter of 1.7 cm and dried under vacuum at 120 °C for 12 h, the working electrode then was obtained, which contained about 3 mg active materials. Two-electrode system with two working electrodes sealed in a button cell and 1 M $LiNO_3$ as the electrolyte was adopted to measure the electrochemical. The two working electrodes with same mass of materials were separated by a polypropylene membrane. The button cell should be kept static for 24 h so that the electrode materials were thoroughly wetted with the electrolyte. The cyclic voltammetry (CV) and galvanostatic charging/discharging (GCD) measurements were conducted on the electrochemical workstation. All the electrochemical tests were performed at room temperature.

Results and Discussion

Chemical components of products were investigated by XRD. Figure 1 shows the typical diffraction patterns of the as-prepared products. It can be seen in the XRD patterns that the apparent peaks at 2θ = 15.4°, 25.3°, 49.3° and 59.6° are well indexed to monoclinic VO_2 (JCPDS no. 31-1438) with standard lattice constants of a = 12.03 Å, b = 3.69 Å, c = 6.42 Å (α = 90.0°, β = 106.6°, γ = 90.0°), while the rest of main peaks are attributed to monoclinic V_6O_{13} (JCPDS no. 43-1050) with standard lattice constants of a = 11.92 Å, b = 3.68 Å, c = 10.14 Å (α = 90.0°, β = 100.9°, γ = 90.0°). Therefore, XRD patterns indicate that the products are mixture of VO_2 and V_6O_{13} crystals.

The morphology of as-prepared products was shown in Fig. 2. It can be seen from Fig. 2a–b that the products are microspheres with diameter of ~5 μm. These

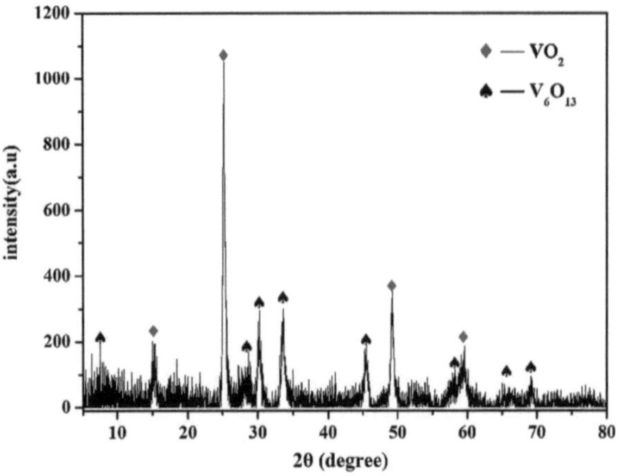

Fig. 1 XRD pattern of the hierarchical VOx microspheres

Fig. 2 SEM images of as-prepared products (**a–b**) and cross-section of the microsphere cut by FIB (**c–d**)

microspheres are made up of ~400 nm long nanoribbons. 4–6 nanoribbons combine together radially, with one end clustering in the centre and the other end dispersing around. To further figure out internal structure of the microsphere, FIB was adopted to cut off a part of microsphere as presented in Fig. 2c–d. It is clear that the interior of microsphere is not completely solid, some radial nanobelts (dark belt-like areas in Fig. 2d) play the roles of backbones, on which the nanoribbons grow. The hyperbranching growth of nanobelts and nanoribbons formed those microspheres. These solid nanobelt backbones contribute to high structure stability of microspheres. The hierarchical structure, resulting from the hyperbranching growth, brings in loads of pores and large surface area, which are beneficial to the easy accessibility of electrolyte ions. It thus indicates the potential high performance of the as-prepared hierarchical microspheres as electrodes of supercapacitors.

To further investigate the microstructure of products, TEM studies were used (Fig. 3). A cluster of hierarchical VOx microspheres was observed in Fig. 3a, as can be seen in the image the deep dark and wide belt-like areas distribute radially and form the backbones of the cluster, which agrees with SEM images in Fig. 2. At the same time, nanoribbons grow around those nanobelts as the light dark areas shown in Fig. 3a. The high resolution TEM images of a part of cluster display the combination of nanobelts and nanoribbons shown in Fig. 3b–c. The deep dark areas are connected to nanobelts and the light dark areas are nanoribbons. The interplanar spacing of 0.27 and 0.32 nm are in agreement with the spacing of V_6O_{13} (310) plane and ($\bar{2}03$) plane, respectively, as indicated in Fig. 2, while the lattice fringes with spacing of 0.16 and 0.19 nm are also found in Fig. 3c, which correspond to the spacing of ($\bar{4}04$) and ($\bar{1}13$) plane, respectively, in VO_2 crystal as shown in XRD. According to TEM images, it further confirms the existence of V_6O_{13} and VO_2.

The electrochemical properties of hierarchical VOx microspheres were investigated as electrodes for supercapacitor using a two-electrode cell configuration. Cyclic voltammetry (CV) and galvanostatic charging/discharging measurements were conducted. Fig 4a exhibits the typical CV curve of hierarchical VOx microspheres at the scan rates of 5 mV/s, 15 mV/s, 20 mV/s, 30 mV/s and

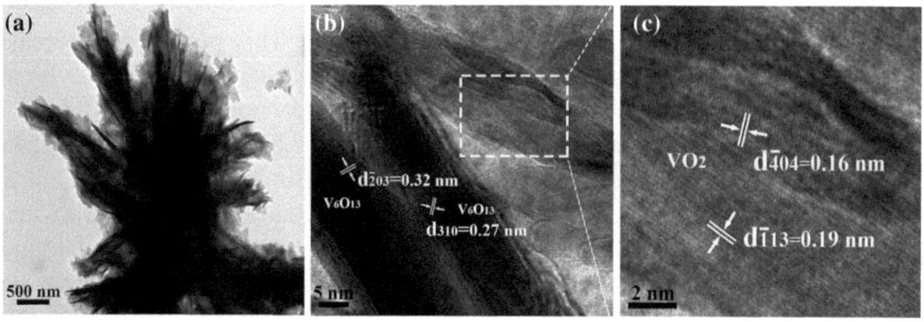

Fig. 3 TEM images of the hierarchical microspheres

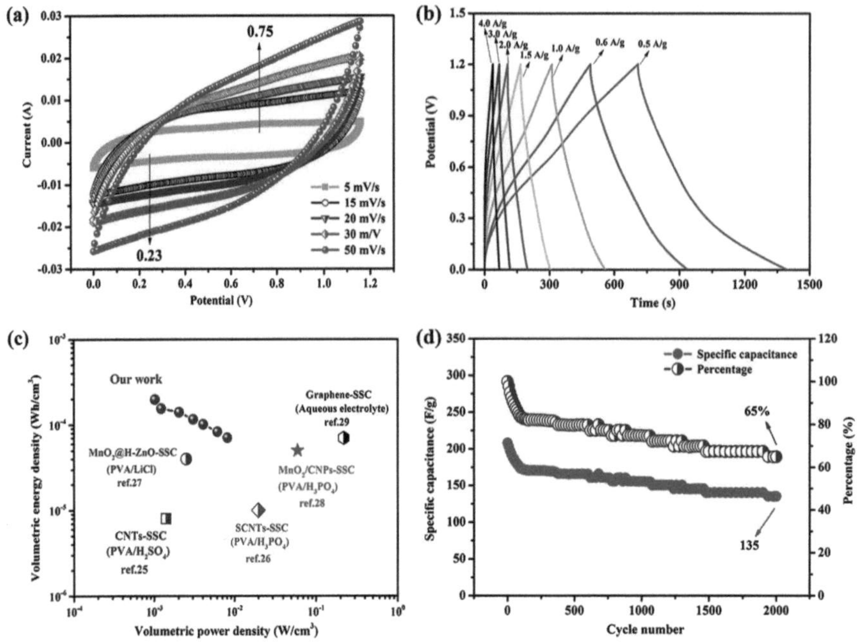

Fig. 4 **a** CV curves of the hierarchical VOx microspheres at different scan rates; **b** Galvanostatic charging/discharging curves of hierarchical VOx microspheres at different current densities; **c** Ragone plot of the symmetric supercapacitor based on hierarchical VOx microspheres; **d** Cycling performance of hierarchical VOx microspheres at current density of 4.0 A/g

50 mV/s, respectively, within the potential range from 0 V to 1.2 V. It is clear that the shape of CV curve at low scan rate is nearly rectangular, following the electric double layer charging/discharging mechanism, which indicates that hierarchical VOx microspheres own perfect charge storage ability and high efficiency. A pair of redox peaks of the CV curves at 5 mV/s advent at 0.23 and 0.75 V, and the electrochemical redox reactions associated with this pair of redox peaks as follows:

$$VO_x + xLi^+ + xe^- \leftrightarrow Li_xVO_x \tag{1}$$

where x is the mole fraction of intercalated Li^+ ions. This pair of redox peaks indicates the charge storage mechanism is pseudocapacitance, which has broad application prospect for electrode materials of supercapacitors [11]. This superior performance of the electrode materials is mainly due to V_6O_{13} crystals with metallic conductivity that contribute to the electron transfer. But at scan rate as high as 50 mV/s, the curve exhibits a deformed shape and the redox peaks are not obvious as low scan rate, the reason is that it is difficult for Li^+ ions to intercalate and deintercalate the mesopores deep inside the microspheres at such high scan rate Fig. 4b shows the galvanostatic charging/discharging curve at different current densities. The specific capacitance of supercapacitors can be calculated through galvanostatic charging/discharging measurement, the corresponding formula as follows:

$$C = 2I\Delta t/m\Delta V \quad (2)$$

where C is the specific capacitance, I is the constant current; Δt is the discharging time; m is the mass of electrode material; ΔV is the potential drop upon discharging. At the current density of 0.5 A/g (i.e. 0.8 mA/cm^2), the gravimetric specific capacitance of hierarchical VOx microspheres is as remarkably high as 581 F/g, while the corresponding volumetric specific capacitance is 3.94 F/cm^3. In terms of the value of gravimetric specific capacitance, it is much higher than most of reported vanadium oxide nanostructures based electrodes such as flower-like V_2O_3 (218 F/g at 0.05 A/g), VO_2 nanobelts (191 F/g at 1 A/g), V_4O_9 microflowers (392 F/g at 0.5 A/g) and polypyrrole coated V_2O_5 nanoribbons (308 F/g at 0.1 A/g), and even vanadium oxide/carbon composite nanostructures including Ti-doped VO_x/CNT composites (310 F/g at 5 mV/s) and V_2O_3 nanoflakes@C composites (205 F/g at 0.05 A/g) [2, 12–16]. It is also higher than those of other transition metal materials such as MnO_2 nanorods (298 F/g at 5 mV/s) and flower-like $Co_3(PO_4)_2$ (350 F/g at 1 A/g) [17, 18]. For volumetric specific capacitance, which is also significantly higher than many other reported supercapacitors, such as the asymmetric supercapacitor based on hydrogen-treated (denoted as H-) TiO_2@MnO_2 and H-TiO_2@C (0.71 F/cm^3 at 10 mV/s) [19] and the symmetric supercapacitor based on MnO_2@carbon fiber (2.5 F/cm^3 at 0.02 A/cm^3) [20]. The curves at all current densities exhibit symmetric triangular shape even at a high current density of 4.0 A/g, which indicates that the supercapacitor based on hierarchical VOx microspheres is an ideal capacitor that charge can transport quickly [21, 22]. The reasons associated with such superior specific capacitance of supercapacitor based hierarchical VOx microspheres are that hyperbranched growth of nanoribbons leads to high specific area that provide a large amount of active surface sites for redox reaction to take place in, and hierarchical structure exists mass mesopores which can facilitate the intercalation/deintercalation of electrolyte ions so that the rate of redox reactions on the surface can be increased, and the high metallic conductivity of V_6O_{13} accelerates the electron transfer inside the bulk microspheres, which is beneficial to increasing the redox reaction rate. It is also worthy to point out that VOx microspheres possess higher mass density compared with carbonaceous materials, which can decrease the volume of electrodes and lead to higher volumetric specific capacitances [19, 20].

The energy density (E) and power density (P) are shown in Fig. 4c. According to the two-electrode configuration, E and P are calculated by the following formula:

$$E = 1/8CV^2 \quad (3)$$

$$P = E/\Delta t \quad (4)$$

where C is the specific capacitance, V is the cell voltage and Δt is the discharge time. The maximum energy density was calculated to be 29 Wh/kg at the power density of 150 W/kg and the maximal power density is 1.2 kW/kg at 4.0 A/g. This energy density is much higher than those of symmetrical supercapacitors based on V_2O_5/carbon nanotubes (16 Wh/kg at power density of 800 W/kg) [23] and V_2O_5 nanotube spherical clusters (11.6 Wh/kg at 0.1 A/g) [24]. The corresponding volumetric energy density as shown is 0.2 mWh/cm^3, which is also substantially higher than values reported for other supercapacitors, such as the symmetric supercapacitors (SSC) based on multiple walled carbon nanotubes (0.008 mWh/cm^3 at 1.4 mW/cm^3) [25], single walled carbon nanotubes (0.01 mWh/cm^3 at 0.02 W/cm^3) [26], MnO_2@H-ZnO (0.04 mWh/cm^3 at 2.5 mW/cm^3) [27], MnO_2/Carbon nanoparticles (0.05 mWh/cm^3 at 0.06 W/cm^3) [28], and graphene (0.07 mWh/cm^3 at 0.22 W/cm^3) [29]. The high energy density, together with high power density, is due to the high specific surface area of the hierarchical microspheres, the high conductivity and high density of V_6O_{13} in the hierarchical microspheres.

Another important criterion for supercapacitor is cycling performance examined at 4.0 A/g as shown in Fig. 4d. Notably, 65% of initial specific capacitance has been maintained after 2000 cycles, even though the specific capacitance decreases sharply by 18% of initial value in the first 130 cycles, after which it decreased slightly. The reason of decay is that V(V)-bearing vanadates was dissolved into electrolyte during the charging/discharging process, which is common in supercapacitors and lithium-ion batteries based on vanadium oxide materials [16, 30–32]. Though the capacitance reduced by 35% after 2000 cycles, the cycling stability of hierarchical VOx microspheres is obviously higher than that of other morphologies based on pure vanadium oxides without carbon, especially those with high initial specific capacitances [7, 33]. This remarkable stability benefits from the hyperbranched growth of alternative VO_2 crystals and V_6O_{13} crystals in the microspheres.

Conclusions

In summary, hierarchical VOx microspheres forming from hyperbranched growth of growth of have been successfully synthesized by solvothermal method. These as-prepared VOx microspheres with diameter of ~5 μm are made up of ~400 nm long nanoribbons. Both XRD and TEM indicate that the products consist of V_6O_{13} crystal with good metallic conductive and VO_2 crystal. The hierarchical VOx microspheres prepared for electrode materials of supercapacitors exhibit superior electrochemical performance with high specific capacity of 581 F/g (3.94 F/cm^3) and outstanding energy density of 29 Wh/kg (0.2 mWh/cm^3). The cycling stability of hierarchical VOx microspheres is also obviously higher than many other reported materials.

Acknowledgements The authors would like to acknowledge the financial supports by National Natural Science Foundation of China (No. 21105134 and No. 51474041), the Natural Science Foundation Project of CQ CSTC (No. cstc2012jjB50011) and Fundamental Research Funds for the Central Universities of China (No. 106112015CDJZR465505).

References

1. J.R. Miller, A.F. Burke, Electrochemical capacitors: challenges and opportunities for real-world applications. Electrochem. Soc. Interface **17**(1), 53–57 (2008)
2. K.-J. Huang et al., One-step preparation of layered molybdenum disulfide/multi-walled carbon nanotube composites for enhanced performance supercapacitor. Energy **67**, 234–240 (2014)
3. V. Augustyn, P. Simon, B. Dunn, Pseudocapacitive oxide materials for high-rate electrochemical energy storage. Energy Environ. Sci. **7**(5), 1597–1614 (2014)
4. N. Bahlawane, D. Lenoble, Vanadium oxide compounds: structure, properties, and growth from the gas phase. Chem. Vapor. Depos. **20**(7–9), 299–311 (2014)
5. D.P. Nair et al., Effect of surfactants on electrochemical properties of vanadium-pentoxide nanoparticles synthesized via hydrothermal method. J. Nanosci. Nanotechnol. **15**(6), 4392–4397 (2015)
6. N.N. Wang et al., Facile hydrothermal synthesis of ultrahigh-aspect-ratio V2O5 nanowires for high-performance supercapacitors. Curr. Appl. Phys. **15**(4), 493–498 (2015)
7. H.W. Wang et al., One-step strategy to three-dimensional graphene/VO2 nanobelt composite hydrogels for high performance supercapacitors. J. Mater. Chem. A **2**(4), 1165–1173 (2014)
8. J. Shao et al., One-step hydrothermal synthesis of hexangular starfruit-like vanadium oxide for high power aqueous supercapacitors. J. Power Sources **219**, 253–257 (2012)
9. J.X. Zhu et al., Building 3D structures of vanadium pentoxide nanosheets and application as electrodes in supercapacitors. Nano Lett. **13**(11), 5408–5413 (2013)
10. P.D. Dernier, Structural investigation of the metal-insulator transition in V_6O_{13}. Mater. Res. Bull. **9**(7), 955–963 (1974)
11. M.H. Yu et al., Valence-optimized vanadium oxide supercapacitor electrodes exhibit ultrahigh capacitance and super-long cyclic durability of 100 000 cycles. Adv. Funct. Mater. **25**(23), 3534–3540 (2015)
12. P.H. Jampani et al., High energy density titanium doped-vanadium oxide-vertically aligned CNT composite electrodes for supercapacitor applications. J. Mater. Chem. A **3**(16), 8413–8432 (2015)
13. H. Wang et al., One-step strategy to three-dimensional graphene/VO_2 nanobelt composite hydrogels for high performance supercapacitors. J. Mater. Chem. A **2**(4), 1165–1173 (2014)
14. H. Pang et al., 2D single-or double-layered vanadium oxide nanosheet assembled 3D microflowers: controlled synthesis, growth mechanism, and applications. Nanoscale **5**(17), 7790–7794 (2013)
15. Q. Qu et al., Core–shell structure of polypyrrole grown on V_2O_5 nanoribbon as high performance anode material for supercapacitors. Adv. Energy Mater. **2**(8), 950–955 (2012)
16. H. Liu et al., Flowerlike vanadium sesquioxide: solvothermal preparation and electrochemical properties. ChemPhysChem **11**(15), 3273–3280 (2010)
17. Z.K. Ghouri et al., High-efficiency super capacitors based on hetero-structured alpha-MnO2 nanorods. J. Alloy. Compd. **642**, 210–215 (2015)
18. H.Y. Li et al., Self-assembled 3D cobalt phosphate octahydrate architecture for supercapacitor electrodes. Mater. Lett. **152**, 25–28 (2015)
19. X.H. Lu et al., H-TiO_2@MnO_2//H-TiO_2@C Core-shell nanowires for high performance and flexible asymmetric supercapacitors. Adv. Mater. **25**(2), 267–272 (2013)

20. X. Xiao et al., Fiber-based all-solid-state flexible supercapacitors for self-powered systems. ACS Nano **6**(10), 9200–9206 (2012)
21. M. Kim, J. Kim, Development of high power and energy density microsphere silicon carbide-MnO_2 nanoneedles and thermally oxidized activated carbon asymmetric electrochemical supercapacitors. Phys. Chem. Chem. Phys. **16**(23), 11323–11336 (2014)
22. Y. Hou et al., Nanoporous metal based flexible asymmetric pseudocapacitors. J. Mater. Chem. A **2**(28), 10910–10916 (2014)
23. I. Shakir et al., Ultra-thin and uniform coating of vanadium oxide on multiwall carbon nanotubes through solution based approach for high-performance electrochemical supercapacitors. Electrochim. Acta **111**, 400–404 (2013)
24. S.D. Perera et al., Vanadium oxide nanotube spherical clusters prepared on carbon fabrics for energy storage applications. ACS Appl. Mater. Interfaces **3**(11), 4512–4517 (2011)
25. Y.J. Kang, et al., All-solid-state flexible supercapacitors based on papers coated with carbon nanotubes and ionic-liquid-based gel electrolytes, *Nanotechnology*, **23**(28) (2012). (vol 23, 065401, 2012)
26. M. Kaempgen et al., Printable thin film supercapacitors using single-walled carbon nanotubes. Nano Lett. **9**(5), 1872–1876 (2009)
27. P.H. Yang et al., Hydrogenated ZnO core-shell nanocables for flexible supercapacitors and self-powered systems. ACS Nano **7**(3), 2617–2626 (2013)
28. L.Y. Yuan et al., Flexible solid-state supercapacitors based on carbon nanoparticles/mno_2 nanorods hybrid structure. ACS Nano **6**(1), 656–661 (2012)
29. M.F. El-Kady et al., Laser scribing of high-performance and flexible graphene-based electrochemical capacitors. Science **335**(6074), 1326–1330 (2012)
30. L.J. Cao et al., Ultrathin single-crystalline vanadium pentoxide nanoribbon constructed 3D networks for superior energy storage. J. Mater. Chem. A **2**(32), 13136–13142 (2014)
31. M.-H. Bai et al., Electrodeposition of vanadium oxide-polyaniline composite nanowire electrodes for high energy density supercapacitors. J. Mater. Chem. A **2**(28), 10882–10888 (2014)
32. S.-Z. Huang et al., Annealed vanadium oxide nanowires and nanotubes as high performance cathode materials for lithium ion batteries. J. Mater. Chem. A **2**(34), 14099–14108 (2014)
33. H. Liu, et al., *Chem. Phys. Chem.* **11** (2010)

Potential of Magnetotactic Bacteria for the Fabrication of Iron Nanoparticles

T. Thuy Minh Nguyen, Manish D. Baviskar and Paul Bernazzani

Abstract Magnetotactic bacteria are typically found in soils rich in iron. These prokaryote bacteria have the property of using ingesting atoms of iron and generating magnetosomes of nanoparticles of either diamagnetic or paramagnetic nature within each cell. We report on the use of magnetotatic bacteria for the production of uniform nanoparticles of iron mineral magnetite (Fe_3O_4). The potential for mass production was investigated along with the molecular, physical, and magnetic properties of the magnetosomes using various growth and microscopy techniques. Results reveal that the magnetic particles are stable and that bacteria growth can be optimized to produce magnetosomes with different magnetic properties, suggesting that the industrial development of these bio manufactures lies in the foreseeable future.

Keywords Magnetosomes · Magnetic bacteria · Nanoparticles · Iron

Introduction

The synthetic production of nanoparticles, the cornerstone of nanotechnology, involves various physical and chemical methods. However, a disadvantage of these methods is the production of toxic byproducts, possibly making them not environmentally safe methods [1]. Nanoparticle synthesis using biological systems would follow green chemistry principles as the reagents are eco-friendly including the reducing agent and the capping agent in the reaction [2, 3].

Biogenic nanoparticles involve natural phenomena that take place in the biological systems. Bacteria are considered as the most potent eco-friendly nanofactories [1]. Magnetotactic bacteria (MTB) are a heterogenous group of aquatic microorganisms that have the ability to orient and migrate along geomagnetic field

T. Thuy Minh Nguyen (✉) · M.D. Baviskar · P. Bernazzani
Department of Chemistry and Biochemistry, Lamar University,
Beaumont, TX 77710, USA
e-mail: ttnguyen15@lamar.edu

The Minerals, Metals & Materials Society, *TMS 2017 146th Annual Meeting & Exhibition Supplemental Proceedings*, The Minerals, Metals & Materials Series,
DOI 10.1007/978-3-319-51493-2_2

lines, a behavior named magnetotaxis. This property is based on specific intracellular structures, the magnetosomes which, in most MTB are nanometer-sized, membrane bound magnetic particles, composed of the iron mineral magnetite (Fe_3O_4) or more rarely, greigite (Fe_3S_4). These magnetosomes are organized in one or more straight chains parallel to the long axis of the cells. Such an arrangement confers a magnetic moment to the cell. In the northern hemisphere these MTB are known to move towards the geographic North Pole, while in the southern hemisphere they are known to migrate towards the geographic south. Magnetotactic bacteria are found in diverse aquatic habitats and include coccoid to ovoid cells, rods, vibrios and spirilla from different water bodies (freshwater, seawater), sediment and soil. Reports of South-seeking magnetotactic bacteria found in Southern hemispheres sediments as well as of samples held in similar magnetic conditions in the Northern Hemisphere which verifies the hypothesis that downward direction is advantageous orientation and upward detrimental for the survival of the magneto tactic bacteria with unidirectional motility [4–6].

Bacterial magnetic nanoparticles have great useful potential in biotechnological and biomedical applications. In this study, a liquid growth medium will be modified for cultivation of a magnetotactic bacterium that has been isolated from sediment sample. These modifications include changes in the type and amounts of vitamin, minerals, carbon sources, etc. Serum bottles and designed air-tight laboratory bottles can be used to create microaerobic conditions in order to develop a method for scale-up experiment.

Magnetotactic bacteria (MTB) synthesize unique nanoparticle structures within magnetosomes. MTB's nanoparticles are coated with a thin organic membrane that results in high and homogeneous dispersions in aqueous solutions compared to artificial magnetic nanoparticles, making them ideal biotechnological materials [7, 8]. It has already been demonstrated, that isolation and axenic cultivation of MTB in pure culture is very difficult. To date only a small number of isolates and an even smaller number of genera and species are available [9, 10]. Recently, pure cultivation of magnetic bacteria in defined medium has provided effective advancement on the application of MTB's nanoparticles. Because of MTB's fastidious culture requirements, growth of most of them on a large scale is extremely difficult. Mass cultivation of MTB for magnetosome production may be one of the most important biotechnological processes in the application of MTB's nanoparticles. We report the investigation of a new approach for large-scale production of magnetotactic bacteria. The main goal of this study was development a method with low temperature, lower energy cost and high yield that may make biogenical magnetic nanoparticles a more economical and energy sustainable process than some of the competing processes.

Methodology

Isolation and Screening of Bacteria

Samples were collected in sterile flasks and plastic containers for bacteriological analysis. The samples were stored in dark cold room at 40 °C for 28 days. The magnetotactic (MTB) bacteria were isolated using the race track method. Isolated bacteria were tested for magnetic particle synthesis ability.

Optimization and Validation of Growth Condition

Optimum growth condition and media for each isolates will be optimized by adjusting pH, temperature and Different chemical concentration and oxygen requirement. 250 ml Erlenmeyer flask will be used where 10 ml of Sterile Luria-Bertani (LB) broth liquid media will be used as baseline growth media to identify mixo-tropic growth of magnetic tactic bacteria. Then it will be incubated in Shaker overnight at 30 °C AT 150 RPM. Optical density of culture will be checked after 8 to 10 h at 600 nm using spectrophotometer to determine growth of bacteria.

Optimization of Nano Magnetic Practical Synthesis Process

Isolated cells were seeded in optimized media incubated at optimum growth condition by period analysis of magnet in different growth condition. Prussian blue assay was used to evaluate the presence of magnetosomes.

Source of Microorganism and Liquid Media Preparation

Magnetotactic bacteria were isolated from a water/sediment microcosm that was collected from the Marquez Crater. Control cells acquired from ATCC were also used. After magnetic collection and isolation methods, LB media was used for cultivation experiments. A modified liquid medium (MLM) was also used which contained Wolfe's vitamin solution, Wolfe's mineral solution, sodium succinate, yeast extract, $MgSO_4 \times 7H_2O$, peptone casein, potassium phosphate buffer (pH 7), NH_4Cl, sodium acetate anhydrous, resazurin, ferric citrate, HCl and the pH was adjusted to 7.0. HCl and vitamin solution were added to the MLM after autoclaving.

Microscopy Studies

Microscopy, isolated cells of the MTB were diluted by sterilized phosphate buffer solution (pH = 7, 10 mM) and then cells were placed on the surface of glass slides. The morphology and configuration of the MTB and their magnetosomes were investigated with different staining techniques.

Isolation and Purification of Magnetosomes

The magnetotactic bacteria were harvested by centrifugation (8000 rpm, 15 min, 4 °C) and washed by sterilized phosphate buffer solution (pH = 7.0). Then the precipitated cells were resuspended in 1 N NaOH and boiled for 30 min to lyse the cells. Magnetosomes from the disrupted cells were collected at a graduated cylinder by magnets for 1 h, then the nonmagnetic fluid was removed by aspiration and magnetic nanoparticles washed with buffer. Finally, the magnetosomes attracted to the magnet were carefully suspended in sterilized phosphate buffer solution (pH = 7.0).

Collection of Magnetic bacteria was done based on the cells' swimming response to a magnetic field. The south pole of a permanent magnet was attached outside a jar containing the water and sediment samples, 1 cm above the sediment surface. After 2–4 h 1–2 ml of the water in the bottle near the wall adjoining the magnet was collected with a pipette and transferred to sterile tubes to be used for further studies. "Race track" purification and enrichment of the MTB The modified capillary "race track" (CRT) method was used to purify and enrich the MTB obtained by magnetic collection. A capillary tube (length, 6–9 cm) sealed at one end in a gas flame was filled with medium by means of a long hypodermic syringe and was fitted to the narrow end of a Pasteur pipette. The sample material (magnetically collected cells) was placed on the top of a sterile, wetted cotton plug in the wide mouth end of the pipette that served as a reservoir. The capillary was exposed to a magnetic field produced along it with a permanent magnet for 5 h. The MTB migrated through the cotton plug towards the closed end of the capillary. The tip containing the accumulated MTB was then broken off and with the help of a sterile hypodermic needle the organisms transferred to the sterile enrichment medium taken in test tubes that were incubated at 30–350 °C for about two days. This method was repeated two more times to purify the MTB fully. Isolation of magnetic bacteria. The purified magnetic cells were then isolated by the streak plate method using a magnetic field and preserved (at 40 °C) on the same medium.

Assessment of Culture Magnetism

The isolates were tested for their magnetotactic response using the 'hanging drop technique' under an optical microscope, with the south pole of a bar magnet being placed some 10 cm distant from the slide. Their magnetic response was also tested in terms of the spreading of their growth on the surface of a semisolid (50:50) LB and MSGM medium [0.8% agar]. The isolates were inoculated in a straight line at the center of the medium in petriplates. The plates were incubated in a magnetic field created by placing the opposite poles of two different bar magnets on either side perpendicular to the line of streaking. The growth pattern after incubation was observed for any spreading towards the magnet poles. The isolated magnetic cultures and a known non-magnetic bacterial culture [*E. coli*] as control were grown in the MSGM medium that contained Ferric Quinate as source of iron. After obtaining sufficient growth (approximately 30–40 mg dry weight), the cell mass was separated by centrifugation at 10,000 g in a research centrifuge. The cell mass thus obtained was dried to constant weight at 105 °C in a hot air oven. Iron content of the cell mass was determined by Infra-red and UV-visible spectroscopy on a Model, using the tri-acid digestion method of iron extraction from the cells.

Results and Discussion

Enrichment and Isolation

Magnetotactic bacteria (MTB) were successfully enriched from sediment samples obtained from the Marquez crater using the magnetic collection method and purified by the capillary racetrack method. The mixed bacterial culture that was obtained was observed under a microscope and seen to include more than one morphological type. The streak plate method of isolation as well as a total viable count of the original samples were performed to verify the efficacy of the magnetic purification methods. About four different morphological forms of bacteria were obtained as pure cultures and different individual colony characteristic were observed. The bacteria forms included gram positive rod shaped bacteria, gram negative slender rod and gram positive coccus.

The original samples were grown under different media conditions. Figure 1 present the images of Petrie dishes containing the samples as received and following 10 days of incubation in standard media concentration. The growth potential of the samples is observed and optimization can be obtained. Following culture growth, magnetotactic bacteria were isolated using the race track method. Figure 2 shows the microscopy images of bacteria where two distinctive types were identified, one north seeking magnetotactic bacteria and another south seeking. These findings lead to the assessment of the magnetism and of the capacity of the samples to produce magnetosomes.

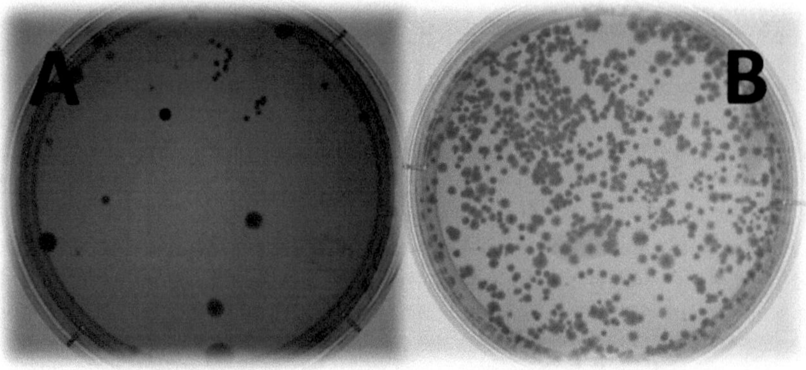

Fig. 1 Images of sample cultures as obtained in different conditions. Original sediment (panel **A**), following 10 days of incubation (panel **B**)

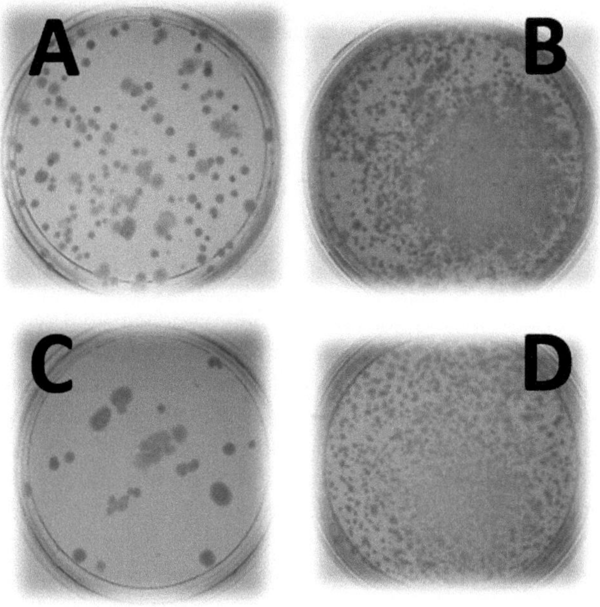

Fig. 2 Images of north (panels **A** and **B**) and south (panels **C** and **D**) seeking bacteria following 2 days (panels **A** and **C**) and 10 days (panels **B** and **D**) incubation times

Assessment of Culture Magnetism and Iron Analysis

The capability of the magnetotactic bacteria to orient themselves while under the influence of a magnetic field was evaluated by exposing the sediment containing the bacteria to both north and south poles of a magnet. Figure 3 compares the incubation

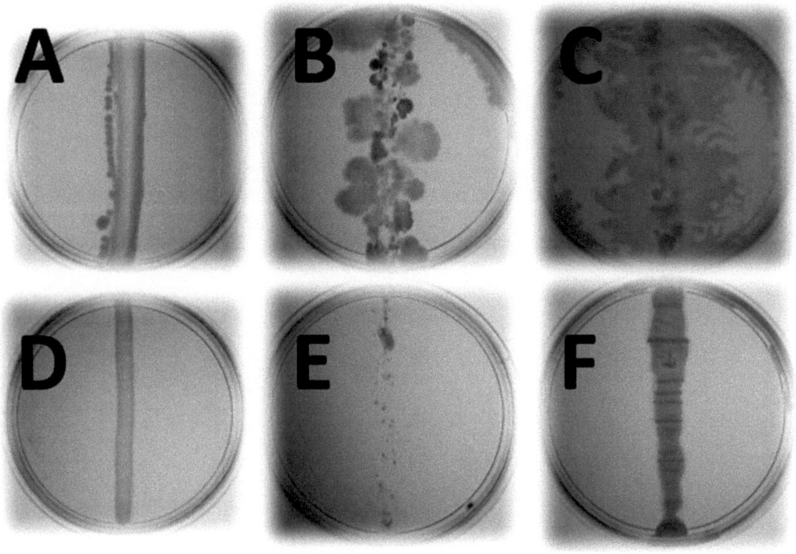

Fig. 3 Images of bacterial growth when exposed to (panels **A**, **B** and **C**) and in the absence of an external magnetic field (panels **D**, **E**, and **F**). Panels **A** and **D** represent the control *E. coli* samples while panels **B** and **E** represent the cultures after 2 days of incubation and panels **C** and **F** represent the cultures after 10 days of incubation

of the bacteria when exposed to a magnetic field and under normal conditions. Bacteria were plated on a line going up and down on the figure, while the magnetic field is in a left to right orientation. Nonmagnetic cells, *E. coli*, were used as a control. The figure reveals that the sample containing both south and north seeking magnetotactic bacteria results in cell growth that expand toward left and right directions compared to control samples and those without exposure to a magnetic field. This result supports the fact that the prominent growth of magnetosome containing cells was achieved.

To further evaluate the magnetic properties of the cultures, the mixture before isolation as well as the individual isolated cultures were tested by the hanging drop technique and the semisolid medium method. Figure 4 presents the images of the hanging drop technique. Both, the mixed culture as well as the isolated organisms showed a response as expected. In the hanging drop observation. The cultures aligned parallel to each other along the magnetic field lines and showed a migration towards the South Pole and north pole of the magnet wherever it was placed. On the semisolid medium too, the MTB when placed in a magnetic field showed pronounced migration towards the South Pole referred to as ‘south seeking’, while in the absence of magnetic field they did not show any migration. The non-magnetic, motile control bacterial culture [*E. coli*] meanwhile did not show any migration towards any pole and grew as a straight line as inoculated but magnetic motile control Magnetospirillum magnetotacticum (ATCC-3163) shows orientation.

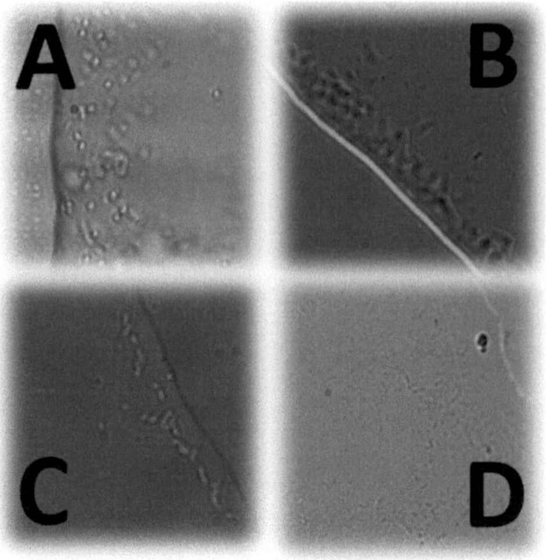

Fig. 4 Images of the magnetic migration by the hanging drop method. The different images represent north (panels **A** and **B**) and south (panels **C** and **D**) seeking bacteria following 2 days (panels **A** and **C**) and 10 days (panels **B** and **D**) incubation times

Table 1 Optimization parameters for growth of magnetosome containing bacteria

	Temperature optimization		Growth on liquid media (600 nm) after 3 h incubation (a.u.)	
Isolates	Range temperature (°C)	Optimal T (°C)	LB broth	Magnetic spirillum growth medium
A1-N	4–45	30	0.78	0.52
B1-N	4–45	30	0.83	0.39
A2-N	21–45	37	0.88	0.59
B2-N	21–45	37	0.83	0.32
A1-S	4–45	30	0.98	0.41
B1-S	4–45	30	0.76	0.29
A2-S	21–45	37	0.90	0.38
B2-S	21–45	37	1.27	0.24

Along with the temperature and media optimization data, these results and the morphological characters of the isolates are presented in Table 1. The last column represents the iron content of each isolated bacterial strain. The unambiguous correlation between the magnetic response and the iron content of the cells is clearly evident. It can also be seen that the magnetic bacteria accumulated up to more iron within their cells as compared to the non-magnetic cell. These results suggest that the growth of magnetosome containing bacteria can be optimized to yield promising amounts of cells. The next step would be the scale up of the optimized growth to yield manageable amount of nanoparticles.

Conclusion

Magnetosome containing magnetotactic bacteria have been isolated from sediment samples. Partial growth optimization revealed that the temperature range, incubation temperature and type of media were critical for the efficient growth, but that these could be tailored. The presence of visible magnetic clusters that were influenced by the presence or absence of a magnetic field confirmed the nature of the nanoparticles that the magnetosome contain.

Acknowledgements The partial support of the Welch Foundation (grant V-0004) is gratefully acknowledged.

References

1. V. Arya, Living systems: eco-friendly nanofactories. Digest. J. Nanomater. Biostruct. **5**, 9–21 (2010)
2. S. Li, Y. Shen, A. Xie, X. Yu, L. Qui, L. Zhang, Q. Zhang, Green synthesis of silver nanoparticles using Capsicum annum L. extract. Green Chem. **9**, 852–858 (2007)
3. T.C. Prathna, T. Lazar Mathew, N. Chandrasekaran, A.M. Raichur, A. Mukherjee, Biomimetic synthesis of nanoparticles: science, technology & applicability, in *Biomimetics learning from Nature*. ed. by A. Mukherjee, (IN- TECH Publications, Vienna, Austria, 2010), pp. 1–20. ISBN number 978-953-307-025-4
4. D. Schüler, Formation of magnetosomes in magnetotactic bacteria. J. Mol. Microbiol. Biotechnol. **1**, 79–86 (1999)
5. D.A. Bazylinski, R.B. Frankel, Magnetosome formation in prokaryotes. Nat. Rev. **2**, 217–230 (2004)
6. D. Schüler, Molecular analysis of a subcellular compartment: the magnetosome membrane in Magnetospirillum gryphiswaldense. Arch. Microbiol. **181**, 1–7 (2004)
7. T. Matsunaga, A. Arakaki, Molecular bioengineering of bacterial magnetic particles for biotechnological applications, in *Magnetoreception and Magnetosomes in Bacteria*, vol. 3, ed. by D. Schüler (Springer, Berlin, 2007), pp. 227–254
8. J. Xie, K. Chen, X. Chen, Production, modification and bioapplications of magnetic nanoparticles gestated by magnetotactic bacteria. Nano. Res. **2**, 261–278 (2009)
9. D.A. Bazylinski, T.J. Williams, Ecophysiology of magnetotactic bacteria, in *Magnetoreception and Magnetosomes in Bacteria*, vol. 3, ed. by D. Schüler (Springer, Berlin, 2007), pp. 37–75
10. D. Faivre, D. Schüler, Magnetotactic Bacteria and Magnetosomes. Chem. Rev. **108**, 4875–4898 (2008)

Facile Green Synthesis and Characterization of Water Soluble Superparamagnetic Iron Oxide-Gold Porphyrin Conjugate for Improved Photodynamic Therapy

Olayemi J. Fakayode, Sandile P. Songca and Oluwatobi S. Oluwafemi

Abstract Porphyrin has been proven to be an efficient photosensitizer for photodynamic therapy (PDT). However, the problem of undesired localization in non-tumour environments has lowered its therapeutic efficacy. To make porphyrin tumour specific, we have successfully conjugated a meso-substituted symmetric 4-hydroxyphenyl porphyrin (THPP) to superparamagnetic iron oxide nanoparticles-gold core-shell nanostructure (SPIONs-Au). The magnetic porphyrin conjugate (MPC) was characterized using UV-Visible spectrophotometry (UV-Vis), Photoluminescence spectroscopy (PL) and Fourier Transform infrared spectroscopy (FT-IR). The as-synthesized MPC exhibited strong attraction towards external magnetic field and high singlet oxygen quantum yield sufficient for the eradication of cancer cells in vitro.

Keywords Porphyrin · SPIONs · Gold · PDT · Magnetic targeting · Breast cancer

---

O.J. Fakayode (✉) · O.S. Oluwafemi (✉)
Department of Applied Chemistry, University of Johannesburg, 17011, Doornfontein, Johannesburg 2028, South Africa
e-mail: olayemifakayode@gmail.com

O.S. Oluwafemi
e-mail: oluwafemi.oluwatobi@gmail.com

O.J. Fakayode · O.S. Oluwafemi
Centre for Nanomaterials Science Research, University of Johannesburg, Johannesburg, South Africa

S.P. Songca
Department of Chemistry, University of Zululand, Private Bag, X1001, Kwadlangezwa 3886, South Africa

The Minerals, Metals & Materials Society, *TMS 2017 146th Annual Meeting & Exhibition Supplemental Proceedings*, The Minerals, Metals & Materials Series,
DOI 10.1007/978-3-319-51493-2_3

Introduction

Cancer occurs in many parts of human body [1–3] and thus represents one of the human diseases very difficult to control. Conventional treatments such as chemotherapy, ionizing radiotherapy and surgery have been inefficient [4] for many years [5] due to many side effects including non-specific therapeutic function [6], resistance to drugs over time, cosmetic disadvantages and psychological depression experienced by patient after surgical operation. Photodynamic therapy (PDT) emerged as a result of a search for better alternatives for cancer treatment. PDT is a non-invasive technique [7–9] that employs the use of a non-toxic light, photosensitizer prodrug (PS) and molecular oxygen for the treatment of cancer cell invasion [10]. The light activates the prodrug within tumour leading to the generation of reactive oxygen species that irreversibly destroy cancer cells. As a result, PDT has been successfully used to treat many cancerous and even non-cancerous diseases [11]. Among the photosensitizers used in PDT, porphyrin PS was the first to be clinically accepted due to its non-toxicity, biocompatibility, high reactive oxygen species generation and efficient PDT efficacy. However, despite these advantages, porphyrin PDT has certain limitations such as prolong photosensitivity, non-specific therapeutic function and inability to treat deep-seated tumour. To this end, the deep-seated tumour and photosensitivity problems have been resolved to some extent by the development of the second generation porphyrins. However, the problem of non-specific therapeutic function as a result of poor whole tumour accumulation strategy still remains unresolved. This has led to the destruction of normal cells within the vicinity of cancer environment and inevitable overdose of the prodrug. To resolve this problem, many third generation porphyrin conjugates are now emerging and their PDT efficacy evaluated [12, 13]. In this study, we have successfully attached a meso-substituted porphyrin to a superparamagnetic nanoparticles-gold core-shell (SPIONs-Au) nanostructure to obtain an external magnetic field—driven porphyrin system.

Materials and Methods

All chemicals were purchased from Sigma Aldrich and used without further purification. The UV-Vis analysis was performed on UV-Visible spectrophotometer (Perkin Elmer (UK), Lambda 25, 1 nm slit width; scan range: 200–900 nm). The FT-IR evaluation was conducted on a UATR-FT-IR spectrometer (Perkin Elmer (UK). The sample holder was precleaned with 50% ethanol and dried before sample analysis. The photoluminescence study was conducted on a spectro- Fluorophotometer (RF-6000; Shimadzu, Japan), wavelength range of 200–900.

The SPIONs were synthesized using indirect coprecipitation method following modification of a previously reported method [14]. This is followed by coating with gold at 73 °C for some time to form the SPION-Au core-shell nanostructure

(SACS). SACS was then interacted with a polyethylene glycol (PEG) matrix together with meso-5, 10, 15, 20-tetrakis (4-hydroxyphenyl) porphyrin to obtain water-soluble SPION-Au-porphyrin conjugate.

Results and Discussion

The as-synthesized SPIONs-Au-porphyrin conjugate exhibits red light absorption at 660 nm and photoluminescence ultraviolet light emission at 334 nm under 670 nm excitation (Fig. 1D). The result from the photoluminescence study suggests that the as-synthesized conjugate may carry out its therapeutic function synergistically via direct release of UV-light energy into tumour in addition to serving as PDT agent. FT-IR (Spectra not shown) reveals the conjugate is encapsulated within the polymer matrix rather than conforming to head-to-tail configuration.

The conjugate is highly attractive to external magnetic field (Fig. 2B) and exhibits singlet oxygen generation potential with 1, 3-diphenylisobenzofuran displaying singlet oxygen quantum yield of about 0.1 in 200 s which is sufficient for the eradication of breast cancer cells in vitro [15]. Furthermore, the conjugate shows

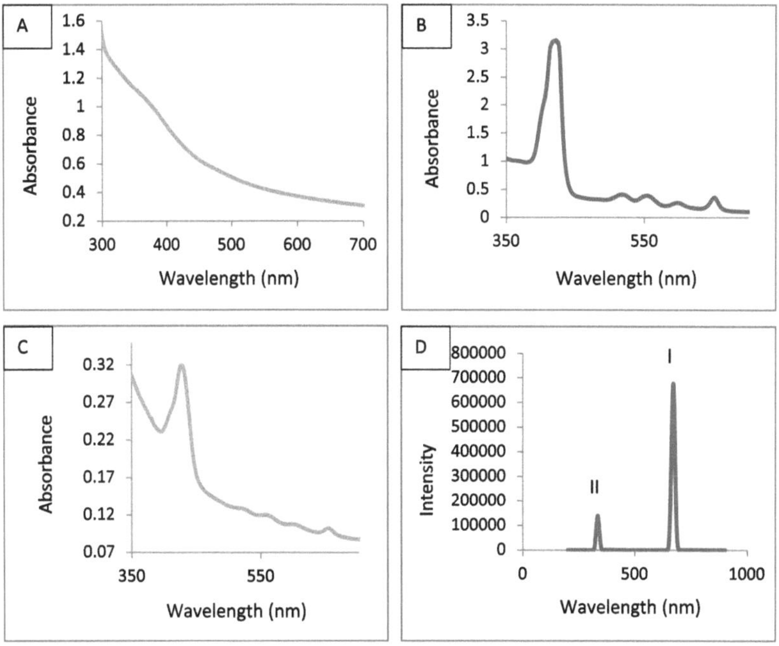

Fig. 1 Characterizations of the as-synthesized conjugate. UV-Vis: SPIONS-Au core-shell nanostructure (**A**), Porphyrin in MeOH (**B**), as-synthesized SPIONs-Au-porphyrin conjugate (**C**); photoluminescence spectra of the conjugate (**D**) (I = excitation peak at 670 nm, II = emission at 334 nm)

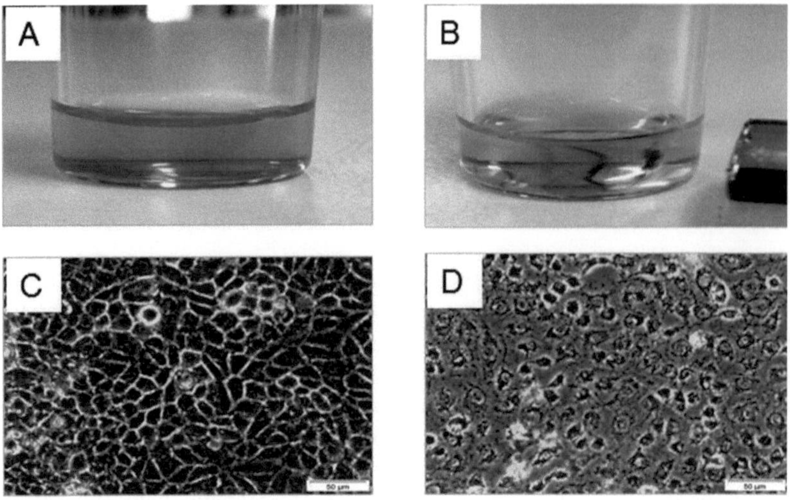

Fig. 2 Digital images of the as-synthesized SPIONs-Au-Porphyrin conjugate: in phosphate buffer saline solution (pH 7.4) in the absence (**A**) and presence (**B**) of external magnetic field; morphology of the cancer cells after incubation for 24 h with the conjugate (1 μg/mL) in the dark (**C**) and after light irradiation (**D**)

no dark cytotoxicity after incubation with breast cancer cells (MCF-7) for 24 h (in vitro). However, it becomes highly cytotoxic following exposure of the cells to 41 mW light at 681.5 nm for 40 min (Fig. 2D). This result shows that the conjugate exhibits high cancer cell uptake and retention even in the absence of external magnetic field. This may be assigned to the intrinsic enhanced permeability and retention effect of the PEG coating [15] on the conjugate. Thus, we hypothesized that applying external magnetic field to drive the as-synthesized SPIONs-Au-porphyrin conjugate into tumour will help in reducing porphyrin dose and time of incubation before light irradiation, thus enhancing PDT efficacy in terms of cost-effectiveness and therapeutic time. These effects are currently being evaluated.

Conclusion

Water-soluble SPIONs-Au-porphyrin conjugate has been synthesized, characterized and demonstrated to possess tumour-specificity property via attractive response to external magnetic field. Further work on the effect of the external magnetic field on the PDT efficacy of the conjugate on breast cancer cells is currently being exploited.

References

1. A.C. Silva, T.R. Oliveira, J.B. Mamani, S.M.F. Malheiros, L. Malavolta, L.F. Pavon, T.T. Sibov, E. Amaro, A. Tannús, E.L.G. Vidoto, M.J. Martins, R.S. Santos, L.F. Gamarra, Application of hyperthermia induced by superparamagnetic iron oxide nanoparticles in glioma treatment. Int. J. Nanomed. **6**, 591–603 (2011)
2. K. Vermaelen, G. Brusselle, Exposing a deadly alliance: Novel insights into the biological links between COPD and lung cancer. Pulm. Pharmacol. Ther. **26**, 544–554 (2013)
3. M. Arbyn, X. Castellsagué, S. de sanjosé, L. Bruni, M. Saraiya, F. Bray, J. Ferlay, Worldwide burden of cervical cancer in 2008. Ann. Oncol. **22**, 2675–2686 (2011)
4. M.H. Aziz, R. Kumar, N. Ahmad, Cancer chemoprevention by resveratrol: in vitro and in vivo studies and the underlying mechanisms (review). Int. J. Oncol. **23**, 17–28 (2003)
5. X. Zhao, Z. Chen, H. Zhao, D. Zhang, L. Tao, M. Lan, Multifunctional magnetic nanoparticles for simultaneous cancer near-infrared imaging and targeting photodynamic therapy. RSC Adv. **4**, 62153–62159 (2014)
6. C. Tang, P.J. Russell, R. Martiniello-Wilks, J.E.J. Rasko, A. Khatri, Concise review: Nanoparticles and cellular carriers-allies in cancer imaging and cellular gene therapy? Stem Cells **28**, 1686–1702 (2010)
7. S. Banfi, E. Caruso, S. Caprioli, L. Mazzagatti, G. Canti, R. Ravizza, M. Gariboldi, E. Monti, Photodynamic effects of porphyrin and chlorin photosensitizers in human colon adenocarcinoma cells. Bioorganic Med. Chem. **12**, 4853–4860 (2004)
8. B. Gao, Y. Liu, H. Yin, Y. Li, Q. Bai, L. Zhang, Water-soluble dendritic polyaspartic porphyrins: potential photosensitizers for photodynamic therapy. New J. Chem. **36**(1), 28 (2012)
9. M.B. Gariboldi, R. Ravizza, P. Baranyai, E. Caruso, S. Banfi, S. Meschini, E. Monti, Photodynamic effects of novel 5,15-diaryl-tetrapyrrole derivatives on human colon carcinoma cells. Bioorganic Med. Chem. **17**(5), 2009–2016 (2009)
10. M. Ethirajan, Y. Chen, P. Joshi, R.K. Pandey, The role of porphyrin chemistry in tumor imaging and photodynamic therapy. Chem. Soc. R ev. **40**, 340–362 (2011)
11. Z. Hu, O. Nancy, M.R. Hamblin, Photodynamic therapy as an emerging treatment modality for cancer and non-cancer diseases. J. Anal. Bioanal. Tech. 1–3 (2014)
12. M. Thandu, V. Rapozzi, L. Xodo, F. Albericio, C. Comuzzi, S. Cavalli, Clicking' porphyrins to magnetic nanoparticles for photodynamic therapy. ChemPlusChem **79**(1), 90–98 (2014)
13. J.R. Mccarthy, J. Bhaumik, R. Weissleder, High-yielding syntheses of hydrophilic conjugatable chlorins and bacteriochlorins. Org. Biomol. Chem. **7**, 3430–3436 (2009)
14. W. Lu, Y. Shen, A. Xie, W. Zhang, Green synthesis and characterization of superparamagnetic Fe_3O_4 nanoparticles. J. Magn. Magn. Mater. **322**(13), 1828–1833 (2010)
15. T. Jing, L. Fu, L. Liu, L. Yan, Polymer chemistry guided photodynamic therapy. Polym. Chem. **7**, 951–957 (2016)

Synthesis of AgNP's from Industrial Waste Using Electrochemical Techniques

J.A. Elizalde Mata, R.F. Islas Hernández, J.L. Navarro Jiménez, P.A. Ramírez Ortega, M.U. Flores Guerrero and L. García Hernández

Abstract In the present word it propose synthesize AgNP's from a industrial solution rich on Ag+ obtained from solid waste, using a typical three-electrode cell; a titanium rod as work electrode, a DSA as counter electrode and a (SCE) satured calomel electrode as reference electrode. The electrochemical techniques used for the synthesis were: ciclic voltamperometry, chronopotentiometry and galvanostatic pulses. The chronopotentiometries allowed to identify the current with greater stability in ion silver Reduction (–282.6 μA), which was applied to the galvanostatic pulses technique. The characterizations by SEM-EDS, AAS y XRD proved the presence of Ag nanoparticles, with sizes among 36 nm and quasi-spherical morphologies. This checks the feasibility of obtaining nanostructured materials from industrial waste for subsequent application in industry, contributing solid waste treatment to reduce and its impact on the environment.

Keywords Wastes · Silver · Electrochemistry · AgNP's

Introduction

The increase in world population has led to a sharp increase in industrial waste generating a problem of environmental pollution. Currently, electrochemical processes are used to recover metals contained in such wastes. Because of its low cost, which opens the possibility to silver (*Ag*) recovery, present in solid wastes, and thus use it as raw material for coating metals, which encourages new application areas [1, 2].

J.A. Elizalde Mata · R.F. Islas Hernández · J.L. Navarro Jiménez ·
P.A. Ramírez Ortega (✉) · M.U. Flores Guerrero · L. García Hernández
Centro de Desarrollo de Nanotecnología, Área de Electromecánica Industrial,
Universidad Tecnológica de Tulancingo, Camino a Ahuehuetitla #301 Col. Las Presas,
43642 Tulancingo, Hidalgo, Mexico
e-mail: pedro.ramirez@utec-tgo.edu.mx

The Minerals, Metals & Materials Society, *TMS 2017 146th Annual Meeting & Exhibition Supplemental Proceedings*, The Minerals, Metals & Materials Series,
DOI 10.1007/978-3-319-51493-2_4

The industrial wastes, e.g. radiographic films, photographic paper, electronic circuits, etc. It contains noble metals such as gold, platinum and silver. The noble metals have applications in medicine, textiles, personal cleanliness, environment, and others. They have a great impact because they possess the characteristics for different applications [3].

Is possible to use electrochemical techniques for the silver (*Ag*) recovery from industrial waste, like radiographic films [3, 4]. The most common radiographic film which is composed of a base on which adheres on both sides an emulsion, which contains silver halides. This emulsion is attached to the base by an adhesive layer and both emulsion layers are protected by a protective layer [5].

Experimental Methodology

Liquid Solution Preparation

A 230 ppm Ag^+ solution was prepared from a solid industrial waste, using HNO_3 2.5% v/v like lixiviant agent.

Electrochemistry Characterization

Electrochemistry System

All the experiments were realized in a typical three electrodes electrochemistry cell, which has a volume of 100 ml; as work electrode a titanium rod was used, as counter electrode a DSA type electrode and as reference electrode a Saturated Calomel Electrode. The work solutions were prepared using analytical grade reactive products and deionized water. All experiments were performed at atmospheric pressure and constant temperature of 25 °C.

Electrodes Preparation

To produce electrochemical responses titanium electrodes were used. 0.8 cm^2 of surface were used as the area of the electrodes, the electrode surface was polished with 0.3 mm AlO_3 powlder/water and rinsed with deionized water followed by ultrasonic bath for 15 min to remove traces of AlO_3 and grease and a final rinse with the deionized water.

Voltamperometric and Chronopotentiometric Studies

A Cyclic Voltamperometry with 100 ml of industrial solution rich in Ag^{+} were realized. Within a cell with a capacity of 100 ml with an experimental setup of two electrodes with an one and a half inches gap. The study window was –0.6 mV versus SCE 1 with a scan rate of 25 mVseg-1.

The chronopotentiometric studies were conducted at different reduction currents obtained from voltammetric study with a duration of 180 min, taking samples of 1 ml every 20 min.

The electrochemical studies were performed out using a potentiostat galvanostat; Brand Princeton Applied Research; Model Versa Stat 4, with capacity of 1 A.

Synthesis of Nanoparticles by the Galvanostatic Pulses Technique

Galvanostatic pulses technique was used with an experimental setup of two electrodes (Fig. 1), as working electrode was a titanium rod (0.8 cm^2 of geometric area) and as counter electrode was a DSA rod. The cell capacity was 100 ml which was immersed in an ultrasonic bath. Galvanostatic pulses were performed for 30 s to 150 cycles, with a time of 7×10^{-5} s between a sonic pulse and 42 kHz of sonic frequency.

Characterization

- The concentration of silver present in the solution was evaluated by atomic absorption spectroscopy.
- The deposits obtained on Ti electrode were characterized by SEM-EDS and XRD.
- The solutions used in the galvanostatic pulse technique were characterized by UV—Visible Spectroscopy.

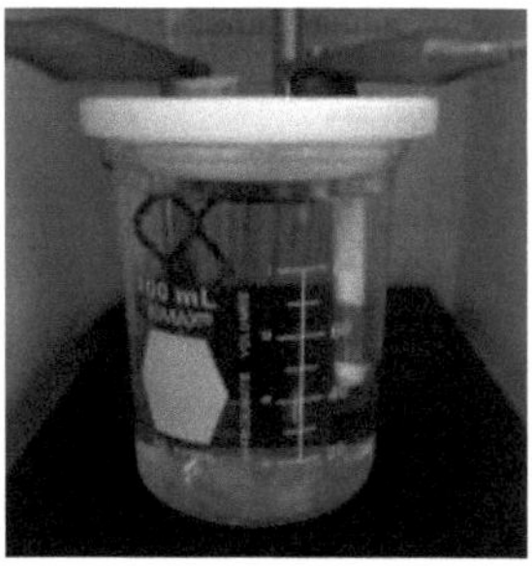

Fig. 1 Typical three-electrode cell, the *left side* the counter electrode DSA and the *right side Ti* working electrode submerged within a CELCA with capacity of 100 ml

Discussion and Results

Figure 2a shows the cyclic voltammetry for free of silver solution (HNO3 2.5% *v/v*). In Fig. 2b voltammogram of industrial solution rich Ag^+ is shown, where two reduction processes on Ti electrode were appreciated, the first process in the currents range of –62.6 to –282.6 μA (333,083–227,279 mV vs. SCE), which is attributed to the reduction of silver ion species on *Ti* electrode, while the second reduction process begins in μA –248.0 (–0.4 mV vs. SCE) which is attributed to the reduction of the medium. In the reverse scan a process it is observed in the range of 261.62 to 339 332 mV potential versus SCE what attributed to oxidize the silver deposited on the forward scan.

From the results of voltammetric study (Fig. 2b) eight different currents corresponding to the process of reduction of silver were selected, which will be used in the chronopotentiometric studies.

In Fig. 3 it is observed the galvanostatic transistors obtained on *Ti* electrode in the currents range of –62.2 to –282.6 μA. it can be seen that as the current takes more cathodic values, the potential becames more negative for all the time. Moreover, all ranges of current have the same slope, indicating that the electroactive spice of Ag has not been finished, due to high concentration in the solution.

In Fig. 4 shows the variation of the Ag normalized concentration respect to the initial concentration C(t)/C(0) function of electrolysis time when is imposed a current of: –62.6, –217.6 y –282.6 μA on *Ti* electrode during electrolysis time of 180 min. It is observed that the variation of the normalized concentration of Ag^+ diminish respect the time of electrolysis, for the three values of current imposed on *Ti* electrode. Also, it is observed that as the current takes more cathodic values, the reduction of *Ag* increases, as it presents a greater decrease in the concentration of the species Ag^+ presents in the solution, in the current of –282.6 μA the ionic silver species was reduced almost entirely, obtaining a 97% recovery on the *Ti* electrode.

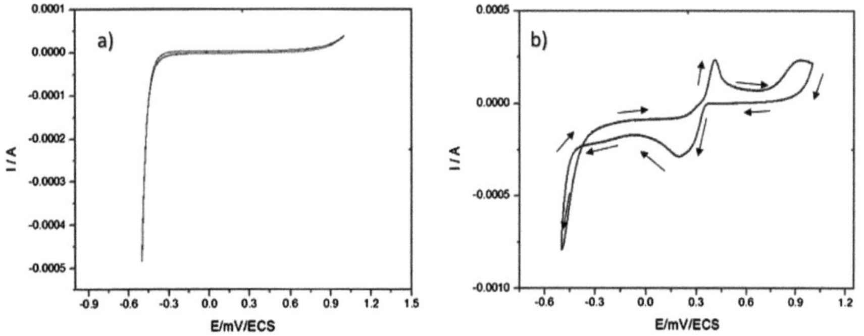

Fig. 2 Voltammograms obtained on the Ti electrode (0.8 cm^2) at a scan rate of 25 $mVseg^{-1}$. **a** Solution free of *Ag* ions and **b** industrial solution with *Ag* (293 ppm)

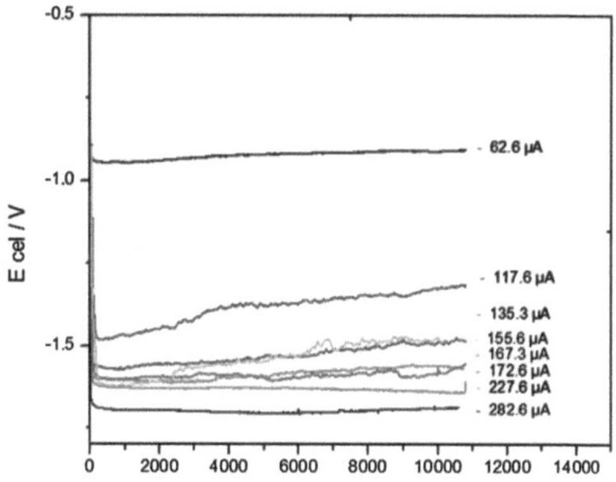

Fig. 3 Chronopotentiograms obtained of the reduction of the ion Ag^{+} on the *Ti* electrode, in function of the macroelectrolysis time with impoused currents in a range of –62.6 to –282.6 μA (geometric area 0.8 cm^2)

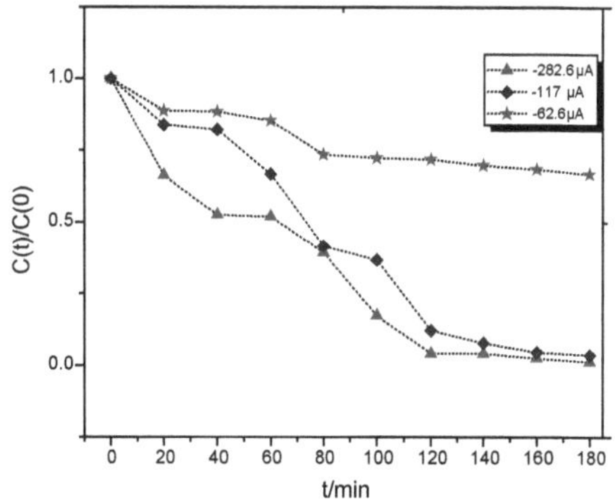

Fig. 4 Variation of the *Ag* normalized concentrations from industrial wastes solution, in function of the macroelectrolysis time with imposed currents in a range of –282.6 to–62.6 μA, on a *Ti* electrode (geometric area 0.8 cm^2)

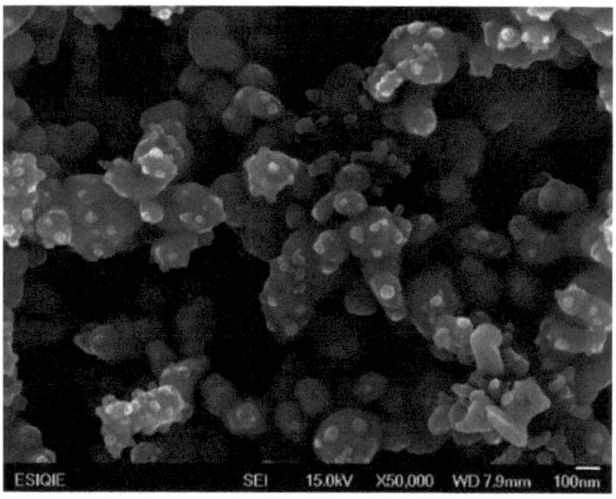

Fig. 5 X50, 000 micrograph obtained of *Ti* electrode after electrochemical time of 180 min using a current of –282.6 μA. (Made on a SEM brand Jeol; JSM-6701F model)

In order to know the composition and morphology of the deposits obtained on *Ti* electrode by imposing a current of –282.6 μA, was performed a characterization by SEM-EDS. In the Fig. 5 shown the micrographs by SEM, is observed the formation of agglomerates of nanoparticles with sizes between 40 and 60 nm with a spherical morphology, which are agglomerated into larger particles.

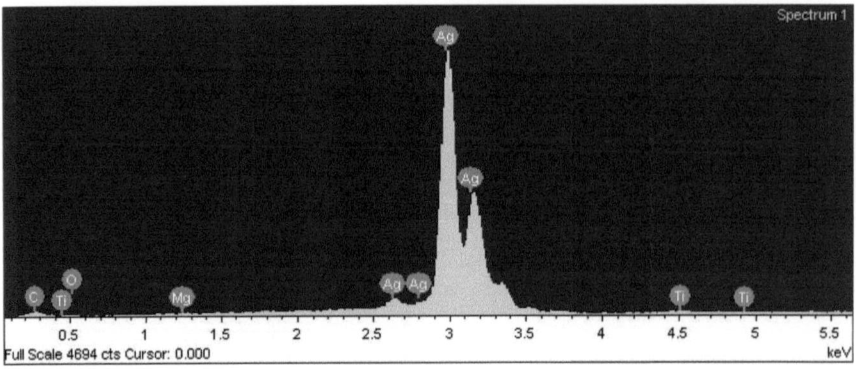

Fig. 6 EDS spectrum obtained from the deposit on the *Ti* electrode, using a current of –282.6 μA for 180 min. (Made on a SEM brand Jeol; JSM-6701F model)

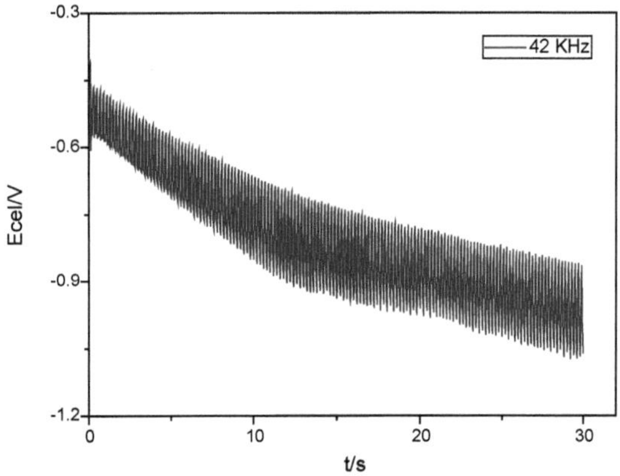

Fig. 7 Galvanostatic pulses, in a current of μA –282.6 with 150 cycles for 30 s. A sonication constant frequency of 42 kHz was used

For know the chemical composition of the deposit was made an EDS (Fig. 6). The results indicate that the deposits obtained on *Ti* electrode is silver.

Galvanostatic pulses technique was used for the synthesis of *AgNP's*, a constant current of –282.6 μA was imposed, for 30 s, to 150 cycles, with a time of separation between 7×10^{-5} s pulse and a sonic frequency of 42 kHz. The results of the study are shown in Fig. 7.

Theoretically the ultrasound helps the dispersion of the particles to be deposited on the electrode, helping the formation of nanoparticles and avoiding agglomerates. Based on this idea it has hypothesized that some nanoparticles deposited on the *Ti* electrode break off and remain immerses in the solution resulting in a *AgNP's* colloidal solution, to check that exist nanometric silver particles in the solution,

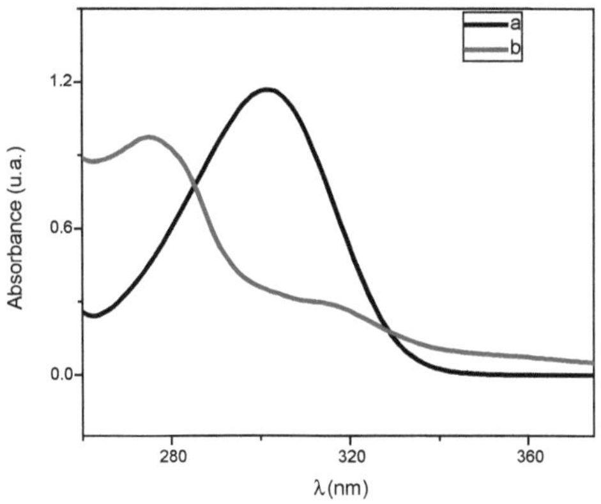

Fig. 8 Spectrum obtained by UV-Visible of the solutions used in the galvanostatic pulse technique with a sonic frequency of 42 kHz. **a** Free Ag ions industrial solution and **b** industrial solution with *Ag* (293 ppm)

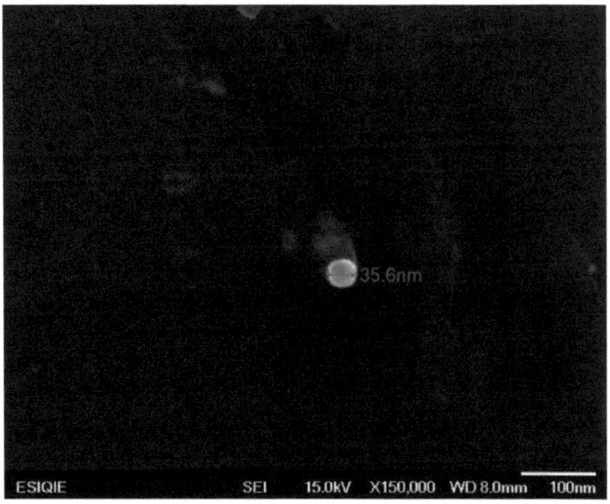

Fig. 9 Micrograph of nanoparticles deposit on the *Ti* electrode, taken with a SEM brand Jeol; JSM-6701F model

characterization was carried by an ultraviolet visible spectroscopy hope to observe the *AgNP's* typical plasmon. In Fig. 8 shown the UV-Vis spectroscopy made to the solution, is present an absorbance curve between 270 and 280 nm, this plasmon can be attributed to the presence of nanoparticles with very small sizes.

In Fig. 9 the micrograph of deposit by Galvanostatic pulses synthesis by at a sonication frequency of 42 kHz, where nanoscale particles are observed and spherical morphology of about 36 nm sizes is presented. It was found that these nanoparticles are *Ag* by EDS (Fig. 10).

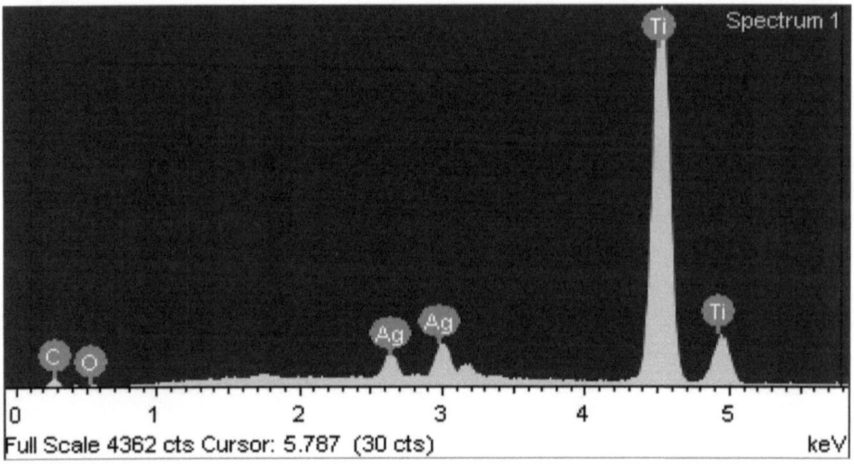

Fig. 10 EDS of the deposited nanoparticles on the *Ti* electrode. Made with SEM brand Jeol; JSM-6701F model

Conclusions

- The voltammetric studies indicate that reduction processes on *Ti* electrode ranging from –62. 6 μA to –282. 6 μA.
- Chronopotentiometry studies, helped identify where most current recovery of silver on the *Ti* electrode is generated (–282.6 μA).
- The evaluation of the change in concentration of the solution rich in silver ions, for AAS, it was determined that a current of –282.6 μA have a 97% reduction of silver over 180 min.
- The micrographs taken by SEM, allowed appreciate the morphology of the deposit obtained on *Ti* electrode by chronopotentiometry technique, using current –282. 6 μA which generated higher deposit on the electrode.
- Characterization by UV-Visible, indicates that there are changes in the solution, obtaining a different plasmon absorbance at the control sample, which can be attributed to the presence of *AgNP's* colloidal in solution.
- It was found that the synthesis of *AgNP's*, by galvanostatic pulses technique with –282 μA current is viable, the micrographs obtained by SEM show obtaining particles with sizes around 36 nm with spherical morphology.
- The EDS analysis, found that deposits by synthesizing *AgNP's* on Ti electrode were formed from *Ag*.

Acknowledgements The authors externalize their gratitude to the Nanotechnology Development Center of Technological University of Tulancingo, Hidalgo Mexico for permission to use the facilities and equipment to carry out this work.

References

1. M.T. Oropeza, *Alternativas a la Recuperación no Contaminante de plata: Disolución y Formación de Electrodepósitos* (Electrocristalización), (Tesis de Maestría, UAM-I, México, D.F., 1991)
2. O.P. Ramírez, *A, Estudio electroquímico preliminar para depositar Ag proveniente de los desechos sólidos de la industria fotográfica y radiográfica* (Tesis de Licenciatura, UAEH, Hidalgo, México, 2006)
3. K. Guenter, Silver Recycling from photographic solutions: review of the methods. Chem. Labor Betr. **32**, 40–48 (1981)
4. A. Alan Melo López, Estudios electroquímicos preliminares en un rector tipo prensa para la recuperación de Ag proveniente de los efluentes de la industria Fotográfica y Radiográfica, (Tesis de Licenciatura, UAEH, México, 2006)
5. Environment Information from Kodak, *The Technology of Silver Recovery for Photographic Processing Facilities*. Eastman Kodak Company, referencia J-212 (1999)

Synthesis of Mn_2O_3 Nanopowders with Urea and Citric Acid by Solution Combustion Route

Esma Yilmaz, M. Seref Sonmez, Bora Derin, Filiz Cinar Sahin and Onuralp Yucel

Abstract In this study, synthesis of Mn_2O_3 powders by solution combustion route was carried out. Urea (CH_4N_2O) and citric acid ($C_6H_8O_7.H_2O$) was used as a fuel material in solution combustion synthesis, while manganese nitrate (Mn $(NO_3)_2.4H_2O$) was used as an oxidant. Surface area of Mn_2O_3 synthesised by urea was calculated as 19,37 m^2/g and 25,07 m^2/g after the calcination at 400 °C and 500 °C, respectively. Surface area of Mn_2O_3 synthesised by citric acid was calculated as 63,77 m^2/g when fuel/oxidizer was 1/1, ignition temperature of 250 °C and calcination temperature at 500 °C. Spherical particles of Mn_2O_3 having average particle size of 50–60 nm was synthesised by using urea, while foam like porous structure of Mn_2O_3 was obtained when citric acid was used as a fuel in solution combustion synthesis. Pore diameters between 27 and 55 nm was determined for this powders.

Keywords Solution combustion synthesis · Mn_2O_3 · Nanopowders · Nanopores

E. Yilmaz (✉) · M.S. Sonmez (✉) · B. Derin · F.C. Sahin · O. Yucel
Faculty of Chemistry & Metallurgy, Department of Metallurgical & Materials Engineering, Istanbul Technical University, Istanbul, Turkey
e-mail: esmayilmazz@gmail.com

M.S. Sonmez
e-mail: ssonmez@itu.edu.tr

B. Derin
e-mail: bderin@itu.edu.tr

F.C. Sahin
e-mail: cinar@itu.edu.tr

O. Yucel
e-mail: yucel@itu.edu.tr

The Minerals, Metals & Materials Society, *TMS 2017 146th Annual Meeting & Exhibition Supplemental Proceedings*, The Minerals, Metals & Materials Series,
DOI 10.1007/978-3-319-51493-2_5

Introduction

Solution combustion synthesis is a simple and reproducible method used to obtain several types of ceramic oxides for a variety of applications [1–3]. It is well recognized that combustion-based techniques, known in the literature as combustion synthesis (CS) or self-propagating high-temperature synthesis (SHS) are an effective energy saving method for the synthesis of a variety of advanced materials [4]. A combination of CS and reactive solution approaches leads to solution combustion (SC) synthesis methods [1]. Typically, SC involves a self-sustained reaction in solutions of metal nitrites and different fuels, which can be classified based on their chemical structure, i.e. the type of reactive groups (e.g. amino, hydroxyl, and carboxyl) bonded to the hydrocarbon chain [5]. The reaction between fuel and oxygen containing species, formed during the decomposition of the nitrite species, provides high-temperature rapid interaction. In a typical scheme, an initial liquid solution of desired reagents, after preheating to a moderate temperature (150–200 °C), self-ignites along the whole volume (VCS mode) leading to the formation of fine solid products with tailored composition. Thus it can be concluded that SC is an effective method for synthesis of nano scale materials and it has been used for the production of a variety (currently more than 1000) of fine complex oxide powders for different advanced applications including catalysts, fuel cells, and biotechnology [2, 3, 6, 7].

Amongst many transition metal oxides being investigated for scientific applications, manganese oxides emerge as potential candidates for many niche applications such as super capacitors, alkaline and rechargeable batteries, sensing devices, catalysts, waste water treatment, in terms of cost and environmentally benign qualities. When the chemical valence states of manganese oxide are concerned, it possesses different oxidation states (2^+, 3^+, 4^+, and 6^+) [8]. As far as manganese oxide phases are concerned, the most common phases of manganese oxide in air at different temperatures are shown below. That is, the following thermal oxidation reactions are possible in manganese oxides [9]. It is well known that manganese exhibits thermochemical changes as shown below:

$$MnO_2 \overset{500^\circ C}{\longrightarrow} Mn_2O_3 \overset{900^\circ C}{\longrightarrow} Mn_3O_4 \overset{1700^\circ C}{\longrightarrow} MnO \qquad (1)$$

Nano-porous materials as an important class of nanostructured materials possess high specific surface area, large pore volume, uniform pore size and rich surface chemistry. These materials present great promises and opportunities for a new generation of functional materials with improved and tailorable properties such as adsorption, membranes, sensors, energy storage, catalysis, etc. Generally porous materials have porosity (volume ratio of pore space to the total volume of the material) between 0.2–0.95. Nanomaterials are subset of porous materials, typically having large porosities (greater than 0.4), and pore diameters between 1–100 nm [10, 11]. It was reported that there are many strategies for the synthesis of metal oxide nanomaterials: for example, sol–gel, chemical co-precipitation, chemical vapor deposition, mechanical alloying, solvothermal synthesis, microwave heating, and reverse micelle synthesis [11–14]. The most obvious advantage of solution combustion synthesis among these techniques is its time- and energy-efficiency. Once the mixture of reagents is ignited, the high self-generated energy can convert precursors into the corresponding oxides without the requirement of an additional external energy input. Hence, the products are formed in several minutes [11].

In this study synthesis of manganese oxide by solution combustion synthesis (SCS) for CO_2 capture was carried out. Urea and citric acid was used as a fuel material in SCS experiments, while manganese nitrate was used as an oxidant. The use of citric acid as a fuel was the originality of the study as there have been no any work carried out before. Several properties of Mn_2O_3 powders were compared synthesised by urea and citric acid as a fuel, respectively.

Experimental

Precursor materials which were manganese nitrate tetrahydrate ($Mn(NO_3)_2.4H_2O$, MERCK, $\geq$ 98.5%) as oxidizer material and citric acid monohydrate ($C_6H_8O_7.H_2O$, MERCK) and urea (CH_4N_2O, MERCK, $\geq$ 99.5%) as fuel materials mixed in various stoichiometric ratios. Fuel to oxidizer ratio, φ, was calculated from following equation [15]:

$$\varphi = \frac{|total\,valency\,of\,fuel|}{|total\,valency\,of\,oxidizer|} \tag{2}$$

Two different φ values, 1:1 and 1:1.5 of CH_4N_2O and $Mn(NO_3)_2.4H_2O$ was used for the solution combustion synthesis of manganese oxide by using urea, while $C_6H_8O_7.H_2O$ and $Mn(NO_3)_2.4H_2O$ raito was kept constant as 1:1 when the citric acid was used as fuel. 50 ml of distilled water was added to all mixtures. Solution combustion synthesis was carried out in an muffle furnace for the solution mixtures prepared by the addition of urea at 350 °C. The ignition temperature was 250 °C for citric acid containing specimens which was realized on hot plate. After the combustion all samples were washed, filtrated and dried at 105 °C for a day, respectively. Thermogravimetric analysis (PERKIN ELMER Diamond TGA) were carried out to the samples to determine calcination conditions and it was experimented for three different temperatures (400, 500 and 600 °C) for 5 h. X-ray diffraction analysis of powders were done by X'Pert Philips PW3040 diffractometer using Cu $K_{\alpha 1}$ radiation (2θ range = 20–80°; step = 0.02°). The specific surface area (SBET) of powders were measured by NOVA 2200e that was previously out gassed to remove water and other atmospheric contaminants. Morphology and the chemical composition of the samples were investigated by the scanning electron microscope (JEOL JIB-4501 Multi Beam Focused Ion Beam—Scanning Electron Microscope) and EDS module.

Results and Discussion

Thermogravimetric analysis performed to two different samples prepared by using urea and citric acid as a fuel material were given in Figs. 1 and 2, respectively.

As seen from the Fig. 1, endothermic peak from at about 150 °C referred to water loss and an endothermic peak at near 500 °C reflected the reduction of MnO_2 to Mn_2O_3 while other endothermic peak at about 780 °C referred to the reduction of Mn_2O_3 to Mn_3O_4. Endothermic curve at the beginning of the reaction referred as vaporization of water in Fig. 2. Reduction of MnO_2 to Mn_2O_3 and immense weight

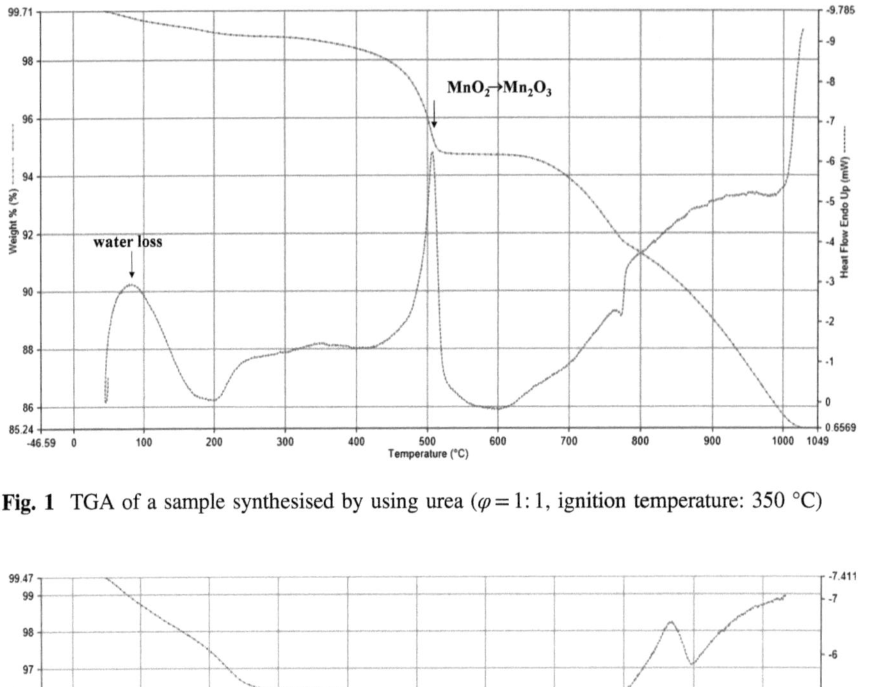

Fig. 1 TGA of a sample synthesised by using urea ($\varphi = 1{:}1$, ignition temperature: 350 °C)

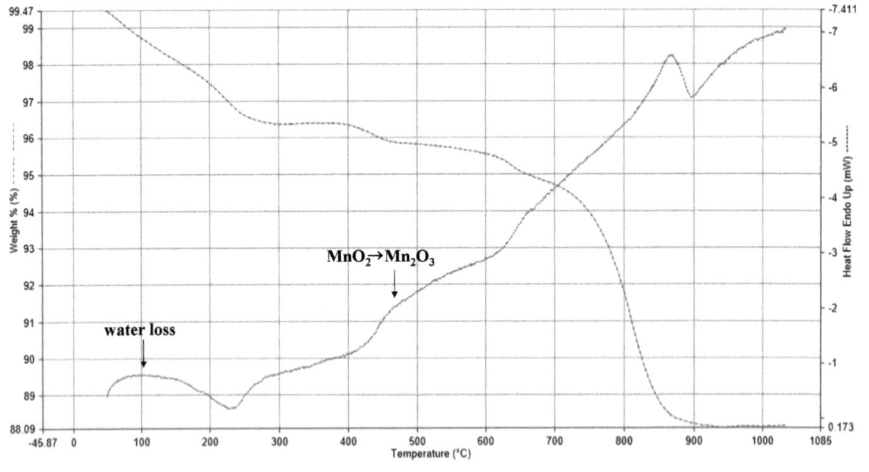

Fig. 2 TGA of a sample synthesised by using citric acid ($\varphi = 1{:}1$, ignition temperature: 250 °C)

loss via reduction takes place at about 500 °C, while endothermic peak seen about 860 °C referred to reduction of Mn_2O_3 to Mn_3O_4. It was assumed that the difference in thermogravimetric behaviour of two samples was due to the different chemical composition of powders obtained after the solution combustion synthesis by using to different fuel materials, urea and citric acid, respectively.

After the calcination experiments that were carried out at temperatures ranging from 400 to 600 °C, x-ray diffraction analysis was performed for the selected powders. It can be seen from the Figs. 3 and 4 that obtained powders are Mn_2O_3 for samples synthesised by using urea and citric acid. Furthermore, specific surface of these two samples calculated as 25.07 m^2/g and 63.77 m^2/g, respectively.

To obtain information about the morphology of the powders obtained by using both urea and citric acid as a fuel material were analysed by SEM (Figs. 5 and 6).

It was seen that the different types of Mn_2O_3 structure was obtained after the solution combustion synthesis and the following calcination process for two different samples. As seen in Fig. 5 spherical particles having between 50 and 70 nm in size was obtained when urea was used as a fuel material during the solution

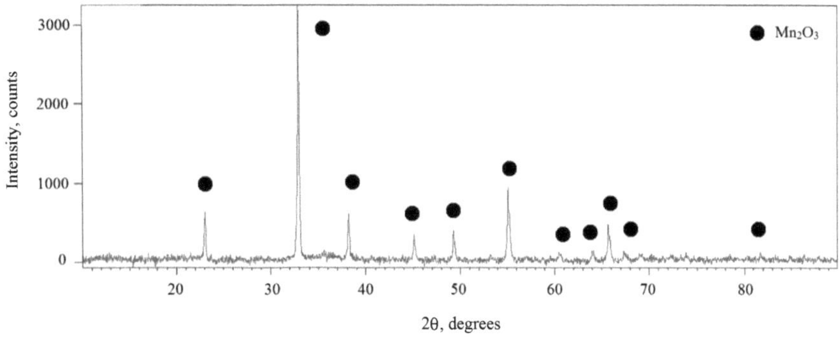

Fig. 3 XRD analysis of powder synthesised by using urea ($\varphi = 1{:}\,1$, calcination temperature: 600 °C)

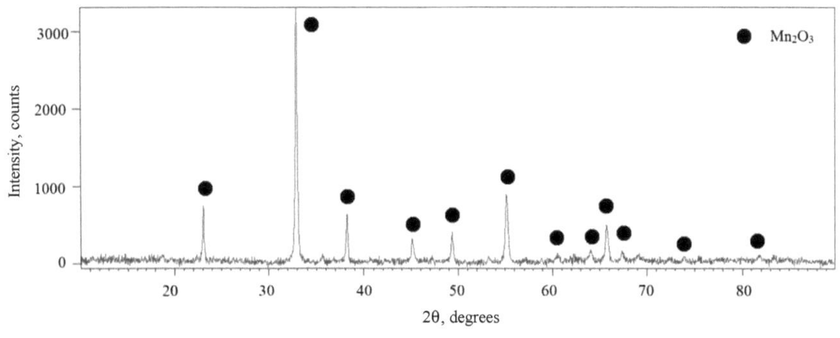

Fig. 4 XRD analysis of powder synthesised by using citric acid ($\varphi = 1{:}\,1$, calcination temperature: 600 °C)

Fig. 5 SEM images of Mn_2O_3 powders synthesised by using urea

Fig. 6 SEM images of Mn_2O_3 powders synthesised by using citric acid

combustion synthesis. On the other hand, foam like porous structure was obtained from the powders synthesised by using citric acid in which minimum pore size was calculated as 27 nm as shown in Fig. 6. According to the EDS analysis of the two powders synthesised by urea and citric acid Mn content was calculated as 73.62% and 61.26%, respectively.

Conclusion

Higher crystallization was obtained when lower amount of fuel was used (fuel/oxidizer = 1/1) for the synthesis of Mn_2O_3 by using urea as a fuel. There is no significant effect of preheating of solution on the final product synthesised by citric acid and manganese nitrate. On the other hand, the reaction time decreases with pre-heating, while reaction time of SCS has no significant effect on the final product. Increasing calcination temperature, decreases crystallization of Mn_2O_3 synthesised by citric acid and manganese nitrate. Surface area of Mn_2O_3 synthesised by urea was calculated as 19,37 m^2/g and 25,07 m^2/g when fuel/oxidizer was 1/1 (calcined at 400 °C) and fuel/oxidizer was (calcined at 500 °C), respectively. Surface area of Mn_2O_3 synthesised by citric acid was calculated as 63,77 m^2/g when fuel/oxidizer was 1/1, ignition temperature of 250 °C and calcined at 500 °C. Spherical particles of Mn_2O_3 having average particle size of 50–60 nm was synthesised by using urea, while foam like porous structure of Mn_2O_3 was obtained when citric acid was used as a fuel in SCS. This structure was determined as a candidate for being CO_2 capture material. Minimum 27 nm pore diameter was calculated from SEM images. Cyclic gas absorption tests will be carried out to these samples to be able to determine CO_2 capture ability.

Acknowledgements The authors would like to thank Anil Uzer, Deniz Albayrak, Oguz Kaan Demet, Sinan Ozdemir, and, Sinan Yalcin for their efforts during the design project in the Department of Metallurgical & Materials Engineering in ITU.

References

1. K.C. Patil, S.T. Aruna, T. Mimani, Combustion synthesis: an update. Curr. Opin. Solid State Mater. Sci. **6**, 507–512 (2002)
2. S.T. Aruna, A.S. Mukasyan, Combustion synthesis and nanomaterials. Curr. Opin. Solid State Mater. Sci. **12**, 44–50 (2008)
3. D.P. Tarrago, C.F. Malfatti, V.C. Sousa, Influence of fuel on morphology of LSM powders obtained by solution combustion synthesis. Powder Technol. **269**, 481–487 (2015)
4. A. Varma, A.S. Mukasyan, Combustion synthesis of advanced materials: fundamentals and applications. Korean J. Chem. Eng. **21**(2), 527–536 (2004)
5. A. Varma, A.S. Mukasyan, K. Deshpande, P. Pranda, P. Erii, Combustion synthesis of nanoscale oxide powders: mechanism, characterization and properties. Mat. Res. Soc. Symp. Proc. **800**, 113–124 (2003)
6. T. Mimani, K.C. Patil, Solution combustion synthesis of nanoscale oxides and their composites. Mater. Phys. Mech. **4**, 134–137 (2001)
7. A.S. Mukasyan, P. Dinka, Novel approaches to solution-combustion synthesis of nanomaterials. Int. J. Self Propag. High Temp. Synth. **16**(1), 23–35 (2007)
8. T. Nathan, M. Cloke, S.R.S. Prabaharan, Electrode properties of Mn_2O_3 nanospheres synthesized by combined sonochemical/ solvothermal method for use in electrochemical capacitors. J. Nanomater., 1–8 (2008)
9. L. Xiaodi, C. Chen, Y. Zhao, B. Jia, A review on the synthesis of manganese oxide nanomaterials and their applications on lithium-ion batteries. J. Nanomater., 1–7 (2013)
10. G.Q. Lu and X.S. Zhao, *Nanoporous Materials: Science and Engineering*, Series on Chemical Engineering. vol. 4 (College Press, London Imperial, 2004), pp. 1–4
11. F. Li, J. Ran, M. Jaroniec, S.Z. Qiao, Solution combustion synthesis of metal oxide nanomaterials for energy storage and conversion. Nanoscale **7**, 17590–17610 (2015)
12. J.A. Rodríguez, M. Fernández-García, *Synthesis, Properties, And Applications Of Oxide Nanomaterials* (JohnWiley & Sons Inc, Hoboken, New Jersey, 2007)

13. T. Guo, M.S. Yao, Y.H. Lin, C.W. Nan, A comprehensive review on synthesis methods for transition-metal oxide nanostructures. CrystEngComm **17**, 3551–35585 (2015)
14. D. Bekermann, D. Barreca, A. Gasparotto, C. Maccato, Multi-component oxide nanosystems by chemical vapor deposition and related routes: challenges and perspectives. CrystEngComm **14**, 6347–6358 (2012)
15. S.R. Jain, K.C. Adiga, V.R.P. Verneker, A new approach to thermochemical calculations of condensed fuel-oxidizer mixtures. Combust. Flame **40**, 71–79 (1981)

Part II
Additive Manufacturing of Metals: Establishing Location-Specific Processing-Microstructure-Property Relationships

Evolution of Aluminum Alloys Structure at Production Phases of 3D Products by Methods of Additive Technologies

Victor Mann, Aleksandr Krokhin, Aleksandr Alabin, Sergey Zmanovskiy, Valentin Konkevich and Ivan Redkin

Abstract The structure and phase structure of an aluminum alloy type Al10SiMg, received in the conditions of classical molding castings and the 3D printing by selective laser melting is investigated. Consistent patterns of redistribution of the alloying elements in structure in the course of powder production, the 3D printing and the subsequent heat treatment are determined. The analysis of mechanical properties level of the samples received on 3D technology in comparison with castings is carried out. On the basis of the received regularities, the new compositions on the basis of aluminum, providing a combination of high level of mechanical and physical characteristics for 3D products are offered. Purpose of new materials for 3D printing—critical parts operating under high loads for application in transport products, in particular the automotive and aircraft industry.

Keywords Additive technology · Aluminium alloys · Spherical powders · Alsi10 mg · Selective laser melting

Introduction

Considering the development of production of construction materials, it can be noted that the development of additive technology has touched the most of materials such as titanium alloys, heat-resistant nickel alloys, alloyed steels. With regard to, then they pay less attention as the cost of parts made of aluminum alloys is relatively low, and the cost of additive production transfer it to another price category. At first glance, it is quite difficult to find parts made of aluminum alloys, production of which would justify the cost of the additive production in the event, if

V. Mann · A. Krokhin · A. Alabin · S. Zmanovskiy · I. Redkin (✉)
Technical Directorate, UC RUSAL, Moscow, Russian Federation
e-mail: ivan.redkin@rusal.com

V. Konkevich
Moscow Aviation Institute (National Research University), Moscow, Russian Federation

The Minerals, Metals & Materials Society, *TMS 2017 146th Annual Meeting & Exhibition Supplemental Proceedings*, The Minerals, Metals & Materials Series, DOI 10.1007/978-3-319-51493-2_6

it had not been able to achieve a significant increase in their physical and mechanical properties compared to conventional technology.

Among cast aluminum alloys most commonly used in industry alloys based on Al–Si system. These alloys have relatively low strength, but practically feasible and inexpensive to manufacture.

A comparative study of the properties of parts made of alloys based on Al–Si system obtained by additive technology and conventional casting techniques is of interest in that the additive technology, specifically, for example, for SLM technology, characterized by passing the process under rapid cooling rate during crystallization. These conditions are extremely favorable precisely for aluminum alloys, as they contribute to increase the solubility in solid solution (formation of supersaturated solid solution), milled structure change morphology excess phases and the formation of metastable phases. Accordingly, the effects of rapid crystallization should help to improve the properties silumin parts of—refinement of eutectic Si, it's sferoidezation, and etc. Therefore, this article is regarded to structure and properties of AlSi10 Mg powders, as well as samples, obtained by 3D printing technology. This alloy is an analogue of the traditional A356 alloy (content of Si slightly lower, ~ 7%). A characteristic feature of these alloys is an additional alloying with small (< 1%) amount of magnesium (Mg). As a result alloy becomes thermostrengthened and after quenching and artificial aging significantly increases durability due to allocation of the Mg_2Si phase.

In article benefits of additive technology, feature of the powders choice for SLM, properties of 3D details, types of defects, regularities of properties change are considered in case of heat treatment of details, etc.

Research of Powders of Alsi10 Mg Alloy for Production of 3D Details

For a research powder samples from aluminum alloy with markings ASP-5AlSi10 Mg, ASP-15AlSi10 Mg, ASP-22AlSi10 Mg, ASP-27AlSi10 Mg have been used.

The conducted researches of ASP showed that the average size of particles of the provided powders made of aluminum alloy:

ASP-5AlSi10 Mg—4 ± 1.4 μ;
ASP-15AlSi10 Mg—11 ± 4.5 μ;
ASP-22AlSi10 Mg—20 ± 7.4 μ;
ASP-27AlSi10 Mg—24 ± 9 μ.

The structural analysis has shown that a form of particles mainly spherical. In powders of large fractions the non-uniform particle size distribution which is characterized by existence of a large number of particles of small fractions against the background of large particles (Fig. 1) is observed. With increase in the size of

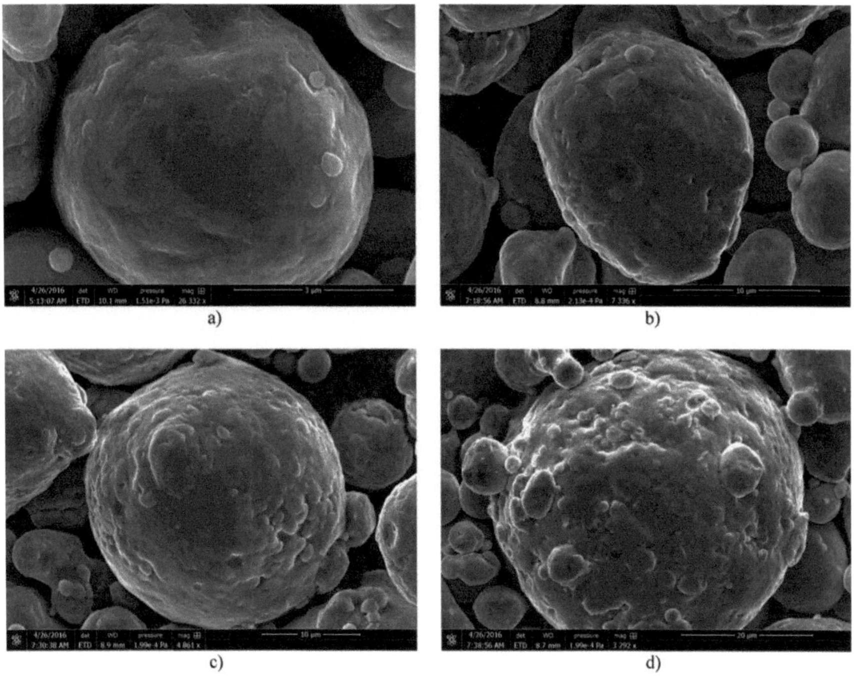

Fig. 1 The characteristic shape of the powder and the topography of the surface of particles: **a** ASP-5AlSi10 Mg; **b** ASP-15AlSi10 Mg; **c** ASP-22AlSi10 Mg, **d** ASP-27AlSi10 Mg

particles their surface becomes more friable and rough, outgrowths appear. The topography of a surface of particles was investigated with use of the ET detector in the mode of secondary electrons. It, obviously, can be explained with a small superficial tension of alloy (about reduction of a superficial tension at introduction of silicon and magnesium it has been told in work [1]).

The chemical composition of ASP (Fig. 2) defined by an electron microscope using energy dispersive spectrometer (EDS-analysis) showed that the composition has a dispersion particle concentrations for silicon and magnesium: 7–14% Si; 0.4–0.7% Mg. However, you must take into account the relative accuracy (a few percent) of the reference-free method for determining the elemental composition of materials.

The study of the powder structure was conducted on a scanning electron microscope FEI Quanta FEG 650 (SEM) with energy dispersive spectrometer X-MaxN SDD (EDS). The microstructure of powder particles is dispersed. The main structural components are the aluminum dendrites and eutectic solid solution (Al) + (Si) (Fig. 1b–d). Research of a microstructure of samples, were carried out on an optical microscope of OLYMPUS GX71 (OM). The microstructure of

a)

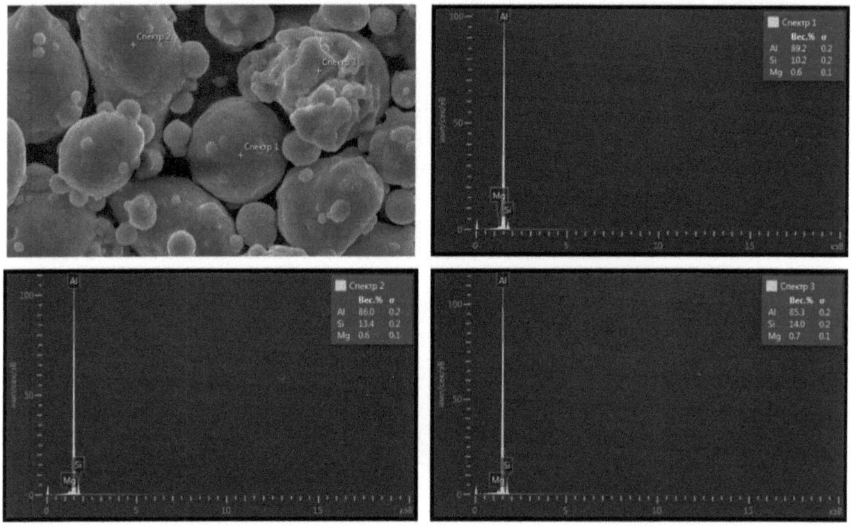

b)

Fig. 2 EDS-analysis of powder samples: **a** ASP-5 AlSi10 Mg; **b** ASP-27AlSi10 Mg

ASP-5AlSi10 Mg powder particles is not detected by methods of optical microscopy at x1000 increase (Fig. 1a).

The analysis of structure of powders allows to draw a conclusion that in case of their production rather high speed of chilling in case of crystallization was reached. Currently, it is generally accepted equation:

$$d = av^{-n}$$

where: d—dendrit parameter, microns; a—cooling rate, K/sec; v and n—constants, depending on coefficient of diffusion, concentration of a component, distribution coefficient at crystallization, etc. [2] (Fig. 3).

According to work [3] for particles with small section cooling speed at crystallization can approximately be determined in size of dendrit parameter:

$$v = \frac{10^5}{d^3},$$

Considering that the size of dendrit parameter of ASP-15AlSi10 Mg, ASP-22AlSi10Mg, ASP-27AlSi10Mg changes from about 1 to 4 microns, it is possible to draw a conclusion that cooling speed at crystallization approximately for these three consignments of powder can be estimated as $1 \times 10^4 \ldots 1 \times 10^5$ K/sec. In the conditions of fast cooling, crystallization of powder goes according to the metastable chart of a state that allows to receive, for example (Fig. 4), alloys of the eutectic type at 17–19% of Si for the speed of 1×10^3–1×10^4 K/s.

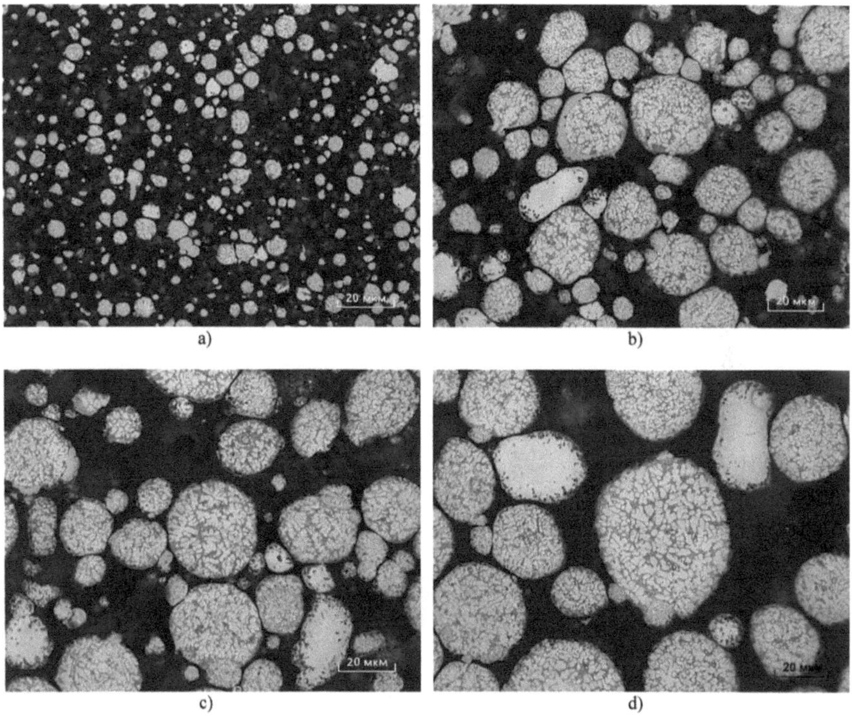

Fig. 3 The microstructure of the powder particles: **a** ASP-5 AlSi10 Mg; **b** ASP-15AlSi10 Mg; in - ASP-22AlSi10 Mg; g - ASP-27AlSi10 Mg

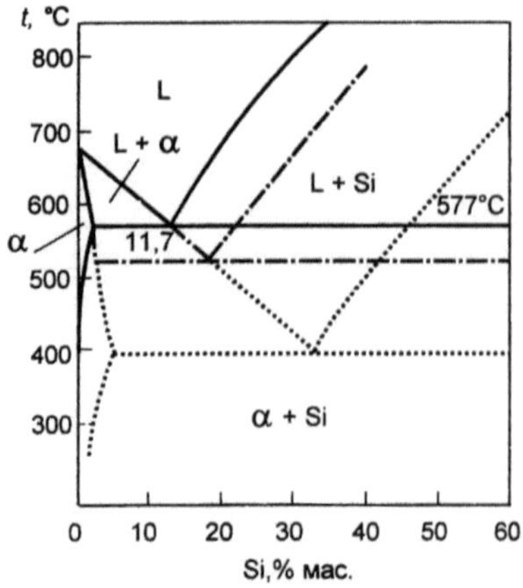

Fig. 4 Non-equilibrium phase diagram of the Al–Si [4]: _._._ Cooling rate during crystallization 1 × 10^3–1 × 10^4 K/s........ Cooling rate during crystallization 1 × 10^6 K/sec

Analyzing the specified chart it is possible to assume that "uncertainty" on an optical microscope of structure of ultrasmall particles ASP-5 AlSi10 Mg, is explained by abnormally high solubility of silicon in conditions over fast crystallization, but not technical imperfection of a method of optical microscopy.

In works [5, 6] it is reported that at a cooling rate 10^5–10^6 K/s, concentration of silicon in solid solution is 10, 7 and 11%.

Research of Samples Obtained by the Additive Technology of 3D-Printing

The samples received by the SLM method, from AlSi10Mg alloy powder, were tested on stretching, subjected to heat treatment then carried out determination of hardness.

Heat treatment of the samples received on additive technology was carried out in Nabertherm TR 120 and N120 furnaces.

Modes of heat treatment:

- stabilizing annealing T = 300 °C; τ = 2 h;
- quenching in water T = 530 °C; τ = 5 h;
- aging T = 160 °C; τ = 7 h.

Measurement of hardness according to Brinell was taken according to DIN EN ISO 6506-1 on the universal hardness gage "EMCOTEST" DuraVision. The research of a microstructure of the samples received on additive technology was conducted on the optical microscope of OLYMPUS GX71 (OM) equipped with the program Thixomet.PRO complex. The research of structural components was conducted on the electronic scanning microscope of FEI Quanta FEG 650 (SEM) with power dispersive spectrometer X-MaxN SDD (EDS).

The microstructure of details fine, assessment of the size of a dendrit cell was carried out on electronic images of a microstructure at increase 30000x. The size of a dendrit cell is 0.5–2 μ. Thus cooling speed at crystallization of the powder particles melted by the laser very high, here it is possible to estimate as10^5–10^6 K/s.

In volume of details the light extended borders with the integrated structure which was created as a result of consecutive crystallization of the melted powder particle on the previous grown-up layer are observed. Actually in a boundary zone there was a repeated fusion of a surface of the put layer under the influence of the melted particle.

Discontinuities at the interfaces between initial particles of powder it isn't revealed. On a microstructure of samples in an initial state and after heat treatment on all carried-out modes, gas porosity which increases after heat treatment is found. Results of assessment of porosity are given in Table 1.

The analysis of influence of heat treatment on a microstructure has shown that after the stabilizing annealing, borders of the melted-off particles remain visible, but more uniform structure is observed. In [1] the author notes that under nonequilibrium conditions (after crystallization with a high cooling rate), a tendency to local segregation appears. At the same time crystals of silicon can appear in alloys where silicon has to enter the Mg_2Si connection. Thanks to a segregation the Mg_2Si and Al_8Mg_5 connection can be present at alloys which in an equilibrium state are single-phase (Fig. 5).

Respectively due to course of diffusive processes and change of morphology of silicon (Fig. 6b) the structure of a sample changes.

Table 1 Results of porosity evaluation

State	Pore volume (%)	Average pore diameter (μ)	Maximum pore diameter (μ)
As built	1.2	3.4	106
Stabilizing annealing	1.2	4	106
Hardening	3.6	4.5	121
Hardening, aging, T6	5	5	136

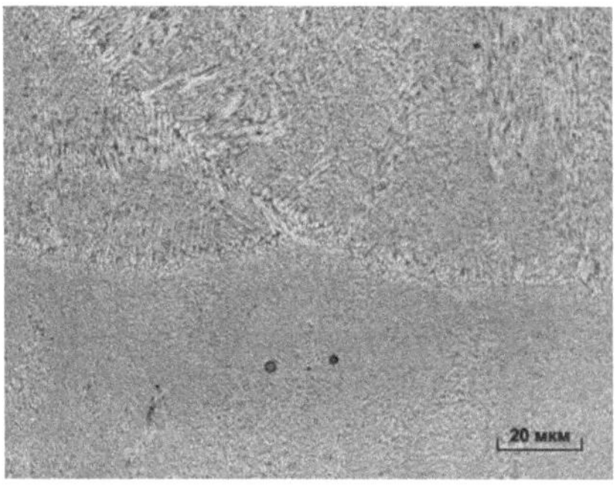

Fig. 5 Microstructure of 3D parts. The boundaries formed by growing layers

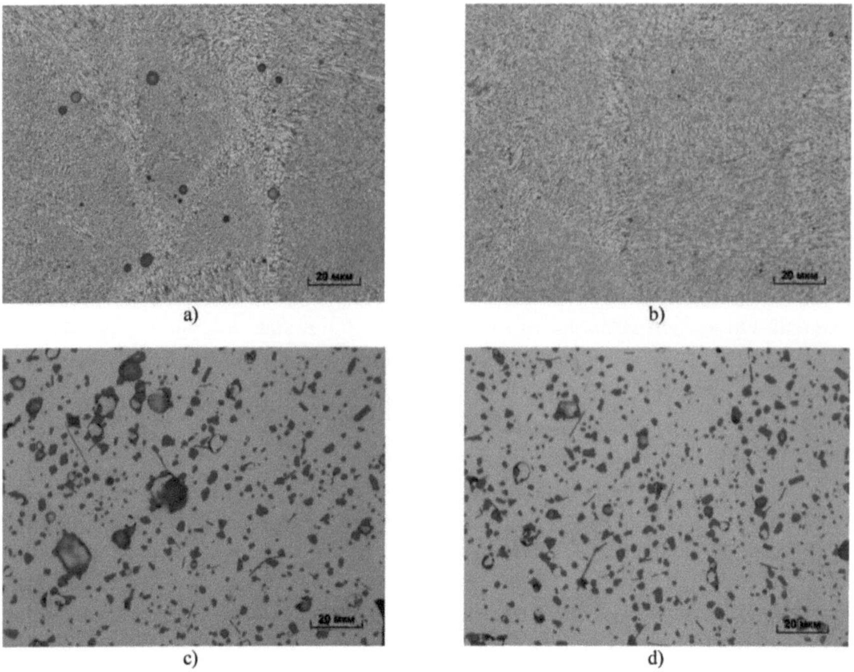

Fig. 6 The microstructure of the samples obtained by the additive technology: **a** as built; **b** stabilizing annealing; **c** hardening; **d** T6

After training and final heat treatment on the T6 mode in structure there have taken place processes of fragmentation, a sferoidization and coagulation of silicon in an eutectic (Fig. 6c, d). Particles of a silicon phase have the globulyarny form, their average size makes 2–3 μ, the maximum size of single particles is 10 μ. In structure the thin lamellar crystals of a ferriferous β-phase located on the former borders the dendrit cells (Fig. 2d) which were not detected in structure of powder and in the samples received on additive technology are found. The reason of the aforesaid structural formations has to be in addition studied as it is known that at ultra-fast crystallization course of processes of formation of the metastable phases containing iron (Al_6Fe instead of Al_3Fe) was possible [2].

Results of hardness measurement on Brinell have shown that in "as built" state the sample is characterized by the highest hardness (Table 2). The stabilizing annealing has led to decrease in hardness on 20 units, as a result of removal of internal tension which has arisen at rapid crystallization in the course of particles melting, leveling diffusions and, perhaps as a result of disintegration of solid solution, phase changes. In state T6 level of hardness in comparison with the hardened state is significantly increased due to the release of the dispersed phase Mg_2Si in aging. Certainly, porosity had to have significant effect on hardness.

Table 2 Results of measurement of hardness on DIN ENISO 6506-1

State	Hardness
As built	103 HBW 2.5/62.5
Stabilizing annealing	80 HBW 2.5/62.5
Hardening	67 HBW 2.5/62.5
T6	95 HBW 2.5/62.5

Table 3 Mechanical properties of the samples obtained by SLM printing technology

Sample number	UTS, MPa	YTS, MPa	Elongation (%)
1	465	279	7.0
2	456	259	5.2
3	456	259	5.2
Average	459	266	5.8

Tests on static stretching carried out by the universal test machine Instron 5982 controlled by the Bluehill 3 software with PC. Computation of mechanical properties was carried out by the Bluehill 3 software in an automatic mode on the basis of the basic data and results received during test. Strength characteristics were determined by the chart of stretching, relative lengthening paid off on a difference between the settlement length of a sample before test (Table 3).

The analysis of the received results shows that during work on production of samples by SLM method from AlSi10 Mg alloy powder, it was succeeded to receive very high strength rates.

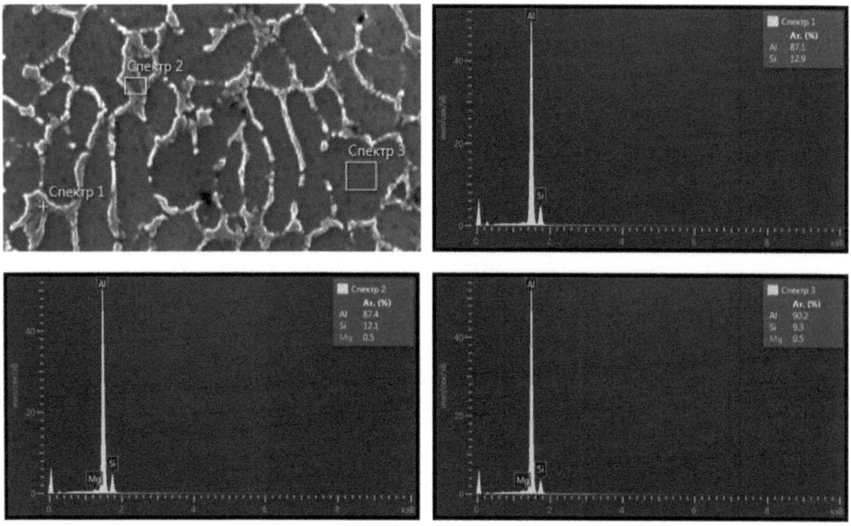

Fig. 7 Structure and results of the EDS analysis of the samples received on additive technology

Table 4 Comparison of properties of A356 (in castings) and AlSi10Mg (in 3D printed samples) alloys

Parameter	A356T6	AlSi10Mg
UTS, MPa	234	459
YTS, MPa	165	266
Elongation (%)	3.5	5.8

The carried-out SEM and EDS analyses have shown that in an initial state samples are characterized by ultradisperse structure with a nanodimensional silicon eutectic, dendrites of solid solution on the basis of aluminum with soluble Mg (about 0.5% of Mg are defined). Considering that the volume fraction of the eutectic phase is much less, than it shall correspond close to the eutectic composition of Al10SiMg alloy, it is possible to assume that in the course of additive production, as a result of an laser overheat and crystallization, essential dissolution of silicon in solid solution was happened (Fig. 7).

Conclusion

The conducted researches have shown that use of 3D printing technology by SLM method allows to reach for AlSi10 Mg alloy durability significantly above, than at the similar A356 alloy produced by casting (Table 4).

Structure analysis of samples obtained by additive technology shows that the temperature-physical parameters which occur during aditive alloy formation, correspond to rapid solidification technology, and thus, it is possible to realize all the advantages of rapid crystallization in 3D printed parts. Therefore, for use in the production of the additive is not advisable to use standard aluminum alloys, and it is necessary to develop special alloys, which can realize the benefits of fast crystallization:

- Increase in the level alloying of transition metals;
- Dispersing excess phases;
- The formation of metastable phases in the finished parts.

Costs for development of additive production can be justified, only provided that the alloys having the high durability, thermal stability and other special properties which can't be received use of traditional technologies will be developed.

References

1. L.F. Mondolfo, Structure and properties of aluminum alloys, M., Metallurgy, (1979)
2. V.I. Dobatkin, V.I. Elagin, V.M. Fedorov, Quickly crystallized aluminum alloys// M., VILS, (1995)
3. H. Jones, Aluminium (BRD), 54(4), pp. 274–281. 67.(1978)

4. B.I. Bondarev, Y. Shmakov, The technology quickly crystallized aluminum alloys. M.: VILS, (1997)
5. M. Itagati, B. Diessen, N.I. Grant, Trans. ASM **61**(2), 330–335 (1968)
6. H. Matyja, K.C. Russell, B.C. Diessen, N.I. Grant, Metallurgical Trans.,6(12), 2249–2252, (1975)

Characterization of Multiperforated Plates Manufactured by SLM and EBM for Aeroengine Applications

M. Thomas, O. Lambert, C. Davoine, F. Popoff, C. Dupuy, P. Peyre and R. Dendievel

Abstract Additive Manufacturing has shown growing interest in the last years compared to traditional techniques for aeronautical complex shaped parts. In this context, the development of multiperforated plates by additive techniques such as Selective Laser Melting and Electron Beam Melting with respect to conventional substractive technique by laser drilling can be promising for combustion chamber applications. Multiperforated plates have then been manufactured by SLM at ENSAM ParisTech and by EBM at SIMaP to provide efficient cooling system within the chamber. The present work has been focused on the detailed analysis of these different plates, including microstructural characterization and permeability measurements. A special attention was paid to the surface roughness since this parameter might be beneficial for an increasing heat transfer surface, but also might be deleterious for the film cooling stability because of the onset of turbulent regimes.

Keywords SLM processing • EBM processing • Permeability • Combustion chamber

Introduction

The Co- and Ni-based superalloys currently used in combustion chambers of jet engines must withstand severe thermomechanical stresses acting in service conditions. Constraints are high in the combustion chambers with a gas temperature up to 1900 K. The lifetime of such high temperature materials is strongly affected by the

M. Thomas (✉) · O. Lambert · C. Davoine · F. Popoff
The French Aerospace Lab, ONERA, 92322 Châtillon, France
e-mail: marc.thomas@onera.fr

C. Dupuy · P. Peyre
ENSAM ParisTech, Paris, France

R. Dendievel
SIMaP, Grenoble, France

The Minerals, Metals & Materials Society, *TMS 2017 146th Annual Meeting & Exhibition Supplemental Proceedings*, The Minerals, Metals & Materials Series,
DOI 10.1007/978-3-319-51493-2_7

number of cyclic thermomechanical loadings. In view of improving the engine thermodynamic cycles, higher temperature levels are required in the combustion chamber. Thus, thermal barriers which are deposited by plasma spraying, are already used to limit the internal wall temperature of combustion chambers of the last generation of jet engines. Now, any further increased engine efficiency is directly related to the ability to raise the temperature and pressure in hot parts of aerobic propulsion systems, i.e. at temperatures beyond the thermomechanical capabilities of existing materials. Nowadays, advanced materials such as thermo-structural composites that can withstand higher temperature levels are investigated to replace superalloys.

However, such an improved engine efficiency may have a detrimental impact on NOx (nitrogen oxide and nitrogen dioxide) emissions. In this regard, the Advisory Council for Aeronautical Research in Europe (ACARE) recommends cutting NOx emissions by 80% in new planes between 1990 and 2020. Therefore, major efforts of engine manufacturers and research institutes deal with the development of innovative technological breakthroughs that will enable the design of low emission combustion chambers operating at high pressure, while preserving high engine efficiency. Among these new technologies, the design of advanced cooling systems can be highly beneficial to preserve the material lifetime. For instance, multi-perforated combustor liners are used to form a film cooling that will protect the wall from the convective heat flux. This cooling technique is based on the derivation of a portion of air flow from the high pressure compressor in a bypassed circuit to be injected into the combustion chamber to contribute to the formation of a film cooling. For structural limitations, the multi-perforated plates exhibit a low void ratio of about 2%. However, a significant amount of air (about 25–30% of the total air flow) entering the compressor is diverted for cooling. It is recalled that the existing multiperforated plates are commonly made by laser drilling, with the possibility to vary the hole size and orientation in order to guarantee the proper cooling efficiency. However, one can imagine other alternative processing techniques to manufacture such multiperforated plates with possibly better cooling properties depending on the material architecture and the chemical composition.

Additive Manufacturing (AM) has shown growing interest in the last years compared to traditional techniques based on near net-shaping and subsequent machining for aeronautical complex shaped parts and small series [1–3]. AM processing techniques, with their versatility and precise control of operating conditions, are suitable to produce every type of structural architectures from a CAD file. Therefore, a comparison is made in the present work between multiperforated plates obtained by layer-by-layer AM and reference plates produced by laser drilling. The challenge with the use of novel cooling materials for combustion chambers is that their geometric characteristics are to be controlled in order to maintain a good balance of mechanical and thermal properties compatible with the desired application. The aim of this work was to establish whether enhanced cooling could be achieved by using AM porous structures with respect to the commonly used multiperforation solution.

Experimental Details

The materials under investigation are a 321 stainless steel for the reference perforated plates, a 316L stainless steel for SLM processing and a Ti64 alloy for EBM processing. A number of porous walls have been manufactured by SLM at ENSAM ParisTech and by EBM at SIMaP to provide comparative cooling systems.

Experimental details of the two AM technologies used for this investigation are provided below:

- The SLM (Selective Laser Melting) process can produce metal parts with different metallic alloys options using 3D-CAD files. The investigated samples were manufactured from a SLM Solutions SLM®125^{HL} machine at ENSAM ParisTech. This system benefits from a single 400 W IPG fiber laser for a maximum scan speed of 10 m/s, and a delivery of 2 l/min of argon during the process in order to prevent melting pool oxidation.
- The EBM (Electron Beam Melting) process can deliver high energy and fast build speed owing to MultiBeam technology for a high productivity. The ARCAM A1 machine at SIMaP was used for this work, allowing solid and porous Ti64 parts to be produced with a controlled microstructure and without residual stress by using a beam power of up to 3500 W and maintaining a vacuum level of 1×10^{-4} Torr.

Furthermore, standard metallographic procedure, including electro-discharge cutting, mechanical polishing, optical and SEM observations, image analysis, was used to characterize the different multiperforated samples. The assessment of surface roughness was obtained by scanning different surfaces for roughness profiles by means of a MarTalk instrument. The influence of several geometrical parameters relative to the multiperforated plates was also studied in permeability tests at room temperature.

Experimental Results

Reference Multiperforated Plates

Four reference 1.2 mm thick plates in 321 stainless steel and of dimensions 98 × 98 mm have been laser drilled in this study for further characterization. First of all, the nominal geometrical characteristics from CAD files are summarized in Table 1. In parallel to the volumic porosity which is often related to the apparent density, the surface area is also a critical parameter because it reflects the number of holes and their inclination in the sense that a more inclined hole offers a larger surface area. It is interesting to note that, by adjusting the geometrical characteristics of the holes in terms of diameter, orientation and spacing, this allows a two-by-two comparison. Indeed, plates P3 and P7 exhibit the same volumic

Table 1 Nominal characteristics of reference multiperforated plates

Reference sample	Hole diameter (mm)	Hole orientation	Hole pattern	Hole spacing (mm)	Volumic porosity (%)	Specific surface area (mm^{-1})
P3	0.5	33	squared	2.8	3	0.24
P4	0.5	60	squared	2.8	5	0.40
P6	0.3	60	squared	2.17	3	0.40
P7	0.3	33	squared	1.68	3	0.40

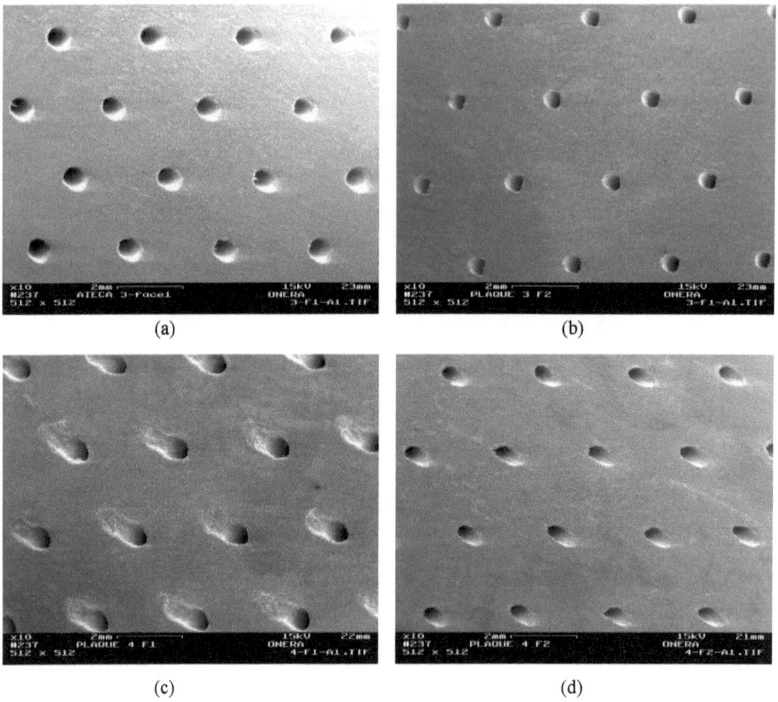

Fig. 1 SEM micrographs of plates P3 and P4 for both sides, with the incident side on the *left hand-side* (**a** and **c**)

porosity rate but different specific surface area whereas plates P4 and P7 exhibit different volumic porosity rate but have the same specific surface area.

In Fig. 1, SEM observations are presented at low magnification for plates P3 and P4. Firstly, a pronounced flare of the holes can be clearly emphasized. Flaring is located on the incident surface of the plates. Secondly, as the holes are more inclined, such as in the case of plate P4, traces of the laser fusion occur at the surface of the plate, as shown in Fig. 1c. For the smallest holes, some additional features can be revealed such as non-cylindrical holes and irregular surfaces.

Table 2 Characteristics of reference multiperforated plates by measuring the average hole diameter

Reference sample	Incident diameter (mm)	Opposite diameter (mm)	Average diameter (mm)	Hole spacing (mm)	Volumic porosity (%)	Specific surface area (mm^{-1})
P3	0.72	0.48	0.60	2.8	4.30	0.29
P4	0.65	0.47	0.56	2.8	6.28	0.45
P6	0.50	0.32	0.41	2.17	5.61	0.55
P7	0.47	0.27	0.37	1.68	4.54	0.49

Table 3 Characteristics of reference multiperforated plates by measuring the density

Reference sample	Bulk density	Measured density	Average diameter (mm)	Hole spacing (mm)	Volumic porosity (%)	Specific surface area (mm^{-1})
P3	7.9	7.524	0.63	2.8	4.75	0.30
P4	7.9	7.386	0.57	2.8	6.51	0.46
P6	7.9	7.420	0.43	2.17	6.07	0.57
P7	7.9	7.440	0.42	1.68	5.83	0.56

Attempts were then made to measure a series of hole diameters on both sides of the different plates in the SEM microscope. Average values have been calculated to estimate the volume porosity and the specific surface area, assuming that the holes were perfectly conical. Results from Table 2 show that the hole diameters are systematically larger than the nominal values. The differences in holes diameter between both sides is roughly constant and equal to 0.2 mm, irrespective of the plates. Unexpectedly, plate P4 exhibits slightly lower differences in holes diameter between both sides while having the more inclined holes which means a longer path for the laser to open at the other end.

Furthermore, the volume porosity and the specific surface area have been determined in some other way, i.e. by measuring the plate density (Table 3). It is expected to yield more precise results than from the measurement of a few hole diameters. Therefore, the present values of volume porosity and of specific surface area have been taken into account to determine any further correlation with the permeability measurements.

Now, with regard to the permeability tests on plates P3, P4, P6 and P7, measurements were made on both sides. In Fig. 2, the pressure drop per unit of length is plotted against the superficial velocity Vs up to 2.5 m/s which is representative of the air flow in jet engines. These non-linear curves can be described by the well-known Darcy-Forchheimer expression.

Quite clearly, the plates appear to be slightly more permeable when the air flow is injected through the incident laser holes, which is the most flared side. This is consistent with the work of Zaragoza et al. investigating the permeability of metal foams with graded pore size [4]. Additional information can be collected for the

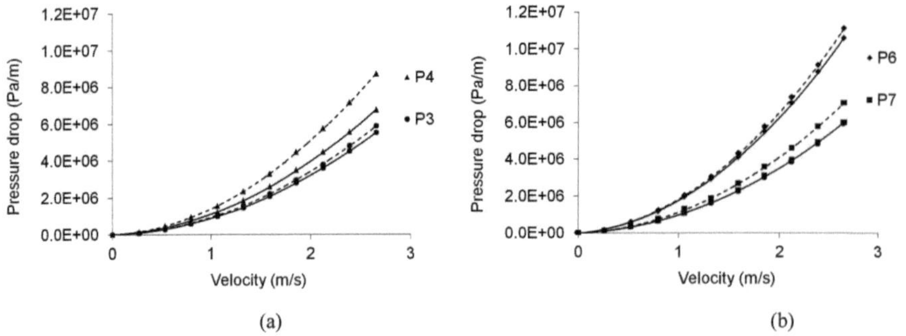

Fig. 2 Permeability curves: **a** plates P3 and P4, **b** plates P6 and P7. Solid lines correspond to the incident side and dotted lines to the opposite side

Fig. 3 Relationship between the geometrical characteristics and permeability for the multiperforated plates

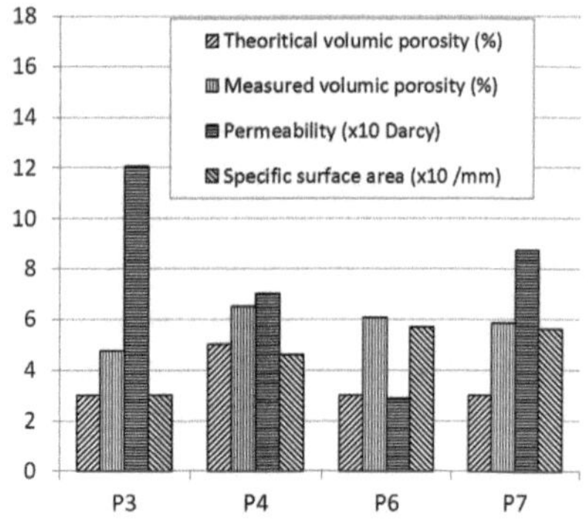

four plates by comparing the Darcy permeability (average value of both sides) which corresponds to the linear part of the curves (Fig. 3). It then appears that plate P4 exhibits a lower permeability than plate P3. Therefore, the higher volumic porosity in P4 does not seem to affect the permeability because this increased porosity is just related to the fact that the holes are more inclined. It is important to note that, in the permeability tests, the air flow blows perpendicularly to the plates, so that the air flow should be more turbulent with hole angle of 60° (P4) than with hole angle of 33° (P3). In addition, holes in plate P3 are found to be slightly larger than the ones in plate P4 (Table 3). It is then interesting to compare plates P3 and P7 having the same hole inclination. For P7, the hole diameter and the hole spacing are smaller which leads to a slightly higher volumic porosity than that of P3. Despite this higher porosity for P7, the permeability is unexpectedly lower. This means that, with an almost twice specific surface area, the air flow through plate P7 can be affected by the surface roughness inside the holes, this effect being more pronounced for smaller holes. Finally, it is instructive to compare P6 and P7, both plates having the same hole diameter, approximately the same volumic porosity and the same specific surface area. Figure 3 unambiguously shows that P7 is considerably more permeable than P6. The reason for this is that the holes in P7 are less inclined than in P6, which is consistent with the first comparison between plates P3 and P4.

Multiperforated Walls Produced by EBM

As illustrated in Fig. 4, the two types of walls either with 0.5 mm holes or with 1 mm holes, exhibit a significant surface roughness. The latter is similar for the internal part and the outer part of the walls. This is consistent with the fact that these walls are built vertically and so both inner and outer parts of the wall have the same orientation in the built chamber.

Moreover, due to these serrated internal surfaces of the holes, it is difficult to appreciate the precise internal diameter of the holes. A rapid visual inspection would probably underestimate the actual hole size. Finally, it appears that the expected 0.5 mm holes as illustrated in the top view are not free from individual molten powder particles. As a consequence, permeability might be strongly affected by such sintered powders obstruction for the air flow within a number of holes. In particular, the most burried sections at mid-thickness are presumably the most obstructed ones. The dilemna is that, regardless of whether an obstruction is narrow or large, the air flow will not be able to pass through.

Surface roughness profiles have been performed along X- and Z-axis. For a surface area without holes, the experimental results are given in Table 4. It appears that Ra values are slightly higher for the Z-axis, presumably due to the powder layer-by-layer interfaces.

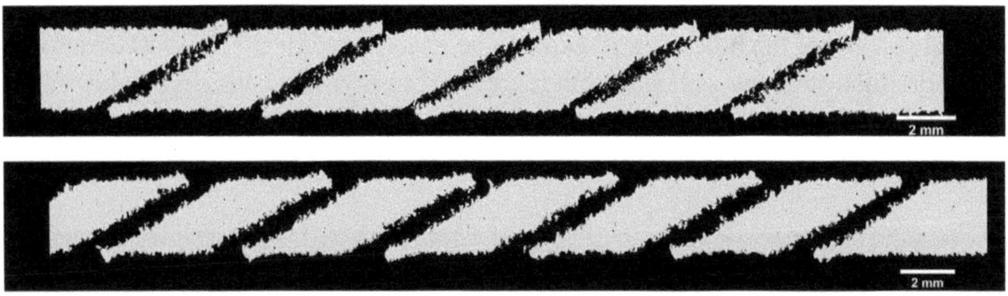

Fig. 4 Illustration of the cross section of EBM walls for two different hole sizes

Table 4 Roughness measurements on both sides of an EBM wall

	Direction	Ra (μm)
Incident side	Along Z-axis	19.7
	Along X-axis	25.6
Opposite side	Along Z-axis	17.9
	Along X-axis	21.2

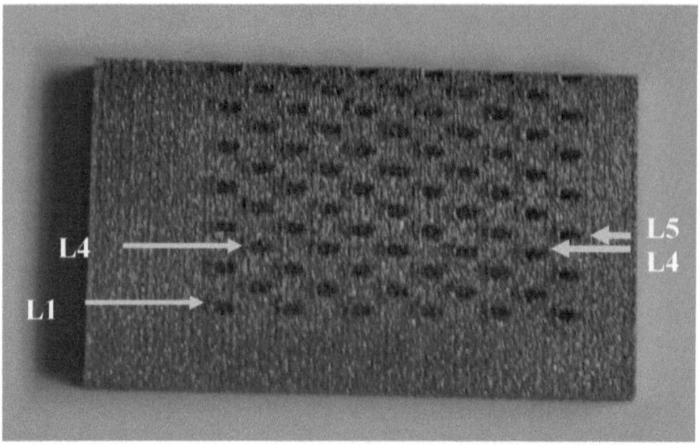

Fig. 5 Illustration of the scan lines for the roughness profiles of an EBM wall

An example of scans made at the surface of the EBM wall to evaluate the effect of surface roughness in the presence of holes is provided in Fig. 5. For theses scans performed along a series of holes, Ra values of approximately 60 μm have been measured. It is believed that such a significant surface roughness may affect the stability of film cooling.

Multiperforated Walls Produced by SLM

The cross section of the two types of SLM walls is illustrated in Fig. 6. The surface roughness of the walls is definitely better than that of the EBM walls, with Ra values between 5 and 8 μm. However, contrary to the EBM walls, the wall surface is significantly smoother than the internal surface of the holes. Moreover, for the internal surface of the holes, there seems to be an asymmetrical behavior, depending on the orientation of the wall side with respect to gravity. Thus, the right hand-side of the holes in Fig. 6a and the left hand-side of the holes in Fig. 6b should be the

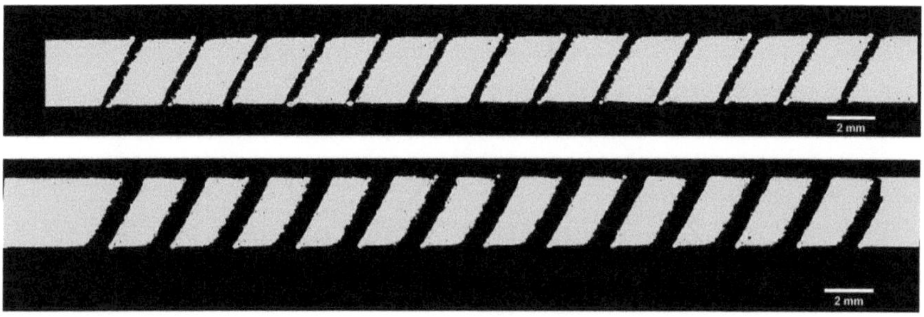

Fig. 6 Illustration of the cross section of SLM walls for the two different hole sizes

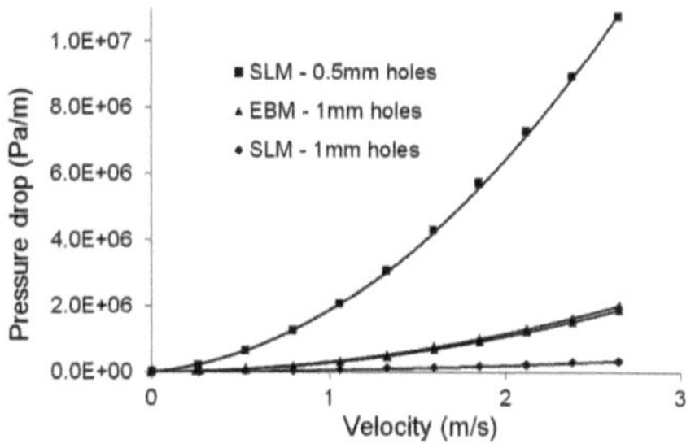

Fig. 7 Permeability curves for three AM samples

over-hanging sides during the built. The fact that EBM walls did not show such an asymmetrical feature deals with the fact that the powders are pre-sintered in the Arcam machine before melting of the successive layers. Finally, a protuding ridge is visible where the holes are opening onto the surface of the SLM walls. This is presumably related to the minimum required size of the melt pool which makes difficult to duplicate sharp outer edges of the wall, as already observed in [5]. What still needs to be established is the effect on film cooling of these protuding ridge which might be compared to small upstream ramps as described in [6].

Permeability Tests for EBM and SLM Samples

Permeability curves for the 3 mm thick EBM and SLM walls are displayed in Fig. 7. A direct comparison of EBM and SLM walls with 1 mm holes reveals that the SLM wall are more permeable than the EBM one, which may be related to the more favorable inclination of the holes (30° vs. 60°) and to a lower surface roughness. Quite expectedly, the SLM plate with smaller walls is less permeable. Finally, permeability tests carried out on both sides of the EBM wall yield the same values, which is indicative of a more symmetrical behavior than in the case of the laser drilled plates inducing a flare of the holes.

Conclusion

Multiperforated plates produced by Selective Laser Melting and Electron Beam Melting have been compared to conventional laser drilled multiperforated plates. The present investigation has shown that it is important to control the geometrical and metallurgical properties of these AM porous structures because of the significant influence of the AM process on the physical properties of the finished parts. AM is capable of providing nice and regular multiperforated structures with a good reproducibility in permeability tests. Based on the geometrical characterization of the different plates and of the related roughness and permeability measurements, these experiments were helpful to demonstrate that both EBM and SLM walls provide cylindrical holes with a good symmetry on both sides of the plate. SLM offers the advantage of a smoother surface than that of EBM but suffers from an asymmetrical roughness of the internal surface of the holes, depending on the built orientation of these samples in the SLM chamber.

However, a higher surface roughness is promoted by both SLM and EBM processes compared to the commonly used multiperforated plates. This increase in roughness should result in more rapid turbulent regimes under air flow. This work also highlights the predominant effect of the direction of the incident air flow toward the multiperforated plates. For comparison, additional permeability experiments are then needed under razing air flow with respect to the surface of the multiperforated plates. Now, if the increased surface roughness for AM samples could be detrimental for the film cooling stability, on the other hand heat transfer efficiency can be favored as to provide a more efficient cooling system. Such a hypothesis will be verified in the near future by complementary rig tests with aerodynamic and high temperature conditions.

Acknowledgements Fundings from Direction Générale de l'Armement (DGA) and Agence National de la Recherche (ANR) under project MOSART and from ONERA under project MOTRICE together with technical support from SAFRAN are greatly acknowledged.

References

1. H. Gong et al., Analysis of defect generation in Ti–6Al–4 V parts made using powder bed fusion additive manufacturing processes. Addit. Manuf. **1–4**, 87–98 (2014)
2. S.S. Babu et al., Additive manufacturing of materials: Opportunities and challenges. MRS Bull. **40**(12), 1154–1161 (2015)
3. X. Zhao et al., Comparison of the microstructures and mechanical properties of Ti–6Al–4 V fabricated by selective laser melting and electron beam melting. Mater. Des. **95**, 21–31 (2016)
4. G. Zaragoza, R. Goodall, Metal foams with graded pore size for heat transfer applications. Adv. Eng. Mater. **15**(3), 123–128 (2013)
5. G.A.O. Adam, D. Zimmer, Design for additive manufacturing—element transitions and aggregated structures. CIRP J. Manuf. Sci. Technol. **7**, 20–28 (2014)
6. S.P. Chen et al., Effects of upstream ramp on the performance of film cooling. Int. J. Therm. Sci. **50**, 1085–1094 (2011)

Direct Laser Metal Deposition of Eutectic Al-Si Alloy for Automotive Applications

Amrinder Singh, Abhishek Ramakrishnan and Guru Prasad Dinda

Abstract Various additive manufacturing techniques have been able to manufacture Al-Si hypo- and hyper-eutectic alloys but practical implementation of additive manufacturing of Al-Si alloys in automotive industry still remains a big challenge. This paper deals with the challenges of building tall wall and cuboid shapes eutectic Al-Si components by direct laser metal deposition technique for automotive applications. Microstructural investigation using optical and scanning electron microscopy revealed a 99.9% dense component with very fine hypoeutectic microstructure. Tensile test showed an impressive elongation of 9% with an ultimate tensile strength of 225 MPa. This investigation revealed that direct laser metal deposition could be successfully implemented on automotive shock tower hood without any distortion or bending. This paper also presents the effect of various laser deposition parameters like laser power, powder flow rate, and scanning speed on the microstructure and mechanical properties distribution from substrate to deposit.

Keywords Additive manufacturing · Al-Si alloy · Laser metal deposition · Laser cladding

Introduction

The Additive Manufacturing (AM) of eutectic Al-Si alloy has received a special attention in last couple of years because of its endless applications in automotive and other sectors. Most of the castings used in automotive industry uses Al-Si alloys, putting necessary emphasis on implementation of deposition of Al-Si alloy using AM. Numerous studies show additive manufacturing techniques been used to successfully deposit Al-12Si but none of them have tried to deposit on existing automobile parts and none of them briefly discussed challenges if AM is applied at

A. Singh (✉) · A. Ramakrishnan · G.P. Dinda
Department of Mechanical Engineering, Wayne State University, Detroit, MI 48202 USA
e-mail: bb8314@wayne.edu

The Minerals, Metals & Materials Society, *TMS 2017 146th Annual Meeting & Exhibition Supplemental Proceedings*, The Minerals, Metals & Materials Series, DOI 10.1007/978-3-319-51493-2_8

commercial level [1–5]. This study specifically deals with challenges involved using Direct Laser Metal Deposition technique (LMD) to deposit Al-12Si on the existing cast parts. LMD combines both laser cladding along with the rapid prototyping, which gives full freedom to manufacture any sophisticated shape. LMD has an advantage over laser sintering technique, it can print on existing parts without requiring the extra process of combining the additively manufacture part to existing part. This study has proven the reliability and consistency of the LMD to achieve the deposition of Al-12Si on existing Al-Si casting. LMD has an upper hand as it deposits highly dense (99.9%) eutectic Al-Si alloy as compared to the maximum density achieved in sintering which is 99.5% via Pulsed Selective Laser Melting (P-SLM) [6]. The practice of casting all components first and then joining them using various welding techniques has long been followed in the automotive industry. The best solution is LMD, which brings us the possibility to print other components on the existing cast part. This provides the designer with a lot of freedom to change design as per customer requirements. This study could be used as a guide for implementation of direct metal deposition technique in automotive sector. LMD has vast application other than manufacturing parts including surface coating, repair, remanufacturing etc. [7–11]. Laser sintering has been used for depositing hyper and hypo eutectic alloys [1–6, 12, 13] but so far mechanical test results has been only provided by Wang [5] using Powder bed technique (Laser Sintering) which is comparable with the results obtained using LMD technique.

Experimental

Robotic Laser Metal Deposition (LMD) technology developed at Wayne State University having four main components 1.2 KW Diode Laser, ABB IRB-1410 six axis Robot, Coaxial nozzle, Powder feeder. LMD technology combines the rapid prototyping with laser cladding into solid freeform fabrication process as shown in Fig. 1. Copper Nozzle shown in Fig. 1c shows all the different openings for powder delivery, Nozzle gas, and shaping gas. This copper nozzle is attached at the end of

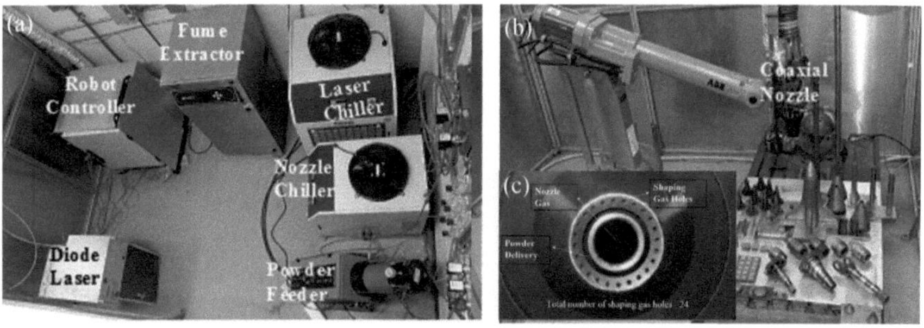

Fig. 1 LMD control units (**a**), coaxial nozzle connected to the end of robotic arm (**b**), and bottom view of coaxial nozzle showing delivery holes for shaping, powder, and nozzle gas (**c**)

Table 1 Composition of Al-Si (S20) alloy received from Valimet Inc

Si	Fe	Al
11.2	0.15	88.65

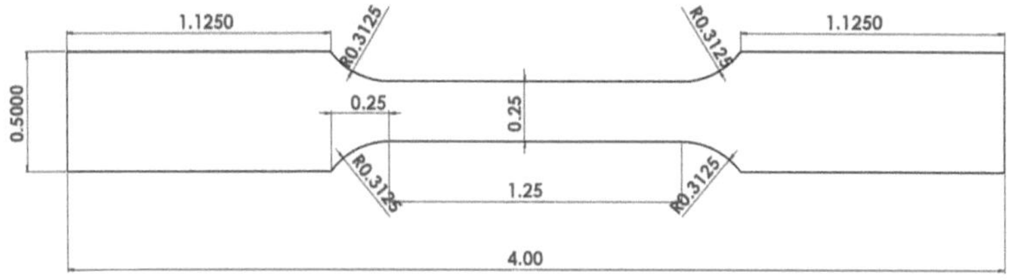

Fig. 2 Schematic of tensile test sample (All dimensions are in inches)

robotic arm to provide a specific tool path motion. The coolant from chiller is circulated inside the nozzle to avoid overheating otherwise nozzle can be damaged under immense heat produced by laser. Nozzle is made from Copper to dissipate the heat at faster rates. Powder hopper delivers the powder to delivery openings in the nozzle, powder is then concentrated on the substrate with the help of shaping gas coming from shaping gas holes. The nozzle gas helps to protect the focusing laser lens. Previously deposited layer is also partially melted while melting the incoming powder, providing very high density and good bonding at layer interface.

Gas atomized Al-Si alloy was supplied by the Valimet Inc within the mesh size of −100/+325 was used for the deposition purposes. The composition was given in as a data sheet shown in Table 1. The deposition was done on 6061 rolled Aluminum plate for initial tests.

To achieve high density of Al-Si deposit, various parameters (Laser Power, Powder Feed Rate, Linear Velocity) had to be optimized as each parameter has significant effect on the build quality of deposit. Laser power has to be more than usual to melt Al-Si alloy, as the thermal conductivity of the Al-Si alloy is substantially higher than the other alloys. This was also stated in the study by R Cottam on cladding process [14]. Powder feed rate has to be high enough to maintain the enough deposition rate. Scanning velocity is also crucial parameter that when optimized will maintain continuous bead. Not only these parameters but also working height from nozzle to build height has to be constant all the time to get the defect free deposition, which is trickier in case of Al-Si alloys. The main objective of this study is to deposit Al-Si alloy on any automotive cast parts to prove its utility. For this purpose, a Shock Tower as shown in Fig. 9 was chosen because there is lot of accessories that are welded on to the shock tower hood. Before proceeding with deposition on the shock hood, it was necessary to do some mechanical testing on the deposited part to prove its reliability. The tensile test sample prepared from a Al-Si block (100 mm × 17 mm × 15 mm) deposited using optimized laser deposition parameters. The block is then machined to a dog bone shape using milling operation with dimensions as shown in Fig. 2. Resulting milled dog bone block is shown in Fig. 3a. Three tensile test samples of 2 mm thickness were cut using wire EDM from above dog bone shaped

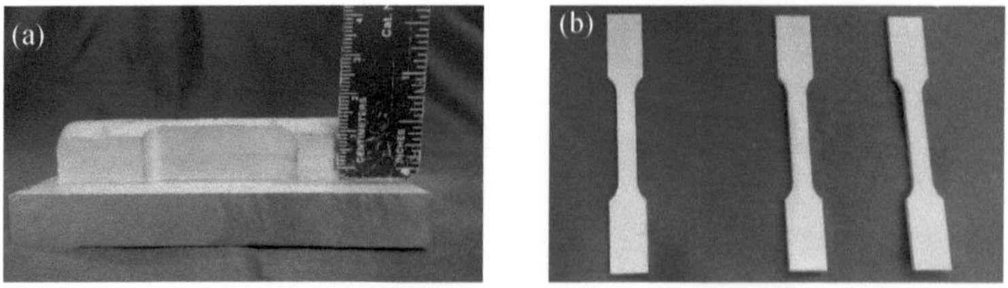

Fig. 3 Al-Si block deposit (**a**), and EDM cut sample from the block (**b**)

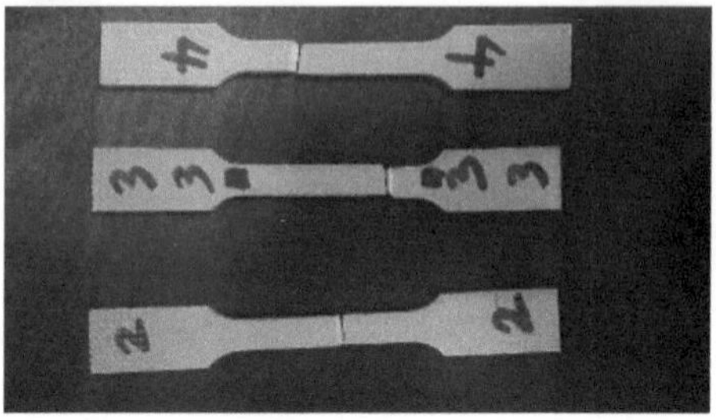

Fig. 4 Position of failure in tensile sample

block as shown in Fig. 3b. Universal tensile testing machine (MTS 810) was used for tensile testing of final samples.

Significant clutching pressure is set so that the sample does not slip while under tension. Tensile testing is performed at ramp up speed of 0.05 mm/s. Three samples were tested and position of failure is shown in Fig. 4. After the conclusive results from tensile test, a single wall was deposited on the substrate to achieve maximum height.

This was done to allow parameter optimization for single wall before printing on the shock tower hood. In addition to the tensile test, microstructural investigation was carried out on both single wall and block deposition to observe the microstructure of single wall and block structure.

The microstructural evolution in case of single wall is well documented by Thijs et al. [6]. The tool path is chosen as hatch style (Fig. 5a) for block and nozzle moves in y-axis only left and right. Powder and laser is never switched off in case of single wall deposition. Consecutive increment in layer height was kept at 0.4 mm for both block and single wall deposition. Metallographic etching was done by dipping samples in solution of 100 ml H_2O, 25 ml HNO_3, 15 ml HCL, and 10 ml Hf for samples for 10–12 s.

Results and Discussion

Block Deposition

Tensile test samples made from the block deposition (using hatch pattern as shown in Fig. 5a) revealed excellent elongation of average 9%, which is comparable with cast

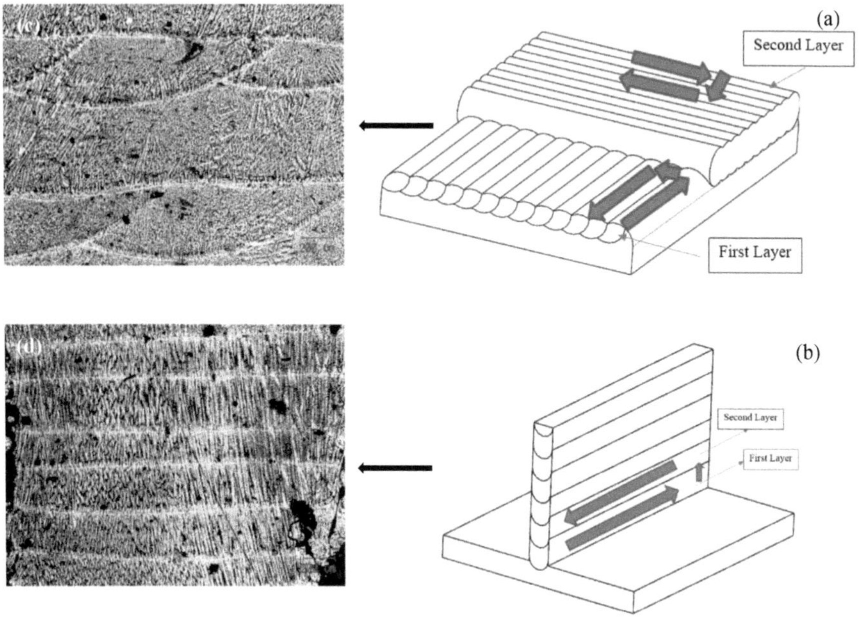

Fig. 5 Block hatch deposition pattern (**a**), single wall deposition pattern (**b**), optical micrograph of the as-deposited block sample (**c**), and optical micrograph of the as-deposited single wall deposit (**d**)

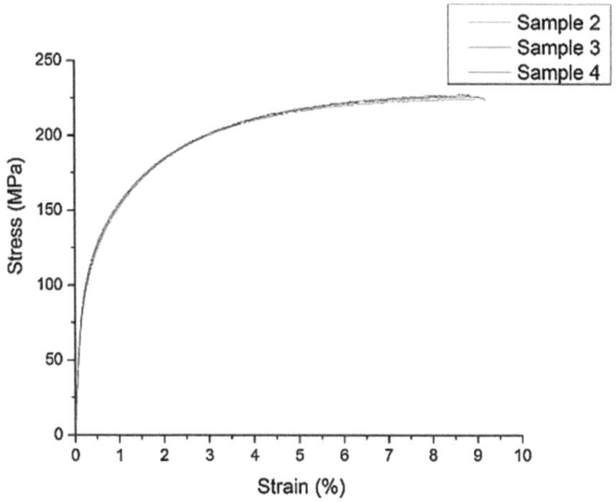

Fig. 6 Stress strain curves of the as-deposited Al-11.2Si samples

properties and ultimate tensile strength was found to be 225.7 MPa as shown in Fig. 6. These results indicate that deposition is of good quality and hence practical implementation is feasible. Optimized laser deposition parameters are given in Table 2.

Table 2 Optimized laser deposition parameters

	Block deposition	Single wall
Laser power (W)	900	900[1]
Powder flow rate (g/min)	3.5	6.3
Scanning speed (mm/min)	720	576

[1]Laser power was reduced by 300 W after every 5 mm layer deposition until final laser power reached to 800 W

In the very first layer of deposition the powder flow rate was kept low (2.6 g/min) to achieve good bonding with the substrate. It was observed that the laser power particularly has to be kept high (900 W) for the aluminum alloys to achieve the continuous melt pool because of high thermal conductivity of aluminum.

Figure 5 clearly shows that microstructure is dependent on the tool path pattern. The microstructure of transverse section of deposited block shows an interesting circular overlaps where the coarse dendrites slowly transformed into fine dendrite microstructure. This circular bead is the liquid/solid interface. There was no major defects like cracks, de-bonding at interface or between layers observed. Figure 7 clearly shows the microstructure variation in same sample. There is dendrite structure above the circular band and gradually transforms to fine microstructure. Figure 7d shows the top left edge of the sample where clearly the equiaxed Al dendrites was observed which indicates that thermal gradient was low and solidification velocity to be high at the top and no re-melting was happened at the top. It was observed low hardness value of 70 HV on average at the coarser dendrite area around circular pattern, which is due to coarsening of silicon particles. The microhardness value at the top found to be 90 on an average. This high value of hardness is due to the presence of fine equiaxed microstructure. This increase in hardness value is about 28% from the soft circular pattern area to the top layer of deposit. Dinda et al. has reported 10% change in hardness while depositing Aluminum 4047 alloy using CO_2 laser [15].

Single Wall Deposition

Laser deposition parameters had to be optimized again for single wall as shown in Table 3. This deposition was crucial as we are trying to replace the welded bracket with deposited bracket. To accomplish the bracket deposition on shock tower hood, we had to demonstrate that LMD can build tall enough deposit to replace welded bracket design. It was decided that a single wall need to be built on the 6061 rolled Aluminum alloy substrate first for testing maximum achievable height. The very first layer was deposited with high laser power (950 W) and less powder flow rate (2.6 g/min) for good bonding between substrate and deposit. Consecutive layers were deposited using reduced laser power. As the wall height increases double layer grows on each side of deposited wall as shown in Fig. 8. When observed closely, it

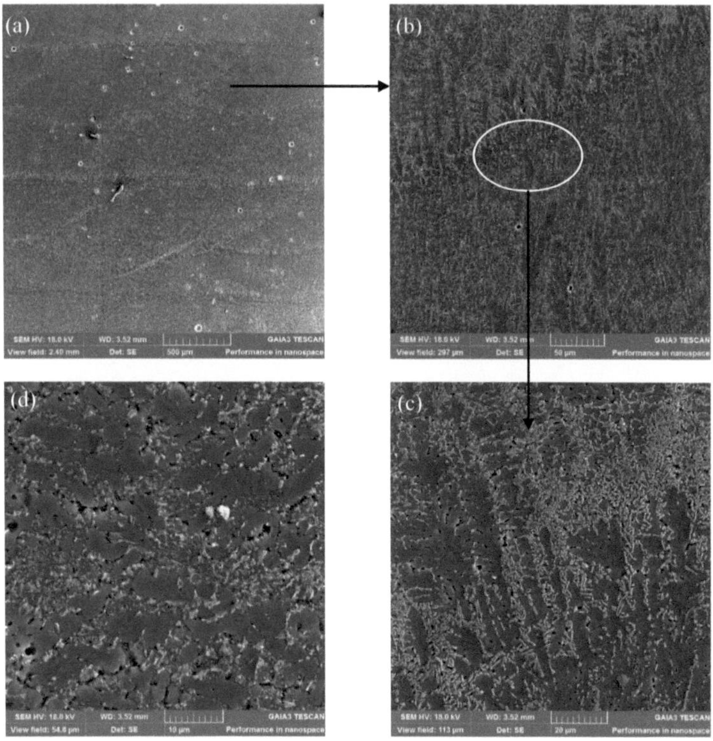

Fig. 7 SEM micrographs (*Block*) showing dendrites around circular pattern (**a–c**), and Equiaxed microstructure at top part of sample (**d**)

Table 3 Mechanical properties of as-deposited Al-12Si alloy

Yield strength (MPa)	122.43
Modulus of elasticity (E) (GPa)	50.802
Ultimate tensile strength (MPa)	225.7
Elongation (%)	9.0

was revealed that some powder was not caught in the melt pool and had in process fallen on either sides of the wall. With the increase in the height of the deposit, heat dissipation rate falls and hence more heat is available in the wall at that time. This extra heat sinters the extra powder falling on sides. Due to sintering issue, the height was limited to 15 mm. To cope with this issue a gradually decreasing laser power after every 5 mm was deployed, alternative solutions to this problem are currently being developed and work is in progress. Besides the sintering issue, top of the deposition was not found to be flat. This can be due to unstable melt pool and because of high thermal conductivity of aluminum. The thickness test was also conducted to standardize the process for industry applications.

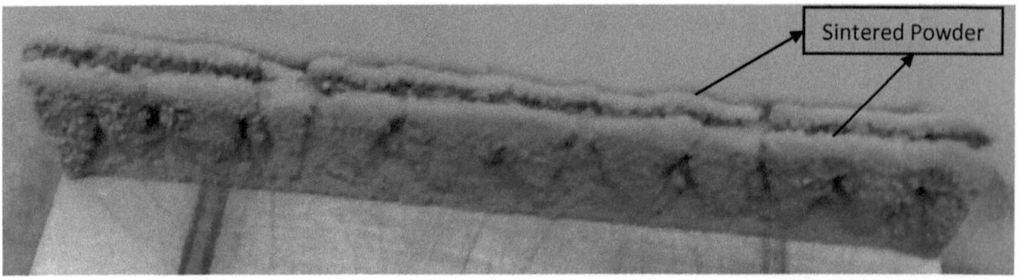

Fig. 8 Single wall deposition of Al-11.2Si alloy with sintered powder on both sides

Fig. 9 Shock Tower Hood with Al-12Si deposition using LMD

Three Al 6061 rolled plate with different thickness (1.5, 2.0, and 3.0 mm) were tested as substrate and single wall was deposited on each substrate. It was found that substrate thickness has to be more than 1.5 mm as the 1.5 mm thin substrate was completely melted. After all these feasibility tests deposition was successfully made on the shock tower hood with the optimized parameters shown in the Fig. 9. Lastly it was checked if single wall deposition had resulted in any deformity beneath the shock tower. Very minor impression of distortion was found on the substrate. Microhardness results shows the maximum value of 98 HV around the edges of the deposition.

The dendritic structure is more prominent in case of single wall as shown in Fig. 10a, since this is single wall and most of the heat is being conducted towards substrate at the bottom. This is well known fact that dendrites grow opposite to the heat flow direction. The gray primary dendrites is the aluminum and bright constituents represents the silicon particles. Inter dendritic region has eutectic structure as shown in Fig. 10b. The average microhardness at top was found to be 82 HV and 72 in the middle (lengthwise) of the wall. Dendritic structure throughout the single

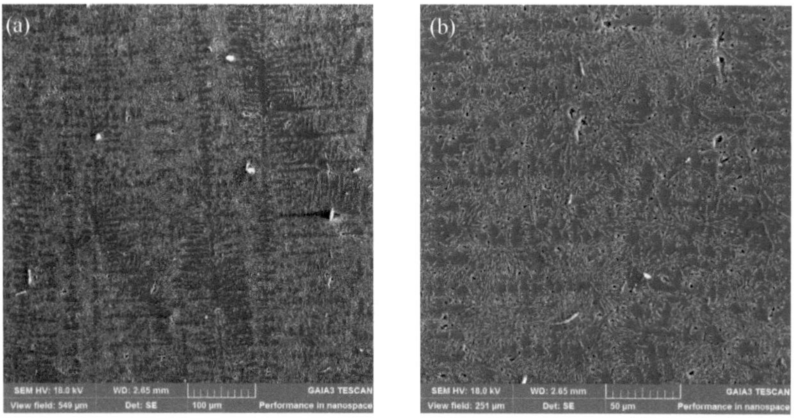

Fig. 10 Micrographs of single wall deposit showing dendritic (**a**), and equiaxed microstructure (**b**)

wall makes it softer but it tends towards the eutectic side at the top of single wall. There is about 12% increase in hardness value from dendritic to eutectic microstructural change in case of single wall deposition.

Conclusion

Current investigation revealed that LMD can be successfully applied at commercial level for deposition on the existing cast aluminum components saving both time and cost of making another part. In addition, welding process can be entirely skipped. Tensile tests showed a good combination of yield strength (122.43 MPa) and tensile elongation (9%). Single wall deposition higher than 15 mm can be achieved by solving the sintering issue. Reducing laser power as the deposition height grows has significant effect only up until 15 mm. After 15 mm, the laser power can not be reduced further as any more reduction will affect the melt pool stability. Columnar Aluminum dendrites develop at bottom of the each layer of block deposition and kind of continuous in case single until the top layer where the tendency of equiaxed microstructure is predominant. Microhardness values showed the dependence on the type of microstructure formation, clearly showed that equiaxed microstructure has higher value of microhardness.

Acknowledgements The author wish to acknowledge the General Motors for providing the Shock Tower Hood for deposition purposes. The authors are indebted to Paul Wolcott and Yiran Tong for their valuable input.

References

1. R. Chou et al., Additive manufacturing of Al-12Si alloy via pulsed selective laser melting. JOM **67**(3), 590–596 (2015)
2. G.P. Dinda, A.K. Dasgupta, J. Mazumder, Evolution of microstructure in laser deposited Al–11.28%Si alloy. Surf. Coat. Technol. **206**(8–9), 2152–2160 (2012)
3. M. Nahmany et al., Electron beam welding of AlSi10 Mg workpieces produced by selected laser melting additive manufacturing technology. Addit. Manuf. **8**, 63–70 (2015)
4. E.O. Olakanmi, K.W. Dalgarno, R.F. Cochrane, Laser sintering of blended Al-Si powders. Rapid Prototyping J. **18**(2), 109–119 (2012)
5. X.J. Wang et al., The effect of atmosphere on the structure and properties of a selective laser melted Al-12Si alloy. Mater. Sci. Eng. A **597**, 370–375 (2014)
6. L. Thijs et al., Fine-structured aluminium products with controllable texture by selective laser melting of pre-alloyed AlSi10 Mg powder. Acta Mater. **61**(5), 1809–1819 (2013)
7. G.P. Dinda, A.K. Dasgupta, J. Mazumder, Laser aided direct metal deposition of Inconel 625 superalloy: microstructural evolution and thermal stability. Mater. Sci. Eng. A **509**(1–2), 98–104 (2009)
8. G.P. Dinda, L. Song, J. Mazumder, Fabrication of Ti-6Al-4 V scaffolds by direct metal deposition. Metall. Mater. Trans. A **39**(12), 2914–2922 (2008)
9. J. Mazumder et al., The direct metal deposition of H13 tool steel for 3-D components. JOM **49** (5), 55–60 (1997)
10. J. Mazumder et al., Closed loop direct metal deposition: art to part. Opt. Lasers Eng. **34**(4–6), 397–414 (2000)
11. W. Zhuang et al., Deep surface rolling for fatigue life enhancement of laser clad aircraft aluminium alloy. Appl. Surf. Sci. **320**, 558–562 (2014)
12. R. Mathew et al., Microstructural refinement of aluminium-zinc-silicon coated steels. Surf. Coat. Technol. (2016)
13. A. Simchi, Direct laser sintering of metal powders: mechanism, kinetics and microstructural features. Mater. Sci. Eng. A **428**(1–2), 148–158 (2006)
14. R. Cottam et al., Laser cladding of high strength aluminium alloy 7075 powder on a 7075 substrate for repair of damaged components
15. G.P. Dinda et al., Microstructural characterization of laser-deposited Al 4047 alloy. Metall. Mater. Trans. A **44**(5), 2233–2242 (2012)

Additive Manufacturing to Produce Standard and Custom Alloy Titanium

James C. Withers and Sion M. Pickard

Abstract Metal additive manufacturing patented in 1925, today uses electron beam or laser beam pool melting as well as using laser or electron beam powder sintering a single layer at a time to build relatively small near net shape parts. Friction Stir Processing (FSP) as a solid state non-melt process has also been demonstrated to build parts layer by layer as well as repair parts. Plasma transferred arc (PTA) power system melt pool processing has low capital cost, very high build rates compared to e-beam or laser and has built larger multiple feet size parts with property values equivalent to superior to wrought material. In the case of titanium, sponge has been used as a low cost feed to produce varieties of alloy compositions with exceptional strength and higher moduli that have also included ceramic particulate with applications in the auto industry. FSP has produced double strengths over analogous wrought alloys.

Keywords Additive manufacturing · 3D printing · Titanium alloys

Introduction

Additive manufacturing (AM) also referred to as 3D printing, has been touted as having the potential to rejuvenate manufacturing in the U.S. Recent AM/3D printing has grown out of rapid prototyping which encompasses building polymer/plastic base compositions layer-by-layer using CAD controlled light beams to react the material or laser beams to sinter or melt form the material into near/net shape products. Within the past few years this technology has reached a degree of maturing that is now being used to production produce componentry.

J.C. Withers (✉) · S.M. Pickard
ATS-MER, LLC, Tucson, USA
e-mail: jwithers@atsmer.com; mercorp@mercorp.com

S.M. Pickard
e-mail: spickard@mercorp.com

The Minerals, Metals & Materials Society, *TMS 2017 146th Annual Meeting & Exhibition Supplemental Proceedings*, The Minerals, Metals & Materials Series,
DOI 10.1007/978-3-319-51493-2_9

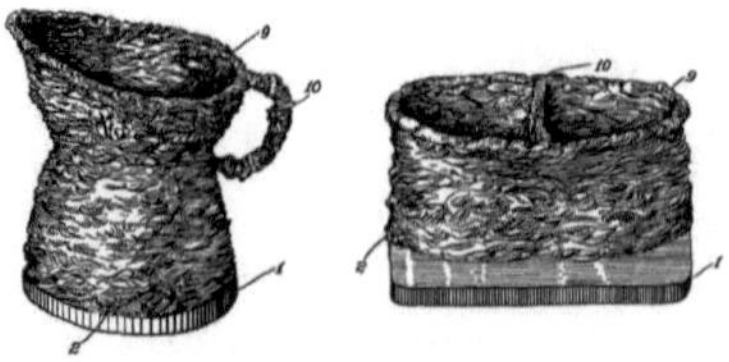
Fig. 1 Figures from Baker's patent [1]

Additive manufacturing (AM) processing technology has been enhanced to the point of using metals in additive metal manufacturing (AMM). However, the routes of additive metal manufacturing (AMM) dates back to at least 1926 when Baker [1] patented the use of an electric arc as a heat source to generate 3D objectives depositing molten metal in superimposed layers as shown from his patent in Fig. 1. In 1971 Mistubishi patented [2, 3] processing utilizing welding of multiwire to produce a functionally graded wall which was followed in the early 80s by Kussmaul [4] manufacturing very large steel parts weighing 79 tons at a build rate of 80 kg/h.

With the advent of computer 3D graphic CAD systems, the capability of fabrication layer-by-layer from a digital design package has generated a paradigm in the capabilities of additive metal manufacturing (AMM) that has taken off exponentially in the mid-2000s. There are a family of processes in which parts are first modeled into a CAD program that slices them into thin layers where specialized machines then build-up parts layer-by-layer according to the pattern in the program by injecting metal powder or wire into a high temperature plasma, laser beam or electron beam where it is melted into a pool that builds the layer; or by laser-sintering a powder within a thin layer of powder onto a substrate or previous built layer. More recently a solid state non-melt/fusion process termed friction stir processing (FSP) has evolved into repair as well as building up layers which has been termed friction stir additive manufacturing (FSAM). The FSP/FSAM and the plasma beam process has been scaled to commercial demonstration of producing parts several feet in dimension in thickness from approximately 0.015 to 7 inches or more. The plasma beam for fusion/melt processing is less expensive to generate than e-beams or laser beams to generate the melt pool as well as the feedstock is melted by direct electrical energy in contrast to the expensive beams. Power systems to generate plasma beams of ionized gas are low cost, require little to no maintenance, are energy efficient and exhibit low operating cost. The plasma beam system often referred to as plasma transferred arc (PTA) accommodates feeds of powder and/or wire. Torch sizes from a few amps to build intricate configurations at surface finishes down to 30 micro-inches as deposited up to hundreds to a thousand amps can build scores to hundreds of pounds per hour that contributes to low cost processing. Multiple torches can be used to increase the very high build rate of the PTA power system which provides a basis for production of 3D printing additive metal manufacturing (AMM) at a cost less than the lower build rate laser and e-beam AMM systems. The process control from the feed through all deposition/build variables can be automated with real-time closed loop control for in situ real time machining that provides the capability to produce tolerances of ±0.001 inch in a defect free predictable property material. In situ machining with

PTA AMM processing was instituted several years ago wherein a defect of any kind or out of tolerance can be machined away in real time and rebuilt providing a net shape capability in a defect-free material. The PTA capability to direct electrical energy melt, provides the capability to melt form very high temperature materials such as tungsten and the refractory metal carbide eutectics that melt at >4000 °C. The PTA can be operated in a closed atmosphere chamber as well as in the open with encompassing inert atmosphere shields similar to welding. The precise control of the cooling rate of the plasma-beam-produced mini-pool provides a capability to control microstructure of the solidifying deposited alloy avoiding all columnar growth patterns. It is possible with the PTA AMM system to build while maintaining the part hot in an annealed state or even in a superplastic state to prevent built-in stresses and warping. Controlled heat management from one build layer to another also assures uniformity of microstructure in the deposit from the first to last layer built. Heat management in built material is also desirable in repair of parts to prevent heat damage in the basic parent part. It is also possible to cryogenically quench cool the solidifying mini-pool to achieve a random oriented nanostructure. This controlled nano-microstructure typically produces strength properties in most metals on the order of 10–25% greater than the same material composition in the wrought state.

The friction stir process FSP-FSAM is a simple straight forward process using a non-consumable tool with a center tip such as shown in Fig. 2. Material in a superplastic state is transferred from the front of the spinning pin under the downward pressure shoulder to the back of the pin in a continuous linear forward plane creating a very fine to nano-structure exhibiting strength well above standard wrought materials. The FSP can be coupled with the PTA fusion/melt process just behind the molten pool that overcomes any tendency to form columnar microstructures. The processing can be performed in a hot annealing build box and/or a preheat such as a torch in front of the PTA which can be followed with the FSP producing materials at very fast build rates exhibiting strengths 2X standard wrought material.

The morphology of the feed material plays a significant role in the quality of geometry produced as well as the properties of the produced material. In the case of titanium; sponge, which is the most basic form and lowest cost of titanium, can be used as the feed with the PTA additive metal manufacturing (AMM) process. Alloying elements can be co-fed with the sponge or electrolytic powder to produce any standard titanium alloy, or custom alloys as well as grading from one composition to another. The capability of grading from one composition to another through the computer control feed systems provides a capability to customize

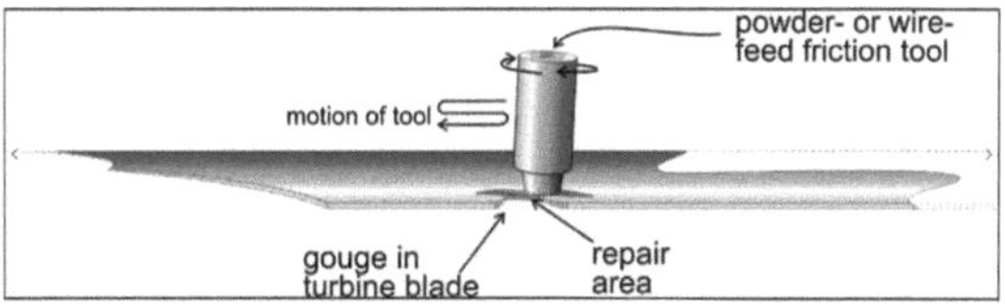

Fig. 2 Illustration of FSP repair of a blade defect

compositions and properties at specific locations within a part. The PTA AMM process when fully instrumented feedback control with in situ real time machining provides the capability to produce parts within a tolerance of ±0.001″ and 30 micro-inch finish, which for many applications is net shape, and with in situ machining produces defect free material in a finished net shape. The PTA AMM process with variable gas composition passed through or into the plasma and control of feed composition can be combined to produce porous parts as well as porosity/foam combined with fully dense material. Porosity volume can be varied from approximately 10 to 80% and fully dense islands or layers can be produced within the porous geometry as well as fully dense skins that may encompass the entire porous body.

Manufacturing of Parts

Three plasma transferred arc (PTA) 3D printing additive metal manufacturing (AMM) systems for producing metals and ceramics is shown in Fig. 3 with the illustration of the PTA process shown in Fig. 4.

Fig. 3 **a** Operational PTA unit, **b** laboratory plasma transferred arc (PTA) system and **c** robotic PTA unit with in situ machining capability at MER Corporation

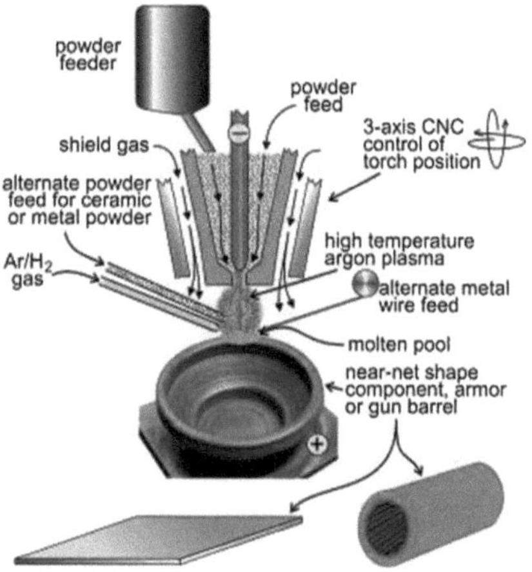

Fig. 4 Illustration of a plasma transferred arc (PTA) additive metal manufacturing (AMM) system in operation

The PTA AMM system has been utilized to produce a wide variety of metals and ceramics. An example of some metal configurations produced by PTA AMM is shown in Fig. 5.

An example of a metal alloy is shown in Fig. 6 and a high temperature ceramic in Fig. 7.

Components are readily produced by PTA AMM processing such as shown in Figs. 8, 9 and 10. Figure 11 shows PTA AMM producing a high modulus (150 GPa) preform in a custom alloy composition forged into connector rods.

An example of producing all and part foam pieces by PTA AMM as shown in Fig. 12.

Fig. 5 A sampling of components produced by the PTA AMM system, bottom right two are 4 × 5 feet and 6 feet

Fig. 6 1:1 Ta-Zr Alloy Produced by PTA AMM

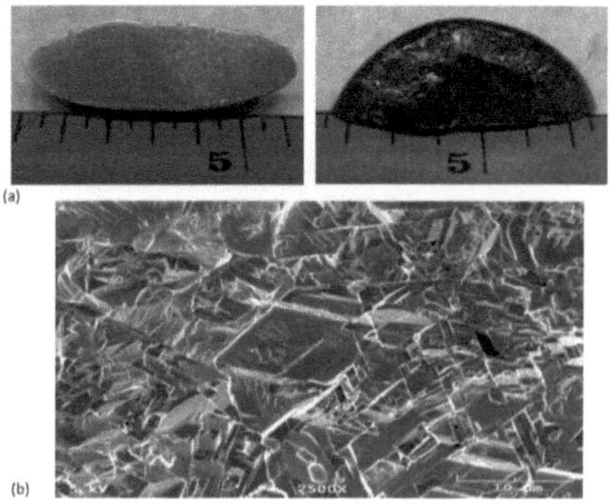

Fig. 7 **a** A boule of HfC-C eutectic cut in half by EDM machinery and **b** SEM of a broken flexure test specimen of HfC-C eutectic

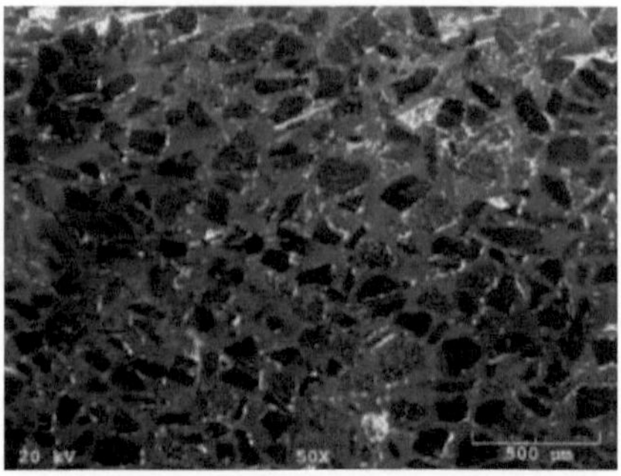

Fig. 8 Al-60 SiC

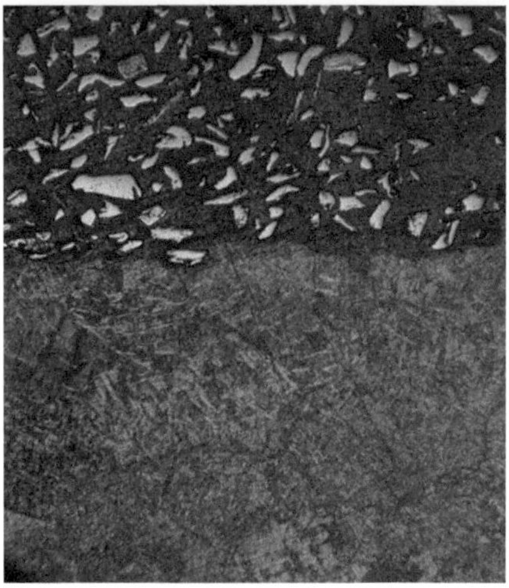

Fig. 9 Al-60 B_4C

Fig. 10 **a** TiB_2/Ti composite integrally built on a Ti-6Al-4 V base by PTA AMM processing; **b** Ti/TiB2 after 2 hits by APM2 at muzzle velocity, no cracking

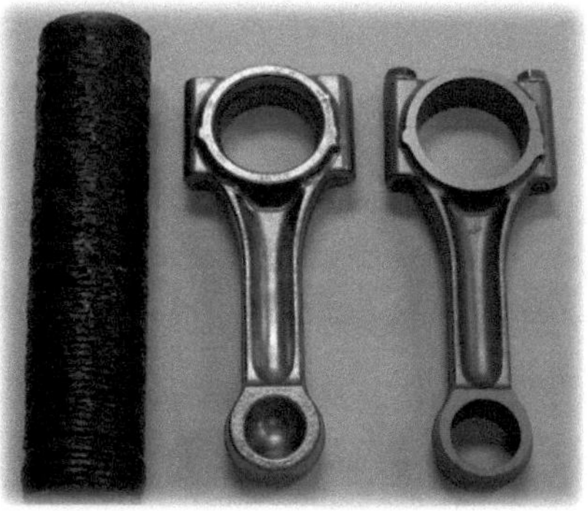

Fig. 11 A PTA high modulus (150 GPa) titanium preform forged into a connecting rod

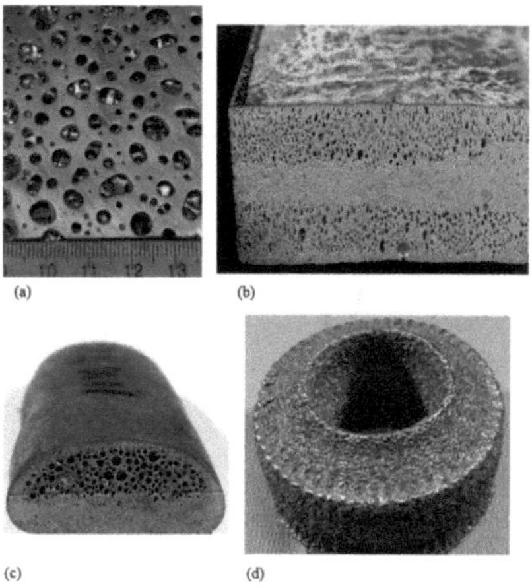

Fig. 12 **a** PTA AMM produced titanium foam, **b** foam with a solid layer, **c** foam with encapsulating skins of different thicknesses and **d** a final foamed part

Table 1 Mechanical properties of PTA AMM prepared Ti-6Al-4 V and proprietary alloys from titanium sponge feed

Ti64 source	UTS (MPa)	Yield strength (MPa)	Modulus (GPa)	Density (gm/cc)	Elongation (%)
Commercial	964	882	117	4.43	12
PTA AMM as deposited	978	882	117	4.43	10
PTA AMM, heat treated	1116	964	117	4.43	12
Proprietary Alloy (1) PTA AMM produced	1316	–	130	4.5	4
Proprietary Alloy (2) PTA AMM produced	2067	–	200	4.5	0.5–2.5

An example of strength of titanium compositions produced using titanium sponge as a feed is as shown in Table 1.

The FSP-FSAM non-melt solid state process in various configurations to process materials is shown in Fig. 13. The FSP-FSAM solid state process eliminates build stresses and warping. The linear speed can be up to scores of inches per minute

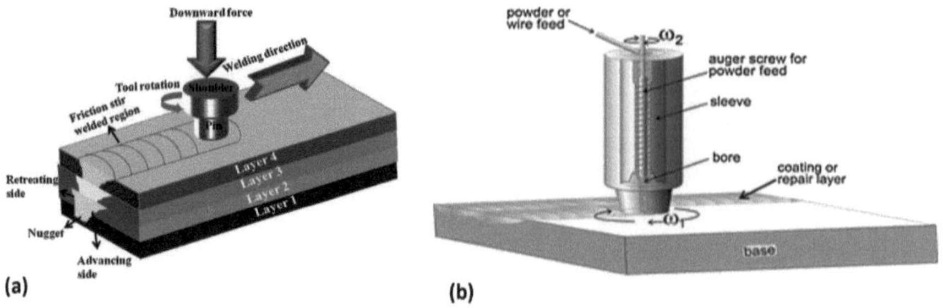

Fig. 13 **a** A schematic illustration of friction stir additive manufacturing process, build direction is perpendicular to the welding direction. **b** FSP/FSAM repairing using a powder or wire feed to base surface

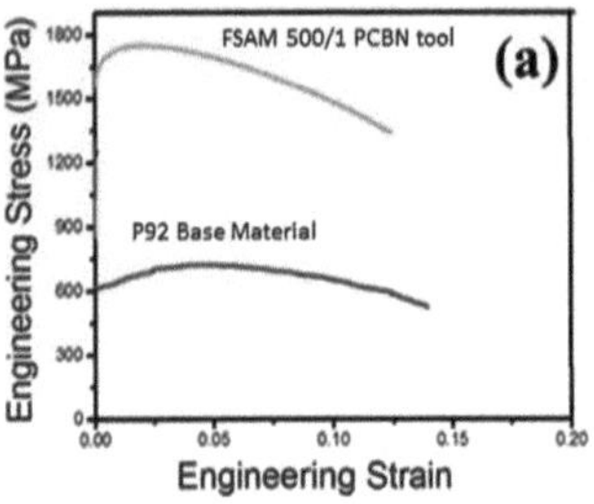

(b)

Condition	Yield Strength (MPa)	Tensile Strength (MPa)	% Elongation
Base	610	720	14
FSAM 500/1 (D2M)	1235	1441	16
FSAM 500/1 (PCBN)	1616	1747	12

Fig. 14 Example of friction stir additive manufacturing (FSAM) of P92 steel showing strength increase of over 250% over base steel produced by standard processing

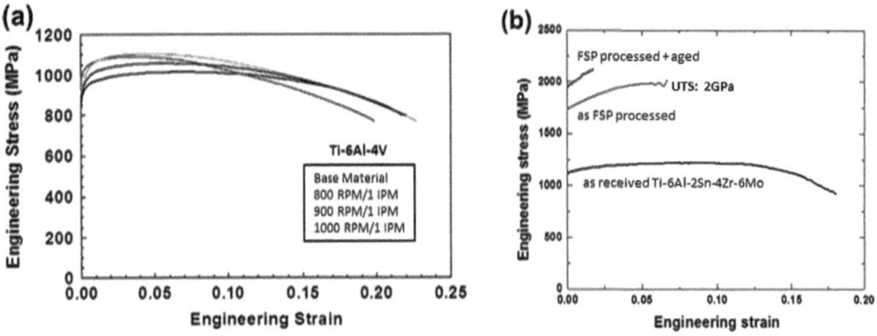

Fig. 15 **a** α + β Ti-6-4 alloy showing strength levels equivalent of β alloys with significant ductility (no effect of processing). **b** Strength levels of near β Ti-6-2-4-6 FSP processed alloy

providing high build rates. Repairs are ideal since there is no melting to damage the parent material being repaired. Very hard and composite claddings can readily be performed utilizing FSP-FSAM. The FSP-FSAM process is ideal to process structural materials such as P92 steel and titanium alloys such as Ti-6Al-2Sn-4Zr-6Mo as shown in Figs. 14 and 15. With such exemplary properties, FSP-FSAM has extraordinary potential to produce enhanced properties in materials treated by this processing.

Conclusions

Layer-by-layer metal additive manufacturing has a plethora of applications to produce standard and custom alloys at lower cost with near immediate turn around which can re-establish a manufacturing base in the U.S. There are multiple additive metal manufacturing processes to melt metal powder and/or wire feed into a pool that builds the part into near/net shape. Plasma transferred arc (PTA) is an energy efficient beam process to low cost melt form a powder and/or wire feed producing standard and custom alloys exhibiting mechanical properties greater than standard processed material producing parts into multiple feet sizes. It is also possible to (PTA) produce porous foam metal alloys that may have solid section within. PTA can be practiced with multiple torches and with friction stir processing (FSP) including building up parts designated as FSAM. The friction stir process is a solid state non-melt process that eliminates stresses and warping, and producing very fine/nano-structures that have demonstrated strengths up to 2X standard wrought material.

References

1. R. Baker, Method of Making Decorative Articles, US Patent 1,533,300, 14 Apr 1925
2. U. Akira, Method of and apparatus for constructing substantially circular cross section vessel by welding, US3558846A, 26 Jan 1971
3. H.R. Edgecomb, Device for Testing Plastic Materials, US118437A, 22 Aug 1913
4. K. Kussmaul, F.W. Schoch, H. Luckow, High quality large components 'Shape Welded' by a SAW process. Weld. J. (1983)

Aiming for Modeling-Assisted Tailored Designs for Additive Manufacturing

Dayalan R. Gunasegaram, Anthony B. Murphy, Sharen J. Cummins, Vincent Lemiale, Gary W. Delaney, Vu Nguyen and Yuqing Feng

Abstract It is well recognized that there are gaps in knowledge on the strongly intertwined process–microstructure–property–performance relationships inherent in the metallic additive manufacturing processes. Computational modeling can assist with filling in some of these gaps by increasing in-depth understanding of these relationships and highlighting cause-and-effect. Additionally, it can capture the knowledge of materials scientists and engineers and apply established physics-based rules to simulate the processes and thus predict the final outcomes. Modeling can also help optimize processes. Some even predict that future generations of additive manufacturing machines will employ 'model-assisted feed forward algorithms' that would leapfrog feedback control methods. In the current article the authors describe the several computational efforts sponsored by CSIRO's 'Lab 22—Australia's Centre for Additive Innovation' aimed at modeling-assisted tailored design. The models in development, e.g. microstructure prediction (both fundamental and empirical), powder bed raking, and residual stress predictions, are described in some detail, and representative results are presented.

Keywords Microstructure model · Powder raking model · Residual stress model · Finite element method · Phase field method · Discrete element method

D.R. Gunasegaram (✉) · A.B. Murphy · V. Nguyen
Manufacturing Business Unit, Commonwealth Scientific and Industrial Research Organisation (CSIRO), Private Bag 10, Clayton (Melbourne), VIC 3169, Australia
e-mail: dayalan.gunasegaram@csiro.au

S.J. Cummins · V. Lemiale · G.W. Delaney
Data61, Commonwealth Scientific and Industrial Research Organisation (CSIRO), Private Bag 10, Clayton (Melbourne), VIC 3169, Australia

Y. Feng
Mineral Resources Business Unit, Commonwealth Scientific and Industrial Research Organisation (CSIRO), Private Bag 10, Clayton (Melbourne), VIC 3169, Australia

The Minerals, Metals & Materials Society, *TMS 2017 146th Annual Meeting & Exhibition Supplemental Proceedings*, The Minerals, Metals & Materials Series, DOI 10.1007/978-3-319-51493-2_10

Introduction

Metallic additive manufacturing (AM) is a disruptive technology that has transformed the design and manufacture of premium-value modern products, most notably in the aerospace and biomedical industries. Additively manufactured parts are designed based on the new design paradigm 'design for performance', rather than the conventional 'design for manufacture' philosophy, since design limitations that are associated with traditional manufacturing routes no longer apply. Thus parts can be optimized with increased freedom for their fitness for use, e.g. reducing the mass of aerospace parts and allowing significant personalization in biomedical applications.

Optimization of part design for performance, however, invariably pushes the boundaries of AM processes; these boundaries arise sometimes through physical limits (e.g. diffusion rates) and at other times owing to restrictions in capabilities of machines. Alternatively, if the capabilities and margins of AM processes and machinery are well understood quantitatively, designers will have a strong understanding of the constraints that they need to impose during the optimization process.

The design of a part is strongly influenced by the chosen material of construction, since the properties of the material often determine its fitness for use, e.g. by affecting strength and rigidity. The AM process provides substantial flexibility to designers in tailoring local properties of the material, allowing the introduction of functionally-graded material properties in a single part. This is because the product is constructed through the layer-by-layer deposition of metal, a process that can be modulated to obtain a pre-determined property at a given location. This is done by taking advantage of the process–microstructure–property–performance relationships, which can be altered through process parameters such as the chosen material and the cooling rate. Hence an in-depth understanding of these relationships is pivotal to consistently achieving desired tailored properties at various locations with a high degree of confidence.

In the popular powder bed systems employed in AM machinery, a 'rake' is used to even out the powders before the heat source fuses the powders to create the part. The raking process affects microstructures indirectly, since the stacking characteristics of the powders on the bed has an influence on the absorption of the incident heat source, and on conductive heat transfer rates during the fusion of powders. Hence a model that mimics the raking process will be able to provide insights into how the process may be optimized to potentially improve fusion characteristics and reduce porosity. The ability to create accurate, tailored designs in microstructure would depend on how well the raking model recreates the process for different powder shapes and sizes.

An issue that often needs to be resolved in AM is the residual stress, which develops in the part as a consequence of the characteristically large thermal gradients that arise during the build stage. The thermal gradients are a result of heat being lost from the already built layers while, at the same time, a thermal load is added to the nascent layers a relatively short distance away. Residual stresses can create distortion and/or cracking in the part as it cools to room temperature. The local microstructure plays a major role in determining how well the material is able to withstand the local stresses. Hence, developing a robust understanding of the interplay between residual stresses and microstructure as well as developing strategies to manage residual stresses are key to making good quality AM parts with pre-determined properties.

Physics-based computational modeling that encapsulates experimental knowledge can assist with addressing the challenges encountered in AM processing by helping to deepen the understanding of the processes and thereby paving the way for tailored designs. For instance, it can assist with investigating the cause of defects by interrogating the cause-and-effect relationships associated with various process parameters, including alloy composition, power of the heat source (e.g. laser beam or electron beam), scan rate, part design, cooling rate, etc. In addition, modeling allows insights into fundamental processes that are not observable in situ; for example, into how the process parameters control the evolution of microstructures. The microstructures govern the properties that ultimately determine the fitness for use, and can be predicted in a virtual domain through validated physics-based models. Such models can also be used for the optimization of AM processes. Frazier [1] even argues that such validated models should be used as an alternative to conventional qualification methods, since part-by-part certification is time consuming, and by extension expensive, in an industry awash with customized products. Looking ahead, Herderick [2] suggested that closed-loop feedback control sensing systems in future AM machines should be integrated with model-based intelligent feed forward schemes to better control the manufacturing cycle. These feed-forward systems typically track certain disturbance variables D that are indicators of an important control variable C that needs to be maintained within set limits. If D at any time points to the possibility of C going out of bounds, the system draws from artificial intelligence derived from model predictions and moves the process by altering a tunable variable T so that C remains within its set limits *before* any damage is done. Whilst models used in feed-forward algorithms are usually those that can be solved quickly (e.g. Artificial Neural Networks or Cellular Automata), physics-based models can still be used to teach such models.

Recognizing that computational modeling is integral to the push to make AM more productive and commercial-ready, workers worldwide have developed several computational models and, in some cases, incorporated these into commercial software packages, e.g. Flow-3D [3]. Recent reviews by King et al. [4] and Gong et al. [5] are excellent sources of information on the metallic-AM-related computational models relating to the powder bed systems; Megahed et al. [6] have, in addition, discussed models relating to blown powder systems as well as residual stress development. In the present article, the authors describe physics-based models being developed by CSIRO's Lab 22—Australia's Centre for Additive Innovation—covering microstructure prediction, raking of powders in powder bed systems and residual stress management.

Microstructure Models

Before tunable microstructures can be achieved, part designers should have access to a strong predictive capability that takes into account the influence of contributing process parameters. Consequently, microstructure modeling becomes an essential part of the tailored design process. Microstructure models can be constructed at the various length scales—from continuum down to the mesoscale and subsequently to the more fundamental atomic or molecular scales. Continuum models based on empirical observations provide rapid predictions using relatively-low computational power for scenarios around which experimental data are available for interpolation. However, they cannot be interrogated from a more fundamental viewpoint. Whilst the atomic and molecular-scale models are capable of lending themselves to such interrogation, the computing power available today can only support modeling of a very small volume of the entire part in this way. However a model constructed at the intermediate micro/mesoscopic scale provides the scope for some useful interrogation, as it aligns with the characteristic length scale of microstructure formation. In addition, it lends itself to simulating a reasonable volume of the part using state-of-the-art parallel computing hardware. These latter models will be able to shed light on phenomena occurring at the microscopic level, e.g. crystallization, crystal growth, and grain boundary formation, as well as account for processes such as diffusion that control the rate of non-equilibrium solidification that characterizes AM. Hence, it is useful to have different microstructure prediction models at the various length scales, since they serve diverse purposes. Here we present a continuum scale model that is being developed and a mesoscale model that is in the planning stages.

Continuum-Scale Semi-empirical Finite Element Analysis (FEA) Model for Ti-6Al-4V

The titanium alloy Ti-6Al-4V is widely used due to its low density, high strength and corrosion resistance. Its mechanical properties are directly affected by the microstructure. The Ti-6Al-4V microstructure is formed during thermomechanical treatment and consists of two elemental phases, a hexagonal-close-packed (hcp) α phase and a body-centered-cubic (bcc) β phase. The microstructure depends strongly on the thermal and mechanical conditions experienced during processing, and different grain morphologies form depending on the thermal history. Figure 1 shows a schematic of the different types of grain morphologies of Ti-6Al-4V that can form based on thermal conditions (thermal spatial gradients, G and cooling rates, R), see [7]. For high cooling rates the β phase transforms into equiaxed martensite α_mart; an example micrograph is shown in Fig. 2. For low cooling rates, the α phase forms Widmanstatten plate-like or basket-weave structures α_wid by nucleation and growth; an example micrograph is shown in Fig. 3.

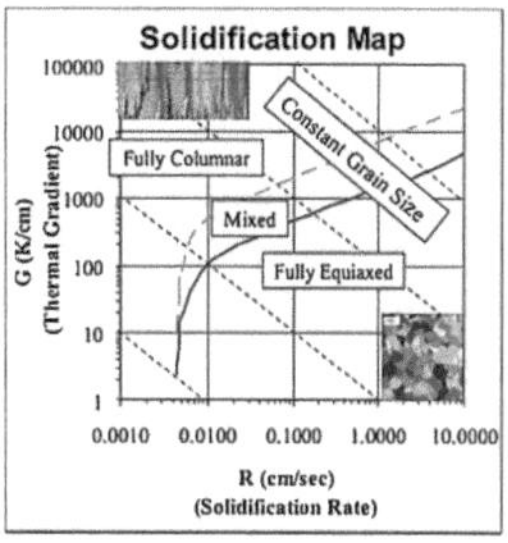

Fig. 1 Schematic showing different Ti-6Al-4V grain morphologies with thermal conditions

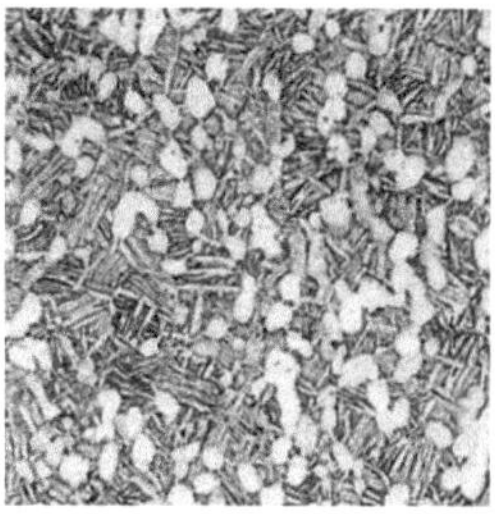

Fig. 2 Micrograph showing Ti-6Al-4V grain morphology after a fast cooling process, taken from [29]

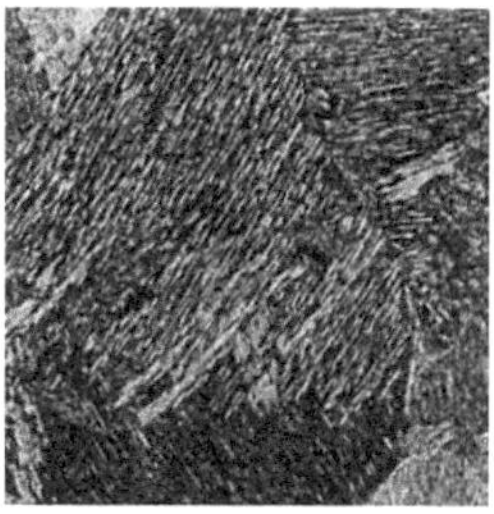

Fig. 3 Micrograph showing Ti-6Al-4V grain morphology after a slow cooling process, taken from [29]

Understanding and controlling the microstructure is important in order to maintain high quality during manufacturing and in the final product. We are currently evaluating a semi-empirical microstructure evolution model for Ti-6Al-4V, proposed in [8], which permits the prediction of the microstructure during heating and cooling cycles as typically imposed in additive manufacturing applications. The approach is semi-empirical in that it does not predict any geometrical quantities in the microstructure, such as grain orientation and shape, but instead represents the microstructure phases, β, α_wid, α_gb and α_mart as volume fractions in a representative volume element (RVE). The transformation from one phase to another is modeled as either instantaneous or time-dependent, depending on which phases are involved. Instantaneous transformations are modeled using the Koistinen–Marburger equation, see [9], an example of which is the β to α_mart transformation:

$$\Delta v_{\alpha_{mart}} = \left(1 - e^{-B(T_{ms} - T)}\right)\left(v_\beta^n + v_{\alpha_{mart}}^n - v_\beta^{eq}\right). \tag{1}$$

In Eq. (1), v_β^n and $v_{\alpha_{mart}}^n$ are the volume fractions of the β and α_{mart} phases at time t^n and temperature T, and T_{ms} is the temperature at which the transformation starts. The equilibrium volume fraction of β is v_β^{eq} and the exponent B is material-dependent.

Time-dependent transformations are modeled using the JMA theory [10], an example of which is the β to α_{gb} transformation:

$$\Delta v_{\alpha_{gb}} = \left(1 - e^{-K(t_c + \Delta t)^N}\right)\left(v_\beta^n + v_{\alpha_{gb}}^n\right) v_{\alpha_{gb}}^{eq} - v_{\alpha_{gb}}^n \tag{2}$$

where $v_{\alpha_{gb}}^n$ is the volume fractions of the α_{gb} phases at current time t^n, N and K are material-dependent JMA constants and

$$t_c = \left[- \ln\left(1 - \frac{v_{\alpha_{gb}}^n / v_{\alpha_{gb}}^{eq}}{v_\beta^n + v_{\alpha_{gb}}^n}\right) / K\right]^{1/N} \tag{3}$$

Further details of the model, including transformations of the other phases, are in [8]. The advantage of this model is its simplicity, low computational cost and ability to cater for a broad range of temperatures. It is, however, a simplification of the microstructure evolution in that it does not model the nucleation and grain growth of the α and β phases as done in the more computationally-intensive microscale models [5–7]. Sensitivities of the model to the choice of JMAK exponents have also been noted in [8]. We plan to validate the model for a laser metal deposition application for which we have experimental data [11].

Mesoscale Phase Field Model (PFM)

We propose to take advantage of the relatively new modeling technique known as the phase field method (PFM), described in detail elsewhere [12, 13], to model microstructures at their characteristic length scales. PFM is ideal for connecting the lower atomic/molecular scales to the continuum scale [13], enabling the development of a multiscale description of AM processes. A knowledge of the microstructure at the microscopic scale is necessary to develop strategies that help control and tune these for tailored outcomes.

PFM has enjoyed increasing popularity in the modeling of microstructures in recent times. This is because it successfully addresses the limitations associated with the classical sharp interface method, where the boundary between phases is assumed to be infinitely thin. In PFM, by contrast, the boundary is considered smeared for mathematical expedience and is assigned a small thickness value,

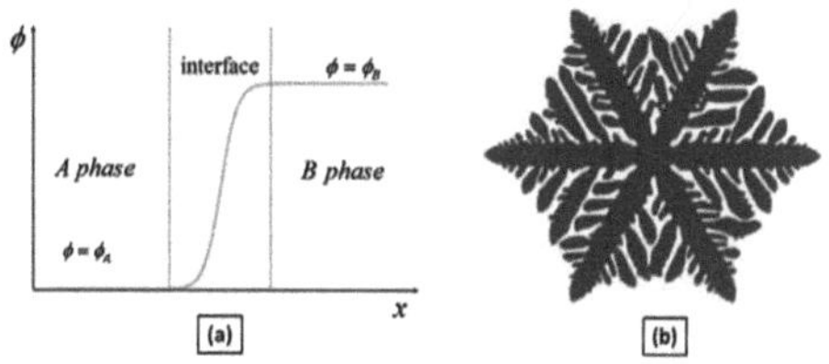

Fig. 4 **a** Schematic plot of the PF variable across an A-B interface. x is the surface normal coordinate [13] and **b** a typical dendrite crystal captured in a PF model [13]

which can be imagined physically as the atomically diffuse nature of the interface. A phase field variable ϕ (e.g. $0 < \phi < 1$) is employed to describe the state of the system at a given location and varies across the phase boundary (Fig. 4a). From the perspective of condensed matter physics, the phase field may be seen as an indicator describing the degree of crystallinity or atomic order or disorder in a system [12]. A function $g(\phi)$ constructed from the thermodynamics that govern the transition between different phases represents the free energy at a given location, the integration of which over the volume gives a complete description of free energy of the system. Kinetics that moves the system to non-equilibrium behavior is then modeled by imposing a condition to minimize the free energy of the system and letting the system evolve accordingly.

Since the model will describe the system at microstructural length scales, it will be able to account for several important phenomena occurring at this scale in AM processing, e.g. crystallization, grain growth (Fig. 4b), grain boundary formation, and solid state transformations limited by diffusion. All such phenomena are influenced by process parameters unique to AM, including the reheating that results from the heat received from newly laid layers, and are yet to be understood well. The predicted AM microstructures will be used to, in turn, estimate local mechanical properties relevant to the continuum scale. The PFM-based modeling work at CSIRO will be reported in future articles.

Discrete Element Method (DEM) Modeling of Powder Raking

A common technique employed in AM powder bed systems is to add successive layers of metal powder by raking a new layer across the existing surface. Understanding this raking process and how the properties of the powder particles (e.g. size, shape, density, interaction properties) and process parameters (e.g. height of powder layer, rake geometry, rake speed) affect the properties of the bed after raking is crucial in optimizing the performance of the system and ensuring the quality of the 3D-printed part.

We have developed a computational model of this raking process using the discrete element method (DEM) [14, 15]. This is a particle-based modeling technique, in which the interactions between individual particles are considered.

This technique has been applied widely in studying industrial particle flows, including screening [16], milling [17], crushing [18] and powder flow in additive manufacturing processes [19]. Extensive details of the DEM model used in this work are given in Refs. [14–18].

The shape of the individual powder particles plays a crucial role in determining their flow properties and the density of the packed powder bed [20]. In our model, we take account of the particle shape by using a distribution of super-quadric particle shapes, which can be matched to the shape distributions of the powder particles being studied. The shapes of our particles are given by:

$$\left(\frac{x}{a}\right)^m + \left(\frac{y}{b}\right)^m + \left(\frac{z}{c}\right)^m = 1 \quad (4)$$

where we refer to m as the shape parameter, and a, b and c are the semi-major axis lengths. For $m = 2$, we recover the general formula for an ellipsoid, while larger values of m generate increasingly cubic shapes. This gives us the ability to investigate a broad range of particle shapes, smoothly transitioning through a range of surface curvatures and aspect ratios [20]. In our model, we set the size and shape distributions of the particles to best match the powder being studied, and calibrate the inter-particle interaction properties using a simulated Hall Flowmeter so as to best match the measured flowability of the powder.

Flow images from our simulations for a reference Ti-6Al-4V metal powder are shown in Fig. 5, with stationary particles colored blue and the fastest moving particles colored red. The raking setup consists of two identical rakes with 0.8 mm wide metal teeth and a gap of 0.2 mm between pairs of teeth. The teeth of the two rakes are offset by 0.5 mm, with a 2.5 mm distance between the 2 rakes. The simulation consists of 875,000 particles ranging in size from 37 to 183 μm, with a median of 67 μm. The shapes of the powder particles were measured from SEM images and found to be very close to spherical, with a maximum aspect ratio of 0.92 used in the simulations to match the particle shapes. The initial powder bed is set up by deposition of the particles from above into a simulation box of length 42 mm by width 8 mm, with periodic boundary conditions employed in the direction normal to the motion of the rake. A small portion of powder is then placed in front of the rake and the rake is moved across the surface of the bed at a speed of 230 mm/s.

Fig. 5 Flow images from a DEM simulation of raking of a reference Ti-6Al-4V metal powder with particles colored according their speed; ranging from 0 m/s (*blue*) to 0.3 m/s (*red*). Images are shown from **a** *side view* and **b** *top view* at a position between the two rakes

Figure 5a shows a side view of the simulation after the rakes have travelled approximately a third of the way along the simulated length. The variations in the speed of the powder in front of the leading rake can clearly be seen, with the highest velocity powder particles flowing down the front of the sloped region. Figure 5b shows a view from above at a position between the two rakes. Here we see the complex flow of the powder particles through the first rake, with moderate speed (yellow-colored) particles flowing through the gaps between the rake teeth. This is followed by another high speed region in front of the second rake (colored red) and a similar reduced flow of powder through the gaps in the second rake. Understanding these fine flow features, and analyzing the surface smoothness and density variations within the bed after raking, allow us to optimize rake geometry and operational parameters such as rake speed to achieve a higher-quality powder bed after raking. Further, as noted in the introduction, the particle distribution strongly affects the thermal conductivity of the powder bed and the absorption of the incident laser or electron-beam energy.

Residual Stress Modeling

An undesirable side-effect of metal AM is the generation of residual stresses and deformation in the component due to the high thermal gradients. Similarly to welding, the quality of the material deposits has a substantial impact on the fatigue life of the structure. In some cases, the post-build distortion may render the component unsuitable for service while the residual stresses may cause issues for subsequent manufacturing processes if untreated. Attempts have been made previously to predict and manage post-build distortion [21].

Residual stress and deformation in AM processes can be modeled at a continuum level by employing the finite element method to carry out coupled thermal-structural analyses, as has been done in, e.g. wire–arc AM [22], wire–electron-beam AM [23] and selective laser melting [24]. In this approach, the transient thermal histories of the material elements obtained from thermal modeling are used for structural analyses, where the mechanical properties of the metal are usually functions of temperature. To model the added material, the deposit elements are deactivated at the beginning of the simulations and gradually activated when the heat source passes through. The material behavior is described by thermo-elasto-plastic constitutive laws, which have been implemented in most commercial finite element solvers.

In our study, the MSC Marc solver [25] was employed for the analyses. The material was assumed isotropic at all temperature ranges and yielding behaviour was assumed to follow the von Mises criterion [26].

The Johnson–Cook constitutive model [27] was initially employed to describe the evolution of yield, since it can simultaneously account for the effects of strain hardening, strain rate and temperature. The flow stress is expressed as

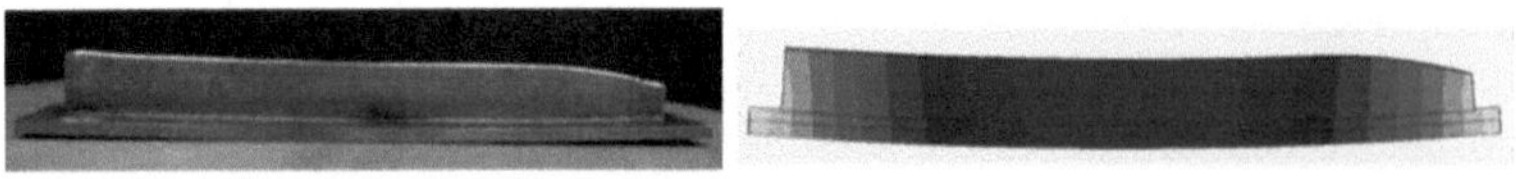

Fig. 6 Post-built distortion ([11]—Boeing approval for public release Ref 14-00402-EOT)

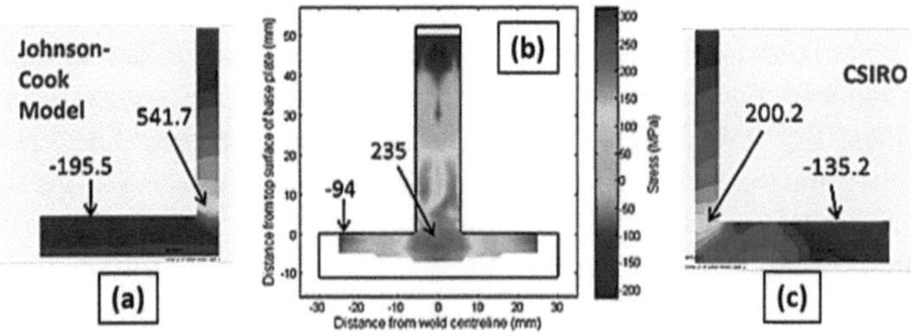

Fig. 7 Residual stress (MPa) at the central section of T-shaped part **a** predicted with Johnson–Cook constitutive model, **b** measured by neutron diffraction method, and **c** predicted with CSIRO constitutive data ([11]—Boeing approval for public release Ref 14-00402-EOT)

$$\sigma_{\mathrm{eq}} = \left(A + B\varepsilon_{\mathrm{eq}}^{n}\right)\left(1 + C\,\ln\dot{\varepsilon}_{\mathrm{eq}}^{*}\right)\left[1 - \left(\frac{T - T_r}{T_m - T_r}\right)^{m}\right] \quad (5)$$

where $\dot{\varepsilon}_{\mathrm{eq}}^{*} = \dot{\varepsilon}_{\mathrm{eq}}/\dot{\varepsilon}_0$ is the non-dimensional plastic strain rate; $\dot{\varepsilon}_{\mathrm{eq}}$ and $\dot{\varepsilon}_0$ are respectively the plastic strain rate and the effective plastic strain rate of the quasistatic test used to determine the yield and hardening parameters A, B and n. T_m is the melting temperature and T_r is room temperature, and A, B, C, n and m are material constants. Appropriate Johnson–Cook parameters for Ti6Al4 V alloy have been discussed in the literature, e.g. [28].

To validate our models, a 600-mm-long, 100-mm-wide and 12.7-mm-thick Ti6Al4 V alloy substrate was additively manufactured using wire-electron beam melting to form a T-shaped part having a 51-mm-high and 11.7-mm-wide deposit, as shown in Fig. 6. It was reported by Nguyen et al. [11] that residual stress measured using the neutron diffraction method at the central section of the built part, shown in Fig. 7b, indicated the model that employed the Johnson–Cook model, shown in Fig. 7a, over predicted the residual stress at the corresponding location. An improved model using CSIRO constitutive data (Fig. 7c) shows a significant improvement in the accurate prediction of residual stress.

Concluding Remarks

Physics-based computational models are a useful tool not only for understanding and optimizing AM processes but, through the application of such improved in-depth understanding, also for developing strategies to control the processes towards achieving tailored designs. As such, they play an integral role in the development of these processes and are likely to form a significant part of their future success. We have discussed in some detail the metallic AM-related models being developed by CSIRO for the prediction of microstructures, raking of powders and the management of residual stresses.

Acknowledgements The authors wish to thank CSIRO colleague Stefan Gulizia for obtaining financial support for the powder raking model development. The contributions of Sri Lathabai, Anthony Miller and John Barnes of CSIRO and Gary Coleman and Amy Helvey of The Boeing Company to the development of the residual stress model are gratefully acknowledged.

References

1. W.E. Frazier, Metal additive manufacturing: a review. J. Mater. Eng. Perform. **23**, 1917–1928 (2014)
2. E. Herderick, Additive manufacturing of metals: a review, in *Proceedings of the Materials Science and Technology (MS&T) Conference*, Columbus, Ohio, 16–20 Oct 2011, pp. 1413–1425
3. https://www.flow3d.com/laser-powder-bed-additive-manufacturing. Accessed 6 Sept 2016
4. W.E. King, A.T. Anderson, R.M. Ferencz, N.E. Hodge, C. Kamath, S.A. Khairallah, A.M. Rubenchik, Laser powder bed fusion additive manufacturing of metals; physics, computational, and materials challenges. Appl. Phys. Rev. **2**, 041304 (2015)
5. X. Gong, T. Anderson, K. Chou, Review on powder-based electron beam additive manufacturing Technology. Manuf. Rev. **1**, 2–11 (2014)
6. M. Megahed, H.-W. Mindt, N. N'Dri, H. Duan, O. Desmaison, Metal additive-manufacturing process and residual stress modeling. Integr. Mater. Manuf. Innovation **5**, 1–33 (2014)
7. J.D. Gockel, J.L. Beuth, Understanding Ti-6Al-4V microstructure control in additive manufacturing via process maps, in *Solid Freeform Fabrication Proceedings*, Austin, Texas, Aug 2013, pp. 666–674
8. M.C. Charles, R. Pederson, L.E. Lindgren, A model for Ti–6Al–4 V microstructure evolution for arbitrary temperature changes. Modell. Simul. Mater. Sci. Eng. **20**, 055006 (2012). doi:10.1088/0965-0393/20/5/055006
9. D.P. Koistinen, R.E. Marburger, A general equation prescribing the extent of the austenite-martensite transformation in pure iron-carbon alloys and plain carbon steels. Acta Metall. **7**, 59–60 (1959)
10. M. Avrami, Kinetics of phase change: I general theory. J. Chem. Phys. **7**, 1103–1112 (1939). doi:10.1063/1.1750380
11. V. Nguyen, S. Lathabai, Y. Feng, A. Miller, J. Barnes, G. Coleman, A.M. Helvey, Further development of a predictive tool for managing distortion in electron beam direct manufacturing. Paper presented at the 25th Advanced Aerospace Materials and Processes (AeroMat) Conference and Exposition, Orlando, Florida, 16–19 June 2014
12. N. Provatas, K. Elder, *Phase-Field Methods in Material Science and Engineering* (WILEY-VCH Verlag GmbH & Co. KGaA, Weinheim, 2010), pp. 1–7
13. N. Wang, L.Q. Chen, Phase field methods, in *Multiscale Paradigms in Integrated Computational Materials Science and Engineering*, ed. by P.A. Deymier et al. (Springer International Publishing, Switzerland, 2016), pp. 195–217
14. P.A. Cundall, O.D.L. Strack, A discrete numerical model for granular assemblies. Geotechnique **29,** 47–65 (1979)
15. P.W. Cleary, Large scale industrial DEM modelling. Eng. Comput. **21**, 169–204 (2004)

16. G.W. Delaney, P.W. Cleary, M. Hilden, R.D. Morrison, Testing the validity of the spherical DEM model in simulating real granular screening processes. Chem. Eng. Sci. **68**, 215–226 (2012)
17. G.W. Delaney, P.W. Cleary, R.D. Morrison, S. Cummins, B. Loveday, Predicting breakage and the evolution of rock size and shape distributions in Ag and SAG mills using DEM. Miner. Eng. **50–51**, 132–139 (2013)
18. G.W. Delaney, R.D. Morrison, M.D. Sinnott, S. Cummins, P.W. Cleary, DEM modelling of non-spherical particle breakage and flow in an industrial scale cone crusher. Miner. Eng. **74**, 112–122 (2015)
19. E.J.R. Parteli, T. Pöschel, Particle-based simulation of powder application in additive manufacturing. Powder Technol. **288**, 96–102 (2016)
20. G.W. Delaney, P.W. Cleary, The packing properties of superellipsoids. Europhys. Lett. **89**, 34002 (2010)
21. V. Nguyen, S. Lathabai, Y. Feng, A. Miller, J. Barnes, G. Coleman, A.M. Helvey, Distortion prediction and minimisation in additive manufacturing, WO/2015/184495 (2015)
22. J. Ding, P. Colegrove, J. Mehnen, S. Williams, F. Wang, P. Sequeira Almeida, A computationally efficient finite element model of wire and arc additive manufacture. Int. J. Adv. Manuf. Technol. **70**, 227–236 (2014)
23. E.R. Denlinger, J. Irwin, P. Michaleris, Thermomechanical modeling of additive manufacturing large parts. J. Manuf. Sci. Eng. **136**, 1–8 (2014)
24. L. Parry, I.A. Ashcroft, R.D. Wildman, Understanding the effect of laser scan strategy on residual stress in selective laser melting through thermo-mechanical simulation. Addit. Manuf. **12**, 1–15 (2016)
25. MSC-Marc *Volume A: Theory and User Information*, MSC Software Corporation (2013)
26. R. von Mises, Mechanik der festen Körper im plastisch deformablen Zustand. Göttin. Nachr. Math. Phys. **1**, 582–592 (1913)
27. G.R. Johnson, W.H. Cook, A constitutive model and data for metals subjected to large strains, high strain rates and high temperatures, in *7th International Symposium on Ballistics* (American Defense Preparedness Association, Koninklijk Instituut van Ingenieurs, The Hague, 1983), pp. 541–547
28. A.S. Khan, Y.S. Suh, R. Kazmi, Quasi-static and dynamic loading responses and constitutive modeling of titanium alloys. Int. J. Plast. **20**, 2233–2248 (2004)
29. B. Gaspar, Microstructural characterization of Ti-6Al-4V and its relationship to sample geometry. Report, Materials Engineering Department, Cal Poly State University-San Luis Obispo, California (2012)

Part III
Additive Manufacturing: Building the Pathway towards Process and Material Qualification

The Use of Laser Ultrasound to Detect Defects in Laser Melted Parts

Sarah Everton, Phill Dickens, Chris Tuck, Ben Dutton and David Wimpenny

Abstract Additive manufacturing (AM) offers a number of benefits over conventional manufacturing, giving an increased design freedom and the opportunity to integrate multiple components, saving weight. The rigorous standard for material integrity set by the Aerospace sector necessitates the development of systems to ensure quality. Laser ultrasonic testing (LU) is a non-contact inspection technique which has been proposed as suitable for in-situ monitoring of additive manufacturing processes, as measurements can be taken at elevated temperatures. This paper will show the capability of this technique for assuring material quality in metal AM parts and compare LU with x-ray computed tomography (XCT).

Keywords Laser ultrasound · Defect characterisation · Non-destructive testing · Additive manufacturing · Laser powder bed fusion

Introduction

The development of additive manufacturing (AM) processes in recent years, due in-part to an increased understanding of key process variables (KPVs), materials and the underlying design principles, has resulted in a rapid uptake of the technology across multiple engineering sectors [1]. With AM offering the opportunity to save on product weight and cost, simplify assemblies and enhance component performance, it is the lack of assurance of AM part quality proving a key barrier for the Aerospace sector [2].

S. Everton · P. Dickens (✉) · C. Tuck
Additive Manufacturing and 3D Printing Research Group, University of Nottingham, University Park, Nottingham NG7 2RD, UK
e-mail: phill.dickens@nottingham.ac.uk

S. Everton · B. Dutton · D. Wimpenny
The Manufacturing Technology Centre, Ansty Business Park, Pilot Way, Coventry, Warwickshire CV7 9JU, UK

The Minerals, Metals & Materials Society, *TMS 2017 146th Annual Meeting & Exhibition Supplemental Proceedings*, The Minerals, Metals & Materials Series,
DOI 10.1007/978-3-319-51493-2_11

The in-situ acquisition of data through process monitoring has been identified as a priority area of research by the US National Institute of Standards and Technology [3] and the UK AM Steering Group [4]. The in-situ detection of defects and use of closed-loop process control will increase AM process robustness and remove the barrier to adoption of AM, for Aerospace. The term AM refers to a number of different processes, characterised by the energy source, feedstock or melting mechanism. A comprehensive introduction to AM is given by Gibson et al. [5]. This paper focusses on the identification of defects generated during manufacturing by laser powder bed fusion (PBF).

Laser PBF has a large number of input parameters which affect the quality of the final component build [6] and therefore, many studies have been carried out to assess the effects of these, sometimes interdependent, parameters [7, 8]. Commonly occurring "defects" include material discontinuities such as pores, inclusions and cracks. Limiting the porosity of AM components is essential to avoid adversely affecting a components material performance [9]. Several different mechanisms are responsible for the generation of pores during laser PBF processing, yielding pores with different characteristics. Selecting parameter combinations that result in over- or under-melting of the material results in increased porosity. The pores can be further distinguished by their shape and can be categorised as spherical [10], keyhole [11] or elongated pores [12]. The different pores vary in size from spherical pores with 20 μm diameter to elongated pores in the region of 500 μm length, sometimes longer.

In time, as understanding of the formation mechanisms and the relationship between input parameters is developed, these defect will be largely eradicated. However, in order to aid this development and to enable the use of AM components where quality assurance is needed, in-situ monitoring can be employed. Many non-destructive monitoring methods have been trialled such as thermographic and visual monitoring systems [13], but these systems are limited to the identification of surface defects. Surface defects can be healed as the subsequent layers are processed so to avoid false positives, inspection systems that allow for identification of sub-surface defects are recommended.

Laser ultrasound (LU) is a good candidate for in-situ inspection of sub-surface defects, as the technique can be utilised on "rough" AM surfaces eliminating the coupling problems associated with conventional ultrasonic methods and can operate at high temperatures. Only a handful of experiments have been carried out to date trialling LU for inspection of AM surfaces [14–16] although capability of detecting AM type defects is more widely studied [17, 18]. Prior to being implemented in-situ, the capability of LU for detection of sub-surface AM defects must be proven, ex situ.

In this study, LU has been used to give material quality assurance for Ti-6Al-4 V AM samples, ex situ. Sample integrity has also been investigated using XCT and validated by destructive testing.

Laser Ultrasound

Background

To generate ultrasonic waves, a pulsed generation laser is directed at the sample surface. In high energy mode, termed the ablative mode, a region of plasma is formed directly above the impact point. When the energy build up is sufficient, the plasma expands and is ejected from the sample surface. The recoil force from this ejection generates an ultrasonic wave which travels into the sample and along the sample surface.

There are several types of ultrasonic waves generated, which are categorised based on their mode of propagation. "Surface" or near-to-surface waves travel along the surface of a sample. A Rayleigh wave is a type of surface wave which propagates with elliptical orbit and can penetrate into a sample approximately the same distance as its wavelength. In this experiment, a pulsed laser with a wavelength of 1064 nm was utilised, so the expected penetration is also in the region of 1064 nm. "Bulk" waves are so named because they penetrate into the bulk of the material from the generation source. These waves will travel until they meet an interface such as the wall, or base of the sample. When a boundary is encountered, the waves can reflect or propagate and energy can be transferred between wave types, termed mode conversion. The two main bulk wave types are longitudinal waves which travel by compression, or transverse waves which have displacements perpendicular to the direction of penetration (also termed Shear waves).

In LU, a laser is also used at the detection point, a fixed distance from the generation laser, to identify surface displacements caused by the arrival of the ultrasonic wave types. An interferometer splits the continuous wavelength laser beam into a reference beam and a beam which is directed at the sample surface. In a fully dense sample, with no defects present, the incident generated and reflected waves arrive at the detection point after an elapsed time. The velocity of each wave type in elastic solids is different, so displacement of the surface over time can be attributed to the arrival of each wave type. The interferometer gives two readings at each point, over the selected time period, an alternating current (AC) and a direct current (DC) reading. The DC monitor gives a signal that is proportional to the power of the signal beam on the detector and gives a measure of the light reflected from the surface. The AC signal output delivers the AC voltage corresponding to the instantaneous out of plane surface displacement at that point, for ultrasonic frequencies. At positions of low light (low DC), the AC signal amplitude will also be low. Dividing through the AC signal by the mean DC value normalises this effect. The compensated AC signal can then be displayed visually for analysis.

If a sub-surface defect is present between the two LU lasers, the tips of defect act as a diffraction source and the detection laser receives diffracted Rayleigh wave signals from the defect, along with direct waves and bulk waves which have bounced back from the side and end walls, or the bottom of the sample.

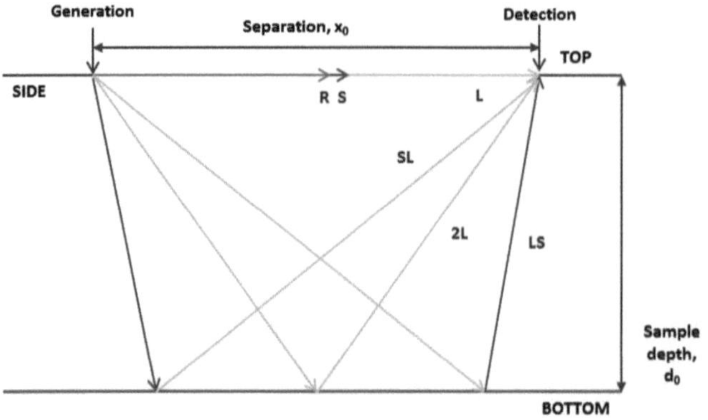

Fig. 1 Schematic showing approximate paths of direct, reflected and mode conversed ultrasonic waves

Fig. 2 A typical A-scan, extracted from LaserScan with surface displacements from wave arrivals indicated, for a defect-free sample

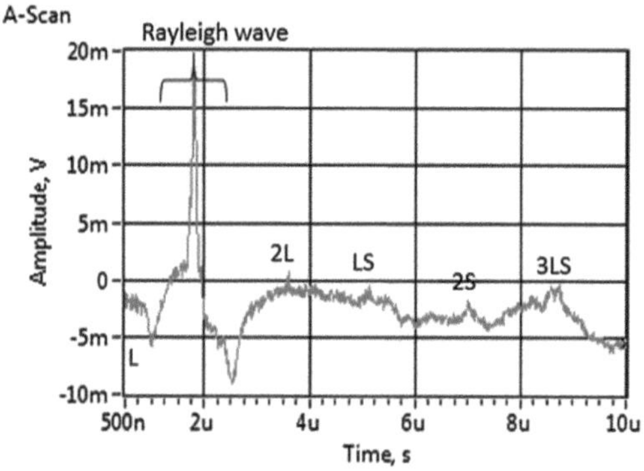

Interrogation of the resulting data in the time-window of the Rayleigh wave arrival will give indication of whether or not defects are present.

Figure 1 shows a schematic of the wave paths for direct longitudinal (L) and shear (S) bulk waves and reflected bulk waves (2L). Mode converted waves reflecting off the bottom interface (LS, SL) are also shown. The surface Rayleigh (R) wave also passes directly between the generation and detection lasers at separation, x_{o}.

Ultrasonic data is typically displayed in two ways: firstly, an A-scan is produced at each scanning location, which displays the received ultrasonic energy at the detection point, as a function of time (Fig. 2). The Rayleigh wave's comparatively large amplitude makes it easier to analyse.

Secondly, a motorised stage is used to move the lasers along and a snapshot is taken at defined distance intervals along the top of the sample. The series of A-scans are then compiled and stacked to create a profile view of the sample, termed a B-scan (Fig. 3). The expected arrival times for each wave and mode-converted wave type are overlaid. The diagonal lines that can be seen are

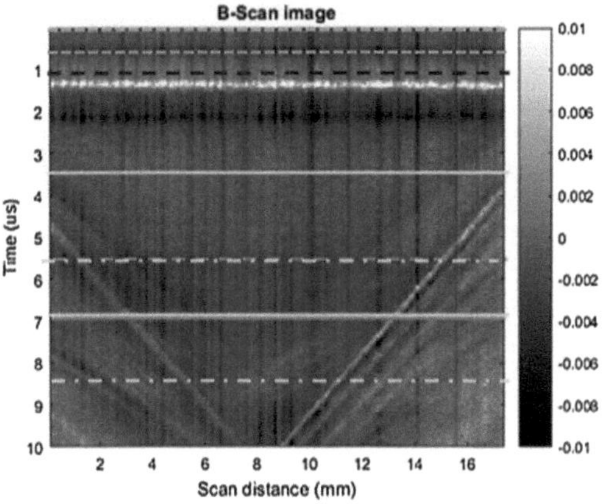

Key	Wave
----	Radio frequency (RF) noise
----	Skimming (L)
- - ·	Rayleigh (R)
——	Reflected longitudinal (2L, 4L)
- - ·	Mode converted (SL, 3SL)
- · -	Mode converted (LS, 3LS)
——	Reflected shear (2S)

Fig. 3 Predicted wave arrival times overlaid on a B-scan for a defect-free sample, with key

reflections from the end walls of the sample, as the laser unit moves farther from one end, the time for the wave to travel to the detection point increases. Similarly, the laser unit moves towards the other end of the sample and the wave travel time decreases.

As the lasers traverse along a sample surface at a fixed separation distance, if a defect is present, any interaction of the Rayleigh wave with the defect shows as a parabolic indication.

Experimental Set-up

The laser ultrasonic testing system uses a pulsed laser (class IV, Q-switched Nd: YAG laser with a wavelength of 1064 nm, capable of delivering 200 mJ energy with each 10 ns pulse at 20 Hz frequency) to generate ultrasonic waves in a sample and a continuous wave detection laser (10 W, 1550 nm ± 10 nm wavelength fibre laser), interferometer, oscilloscope and computer.

The Ti6Al4V cubes were built using a Renishaw AM250 at a power of 200 W, 600 mm/s scanning speed at 150 μm hatch spacing. The build pattern was rotated by an angle increment of 67° at each 40 μm layer. The larger cubes were sliced into 10 × 10 × 30 mm blocks and through-holes with various diameters were manufactured by EDM at a variety of depths. The top surface was left in the as-built condition (Fig. 4).

The LU system was set up such that in-line measurements were taken along the x-direction of the sample, with the generation laser line following the detection laser spot along the length of the sample. The sample was mounted on a motorised stage so that the sample could be moved in the y-direction to examine alternative scanning lines, without the need to relocate the sample (Fig. 5).

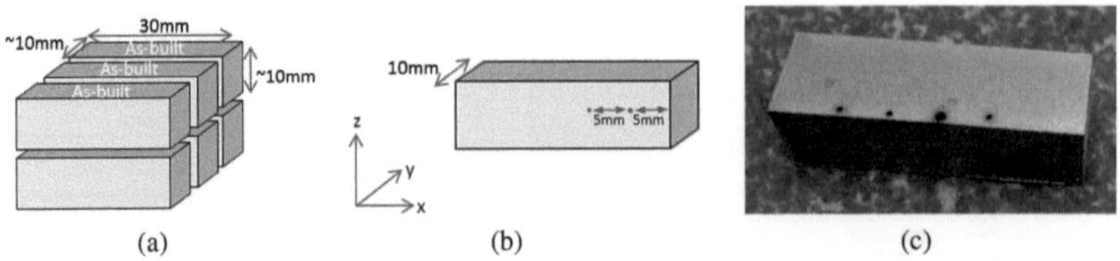

Fig. 4 Schematic showing **a** slicing pattern and block size, **b** targeted hole spacing and **c** a photograph of the sample block

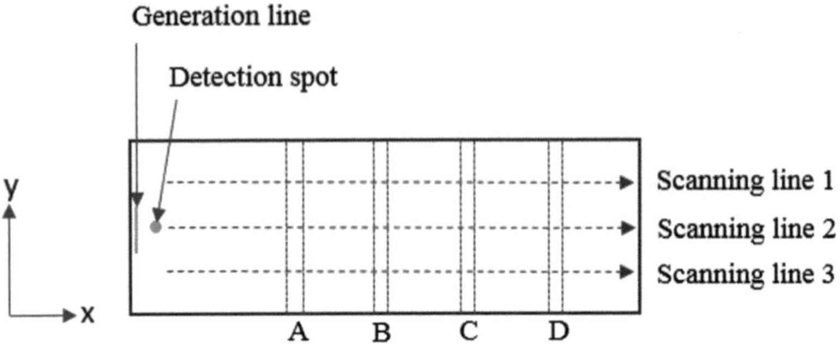

Fig. 5 Schematic showing laser starting positions, the four hole positions and the three laser scanning lines

A laser spacing of 3 mm was calculated as necessary to avoid interference of the reflected bulk waves with the surface wave arrival. The signal was measured over 10 μs at each acquisition point and the average of 64 shots recorded. The LU data generated in LaserScan software was exported for signal processing and analysis in Matlab.

Results

LU

The samples were scanned along three scanning lines indicated in Fig. 5 and a B-scan generated for each dataset (Figs. 6, 7 and 8). The data has been processed to eliminate the laser shot noise, reduce background noise and to remove the DC offset, then scaled by the maximum signal amplitude to allow a direct comparison between B-scans to be made.

The four large through-holes are indicated by four ‘triangles’ in the above B-scans, peaking at roughly 7, 12, 17 and 22 mm scan distance. The reflected and mode-converted waves appear as “trails” after the arrival of the Rayleigh wave

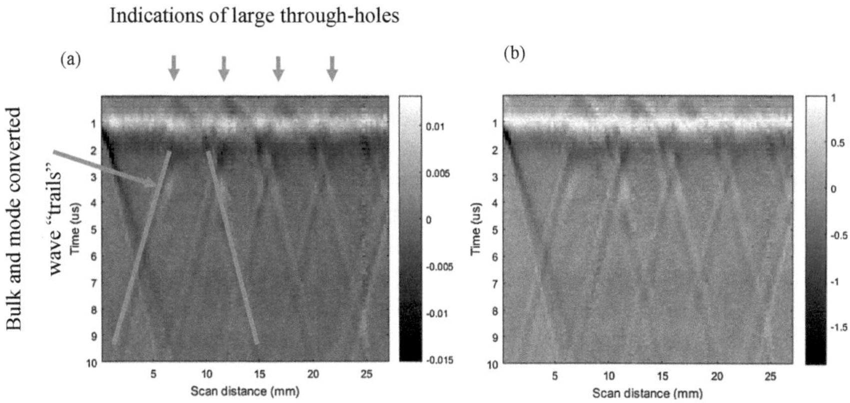

Fig. 6 Scanning line 1 B-scans showing amplitude variation (mV) and elapsed time **a** filtered and **b** filtered and scaled

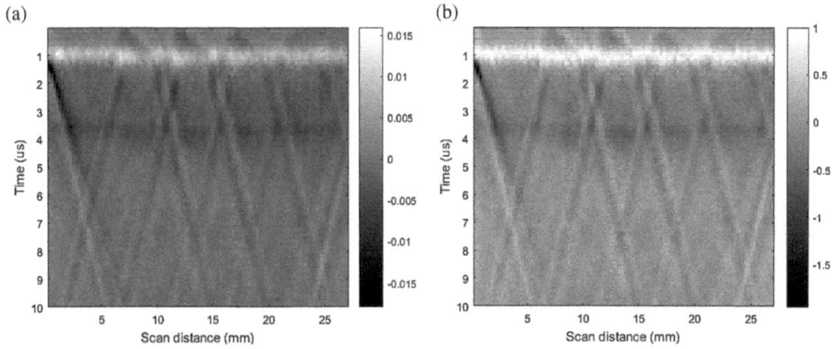

Fig. 7 Scanning line 2 B-scans showing amplitude variation (mV) and elapsed time **a** filtered and **b** filtered and scaled

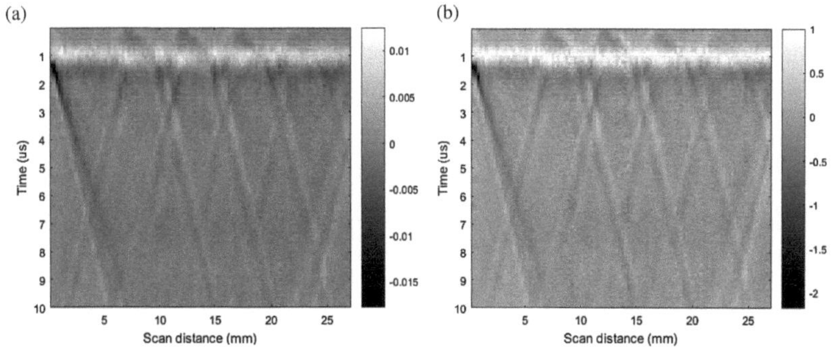

Fig. 8 Scanning line 3 B-scans showing amplitude variation (mV) and elapsed time **a** filtered and **b** filtered and scaled

which arrives at approximately 1 μs. No further defects are indicated. The B-scans generated from different scanning lines can be directly compared, highlighting any changes in wave arrival times. Very small changes are seen in the arrival time of the Rayleigh and reflected shear waves from scan to scan; this is thought to be due to the geometry of the sample.

If the top surface and bottom surface were not cut parallel the laser spacing would be altered when moving between channels. Similarly, if the samples were not a uniform thickness across the y-direction and thus, the reflected bulk wave arrival times would be different, as is the case here. Differences in wave arrival times seen in the "trail" sections are indicative of a variation in hole depth across the channels (y-direction). This was validated using focus variation microscopy (FVM) to measure the hole exit and entry locations. Due to the manufacturing method used to generate the through-holes, a constant cross section, and axes parallel to the surface were not achieved. No other naturally occurring "defects" were obvious from analysis of the LU data carried out, in the areas scanned.

XCT

To check that no naturally occurring defects were in fact present, the samples were analysed using x-ray computed tomography (XCT) and a comparison with the LU results then made. A Nikon XT H 225 ST was used to take 1600 shots with approximately 10.6 μm voxel size. Settings of 186 kV, 220 μA, copper 1 mm filter and 0.08 exposure were selected, resulting in a scan time of 30 min. The scans were reconstructed using 'Volume Graphics' software for interrogation. Following manual alignment, a region of interest was created for the top 3 mm (z-direction) of the sample and the automated defect recognition (ADR) algorithm run to identify any voids present. Thresholds were applied at a probability threshold 0.1 and a minimum void diameter of 0.1 mm and a total of 11 voids were identified in the region. Only two voids was identified in LU scanning line regions. One in scanning line 2, and one in scanning line 3. Both indications were of large voids, greater than 700 μm diameter but were at 2 and 2.8 mm below the surface, respectively; beyond the anticipated 1064 nm limit of detection for the LU system (Fig. 9).

Destructive

In order to validate the LU and XCT results which suggested that no naturally occurring defects were present in the LU scanning line regions, destructive evaluation was carried out. A Robo-Met.3D automated serial sectioning machine was used to generate reconstructed microscope images over the area indicated in Fig. 10.

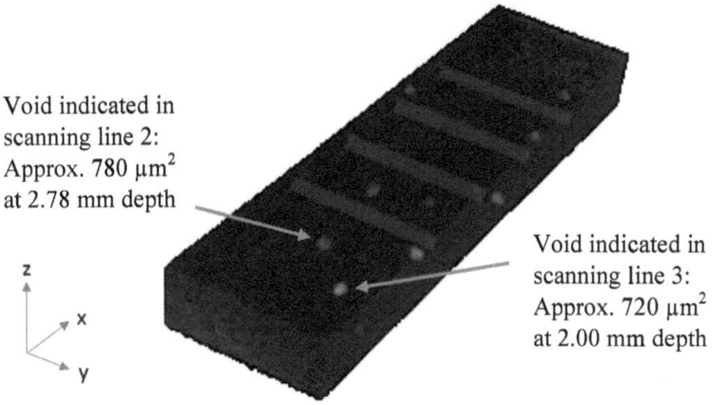

Fig. 9 Extract from 'Volume Graphics' showing voids identified in region of interest by automated defect recognition software

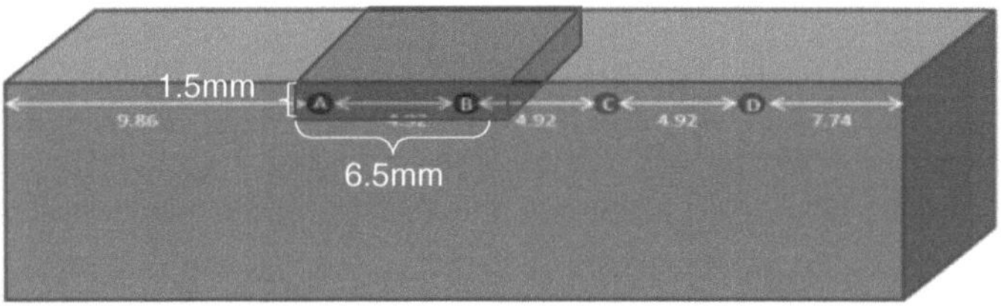

Fig. 10 Schematic (*left*) showing area of interest (*red*) for destructive evaluation

Fig. 11 Images showing good correlation between micrograph (*left*) and XCT image (*right*) at 643 μm depth

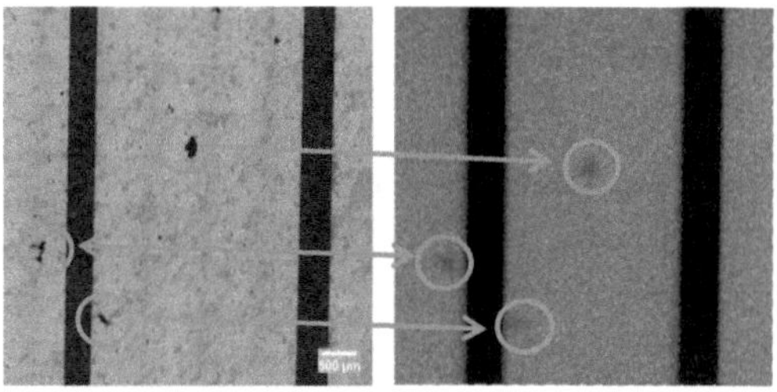

The micrograph images were compared directly with the XCT reconstructions to identify any naturally occurring defects in this area and a reasonable correlation found, albeit with poor contrast. Obvious material discontinuities identified on the micrographs were linked to indications of defects (as deviations in the greyscale) in the XCT reconstructions (Fig. 11). It can be concluded that the XCT gives a good representation of the sample integrity. However, not all of the naturally occurring defects found by destructive analysis were found by the ADR algorithm; this was probably due to the very small changes in greyscaled value. Although a time consuming activity with room for human error, it is recommended that an additional manual analysis of XCT reconstructions is carried out in future analyses.

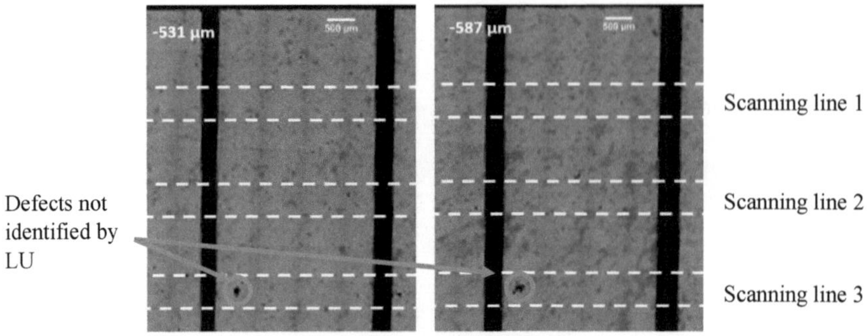

Fig. 12 Micrograph images at depths of 531 and 587 µm, showing the scanning line regions. Previously unidentified (by LU) defects are highlighted

After overlaying the LU scanning lines on the micrographs, it was found that there was one void present in the region destructively analysed, which had not been indicated from the LU B-scan (Fig. 12). The unidentified defect has dimensions of approximately 200 × 150 × 100 µm and is located 250 µm from the EDM'd "hole A" at a depth of roughly 530 µm from the top surface. For reference, "hole A" has a diameter of 580 µm and is 417 µm below the sample surface. It is thought that the presence of such a large, manufactured defect in the vicinity of the naturally occurring defect is having a masking effect.

Conclusions

Laser ultrasound has been used to identify the presence of four large (580–670 µm at 400–800 µm depth), manufactured though-holes in a Ti-6Al-4 V AM sample with an as-built surface finish. No further defects have been identified using LU. Further interrogation of the sample using XCT with ADR identified eleven possible naturally occurring defects. Two of these possible defects were located in the LU scanning line regions, although both were at a depth exceeding the anticipated capability of the LU system with current settings. No false positives were identified.

Destructive evaluation of a small region of the sample revealed a defect with dimensions of approximately 200 × 150 × 100 µm at a depth of 530 µm from the top surface, which had been 'missed' by LU and XCT ADR. It is possible that the close proximity of this naturally occurring void to the manufactured through-hole is masking any change in the Rayleigh wave arrive time in the LU signal. The greyscale settings and threshold values could be responsible for the defect not appearing with ADR. In future, manual analysis of the XCT reconstructions is recommended.

Additional work is required before it can be concluded that LU shows promise as an in-situ inspection system for AM. Multiple build layers could be assessed in a single scan, reducing the area interrogated with each build layer. The laser spot size could be altered to increase the wave penetration. Further work is required to assess the capability of the system to detect naturally occurring defects and to investigate the effect of defect density.

Acknowledgements The primary author is working within the Manufacturing Technology Engineering Doctorate Centre, supported by the Manufacturing Technology Centre, University of Nottingham and the Engineering and Physical Sciences Research Council under Award No. 1361477. Work has been carried out in conjunction with FP7-2012-NMO-ICT-FoF project 313781, AMAZE.

References

1. T.T. Wohlers, *Wohlers Report 2015*, vol. 22. Wohlers Associates (2015)
2. A.M. Uriondo, M. Esperon-Miguez, S. Perinpanayagam, The present and future of additive manufacturing in the aerospace sector: a review of important aspects, in *Proceedings of the Institution of Mechanical Engineers, Part G: Journal of Aerospace Engineering* (2015)
3. Energetics Incorporated, *Measurement science roadmap for metal-based additive manufacturing*, National Institute of Standards and Technology, Maryland, US (2013)
4. AM Steering Group UK. *UK national strategy for additive manufacturing/3D printing* (2015), http://www.amnationalstrategy.uk/steering-group/
5. I. Gibson, D.W. Rosen, B. Stucker, *Additive Manufacturing Technologies: Rapid Prototyping to Direct Digital Manufacturing* (Springer, 2009)
6. M. Van Elsen, *Complexity of Selective Laser Melting: a new optimisation approach,* PhD Thesis published, Katholieke Universiteit Leuven, Belgium (2007), pp. 31–42
7. H. Gong et al., Influence of defects on mechanical properties of Ti–6Al–4 V components produced by selective laser melting and electron beam melting. Mater. Des. **86**, 545–554 (2015)
8. S. Tammas-Williams et al., XCT analysis of the influence of melt strategies on defect population in Ti–6Al–4 V components manufactured by selective electron beam melting. Mater. Charact. **102**(4), 47–61 (2015)
9. K. Monroy, J. Delgado, J. Ciurana, Study of the pore formation on CoCrMo alloys by selective laser melting manufacturing process. Manuf. Eng. Soc. Int. Conf. MESIC Procedia Eng. **63**, 361–369 (2013)
10. C. Weingarten et al., Formation and reduction of hydrogen porosity during selective laser melting of AlSi10Mg. J. Mater. Process. Technol. **221**, 112–120 (2015)
11. W.E. King et al., Observation of keyhole-mode laser melting in laser powder-bed fusion additive manufacturing. J. Mater. Process. Technol. **214**(12), 2915–2925 (2014)
12. H. Gong et al., Defect morphology in Ti-6Al-4 V parks fabricated by selective laser melting and electron beam melting, in *Solid Freeform Fabrication Symposium*. Austin, TX (2013)
13. S.K. Everton et al., Review of in-situ process monitoring and in-situ metrology for metal additive manufacturing. Mater. Des. **95**, 431–445 (2016)
14. S.K. Everton et al., Evaluation of laser ultrasonic testing for inspection of metal additive manufacturing, in *SPIE*, San Francisco, CA, US (2015)
15. S.P Santospirito et al., Defect detection in laser powder deposition components by laser thermography and laser ultrasonic inspections in *SPIE*, San Diego, CA (2013)

16. D. Cerniglia et al., Inspection of additive-manufactured layered components. Ultrasonics **62**, 292–298 (2015)
17. R.S. Edwards et al., Scanning laser source and scanning laser detection techniques for different surface crack geometries. AIP Conf. Proc. **1430**(1), 251–258 (2012)
18. M. Klein, T. Sienicki, J. Eichenbergeer, *Laser-ultrasonic detection of subsurface defects in processed metals.* Patent. 2007: United States

SLM (Near)-Net-Shape Part Design Optimization Based on Numerical Prediction of Process Induced Distortions

Maria San Sebastian, Iñaki Setien, Ane Miren Mancisidor and Alberto Echeverria

Abstract Distortions are critical in metal additive manufacturing, since they increase manufacturing costs, times and generate wastes and scraps due to dimensional inaccuracies in newly developed parts. Currently, corrective measures are often not taken until production step and they are mainly based on trial an error approach. Alternatively, Design Against Distortion paradigm is focused on the development of numerical modelling strategies which can anticipate distortions even from the design stage. This work is focused on the development of a rapid distortion prediction numerical methodology applicable to selective laser melting (SLM) process, taking into account the influence of different process variables, like scanning parameters and scanning strategy. This methodology is applicable to predict distortion behaviour of a real component manufactured by SLM, in order to determine the best supporting strategy and build up orientation. Therefore, current developments entail an innovative way of designing and manufacturing SLM manufactured parts which are more robust against distortions.

Keywords FEM · Distortion · Inherent shrinkage · SLM

Introduction

Distortions are critical during additive manufacturing of metallic parts, since they increase manufacturing costs, times and generate wastes and scraps due to dimensional inaccuracies in newly developed parts. Currently, corrective measures are often not taken until production step and they are mainly based on trial an error approach. Alternatively, Design Against Distortion paradigm is focused on the development of numerical modelling methodologies which can anticipate distortions even from the design stage. However, this new paradigm requires the

M.S. Sebastian (✉) · I. Setien · A.M. Mancisidor · A. Echeverria
IK4 LORTEK, Arranomendia kalea 4A, 20240 Ordizia, Spain
e-mail: msansebastian@lortek.es

The Minerals, Metals & Materials Society, *TMS 2017 146th Annual Meeting & Exhibition Supplemental Proceedings*, The Minerals, Metals & Materials Series,
DOI 10.1007/978-3-319-51493-2_12

development of new modelling strategies that can be efficiently applicable to real parts.

In selective laser melting (SLM) process, the part is constructed on a layer by layer basis. The material is locally and rapidly heated above its melting temperature and then allowed to solidify and cool in order to form a dense geometry. This process involves large thermal gradients. Thereby, residual stresses and distortions are generated due to its local inherent nature. Moreover, post-manufacturing processes, such as cutting of support structures, lead to redistribution of internal stresses and strain modifying as-build distortions.

SLM can be considered as a multipass micro welding process [1] and many of its modelling challenges are related to the numerical modelling of multipass welding [2]. However, despite there are several validated finite element (FEM) based modelling strategies for multipass welding, their direct implementation to SLM process is not straightforward due to computational time limitations. It must be noted that the manufacturing of a real part by SLM will normally require hundreds of layers and thousands of weld seams. Thus, conventional thermo-mechanical FEM simulation is not feasible and adapted simplified methodologies must be developed in order to model this manufacturing process.

Already developed SLM distortion prediction models are based on the activation of full layer or combination of layers at once [3–5]. In these models, a uniform heat input is introduced on the whole layer within each load step [3, 5–7]. This simplification neglects temperature gradients in X-Y direction and does not take into account the effect of scan pattern (stripes, chess-board). Nevertheless, these simplified models have demonstrated good predictive capability in geometrically simple parts [8].

Recently, alternative modelling approaches have been developed with the idea of capturing the effect of scanning strategy. One innovative approach is based on the introduction of non-uniform heat input on layers [9] Different methods for modelling non-uniform heat input have been proposed which increase the calculation time but are able to reproduce the thermal cycling in one layer.

Another significant approach is based on a multi-scale method which includes the calibration of the heat source at microscopic scale, the analysis of scanning strategy at mesoscopic scale and the mapping of resulting inherent strains (eigenstrain) into a macroscopic layer model [2, 10]. This allows replacing computationally costly thermo-mechanical simulations by a mechanical one.

The work presented in this paper is focused on the development and application of an alternative simplified SLM distortion prediction methodology based on inherent shrinkage approach. This is a mechanical approach where no transient thermal analysis is involved. It assumes that the main driving force for distortion is the thermal contraction of the weld metal during cooling to room temperature, and distortion are generated as a consequence of the accommodation by the structure of this contraction. This approach involves that elements in one layer undergo a shrinkage that is equal and opposite to the thermal expansion when the metal is heated up to its melting temperature. It has been previously developed for the calculation of welding distortion of large parts and it requires to adjust the volume of the shrinking elements and their initial temperature [11]. Current work addresses the application of this method and its evaluation in simple geometry part (cantilever) by comparison with experimental distortion measurements. Both isotropic and non-isotropic thermal expansion coefficients (α) have been considered for the model development.

Experimental Work

Cantilever shape specimens were fabricated by SLM as testing geometry. The specimens were made of Inconel 718 and their geometry and dimensions are shown in Fig. 1. Dimensions match with specimens studied by [2].

The specimens were manufactured in a SLM Solutions 280 HL machine equipped with 400 W fiber laser. Samples were manufactured using a layer thickness of 30 μm, and process parameters included in Table 1 for hatch, boundary and contour zones. Preheating temperature was 25 °C.

Cantilever specimens were manufactured over thick baseplate (80 × 80 × 11 mm). The baseplate contained 3 specimens with the same process parameters but different scanning strategies (Fig. 2). Three different scanning strategies were investigated including chess-board pattern, transversal stripes and longitudinal stripes. Squares of chess-board pattern were 7 mm size and the scanning direction was rotated 90° from one square to another. Stripes had a meandering pattern. From layer to layer, scanning strategy was kept constant.

After manufacturing, cantilever specimens were partially cut from the baseplate by saw cutting. The initial leg was left attached to the baseplate. After cutting residual stresses of the parts are redistributed leading to plastic deformation. The final vertical distortions (Z direction) were measured on the top-surface of cantilevers. Measuring points were located in the central longitudinal line along

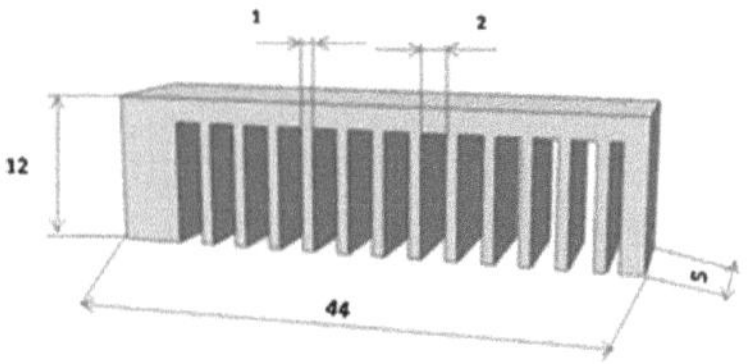

Fig. 1 Cantilever shape specimen with dimensions

Table 1 SLM process parameters

	Hatch	Boundary	Offset hatch
Laser power (W)	200	200	200
Scanning speed (mm/s)	800	1100	1100
Hatch distance (μm)	120	–	–

cantilever's length and keeping a constant distance of 5 mm between points. Distortions were measured on metrology flat table using dial gauge with 10 μm accuracy.

Numerical Model Definition

The inherent shrinkage method was applied to the cantilever specimen geometry. The geometry was meshed using 182098 hexahedral C3D8R elements. A layer-based meshing procedure was applied which gave rise to a mesh composed by parallel layers in the buildup direction one of top of each other. The mesh of the cantilever specimen had in total 100 layers (0.12 mm thickness), each of them containing two elements in the thickness direction.

Inconel 718 material properties at room temperature were introduced in the FE-model (Table 2). Properties were independent of the temperature and an elasto-plastic behaviour was selected.

Concerning the thermal expansion coefficient, three different approaches were studied including isotropic and non-isotropic (orthotropic) behaviour in vertical and transversal directions. Next table summarizes the 3D components of the thermal expansion coefficient that were used in different models. Vertical component is parallel to building direction (Z), longitudinal to the length of the cantilever (X) and transversal to the width of the cantilever (Y) (Table 3).

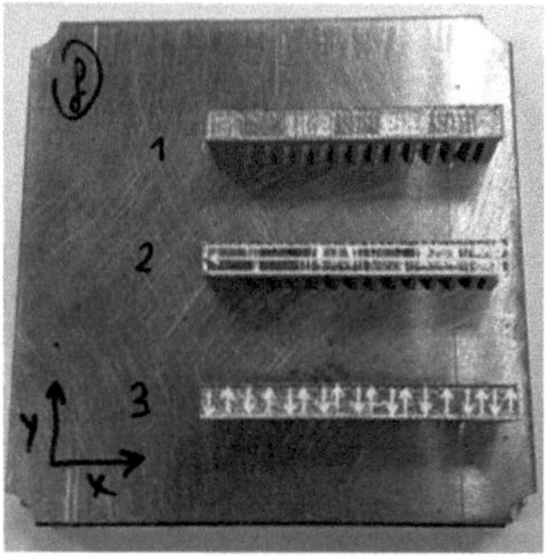

Fig. 2 Scanning strategies followed for the manufacturing of cantilever specimens

Table 2 Inconel 718 material properties

Mechanical properties	Value
Density (kg/m^3)	8146
Young modulus (MPa)	210000
Poisson coefficient	0.38
Yield strength (MPa)	700
Ultimate strength (MPa)	900

The element and layer activation strategy was based on the so called "birth and death" or model change technique. According to this activation strategy, elements are not part of the model until they are activated [12].

Inherent shrinkage was implemented by introducing equivalent thermal loads in the FE-model in selected stacks of meshing layers at the same time. Different number or combinations of layers of FEM mesh were activated in each step with the aim of studying the effect of this activation procedure on the distortion prediction capability and computational cost. It must be noted that real cantilever specimens were manufactured in 400 layers. The simulation of all these layers would require very small size elements in the buildup direction, a huge amount of elements and long computational times. Alternatively, if model layers are thicker, the computational time can be reduced. Thus, the sequential activation of layers in the FE-model was discretized in different number of layers in the building direction, i.e., 5, 10, 50 and 100 layer activation steps. That means that different models with different layer activation steps were developed comprising simultaneous layer activation from 1 to 20 meshing layers. Figure 3 shows a representation of the activation step of first 20 meshing layers in the same step.

Moreover, the effect of the initial melting temperature defined for the introduction of equivalent thermal loads was studied. Initial temperatures that were selected for the activation of new layers were between 1000 and 2000 °C.

Residual stresses were analysed after completing the activation of every model layer and final distortions were calculated after simulating the cutting of the legs from the baseplate.

Abaqus 6.14-2 software was employed for the implementation of new numerical methodology.

Table 3 Components of thermal expansion coefficients used in different models

	Vertical	Longitudinal	Transversal
Model 1—isotropic	1.28×10^{-5}	1.28×10^{-5}	1.28×10^{-5}
Model II—orthotropic (vertical)	2.56×10^{-5}	1.28×10^{-5}	1.28×10^{-5}
Model III—orthotropic (transversal)	1.28×10^{-5}	1.28×10^{-5}	2.56×10^{-5}

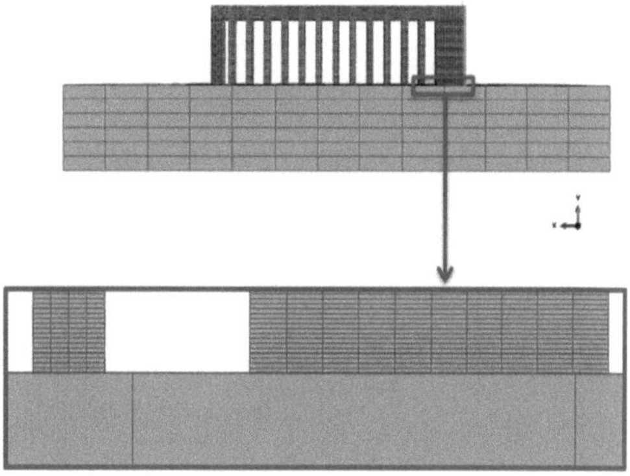

Fig. 3 Combination of meshing layers activated in one step

Results and Discussion

Experimental Distortions

Figure 4 shows experimental distortions along vertical direction that were measured in manufactured cantilever specimens after cutting. As it can be observed, there was a clear influence of the scanning strategy on final distortions. Longitudinal stripes led to the highest distortions, whereas transversal stripes reduced them. These experimental measurements were considered for the validation of new simplified distortion prediction model.

Results of FE-Models

Figure 5 represents the distortion predicted by isotropic FE-model (Model I) using different discretization or number of meshing layer activation steps.

The results show that at least 50 layers are required in the FE-model in order to obtain a good matching between numerical and experimental results. The distortion

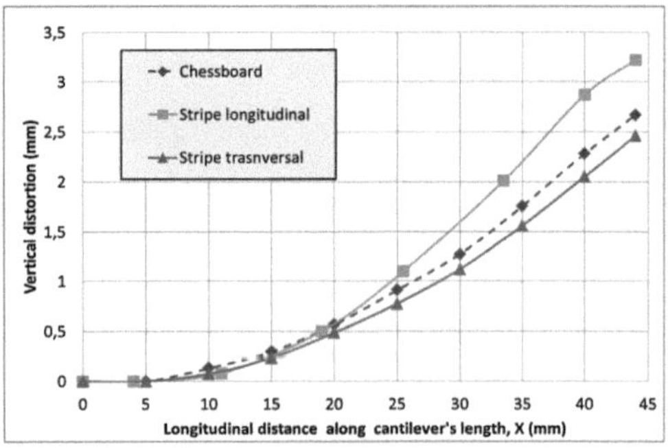

Fig. 4 Experimental vertical distortions of cantilever specimens manufactured with different scanning strategies

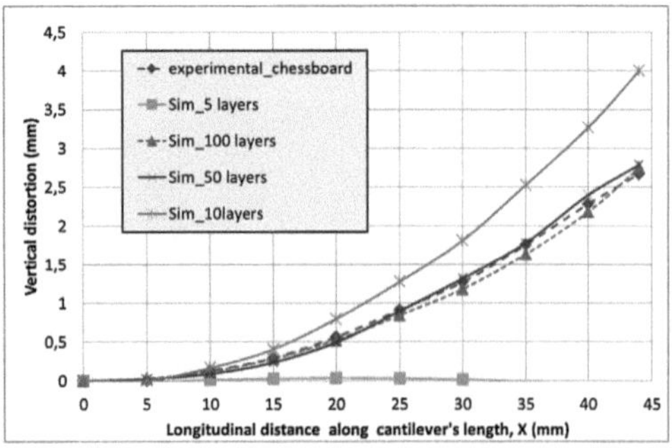

Fig. 5 Comparison of predicted vertical distortion along cantilever length with experimental measurements (chess-board scanning strategy)

prediction of FE-models with 50 and 100 layer activation steps was in good correlation with experimental results obtained using the chess-board scanning strategy.

In terms of computational cost, the time required for model resolution was 2.3 and 8.6 h for the 50 and 100 layers models, respectively. This time was recorded in an Intel Core i7-3770 3.40 GHz microprocessor with 4 CPUs (Figs. 6 and 7).

Concerning the influence of the initial temperature defined in the isotropic model (Model I), no differences were observed between 1000 and 2000 °C in final distortions (Fig. 8). Surprisingly, the distortion predicted by 3 models with different initial temperatures was identical. This can be explained taking into account that

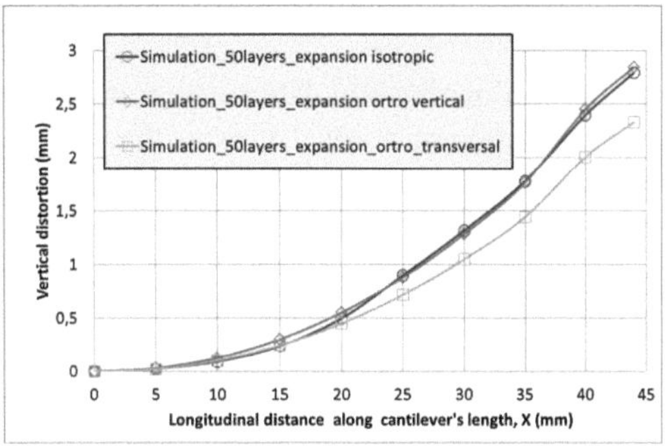

Fig. 6 Comparison of predicted vertical distortion with different thermal expansion coefficient

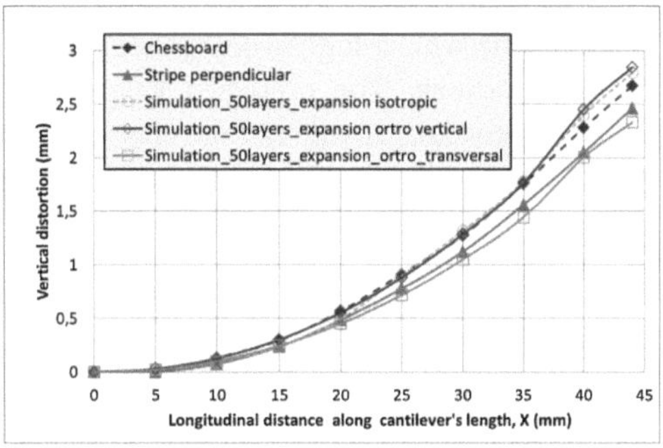

Fig. 7 Comparison of experimental versus simulation distortions with different thermal expansion coefficient

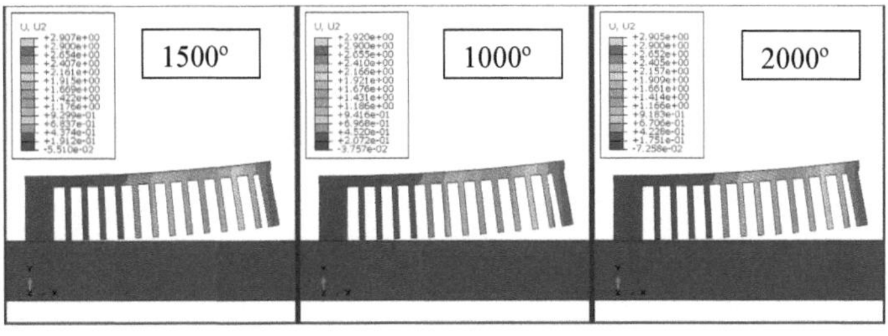

Fig. 8 Vertical distortion predicted with different initial temperatures

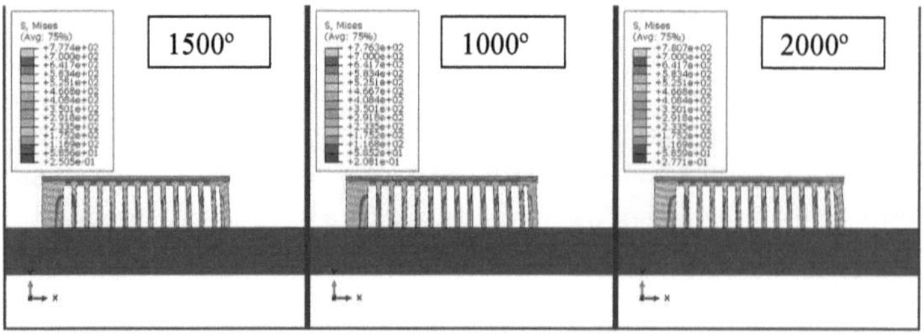

Fig. 9 Residual stresses before cutting process predicted by isotropic numerical model for different initial temperatures

final distortions are simulated after cutting of cantilever's leg. The cutting generates a redistribution of residual stresses that were stored in the part during SLM process. The analysis of the residual stresses and plastic strains before cutting step shows that the distribution and magnitude of residual stresses were comparable for the three initial temperatures (Fig. 9). Therefore, final distortions after residual stress redistribution will be similar.

As it is shown in Fig. 4, experimental distortions mainly depend on selected scanning strategy. Results of predicted distortion considering orthotropic thermal expansion coefficients are show in Fig. 6. Model II with a higher vertical component, yielded comparable results with the isotropic model (Model I) and only a slight increase of the vertical distortions at the edge of the cantilever was observed.

Model III with a higher component in the transversal Y direction gave rise to lower distortions. The predicted shape and magnitude of the curve describing vertical distortions along cantilever's length was in good agreement with experimental distortions measured for transversal stripe scanning strategy (Fig. 7).

Conclusions

The application of a new simplified numerical methodology based on inherent shrinkage method for fast prediction of SLM process induced distortions has been studied in a real geometry (cantilever).

The predictive capability of this simplified methodology has been demonstrated for the studied geometry by comparing model results with experimentally measured distortions in Inconel 718 parts manufactured by SLM.

The following practical conclusions were drawn for the analysis and comparison of results:

- The model application requires a layer-based meshing procedure, i.e., mesh must be composed by parallel layers in the building direction.
- The model is based on a sequential activation of layers ("birth and death" or model change technique). A minimum number of model layer activation steps is

required in order to ensure good prediction. For the geometry studied in this work a good prediction was achieved for a ratio between real manufacturing layers and model layer activation steps of 8 (400 real layers, activated in 50 steps). Above this ratio, the predictive capability decreases, whereas the increase of this ratio does not improve the accuracy of the results, but increases the computational cost.

- Experimental distortions mainly depend on scanning strategy (chess-board, transversal stripes, longitudinal stripes). Despite the simplicity of the model, it was feasible to predict the distortions using different scanning strategies. Good correlation with chess-board scanning strategy was obtained by applying isotropic model (no differences in the 3D components of the thermal expansion coefficients), whereas, outstanding matching with transversal stripe scanning strategy was attained by orthotropic model (higher thermal expansion coefficient in transversal direction).
- The effect of the initial temperature for the layer activation which determines the associated shrinkage was much lower than expected for this material and geometry.
- Current promising results must be validated in more complex parts before making a final conclusion about real contribution of this methodology to the Design Against Distortion paradigm which includes the analysis of distortions of SLM manufactured parts from the design stage.

References

1. B. Baufeld, O. Van der Biest, R. Gault, Additive manufacturing of Ti–6Al–4 V components by shaped metal deposition: microstructure and mechanical properties. Mater. Des. **31** (Supplement 1), S106–S111 (2010)
2. N. Keller, V. Ploshikhin, New method for fast predictions of residual stress and distortion of AM parts, Solid Freeform Fabrication Symposium, 2014, Austin, Texas, Conference paper, pp. 1229–1237
3. G. Branner, Modellierung transienter Effekte in der Struktursimulation von Schichtbauverfahren (2010)
4. G. Branner, J. Schilp, T.A. Krol, Modelle zur thermomechanischen Simulation metallverarbeitender Strahlschmelzprozesse (2009)
5. T.A. Krol, S. Westhäuser, M.F. Zäh, J. Schilp, G. Groth, Development of a simulation-based process chain—strategy for different levels of detail for the preprocessing definitions, vol. 21 (2011), pp. 135–140
6. C. Seidel, T.A. Krol, J. Schilp, M.F. Zaeh, C. Groth, Ansaetze zur rechenzeiteffizienten Struktursimulation additiv gefertigter, filigraner Bauteile. *CADFEM GmbH (ed.): Ansys Conference & 30. CADFEM Users' Meeting.* (Kassel;2012)
7. N. Keller, F. Neugebauer, H. Xu, V. Ploshikhin, Thermo-mechanical simulation of additive layer manufacturing of titanium aerospace structures, no. figure 1 (2013)
8. C. Seidel, M.F. Zaeh, J. Weirather , T.A. Krol, J. Schilp, D. Schmid, Prozessnahe Modellierung des Materialverhaltens beim Laserstrahlschmelzen als Grundlage für die Ergebnisgenauigkeit hinsichtlich der Bauteilmaßhaltigkeit und des Eigenspannungszustandes.

In: *CADFEM GmbH (ed.): ANSYS Conference & 31th CADFEM Users' Meeting.* (Mannheim; 2013)

9. C. Seidel, M.F. Zaeh, J. Weirather, M. Wunderer, T.A. Krol, J. Schilp, C. Groth, Simulation of the laser beam melting process—an approach for an efficient geometry modelling of complex lightweight parts, no. c
10. F. Neugebauer, N. Keller, V. Ploshikhin, F. Feuerhahn, H. Köhler, Multi scale FEM simulation for distortion calculation in additive manufacturing of hardening stainless steel, in *Proceedings of Fraunhofer Direct Digital Manufacturing Conference* (2014), pp. 13–23
11. A. Bachorski, M. Painter, A. Smailes, M. Wahab, Finite-element prediction of distortion during gas metal arc welding using the shrinkage volume approach. J. Mater. Process. Technol. **92–93**, 405–409 (1999)
12. A. Lundbäck, Modelling and simulation of welding and metal deposition. Aerosp. Eng.

Optimizing, Fabricating and Characterizing Additively Manufactured Heat Exchanger Tubing

Paul Korinko, John Bobbitt, Haley McKee, Frederick List and S.S. Babu

Abstract Additive Manufacturing brings unique opportunities to the fabrication world, especially for complex, high value added components that are challenging if not impossible to fabricate using traditional technologies. This project was developed to design, build and test a variety of shapes, internal geometries, and lengths of process tubing for a specialized heat exchanger that operates from nominally 140 to −40 °C. A preliminary design was fabricated and based on the knowledge gained from this component, several design iterations were made to optimize the heat transfer while not adversely affecting the thermal strains. Several additional components were prepared and the parts were characterized for burst strength, tensile properties, microstructure, and chemistry. In addition, the build data were analyzed to determine if a correlation could be made with the properties. This presentation will describe the application for the heat exchanger, the goals of the project, the results for the optimization, fabrication, and characterization.

Keywords Additive manufacturing · Stainless steel · Characterization

P. Korinko (✉) · J. Bobbitt
Savannah River National Laboratory, Aiken, SC, USA
e-mail: paul.korinko@srnl.doe.gov

H. McKee
Kansas City National Security Campus, Kansas City, MO, USA

F. List
Oak Ridge National Laboratory, Manufacturing Development Facility, Knoxville, TN, USA

S.S. Babu
Univeristy of Tennessee Knoxville, Knoxville, TN, USA

The Minerals, Metals & Materials Society, *TMS 2017 146th Annual Meeting & Exhibition Supplemental Proceedings*, The Minerals, Metals & Materials Series,
DOI 10.1007/978-3-319-51493-2_13

Introduction

Freeform fabrication, additive manufacturing (AM), three dimensional printing all describe a common process that produces new materials from starting stock that can be metal, plastic, or ceramic. The heat sources typically used in AM include electron beam and lasers. There are a number of reviews that describe these processes in detail [1–3]. The laser processes can use directed energy and powder injection into the molten pool or localized melting in a powder bed process. The directed energy systems can build on existing components or be used to repair [4–6] while the powder bed systems require a build plate and are not amenable for repair processes but can be used for component manufacturing.

In general, the laser powder bed process accepts a CAD drawing from the designer. This drawing is then sliced into sections for the machine control. Proprietary software is then used to fabricate the part [7]. The parts can then be either used or tested depending on the purpose of the build. There is currently a significant level of effort to define what testing is required for acceptance, with the holy grail being "born qualified" so that all the science and engineering understanding is completed before the parts are made so that destructive testing is not needed. However, to reach this ideal, preliminary work must be accomplished to better understand limitations and design criterial.

This article describes and compares two different type of process tubing. These tubes are intended to replace process tubing that was fabricated by mechanically forming tubing and brazing it together.

Process Tubing Design

A process at The Savannah River Site (SRS) uses process tubing in a flat spiral in a gas chromatography application. This assembly is comprised of two layers of tubing that are coiled and brazed into the desired configuration. This forming and joining process has not been optimized and often results in gaps in the assembly between the electrical heating coils and the product gas coils and between the product gas and the cooling coils. In addition, the contact points between the tubes is limited to the amount of braze that will capillate into the gaps. The end result is shown schematically in Fig. 1. Note that even in the as-designed configuration there are gaps between the tubes. Since the tubes are enclosed in a vacuum enclosure, these gaps further impede heat flow.

Additive manufacturing offers the opportunity to eliminate the gaps between the coils by printing the walls concurrently. This attribute can eliminate defects caused by tubing movement during the brazing operation; the tubes may shift while hot due to stress relief. Additive manufacturing can also eliminate the stacking voids by depositing material in the tube to tube interstices. Consequently, additive manufacturing of Type 316L stainless steel on a Renishaw was used to produce two

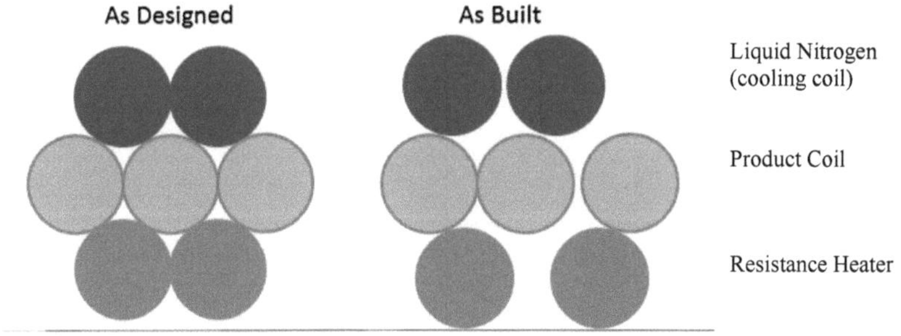

Fig. 1 Schematic of the traditionally designed and fabricated process tubing

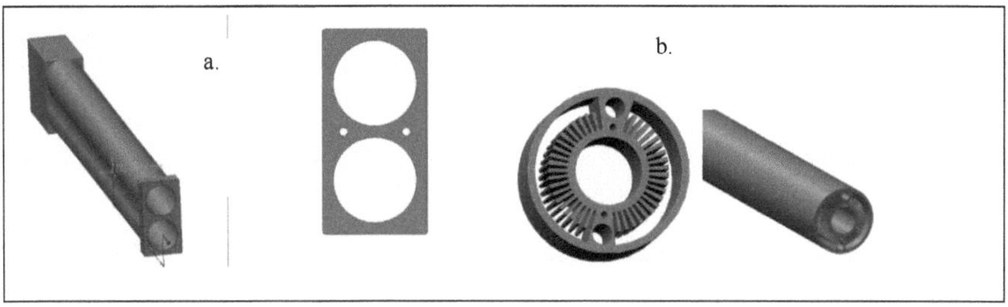

Fig. 2 Initial designs for process tubing to improve heat transfer, **a** "shotgun" and **b** finned tube

different process tubes that were intended to overcome some of the limitations of the traditional fabrication methodology. The process tubing cross-sections are shown in Fig. 2. Note the "shotgun" appearance allows for better contact between the tubes and is also configured to allow the resistance heater to more closely nest between the tubes. The finned tube is somewhat of a deviation from the initial designs with integral resistance heater locations, fins that extend into the cooling coil, and integral thermocouple tubes, as indicated. The parts were printed as described later with the known deficiencies AM which includes residual porosity, poor surface finish, and poor fin formation [8].

Based on the initial experiments for printing, finite element modeling was used to assess these deficiencies, to evaluate the stress levels, and to produce a more optimized solution. A series of process tubing optimization finite element models were conducted using the "as-designed" system as a basis. Simplifications were made to account for the liquid nitrogen flashing as well as the heating of the tubing using the resistance heaters. This methodology was applied uniformly across the design modifications.

The potential improvements in heat transfer rate, based on heating rates, can be seen in Fig. 3. There is an increase in the slopes of the curve for the improved designs compared to the baseline. This increased rate shows that the "shotgun" design is inferior to the improved designs, patent application under development.

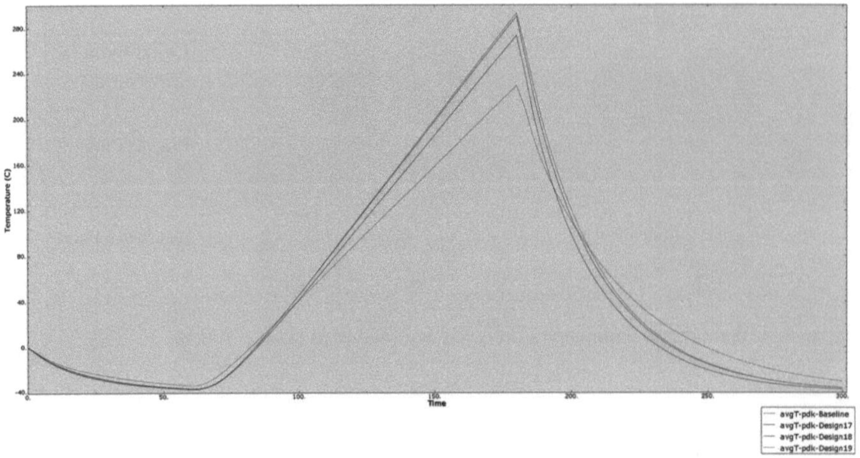

Fig. 3 Potential heat transfer improvements based on FEM

Table 1 Printing conditions used for the Builds described in this document

Parameter	Renishaw 316L SS Build 1	Renishaw 316L SS Builds 2&3
Powder size (μm)	25–44	25–44
Laser power (W)	200	200
Layer thickness (μm)	50	50
Scan pattern	Meander	Stripe—5 mm
Beam size (μm)	75	75
Point distance (μm)	50	60
Point exposure (μs)	90	80
Effective velocity (m/s)	0.56	0.64
Hatch spacing (μm)	150	110

Experimental Approach

Process tube segments, approximately 150 (Builds 2 & 3) and 200 (Build 1) mm tall were printed on a Renishaw AM 250 SLM system. The nominal parameters for the three builds are listed in Table 1. Note the relatively large hatch spacing for Build 1. The hatch spacing was reduced for Builds 2 and 3 to determine if the porosity levels would be decreased. In addition, the scan pattern was modified to improve the sample density and surface roughness.

Samples were visually examined and photographed to show the as printed characteristics. Select samples were metallographically examined and others were burst tested. Burst testing was conducted by welding high pressure fittings onto the tubes and then filling them with water and then pressurizing the tube to failure by opening the valve and bleeding in pressure at a nominal rate. The same initial nominal valve setting was used for each test; the valve was opened further if the pressurization rate decreased during tube expansion. The pressure vs time data were recorded digitally and plotted in Excel.

The fracture surfaces of selected samples were examined on the scanning electron microscope (SEM). The fracture surface was cut from the tubes and opened up to permit better viewing of the fracture.

Results and Discussion

The Build 1 exhibits a matte surface appearance due to the high surface roughness but are uniform metallic gray, as shown in Fig. 4. The Build 2 exhibits a heat tint discoloration that starts at approximately 20 mm from the base plate and gradually darkens as it extends to approximately 130 mm where a clear line of demarcation is observed. Based on the process data, this delineation was not associated with any process upsets. Build 3 samples were printed in both the horizontal and vertical orientations. The samples, Fig. 5, that were printed in the horizontal orientation all exhibit a shiny metallic appearance while the two samples printed in the vertical orientation are more discolored than the Build 2 samples. In addition, the discoloration near the top is significantly darker than Build 2 samples. A review of the process data for builds 2 and 3 does not indicate any obvious reason, e.g., increased bed temperature, increased oxygen content, etc. that might account for these differences. Locally the samples themselves may have been exposed to higher temperatures than the Build 2 samples since the "on-beam time" would have been higher for this build since there were fewer areas being melted.

Samples were burst tested in the as-printed condition after welding high pressure tubing fittings into place. The welder indicated that the Build 1 samples exhibited a significant amount of outgassing during welding, that is attributed to higher surface roughness while the Builds 2 & 3 samples welded very similarly to conventionally produced, e.g., forged, tubing, pipe, etc., austenitic stainless steel and were significantly smoother. The extent of oxidation can be seen in Fig. 6 for the tubes that

Fig. 4 Surface appearance of the Build 1 finned tube samples exhibiting a uniform matte appearance

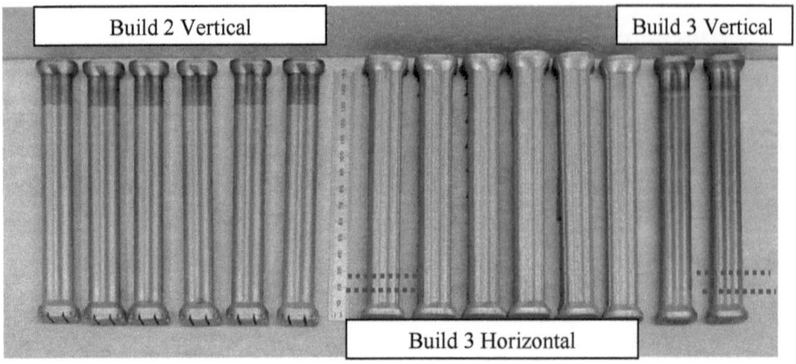

Fig. 5 Appearance of Builds 2 and 3 after removal from the baseplate. Note the *gray* coloration of the horizontally printed parts and the mild oxidation of the vertical built parts and the sheen of the parts rather than the matte surface of the Build 1 samples. The *dashed lines* indicate the orientation for the metallographic samples in Fig. 10

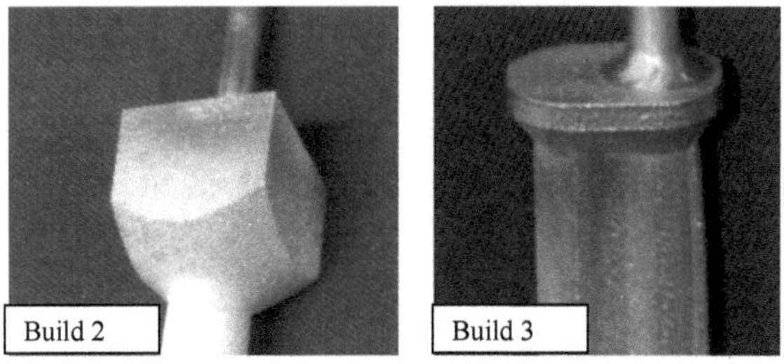

Fig. 6 Surface appearance of welded samples in preparation for burst testing. Note more extensive heat tinting for Build 1

have been welded. The discoloration extends further down Build 1 samples than the Build 2.

Typical burst test results are shown in Fig. 7. Note that the build 1 sample exhibits a low burst result it is consistent in terms of equivalent ultimate tensile strength to that of the horizontally fabricated Build 2 sample but only about 20% of the strength of the vertically fabricated Build 2 sample. All of the Build 2 samples had surface features that may have acted as stress concentrators; grooves for externally attached thermocouples were fabricated in situ while maintaining the desired minimum wall thickness. The typical failure appearance for the Build 2 specimens is shown in Fig. 8 for the horizontal (a) and vertical build (b) orientations. The samples failed at different axial locations with the Build 2, vertically fabricated samples, rupturing in the top half to third of the sample, the Build 3 horizontally fabricated samples failing at the lower third, and Build 3 vertical sample failing near the center of the tube.

The fracture surfaces were examined using scanning electron microscopy. All of the samples exhibited ductile failure modes consisting of dimple rupture with the presence of some larger porosity on the fracture surface. Low magnification images of the fracture surface for horizontally and vertically fabricated samples are shown in Fig. 9. The arrows indicate some of the larger pores. Note that many of the pores

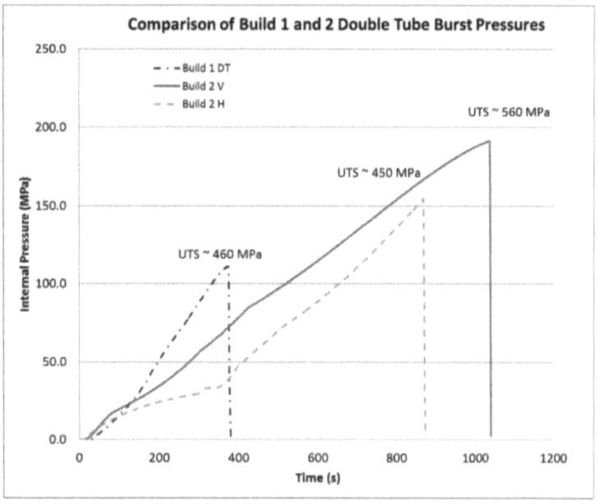

Fig. 7 Burst test results for Build 1 and 2 samples, with corresponding estimated UTS using thin wall pressure vessel calculations for comparison

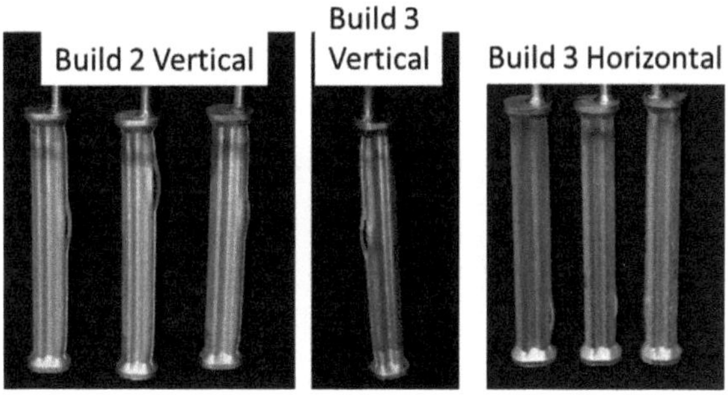

Fig. 8 Photographs of burst tested samples showing failure at the *upper half* for the Build 2, *center* for build 3 vertical and *lower* for the Build 3 horizontal samples

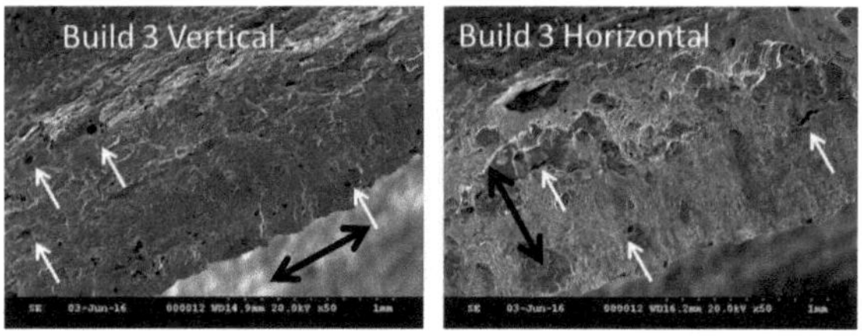

Fig. 9 Fractographs of the burst test samples. *White arrows* indicate pores and *black* general build direction

in the vertical sample are in the outer layer and tend to be spherical in this cross-section while the pores in the horizontal sample are more elongated and likely due to lack of fusion.

The presence of the porosity in the outer layers of the vertical sample can also be seen from the metallographic examination that was completed. The samples were sectioned transversely as indicated in Fig. 10, mounted, polished and electrolytically etched with oxalic acid. Figure 11 shows the microstructure of two samples one in horizontally and one in the vertically printed orientation. The horizontal sample exhibits a fish-scale structure with random pores that are irregularly shaped. There is also some material of less than full density on the lower portion of the tube

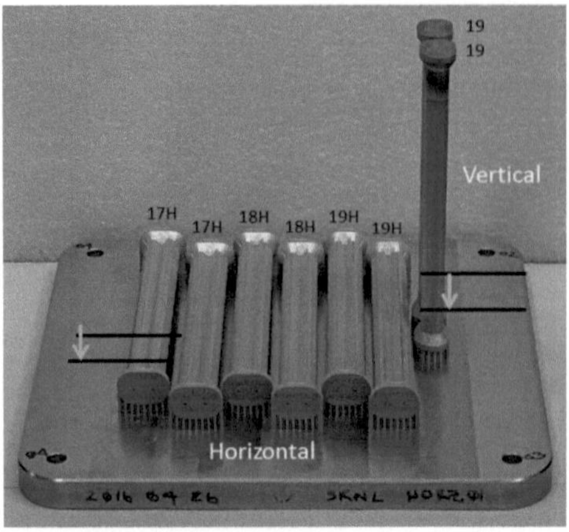

Fig. 10 Arrangement of the vertically and horizontally fabricated parts on a build plate with the orientation for the metallographic samples indicated

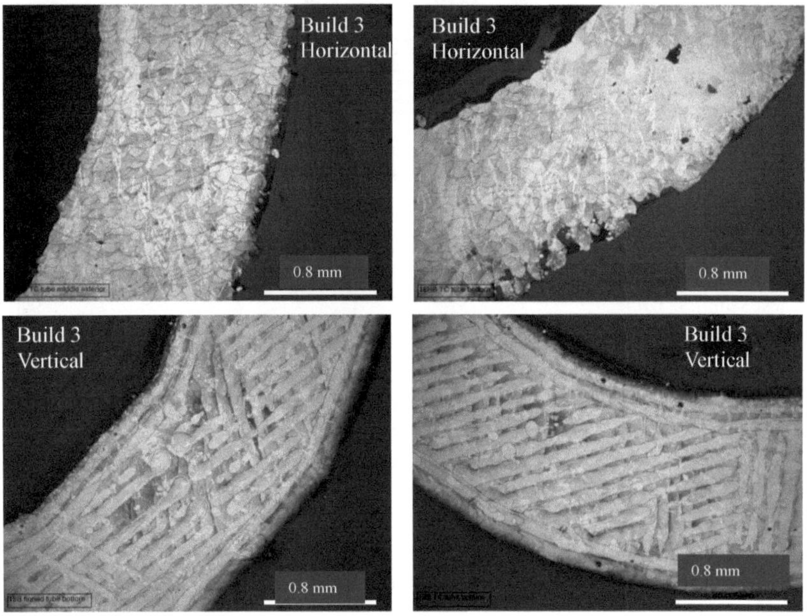

Fig. 11 Cross-section metallographic images from the Build 3 samples in the vertical and horizontal orientations near the bottom of the test sample

where the support structure was attached to the sample. Some of the particles in this area were not fully melted based on the observation of fine dendrites present in th e spherical particles. The vertically fabricated samples exhibit the "basket weave" structure expected on the inner passes from samples fabricated by the laser powder bed process and also due to sectioning effect that intersects successive layers. The perimeter laser scans resulted in random porosity in these layers. This location for pores was also observed in the fractographs from the burst test samples.

Conclusions and Summary

Heat exchanger tubing can be printed using laser bed selective melting. Finite element modeling was successfully used to compare different tubing designs for improved heat transfer.

The quality of the materials produced is related to the printing orientation and parameters; a smaller hatch spacing results in a fewer pores. The burst pressure varied with the build orientation and build number. A few scattered pores were evident on the fracture surfaces. These pores are also visible in cross-section.

Additional work is planned to more fully characterize these materials. The future work includes tensile testing, thermal testing, and more detailed metallographic examinations.

Acknowledgements The authors would like to acknowledge Edward A. Stein for his help with SEM, Metallography, burst testing, and general lab support, George Crow for conducting burst tests, David Worrall for gas tungsten arc welding of the high pressure tubes to the am pipe segments, Robert (Bob) Snyder, Jr. for programmatic support, and Keith Carver for operation of the Renishaw AM250 SLM. This document was prepared for the U.S. Department of Energy under contract number DE-AC09-08SR22470.

References

1. L.E. Murr et al., Metal fabrication by additive manufacturing using laser and electron beam melting technologies. J. Mater. Sci. Technol. **28**(1), 1–14 (2012)
2. B. Vayrea, F. Vignata, F. Villeneuvea, Designing for Additive Manufacturing, in *45th CIRP Conference on Manufacturing Systems 2012, Procedia CIRP 3* (2012), pp. 632–637
3. K.F. Wong, A. Hernandez, A review of additive manufacturing. Int. Sch. Res. Netw., ISRN Mech. Eng. **2012**, Article ID 208760
4. S.S. Babu, et al., Additive manufacturing of materials: opportunities and challenges. MRS Bull. **40** (2015)
5. S.S. Babu, R. Goodridge, Additive manufacturing. Mater. Sci. Technol. **31**(8) (2015)
6. P.S. Korinko, Development of Laser Engineered Net Shape (LENS) additive manufacturing for repair, in *Proceedings of the ASME 2015 Pressure Vessel and Piping Conference*, Boston, MA, 19–23 July 2015
7. D.D. Gu, W. Meiners, K. Wissenbach, R. Poprawe, Laser additive manufacturing of metallic components: materials, processes and mechanisms. Int. Mater. Rev. **57**(3), 133–164 (2012)
8. P.S. Korinko et al., Characterization of Additive Manufacturing for Process Tubing, submitted to Additive Manufacturing, Sept 2016

Scaling Relationships for Direct Ink Writing with Acoustic Focusing

Leanne Friedrich, Rachel Collino, Tyler Ray and Matthew Begley

Abstract Novel methods to control the microstructures of multiphase extruded materials are needed to control spatial variations in structural and functional properties in additively manufactured components. One promising method is acoustic focusing, wherein microparticles suspended in fluids flowing through a silicon microfluidic channel are manipulated by establishing an acoustic wave in the channel. Acoustic focusing quality depends on factors including nozzle and particle geometries, acoustic pressure, matrix rheology, and flow characteristics. Using these factors, we experimentally verified scaling relationships governing acoustic focusing based on classical acoustic forces, using ceramic microspheres in epoxy resin. Further experiments characterized focusing behavior in more complex shear-thinning matrices composed of epoxy resin, fumed silica, and acetone. We designed image processing methods to quantify particle focusing and form holding (the ink's ability to maintain shapes), and we used those methods to identify the trade-offs associated with matrix formulation, printing speed, and acoustic pressure.

Keywords 3D printing · Additive manufacturing · Composites · Acoustic waves

Introduction

3D printing has expanded the manufacturing landscape to include components with intricate shapes of unprecedented complexity. Two-phase printing techniques that allow users to design not only the shape but also spatial variations in particle distribution can introduce additional complexity and opportunities for material reduction. Controlling particles with specific optical, thermal, and electrical properties can impose functionalities like photonic band gaps, coupled electro-mechanical

L. Friedrich (✉) · R. Collino · T. Ray · M. Begley (✉)
University of California, Santa Barbara, USA
e-mail: leanne@umail.ucsb.edu

M. Begley
e-mail: begley@engr.ucsb.edu

The Minerals, Metals & Materials Society, *TMS 2017 146th Annual Meeting & Exhibition Supplemental Proceedings*, The Minerals, Metals & Materials Series,
DOI 10.1007/978-3-319-51493-2_14

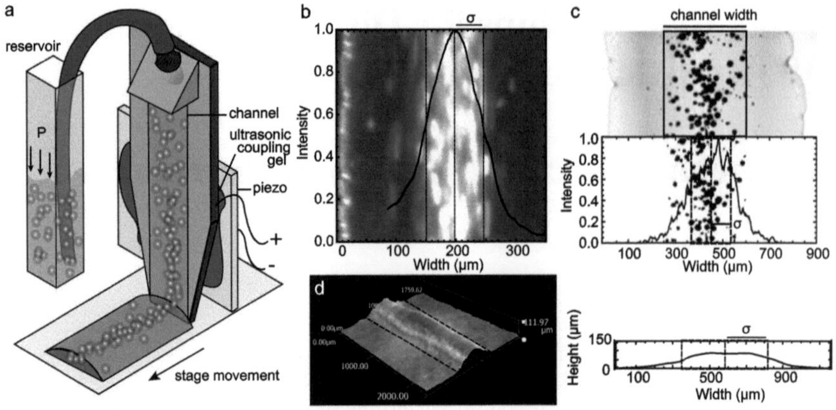

Fig. 1 **a** Diagram of print nozzle. **b–d** Examples of image analysis. **b** Still from video of particles (*light*) in channel (*dark*). *Overlay* average intensity profile throughout channel. **c** Still of particles (*dark*) in printed line (*light*), segmented image, and average intensity profile across line. **d** Height profile of cured line and average height profile across line. σ is the standard deviation

behaviors, and coupled thermo-mechanical behaviors [1–4]. One pathway for decreasing the complexity and cost of fabricating two-phase materials with tailored functionalities is direct ink writing with acoustic focusing.

Direct ink writing, also known as extrusion freeform fabrication, is a method wherein fluid inks are extruded through a nozzle and subsequently cured. This technique has previously been used to fabricate polymer matrix composites with uniform ceramic phase distributions [5–8]. Our group has fabricated SiC fiber-epoxy composites with designed fiber packing densities using acoustic focusing, a technique whereby acoustic waves manipulate particles [9]. With our method, a microparticle-laden epoxy flows through a rectangular glass-on-silicon microfluidic channel that functions as a 3D printing nozzle (Fig. 1a). A piezoelectric actuator establishes a standing wave at the lowest harmonic frequency of the channel, pushing the dense microparticles toward the node of the wave. The focused assembly is then extruded onto a glass slide.

Here, we first examine the theoretical scaling relationships between channel dimensions, particle dimensions, acoustic pressures, focusing zone size, and flow characteristics for material deposition and compare these models to experimental results for ceramic spheres in an epoxy resin. Next, we experimentally explore how non-Newtonian ink formulation, acoustic pressures, and printing speed impact particle packing, distribution preservation, and form holding, or the ink's ability to maintain shapes.

Experimental

Base inks were prepared by mixing epoxy (Miller-Stephenson EPON 828) and fumed silica (Evonik Aerosil R106) at designated weight fractions, then adding a heat-curing agent (Basionics VS03 1-ethyl-3-methylimidazolium dicyanimide) in a 20:1 epoxy:curing agent weight ratio. For printed inks, 14–31 μm silver-coated glass spheres (Cospheric SLGMS-AG-2.74) were mixed with base inks in designated sphere:base weight ratios and adding designated weight percentages of acetone. Inks were mixed between additions for 3 min at 2000 rpm in a planetary mixer (Thinky ARE-310). Weights were measured on a micro-balance (Mettler AE163). Rheological measurements were conducted on a TA Instrument Company ARES-LS1 rheometer, using 25 mm parallel plates with a 2 mm gap. Log frequency sweeps were conducted at 15% shear strain, with a 60 s steady preshear at 1 Hz followed by a 10 min relaxation time. Tests were conducted at room temperature.

Print nozzles consisted of straight etched silicon 150 μm x 350 μm x 35 mm channels anodically bonded to glass capping wafers, fabricated as described in Ref. [10]. Fluidic inlets were drilled into the glass capping layers using a CNC diamond drill. Triangular inlets were fabricated from poly(methyl methacrylate) sheets using a laser cutter (Trotec Speedy 100). Channels were drilled in the inlets using a Dremel drill. Wire ferrules were bonded to the inlets using cyanoacrylate adhesive and epoxy (Devcon HP250). For printing experiments, inks were driven with a high-pressure mass flow controller (Fluigent MFCS-EZ). Nozzles were oriented vertically, and inks were extruded onto a glass slide mounted on a screw-driven linear stage (Zaber T-LSR150B). Inks were cured on a hot plate at 180 °C (Fig. 1a). Extrusion pressures were calculated to match stage speeds as described in Ref. [11], then calibrated manually. For particles in resin, the inks were driven with a syringe pump (Harvard PhD) at 1 μL/min. Particle motion was recorded using an inverted light microscope (Nikon Ti-U Eclipse) and camera (Point Grey Grasshopper3).

10 mm square by 1 mm piezoelectric actuators (Navy I, American Piezo) were attached to the circuits using ultrasonic coupling gel and driven using a sinusoidal function generator (Fluke 294) run through an amplifier (Mini-Circuits LZY-22+). Signals were generated at 2.09 MHz and measured using an oscilloscope (Agilent DSO-X 2024A). For printing experiments, the piezoelectric actuator was cooled using a water-cooled copper stage and thermal joint compound (Wakefield Type 120) and maintained at 20 °C.

Videos of the channel near the exit were captured for each print using an Infinitube FM-200 objective with a Guppy Pro F-503B camera. Wolfram Mathematica 10.1 was used to process the videos. The left edge of the channel was detected automatically using a Shen Castan edge detection algorithm and a Hough transform with a threshold of 0.25. 35 normalized intensity profiles were collected across the width of the channel for 8 frames throughout the print. An average profile was calculated, and the standard deviation of the profile normalized by the channel width was taken to be the focusing width (Fig. 1b).

Using a Keyence VHX-5000 microscope, three-dimensional surface profiles were collected from cured lines coated with a thin layer of Helling 3D laser scanning anti-glare spray. 3D tilt correction was executed. Profiles were analyzed using Wolfram Mathematica 10.1. Surface profiles of two regions in each line were collected, and cross-sectional profiles were averaged across the line. The standard deviation of the cross-sectional profile divided by the height serves as a form holding performance metric. For measurements of focused lines, only samples with cross-sections comprising between 80 and 150% of the channel area were included (Fig. 1d).

Images of cured lines were captured using an inverted light microscope (Nikon Ti-U Eclipse). Using Wolfram Mathematica 10.1, images were thresholded to only detect particles using morphological binarization (top threshold = 0.85, bottom threshold = 0.3). Intensity profiles were captured along the width of the line and averaged to create a composite intensity profile. The standard deviation of this profile divided by the standard deviation of the height profile serve as a focused width metric (Fig. 1c).

Scaling Relationships

Classical solutions for acoustic forces acting on small, isolated particles provide significant insight regarding the time and length scales governing the motion of particles [12]. Here, we summarize some of the scaling relationships that are outlined in detail elsewhere and use them to rationalize experimental observations described in the next section [12–15].

To begin, consider the primary focusing forces that act to drive particles to the center of the channel (provided they are less compressible than the fluid). Let F_o be the peak value of the primary focusing force, and t_o be the time required to move a particle a distance equal to the channel width, when acted upon by F_o and resisted only by laminar drag forces. (I.e., inertia of the particles is neglected.) Then, the characteristic force and time scale associated with particle motion are given by:

$$F_o = \frac{\pi^2 a^3 \beta_f p_o^2 \phi}{3W}, \text{ with } \phi(\beta, \rho) = \frac{5\rho_p - 2\rho_f}{2\rho_p + \rho_f} - \frac{\beta_p}{\beta_f}; \quad t_o = \frac{6\pi\eta a W}{F_o} = \frac{18\eta W^2}{\pi a^2 \beta_f \phi p_o^2} \tag{1}$$

where a is the particle radius, (β_f, β_p) are the fluid and particle compressibilities, respectively, while (ρ_f, ρ_p) are the fluid and particle densities, respectively. W is the width of the channel, or half wave-length; η is the fluid viscosity and p_o is the amplitude of the acoustic pressure wave.

Strictly speaking, the time required to move an isolated particle from one position to its final position is different from t_o, due to the fact that the primary focusing force varies with position, and scattering with other particles may affect is final location. However, despite these complications, the time required to focus nearly any particle in the transverse direction, t_T, is approximately $t_T = t_o/2$ [16]. Once the time to focus the particles is known, the relationship between the length of the channel, the flow velocity and the degree of focusing can be determined. A detailed analysis of parabolic flow conditions reveals that complete focusing occurs when $L \approx 0.3 v_s t_o$, which corresponds to $t_T = t_o/2$. Neglecting pre-factors close to unity, the implication is that the dimensionless nozzle length, $\tilde{L} = L/(v_s t_o)$, and dimensionless transverse focusing time, $\tau_T = t_T/t_o$, are about equal: $\tilde{L} = \tau_T$.

Next, consider the acoustic forces that act to aggregate particles in the flow direction. Attractive secondary forces between particles aligned perpendicular to the wave direction (along the channel) arise due to wave scattering. Using classical acoustic solutions [16], the time required to aggregate two particles separated by distance D_o is given by:

$$\tau_A \equiv \frac{t_A}{t_o} = \frac{16a^2}{5\tilde{\alpha}_\rho W^2}\left(\left(\frac{D_o}{W}\right)^5 - 1\right), \text{ with } \tilde{\alpha}_\rho = \frac{2}{\pi\phi}\left(\frac{\rho_p}{\rho_f} - 1\right)^2 \quad (2)$$

For moderate volume fractions, the spacing between particles is given by $D_o/W \sim 1/f_V^{1/3}$, where f_V is the volume fraction of particles. This is the result for two isolated particles to aggregate: the time required for an isolated particle to join a clump of particles will be slightly smaller, due to large aggregate volumes (which lead to larger acoustic forces). However, the effect is slight because attractive scattering forces fall off quickly with separation distance. Hence, after a clump forms and starves its neighborhood of particles, it is no longer effective in driving aggregation. The upshot is that one can expect the time scale for aggregation to be most accurately described by the two particle result given above.

For printing applications, it is reasonable to expect that ideal conditions are associated with a balance of transverse and axial printing time scales, i.e. $t_T \sim t_A$. If transverse focusing is faster than axial aggregation, one anticipates printing clumps of particles that have not had time to assemble in the flow direction. Conversely, if axial aggregation is much faster, one can anticipate clumping in the axial direction printer to focusing, which promotes jamming. Therefore, a useful printing process parameter is $P = t_A/t_T = \tau_A/\tau_T$; ideal conditions are associated with $P \sim 1$. Noting that $\tau_T \approx L/(v_s t_o) = \tilde{L}$,

$$P \equiv \frac{16a^2}{5W^2\tilde{\alpha}_\rho \tilde{L} f_V^{5/3}} \quad (3)$$

This expression leads immediately to the printing map shown in Fig. 2, which plots contours of P as a function of normalized nozzle length, $\tilde{L}$, and particle volume fraction, f_V. For points where $\tilde{L} < 0.3$, the nozzle is not long enough to achieve full transverse focusing. This can be changed by using a longer nozzle, slower flow speed, or less viscous fluid.

The experiments described in the previous section provide qualitative validation of the printing regimes implied by the map shown in Fig. 2. Representative optical micrographs are are shown; points A and C correspond to printing conditions where complete transverse focusing has not occurred, as confirmed by the micrographs.

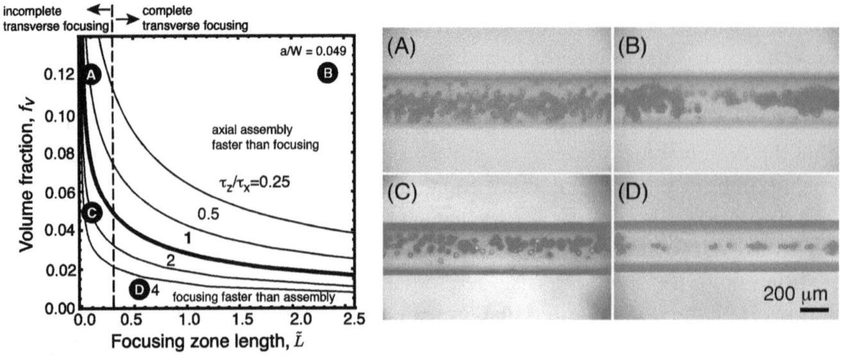

Fig. 2 Contours of assembly parameter P_A^o. Where $P_A^o > 1$, axial assembly is faster than focusing, and where $\tilde{L} > 0.3$, focusing is complete. Images are of barium titanate microspheres of diameter 34.5 μm in Epon 828 resin for each focusing domain

Point C has less axial aggregation than Point A, as indicated by the map and the more frequent axial gaps shown in (C). Similarly, Point B and micrograph (B) show large axial aggregations, due to the fact that extrusion rates are so low. In (D), complete focusing occurs, such that the particles are mostly at the center line, but there is not sufficient time for axial aggregation, leading to large gaps between focused particles.

Focused Microstructures in Printed Lines

While the previously discussed theory considers only Newtonian matrices, direct ink writing requires shear thinning matrices, such that during extrusion, the viscosity is low, but after deposition, the viscosity is high. Although this theory only approximates focusing behavior in shear thinning fluids, we demonstrate that it nonetheless is useful in predicting trends for shear thinning Epon 828 nanocomposite resins. Balancing primary acoustic forces against drag forces as in Eq. 1, we can determine an approximation for the focusing width w_F, where η is the viscosity of the fluid at the flow speed.

$$w_F = W - \frac{\pi a^2 \beta_f p_o^2 \phi L}{9 W \eta v_s} \tag{4}$$

Focusing width scales with the inverse of the calculated flowing viscosity, affirming the primary force model in Eq. 4 (Fig. 3a). When focusing is not employed, there is no correlation between focusing width and viscosity or focusing width and silica, because p_o is 0. When focusing is employed, the focusing width increases quickly with increasing viscosity. The cutoff silica loading, where there is no difference between the focused and unfocused states, is at 10 w% silica.

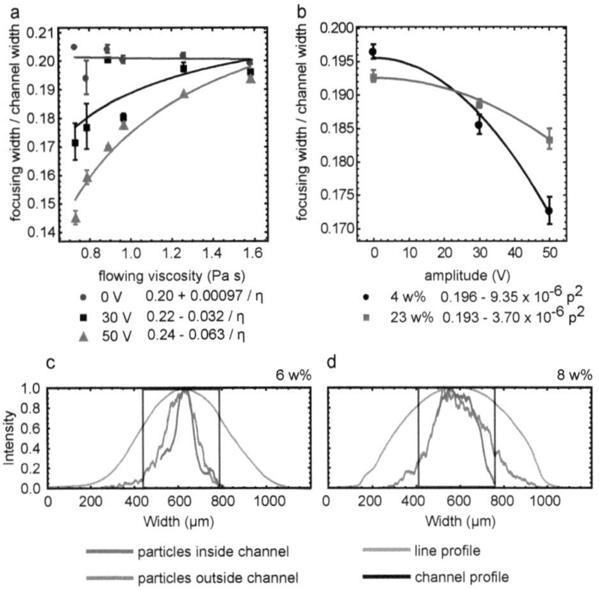

Fig. 3 Experimental data fit to theoretical scaling relationships. **a** Focusing width as a function of calculated viscosity in the flowing state for 1.7 v% particles, 12 v% acetone, and a speed of 1 mm/s. **b** Focusing width as a function of acoustic wave amplitude for 12 v% acetone. **c–d** Normalized intensity and line profiles for 1.7 v% particles in the printed line, 50 V, 1 mm/s, and 12 v% acetone, with varying silica loadings

The focusing distance scales with the square of the piezo input voltage, which scales with the acoustic pressure wave amplitude (Fig. 3b). This also fits with Eq. 4. At higher particle loadings, the focusing width is larger. This may be because secondary scattering increases the focusing width.

The relationship between speed and focusing width is not as simple. As speed increases, the focusing width increases because the particles spend less time in the focusing zone. However, increasing the speed also decreases the fluid viscosity, so the focusing width follows more of a linear than an inverse relationship with speed, so the relationship between speed and focusing width does not match Eq. 4.

Ink formulations also impact form holding. As silica loading decreases, the line width/height increases (Fig. 3c–d). This spreading makes it difficult to print shapes that maintain their form after deposition without support material. Fortunately, the spreading does not impact the particle distribution. Between the channel and the printed line, intensity profiles tend to remain similar widths (Fig. 3c–d). Only the epoxy matrix spreads out. Thus, if forming a narrow distribution of particles is the only priority, then using an inviscid matrix may be sufficient. In contrast, if holding form is critical, it is necessary to use a higher viscosity matrix.

Conclusion

We have formulated a 'focusing zone' model for acoustic focusing combined with direct ink writing that can be used to design the microstructure of multiphase inks, with the opportunity to assemble particles along the length of the nozzle during deposition. This model considers time scales for primary acoustic focusing and secondary scattering and how they depend on ink formulation and printing parameters. These relationships align with experimental results. These principles can be applied to acoustic focusing with direct ink writing of other material systems, and have been demonstrated in both Newtonian and non-Newtonian matrices. Thus, because acoustic focusing only depends on physical parameters like compressibility and size, it can be paired with other printing modalities like hydrodynamic, electrical and magnetic focusing.

Acknowledgements R.R.C., T.R.R., and M.R.B. gratefully acknowledge the support of the Institute for Collaborative Biotechnologies through contract no. W911NF-09-D-0001 from the U.S. Army Research Office, use of the Microfluidics Laboratory and the NanoStructures Cleanroom Facility at the California Nanosystems Institute, as well as the Nanofabrication Facility (a part of the NSF-funded National Nanotechnology Infrastructure Network) at UCSB. L.M.F. was supported in part by a UCSB Chancellor's Fellowship.

References

1. S.H. Kim, S.Y. Lee, S.M. Yang, G.R. Yi, Self-assembled colloidal structures for photonics. NPG Asia Mater. **3**(1), 25–33 (2011)
2. A. O'Halloran, F. O'Malley, P. McHugh, A review on dielectric elastomer actuators, technology, applications, and challenges. J. Appl. Phys. **104**(7), 071101 (2008)
3. Q. Ge, H.J. Qi, M.L. Dunn, Active materials by four-dimension printing. Appl. Phys. Lett. **103**(13), 131901 (2013)
4. S.E. Bakarich, R. Gorkin, M. Panhuis, G.M. Spinks, 4d Printing with mechanically robust, thermally actuating hydrogels. Macromol. Rapid Commun. **36**(12), 1211–1217 (2015)
5. P. Calvert, T.L. Lin, H. Martin, Extrusion freeform fabrication of chopped-fibre reinforced composites. High Perform. Polym. **9**(4), 449–456 (1997)
6. J. Peng, T.L. Lin, P. Calvert, Orientation effects in freeformed short-fiber composites. Compos. Part A: Appl. Sci. Manuf. **30**(2), 133–138 (1999)
7. B.G. Compton, J.A. Lewis, 3d-printing of lightweight cellular composites. Adv. Mater. **26**(34), 5930–5935 (2014)
8. J. Ebert, E. Ozkol, A. Zeichner, K. Uibel, O. Weiss, U. Koops, R. Telle, H. Fischer, Direct inkjet printing of dental prostheses made of zirconia. J. Dent. Res. **88**(7), 673–676 (2009)
9. R.R. Collino, T.R. Ray, R.C. Fleming, J.D. Cornell, B.G. Compton, M.R. Begley, Deposition of ordered two-phase materials using microfluidic print nozzles with acoustic focusing. Extrem. Mech. Lett. (2016)
10. R.R. Collino, T.R. Ray, R.C. Fleming, C.H. Sasaki, H. Haj-Hariri, M.R. Begley, Acoustic field controlled patterning and assembly of anisotropic particles. Extrem. Mech. Lett. (2015)
11. Y. Son, Determination of shear viscosity and shear rate from pressure drop and flow rate relationship in a rectangular channel. Polymer **48**(2), 632–637 (2007)
12. T. Laurell, F. Petersson, A. Nilsson, Chip integrated strategies for acoustic separation and manipulation of cells and particles. Chem. Soc. Rev. **36**(3), 492–506 (2007)

13. L.V. King, On the acoustic radiation pressure on spheres. Proc. Roy. Soc. A: Math. Phys. Eng. Sci. **147**(861), 212–240 (1934)
14. W.L. Nyborg, Radiation pressure on a small rigid sphere. J. Acoust. Soc. Am. **42**, 947–952 (1967)
15. K. Yosioka, Y. Kawasima, Acoustic radiaion pressure on a compressible sphere. Chem. Soc. Rev. **36**, 492–506 (2007)
16. R.R. Collino, T.R. Ray, L.M. Friedrich, J.D. Cornell, C.D. Meinhart, M.R. Begley, Scaling relationships for directed assembly of two-phase materials using acoustic focusing. Manuscript submitted for publication (2016)

The Influence of Gas Cooling in Context of Wire Arc Additive Manufacturing—A Novel Strategy of Affecting Grain Structure and Size

Philipp Henckell, Karsten Günther, Yarop Ali, Jean Pierre Bergmann, Jürgen Scholz and Pierre Forêt

Abstract The continuous building process in additive manufacturing with gas metal arc welding (GMAW) provides the main advantage of a decreased processing time but leads to a high heat input in the built work piece. Especially geometrically small parts are affected by a coarse and constantly growing grain structure throughout the continuous reheating process. A novel approach of influencing the temperature-time regime during the additive manufacturing process is an application of additional cooling gas. Experimental trials with argon, hydrogen and nitrogen were carried out and analyzed by means of thermal imaging, hardness measurement and microscopy. The experimental results showed a significant influence of cooling gases on the temperature during the building process. Hence, grain structure and size can be modeled to a homogeneous microstructure by the composition of the gas.

Keywords Additive manufacturing · GMAW · Gas cooling · Cross section · Microstructure · Homogeneous properties · Grain size · Hardness measurements

Introduction

Wire arc additive manufacturing methods (WAAM) have become more important over the last years. The reasons are not only higher deposition rates compared to laser based additive manufacturing processes, but also the easy and uncomplicated handling as well as the relatively low investment costs of the welding equipment. Thus, large individual and specially designed parts can be produced in a cost effective way, offering a strategic key role for special demands and applications.

P. Henckell (✉) · K. Günther · Y. Ali · J.P. Bergmann
Department of Production Technology, Ilmenau University of Technology, 98693 Ilmenau, Germany
e-mail: philipp.henckell@tu-ilmenau.de

J. Scholz · P. Forêt
Linde Gases Division, Linde AG, 85716 Unterschleissheim, Germany

The Minerals, Metals & Materials Society, *TMS 2017 146th Annual Meeting & Exhibition Supplemental Proceedings*, The Minerals, Metals & Materials Series, DOI 10.1007/978-3-319-51493-2_15

Especially the aerospace industry shows a great interest in this technology. Within the scope of the EU—project "RAPOLAC", a robotized additive manufacturing method was developed, based on the TIG and Plasma welding process [1–10]. The production of components based on Ti—and Ni—alloys as well as unalloyed steel was in the center of interest, along with detailed examinations of the process and materials characteristics.

Compared to the relatively low deposition rates of TIG—and Plasma based welding processes, GMAW is a highly productive additive manufacturing method, enabling deposition rates of more than 4 kg/h. Profound investigations of process technological features were carried out by Cranfield University by means of experimental determination of residual stresses and their prediction by simulation modeling [12–19]. Further investigations of Kah et al. [11] showed that the use of controlled short arc processes improves the weld bead stability in constraint positions. Thus, near net shaped components with high geometric complexity can be manufactured with a maximum difference in layer width of 0.2 mm [20–23].

So far, only few examinations are dedicated to material scientific characterizations of the additive manufactured components. Hartke et al. [24, 25] investigated the energy input for high alloyed austenitic stainless steels (1.4842 and 1.4430) in the manufacturing of cylindrical parts using the energy reduced short arc technology (EWM ColdArc). The results showed that the increase of arc energy (increase of the wire feed or decrease of welding speed) affected the sample width negatively and increased the finishing work, which is undesired. The metallurgical results showed that homogenous properties over the length of the part can only be achieved with low arc energy. Furthermore, it was noticed that the tensile strength decreases with increased arc energy.

However, the influence of the temperature-time regime on the microstructure of materials (especially steels) is rarely observed in scientific literature, whereas most of the investigations are dedicated to aerospace materials [26, 27]. The reliable adjustment of material's properties during additive manufacturing is of high importance for later applications. Thus, Suryakumar et al. [28] discovered that a periodic thermal load during GMAW additive manufacturing causes inhomogeneous hardness properties with increasing layer height. This is due to the formation of a coarse grained microstructure depending on the number of cycles and temperature loads during the buildup process. In this context, unalloyed as well as low alloyed steels were examined. These results were also confirmed for additive GTAW technologies [29].

In this case study, a novel strategy of influencing the temperature-time regime by gas cooling was examined. In order to reduce the layer temperatures, additional gas was applied onto the solidified top layer during the buildup process. The investigations are based on low alloyed steel and were processed with an energy reduced short arc technology (EWM—ColdArc). The main focus of this investigation is the comparison between the current WAAM process and the buildup process with additional gas cooling. The process temperature by means of contactless measurement, hardness and of the produced components were analyzed, showing the influence of different types of cooling gases.

Experimental Setup

The experimental trials were carried out on cylindrical pipe structures built layer-wise by an energy reduced short arc GMAW process with a neutral torch position. As a welding power source, the EWM Alpha Q 552 was used in combination with a Binzel Robo water cooled welding torch. The handling device was a six-axis industrial robot KUKA type KR15-2, moving the welding torch in z-direction. The substrate of S355 steel was mounted onto a rotary table with an offset to the welding torch in order to manufacture a component with a diameter of 75 mm. The following Fig. 1 shows the experimental setup.

The following Fig. 2 visualizes the circular arrangement of GMAW torch and additional gas nozzles. The position of the gas nozzles was assumed to affect the buildup process as well as the microstructure of the work piece significantly. Therefore, the position of the additional nozzles was varied and analyzed systematically (90°, 180°, 270°) in order to affect different areas of the cylindrical structure.

Fig. 1 Experimental setup with rotary table and 6-axis industrial robot

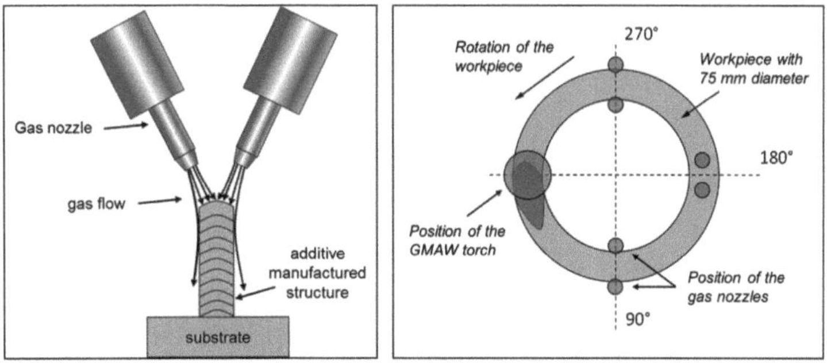

Fig. 2 Position of the cooling gas nozzles (*front view left, top view right*)

Table 1 Properties and parameters of the welding trails

Property	Value
Base material	S355; ferrite steel; 5 mm thickness
Wire material	G4Si1; 1.0 mm diameter; massive wire
Welding voltage	19 V
Welding velocity	0.3 m/min
Stick out	12 mm
Wire feeding speed	6 m/min
Shielding gas (GMAW)	15 l/min – Ar 4.6
Cooling gas	30 l/min – Ar 4.6; Ar + 5% H_2; N_2 + 5% H_2

The experimental trials with different types of gases were carried out with constant process parameters in order to qualify the effects of the additional cooling process. The following Table 1 shows the correlating values of the process parameters used. The trials with additional gas cooling were executed with a gas flow starting from the sixth layer in order to reduce the heat accumulation in the upper parts of the work piece. The measurement of the layer temperature was carried out with a contactless infra-red (IR) camera system type Infratec VarioCam HD research 980 S, set up with a 60 mm Zeiss VarioCAM telephoto objective. After cutting, samples for microstructural cross sections were grinded, polished and etched with 5% nitric acid. The inspection was undertaken with an AxioScope.A1 microscope and an AxioCam ICc3 camera from Zeiss.

Results and Discussion

Additive Manufacturing Without Gas Cooling

This case study was aimed to clarify the influence of additional gas cooling during additive manufacturing with GMAW. The cylindrical geometry of the sample work pieces was chosen to be built continuously and in order to visualize the total heat input and the heat accumulation in the upper layer heights of the symmetrical structure. As an abortion criterion for the welding process, driven out liquid metal was set.

The welding parameters were chosen in that way, a continuous buildup process cannot be accomplished due to excessive heat accumulation in the top layers. Therefore, the adjustment of the layer temperature during the welding process is required to reach a stable temperature time regime. The following Fig. 3 illustrates the resulting structure of the actual buildup condition without additional gas cooling. A process abortion after 40 layers can be seen due to driven out cast. Additionally, a correlating IR-camera image shows the temperature distribution during the additive manufacturing process. A schematically drawn ellipsis in this image points out the rotatory extension of the temperature field with high heat input above 700 °C.

Fig. 3 Resulting work piece (*left*) and IR-thermal image without gas cooling (*right*)

Influence of the Gas Nozzle Positioning

The positioning of the cooling gas nozzles in rotary welding direction (compare Fig. 2, right) was of high interest and carried out with $N_2 + 5\%\ H_2$ gas. Experimental trials with differing gas types were carried out with the gas nozzle arrangement showing the most effective cooling afterwards. The resulting samples are visualized in Fig. 4. The arrangement of the gas nozzles in 90° position behind the welding torch led to the highest cooling effect due to the instant penetration of the mushy weld bead. With an increasing layer number, the temperature in the buildup structure augmented to a critical point where the weld bead was blown out by the gas stream due to a more liquid bead condition.

The 180° position on one hand was determined to analyze the angular position itself. On the other hand the arrangement of the gas nozzles in this position was set directly onto the top layer in order to receive a higher cooling effect (see Fig. 2, right). Though, the resulting additively manufactured sample shows a process abortion directly after turning on the additional cooling gas in the sixth layer. The arrangement of the gas nozzles led to high turbulences in the weld bead and finally a process abortion occurred.

The gas nozzle arrangement in 270° position led to a homogeneous cooling of each layer in the circular structure. A continuous buildup process was possible and a near net shape geometry achieved. All further experimental trails were carried out with this arrangement.

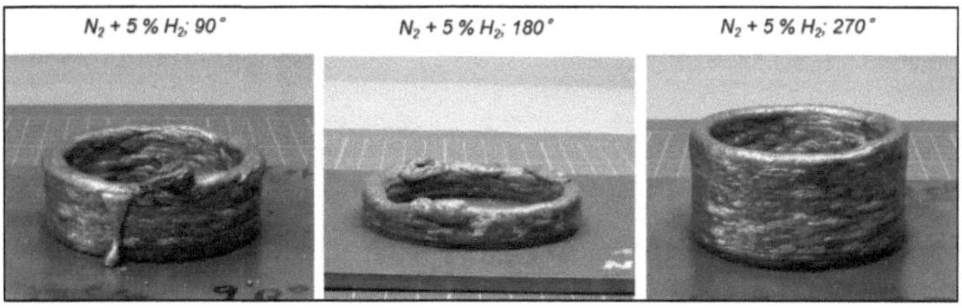

Fig. 4 Effects of the cooling gas nozzle positioning (90°, 180°, 270°)

Additive Manufacturing with Cooling Gas—Affecting Process Temperature

Another main focus of this case study was to understand the effect of different types of cooling gases on the processing temperature. Trials were carried out with argon 4.6, argon including 5% hydrogen and nitrogen with 5% hydrogen. The heat affected area throughout the welding process by IR-camera images can be seen in Fig. 5. Argon 4.6 and argon with 5% hydrogen showed similar results in terms of the temperature field and process behavior. Therefore, the results of argon 4.6 will not be illustrated.

The images also show the schematically added ellipsis, pointing out the area of the top layers with excessive heat input. In comparison to the current condition (Fig. 3, right) a significant reduction of the temperature field with high heat input can be seen in both images. Looking at the IR-camera image of argon with 5% hydrogen a reduction of the described temperature field by nearly 20% throughout the additional gas nozzles (270° position, compare Fig. 2) can be seen. The use of nitrogen with 5% of hydrogen even increased the cooling effect. Also positioned at 270° behind the welding torch, this type of cooling gas affects a larger area of the top layers, leading to a further decrease of the temperature field of about 30% compared to the current condition. The heat accumulation during the additive buildup process can be reduced, respectively controlled throughout the type of cooling gas used during the process.

Comparing the properties of different types of gases, an explanation for such differences in cooling effects can be found at the specific heat capacity and the thermal conductivity. The values for nitrogen and hydrogen exceed those of argon by a minimum factor of two. This leads to a more effective absorption and dissipation of the heat, reducing the temperature of the top layers in additive manufactured structures. The following Table 2 shows the properties of the gases described above.

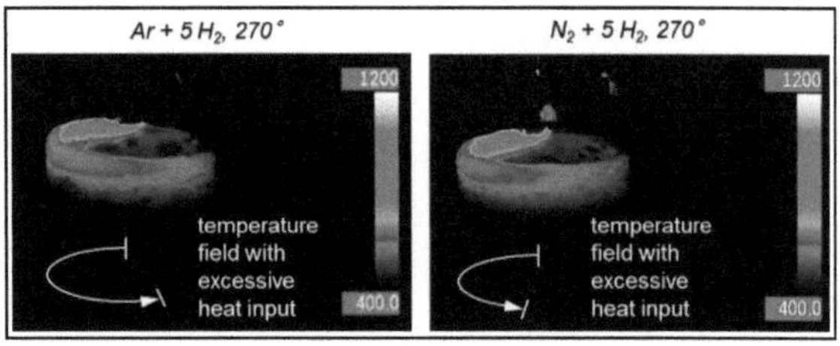

Fig. 5 Effects of different cooling gas types on thermal imaging during additive manufacturing

Table 2 Properties of different types of gases at room temperature [30, 31]

Type of gas	Specific heat capacity [J/g * K]	Thermal conductivity [W/m * K]
Argon	0.52	0.018
Nitrogen	1.04	0.026
Hydrogen	14.3	0.186

Additive Manufacturing with Cooling Gas—Affecting Microstructure

In order to qualify the process of gas cooling during WAAM, cross sections were prepared, showing the microstructure from substrate to top. Geometrical deviations towards a near net shape structure can be determined. Microstructural images and Vickers hardness measurements validate the additive manufacturing process.

The following Fig. 6 describes the properties of the additively built work piece without additional gas cooling. Due to the excessive heat input during the process, the cross section shows layer geometries with large width and low height towards the top layer. During the unstable buildup process, a near net shape wall thickness over all layers could not be accomplished. The microstructural analysis was carried out in the middle of the additively manufactured sample that is characterized by the rectangle in the cross section image. A coarse grain structure consisting of ferrite and perlite due to cyclic reheating and heat accumulation in the top layers can be seen.

Hardness measurements over the different layers show an average value of 160 HV1 with strongly varying values up to 50 HV1. As a result, the buildup process without additional cooling gas leads to inhomogeneous material properties throughout the coarse grain structure.

Using the additional gas cooling with nitrogen including 5% of hydrogen, a near net shape structure can be seen in the cross section (see Fig. 7). Width and height

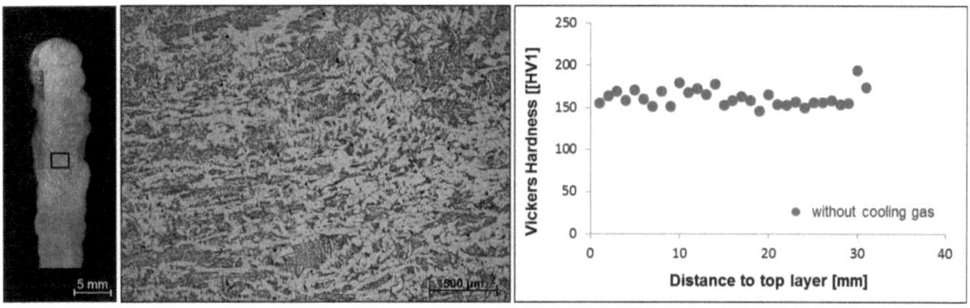

Fig. 6 Cross section, microstructure and Vickers hardness of a probe without gas cooling

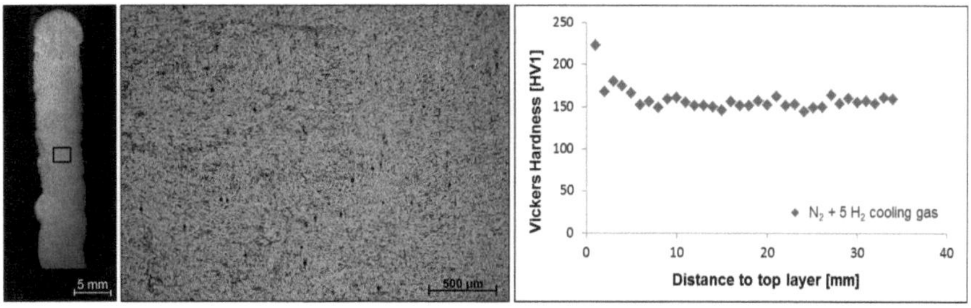

Fig. 7 Cross section, microstructure and Vickers hardness of a probe with additional gas cooling (N2 + 5% H2–270°)

of the different layers conform in geometry throughout the entire sample, leading to low rework during machine finishing. Furthermore, the lower cyclic reheating temperature decreases grain growth during the WAAM process, resulting in a fine grain microstructure. As expected, the according measurement of the Vickers hardness shows homogeneous values of about 150 HV1 with little variation.

Conclusion and Outlook

Additional gas cooling during additive manufacturing with GMAW is a novel way of setting a desired temperature time regime by means of material, built geometry and welding parameters. In this work, all employed gas types improved the current buildup condition without gas cooling, enabling a continuous manufacturing process without heat accumulation in the top layers. The best results were achieved with the additional use of nitrogen with 5% of hydrogen. The high values of specific heat capacity and thermal conductivity of nitrogen and hydrogen in comparison to argon led to a highly effective cooling process. The reduction of the temperature in the top layers also changed the additive manufacturing process towards a decreased cyclic reheating of the lower layers. A near net shape layer geometry could be accomplished throughout the entire height of the structure. Analyzing the microstructures of the sample work pieces, the adapted temperature time regime led to a fine grain structure with homogeneous values in hardness over all layers.

Further investigations will analyze different types of gases such as carbon dioxide or pure nitrogen in order to validate specific heat capacity and thermal conductivity as the determining factor of gas cooling. Quantitative temperature measurements with contacting thermo couples will be carried out to confirm the IR-camera images and qualify the process of additive manufacturing with GMAW further.

Acknowledgements The investigations were carried out in close cooperation between the Linde AG and the production technology group of Ilmenau University of Technology. We would like to thank the organization, especially Mr. Jürgen Scholz as well as Mr. Pierre Forêt for the opportunity and the great support.

References

1. B. Baufeld, O. van der Biest, Mechanical properties of Ti-6Al-4 V specimens produced by shaped metal deposition. Sci. Technol. Adv. Mater. **10** (2009)
2. B. Baufeld, O. van der Biest, R. Gault, Additive manufacturing of Ti–6Al–4 V components by shaped metal deposition. Mater. Des. **31** (2010)

3. B. Baufeld, Effect of deposition parameters on mechanical properties of shaped metal deposition parts. J. Eng. Manuf. **226** (2012)
4. B. Baufeld, Mechanical properties of Inconel 718 parts manufactured by shaped metal deposition (SMD). J. Mater. Eng. Perform. **12** (2012)
5. V.D. Fachinotti, A. Cardona, B. Baufeld, O. van der Biest, Evolution of microstructure during shaped metal deposition. Mechanica Computacional, CILAMCE (2010)
6. E. Brandl, B. Baufeld, C. Leyens, R. Gault, Additive manufactured Ti-6Al-4 V using welding wire: comparison of laser and arc beam deposition and evaluation with respect to aerospace material specifications. Phys. Proc. **5** (2010)
7. F. Martina, J. Mehnen, S.W. Williams, P. Colegrove, F. Wang, Investigation of the benefits of plasma deposition for the additive layer manufacture of Ti–6Al–4 V. J. Mater. Process. Technol. **212** (2012)
8. T. Skiba, B. Baufeld, O. van der Biest, Microstructure and mechanical properties of stainless steel component manufactured by shaped metal deposition. J. Iron Steel Inst. Jpn. **49** (2009)
9. Informationen zum EU-Projekt "Rapolac". http://www.2020-horizon.com/RAPOLAC-Rapid-Production-of-Large-Aerospace-Components(RAPOLAC)-s26465.html
10. V.D. Fachinotti, A. Cardona, B. Baufeld, O. van der Biest, Finite-element modelling of heat transfer in shaped metal deposition and experimental validation. Acta Mater. **60** (2012)
11. P. Kah, R. Suoranta, J. Martikainen, Advanced gas metal arc welding processes. Int. J. Adv. Manuf. Technol. **67** (2013)
12. J.D. Spencer, P.M. Dickens, C.M. Wykes, Rapid prototyping of metal parts by three-dimensional welding. J. Eng. Manuf. **212** (1998)
13. M.P. Mughal, H. Fawad, R.A. Mufti, Three-dimensional finite-element modelling of deformation in weld-based rapid prototyping. J. Mech. Eng. Sci. **220** (2006)
14. H. Fawad, Heat transfer modeling of metal deposition employing welding heat source. Appl. Mech. Mater. **315** (2013)
15. P. Michaleris, Modelling metal deposition in heat transfer analyses of additive manufacturing processes. Fin. Elem. Anal. Des. **86** (2014)
16. J. Ding, Thermo-mechanical analysis of wire and arc additive manufacturing process, Dissertation, Cranfield University (2012)
17. J. Ding, P. Colegrove, J. Mehnen, S. Williams, F. Wang, P.S. Almeida, A computationally efficient finite element model of wire and arc additive manufacture. Int. J. Adv. Manuf. Technol. **70** (2014)
18. R.K. Chin, J.L. Beuth, Successive deposition of metals in solid freeform fabrication processes, Part 1: thermomechanical models of layers and droplet columns. J. Manuf. Sci. Eng. **123** (2000)
19. R.K. Chin, J.L. Beuth, Successive deposition of metals in solid freeform fabrication processes, Part 2: Thermomechanical models of adjacent droplets. J. Manuf. Sci. Eng. **123** (2000)
20. J. Mehnen, J. Ding, H. Lockett, P. Kazanas, Design study for wire and arc additive manufacture. Int. J. Prod. Dev. **19** (2014)
21. P.M.S. Almeida, Process control and development in wire and arc additive manufacturing, Dissertation, Cranfield University (2012)
22. A. Adebayo, J. Mehnen, X. Tonnellier, Limiting travel speed in additive layer manufacturing, in *Konferenzbeitrag zur 9th International Conference on Trends in Welding Research*, Chicago (2012)
23. P. Kazanas, P. Deherkar, P. Almeida, H. Lockett, S. Williams, Fabrication of geometrical features using wire and arc additive manufacture. J. Eng. Manuf. **226** (2012)
24. M.A. Hartke, K. Günther, J.P. Bergmann, Untersuchung zur geregelten, energiereduzierten Kurzlichtbogentechnik als generatives Fertigungsverfahren, in Fachbeitrag zur „Großen Schweißtechnischen Tagung", Berlin (2014)
25. M. Hartke, K. Günther, Y. Ali, J.P. Bergmann, Schweißtechnische Herstellung komplexer 3D-Freiformflächen mittels geregelter MSG-Kurzlichtbogentechnik, in Tagungsband zur 4. Fachtagung „Verarbeitung von Stählen im Kraftwerksbau", Duisburg (2014)

26. P.M.S. Almeida, S. Williams, Innovative process model of Ti–6Al–4 V additive layer manufacturing using cold metal transfer (CMT), in *Fachbeitrag zum 21th Annual International Solid Freeform Fabrication Symposium*, Texas (2010)
27. D. Clark, M.R. Bache, M.T. Whittaker, Shaped metal deposition of a nickel alloy for aero engine applications. J. Mater. Process. Technol. **203** (2008)
28. S. Suryakumar, K. Karunakaran, A study of the mechanical properties of objects built through weld deposition. J. Eng. Manuf. **227** (2013)
29. Z. Jandric, M. Labudovic, R. Kovacevic, Effect of heat sink on microstructure of three-dimensional parts built by welding-based deposition. Int. J. Mach. Tools Manuf. **44** (2004)
30. N. Vargaftik, L. Filippov, A. Tarzimanov, *Handbook of Thermal Conductivity of Liquids and Gases* (CRC-Press Inc., Boca Raton, FL, 1994)
31. Compressed Gas Association Inc, *Handbook of Compressed Gases*, 3rd edn. (Chapman & Hall, New York, NY, 1990)

Microstructure and Mechanical Properties of Laser Deposited Ni/WC Metal Matrix Composite Coatings

Abhishek Ramakrishnan, Amrinder Singh and Guru Prasad Dinda

Abstract In the present study, the feasibility of laser metal deposition of WC reinforced Nickel composite coatings was investigated for the application of hardfacing coating on copper. Ni/60WC MMC were fabricated on the surface of copper with a buffer layer of hastelloy C-276. Additional sets of samples were deposited on a 1020 mild steel substrate. Three principal experiments were conducted in a systematic order to understand the microstructural evolution of Ni/60WC MMC. A multi-track, single-track deposit, and remelting of the as-deposited surface were systematically investigated by optical and scanning electron microscopy. The results revealed microstructural element such as dendrites, equiaxed structure, eutectic structure, and several other complex carbide structures. It was observed that the presence of three principle phases in the deposit consisted of the primary WC, dendritic Ni, and a eutectic structure. Microhardness of the laser deposited samples indicated the gradual variation in hardness as the deposition transits from a highly conductive soft material to a hard facing coating. Homogenization of WC in Ni matrix was well realized in this study.

Keywords Ni-60WC · MMC · Laser metal deposition · Additive manufacturing

Introduction

Laser Metal Deposition (LMD) process has many potential applications, including production of near net shaped components, surface coatings, part repairing, and fabricating functionally graded materials. Material parts that are required to perform under severe environmental conditions often have to display high hardness, high resistance to wear, creep, erosion and corrosion. Some of the examples of these applications are cutting tools, mining parts, oil & gas industry and high voltage

A. Ramakrishnan (✉) · A. Singh · G.P. Dinda
Department of Mechanical Engineering, Wayne State University, Detroit, MI 48202, USA
e-mail: fl9494@wayne.edu

The Minerals, Metals & Materials Society, *TMS 2017 146th Annual Meeting & Exhibition Supplemental Proceedings*, The Minerals, Metals & Materials Series,
DOI 10.1007/978-3-319-51493-2_16

electrical applications [1–6]. Nickel superalloy is used as binder of WC particles in this current investigation. Nickel superalloy as a binder material shows better mechanical properties, resistance to wear, thermal resistance and is cost effective compared to copper, cobalt and iron based superalloys [7–13]. A pre mixed powder of Ni/60WC was used to study the microstructural evolution of WC/Ni MMC considering it to be the most favorable hardfacing material used in the industrial applications. The motivation of this work is to obtain parts with a spatial variation in their structure and thermal properties. It is envisaged that the outer shell would be from Ni/WC to provide the structural strength and the core would be from copper to reduce the thermal resistance, thus encouraging the flow of heat from the cavity to the cooling channels in case of high pressure die casting (HPDC) and hardfacing coating in the oil and gas industry. Copper tungsten alloys are used where the combination of high heat resistance and thermal conductivity, and low thermal expansions are needed. Some of the applications are in electrical resistance welding, as electrical contactors. The primary objective of this investigation is to develop a hardfacing coating on copper alloy for new applications. One knows that depositing WC directly on copper would not yield a good bond, hence it would be good if one deposits a nickel alloy in between because nickel and copper make a good bond because both are face centered cubic and they make a complete solid solution over the entire phase diagram.

Experimental Procedure

A 1.2 KW high power diode laser (LDM 1200–40) is used as a heat source. The laser beam has a circular spot size of 2 mm in diameter. An Inert gas (Ar) is used as a medium to transport the metal powder. The metal powder is delivered such that the powder stream converges at the same point as the focused laser beam. Commercially available gas atomized Ni alloy powder and milled tungsten carbide powder (CARPENTER Powder Products) were premixed. Table 1 shows the chemical composition of the premixed Ni/60WC powder. Image analysis of the powder revealed an average size ranging between 40–150 μ. Figure 1 shows Nickel particles having a spherical structure with a dendritic morphology and WC particles having an irregular morphology. Commercially available pure copper was used as the substrate. C-276 (hastelloy) was used as a buffer layer between Ni/60WC and copper. A set of experiments were conducted to demonstrate the extent of application of joining of dissimilar materials and to show that one can have numerous number of layers with different morphologies and structures in the LMD process.

Table 1 Nominal composition of Ni/60WC powder

Element	C	Si	Mn	P	S	Cr	Mo	Cu	Co	Fe	B	W	Ni
wt%	3.1	1.4	<0.01	<0.01	<0.01	0.004	0.02	<0.01	0.08	0.3	1	56	Bal.

Fig. 1 Ni/60WC pre-mixed metal powder morphology

Table 2 Nominal composition of C-276 powder

Element	C	Si	Mn	P	S	Cr	Mo	W	Fe	V	Ni
wt%	0.006	0.1	0.1	0.01	15.5	0.004	15.5	3.5	6.3	0.01	58.7

An additional set of experiments was conducted on 1020 steel substrates with no buffer layer. Ni/60WC was directly deposited on the 1020 steel substrate (Table 2).

In the present study a systematic analysis is conducted to understand the microstructural evolution and bond between the substrate and the deposit by varying two parameters, the traverse speed between the range of 720 to 90 mm/min and the design of the geometry from multi-track (MT)—single wall (SW)—laser glazing or laser re-melting (LG). A set of multi-track blocks of 25 mm × 25 mm × 6 mm was deposited on a 12.5 mm thick Cu substrate. The bead offset was 1 mm so there was 50% overlapping after each track in the multi-track metal deposit. The direction of the tracks after each layer was perpendicular to the direction of the previous track. The Z increment between layers in all the samples examined was 0.5 mm. Conducting the experiment perpendicular to copper substrate is not recommended. Copper can reflect the laser light in the direction opposite to laser beam and impair the laser unit. Considering the harmful effect to the laser unit, the nozzle was tilted at an angle of 15° prior to the experiment. The metal deposition took place in 3 stages. The first layer was laser glazed to heat the substrate at 180 mm/min followed by deposition on only MT sample deposited on copper substrate. After laser glazing a buffer layer of C-276 was deposited at 360 mm/min, so a good bond is made between copper and C-276. After the buffer layer continuous deposition of Ni/60WC took place at a scanning speed of 720 mm/min perpendicular to the copper substrate. Sample 2, 3, 4 and 5 show the parameters used for a single-track thin layer deposit of Ni/60WC. The length of the track was 20 mm and width of the track was 2 mm and 10 layers were deposited at 0.5 mm Z increments. The substrate used was 1020 steel of 150 × 100 × 12.7 mm. Samples 6, 7, 8, and 9 were only laser glazed for better understanding the microstructural evolution of Ni/60WC. Laser glazing is a term used when there is no powder injected and by means of only the laser beam to re-melt the as deposited surface. The microstructure of the as deposited single-wall (SW), multi-track (MT), and laser glazed (LG) samples were investigated by optical microscopy (Olympus BX51) and scanning electron microscopy (JSM-7600 FE SEM). The samples were etched using a mixture of 10 ml HNO_3, 15 ml HCl & 10 ml CH_3–COOH for optical and scanning electron microscopy. The hardness of the sample was measured using a Vickers microhardness tester (LECO AMH 43) using a load of 100, 500, and 1000 gm for Cu, C-276, and Ni/WC MMC, respectively, for a dwell time of 10 s.

Results and Discussion

With the optimum laser deposition parameter as listed in Table 3 a set of samples were deposited on the copper substrate. However, it should be noted that all combination of deposition parameters selected in this study (Laser power: 500–900 W, speed: 90–720 mm/min, and feed rate: 9.5–17.2 g/min), produced samples with no relevant defects such as cracks, bonding error or porosities at the interface between deposit, buffer layer, and the copper substrate. It is well known that microstructure defines the material properties. Initially when one looks at the optical micrographs it looks like there were some unmelted WC in the nickel matrix. Due to the limited resolution of the optical microscope one could not see in more detail whether WC was dissolved or not, if it's dissolved one cannot really tell to what extent it has been dissolved. It was not clear under the optical microscope so a scanning electron microscope was used to disclose the microstructure, which was not revealed in optical.

Table 3 Systematical DOE of multi-track–single track–laser glazed investigation

Sample #	Sample geometry	Laser power (Watt)	Scanning speed (mm/min)	Powder feed rate (g/min)	Layer thickness (mm)	Material	Deposition rate (g/min)
1	Multi track (square block)	900	720	17.2	0.75	Ni/60WC on C-276	5.75
2	Single wall	900	720	17.2	0.626	Ni/60WC on 1020	3.822
3	Single wall	700	360	14.7	0.767	Ni/60WC on 1020	4.525
4	Single wall	600	180	11.9	0.676	Ni/60WC on 1020	4
5	Single wall	500	90	9.5	0.575	Ni/60WC on 1020	3.45
6	Remelting 10 times	1000	720	0	Nil	Ni/60WC	NA
7	Remelting 3 times	1000	720	0	Nil	Ni/60WC	NA
8	Remelting 10 times	500	90	0	Nil	Ni/60WC	NA
9	Remelting 3 times	500	90	0	Nil	Ni/60WC	NA

Since WC is an extremely brittle material a lot of crack formation in the deposit have been reported in the literature, but we don't see that in our case. Figure 2b shows the distribution of WC particles, which is more or less homogeneous. Anandan et al. [14] and Bartkowski et al. [15] reported that WC particles sank to the bottom of the melt pool due to its high density. But we have not seen that in our case. Fernandez et al. [2] reported some non-homogeneous distribution of tungsten carbide particles but that was not the case in our deposit and more or less uniform distribution of WC particles was observed. Guo et al. [3] reported homogeneous distribution of WC particles and also reported that there were cracks and porosities formed by increasing the WC content in the matrix, but we have not seen any of these phenomenon in our investigation.

Results from Multi-track (MT) Deposit

Figure 2c shows that there was some unmelted WC and some of the boundary of the unmelted carbide was very clean but some of the other carbide particles had fragmented particles. Fragmented particles had precipitated or broken which were smaller and clustered around the parent carbide particles were observed. It looked like more dissolution of the WC particles had occurred to cause those smaller particles. In Fig. 3a the surface was not completely fragmented as well but they were cracked surface indicated by a different contrast. Around the fragmented particle there was different kind of morphology like feathery, rod like and blocky shaped. These were fragmented or partially melted carbide with different kind of morphology. Figure 3a shows distinguished polygonal shape carbide which are rectangular and square shaped in nature drifted away from the parent carbide on the matrix and one can see some dendrites also. It can be distinguished in Fig. 3a as a dendrite and a blocky carbide. There is substantial work reported on this material. However, a clear or a well-established mechanism has not been reported behind the formation of all those particles. For better understanding of the microstructural evolution of these morphologies a systematic study was conducted. One can

Fig. 2 **a** Represents the image of Cu-276-Ni/60WC multi track deposit. **b** Homogeneous distribution of WC particles. **c** Fragmented WC particles about to separate from the parent carbide particle

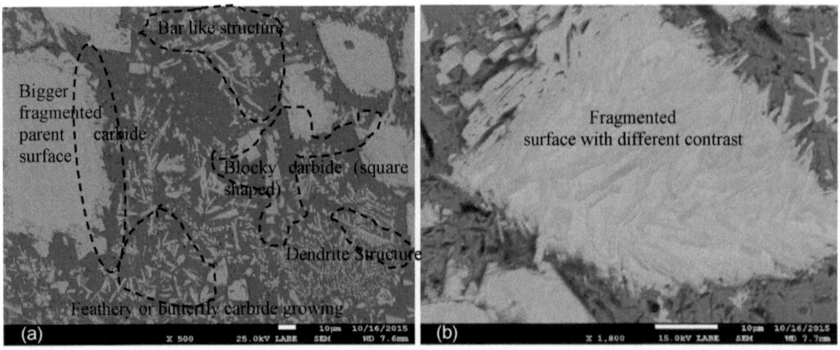

Fig. 3 **a** Shows different microstructures observed in MT deposit. **b** Fragmented particle surface

witness several complex structures in one frame and it is difficult to explain how it was developed because of the transition of structures in different phases.

The particular blocky structure tells us that the carbides are actually fragmented pieces. These define that they are not precipitated from the liquid but they are fragmented particles due to the heat damage. Heat damage occurs due to the rapid heating of the WC. The WC particles are falling on the liquid nickel and the surface becomes hot very quickly and since WC is an extremely brittle material, and it's know that when a material becomes hot it tries to expand. We believe that the center part of the bigger WC particle is still cold and not expanding at the same rate as the surface is trying to expand. Hence the surface becomes too hot, and due to the rapid expansion the surface fragments which develops into a crack. In Fig. 3b surface crack developed profile is seen. If it was a ductile material it can expand a little initially but since WC is a brittle material there is a chance that it might break and it might fracture. That seems to be the reason in Figs. 2c and 3 one can see that the parent carbides crack and some of these cracked particles get separated. However it was observed that the fragmented particles are mostly blocky carbides. The reason for it be drifted away from the parent carbide is that they are basically fragmented and seems that due to the turbulence in the liquid metal to some extent the smaller particles are separated from the parent surface and in some cases to some extent to the matrix. For better understanding the microstructural evolution of butterfly, bar like, rod like and dendritic structure, a SW was deposited with 4 different parameters as shown in Table 3.

Results of the SW Deposit

The tongue in Fig. 4a is the center of the beam, the laser used is a Gaussian beam and in the center there is little bit more intensity so that it melts more at lower scanning speed, which gradually decreases at 180 mm/min and no presence of it at

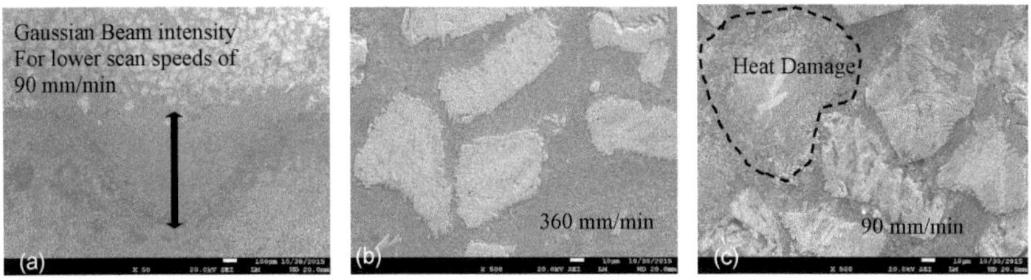

Fig. 4 **a** SW (single wall) deposit fabricated with 90 mm/min. **b** SW at 360 mm/min with very little heat damage and very few or sometimes no signs of blocky carbides. **c** Completely fragmented WC particle

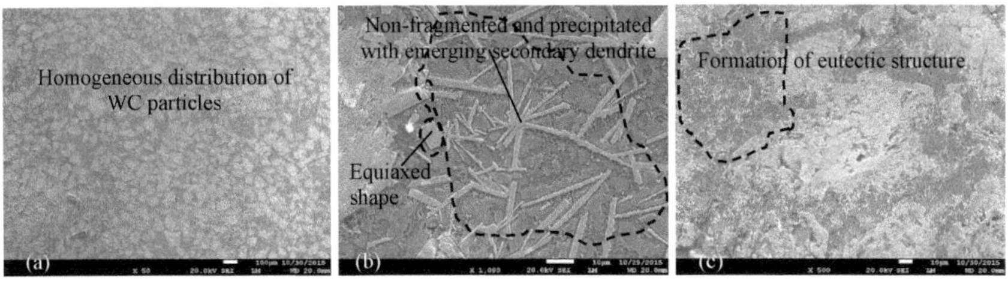

Fig. 5 **a** Homogeneous distribution of WC particles at a deposition speed of 90 mm/min. **b** Fragmented and non-fragmented primary carbides. **c** Formation of eutectic structure at 90 mm/min

720 mm/min. The particles do not sink to the bottom of the melt pool and it stays uniformly at the same place. SW deposit technique was used to develop the material quickly. Uniform homogeneous distribution of WC particles is observed in all the SW deposits. At extremely low scanning speed of 90 mm/min, the fragmented pieces are more. Blocky carbides are more because it was scanned at lower speed and it got more heat and it fragmented to more pieces.

At the same time some particles after fragmentation was dissolved, also because we have smaller particles in the powder itself. Initially the fragmented pieces were a bigger particle and since it was in the liquid of the nickel alloy matrix for a longer time it dissolved to some extent, and it became smaller. Also not all particles are fragmented perhaps due to the crystallographic orientation. This it is just a hypothesis based on the observation also some particles are easy to be broken and some particles may not. Fragmentation patterns were not quite sure on why all particles do not fragment in the same manner. One can see more fragmentation at 90 mm/min due to more heat input compared to 720 mm/min. In 720 mm/min deposition speed most of the particle is intact and only the surface is fragmented, but here in 90 mm/min there are lot of blocky carbides and more broken particles due to heat damage shown in Fig. 5. Particles break down to many different shapes during LMD and that is expected due to longer exposure time in the melt pool at slower scan rate. In Fig. 5b in case of the fragmented pieces we can see that it is one piece, and fragmented particles cannot be joined. The precipitated carbides are seen and everything here is growing from a joint. Thus they are not the fragmented pieces, and all the fragmented pieces until now which were noticed was not that long, it was not two dimensional it was more equiaxed shaped. The longer rod shape is not the fragmented particles it looks more like it re-precipitated during solidification and is developed. It is still not known whether these are the primary carbides or not and on which side of the phase diagram it precipitated or at what temperature. In the precipitated rod like longer particle in Fig. 5b it can be observed that a secondary dendrite emerging are at an angle in all the rod like structures and it further proved it was not the fragmented pieces and that they are precipitated during solidification. In 180 mm/min the trend is the same it's fragmented but is not as severe as 90. In 360 mm/min from Fig. 4b one can see fragmentation with very few or no traces of the blocky carbides. Now we can understand the evolution of blocky carbides. In 90 mm/min during solidification one may see some eutectic structure shown if Fig. 5c. We have two morphologies, rod kind with emerging smaller secondary dendrites and eutectic structure, and now we have to see how these developed.

Results of Laser Glazing (LG)

The Ni/60WC deposits fabricated with a scanning speed of 720 mm/min were laser glazed (remelted) with a scanning speed of 720 and 90 mm/min. Note that the scanning speed is about same as the solidification velocity. As the melts pool moves the solid liquid interface moves at the same speed. Laser glazed samples scanned at 720 mm/min for 3 times and 10 times are shown in Fig. 6. Some un-melted carbide particles was seen Fig. 6a. In Fig. 6b most of the area is dissolved and only one thing totally different from MT and SW is the very well defined dendrite structure. The dendrites formed even in the regular MT 720 mm/min structure, but the dendrites were only in few places. To form dendrites the particles has to go to the liquid and during solidification they will form the dendrites. In 720 mm/min every time when we deposit the next layer we re-melt the previous layer a little bit and at

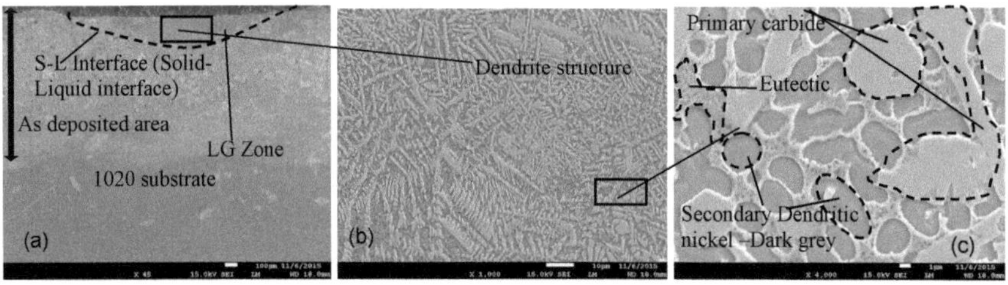

Fig. 6 **a** Overview structure of 720 mm/min LG region remelted for 10 times. **b** Well defined dendritic structure at the center of the remelted area. **c** Primary carbide phase- sec, dendritic nickel and eutectic structure

that instance there was more dissolution and the same layer re-precipitated which may form some dendrites in MT samples. No heat-affected particles here were observed in the re-melted area because the heat affected particles are already dissolved. Upon 3 and 10 times LG at 720 mm/min all the particles dissolve and re-precipitate to form dendrites and the undissolved particles are re-melted in small areas in 3 times re-melted surface and mostly dissolved in 10 times re-melted region. The microstructural morphology and amount of various phases strongly depend on the solidification velocity and temperature gradient. The microstructural anisotropy in the deposit can be explained in terms of the local solidification condition and thermal excursion at each point of the deposit. In laser deposition, the substrate acts as a heat sink and solidification is usually directional. During solidification of the melt pool, columnar grains grow opposite to the heat flow direction [14]. Usually, high thermal gradient and low solidification velocity produce a columnar structure, whereas low thermal gradient and high solidification velocity result in an equiaxed structure [16].

In LG of Ni/60WC at 720 mm/min for 3 and 10 times it was observed that at higher cooling rates dendrite structure and columnar bar like structure, butterfly and rod like structures was formed as shown in Fig. 6c. It was found that the primary carbides which has dendrite structure was not nickel, it is WC. It formed first in the hyper eutectic region. So here the first phase to form is tungsten carbide and the remaining was the liquid. It is seen here that they are the primary phase because they developed first in the phase diagram. It is a three-phase structure where we have the primary carbide and pockets of dendritic Nickel and the eutectic structure. By looking at the microstructure we can explain that the first phase to form was the primary WC second to form was the Ni and finally eutectic phase (Ni + WC) formed. Primary carbide particles have some hexagonal shape shown in Fig. 6c and this is still the primary carbides. It seems as if they got more time to cool to form an equiaxed structure. In 720 mm/min 10 and 3 times we see the same primary carbides and dendritic nickel and a eutectic structure. In 720 mm/min 3 times there was more undissolved WC particles and in 10 times we see mostly dissolved WC particles, which precipitated during solidification.

Figure 7a, b show that all the particles were melted in 90 mm/min LG samples due to the effect of slower scan rate. There is a different structure, which has a

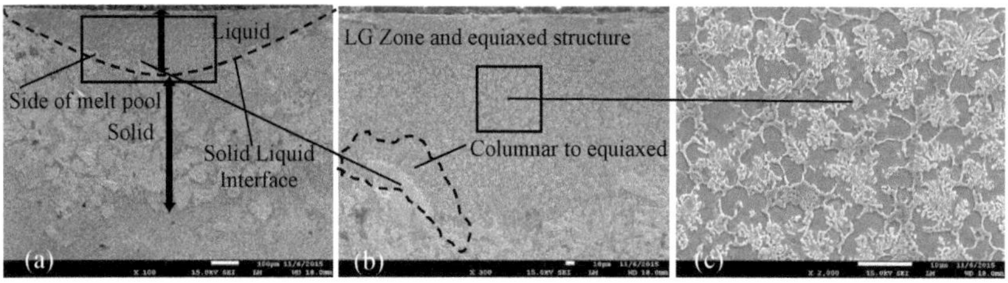

Fig. 7 **a** Overview of 90 mm/min LG/re-melted surface. **b** Overview of the equiaxed eutectic structure and longer rod like structure at the extreme edges. **c** Equiaxed and eutectic structure

substantial difference from 720 LG structure. In the edges of the melt pool or near the solid liquid interface we have structure similar to the 720 LG structure with longer rod like structure transition into an equiaxed and eutectic structure. But mostly there is only the two-phase region as shown in Fig. 7c full of equiaxed primary alpha nickel and eutectic structure.

The primary phase was not WC. Here the first phase formed is nickel alpha primary phase and surrounded by all eutectic. Since the heating is so slow we expect to see more tungsten carbide to be dissolved. In the solid liquid interface the solidification velocity is low at the side of the melt pool. The sides gets more time to cool compared to the trailing side of the melt pool and that is why solidification velocity is relatively low. In the center of the melt pool the solidification velocity is very high and the thermal gradient is relatively low. If it cools fast we're supposed to see primary carbide phase in the center of 90 mm/min sample. Rod kind of structure is nothing but the primary WC, which is formed at lower solidification velocity and higher thermal gradient. Solidification rate actually changes the effective composition in terms of microstructure. Equiaxed primary phase is seen in 90 mm/min 3 and 10 times remelt regions, they take the shape first that is a typical hypo eutectic structure. Solidification velocity can change the effective composition in terms of the phase diagram location. Dinda et al. described that higher cooling rate shifts the eutectic point to the hypo eutectic region [16]. Composition shift takes place depending on which type of material is used. We don't know if the effective composition moves to the higher side or the lower side but the solidification velocity does change the effective composition location where our primary phase is nickel in the 90 mm/min condition. When the samples were made by re-melting the surface several times some of the WC fragmented to blocky and over time since we are moving at low scan speed the WC particles basically sank to the bottom because the WC particles are heavy. The density is more in 90 mm/min and more of a homogeneous distribution of WC. But in 720 one might not observe the same density of WC in 90 mm/min 10 times accumulated at the lower side of the melt pool. During LG, the bottom side of the melt pool only the nickel is melted but on the top surface both WC and Ni was 100% melted. WC does not melt completely on the bottom surface but Ni melts completely. So we observed more eutectic, dendritic and blocky structures at the bottom of the laser glazed surface. By changing the process parameter one can change the microstructure of the deposit, and to further change the microstructure one can re-melt it, and if you re-melt it completely then due to the solidification rate different types of microstructures are formed with different mechanical properties.

Microhardness Analysis

The average hardness distribution was investigated to understand the influence of laser deposition parameters. At the lower end of the copper substrate the hardness was low which is expected and as we move towards the interface between the lower sides of the copper substrate and the C-276 hastelloy buffer layer the value of the hardness increases from 90 to 127 HV. At the interface between the buffer layer and the Ni-60WC deposit the hardness increased to around 300 HV and then suddenly a sporadic jump in hardness to 581 HV in the deposit. The scanning speed does not change the hardness of the uniform matrix structure, so the range of hardness in all the deposits are more or less the same. The hardness of the eutectic matrix in the remelted region was around 656 ± 100 HV. The hardness of the fragmented particles were around 959 ± 100 HV and undissolved crack free WC particles had a hardness value of 2050–2500 HV.

Conclusion

A series of experiments have been successfully conducted to develop a hardfacing coating on a highly conductive material (copper) and on steel substrate that was free of porosities, cracks, lack of fusion, and bonding errors. The as-deposited and laser glazed samples consisted of many different microstructures. LMD processing parameter plays an important role in the evolution of its microstructure. Systematic study revealed the mechanisms behind the formation of fragmented carbide particles. Consequently the fragmentation was due to heat damage and the brittle nature of the WC particles. The fragmented blocky particles drifted away from its parent surface due to the turbulence of the liquid melt pool. Finally the mechanism behind the growth of the two phase (primary Nickel + eutectic structure), and three phase (primary WC + dendritc Nickel + eutectic structure) have been revealed. In conclusion, it is understood that the processing parameters in LMD plays a vital role in the resulting microstructure of Ni-60WC deposit.

References

1. P. Farahmand, R. Kovacevic, Corrosion and wear behavior of laser cladded Ni–WC coatings. Surf. Coat. Technol. **276**, 121–135 (2015)
2. M.R. Fernández, A. García, J.M. Cuetos, R. González, A. Noriega, M. Cadenas, Effect of actual WC content on the reciprocating wear of a laser cladding NiCrBSi alloy reinforced with WC. Wear **324–325**, 80–89 (2015)
3. C. Guo, J. Chen, J. Zhou, J. Zhao, L. Wang, Y. Yu et al., Effects of WC–Ni content on microstructure and wear resistance of laser cladding Ni-based alloys coating. Surf. Coat. Technol. **206**(8–9), 2064–2071 (2012)
4. X. Luo, J. Li, G.J. Li, Effect of NiCrBSi content on microstructural evolution, cracking susceptibility and wear behaviors of laser cladding WC/Ni–NiCrBSi composite coatings. J. Alloy. Compd. **626**, 102–111 (2015)
5. S.F. Moustafa, Z. Abdel-Hamid, O.G. Baheig, A. Hussien, Synthesis of WC hard materials using coated powders. Adv. Powder Technol. **22**(5), 596–601 (2011)
6. J.-S. Xu, X.-C. Zhang, F.-Z. Xuan, Z.-D. Wang, S.-T. Tu, Rolling contact fatigue behavior of laser cladded WC/Ni composite coating. Surf. Coat. Technol. **239**, 7–15 (2014)

7. M. Zhong, W. Liu, Y. Zhang, X. Zhu, Formation of WC/Ni hard alloy coating by laser cladding of W/C/Ni pure element powder blend. Int. J. Refract Metal Hard Mater. **24**(6), 453–460 (2006)
8. P. Wu, H.M. Du, X.L. Chen, Z.Q. Li, H.L. Bai, E.Y. Jiang, Influence of WC particle behavior on the wear resistance properties of Ni–WC composite coatings. Wear **257**(1–2), 142–147 (2004)
9. M. Riabkina-Fishman, A. Weisheit, P. Levin, R. Galun, N. Frage, B.L. Mordike, Laser produced functionally graded tungsten carbide coatings on M2 high-speed tool steel
10. P. Wu, X.N. Tang. Microstructural characterization and wear behavior of laser cladded nickel-based and tungsten carbide composite coatings
11. S. Zhou, Y. Huang, X. Zeng, A study of Ni-based WC composite coatings by laser induction hybrid rapid cladding with elliptical spot. Appl. Surf. Sci. **254**(10), 3110–3119 (2008)
12. S. Zhou, X. Zeng, Q. Hu, Y. Huang, Analysis of crack behavior for Ni-based WC composite coatings by laser cladding and crack-free realization. Appl. Surf. Sci. **255**(5), 1646–1653 (2008)
13. D. Dai, D. Gu, Thermal behavior and densification mechanism during selective laser melting of copper matrix composites: Simulation and experiments. Mater. Des. **55**, 482–491 (2014)
14. S. Anandan, S. Pityana, Dutta Majumdar J. Structure–property-correlation in laser surface alloyed AISI 304 stainless steel with WC + Ni + NiCr. Mater. Sci. Eng. A **536**, 159–169 (2012)
15. D. Bartkowski, A. Młynarczak, A. Piasecki, B. Dudziak, M. Gościański, A. Bartkowska, Microstructure, microhardness and corrosion resistance of Stellite-6 coatings reinforced with WC particles using laser cladding. Opt. Laser Technol. **68**, 191–201 (2015)
16. G.P. Dinda, A.K. Dasgupta, S. Bhattacharya, H. Natu, B. Dutta, J. Mazumder, Microstructural characterization of laser-deposited Al 4047 alloy. Metall. Mater. Trans. A **44**(5), 2233–2242 (2012)

Composite Powder Consolidation Using Selective Laser Melting: Input Energy/Porosity Morphology/Balling Effect Relation

H. Salem, H.G. Salem and M.M. Attallah

Abstract In this study AlSi10 Mg/Al_2O_3 composite was processed via selective laser melting. The influence of different process parameters namely, laser power, scan speed and hatch spacing was investigated to determine their effect on the physical and mechanical properties of the consolidated powders. Input energy and balling effect, had a major influence on the developed pore morphology. Poor consolidation associated with poor heat diffusion and insufficient melting occurred at the low input energy. Intermediate input energy resulted in the formation of continuous porosity along the building direction separated by consolidated zones of equal width and associated with uniform balling effect. High input energy developed distorted porosity with non-uniform distribution associated with irregular course balling. The formed continuous porosity is attributed to the segregation and migration of alumina particles to the spacing between the deposited successive powder layers followed by their selective melting forming oxide films that acts as a barrier to the complete solidification.

Keywords Selective laser melting · AlSi10 Mg/Al_2O_3 composite · Porosity analysis

H. Salem
Nanotechnology, American University in Cairo, New Cairo, Egypt

H.G. Salem (✉)
Mechanical Engineering Department, American University in Cairo, AUC Avenue, PO. Box 74, New Cairo 11835, Egypt
e-mail: hgsalem@aucegypt.edu

M.M. Attallah
The Advanced Materials Processing Lab. (AMPLAB), IRC in Materials Processing School of Metallurgy & Materials, University of Birmingham, Birmingham, UK

The Minerals, Metals & Materials Society, *TMS 2017 146th Annual Meeting & Exhibition Supplemental Proceedings*, The Minerals, Metals & Materials Series, DOI 10.1007/978-3-319-51493-2_17

Introduction

Additive manufacturing (AM) is a technology that depends on manufacturing parts in the final shapes, from loose powders directly to the bulk form. This technology is regarded as the next predominant technology in the industry for the various advantages it is adding, including less design constrains compared to conventional manufacturing methods providing space for innovation and product customization, and material saving since AM allows reusing the non-consolidated raw material in the following building processes, which reduces the manufacturing cost, and more importantly the short lead-time by which AM processes can produce the final required product [1, 2]. AM includes different processes that are classified based on the heating source and the form of supplied material for fabrication [3]. One of which is selective laser melting (SLM), which has a broad range of applications in different fields such as aerospace, military and implants sectors [1]. SLM depends on the interaction between the laser and material powder. Laser is used to melt the powder layer based on the CAD data provided. The input energy sufficiently heats the powder until it melts creating the melt pool. Moving through the powder layers, while overlapping the heat between the adjacent melt pools, creates the final structure. Layer by layer deposition is how a complete product of interest is build [4].

Various alloys are processable by SLM, such as Ti-based alloys, stainless steels, Cobalt-gold alloys and Aluminum alloys. Al-alloys are considered one of the most interesting engineering alloys, due to the ability of SLM structures to meet the required performance, with generally high mechanical properties [5]. However, Aluminum alloy powders have low absorptivity and high thermal conductivity, which requires supplying higher input energy levels to compensate the heat loss and reflection, in addition to the high oxygen affinity of Aluminum [6]. AlSi10 Mg alloy is a widely used alloy in casting, and has been one of the main candidates for the SLM. It has been extensively studied from mechanical and densification behavior point of view [7–9]. The presence of silicon increases the fluidity of the alloy in addition to decreasing the thermal expansion coefficient. However, it is hard to machine Al–Si alloy for the existence of hard silicon particles. This problem is mitigated by the presence of magnesium, which enhances the machinability of the alloy and improves the corrosion resistance [10]. Several factors affect the output from the SLM process and they are mainly the laser power, scan speed and hatch spacing. All the mentioned parameters are usually integrated in a single function (the input energy density function) as shown in Eq. (1)

$$\mathrm{E}_{\mathrm{density}} = \frac{p}{\mathrm{h} * \nu * \mathrm{t}} \tag{1}$$

where p represents the laser power, h is the hatch spacing between the adjacent laser tracks, ν is the scan speed and t the layer thickness [8].

Input energy has a great influence on the densification behavior of the AM materials as will be illustrated later. With the continuous development in the industry, alloys with their limited properties became insufficient for fulfilling the challenging material requirements. For that reason it has become mandatory to introduce new materials that can cope with the industry needs, and that was when metallic matrix composites were introduced to be fabricated using SLM. Aluminum matrix composites combine lightweight and high strength, which is needed in various fields due to the improvement in mechanical properties, which is induced by the ceramic particles. Studies focusing on the densification and mechanical behavior of different AMC are limited [5]. For example, a study was performed by Gu [6] investigating SLM of TiC/AlSi10 Mg nanocomposite using process parameters of 110 W, 50 μm layer thickness, and scan speed ranging from 100 to 350 mm/s. The study proved that the input energy directly affects the densification behavior, at low (volumetric) input energy of 126 J/mm^3, the process produces bulk-form composite of density 91.7%, which was attributed to insufficient input energy. Near full dense samples were achieved upon increasing the input energy above 293 J/mm^3, as the densification behavior is directly related to the amount of liquid phase formed. According to Dadbakhsh et al. [11], who performed a comparative study between different Al-alloys namely (Al, AlMg1SiCu and AlSi10 Mg), which were used as matrices for reinforcement using 15 wt% Fe_2O_3, AlSi10 Mg showed best results compared to other candidates. This was attributed to the high percentage of alloying elements, which induces changes in the thermal characteristics of the alloy, such as reducing the melting temperature which facilitates the melting and heat diffusion during SLM process, and accordingly improved density results were achieved. In a study for the densification behavior of AlSi10 Mg/SiC, Chang et al. [12] showed that the SiC particle size significantly affects the SLM-processed samples, whereby using SiC with a range of 5–50 μm, fine SiC particles (5 μm) created a high specific surface area which enhanced the powder absorptivity compared to coarse particles (50 μm) this leads to improving the wettability between the SiC particles and Al. Moreover, this resulted in increasing the in situ reaction between the matrix and reinforcement resulted in the formation of $AlSiC_4$ phase that acts as an interface between SiC particles and the aluminum melt.

This paper focuses on studying the densification behavior of AlSi10 Mg/Al_2O_3 composite using SLM process through analyzing the porosity morphology of sectioned samples build within a range of energy density of 49–309 J/mm^3, and investigating the influence of input energy on the balling phenomenon of the samples.

Processing and Experimental Procedure

Material

Aluminum powder was supplied by LPW Ltd, with an average particle size ranges between 12–56 μm (Fig. 1a). Chemical composition and further characterisation of the powder was reported elsewhere [8]. The powder showed a generally irregular morphology, as shown in Fig. 1a. This might result in some agglomeration of the powders during processing, as suggested by Manfredi et al. [13] who found that spherical powders enhance the follow ability of the powder and ensure homogenously distributed layers. Alumina particles ranges between 0.16 and 0.38 μm as shown in Fig. 1b. The alumina powders were supplied by Sigma Aldrich. Figure 1c shows SEM micrograph for the composite powders (2.5wt% Al_2O_3 mixed with AlSi10 Mg alloy powders) using a tubular mixer.

Processing

19 Cubic samples were fabricated by SLM with dimensions 10 mm × 10 mm × 10 mm using Concept Laser M2 CUSING®, equipped with a Yt-fiber laser beam, and a maximum laser power capability of 400 W, scan speed up to 4000 mm/s, and 150 μm is the laser track width. The process was performed in Argon atmosphere with oxygen content less than 0.1%. A simple scanning strategy was used in building the samples as shown in Fig. 2, with powder layer thickness of 20 μm. The main parameters of the building processes that were varied in the experiments are the laser power, scan speed and hatch spacing. Statistical experimental design was used to develop the building parameters matrix upon which building parameters were chosen. The used process parameters as shown in Table 1.

Fig. 1 SEM image for powder morphology **a** AlSi10 Mg, **b** Aluminum oxide powder, **c** powder mixture (*arrows* point at Al_2O_3 particles)

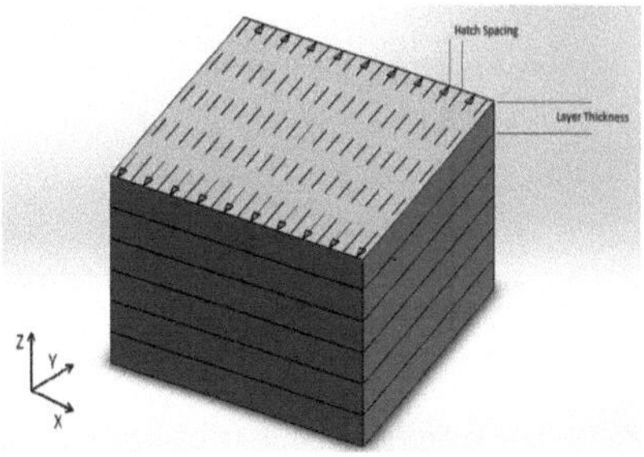

Fig. 2 schematic presentation for SLM building strategy

Table 1 Building parameters matrix

Sample no	Power (W)	Scan speed (mm/s)	al
1	175	1025	0.35
2	175	1675	0.65
3	400	1350	0.5
4	250	1350	0.5
5	100	1350	0.5
6	250	700	0.5
7	325	1675	0.35
8	325	1025	0.35
9	250	1350	0.2
10	175	1675	0.35
11	250	1350	0.8
12	250	1350	0.5
13	250	2000	0.5
14	325	1675	0.65
15	325	1025	0.65
16	250	1350	0.5
17	175	1025	0.65
18	250	1350	0.5
19	250	1350	0.5

Characterization

Samples were sectioned through their X-Z direction, mounted, ground, polished, and etched using Keller's etchant for microstructural analysis. Polished samples were investigated using optical microscope (Leica DM IRM) for producing panoramic images to give a clear overview of the porosity morphology. Density measurement (porosity percent) was conducted using Image J software [14], through measuring the area fraction of the porosity. A field emission scanning electron microscope (FEG-SEM) Leo Supra 55—Zeiss Inc., was used for powder size investigation at accelerating voltage of 20 kV and for investigation of the SLM cubes surface morphology. Energy dispersive X-ray (EDX) detector attached to the

SEM was employed for phase analysis of the sectioned samples. For hardness test, Vickers micro-hardness (Hv) testing machine (Mitutoyo HM-112) was used with a load of 0.5 kg for 30 s.

Results and Discussion

The densification behavior of the AlSi10 Mg/Al_2O_3 composite was extensively studied using electron microscopy and image analysis. Addition of ceramic particles (Al_2O_3) was expected to enhance the mechanical properties. Nonetheless, Alumina particles addition to the Al-alloy powders significantly influenced the consolidation behavior of the processed cubes producing high porosity. This was directly reflected upon the mechanical properties of the SLM cubes. The HV-values reported in the current study for the composite cubes was 109 Hv, which represent 14% reduction in the Hv-values measured for the plain Al-alloy, which was about 127 ± 3 Hv [15, 16]. The low Hv-values were attributed to the combined effect of the metallurgical defects induced by the SLM process along with the formed porosity associated with the Al_2O_3 addition. In the following sections, the metallurgical defects and the Al_2O_3 induced porosity are demonstrated.

Porosity Analysis

Figure 3 presents a bar chart for the variation of the processed samples relative densities percent as a function of input energy, according to the processed parameters shown in Table 1. In addition, demonstration of panoramic images for samples 5, 17, 16 and 9, representing the four consolidation mechanisms of the Al/Al_2O_3 obtained over different ranges of input energies (Fig. 3). The four main mechanism which were observed are classified as follows: (1) First mechanism, occurred over energy density range of 49–53 J/mm^3 at which isolated porosity was formed; (2) Second mechanisms occurred over energy density range of 77–100 J/mm^3, where interrupted porosity was formed; (3) Third mechanism, occurred over energy density range of 123–163 J/mm^3, where longitudinal connected porosity (gaps) was formed along the building direction; (4) Fourth mechanism occurred over energy density range of 185–309 J/mm^3 where disrupted non-uniform porosity was formed.

Figure 3a shows the first mechanism in which the input energy was not insufficient to allow for consolidation between the adjacent melt pools. This was manifested in the form of isolated porosity across the whole section [8]. Second mechanism (Fig. 3b) showed improved consolidation due to the overlap between melt pools, which was observed in the increased relative density value. However, the input energy was not high enough to create fully dense bulk samples, which resulted in forming longitudinally aligned interrupted porosity along the building

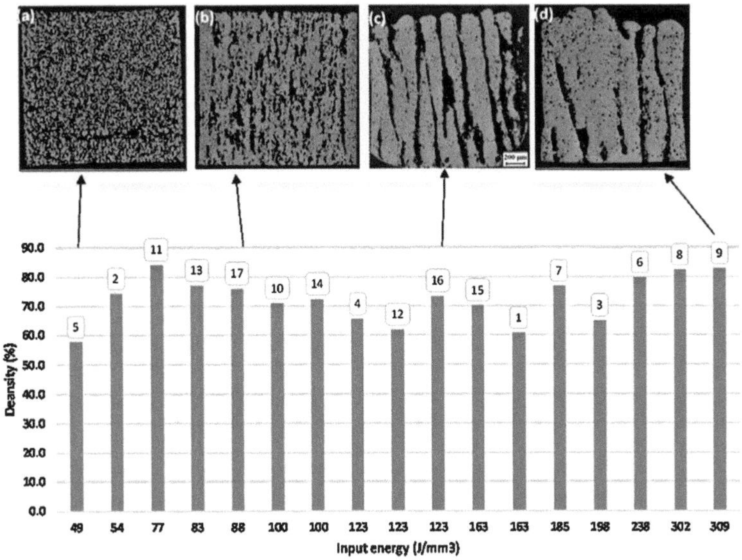

Fig. 3 Density variation as a function of input energy ranges representing the 4-mechanisms **a** First, **b** Second, **c** Third and **d** Fourth

Fig. 4 Effect of marangoni flow on melt pool shape [18]

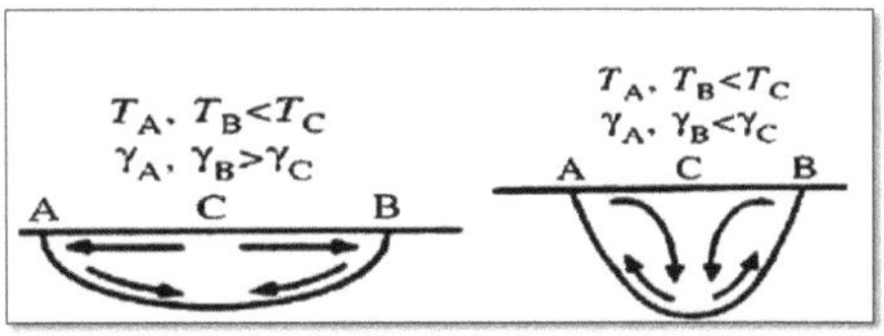

direction. Nevertheless, the second mechanism input energy resulted in the increase in density compared to the first mechanism. It is worth noting that increasing the input energy (100 J/mm^3) within the second mechanism resulted in a gradual decrease in the relative density due to change in the porosity volume fraction. Increasing the input energy over the range of 123–163 J/mm^3 (third mechanism, Fig. 3c) produced tracks almost of equal width separated by connected porosity along the building direction forming gaps between the adjacent tracks. At relatively higher energy input density, it seems that steep temperature gradients developed within the melt pool, which promoted surface tension gradient between their centers and edges. This initiates Marangoni flow, which is the flow of the fluid outwards from the region of high temperature (melt pool center) to that of low temperature (melt pool edges) [6, 17]. Olakanmi [18] suggested that increasing the oxygen content causes a change in the nature of the Marangoni flow form flowing outwards creating shallow and wide melt pools to inwards melt flow forming narrow and deep melt pools as shown (Fig. 4). The change in the melt pool dimensions has a direct impact on the scan track dimensions leading to reducing the scan track width [19].

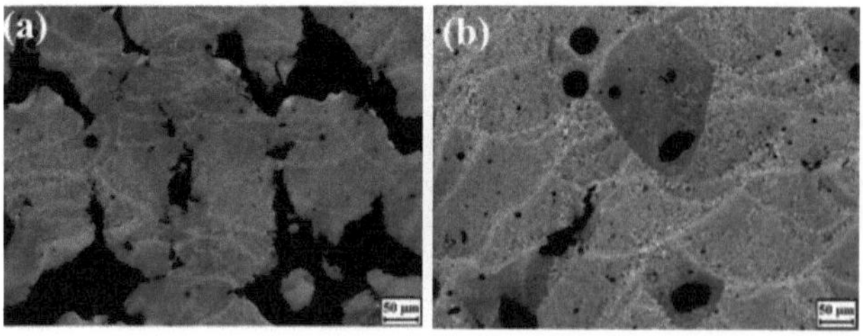

Fig. 5 OM images for samples representing **a** First mechanism (Sample 5), **b** third mechanism (sample 4)

The formation of Oxide films within the aligned gaps along the building direction were evident by EDX analysis (shown in section "Microstructure and EDX Analysis"). It is suggested that the formed oxide films between the scan tracks are responsible for hindering the consolidation creating the third mechanism [20].

Although, it is hard to visualize the Marangoni flow nature transition through the dimensional changes of the melt pool since all samples have the same Al_2O_3 content, a comparison was made for the evaluated melt pool shapes of representative samples for: (1) the first mechanism at which input energy was insufficient for initiating the Marangoni flow Fig. 5a and (3) the third mechanism Fig. 5b at which high energy density and high oxygen content the two main conditions for changing the Marangoni flow are present. As shown in Fig. 5, the difference in melt pool depth between both mechanisms is clear. Measuring the depth of the melt pools of the representative conditions, it was found that the first mechanism produced an average melt pool depth of 54 μm (Fig. 5a), while that for the third mechanism was 90 μm (Fig. 5b). These findings clearly manifested the influence of Marangoni flow nature on the melt pool size and accordingly on the scan track width.

For the fourth mechanism based on the porosity morphology analysis, the disrupted non-uniform porosity as shown in Fig. 3d. Unlike the lower input energy mechanism, input energy range of 185–309 J/mm^3 resulted in liquid phase with low viscosity and higher temperature, leading to the localized tearing of the oxide films [11], which allowed partial consolidation between the scan tracks. Accordingly, the porosity morphology changed from continuous connected longitudinal gaps between scan tracks to the distorted porosity.

Microstructure and EDX Analysis

The etched cross sections cut parallel to the building direction revealed the melt pool microstructure as shown in Fig. 6. It was observed that spherical metallurgical pores were formed within the melt pools, which size varied as a function of increasing the input energy. Shows the relation between energy density and the metallurgical pore size. It is well known that the spherical shaped pores are formed during the solidification process of the melt pool, which occur by forming gas

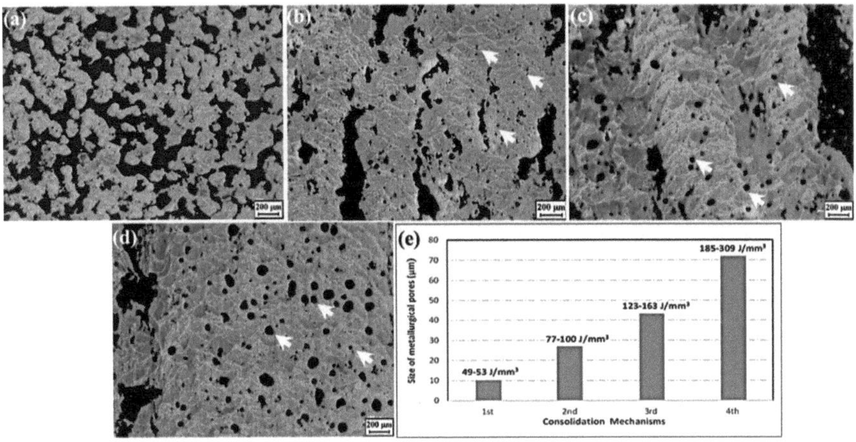

Fig. 6 OM images showing influence of input energy range on the metallurgical pores size for the **a** First, **b** Second, **c** Third, **d** Fourth mechanisms and **e** A bar chart for the variation of Pore size with Energy for the 4-mechanisms. *Arrows* point at Metallurgical pores

bubble surrounded by liquid state [21]. Accordingly, metallurgical pores are formed when high input energy is employed for consolidation [22]. When comparing the metallurgical pores size generated in different energy ranges it was concluded that the pores size increased from 10–72 μm for the first (Low energy density) and fourth mechanisms (highest energy density), respectively. It is clear that for low input energy range (first mechanism) the metallurgical pored are too small to be observed at the employed magnification as shown in Fig. 6a. For higher input energies (second, third and fourth mechanisms), the metallurgical pores can be easily observed as pointed out by arrows in Fig. 6b–d, respectively.

SEM imaging and EDX analysis were employed on representative samples for the four-mechanism as a supportive evidence for the proposed scenario. Figure 7a is representing the first mechanism. Although, no oxide films were observed clearly in Fig. 7a oxides were evident in Point 2 at the boundaries of the pores in EDX analysis. It is expected that Alumina did not melt when low input energy is employed. Figure 7b represents the second mechanism, oxygen content is higher in the interrupted porosity this may be attributed to the high oxygen affinity of aluminum powders in addition to the presence of Al_2O_3 particles. The third mechanism image and EDX analysis (Fig. 7c) showed the formation of a continuous oxide layer on the peripheries of the longitudinal gaps along the building direction. This could be attributed to the melting of alumina particles, which was associated with the relatively high input energy. The direct mixing of both powders rather than using ball milling deprived the mixture from having the alumina particles impingement in the soft Al-powder. Instead the Al_2O_3 particles stuck on the surface of the Al-particles as shown in Fig. 1c, which facilitate segregation of the Alumina particles within the gaps creating oxide film along the formed gaps. Figure 7d represents the fourth mechanism at which distorted non-uniform porosity associated

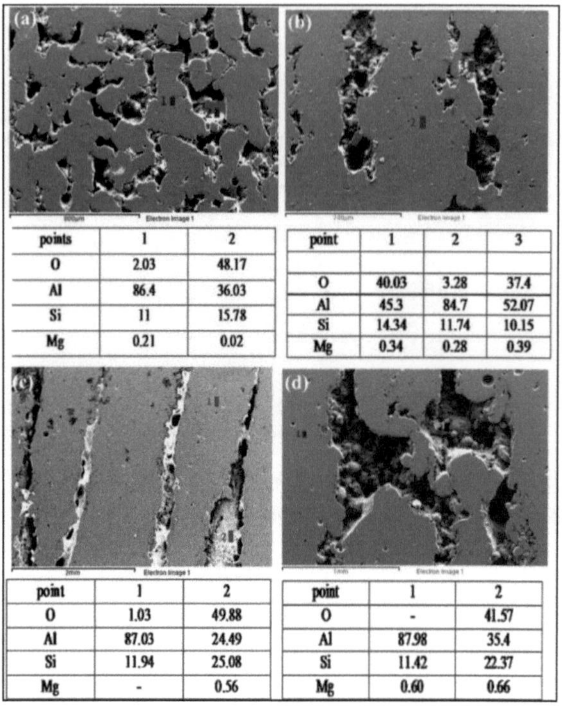

points	1	2
O	2.03	48.17
Al	86.4	36.03
Si	11	15.78
Mg	0.21	0.02

point	1	2	3
O	40.03	3.28	37.4
Al	45.3	84.7	52.07
Si	14.34	11.74	10.15
Mg	0.34	0.28	0.39

point	1	2
O	1.03	49.88
Al	87.03	24.49
Si	11.94	25.08
Mg	-	0.56

point	1	2
O	-	41.57
Al	87.98	35.4
Si	11.42	22.37
Mg	0.60	0.66

Fig. 7 SEM images for cross section of representative sample for each mechanism **a** First, **b** Second, **c** Third and **d** Fourth with their corresponding EDX analysis for the regions listed in tables below for each

with high oxygen content can be observed within which overlapping between separated scan tracks was formed. Moreover, evidence for loose unconsolidated powders was observed within the distorted porosity.

Balling Effect and Oxide Film Formation

The input energy showed significant influence on the surface morphology of the SLM cubes. The combined effect of Marangoni flow, which resulted in the melt pool instability and the low wettability induced by the addition of Alumina particles resulted in the liquid spheroidization, this phenomena is known as balling effect. The uneven surface created by the balling acts as an impediment to having homogenous layer of deposited powder [17, 23]. Two types of balling phenomena were formed under the selected SLM conditions in the current research, one of which is the formation of regular ball shapes of ~0.8–1 mm in average size as shown in Fig. 8a, this morphology appeared in the third consolidation mechanism range. And the second balling type appeared at the fourth consolidation mechanism range at which large irregular agglomerates with inter-connected porosity between them as in Fig. 8b. Deposition of successive powder layers the presence of balling

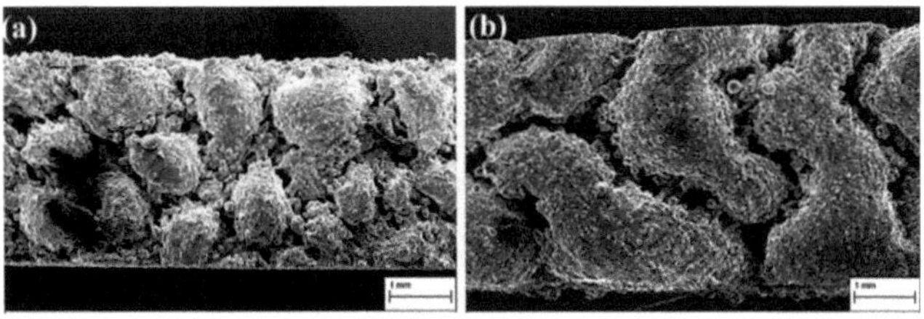

Fig. 8 SEM for surface morphology showing types of balling phenomena **a** Third mechanism, **b** Fourth mechanism

phenomena, Al_2O_3 particles with their relatively small size compared to Al-particles. Al_2O_3 particles segregates and agglomerate inside the connected porosity creating the oxide film upon melting.

Conclusions

The input energy is a major factor influencing the densification behavior of AlSi10 Mg/Al_2O_3 SLM-processed composite. Within the energy densities range employed in the current work, 4-mechanisms were identified based on the porosity morphology/input energy. The porosity morphology changes depends on the combined effect of Al_2O_3 and the input energy. The size of metallurgical pores increases with increasing input energy density. The surface morphology reflects the change in energy through the changes in the type of balling phenomena. Ball milling of the reinforced particles is essential for eliminating the probability of having Al_2O_3 agglomerations rather than homogenous dispersion.

References

1. D. Gu, *Laser Additive Manufacturing (AM): Classification, Processing Philosophy, and Metallurgical Mechanisms* (Springer, Berlin, Heidelberg, 2015)
2. L. Hao, D. Raymond, G. Strano, S. Dadbakhsh, Enhancing the sustainability of additive manufacturing, in *5th International Conference on Responsive Manufacturing—Green Manufacturing (ICRM 2010)* (2010), pp. 390–395
3. N.T. Aboulkhair, Additive manufacture of an aluminium alloy : processing, microstructure, and mechanical properties, Ph.D. thesis, Nottingham University (2015)
4. L. Thijs et al., A study of the microstructural evolution during selective laser melting of Ti-6Al-4 V. Acta Mater. 3303–3312 (2010)
5. D. Manfredi et al., Additive manufacturing of Al alloys and aluminium matrix composites (AMCs). Light Met. Alloy. Appl. 3–34 (2014)
6. D. Gu, Selective laser melting additive manufacturing of TiC/AlSi10 Mg bulk-form nanocomposites with tailored microstructures and properties. Phys. Proc. 108–116 (2014)
7. N.T. Aboulkhair, N.M. Everitt, I. Ashcroft, C. Tuck, Reducing porosity in AlSi10 Mg parts processed by selective laser melting. Addit. Manuf. 77–86 (2014)

8. N. Read, W. Wang, K. Essa, M.M. Attallah, Selective laser melting of AlSi10 Mg alloy: process optimisation and mechanical properties development. Mater. Des. 417–424 (2015)
9. D. Buchbinder et al., High power selective laser melting (HP SLM) of aluminum parts. Phys. Proc. 271–278 (2011)
10. I.J. Polmear, *Light Alloys Metallurgy of the Light Metals* (British Library Cataloguing in Publication Data, Britain, 1981)
11. S. Dadbakhsh, L. Hao, Effect of Al alloys on selective laser melting behaviour and microstructure of in situ formed particle reinforced composites. J. Alloys Compd. 328–334 (2012)
12. F. Chang, D. Gu, D. Dai, P. Yuan, Selective laser melting of in-situ Al4SiC4 + SiC hybrid reinforced Al matrix composites: influence of starting SiC particle size. Surf. Coat. Technol. 15–24 (2015)
13. D. Manfredi et al., Direct metal laser sintering: an additive manufacturing technology ready to produce lightweight structural parts for robotic applications. Metall. Ital. 15–24 (2013)
14. M.D. Abràmoff, P.J. Magalhães, S.J. Ram, Image processing with image. J. Biophotonics Int. 36–41 (2004)
15. L. Thijs et al., Fine-structured aluminium products with controllable texture by selective laser melting of pre-alloyed AlSi10 Mg powder. Acta Mater. 1809–1819 (2013)
16. K. Kempen, L. Thijs, J. Van Humbeeck, J.-P. Kruth, Mechanical properties of AlSi10 Mg produced by selective laser melting. Phys. Proc. 439–446 (2012)
17. D. Gu, H. Wang, G. Zhang, Selective laser melting additive manufacturing of Ti-based nanocomposites: the role of nanopowder. Metall. Mater. Trans. A Phys. Metall. Mater. Sci. 464–476 (2014)
18. E.O. Olakanmi, Selective laser sintering/melting (SLS/SLM) of pure Al, Al–Mg, and Al–Si powders: effect of processing conditions and powder properties. J. Mater. Process. Technol. 1387–1405 (2013)
19. H.J. Niu, I.T.H. Chang, Selective laser sintering of gas atomized M2 high speed steel powder. J. Mater. Sci. 31–38 (2000)
20. E. Louvis, P. Fox, C.J. Sutcliffe, Selective laser melting of aluminium components. J. Mater. Process. Technol. 275–284 (2011)
21. L. Li, Repair of directionally solidified superalloy GTD-111 by laser-engineered net shaping. J. Mater. Sci. 7886–7893 (2006)
22. H. Gong, K. Rafi, H. Gu, T. Starr, B. Stucker, Analysis of defect generation in Ti-6Al-4 V parts made using powder bed fusion additive manufacturing processes. Addit. Manuf. 87–98 (2014)
23. D. Gu, Y.C. Hagedorn, W. Meiners, G. Meng, R.J.S. Batista, K. Wissenbach, R. Poprawe, Densification behavior, microstructure evolution, and wear performance of selective laser melting processed commercially pure titanium. Acta Mater. 3849–3860 (2012)

Part IV
Advanced Characterization Techniques for Quantifying and Modeling Deformation Mechanisms

Strain Localization Structures in Textured Magnesium AZ31 Under Reversed Loading via Microscopic Digital Image Correlation

Enver Kapan, Nima Shafaghi, Sevinç Uçar and C. Can Aydıner

Abstract In favorable crystallographic texture and loading path combinations, Magnesium AZ31 deformation is dominated by twinning. A marked component of the material behavior in this regime is the spatial distribution of the accommodated strain: there is sharp strain localization at every length scale including macroscopic, stemming from an abrupt propagation of twinning events across neighboring grains (e.g., Aydiner and Telemez, Int. J. Plas. 2014(56), p. 203). Strain mapping with digital image correlation (DIC) has been instrumental in bridging length scales and highlighting strain heterogeneity levels in this material. With a multiple-optical-axis DIC apparatus, strain maps are obtained with macroscopic and microscopic resolution for the full field. For the sharp rolling texture, strain heterogeneity patterns can be directly interpreted on the micro-mechanisms of the abundant orientation. These patterns are presented at representative points of the reverse-loading cycle and interpreted with respect to operative twin and slip mechanisms.

Keywords Magnesium alloys · Twinning · Strain heterogeneity · Cyclic loading

Introduction

Crystallites of Magnesium alloys are characterized by an odd combination of approximate elastic isotropy and severe levels of plastic anisotropy. The unbalanced distribution of critical resolved shear stress among operative slip systems (basal, prismatic, pyramidal <a>, pyramidal <c + a>) is partly responsible for the latter. Most notably in this respect, the basal slip is cited to be an order easier than the rest [1, 2]. The other underlying factor is famously the profuse activation of the

E. Kapan · N. Shafaghi · S. Uçar · C.C. Aydıner (✉)
Department of Mechanical Engineering, Bogaziçi University, 34342 Bebek, Istanbul, Turkey
e-mail: can.aydiner@boun.edu.tr

The Minerals, Metals & Materials Society, *TMS 2017 146th Annual Meeting & Exhibition Supplemental Proceedings*, The Minerals, Metals & Materials Series,
DOI 10.1007/978-3-319-51493-2_18

$\{10\bar{1}2\} < \bar{1}011 >$ tensile twin [3], which also made these materials ideal systems to study the physics of deformation twinning. The effect of the tensile twin is compound on wrought Magnesium alloys. First, it is the reason why straightforward and sharp textures form during the manufacturing processes. (Extrusion texture is simply characterized by c-axes distributed with polar symmetry about the extrusion axis; rolling texture is even sharper with c-axes aligned with the normal direction of a rolled plate.) Second, it causes extreme tension-compression anisotropy in the resulting sharply-textured polycrystalline aggregates since twinning activation is unipolar. In the direction that twinning is favored, strain is predominantly accommodated by the tensile twin. This results in a near flat region on the stress strain curve, called the 'twin plateau'. Loading in the reverse sense, however has to resort to harder plastic mechanisms. Consequently, when these samples are subjected to cyclic loading, highly distorted hysteresis loops are revealed [4–6].

Another factor that is heavily affected by the direction of load is strain heterogeneity. As with any material, whether strain is accommodated with heavy localization into bands, or homogenously over the entire sample/component, is critical regarding failure and formability of Magnesium alloys. The fact that tensile twin aggravates strain heterogeneity levels are shown by high-spatial resolution strain measurements with, e.g., digital image correlation [7, 8]. Aydıner and Telemez [8] presented precisely that rolled Magnesium AZ31 deformed via sharp macroscopic shear-banding over the twin plateau using multi-scale, in situ DIC. These bands were manifested by coordinated twinning events as deduced by the micro-scale component of this work. Further, simultaneous measurement over two orthogonal surfaces of the sample allowed the 3-D conclusion that the macroscopic bands were volumetric bands; they were extruded $\pm 45^{\circ}$ cross-patterns that go through the entire sample. The fixed orientation of these bands approximately conformed to the cross-pattern of the two highest resolved-shear-stress $\{10\bar{1}2\} < \bar{1}011 >$ tensile twin variants inside the *abundant* crystallite orientations. In general terms, the localizing nature of the tensile twin is consistent with studies of twin transmission from one grain to the next (note, e.g., Beyerlein et al. [9], Barnett et al. [10] and Wang et al. [11]).

The current in situ DIC study again considers the sharp rolling texture with Magnesium AZ31. However, the load is reversed twice in a compression-tension-compression experiment to study the strain heterogeneity levels during deformation stages with distinct strain accommodation mechanisms. The predominant mechanisms in the respective stages are as follows in the order of activation: twinning of the virgin material (first compression), detwinning (load reversed to tension), slip plasticity (further tension), and twinning after slip plasticity (load reversed to second compression). For a confined treatise, the results presented here focus on strain fields with macroscopic resolution.

Materials and Methods

Rolled stock material of the AZ31 Magnesium alloy (3wt% Al, 1wt% Zn) is procured from Alfa Aesar, USA, in the form of a 100 × 150 × 9.52 mm^3 hot-rolled plate. In house metallography yields equiaxed grains of 11 μm average size, determined with the lineal intercept method. Dog-bone samples have been cut in with a *nominal* 3 × 3 mm^2 square cross-section shown in Fig. 1. This is a hefty cross-section given the 8 mm gage length, chosen to preclude buckling about both lateral axes.

The crystallographic texture of the as-received plate is measured with a Panalytical X-Pert Pro MPD four-cycle diffractometer. The pole figures provided in Fig. 2 conform to the typical rolling texture of Magnesium AZ31. The alignment level of (00.1) c-axes with plate normal direction (ND) is about 6 mrd (multiples of

Fig. 1 Dog-bone sample geometry utilized for reversed loading

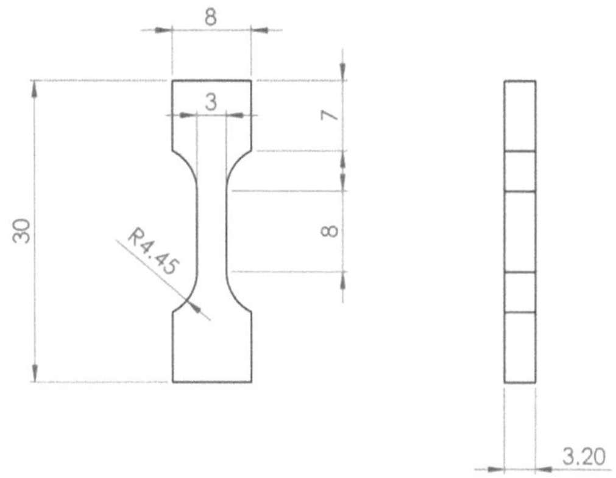

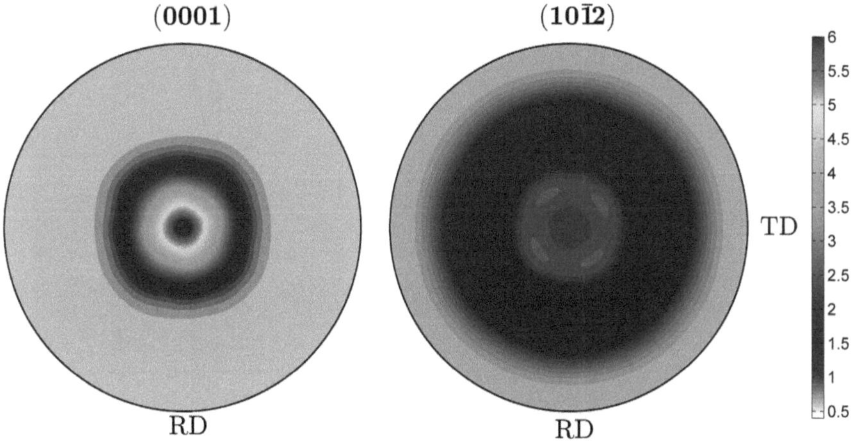

Fig. 2 (00.1) and (10.2) pole figures for the Magnesium AZ31 plate. RD, TD stand for rolling and transverse directions. Contours are provided in multiples of random (mrd)

random), presenting a very sharp texture. Near polar symmetry about the ND axis is noted, also visible on the complementary (10.2) pole figure. In this study, the dog-bone sample's longitudinal axis has been coincided with rolling direction (RD) and the top face is normal to transverse direction (TD).

A small form-factor Kammrath and Weiss load frame, with grips and structure specialized for reversed loading, is utilized. Load is applied incrementally in displacement control. Images are recorded with a 5 Megapixel AVT camera, fitted with an Edmund Optics 0.5 × telecentric lens. The optical resolution in this setup is 6.8 μm/pixel. The spatially-resolved strain data presented in this manuscript is obtained by applying the digital image correlation (DIC) technique on the top surface. First-order 40 × 40 pixel subsets are utilized with rectangular DIC grids of roughly 15 pixel spacing (~0.1 mm). Here, *y*-coordinate is aligned with the loading direction RD. The reported axial strains, ε_{yy}, are obtained by numerically differentiating v (*y*-direction displacement) with respect to *y* over the DIC grid. For this, the central difference formula is used. Further details of the DIC implementation is provided elsewhere [8].

Results

Average Behavior

Figure 3 shows the stress-strain curve of this experiment. Here, the stress and strain axes correspond to engineering stress and the average of the DIC-determined ε_{yy}

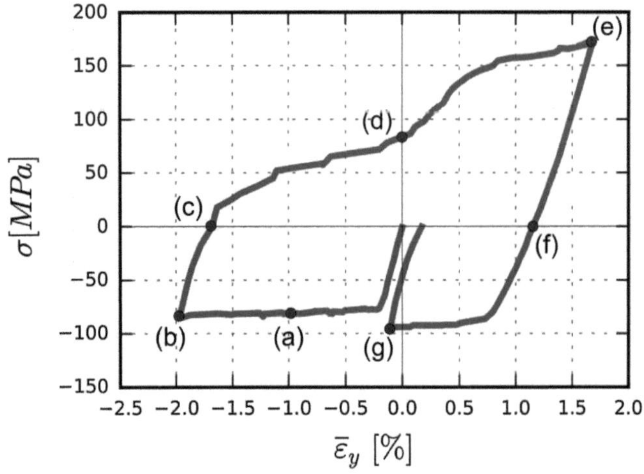

Fig. 3 Engineering stress versus DIC-average axial strain, presenting the load path of the compression-tension-compression cyclic loading experiment. Points *a*–*g* are critically selected points where spatially-resolved strain data is provided

field, respectively. The experiment commences in compression and its path can be followed by following points *a–g* in their alphabetic order. Points *a* and *b* are on the initial twin plateau. Point *c* depicts the moment of the first unload. Then reverse load is applied in tension. Point *d* is roughly the first strain zero-crossing, where the average strain is recovered entirely for the first time. The stage between *c* and *d* that exhibits a huge Bauschinger effect is associated with detwinning. Detwinning is simply the reversal of the twins formed in the initial plateau. After twin-transformed regions are recovered (past point *d*), the slope of the stress-strain curve increases sharply primarily with elastic straining. After a certain stress level, slip plasticity drops the slope appreciably. In this regime, point *e* is reached. Point *f* is the point of second unload, this time from tension. A final compressive stage is commenced until the material is pushed to its second zero-average-strain crossing. Exact average strains for the points *d* and *g* are 0.07% and −0.1%, respectively. With these low values, these points will be coined first and second zero-strain-crossing for brevity.

Macroscopic Strain Heterogeneity

The geometric structure of the deformation is depicted best with contour maps of axial strain in Fig. 4, for marked points *a–g*. Yet, these are not precise in providing quantitative information. Quantitative distribution of the strain data at exactly the same points is complemented with the strain histograms of Fig. 5.

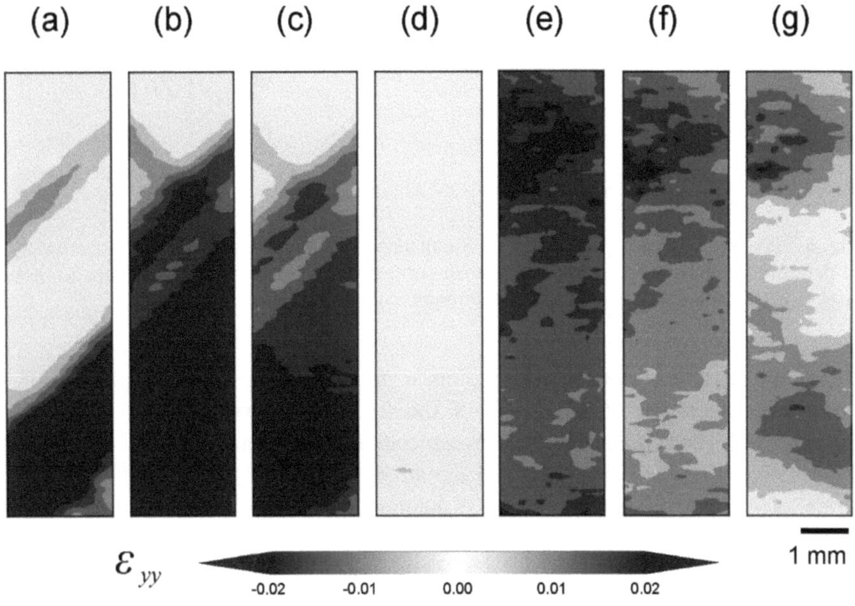

Fig. 4 Axial strain maps on the 8 mm gage section of the sample at consecutive stages of deformation marked *a–g* on the stress-strain curve of Fig. 3

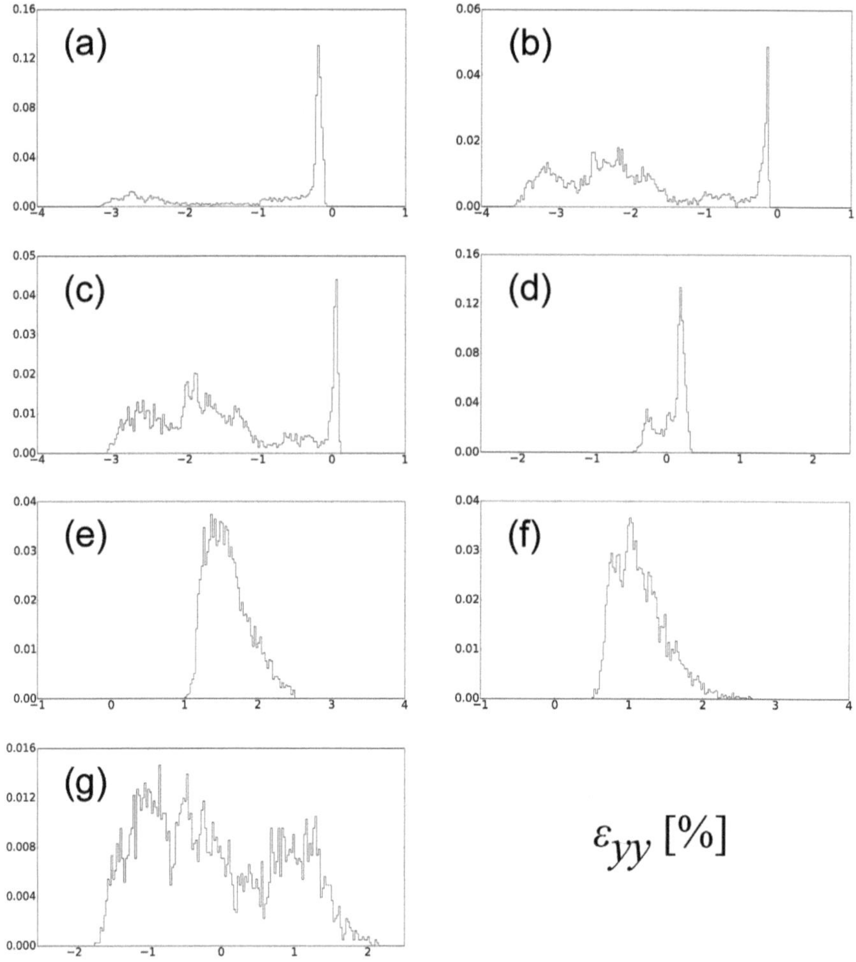

Fig. 5 Axial strain (εyy) histograms on the 8 mm gage section of the sample at consecutive stages of deformation marked *a*–*g* on the stress-strain curve of Fig. 3. Vertical axis is frequency. Strain span is kept at 5% for comparability at each point. Number of bins is 200

On the first twin plateau, two points *a* and *b* are provided. A very detailed multi-scale depiction of this stage of the loading is provided in Aydıner and Telemez [8]. There is extreme macroscopic strain heterogeneity in the sample, and strain accommodation proceeds with the advance of strictly structured 45° degree strain localization bands. Point *a* shows the main band approaching from the bottom and another distinct localization structure forming at the top half of the sample. At +2% strain (point *b*), strain has been accommodated by most of the sample, yet still a dormant region remains at the top. The dormant regions correspond to the sharp central peaks in the histograms of Fig. 5a, b. As expected, the dormant population has a much higher frequency at point *a* (peak frequency $\sim$ 0.13) compared to point b (peak frequency ~0.05). Both points indicate intense strain heterogeneity, namely, strains outside the dormant peak are distributed up to near −4%.

Point *c* is the point of first unload. The strain map shows noticeable recovery with respect to point *b* upon unload, evident from the comparison of Fig. 4b, c. Corresponding histogram comparison (Fig. 5b, c) shows that the highest strain zones beyond −3% recover to lower values. The dormant frequency is virtually unaffected though. Only, a positive shift in the location of this peak is apparent, expected from the elastic unload.

Point *d* is the first-zero strain crossing and it is the point where the initial compressive macroscopic strain in the sample is entirely stretched back. The level of strain recovery of the local strain fields upon detwinning is remarkable. The histogram in Fig. 5d points to a strain span of under 1%, yet with a two peak composition. The tall peak on the right is the dormant peak that swallowed the recovered regions and shifted right incurring low levels of likely elastic strain. This corresponds to the large region on the top in Fig. 4d that has a red hue due to the small positive strain. The most heavily deformed region on the twin plateau (the bottom part of the sample) is following the detwinning procedure with a delay, evident with its blue hue (persisting compressive strain). This region constitutes the smaller peak on the left in the histogram.

Going into plasticity with positive strain increments and larger tensile stress levels, the top region preferentially incurs larger strains. More clearly, what remained dormant during the compressive stage, incurs elastic tension first and then runs into tensile plasticity first. Nevertheless, the strain span at point *e* of average 1.6% strain is certainly very limited compared to the twin plateau. Figure 5e histogram points to a span of under 1.5%. Further, there is no distinction of activated and dormant regions (no dormant peak in the histogram). 45° macroscopic patterns are also completely absent in Fig. 4e. These can be interpreted as distinctions between twin-mediated localized deformation (up to point *b*) and the more regularized behavior of slip plasticity (up to point *e*).

Unload from point *e* marks the location of point *f*. Noting that Magnesium is an elastically compliant material (Young's moduli around 45 MPa), large unload strains are reasonable; unloading from 160 MPa would point to an elastic recovery of close to 0.4%. This is visible on the strain maps, comparing Fig. 5e, f. The corresponding histogram comparison also points to a mere shift of the large strain peak. Note peak shift without shape change is the statistical signature of elastic recovery.

Reverse loading to point *g*, which we might virtually call the second zero-average-strain-crossing exacerbated the strain heterogeneity that is the basically remnant from point *b*. Now, the bottom section that was late to get tensile strain recovers tension first and goes to compression patterns before the top section of the sample. Further, in the bottom the tensile twin becomes active again, and as an explosive mechanism, pushes the left edge of the histogram (Fig. 5g) to around −2%. The top part of the sample, as observed from the histogram, has actually hardly changed in terms of strain distribution since the unload point *f*. The summary is as follows: Before the top section could recover its positive strain, the tensile twin mechanism jumped in at the bottom section and provided localized strain accommodation for the imposed compressive strain steps. The end result is an average zero strain whose composition has a 4% strain span (−2 to 2% strain), fundamentally different from the first zero strain crossing (point *d*) that had very limited strain heterogeneity.

Hence, this study clearly shows the remnant complications of the first strain heterogeneity left in twinned Magnesium AZ31 upon cyclic loading. It is likely not a stretch to predict that the shown cyclic strain heterogeneity development plays important role in the failure of these materials.

Acknowledgements This work was supported by the Scientific and Technological Research Council of Turkey, TÜBİTAK, Grant No: 114M215.

References

1. G. Proust, C.N. Tomé, A. Jain, S.R. Agnew, Modeling the effect of twinning and detwinning during strain-path changes of magnesium alloy AZ31. Int. J. Plast. **25**, 861–880 (2009)
2. B. Clausen, C.N. Tomé, D.W. Brown, S.R. Agnew, Reorientation and stress relaxation due to twinning: modeling and experimental characterization for Mg. Acta Mater. **56**, 2456–2468 (2008)
3. J.W. Christian, S. Mahajan, Deformation twinning. Prog. Mater Sci. **39**, 1–157 (1995)
4. C.H. Cáceres, T. Sumitomo, M. Veid, Pseudoelastic behaviour of cast magnesium AZ91 alloy under cyclic loading–unloading. Acta Mater. **51**, 6211–6218 (2003)
5. G.E. Mann, T. Sumitomo, C.H. Cáceres, J.R. Griffiths, Reversible plastic strain during cyclic loading–unloading of Mg and Mg–Zn alloys. Mater. Sci. Eng. A **456**, 138–146 (2007)
6. T. Hama, N. Kitamura, H. Takuda, Effect of twinning and detwinning on inelastic behavior during unloading in a magnesium alloy sheet. Mater. Sci. Eng. A **583**, 232–241 (2013)
7. K. Hazeli, J. Cuadra, P. Vanniamparambil, A. Kontsos, In situ identification of twin-related bands near yielding in a magnesium alloy. Scr. Mater. **68**, 83–86 (2013)
8. C.C. Aydıner, M.A. Telemez, Multiscale deformation heterogeneity in twinning magnesium investigated with in situ image correlation. Int. J. Plast. **56**, 203–218 (2014)
9. I.J. Beyerlein, L. Capolungo, P.E. Marshall, R.J. McCabe, C.N. Tomé, Statistical analyses of deformation twinning in magnesium. Philosoph. Mag. **90**, 2161–2190 (2010)
10. M.R. Barnett, M.D. Nave, A. Ghaderi, Yield point elongation due to twinning in a magnesium alloy. Acta Mater. **60**, 1433–1443 (2012)
11. F. Wang, S. Sandlöbes, M. Diehl, L. Sharma, F. Roters, D. Raabe, In situ observation of collective grain-scale mechanics in Mg and Mg–rare earth alloys. Acta Mater. **80**, 77–93 (2014)

Deformation and Strengthening Mechanisms in AISI 321 Austenitic Stainless Steel Under Both Dynamic and Quasi-Static Loading Conditions

A.A. Tiamiyu, A.G. Odeshi and J.A. Szpunar

Abstract The mechanical response of AISI 321 austenitic stainless steel under compressive loads at strain rates of 6600 s^{-1} and 4.2×10^{-3} s^{-1} were studied using the split Hopkinson pressure bar and Instron R5500 mechanical testing system respectively. Specimens subjected to quasi-static compression showed lower yield strength and higher strain hardening capacity than the dynamically impacted specimen. High-resolution electron backscattered diffraction (HR-EBSD) study revealed that precipitation of nano-sized carbide and evolution of strain-induced martensite contributed to strengthening while plastic deformation mechanisms occurred in the specimens by slip and mechanical twinning during deformation under both quasi-static and dynamic loading conditions. The strain-induced phase transformation follows the FCC γ-austenite → BCC ά-martensite kinetic path with both phases maintaining the Kurdjumov-Sachs' $\{(111)_{\gamma} \| (110)_{\acute{\alpha}}$ and $<-101>_{\gamma} \| <1-11>_{\acute{\alpha}}\}$ orientation relationship. A transformed shear band consisting of nano-grains with an average size of 0.28 μm was one of the microstructural features of the dynamically impacted specimen. HR-EBSD analysis revealed that the equiaxed ultra-fine grain structure in the TSB developed by rotational dynamic recrystallization mechanism while dynamic recovery occurred at the interface between the inside and outside of the band. During the deformation under both loading conditions, volume fraction of compression direction (CD)//{110} and CD//{111} increases substantially and slightly, respectively at the expense of CD//{100} fibre texture for the austenitic phase.

Keywords Dynamic impact · Quasi-static compression · EBSD · Shear band · Texture

---

A.A. Tiamiyu (✉) · A.G. Odeshi · J.A. Szpunar
Department of Mechanical Engineering, University of Saskatchewan,
Saskatoon, SK, Canada
e-mail: ahmed.tiamiyu@usask.ca

A.G. Odeshi
e-mail: ago145@mail.usask.ca

J.A. Szpunar
e-mail: jerzy.szpunar@usask.ca

The Minerals, Metals & Materials Society, *TMS 2017 146th Annual Meeting & Exhibition Supplemental Proceedings*, The Minerals, Metals & Materials Series,
DOI 10.1007/978-3-319-51493-2_19

Introduction

AISI 321 is an austenitic stainless steel (ASS) stabilized with titanium to reduce grain boundary sensitization, yet they possess relatively low yield strength that hinders their structural applications [1]. They are attractive material that finds applications in load-bearing structures such as nuclear power reactors, boilers, chemical reactors, automobile exhaust systems, aircraft exhaust stacks and exhaust manifolds, welding, pressure vessels, mufflers for engines, carburetors, expansion bellows, stack liners, fire walls and furnace parts [2, 3]. With such applications, it becomes very critical to decipher the deformation and strengthening mechanisms under high and low strain rates.

AISI 321 has a face-centered cubic (FCC) structure [4] with low stacking fault energy (SFE) [5] but the metastable austenitic microstructure can undergo partial transformation to a martensitic microstructure (strain-induced martensite) during deformation. Strain-induced martensite (ά-martensite) are, however harder than the austenite culminating in a composite strengthening and a high strain hardening effect. This strain hardening is attributed to TRIP (Transformation-Induced Plasticity) effect; a phenomenon that promotes an excellent combination of high strength and ductility during deformation of stainless steel in the annealed condition [6]. Under a high strain rate loading condition, common microstructural feature of deformed specimens is the development of adiabatic shear bands (ASBs). These are the paths of intense shear strain localization due to loss of load carrying capacity from excessive thermal softening in the region of intense adiabatic heating [7]. Zener and Hollomon [8] accredited the occurrence of ASB to the competition between strain-rate hardening and thermal softening such that localized deformation occurs when thermal softening dominates strain-hardening.

The objectives of this work are therefore to characterize the microstructural and textural evolution in AISI 321 austenitic stainless steel under both quasi-static and dynamic impact loading using high-resolution EBSD techniques.

Materials and Methods

The material used in this study is a hot-rolled AISI 321 austenitic stainless steel plate with nominal composition of 17.61 Cr, 9.17 Ni, 1.56 Mn, 0.42 Mo, 0.40 Si, 0.36 Ti, 0.30 Cu, 0.15 Co and 0.044 C in weight percent. Cylindrical test samples with length-diameter ratio of 1 (4 mm in diameter and 4 mm in length) was machined along the normal direction of the as-received plate. While the dynamic impact test was conducted at a strain rate of 6600 s^{-1} using the split Hopkinson pressure bar (SHPB), quasi-static compression was conducted using Instron R5500 mechanical testing system at a strain rate of 4.2×10^{-3} s^{-1}. Metallographic sample preparation involved initial mechanical grinding with SiC papers followed by electrolytic polishing. High-resolution EBSD measurements were conducted using a SU 6600 Hitachi field emission scanning electron microscope coupled with an Oxford Instruments Nordlys Nano EBSD detector. A step size of 120 nm was used for obtaining large area scans and 60 nm was used for obtaining area scans on the shear band region. X-ray measurement was conducted using the Bruker D8 advance diffractometer with Cr Kα radiation. All microstructural analysis and measurements in this study were conducted on the compression plane.

Results and Discussion

Microstructure Before Mechanical Loading

The optical micrograph (Fig. 1a) shows the austenitic phase of the as-received alloy with an average grain diameter of 45 ± 5 microns. EBSD inverse pole figure (IPF) and phase color maps (Fig. 1b, c) indicate a near-random orientation with a substantial amount of annealing twin boundaries. The presence of a very low fraction of TiC particles is observed, predominantly at grain boundaries.

Mechanical Behavior Under Dynamic and Quasi-Static Loading

Figure 2a, b represents the engineering stress-strain and hardening curves of specimens subjected to dynamic impact and quasi-static compression, respectively. The alloy exhibit good strength-ductility combination (above 1380 MPa at 0.41 engineering strain). The yield strength (σ_{YS}) of the alloy under both loading conditions was determined to be 506 and 300 MPa for high and low-strain rate loading conditions, respectively. As deformation proceeds beyond the yield point, differences exist in the deformation process in the plastic region for both loading conditions. Hardening curve (Fig. 2a) shows a distinct competition between the strain hardening and thermal softening of the impacted specimen. While strain hardening

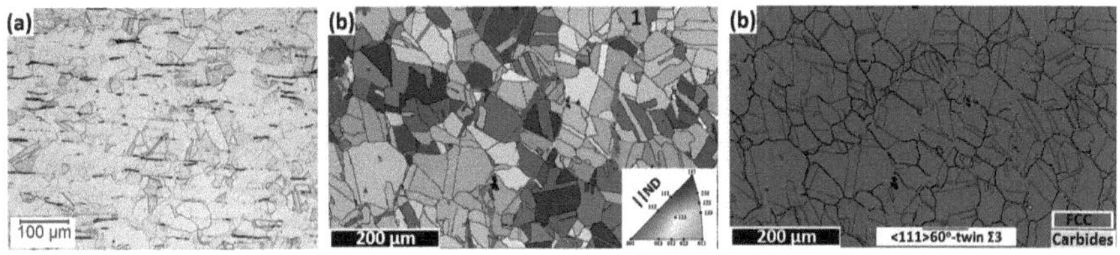

Fig. 1 **a** Optical micrograph, EBSD, **b** inverse pole figure map and **c** phase color map showing annealing twins and TiC precipitates on plane perpendicular to the normal direction (ND) in undeformed AISI 321 austenitic stainless steel

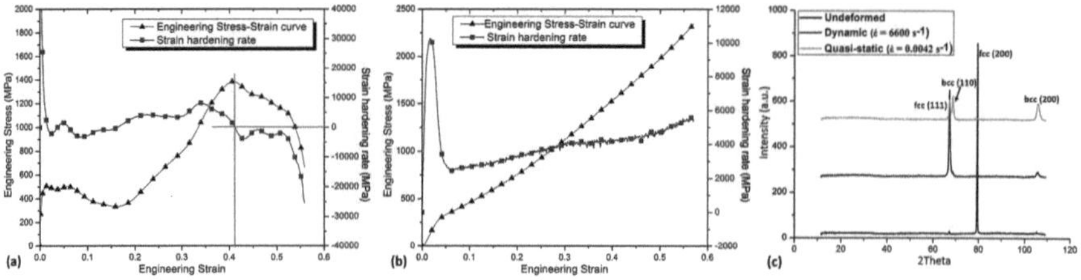

Fig. 2 Engineering stress-strain and hardening curves of **a** dynamically impacted and **b** quasi-statically compressed specimens, **c** X-ray diffractograms of deformed and undeformed specimens

dominated the plastic region up to 0.41 strain, thermal softening gained dominance beyond the 0.41 strain as indicated by the drop in the hardening curve below zero (*intersection of red lines in* Fig. 2a). This phenomenon leads to a thermally-induced mechanical instability and eventual drop in flow stress. Thermal softening has previously been attributed to the conversion of 90% of the kinetic energy of the impact projectile into thermal energy that raises the temperature of specimen under high strain rate loading conditions [9]. The onset of thermal softening corresponds to the ultimate compressive strength (σ_{UCS}) of AISI 321 stainless steel at 1380 MPa under the dynamic loading condition.

On the other hand, Fig. 2b shows that strain hardening dominated the deformation process under the quasi-static loading with no observed drop in the hardening curve below zero. The consistent dominance of strain hardening is not unconnected from the little or no temperature rise during quasi-static compression. Strain-induced transformation occurred in the investigated stainless steel under both loading conditions as indicated in the X-ray diffractograms (Fig. 2c). In comparison with impacted specimen, quasi-statically compressed specimen showed higher BCC ά-martensite peaks. The first BCC ά-martensite peak [ά(110)] appeared on the shoulder of an FCC ɣ-austenite peak [ɣ(111)] while the second peak is the ά(200) as shown in Fig. 2c. This suggests that the mechanical response of AISI 321 ASS is highly rate-dependent.

Microstructural Evolution

The onset of the thermo-mechanical instability as highlighted in previous section, relates microstructurally to the instance of extreme thermal softening and strain localization along narrow regions referred to as adiabatic shear band (ASB) [5]. Deformed shear band (DSB) and transformed shear band (TSB) are the two classifications of ASB with the former consisting of distorted elongated grains and the latter consisting of equiaxed ultrafine grains [10].

Micro-Texture Measurement Near the Edge/ASB Region of the Deformed Specimens

Figures 3 and 4 provides the result of the detailed EBSD analysis near the edge of the compression plane of the specimen. Only the impacted specimen showed evidence of a TSB (Fig. 3a–c); an observation linked to the combined effect temperature rise/thermal softening and strain localization during high strain rate loading. Other authors have shown that the temperature rise could be as high as 943 K in Fe–Cr–Ni austenitic stainless steel [11], 1409 K in β-Ti alloy [12] and $0.5T_m$ (T_m melting point of steel) in 304L ASS [13].

While the TSB region appears dark at a low magnification in Fig. 3a, <110> || CD fibre are the major grain orientation of the austenitic phase outside the TSB region. This is similar to the observed orientation in the austenitic phase of the quasi-statically compressed specimen (Fig. 3d). Similarly, <111> ||CD and <100> ||CD fibres are the major orientations in the martensitic phase in specimens under both high (Fig. 3a) and low (Fig. 3d) strain-rate loading conditions respectively. These orientations agree with compression textures in FCC and BCC metals [14]. Phase maps (Fig. 3b, e) revealed the martensitic transformation of the parent austenitic phase with more fraction of BCC ά-martensite in specimens compressed at low strain rate. This is quite in agreement with the X-ray diffraction result in Fig. 2c. The implication of the aforementioned observation is that the stability of austenitic phase is higher in the specimens under dynamic impact owing to an in increase in SFE (brought about by temperature rise) than those under quasi-static loading condition. The stability of austenite and stacking fault energy (SFE) which depend on the chemical composition, temperature and austenite grain orientation were

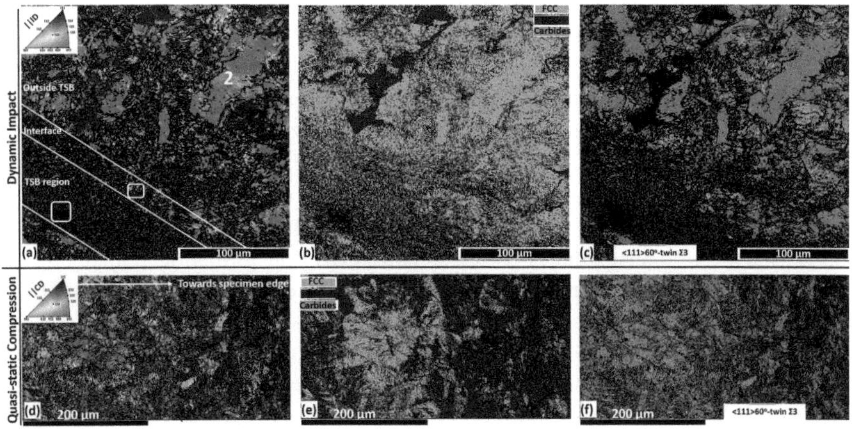

Fig. 3 EBSD maps: **a** and **d** IPF-Z maps of both FCC γ-austenite and BCC ά-martensite, **b** and **e** phase maps showing the FCC γ-austenite, BCC ά-martensite and TiC and **c** and **f** coincidence site lattice (<111> 60°-twin Σ3) of the **a–c** dynamically impacted and **d–f** quasi-statically compressed specimens (*For interpretation of the references to color in this figure legend, the reader is referred to the web version of this article*)

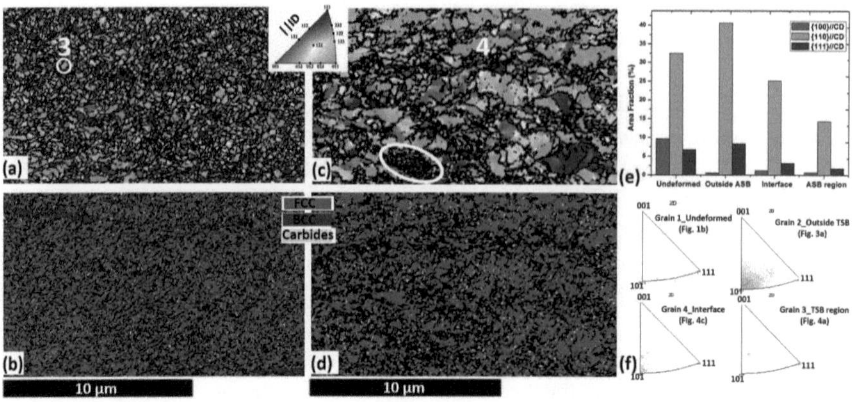

Fig. 4 HR-EBSD map analysis of TSB and interface regions (*marked in* Fig. 3a: **a** IPF-Z map of TSB region, **c** IPF-Z map of interface region, **b** and **d** phase color maps of **a** and **c** respectively, **e** area fraction plot of selected fibers in FCC ɣ-austenite, **f** inverse pole figures of selected FCC ɣ-austenite grains in the undeformed specimen, outside, interface and TSB regions (*For interpretation of the references to color in this figure legend, the reader is referred to the web version of this article*)

earlier reported to influence the extent of phase transformation in austenitic steels [15]. The higher stability of austenite in the specimen deformed under high strain rate loading condition contributed to the lower strain hardening capability in Fig. 2a in comparison with the strain hardening in Fig. 2b. In a separate study, no substantial phase transformation occurred in the quasi-static tensile loading of AISI 304 stainless steel at 75°C whereas substantial phase transformation occurred for loading at room temperature [16]. The martensitic reversion during the plastic deformation of AISI 321 steel leads to the phenomenon termed transformation induced plasticity (TRIP) [17]. Therefore, TRIP effect contributed to an enhanced strength and ductility in AISI 321 steel as observed in Fig. 2 a, b [18]; with more contribution to higher work hardening rate observed in specimens compressed at low strain-rate.

In agreement with the X-ray diffractograms (Fig. 2c), EBSD phase maps (Fig. 3b, e) showed that ε-martensite is absent in the investigated steel under both loading conditions. Hence, the sequence of strain-induced transformation tends to follow the FCC ɣ-austenite → BCC ά-martensite transformation. Other authors [19, 20] did not observe ε-martensite in the sequence of transformation in stainless steels while some [4, 15] did. The difference in test conditions could be responsible for the absence of ε-martensite. It is therefore a possibility that ε-martensite is only a step of the austenite-martensite transformation such that the reaction γ-ε-ά is already completed. Worth mentioning in the phase maps (Fig. 3b, e) is the increase in area fraction of nano-sized (*65-70 nm*) TiC particles to about 6% in both loading conditions in comparison to the near-zero fraction in the as-received condition (Fig. 1c). The precipitated TiCs were observed at grain boundaries and dark regions in both the austenitic and martensitic phases. The dark regions around indexed TiC are arrays of dislocations since an unresolvable measure of orientations in dark regions of a deformed specimen is usually attributed to high dislocation density [21]. The in-grain TiC precipitates therefore contribute to the strengthening as they serve as a barrier to dislocation motion [22]. Other deformation mechanisms or perhaps the source of strengthening is the presence of deformation twins in specimens under both loading conditions (Fig. 3c, f). Deformation twins contributes to the strain hardening capability of the steel since they act as barriers to dislocation glide during plastic deformation. A near-zero increase in temperature during compression that increases the stability of martensite phase is expected to promote more twinning activities in the specimens subjected to quasi-static compression. Therefore, the driving force for TWIP in specimen under high strain rate loading conditions should be lower than during quasi-static compression.

Figure 4 shows the HR-EBSD analysis carried out in the TSB and interface regions (*marked as white rectangles in* Fig. 3a) of the impacted specimen at step size of 60 nm. Equiaxed ultra-fine grains with an average diameter of 0.28 μm were the observed features in the TSB region (Fig. 4a). In a separate study, HR-EBSD technique was used to observe DRX grains with diameter in the range of 0.1–0.3 μm in the ASB developed in AISI 304 ASS [11]. The interface (Fig. 4c) between inside and outside TSB region is a combination of equiaxed grains (*a section is enclosed in white circles*) and lamellar grains elongated along the shear direction. The lamellar grains are reported to be the source of the equiaxed grains [23]. The ultra-fine grains in the TSB are believed to develop by dynamic recrystallization which is the result of the gradual rotation of subgrains coupled with dislocation annihilation [24]. The onset of DRX is preceded by dynamic recovery prior to the formation of new grains. Controversy exists over the effects of dynamic recovery on dynamic recrystallization but there is a general notion that these two processes are competing [25]. DRX mechanisms are classified into migrational (MDRX) and rotational (RDRX) [26]. The migrational dynamic recrystallization mechanism is ruled out in explaining the fine grain structure observed in ASB in the investigated steel because the formation time of ASB is extremely short. At the final engineering strain of 0.56 in this study, it took 135 μs and 136.2 s to complete the deformation under impact and quasi-static loading, respectively. Nesterenko et al. [24] proposed that the RDRX mechanism involves concurrent plastic deformation and thus, the model became one of the most accepted mechanisms to explain developed ultrafine grain structure inside TSBs. The first stage of the proposed mechanism is the accumulation of random, but homogeneously distributed dislocations that re-arranges to elongated dislocation cells culminating to the second stage of the mechanism. The elongated dislocation cells become elongated subgrains in the third stage as deformation progresses. Further increase in deformation results in subgrains break up into equiaxed subgrains due to the minimization of interfacial energy in the fourth stage. Rotation of the subgrains resulting in the evolution of equiaxed dynamically recrystallized grains occurs in the fifth and final stage. While the observed equiaxed DRX grains in the TSB region (Fig. 4a) of the investigated steel are features of the final stage of the schematic model, the interface region (Fig. 4c) shows a combined feature of the third to fifth stages of this mechanism proposed by Nesterenko. The observed weak-textured ultrafine grains and elongated subgrains with lamellar boundaries in Fig. 4c suggests that the interface region underwent dynamic recovery [23].

Phase maps confirmed the existence of strain-induced phase transformation inside TSB (Fig. 4b) and interface region between the TSB and the rest of the material (Fig. 4d). In the study of the dynamic impact response of Ni-Cr steel, XRD techniques was used to observe the co-existence of both austenite and martensite phases in the TSB region with the former suspected to be the retained austenite [27]. Under high strain rate loading conditions, other alloys such as Titanium alloy [28] and β-Ti alloy [12] also confirmed the co-existence of both austenite and martensite phases in ASB region. In the event of fracture of the investigated steel, the presence of BCC ά-martensites in the TSB could promote the initiation and propagation of cracks.

The area fraction of <100> ||CD, <110> ||CD and <111> ||CD fibers in FCC γ-austenite phase in the undeformed specimen, outside, interface and TSB regions is presented in Fig. 4e. In comparison with the undeformed specimen, dynamic impact loading condition lead to an increase in area fraction of <110> ||CD and <111> ||CD texture fibers in the region outside the shear band. From the outside TSB towards the core of the TSB, the area fraction of the aforementioned fibers decreases. However, dynamic impact loading led to a substantial decrease in the area fraction of <100> ||CD fiber in all regions of the specimen. Although the <110> ||CD fiber was strong in all regions, the area fraction of <110> ||CD fiber in TSB region was the lowest which can be attributed to the lattice rotation caused by severe shear strain localization in the region. Therefore, large strain localization in the TSB leads to off- <110> ||CD texture in FCC austenite.

Slip is another plastic deformation mechanism in the region outside the TSB. This is explained by considering the spread of orientation in the inverse pole figures of selected grains of FCC γ-austenitic phase in the undeformed and deformed specimens (Fig. 4f). The spread of orientation in an IPF triangle can hint the extent of deformation in a grain [29]. The orientation spread of grain 1 (*from an undeformed specimen in* Fig. 1b) and grain 3 (*within the shear band in* Fig. 4a) converges to a point in the IPF triangle; suggesting strain free grains. In the same way, grains outside the TSB (*grain 2 in* Fig. 3a) show the widest orientation spread; an indication of large plastic deformation by slip. It is worth noting that large plastic strains are predominantly by slip in spite the abrupt change in orientation of grains in the investigated steel by mechanical twinning. During the slip process, crystal lattice tends to rotate so that the normal to the active slip-plane align with the applied stress direction [14]. Hence, during deformation, an initially random-oriented crystals realigned into a preferred orientation such as the <110> ||CD texture fibers.

Macro-Texture Measurement at the Center of the Deformed Specimens

The summary of the macrotexture measurement conducted at the center of the deformed specimens under high and low strain rate of deformation is presented in Table 1. During the deformation under both loading conditions, volume fraction of CD//{110} and CD//{111} substantially and slightly increases, respectively at the expense of CD//{100} fibre texture for the austenitic phase. Similarly, Brass ({110}//<112>) and S ({123}//<634>) texture components developed substantially with a slight increase in Goss ({110}//<001>) and Copper ({112}//<111>) at the expense of the Cube ({001}//<100>) and R cube ({001}//<110>). The observed data (Table 1) from the macrotexture measurement also confirm the occurrence of strain-induced phase transformation of austenite to martensite under both loading conditions. The pole figures of the FCC ɣ-austenite (Fig. 5a) and BCC

Table 1 Volume fraction (%) of major texture components

		CD//{100}	CD//{110}	CD//{111}	Cube	Goss	Brass	S	Copper	R cube
FCC	Undeformed	25.1	26.9	6.0	4.9	3.2	6.0	7.0	2.0	6.7
	Dynamic impact	2.8	40.8	13.6	0.5	4.8	10.3	10.2	3.8	0.5
	Quasi-static	5.5	36.7	8.3	1.0	4.1	8.9	11.5	4.4	1.0
BCC	Undeformed	–	–	–	–	–	–	–	–	–
	Dynamic impact	16.2	16.2	13.0	3.2	1.8	3.4	8.7	4.5	3.3
	Quasi-static	19.0	23.4	8.8	3.8	3.0	5.5	7.1	2.4	4.2

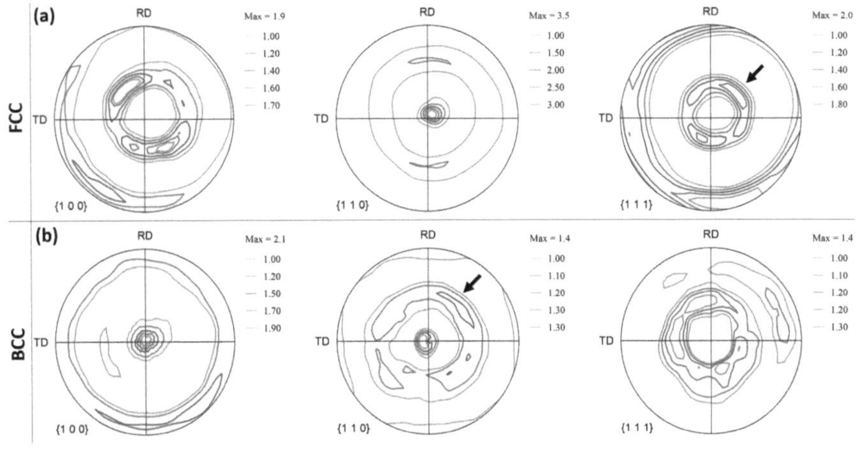

Fig. 5 Macro-texture pole figure measurement of **a** FCC ɣ-austenite and **b** BCC ά-martensite at the center of the quasi-statically compressed specimen

$\acute{\alpha}$-martensite (Fig. 5b) confirmed that the orientation relationships between two phases as indicated by black arrows follow Kurdjumov-Sachs' $\{(111)_{\gamma} \| (110)_{\acute{\alpha}}$ and $<-101>_{\gamma} \| <1-11>_{\acute{\alpha}}\}$.

Conclusion

The mechanical response of AISI 321 austenitic stainless steel under dynamic impact and quasi-static loading conditions were studied and compared. The result showed that specimens subjected to quasi-static compression exhibits lower yield strength and higher strain hardening capacity than the impacted specimen. The high resolution EBSD study revealed that precipitation of nano-sized carbide and evolution of strain-induced martensite contributed to strengthening while plastic deformation mechanisms occurred in the specimens by slip and mechanical twinning during deformation in both quasi-static and dynamic loading conditions. The strain-induced phase transformation follows the FCC γ-austenite → BCC $\acute{\alpha}$-martensite kinetic path with both phases maintaining the Kurdjumov-Sachs' $\{(111)_{\gamma} \| (110)_{\acute{\alpha}}$ and $<-101>_{\gamma} \| <1-11>_{\acute{\alpha}}\}$ orientation relationship. A transformed shear band consisting of nano-grains with an average size of 0.28 μm was the microstructural features of dynamically impacted specimen. High resolution EBSD analysis of the TSB revealed that the equiaxed ultra-fine grain structure developed by rotational dynamic recrystallization mechanism while dynamic recovery occurred at the interface between the inside and outside of the band. During the deformation under both loading conditions, volume fraction of CD//{110} and CD//{111} increases substantially and slightly, respectively at the expense of CD//{100} fibre texture for the austenitic phase.

Acknowledgements The authors wish to acknowledge the financial support of Natural Sciences and Engineering Research Council of Canada (NSERC) for the financial support of this study.

References

1. M. Eskandari, A. Kermanpur, A. Najafizadeh, Formation of nanocrystalline structure in 301 stainless steel produced by martensite treatment. Metall. Mater. Trans. A **40**, 2241–2249 (2009)
2. K. Guan, X. Xu, H. Xu, Z. Wang, Effect of aging at 700°C on precipitation and toughness of AISI 321 and AISI 347 austenitic stainless steel welds. Nucl. Eng. Des. **235**, 2485–2494 (2005)
3. R.K.C. Nkhoma, C.W. Siyasiya, W.E. Stumpf, Hot workability of AISI 321 and AISI 304 austenitic stainless steels. J. Alloys Compd. **595**, 103–112 (2014)
4. J. Talonen, H. Hänninen, Formation of shear bands and strain-induced martensite during plastic deformation of metastable austenitic stainless steels. Acta Mater. **55**, 6108–6118 (2007)

5. W. Zhang, J. Wu, Y. Wen, J. Ye, N. Li, Characterization of different work hardening behavior in AISI 321 stainless steel and hadfield steel. J. Mater. Sci. **45**, 3433–3437 (2010)
6. P. Hedström, Deformation induced martensitic transformation of metastable stainless steel AISI 301 (2005)
7. A.A. Tiamiyu, A.Y. Badmos, A.G. Odeshi, Effects of temper condition on high strain-rate deformation of AA 2017 aluminum alloy in compression. Mater. Des. **89**, 872–883 (2016)
8. D. Yang, Y. An, P. Cizek, P. Hodgson, Development of adiabatic shear band in cold-rolled titanium. Mater. Sci. Eng. A **528**(12), 3990–3997 (2011)
9. S. Nemat-Naser, Introduction to high strain-rate testing, high strain rate tension and compression tests. in *ASM Handbook*, vol. 8 (2000), pp. 942–955
10. W.Q. Song, S. Sun, S. Zhu, G. Wang, J. Wang, M.S. Dargusch, Compressive deformation behavior of a near-beta titanium alloy. Mater. Des. **34**, 739–745 (2012)
11. B.F. Wang, Z.L. Liu, X.Y. Wang, Z.Z. Li, An EBSD investigation on deformation-induced shear bands in a low nickel austenitic stainless steel under controlled shock-loading conditions. Mater. Sci. Eng. A **610**, 301–308 (2014)
12. Y. Yang, F. Jiang, B.M. Zhou, X.M. Li, H.G. Zheng, Q.M. Zhang, Microstructural characterization and evolution mechanism of adiabatic shear band in a near beta-Ti alloy. Mater. Sci. Eng. A **528**(6), 2787–2794 (2011)
13. M.A. Meyers, Y.B. Xu, Q. Xue, M.T. Pérez-Prado, T.R. McNelley, Microstructural evolution in adiabatic shear localization in stainless steel. Acta Mater. **51**(5), 1307–1325 (2003)
14. H. Hu, Texture of metals. Texture **1**(4), 233–258 (1974)
15. C. Herrera, D. Ponge, D. Raabe, Design of a novel Mn-based 1 GPa duplex stainless TRIP steel with 60% ductility by a reduction of austenite stability. Acta Mater. **59**(11), 4653–4664 (2011)
16. H. Wang, Y. Jeong, B. Clausen, Y. Liu, R.J. Mccabe, F. Barlat, C.N. Tomé, Effect of martensitic phase transformation on the behavior of 304 austenitic stainless steel under tension. Mater. Sci. Eng. A **649**, 174–183 (2015)
17. M. Zhang, L. Li, R.Y. Fu, D. Krizan, B.C. De Cooman, Continuous cooling transformation diagrams and properties of micro-alloyed TRIP steels. Mater. Sci. Eng. A. **438–440**, 296–299 (2006)
18. D. Mohr, J. Jacquemin, Large deformation of anisotropic austenitic stainless steel sheets at room temperature: multi-axial experiments and phenomenological modeling. J. Mech. Phys. Solids **56**(10), 2935–2956 (2008)
19. B.R. Kumar, A.K. Singh, B. Mahato, P.K. De, N.R. Bandyopadhyay, D.K. Bhattacharya, Deformation-induced transformation textures in metastable austenitic stainless steel. Mater. Sci. Eng. A **429**, 205–211 (2006)
20. H.F.G. De Abreu, S.S. De Carvalho, P. De Lima Neto, R.P. Dos Santos, V.N. Freire, P.M.D. O. Silva, S.S.M. Tavares, Deformation induced martensite in an AISI 301LN stainless steel: characterization and influence on pitting corrosion resistance. Mater. Res. **10**(4), 359–366 (2007)
21. M.T. Pérez-Prado, J.A. Hines, K.S. Vecchio, Microstructural evolution in adiabatic shear bands in Ta and Ta-W alloys. Acta Mater. **49**, 2905–2917 (2001)
22. M. Chabaud-Reytier, L. Allais, C. Caes, P. Dubuisson, A. Pineau, Mechanisms of stress relief cracking in titanium stabilised austenitic stainless steel. J. Nucl. Mater. **323**(1), 123–137 (2003)
23. J.F.C. Lins, H.R.Z. Sandim, H.-J. Kestenbach, D. Raabe, K.S. Vecchio, A microstructural investigation of adiabatic shear bands in an interstitial free steel. Mater. Sci. Eng. A **457**(1–2), 205–218 (2007)
24. V.F. Nesterenko, M.A. Meyers, J.C. LaSalvia, M.P. Bondar, Y.J. Chen, Y.L. Lukyanov, Shear localization and recrystallization in high-strain, high-strain-rate deformation of tantalum. Mater. Sci. Eng. A **229**(1–2), 23–41 (1997)
25. J. Hines, K. Vecchio, S. Ahzi, A model for microstructure evolution in adiabatic shear bands. Metall. Mater. Trans. A **29**, 1998 (1998)

26. B. Derby, The dependence of grain size on stress during dynamic recrystallisation. Acta Metall. Mater. **39**(5), 955–962 (1991)
27. Y. Xu, J. Zhang, Y. Bai, M.A. Meyers, Shear localization in dynamic deformation: microstructural evolution. Metall. Mater. Trans. A **39**(4), 811–843 (2008)
28. Z.P. Wan, Y.E. Zhu, H.W. Liu, Y. Tang, Microstructure evolution of adiabatic shear bands and mechanisms of saw-tooth chip formation in machining Ti6Al4V. Mater. Sci. Eng. A **531**, 155–163 (2012)
29. A.A. Tiamiyu, R. Basu, A.G. Odeshi, J.A. Szpunar, Plastic deformation in relation to microstructure and texture evolution in AA 2017-T451 and AA 2624-T351 aluminum alloys under dynamic impact loading. Mater. Sci. Eng.A **636**, 379–388 (2015)

Analysis of Dislocation Structures in Ferritic and Dual Phase Steels Regarding Continuous and Discontinuous Loading Paths

Gregory Gerstein, Till Clausmeyer, Florian Gutknecht, A. Erman Tekkaya and Florian Nürnberger

Abstract In sheet-bulk metal forming processes the hardening behavior of the material depends on the sequence of deformation steps and the type of deformation. Loading path changes induce transient hardening phenomena. These phenomena are linked to the formation and interaction of oriented dislocation structures. The aim of this study is to investigate the effect of continuous and discontinuous loading path changes on the dislocation microstructure in ferritic and ferritic-martensitic dual-phase steel, respectively. For the experiments a biaxial test stand was used, which permits to continuously change the load from tension to shear. In the ferrite single-phase steel transmission-electron microscopy reveals a reduced evolution of oriented dislocation structures for continuous loading path changes compared to discontinuous loading path changes. This evolution is further decreased in dual-phase steel compared to the ferritic steel. Microstructural results for the ferritic steel are accompanied by simulation results with a transient hardening model.

Keywords Sheet-bulk metal forming · Dislocation structures · Material modeling · Parameter identification

---

G. Gerstein (✉) · F. Nürnberger
Institut für Werkstoffkunde (Materials Science), Leibniz Universität Hannover, Hannover, Germany
e-mail: gerstein@iw.uni-hannover.de

F. Nürnberger
e-mail: nuernberger@iw.uni-hannover.de

T. Clausmeyer · F. Gutknecht · A.E. Tekkaya
Institute of Forming Technology and Lightweight Construction, TU Dortmund University, Dortmund, Germany
e-mail: Till.Clausmeyer@iul.tu-dortmund.de

F. Gutknecht
e-mail: Florian.Gutknecht@iul.tu-dortmund.de

A.E. Tekkaya
e-mail: Erman.Tekkaya@iul.tu-dortmund.de

The Minerals, Metals & Materials Society, *TMS 2017 146th Annual Meeting & Exhibition Supplemental Proceedings*, The Minerals, Metals & Materials Series,
DOI 10.1007/978-3-319-51493-2_20

Introduction

In sheet-bulk metal forming processes a certain degree of pre-damage is induced into the sheet material during processing which depends on the sequence of deformation steps and the type of deformation. Controlling the direction and type of loading allows avoiding critical damage values. Since suited modifications in the dislocation structure formed during deformation prevent or at least hinder ductile damage, knowledge about load path dependencies is of current interest. Thus, a ferritic-martensitic dual-phase steel DP600 and two single-phase ferritic steels were investigated regarding both a continuous and discontinuous change in the loading direction. The discontinuous change in the load direction was carried out by plane strain tension and shear tests. The results were used to parametrize a micromechanical motivated transient hardening model. The investigations aimed to understand the active physical deformation mechanisms at the micro scale regarding the response on the macroscopic scale. Currently, in order to fully exploit these materials in sheet-bulk metal forming, the process is simulated by means of the finite-element method. Constitutive models relating mechanical stress to mechanical strain are required for this purpose. Commonly used material models describe the material behavior at the macroscopic scale in a phenomenological manner. Central to these models is that the physical motivation originates in the interpretation of stress-strain data of mechanical tests, such as tensile or shear tests [1]. Several studies on the strain path-dependent behavior of steels reported an increasing yield stress for certain sequences of biaxial stretching [2] and the formation of new Dislocation structures [3–5]. According to [6] the formed dislocation structures mainly determine the hardening behavior under load path changes. Structural properties, such as grain size, grain morphology and texture, are of minor importance. The mechanical behavior under strain path changes was studied by van Riel regarding the ferritic interstitial free steel DC06, the dual-phase steel DP600, the high strength steel H340LAD and the aluminum alloy AA5182 [7].

A detailed model requires consideration of the dislocation interactions on a microscale [8–11] though such simulations are computationally costly and rather complex. Efficient phenomenological models do not rely on crystallographic data for describing plastic anisotropy resulting from texture [12] A phenomenological modeling approach was successfully used by Teodosiu et al. based on transmission-electron microscopical investigations of metals subjected to mechanical loading after changing the strain path [13].

Experiments

The single- and the dual-phase steels were subjected to plane strain tension and simple shear loading. Simple shear tests were performed for the steels DC04 and DP600 with the shear direction parallel and orthogonal to the rolling direction,

Table 1 Chemical composition (mass%) of the investigated steels (Fe: balance)

Alloying component	C	Si	Mn	P	S	Al	N	Ti	B	Cr, Mo, Ni
DC04	0.08		0.4	0.03	0.03					
DC06	0.003	0.018	0.137	0.013	0.01	0.035	0.0027	0.079		
DP600	0.14	1.5	2.0	0.07	0.015	0.015			0.005	1.00

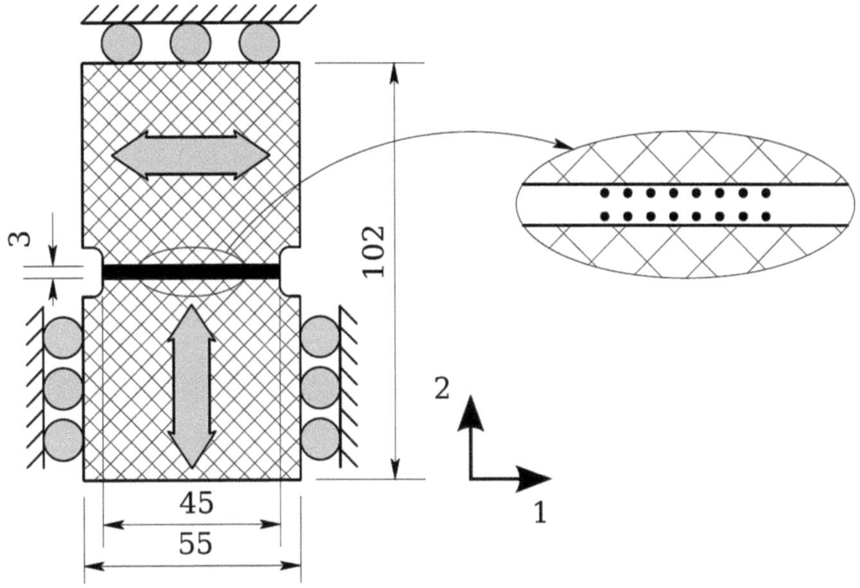

Fig. 1 Biaxial test set-up. Geometry of the tension-shear specimen and the measurement region of height 3.0 mm and width 45.0 mm. The *checkered region* indicates the actual specimen and the *black area* marks the actual deformation zone. The tension direction is direction 2 and the shear direction is direction 1 [14]

where the rolling direction is parallel to the tensile direction. The continuous tests were performed at the Department of Engineering Technology at Twente University, Netherlands, for the steel DC06. A biaxial testing device which is capable of loading a specimen in plane strain tension and in simple shear was used for this purpose. The normal strain and the shear are determined in terms of the components F_{22} and F_{21} of the deformation gradient $\boldsymbol{F}$ by optical strain measurement. The horizontal and vertical forces are measured with load cells in order to compute the Cauchy stress tensor $\boldsymbol{\sigma}$. Table 1 states the chemical composition of each type of steel.

The specimens for biaxial testing (Fig. 1) were obtained by EDM in order to minimize the influence of residual stresses at the edges. The tests were performed at constant nominal strain rates of $10^{-3}\ \mathrm{s}^{-1}$. The stress-strain responses for the

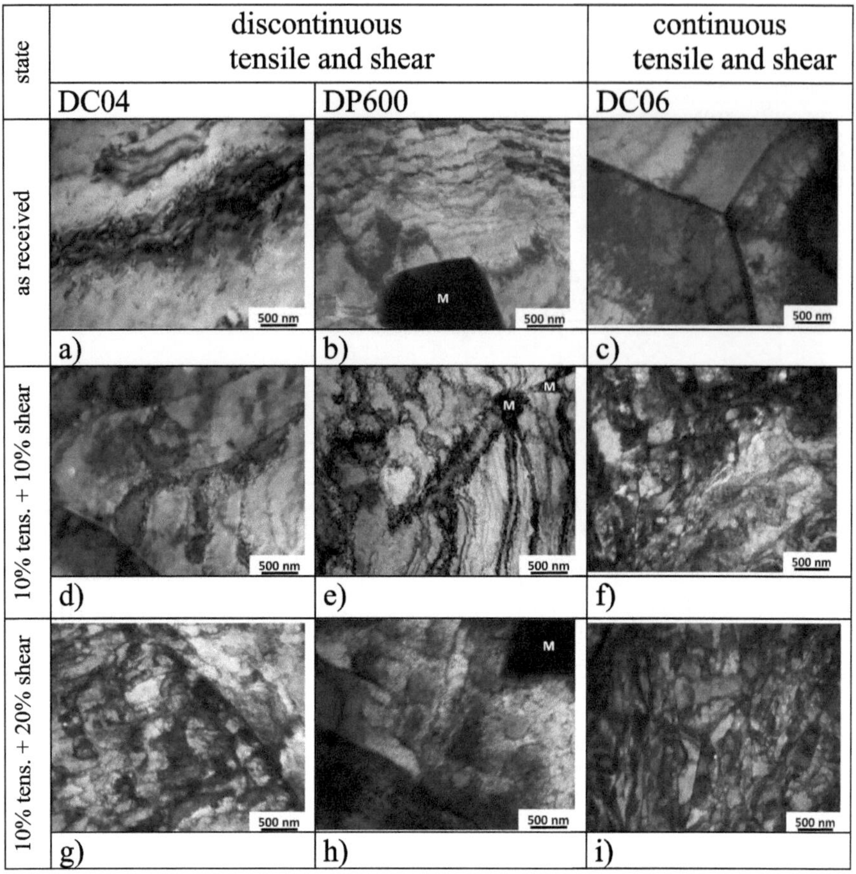

Fig. 2 TEM images of dislocation structures: **a–c** microstructure as-received; **d–f** after plane strain tension of 10 and 10% simple shear **g–i** after plane strain tension of 10 and 20% simple shear Tension in the rolling direction (RD); shear (SD) orthogonal to the RD

interstitial free steel DC06 and for DP600 were reported previously in [15] (Fig. 2) and [16] (Fig. 3), respectively. The curves show a typical monotonic increasing stress-strain behavior.

The deformation states and the corresponding normal or shear strain values, at which transmission electron microscopy samples were taken, are given in Table 2. In order to obtain TEM foils, flat disks with a diameter of 3.0 mm and a thickness of approximately 400 μm were cut from the area of deformation by EDM. These disks were mechanically thinned and polished from both sides to a thickness of 80 μm. The TEM-foils were electrolytic polished with an electrolyte consisting of 120 ml 40% perchloric acid, 440 ml butoxy ethanol and 440 ml 100% acetic acid in a Tenupol Electropolish 5 (Fa. Struers). The microstructure investigations were performed on a JEOL JEM2010 transmission electron microscope with a 200 kV electron gun.

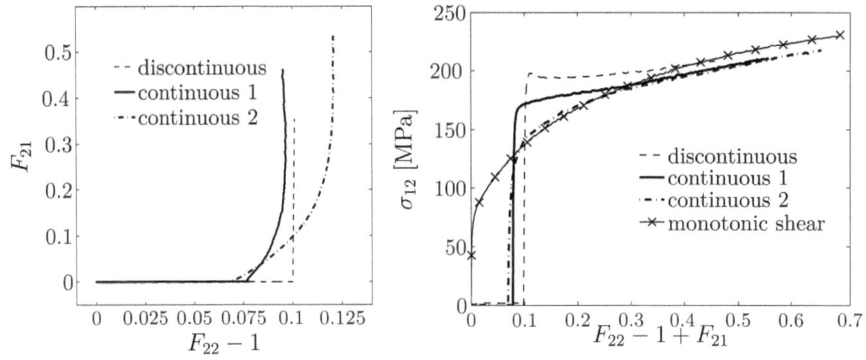

Fig. 3 Test results for continuous transitions from plane strain tension loading to simple shear loading (continuous 1). For comparison, the results for the orthogonal loading path change (discontinuous) is shown along with the curve of monotonic shear loading (monotonic) [17]

Table 2 Deformation states of the TEM-samples

Material	Loading path (strain ε or γ (−))			
DC04	Discontinuously		Continuously	
DC06	Plain-strain (tension)	0.1	Plain-strain (tension) + shear	0.1ε + 0.1γ
DP600		0.2		0.2ε + 0.1γ
	Shear	0.1		0.1ε + 0.2γ
		0.2		
		0.3		

Results and Discussion

The change of the dislocation structures in the single- and dual-phase steels in dependence on the nature and magnitude of the applied load are depicted in Fig. 2. At total deformations above 30% the dislocation structures saturate and the morphology does not evolve further with increasing strain values as previously reported by Fernandes and Schmitt [3]. Structures of the dislocation cells are smaller in the single phase steels in comparison to structures of cells in the dual-phase steel (Fig. 2 a–c). The aspect-factor of the dislocation cells (ratio of length of a cell to its width) at total deformation of combined tension and shear of 20% in the DC04 and DC06 steels is within the range of 0.5-0.7. In comparison, the aspect-factor for dislocation cells of the dual-phase steel DP600 is 0.7-0.9 (Fig. 2 d– f). At total deformations of 30% the nature of the deformation process changes in the dual-phase steel. In addition to dislocation sliding in the ferrite matrix nteracttions between ferrite and martensite become more and more dominant (Fig. 2h). Continuous changes in the loading path result in a considerable crushing of dislocation cell structures by reducing the degree of disorientation with respect to neighboring cells (i.e. a decrease of texture). This is evidenced by the fact that

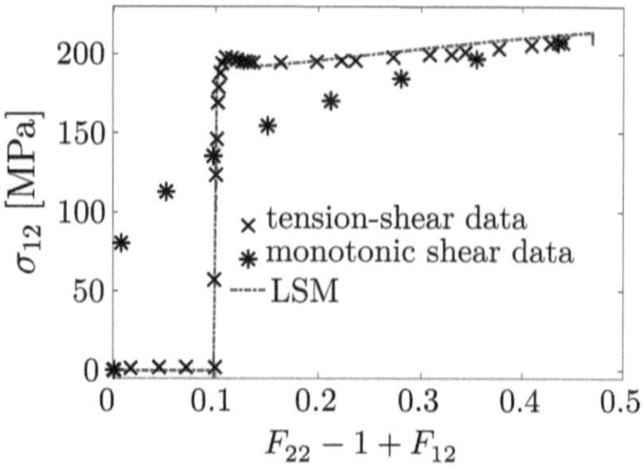

Fig. 4 Comparison of model and experimental results for shear stress as a function of deformation in the interstitial free steel DC. The material was subjected to plane-strain tension in the rolling direction (up to 10%) followed by an orthogonal loading path change to simple shear in the transverse direction [18]

dislocation cells can be simultaneously observed in adjacent grains irrespective of their orientation (Fig. 2i). In general, the dislocation patterning is more homogeneous for adjacent grains after continuous loading path changes compared to discontinuous loading path changes. This can be explained by the fact that the changes in active glide systems are smaller for the continuous change. Such change of microstructure is in good correlation with results of mechanical tests of the discontinuously and continuously loaded samples (Fig. 3).

The aforementioned change in the value of dislocation cells size in grains which are having a completely different crystal geometry orientation can be explained by a step thresholdwise change of the stress values by tension-shear loading (Fig. 4).

The DC06 exhibits a distinct increase (>60 Mpa) in shear stress after the discontinuous change from plane strain tension to shear loading (see Fig. 3). This phenomenon is referred to as cross-hardening. For the observation of cross-hardening the stress after a loading path change is compared to the stress of a monotonic test. Here, the sum of the normal strain $F_{22} - 1$ and shear F_{21} is chosen as strain measure. The increase in stress is smaller (≈50 Mpa) for the continuous transition (continuous 1). Moreover, no change in curvature immediately after the transition is observed. In the continuous loading path change with additional super-imposed tension (continuous 2), the stress-strain curve after the change in loading path coincides with the monotonic curve.

During monotonic loading persistent dislocation structures form parallel to the most active glide systems (see Fig. 2). After a continuous loading path change new glide systems are activated. The dislocations, which move on these systems during the subsequent loading eventually, have to overcome the persistent dislocation structures. The interaction of these two dislocation populations leads to the observed cross-hardening. With ongoing deformation in the new direction micro shear bands form. This results in a hardening stagnation. With further continued loading hardening due to glide activity on the now active set of glide systems resumes. This is marked by the saturation of the hardening similar to a monotonic test. The mechanisms leading to cross-hardening and hardening stagnation have been described by Rauch and Schmitt [4] for discontinuous loading path changes.

The TEM analysis of the continuous loading path changes indicates that changes in the size, the nature of distribution of dislocation cells and the density of dislocations in cells is similar in adjacent grains. For this purpose grains attached to triple points were investigated in particular. During discontinuous loading path changes limited distinct sets of glide systems are activated. During continuous loading path changes, the sets of glide systems change gradually. Therefore glide activity is distributed more heterogeneously on different systems. Consequently, the dislocation structures observed in a single plane of foil are distributed more heterogeneously.

Due to the fact that for a continuous change the changes in the active set of glide systems are smaller, meaning that some systems which have been active during pre-straining might also be active in the subsequent stage of deformation, the interactions of different populations of dislocations are lower. This is reflected by less cross-hardening in the mechanical tests.

The observed mechanical behavior can be modelled with continuum mechanical models which consider the effect of dislocation interaction in an efficient way. For the prediction of the mechanical behavior in sheet-bulk forming an engineering scale model is required. A model based on J2-plasticity, which accounts for the interaction of dislocation populations by introducing an evolving fourth order tensor in the yield function is presented in [11]. Central to the model formulation is the yield function

$$\phi = \sqrt{[M - X] : [\mathscr{A} : [M - X]]} - \sigma_Y(\varepsilon_p)$$

where $\boldsymbol{M}$, $\boldsymbol{X}$, and $\mathscr{A}$ represent the Mandel stress, the backstress and the fourth order anisotropy tensor. Isotropic hardening is considered by the hardening function $\sigma_Y(\varepsilon_p)$ in terms of the equivalent plastic strain ε_p. Material parameters for the model are identified on the basis of rheological tests [11]. The evolution relation for the anisotropy tensor distinguishes between loading parallel to previous plastic deformation and loading orthogonal to this direction. The model is capable to capture the transient hardening behavior observed in the mechanical tests as shown in Fig. 4.

Conclusion

On the basis of discontinuous and continuous tension to shear tests the evolution of the dislocation structures has been investigated for single phase interstitial free steels (DC04 and DC06) and a dual phase steel (DP600). The development of dislocation structures was characterized and compared for different load paths to provide information on the mechanism at the micro scale governing the macroscopic response. The observations suggest that the plastic deformation mechanisms are defined by the form and distribution of dislocation cell structures rather than by the glide mechanism only. The characterization of dislocation cell structures revealed that the deformation by a continuous tension-shear load path in the steel DC06 is more homogeneous in adjacent grains than for a discontinuous load path.

Investigations for different loading histories provide a basis to conceptualize a micromechanical-based numerical model.

Acknowledgements Funding by the German Research Foundation (DFG) within the scope of the Transregional Collaborative Research Centre on sheet-bulk metal forming (SFB/TR 73) in the subproject C4 'Analysis of load history dependent evolution of damage and microstructure for the numerical design of sheet-bulk metal forming processes' is highly acknowledged.

References

1. Barlat et al., An alternative to kinematic hardening in classical plasticity. Int. J. Plast **27**, 1309–1327 (2011)
2. A.K. Ghosh, W.A. Backofen, Strain hardening and instability in biaxially stretched sheets. Metall. Trans. **4**, 1113–1123 (1973)
3. J.V. Fernandes, J.H. Schmitt, Dislocation microstructures in steel during deep drawing. Philos. Mag. A **48**, 841–870 (1983)
4. E.F. Rauch, J.H. Schmitt, Dislocation substructures in mild steel deformed in simple shear. Mater. Sci. Eng., A **113**, 441–448 (1989)
5. S. Thuillier, E. Rauch, Development of micro-bands in mild steel during cross loading. Acta Metall. Mater. **42**, 1973–1983 (1994)
6. D. Wilson, P. Bate, Influences of cell walls and grain boundaries on transient responses of an IF steel to changes in strain path. Acta Metall. Mater. **42**, 1099–1111 (1994)
7. M. van Riel, Strain path dependency in sheet metal-experiments and models. Ph.D. Thesis, Twente University, Enschede, The Netherlands, 2009
8. L. Bardella, A deformation theory of strain gradient crystal plasticity that accounts for geometrically necessary dislocations. J. Mech. Phys. Solids **54**, 128–160 (2006)
9. L.P. Evers, W.A.M. Brekelmans, M.G.D. Geers, Scale dependent crystal plasticity framework with dislocation density and grain boundary effects. Int. J. Solids Struct. **41**, 5209–5230 (2004)
10. B. Holmedal, P. van Houtte, Y. An, A crystal plasticity model for strain-path changes in metals. Int. J. Plast **24**, 702–707 (2008)
11. T. Clausmeyer et al., Phenomenological modeling of anisotropy induced by evolution of the dislocationstructure on the macroscopic and microscopic scale. Int. J. Mater. Form. **4**, 141–154 (2011)
12. Behrouzi et al., Inherent and induced anisotropic finite visco-plasticity with applications to the forming of DC06 sheets. Int. J. Mech. Sci. **89**, 101–111 (2014)
13. E. Nesterova, B. Bacroix, C. Teodosiu, Experimental observation of microstructure evolution under strain-path changes in low-carbon IF steel. Mater. Sci. Eng. A **309**, 495–499 (2001)
14. M. van Riel, A.H. van den Boogaard, Stress-strain responses for continuous orthogonal strain path changes with increasing sharpness. Scripta Mater. **57**(5), 381–384 (2007)
15. T. Clausmeyer et al., Experimental characterization of microstructure development during loading path changes in bcc sheet steels. J. Mater. Sci. **48**, 674–689 (2013)
16. K. Isik et al., Evaluation of Void Nucleation and Development during Plastic Deformation of Dual-Phase Steel DP600. Steel Res. Int. **87**, (2016). doi:10.1002/srin.201500483
17. T. Clausmeyer, Evolution of plastic anisotropy in metals: Material models, experiments and applications. Diss., Fakultät Maschinenbau. Technische Universitaet Dortmund (2016)
18. T. Clausmeyer, B. Svendsen, Comparison of two models for anisotropic hardening and yield surface evolution in bcc sheet steels. Eur. J. Mech. A/Solids **54**, 120–131 (2015)

Microstructural Characterization of Inconel 600 Tubes After Tensile Tests at Various Temperatures and Strain Rates

Cécile Davoine, Vincent Marcadon, David Lévêque, Fabienne Popoff, Nicolas Horezan, Denis Boivin and Gérald Portemont

Abstract As a first step of studying the mechanical behavior of cellular material made with tubes assembled by brazing, this work aims at characterizing the elasto-viscoplastic behavior of Inconel 600 tubes subjected to annealing pre-treatments representative of brazing process. Tensile tests highlighted rate sensitivity at room temperature and at 800 °C that can be attributed to microstructural evolutions occurring both during annealing and during the tests at 800 °C (dynamic recrystallization, recovery, grain growth and coarsening of precipitates). Post-mortem analysis based on Electron BackScattered Diffraction (EBSD) was carried out on broken samples. Local misorientations have been extracted from crystallographic orientation data in order to evaluate plastic residual strain. Kernel Average Misorientation (KAM) maps revealed high-strained area near nitrides precipitates for the annealed tubes loaded at room temperature and high strain rate. Comparisons of averaged misorientation are also discussed regarding microstructural evolutions and strain rate sensitivity.

Keywords Inconel 600 · Electron backscattered diffraction (EBSD) · Strain rate sensitivity · Residual strain

Introduction

Thanks to their large densification strain at the plateau stress during the plastic compression, cellular materials are used as impact energy absorbers. The constitutive material of the cells is thus subjected to severe plastic deformations. Previous work highlighted the influence of the constitutive material behavior on effective

C. Davoine (✉) · V. Marcadon · D. Lévêque · F. Popoff · N. Horezan · D. Boivin
ONERA, The French Aerospace Lab, 29 Avenue de la Division Leclerc, 92320 Châtillon, France
e-mail: cecile.davoine@onera.fr

G. Portemont
ONERA, The French Aerospace Lab, 5 Boulevard Paul Painlevé, 59800 Lille, France

The Minerals, Metals & Materials Society, *TMS 2017 146th Annual Meeting & Exhibition Supplemental Proceedings*, The Minerals, Metals & Materials Series,
DOI 10.1007/978-3-319-51493-2_21

mechanical properties for this kind of materials [1], and the fact that the main mechanism governing the mechanical behavior is the localized plasticity in the walls of the cells [2]. Simple cellular materials can be created by brazing hollow spheres [3], corrugated (prismatic) cellular structures, corrugated sheets or thin-wall tubes. Tubes stacking cellular materials are interesting for high temperature applications because they can be cooled by coolant flowing inside. For instance, in the field of gas turbine applications, tubes can be coiled around the engine to provide efficient containment in case of blade-shedding. Blade-shedding is part of a safety process to preserve turbines disks from burst in the event of over-speed regime and consists in generating controlled blades ruptures in order to decrease centrifugal efforts sustained by the disk. This function provides the ultimate protection in the event of a critical overspeed situation. In our work, the commercially available thin-wall tubes are made in Inconel 600 which is a Ni–Cr–Fe alloy that exhibits a good combination of strength, formability and environmental resistance at elevated temperatures. High temperature brazing is needed to avoid the melting of the joint in use. Nickel-base fillers metals usually contain several "melting point depressant" which can form eutectic. A sufficient amount of time at the brazing temperature could prevent the formation of intermetallics during cooling by alloying a complete isothermal solidification to take place at the brazing temperature. During brazing step and during the use of the cellular material, several microstructural evolutions influencing mechanical properties may occur: grain growth, recrystallization, recovery and coarsening/dissolution of precipitates. In previous studies [4–6], tensile tests have been performed on tubes at various strain rates and temperature until fracture to investigate their elasto-viscoplastic properties. The effects of the heat treatment and tests temperature on strain rate sensitivity have been highlighted. In the present work, these effects will be discussed in relation to microstructural evolutions. Post-mortem analysis based on Electron BackScattered Diffraction (EBSD) was carried out on broken samples. Recent advances in the development of EBSD characterization tools have enabled to study deformed materials in order to highlight the dislocation density. These techniques have greatly improved our ability to quantify deformation mechanisms such as dislocations, twinning and microstructural changes accompanying deformation (i.e. texture evolution, grain morphology changes, and localized strain). EBSD technique enables analyses with high spatial resolution and sensitive examination of plastic strain that are otherwise difficult because of spatial/angular resolution and sampling limitations of X-ray Diffraction (XRD) and Transmission Electron Microscopy (TEM), respectively. First, qualitative information given by the local misorientation through Kernel Average Misorientation (KAM) measurements performed at different step sizes is discussed. Then a comparative study based on quantitative information extracted from both KAM and Grain Orientation Spread (GOS) leads to identify the dislocation density as the major microstructural parameter governing the mechanical behavior of the studied material.

Experimental Procedure

A first set of uni-axial high velocity tensile tests has been performed on tubular samples at various strain rates for different initial material configurations. In particular, three initial states have been compared: as-received, simple heat treated state (TT_1) and double heat treated state (TT_1+TT_2). The samples were machined from 75 mm length tubes with external and internal diameters equals to 6 mm and 4 mm respectively, and the external diameter in the center of the specimen has been reduced by eletromachining at 5 mm. The heat treatment under vacuum at 1000 °C during 15 min (TT_1) has been chosen to be representative of the brazing step of 'model' cellular structure elaboration consisting of regularly stacked circular tubes. The second step of the heat treatment is an annealing at 1050 °C during 16 h (TT_2). This annealing treatment allows consolidating tubes stacking by diffusion of braze material. More details regarding geometry and tests campaigns are given in [4, 5]. The tests revealed that for the as-received samples a strain rate effect appears from the tests performed at 0.7 s^{-1}, whereas few or no strain rate sensitivity was measured for the samples having simple (TT_1) or double heat treated state (TT_1+TT_2). The strain rate sensitivity has been attributed to the initial amount of dislocation. The results of this first set of tests performed at room temperature and high speed loading are summarized in Table 1.

A second set of low strain rate tests has been performed on tubes at various strain rates at room temperature, 600 °C and at 800 °C [6]. In that study, their outer diameter and thickness equaled 4 mm and 250 μm, respectively. Although no viscosity effect was observed at room temperature and 600 °C, a significant one appeared at 800 °C. The results of strain rate sensitivity at room temperature and at 800 °C and the corresponding analyzed samples are summarized in Table 1.

Table 1 Results of tensile tests performed at room temperature and at 800 °C in our previous work [4–6], and corresponding analyzed samples

Results of tensile tests					Analyzed samples		
Initial states	Tests temperature (°C)	Strain rate range (s^{-1})	Strain rate sensitivity	Ref.	Strain rate (s^{-1})	Strain to fracture (%)	Ref.
As-received	25	1.3×10^{-2}–211.0	Yes	[4, 5]	211.0	16.4	A
TT_1	25	7.0×10^{-3}–74.0	No	[4, 5]	66.2	30.0	B
TT_1+TT_2	25	1.0×10^{-3}–206.0	Small	[5]	46.6	43.0	C
As-received	25	5.0×10^{-5}–5.7×10^{-3}	No	[6]	5.0×10^{-5}	8.6	D
As-received	800	1.2×10^{-4}–8.5×10^{-3}	Yes	[6]	8.5×10^{-3}	10.3	E

In order to investigate the evolution of the microstructure resulting from tensile tests and microstructural changes accompanying deformation, post-mortem EBSD analysis were carried out on broken samples. In addition of the crystallographic informations, Grain Orientation Spread (GOS) and Kernel Average Misorientation (KAM) analysis provide information on residual plastic strain. One of the two parts of the broken tubes was cut in a plane that passes through the tube axis. The parts were cold-mounted in epoxy and polished flat in a plane parallel to the tubes axis. All the samples were prepared using the same standard metallographic procedures that incorporated a final polish using 0.1 μm colloidal silica. It is well known that plastic strain degrades the quality of diffraction patterns. This is one of the reasons why careful preparation of sample surface is critical to successful EBSD analyses. The scanning was performed near the fracture areas. The EBSD patterns were indexed in the face-centered-cubic lattice corresponding to nickel crystallographic system (space group: Fm3m) using a 0.5 and 0.1 μm step sizes. The biggest precipitates are titanium nitrides TiN which presents a crystalline structure similar to that of Inconel 600. The same grain dilation cleanup was applied for all the scans. A grain was defined as a group of at least 30 neighboring indexed points having the same crystal orientation within a tolerated angular deviation of 5°. Twins were considered as grain boundaries in order to avoid overestimation of GOS. Grains are outlined with white boundaries. The same color scales, from 0° to 5° misorientation for KAM, are used for all the maps.

Local Misorientation (KAM Analysis)

KAM is used to represent the average misorientation between a given point of measure and its nearest neighbors which belongs to the same grain. Qualitative maps describing the organizing of the defect density have been established through KAM analysis for all the samples (A, B, C, D and E). Examples are given for the samples C and D in Figs. 1 and 2. The blue color in all the maps indicates very little misorientation, whereas the green, yellow, orange and red colors represent progressively increasing levels of local misorientation. Large scale KAM maps clearly show the most strained area of samples C and D (Fig. 1 on the left and Fig. 2 on the left). The crystallographic defect density is higher at the internal and external surfaces of tubular sample D than in the middle. Such inhomogeneity of local misorientation through the wall thickness has been also found for samples A and B. This inhomogeneity can be attributed to a local work hardening occurring during the drawing process of the initial tubes. Indeed, fine wall tubes are drawn through a mold using a guide bar to keep the tube straight and give the inner diameter after drawing. A significant plastic deformation occurs and still remains in these areas in the as-received state and after the tests. Residual strain is more homogeneous for sample C through its wall thickness because the double heat treatment restores the

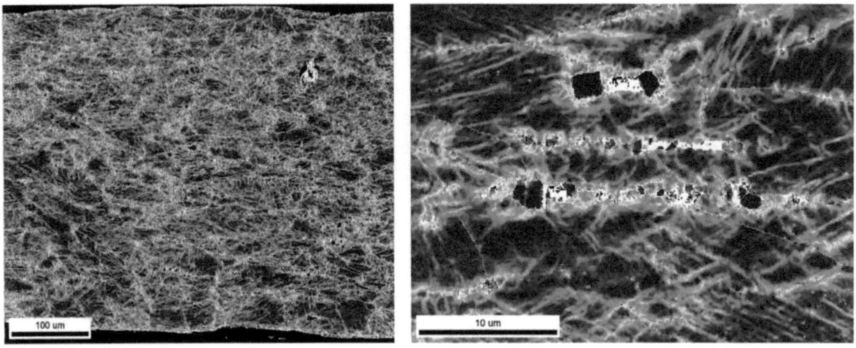

Fig. 1 Maps of KAM for sample C; step size = 0.5 μm on the *left*, step size = 0.1 μm on the *right*

material and recovery results in annihilation of dislocations. The hardening behavior is mainly modified by heat treatments which decreases the dislocation density. A lower dislocation density can promote a dynamic strain aging which manifested by strain rate effect at room temperature. These results are consistent with high velocity tensile tests which show that strain rate sensitivity disappeared for the samples having either simple or double heat treated states compared to the as-received state.

Moreover, KAM maps at a smaller scale were interesting to characterize the area surrounding precipitates, as shown in Fig. 1 on the right and Fig. 2 on the right. The bigger one (few μm) with angular edges are nitrides. In all cases, areas surrounding precipitates are high-strained regions as shown by the high value of KAM, whereas hard precipitates are very low- strained regions. When the strain rate is high (specimens A, B, and C), the microstructure is quite different from those resulting from lower strain rates (specimens D and E). Non-indexed areas are large cracks apparent in white in the maps. During heat treatment T_2, the amount of precipitates increased significantly. The precipitation of grain boundary carbides at

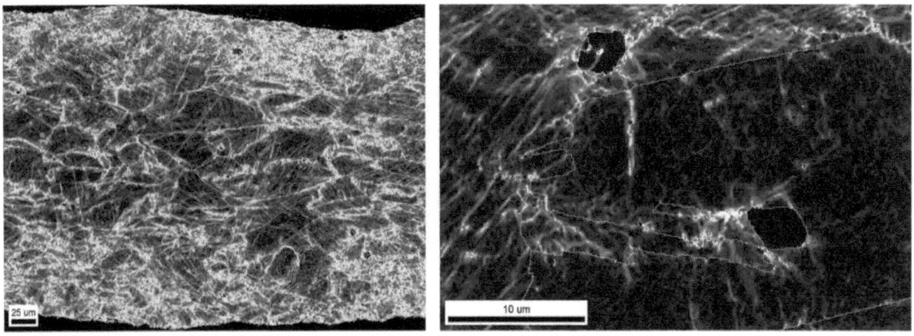

Fig. 2 Maps of KAM for sample D; step size = 0.5 μm on the *left*, step size = 0.1 μm on the *right*

high temperature is a common phenomenon in Ni-base alloys. In this austenitic solid-solution alloy, the precipitated phases present in the microstructure are titanium nitrides, titanium carbides and chromium carbides. Indeed, the annealing is followed by the natural cooling of the furnace. Our specimens are thus not rapidly cooled in the 540–760 °C temperature range and carbides precipitate in that range.

Quantitative Analysis

A comparative study based on quantitative information has been performed. Grain Orientation Spread, determined in each grain, is defined as the average orientation deviation between each point in a grain and the average orientation of the grain. Average KAM and GOS values for the large scale scans are summarized in Table 2 for the five specimens (A, B, C, D and E). Although the strain to fracture was higher for sample C than for samples A and B, average KAM and GOS value were lower. It can be explained by the recovery occurring during the double heat treatment before the tensile tests. The final resulting work hardening is lower, although the ductility was increased.

The results of GOS measurements for the specimen D, loaded at room temperature at $5.0 \times 10^{-5}\ s^{-1}$ to a strain of 8.6%, and the specimen E, loaded at 800 °C at $8.5 \times 10^{-3}\ s^{-1}$ to a strain of 10.3% are compared in Fig. 3. GOS distributions correspond to the fraction of the indexed area for each range of GOS angle in degrees. The low values of GOS for specimen E (between 0° and 2° of GOS) indicate that the resulting microstructure was partially recrystallized. Dynamic recrystallization occurred during tensile tests performed at 800 °C. Smaller average GOS and KAM values for sample E compared to sample D can be attributed to dynamic recovery and recrystallization at 800 °C during the tensile tests. These results are consistent with viscosity effect: a competition between work-hardening on the one hand, and dynamic recrystallization and recovery on the other hand, took place during the tests.

Table 2 Results of average KAM and GOS (°) values and the corresponding value of strain to fracture (%) for the five analyzed samples (large scale scans, step size = 0.5 μm)

Ref.	Average KAM value (°)	Average GOS value (°)	Strain to fracture (%)
A	1.13	6.81	16.4
B	1.18	7.22	30.0
C	0.99	6.74	43.0
D	1.63	7.12	8.6
E	1.13	5.16	10.3

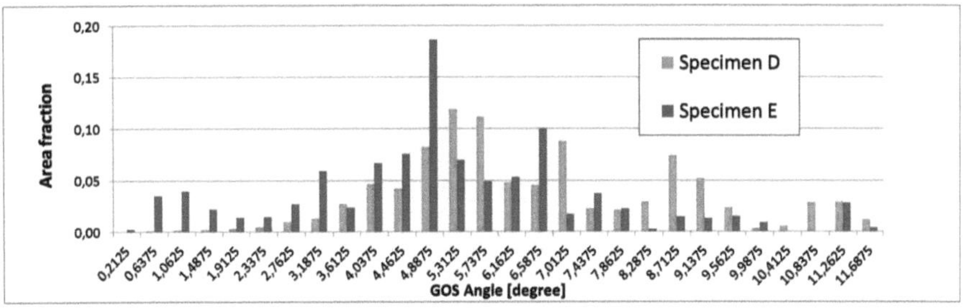

Fig. 3 Comparison of GOS distribution for samples D and E

Conclusions

Microstructural changes that occur when Inconel 600 is heat treated affect its mechanical behavior. Tensile tests performed at room temperature on samples in different initial states indicate that the loss of strain rate effect can be attributed to recovery occurring during heat treatments. The lower dislocation density can promote dynamic strain aging inducing strain rate effects at room temperature. The comparison of post-mortem EBSD analyses performed on samples with different initial states showed similar levels of resulting strain hardening, although their elongation at fracture were very different. Qualitative KAM analyses at different scales revealed high-strained areas around nitrides precipitates. Large cracks were observed between nitrides precipitates for the annealed tubes (TT_1+TT_2) loaded at high strain rate at room temperature. These cracks did not lead to premature fracture probably thanks to the enhanced ductility. Qualitative and quantitative analyses of KAM and GOS measurements highlighted dynamic recrystallization that occurred during tensile tests performed at 800 °C. A competition between work-hardening on one side, and dynamic recrystallization and recovery on the other side, took place leading to viscosity effect highlighted by tensile tests at low strain rate.

References

1. V. Marcadon, C. Davoine, B. Passilly, D. Boivin, F. Popoff, A. Rafray, S. Kruch, Mechanical behaviour of hollow-tube stackings: experimental characterization and modelling of the role of their constitutive material behaviour. Acta Mater. **60**, 5626–5644 (2012)
2. V. Marcadon, B. Passilly, C. Davoine, A. Götzfried, F.H. Leroy, About the use of microindentation to determine the constitutive material's mechanical behaviour of cellular structures. EPJ Web Conf. **6**, 04003 (2010)
3. C. Davoine, A. Götzfried, S. Mercier, F. Popoff, A. Rafray, M. Thomas, V. Marcadon, Metallic hollow sphere structures manufacturing process. In: MRS Proceedings, vol. 1188, Jan 2009, 1188-LL05-10

4. G. Portemont, C. Davoine, B. Langrand, V. Marcadon, Cellular materials made of stacked tubes: influence of the manufacturing process on the dynamic behaviour of the constitutive material. Part II: mechanical behavior, ODAS 2014
5. C. Davoine, G. Portemont, N. Horezan, B. Langrand, V. Marcadon, F. Popoff, Mechanical behaviour of hollow tube stackings: cellular materials made of stacked tubes: influence of the manufacturing process on the dynamic behavior of the constitutive material. Part I: microstructure, ODAS 2014
6. V. Marcadon, C. Davoine, D. Leveque, A. Rafray, F. Popoff, N. Horezan, D. Boivin, High temperature mechanical behavior of tubes stackings—Part I: microstructural and mechanical characterization of Inconel 600 constitutive material. Mater. Sci. Eng.: A **676** (In Progress) (31 Oct 2016)

Part V
Advanced High-Strength Steels

Influences of Thermomechanical Treatments on the Microcstructure Evolution and Mechanical Properties of Nano-precipitates Strengthened Steels

Yu Zhao, Songsong Xu, Hao Guo, Junpeng Li and Z.W. Zhang

Abstract Nano-precipitates strengthened steels exhibit excellent combination of high strength, high toughness and good weldability. In this study, the optimized thermomechanical processing and aging treatments are used to control simultaneously the matrix microstructure and nano scale precipitates. It is found that the mechanical properties are significantly influenced by thermomechanical processing. The aging temperature and aging time play a key role to control the strength of steel. XRD results reveal the transformation of phases after aging treatment was investigated. Optical microscope (OM) and scanning electron microscope (SEM) were used characterized microstructure and fracture modes. The effects of aging treatment and thermomechanical processing on the microstructure evolution and mechanical properties are also discussed.

Keywords Microstructure · Ferritic steel · Mechanical properties · Nano-precipitates

Introduction

Nano-cluster-strengthened ferritic steel with high strength, high toughness, good corrosion resistance and weldability, can be widely used in automobile, construction, bridging industry, etc. [1]. The ways of strengthen of nanoprecipitates strengthened ferritic steel were different from the martensitic steels with carbon hardness. The steel with high carbon content could form a brittle heat affected area next to the weld, which would lead to decrease in weldability property. The nano-precipitation strengthened steel with a low carbon content, and the strength can be greatly improved by the precipitation of nano-phases [2–5]. At the same

Y. Zhao · S. Xu · H. Guo · J. Li · Z.W. Zhang (✉)
Key Laboratory of Superlight Materials and Surface Technology, Ministry of Education, College of Materials Science and Chemical Engineering, Harbin Engineering University, Harbin 150001, People's Republic of China
e-mail: 356686211@qq.com

The Minerals, Metals & Materials Society, *TMS 2017 146th Annual Meeting & Exhibition Supplemental Proceedings*, The Minerals, Metals & Materials Series,
DOI 10.1007/978-3-319-51493-2_22

time, the maximum combination of the strength, plastic, weldability and impacting property of the material can also be guarantied [6]. As we all know, the Mechanical behaviors of studied steel were controlled by aging treatment. Different aging time and temperature have a significant effect on size and number density of nanophase precipitates [7].

In 1930s, researchers have found that the strength of steel could be improved by adding copper. When heat treatment temperature was set in the range of 450–600 °C, it could lead to the precipitation of Cu element, and the yield strength could be increased by 100–200 MPa [8, 9] .It is shows that when the size of the nano-phase is 2.5–3.0 nm, the precipitation hardening peak could be obtained [8]. In general, when the nano-particles grow to a larger size (>10 nm), the crystal structure of the particles transforms form the body-centered cubic (BBC) structure to the face centered cubic (FCC) structure [9, 10]. Some scholars also found that by optimizing the content of alloy elements and thermomechanical processing, a proper copper rich precipitates and matrix microstructure can be obtained, leading to excellent mechanical properties [11]. For this ferritic steel, the main deformation mechanism is dislocation glide and climb. The hardening and strengthening of metals were obtained result from precipitates in which hinder the movement of dislocation. Meanwhile, excellent plasticity and fracture toughness were illustrated by the interaction of precipitates and dislocations in the body-centered cubic (bcc) ferritic matrix [12–14]. When we enhance the aging temperature, the energy of dislocation increase and fast move, while internal stress enhanced and dislocation sources started to move [15]. Then the frequency of the screw dislocation crossing sliding was increased, then some of screw dislocations attracted each other and disappeared automatically. At the same time, the frequency of Edge dislocation climbing was also improve, it will give rise to the dissolve rate of unlike dislocations as well. At different temperature, the dislocation configuration will progress from dislocation wall to dislocation cell during a certain time creep. Furthermore, dislocation cell trend is more obvious with increasing temperature and time [16–18]. This shows that the influence of thermal effect on dislocation configuration is significant. Secondly, temperature increases reduce fault energy, and under the external force, the local stress gradient makes the whole dislocation decompose and grow a large number of faults [19].

In this study, the aging temperature and time were controlled to regulate the size and number density of nanophase. The mechanical properties was evaluated mainly in quasi static tensile and microhardness test. XRD was used analysis the change of phases, optical microscope (OM) were reveal the microstructure, and scanning electron microscope (SEM) showed the appearance of fracture.

Table 1 Alloy compositions of the steels studied (wt%)

Fe	Cu	Mn	Ni	Al	B	Ti	Si	Mo	Cr
Bal	2.5	2	4.0	1.0	0.02	0.05	0.2	1	1

Experimental

High purity alloying elements of the steel were melted and cast in an induction melting furnace, resulting in an ingot of 200 kg. The alloy compositions of this steel are listed in Table 1.

The as-cast were hot rolled at 1000 °C for multiple passes after cutting the front and back of the ingot. The final thickness of the steel plate was 12 mm and the total reduction of 80%, following were fastly quenched by water. The as-rolled samples were directly aged at 500 °C, 550 °C and 600 °C for 1 h, 2 h, 5 h and 10 h, respectively. Then the aged samples were quenched by water.

The average Vickers hardness measurement was conducted with a load of 1000 g for 15 s, and every sample was test over 10 different locations. The rod tensile samples were processed into the specimens with 5 mm in diameter and 25.4 in gauge length by numerically controlled lathe. The tensile tests were conducted on INSTRON 5565 at room temperature with a strain rate of $\sim 1 \times 10^{-3}$/s. The yield strength was determined with the 0.2% off set plastic strain method.

The phase components of the specimens were determined by X-ray diffraction (XRD) with the following experimental parameters: a Cu Target K alpha ray, tube pressure is 40 kV tube pressure, 20°–90° scanning angle, and 5°/min of scanning speed. Optical microscopy (OM) and scanning electron microscope (SEM) were also used to characterize the microstructure and fracture morphology of the specimens. The OM and SEM samples were polished to a final surface finish of 0.2 μm by standard mechanical polishing procedures. Then the samples were etched for approximately 5–15 s with a 4 vol.% Nital solution.

Results and Discussion

Microstructure Characterization

The microstructure of the as-rolled is shown in Fig. 1 After rolling, the dreich ferritic grains were emerged along the rolling direction. There was not clear the second phase in the matrix.

Figure 2 shows the microstructures of the samples aged at 500 °C for 1 h, 2 h, 5 h and 10 h, respectively. Compared with the metallographs of as-rolled sample, the number of short-range rolling layers and broken grains significantly decrease. Then it can be seen that the metallographs is mainly composed of dreich and polygonal ferrite grain with different size. Along the aging time, the microstructures of the matrix do not change clearly. At the same time, the grain size of samples was increased slightly.

Fig. 1 Microstructures of the as-rolled samples

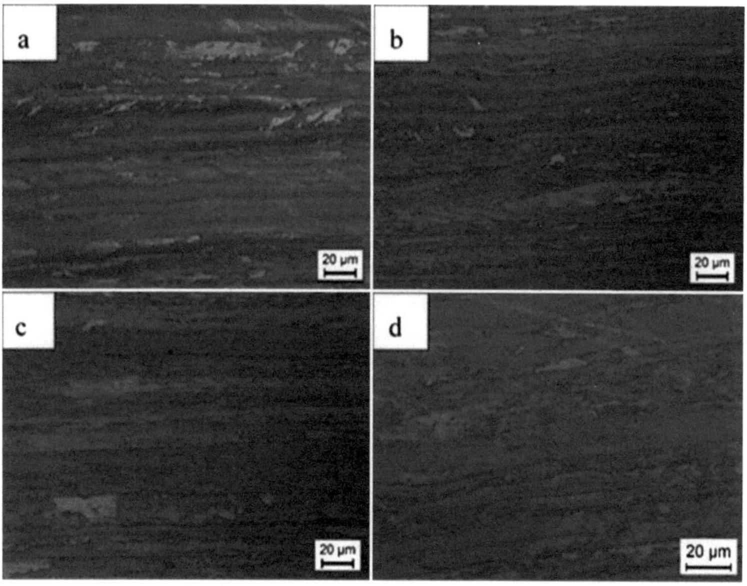

Fig. 2 Microstructures of the samples after aging at 500 °C: **a** 1 h, **b** 2 h, **c** 5 h and **d** 10 h

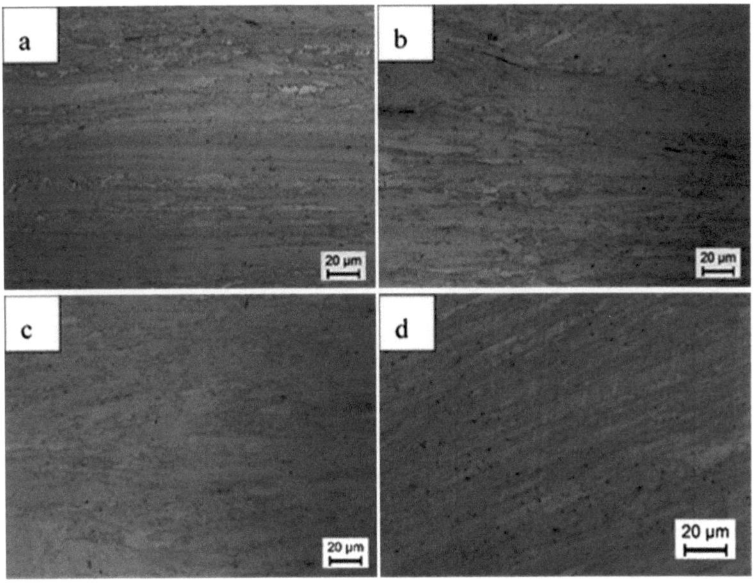

Fig. 3 Microstructures of the samples after aging at 550 °C: **a** 1 h, **b** 2 h, **c** 5 h and **d** 10 h

Figures 3 and 4 shows the microstructures of the samples aged at 550 °C and 600 °C for 1 h, 2 h, 5 h and 10 h, respectively. Increasing the aging temperature to 550 °C or 600 °C, the microstructures are similar to that after aging at 500°C, while

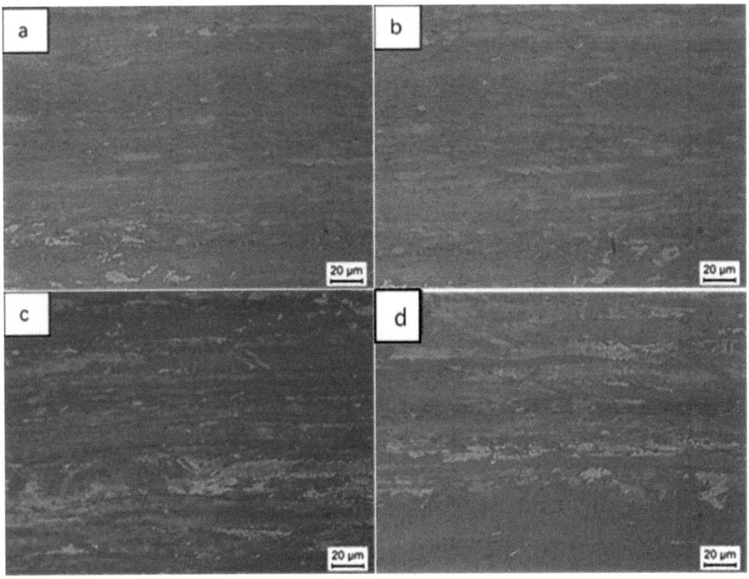

Fig. 4 Microstructures of the samples after aging at 600 °C: **a** 1 h, **b** 2 h, **c** 5 h and **d** 10 h

the grain size grows slightly to ~10 μm and ~20 μm for 550 °C and 600 °C, respectively. It can be concluded that aging treatment at different temperatures do not change the microstructures significantly. The microstructures of aging samples keep the same pattern to the as-rolled samples.

XRD Analysis

Figure 5 shows the diffraction spectra of this studied steel after various aging treatments. From the results, it can be conclude that increasing the aging time and aging temperature, the diffraction spectra remain unchanged. Only three peaks representing ferrite are found in the matrix, there no another phase was found. Due to the nanophase is very small, its size just 2–10 nm, so it is hard to characterize by XRD. Different aging treatments were compared under the same matrix phase.

Mechanical Properties Characterization

The microhardness testing results are shown in Fig. 6. The microhardness of the as-rolled samples is about 275 HV. After aging treatment at different temperature, the microhardness of the steel increases significantly. So the presence of nano-phase

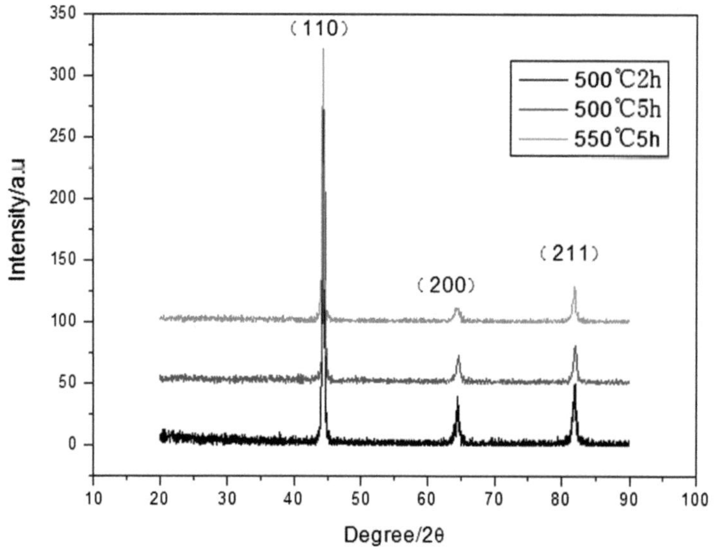

Fig. 5 X-ray diffraction spectra for different aging treatment of the nanophase strengthen ferritic steel

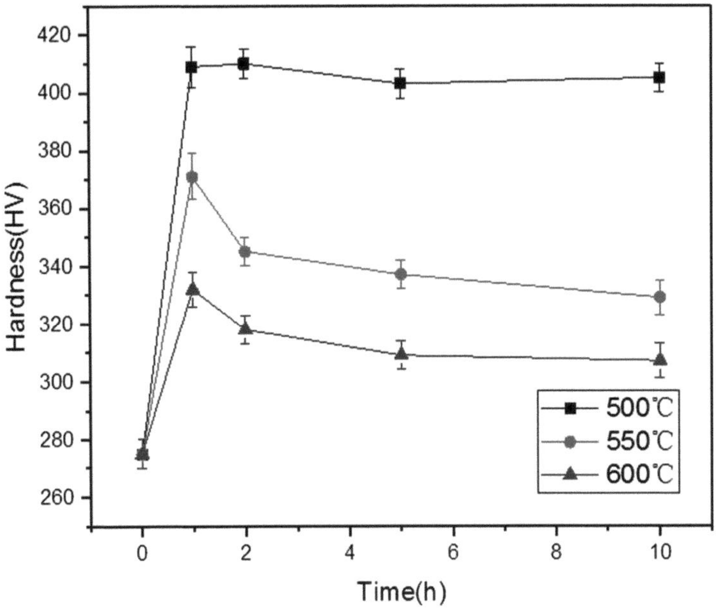

Fig. 6 Microhardness as a function of aging time at different temperature (the time 0 h is solid solution sample)

is the mainly result of improving hardness. For aging at 500 °C, the highest microhardness of specimens was appeared at 2 h, and it increase to 412 HV. With enlarging the aging time, the microhardness decreased slightly. But the hardness was higher than that of as-rolled sample. When the aging temperature was improved to 550 and 600 °C, the HVN was reduced clearly. However, the peak of hardness was taken forward to 1 h. These phenomenas are similar to the traditional aging processes in aged-strengthened alloys. Although the temperature was increased to 600 °C, and the aging time was along to 10 h, the microhardness of aging samples keep higher than the as-rolled samples. Therefore, in this steel, the hardening effect of nanophases is remarkable.

The effect of aging time and temperature on tensile properties is shown in Figs. 7 and 8, respectively. In Fig. 7a, the tensile properties of samples aged at 500 °C for 1 h, 5 h, 10 h. The highest tensile strength of 1206 MPa can be obtained after aging for 2 h, and the elongation was 12%. When improving the time, the strength decrease while elongation increase. When aging treatment at 500 °C, the strength decrease slowly with prolong the aging time. Compare with the samples of aging at 550 °C in Fig. 7b, the peak strength was obtained at 1 h, and the stress declined to 1089 MPa. Different from aging at 500 °C, the stress of samples aging at 550 °C was deduced observably with aging time. The strength of the sample aging 10 h

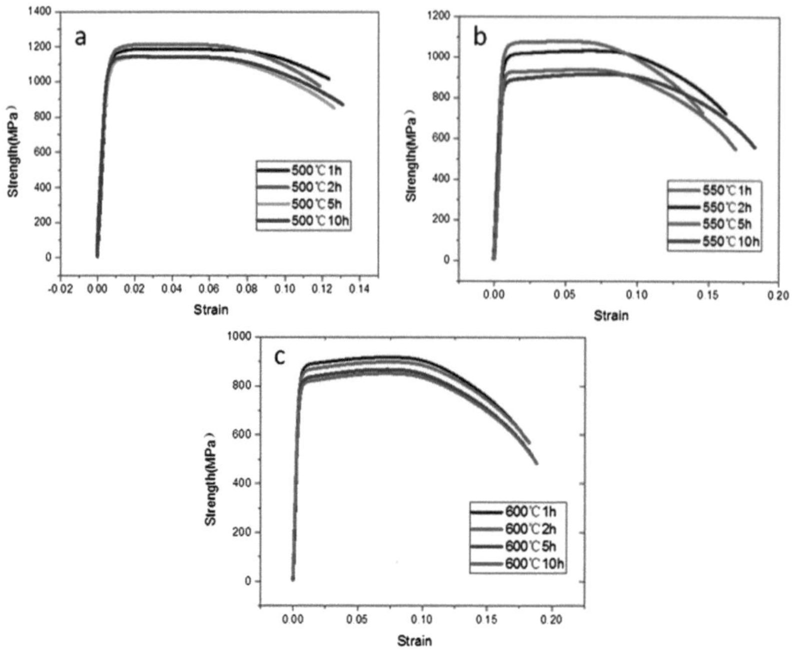

Fig. 7 Tensile properties of the investigated steels

brings down to 880 MPa and the ductility improved to 18%. So the effect of aging time on tensile property of the steels aging at 550 °C was remarkabler than that of aging at 500 °C. When increasing the temperature to 600 °C, the strength will decrease ulteriorly. And the effect of time on strength is similar to the samples aging at 500 °C. Enlarging the aging time at 600 °C, the strength reduced lightly. Comprehensive analysis above result, it can be conclude that the aging temperature and aging time have an important affect on mechanical properties of nano-precipitates strengthened steels. Control the temperature and time of aging treatment, it means control the nucleation and grown of nano-phase. Therefore, the optimal matching of size and number density of nano-precipitates can obtain the best mechanical properties.

It is known that the aging process leads to the formation of nano-scale Cu-rich precipitates, which can enhance the strength significantly by pinning the dislocations during deformation. The increasement and interaction of dislocations improved the strength of steel. While due to the core of the nano-phase is riched Cu cluster, and the dislocations can cut through the nano-phase. So the strength increased without sacrificing the elongation.

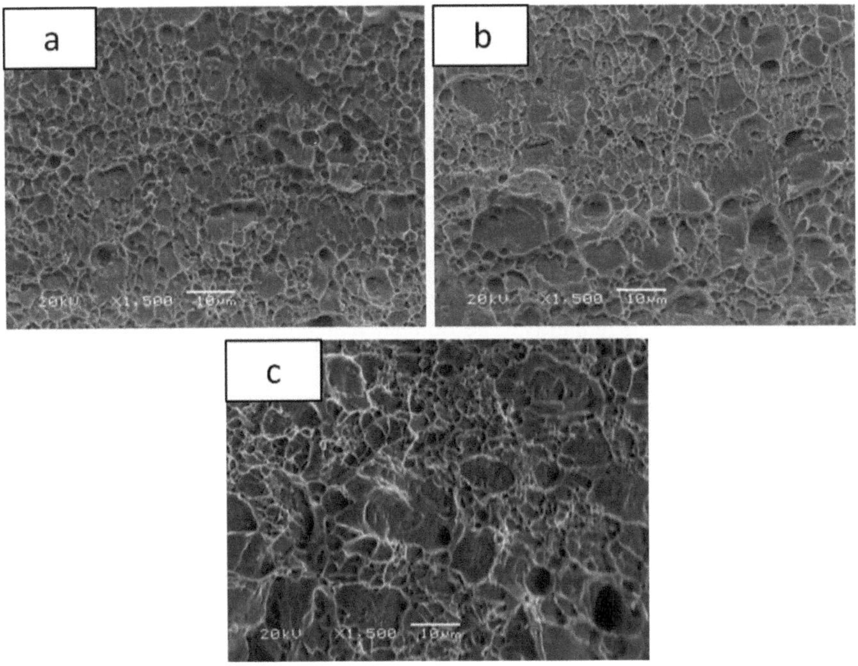

Fig. 8 Tensile fracture morphologies of the samples aged at different temperature for 5 h: **a** 500 °C, **b** 550 °C, **c** 600 °C

SEM Analysis of Tensile Fracture

Figure 8 shows the fracture surface of the sample after aging treatment at diverse temperature and same holding time (5 h). It can be seen that the fracture morphology of the studied steel is a dimple pattern. The fracture morphology remains the same with increasing the aging temperature. However, the size of dimples was improved with temperature. As we know, the mainly micro-morphology of the ductile fracture metal is the existence of the dimples which are formed by the aggregation of the holes. These cavities went through nucleating, growing up and then gather to the fracture under extra loading. Therefore, this steel kept microvoid coalescence fracture, which belongs to plastic fracture.

Conclusions

Nano-precipitates strengthened steels as important structural material, and it have high application prospect. In this study, the effect of aging temperature and aging time on mechanical behavior was investigated. The present work has revealed that:

1. The aging treatment has no significant effect on microstructure. Increasing the aging temperature and time do not cause the change of microstructure, and just enlarge the size of ferritic. The matrix phase do not transform with changing the aging process, it remain ferritic phase.
2. The microhardness was effect significantly by aging treatment. The hardness of all aged samples was higher than as-rolled samples. There will be a peak of hardness with increasing aging time under same temperature.
3. The variation of strength was similar to the transformation of microhardness. When aging time is the same, the tensile strength decreased and the elongation increased with the increase of aging temperature. Under 550 °C, the aging time impact the strength remarkly. The optimal matching of strength and ductility was obtained at 550 °C for 2 h, and the strength was 1008 MPa, elongation was 17%.
4. All fracture mode of aging samples were microvoid coalescence fracture, there were plentiful dimples on the surface of fractures. Then the average size of dimples was increased with improving the temperature.

Acknowledgements This work was supported by the NSFC Funding (51371062 and U1460102), NSFHLJ (ZD201411), the Scientific Research Foundation for the Returned Overseas Chinese Scholars (Heilongjiang Province), the Project for Innovative Talents of Science and Technology of Harbin (2014RFXXJ006) and the Project of High-Tech Ships of Ministry of Industry and Information Technology of the People's Republic of China. Yu Zhao is benefited from the International Exchange Program of Harbin Engineering University for Innovation-oriented Talents Cultivation.

References

1. Z.W. Zhang, C.T. Liu, Y.R. Wen, A. Hirata, S. Guo, G. Chen, M.W. Chen, B.A. Chin, Metall. Mater. Trans. A **43**, 351–359 (2011)
2. Z.B. Jiao, J.H. Luan, M.K. Miller, C.T. Liu, Acta Mater. **97**, 58–67 (2015)
3. X.P. Wang, A.M. Zhao, Z.Z. Zhao, Y. Huang, L. Li, Q. He, Int J. Minerals Metall. Mater. **21**, 266–272 (2014)
4. W. Xu, P.E.J. Rivera-Díaz-Del-Castillo, W. Yan, K. Yang, D.S. Martín, L.A.I. Kestens, S.V.D. Zwaag, Acta Mater. **58**, 4067–4075 (2010)
5. Z.W. Zhang, C.T. Liu, S. Guo, J.L. Cheng, G. Chen, T. Fujita, M.W. Chen, Y.-W. Chung, S. Vaynman, M.E. Fine, B.A. Chin, Mater. Sci. Eng. A **528**, 855–859 (2011)
6. M. Zrilic, V. Grabulov, Z. Burzic, M. Arsic, S. Sedmak, Int. J. Press. Vessels Pip. **84**, 139–150 (2007)
7. L. D'Angelo, L. D'Onofrio, G. Gonzalez, J. Alloy. Compd. **483**, 154–158 (2009)
8. S.K. Dhua, S.K. Sen, Mater. Sci. Eng. A **528**, 6356–6365 (2011)
9. S.P. Fitzgerald, S.L. Dudarev, J. Nucl. Mater. 386–388, 67–70 (2009)
10. Z. Zhang, C.T. Liu, M.K. Miller, X.L. Wang, Y. Wen, T. Fujita, A. Hirata, M. Chen, G. Chen, B.A. Chin, Sci. Rep. **3**, 975–978 (2013)
11. T. Harry, D.J. Bacon, Mrs Online Proc. Libr. **439**(2011)
12. V. Jablokov, D.M. Goto, D.A. Koss, J.B. Mckirgan, Mater. Sci. Eng. A **302**, 197–205 (2001)
13. M. Kapoor, D. Isheim, G. Ghosh, S. Vaynman, M.E. Fine, Y.-W. Chung, Acta Mater. **73**, 56–74 (2014)

14. S. Vaynman, D. Isheim, R.P. Kolli, S.P. Bhat, D.N. Seidman, M.E. Fine, Metall. Mater. Trans. Part A **39**, 363–373 (2008)
15. S.K. Mishra, S. Das, S. Ranganathan, Mater. Sci. Eng. A **323**, 285–292 (2002)
16. R.D.K. Misra, P. Venkatsurya, K.M. Wu, L.P. Karjalainen, Mater. Sci. Eng. A **560**, 693–699 (2013)
17. M.S. Rashid, Science **208**, 862–869 (1980)
18. P.K. Ray, R.I. Ganguly, A.K. Panda, Trans. Indian Inst. Met. (2003)
19. Z. Xiong, S. Liu, X. Wang, C. Shang, X. Li, R.D.K. Misra, Mater. Sci. Eng. A **636**, 117–123 (2015)

Effects of Solid Solution Treatment on the Microstructure and Mechanical Properties in the Ultra-High Strength Steel Strengthened by Nanoscale Particles

Songsong Xu, Yu Zhao, Hao Guo, Mingxing Qiu, Jing Zhang, Junpeng Li and Zhongwu Zhang

Abstract The microstructure and mechanical properties of the ultra-high strength steel by nanoscale particles are studied in detail with different solid solution treatment. The results indicated that the microstructure has been transformed from ferrite bands to sorbite and polygonal ferrite with the solid solution's temperature increasing from 600 to 950 °C. The microhardness has three part which are ascent, descent stage and peak value, what is more, the tensile strength is corresponding to it. The relationship of the microstructure and tensile strength has been investigated and discussion with solid solution in this paper.

Keywords Solid solution treatment • Ultra-high strength steel • Nanoscale precipitation

Introduction

With the requirement of the economic development and national defence industry, the ultrahigh-strength steels with high strength, good weldability and corrosion resistance are urgent to develop [1–4]. Carbon is the main element in the traditional ultrahigh-strength steels (such as HY-80), leading to high strength. But the steels are difficult to weld and require preheating before welding [5, 6]. The problem can be solved by low carbon content and enhancing the strength by nanoscale Cu-rich precipitates [7–11]. The high strength low allow (HSLA) steels by nanoscale Cu-rich precipitates with increasingly developed have been applied widely to shipbuilding and marine engineering, and infrastructure construction, etc. [6, 12, 13]. It would

S. Xu · Y. Zhao · H. Guo · M. Qiu · J. Zhang · J. Li · Z. Zhang (✉)
Key Laboratory of Superlight Materials and Surface Technology, Ministry of Education, College of Materials Science and Chemical Engineering, Harbin Engineering University, Harbin 150001, China
e-mail: zwzhang@hrbeu.edu.cn

S. Xu
e-mail: 554059942@qq.com

The Minerals, Metals & Materials Society, *TMS 2017 146th Annual Meeting & Exhibition Supplemental Proceedings*, The Minerals, Metals & Materials Series,
DOI 10.1007/978-3-319-51493-2_23

cause profound effects on mechanical properties of this HSLA steels that the nanoscale Cu-rich precipitates' size, distribution and the morphology. Z.W. Zhang has used the atom probe tomography and small-angle neutron scattering to represent the individual embryos and the statistical averages of nanoscale Cu-rich precipitation, and find out that how the nanoscale Cu-rich precipitates nucleate from quenche-in Cu embryos [14]. By utilized the atom-probe tomography and X-ray diffraction, Michael D. Mulholland found that co-located Cu precipitates and M_2C carbide strengthening precipitates have been quantified, and the nucleation and growth of M_2C carbide precipitate can offset coarsening of Cu precipitate [15].

Although the nanoscale precipitation appears in the HSLA steels and can solve this problem, the demand for the matrix microstructure in it is accordingly strict. The HSLA steels' matrix microstructure will be changed with the variation of cooling rates, austenitizing temperature and alloy chemistry [16–21]. By austenite decomposing during continuous cooling rate, S.W. Thompson [22] has made the CCT diagram for the HSLA-80 steel (in Fig. 1) and found that the austenite would transform to polygonal ferrite, widmanstatten ferrite, granular ferrite, acicular ferrite, upper bainite, lower bainite and martesite. By varying the continuous cooling rates and using three-dimensional electron backscatter diffraction and transmission electron microscopy, J.S. Kang [23] investigated that granular bainite, acicular ferrite and lath-type bainite will appear in the HSLA steel at the different cooling rates (in Fig. 2). In the present study, by comparably evaluating the microstructure and mechanical properties of the ultra-high strength steel, solid solution effects were investigation.

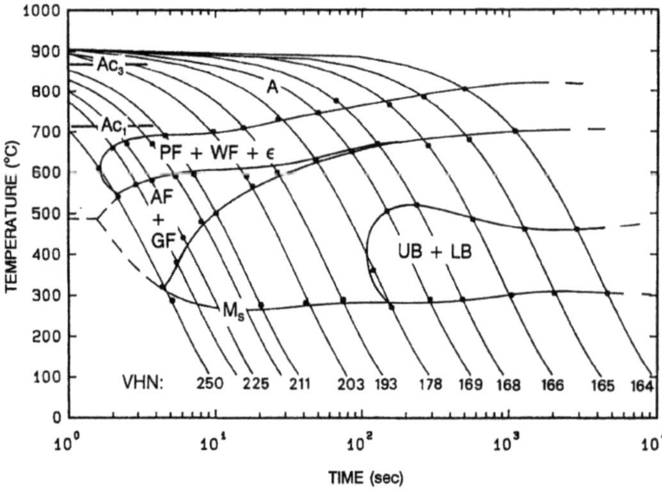

Fig. 1 [22] CCT diagram for the HSLA-80 steel

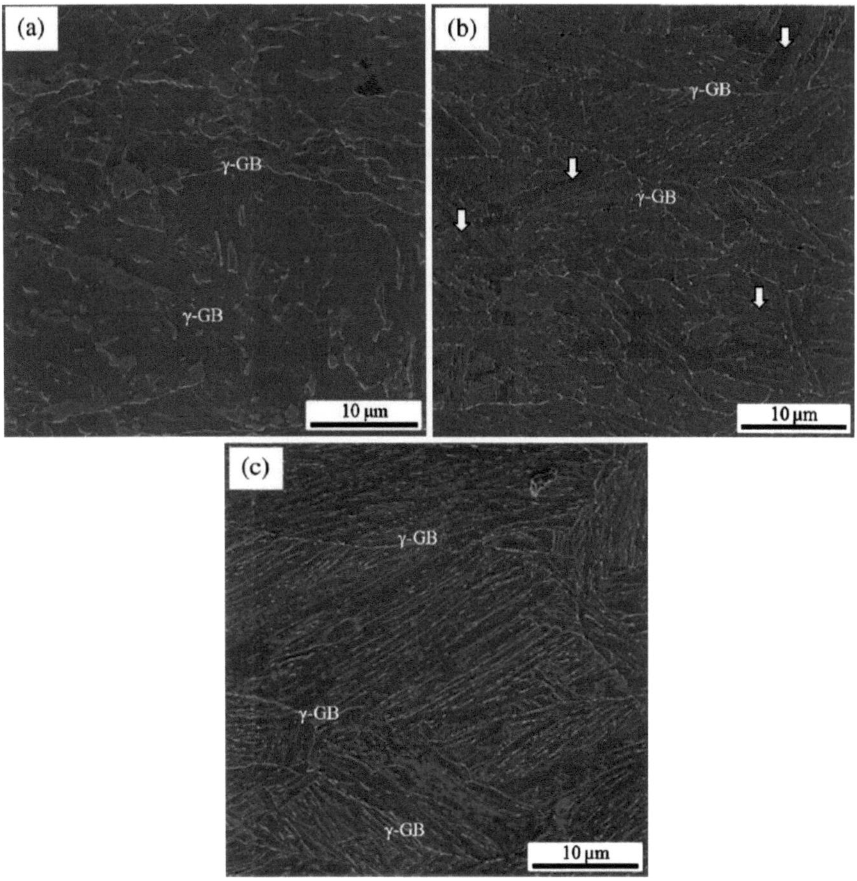

Fig. 2 [23] SEM micrographs of bainitic matrix in the present steel subjected to different continuous cooling rates; **a** granular bainite, **b** mixture of granular bainite and needle-shaped grains and **c** lath-type bainitic grains

Experimental Methods

The USS, with compositions of Fe-1.5Mn-2.5Cu-4Ni-1Al-0.005B-1.5Mo-0.05Nb-0.1Ti-1.5W-0.08C-0.5Si, were prepared by arc melting high-purity elemental metals (99.95 Fe, 99.8 Mn, 99.99 Ni, 99.99 Cu, 99.99 Al wt%). The ingots were remelted five times to ensure chemical homogeneity and drop-cast into 40 kg slabs. The results plates were bloomed at 1000 °C and then finish rolled at 800 °C by multiple passes with a total reduction of ~70% (hot-rolled). To identify the effects of solid solution treatment on microstructure and mechanical properties of the ultra-high strength steel, the samples were solid solution at a range temperature form 700 to 950 °C with 1 h. And the samples were labeled as 700-1, 750-1, 800-1,

850-1, 900-1, 950-1, respectively. Then the samples with solid solution would all age at 500 °C for 5 h and respectively labeled as hot-rolled AG, 700-1 AG, 850-1 AG, 950-1 AG.

Hardness measurements were conducted on a Vickers tester with a load of 1 kg for 15 s. For each specimen, at least ten indents were measured to obtain an average value. Tensile tests along the rolling direction of the samples were conducted on an Instron 5565 testing machine at a strain rate of 10^{-3} s^{-1}. Three specimens were tested in each condition and the average values were reported. The gauge length, width and thickness of the tensile specimens are 12.5, 4 and 1 mm, respectively. The yield strength was determined with the 0.2% offset plastic strain method. The matrix microstructures were characterized using optical microscopy (OM).

Results and Discussion

Microstructure

Light micrographs of experimental steels are present in Fig. 3. For the hot-rolled steel (in the Fig. 3a), the microstructure consisted of band structure and island grain. The phase has been confirmed to be ferrite. After solid solution at 700 °C for 1 h, the band structure was the most of microstructure and it has no island grain in the Fig. 3b. However, the ferrite has transformed to the sorbite, which is extraordinary homogeneous and fine. As we can see in the Fig. 3, the grain size of the sorbite was gradually increased when the temperature was changed from 700 to 850 °C. In the Fig. 3f, the polygonal ferrite was obtained in the steel after solid solution at 900 °C for 1 h. polygonal ferrite has grown with the incense of solid solution temperature.

The deformation temperature was in the austenite non-recrystallization zone. The austenite grains were compressed during rolling, so band structure with island grain were obtained. Based on the iron carbon metal alloy equilibrium diagram, the austenitizing temperature is about 900 °C. So the solid solution from 700 to 850 °C was similar to the high temperature tempering, the ferrite would transform to the sorbite. When the temperature was over 900 °C, the steel would have been austenitizing and formed the polygonal ferrite.

Mechanical Properties

The microhardness of the steel with various solid solution treatment is shown in Fig. 4 as a function of solid solution temperature. The hardness of the steel decreases from ~390 to ~290 HV when the hot-rolled steel has been solid solution at 700 °C for 1 h. And the hardness gradually reaches peak values of ~340HV with

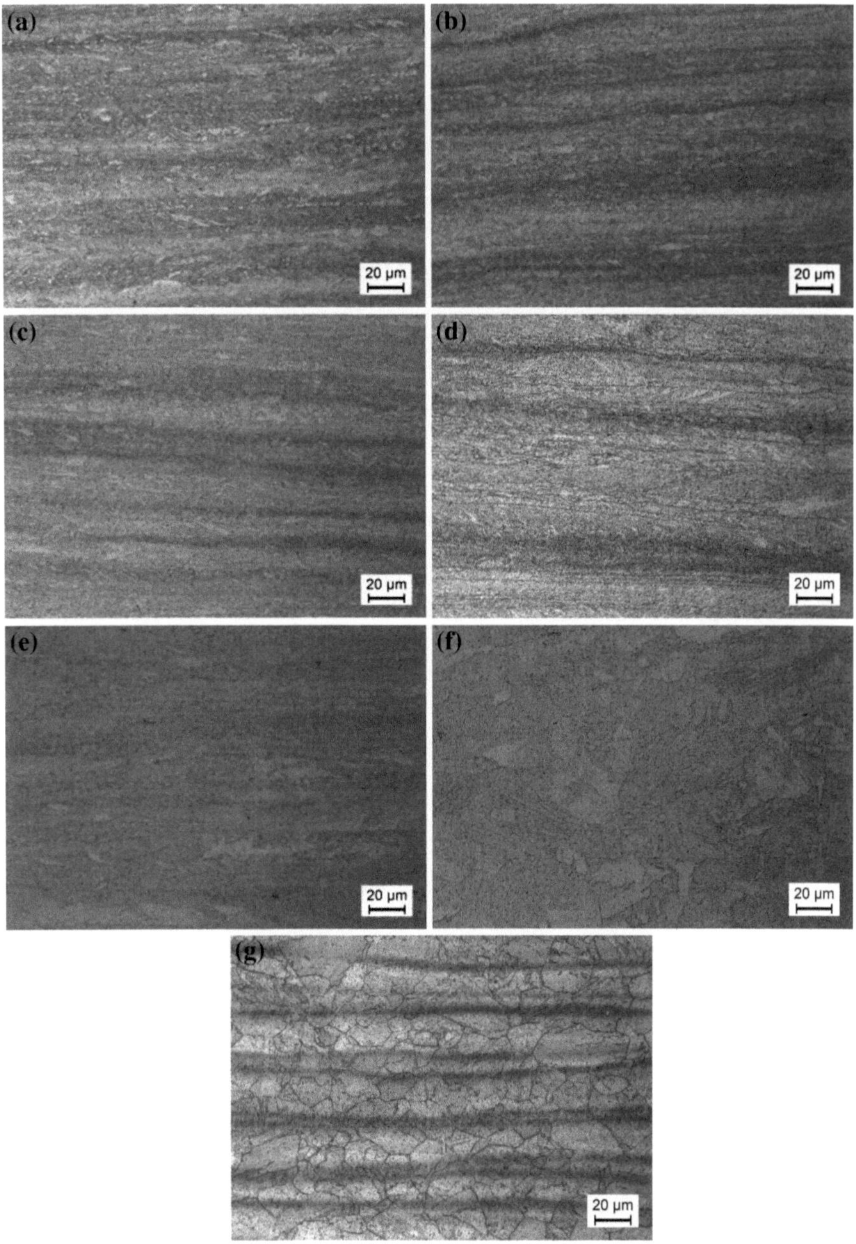

Fig. 3 Microstructures of **a** Hot-rolled steel; **b** 700-1 steel; **c** 750-1 steel; **d** 800-1 steel; **e** 850-1 steel; **f** 900-1 steel; **g** 950-1 steel

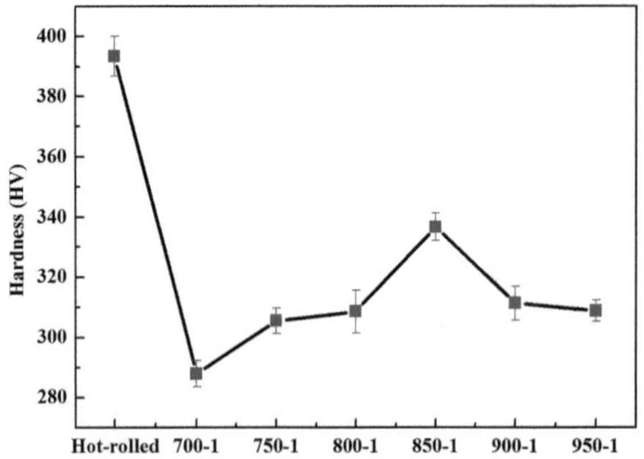

Fig. 4 Hardness as a function of solid solution treatment for the steel

solid solution at 850 °C for 1 h. The temperature surpass 850 °C and leads to the decrease of the hardness to ~300 HV.

Room-temperature tensile tests were performed to further investigate the mechanical behavior of the steel. The engineering stress-strain curve of the steel is shown in Fig. 5. Comparing the steels with various solid solution treatment, the hot-rolled steel have the highest yield strength of ~1100 MPa. The 700-1 steel with the sorbite has yield strength of ~800 MPa, however, the elongation-to-failure of ~26%. While the 850-1 steel has the maximum hardness, it also has a high yield strength of ~1050 MPa. The steel after solid solution at 950 °C for 1 h shows a yield strength of ~1000 MPa. For the solid solution samples with aging at 500 °C for 5 h,

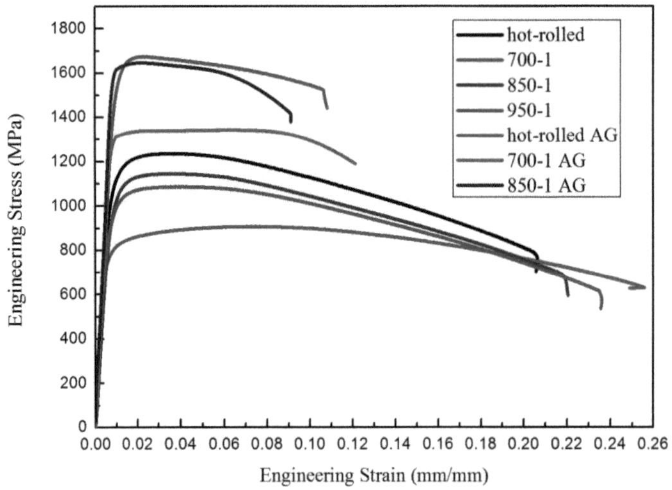

Fig. 5 Room-temperature tensile stress-strain curves of the steel

the yield strength all show a increase of ~500 MPa, on account to the nanoscale precipitation.

After solid solution at 700 °C, the hot-rolled steel's microstructure has changed form ferrite to sorbite. And the sorbite were more uniform and fine than the ferrite, so the 700-1 steel shows lower hardness and yield strength, but has better elongation. With the increase of solid temperature, the grain size of sorbite was bigger. Therefore, the 850-1 steel has higher hardness and yield strength. When the temperature was over 900 °C, the steel would be austenitized and transformed to polygonal ferrite. This is the reason why the 950-1 steel have lower yield strength.

Fracture

The fracture surfaces of the steel are shown in Fig. 6. The hot-rolled show a microviod coalescence fracture mode with bimodal dimples, i.e. both large and small dimples, indicating a ductile fracture mode. The other steels exhibit a similar fracture mode, but the proportion of large dimples is reduced.

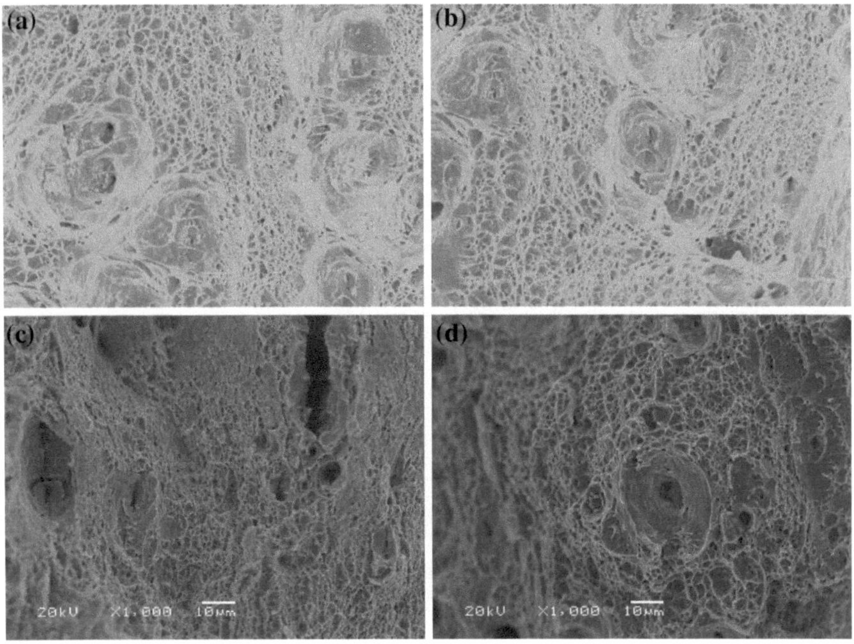

Fig. 6 Room-temperature fracture surfaces of **a** Hot-rolled steel; **b** 700-1 steel; **c** 850-1 steel; **d** 950-1 steel

Conclusion

Based on the systematic study of the effects of solid solution treatment on microstructure and mechanical properties in the ultra-high strength steel strengthened by nanoscale particles, the following conclusion are drawn: With the solid solution temperature's changing from 700 to 950 °C, the lath ferrite has transformed to the sorbite, and finally to the polygonal ferrite. And the steel shows better mechanical properties after solid solution treatment at 850 °C for 1 h, which has the sorbite phase, precipitation strengthening in the steel after aging could contribute ~500 MPa to the yield strength. Therefore, the ultra-high strength steel strengthen by nanoscale precipitations with the sorbite microstructure would have better mechanical properties after aging.

Acknowledgements This work was supported by the NSFC Funding (51371062 and U1460102), NSFHLJ (ZD201411), the Scientific Research Foundation for the Returned Overseas Chinese Scholars (Heilongjiang Province), the Project for Innovative Talents of Science and Technology of Harbin (2014RFXXJ006) and the Project of High-Tech Ships of Ministry of Industry and Information Technology of the People's Republic of China. Xu Songsong is benefited from the International Exchange Program of Harbin Engineering University for Innovation-oriented Talents Cultivation.

References

1. W. Xu, P.E.J. River-Diaz-del-Castillo et al., A new ultrahigh-strength stainless steel strengthened by various coexisting nanoprecipitates. Acta Mater. **58**, 4067–4075(2010)
2. C.S. Pande, M.A. Imam, Nucleation and growth kinetics in high strength low carbon ferrous alloys. Mater. Sci. Eng. A **457**, 69–76 (2007)
3. D. Raabe, D. Ponge, O. Dmitrieva, B. Sander, Nanoprecipitate-hardened 1.5 GPa steels with unexpected high ductility. Scr. Mater. **60**, 1141–1144 (2009)
4. J.H. Schneibel, C.T. Liu, M.K. Miller, M.J. Mills, P. Sarosi, M. Heilmaier, D. Sturm, Ultrafine-grained nanocluster-strengthened alloys with unusually high creep strength. Scr. Mater. **61**, 793–796 (2009)
5. Z.W. Zhang, C.T. Liu, S. Guo et al., Boron effects on the ductility of a nano-cluster-strengthened ferritic steel. Mater. Sci. Eng. A **528**, 855–859(2011)
6. S.K. Dhua, D. Mukerjee, D.S. Sarma, Influence of tempering on the microstructure and mechanical properties of HSLA-100 steel plates. Metall. Mater. Trans. A **32**, 2270
7. H.R. Habibi-Baguirani, Mater. Sci. Eng. A **338**, 142 (2002)
8. S. Vaynman, D. Isheim, R.P. Kolli, S.P. Bhat, D.N. Seidman, M.E. Fine, Metall. Mater. Trans. A **39**, 363 (2008)
9. S.K. Dhua, D. Mukerjee, D.S. Sarma, Metall. Mater. Trans. **34A**, 241 (2003)
10. M.E. Fine, D. Isheim, Scripta Mater. **53**, 115 (2005)
11. D. Isheim, M.S. Gagliano, M.E. Fine, D.N. Seidman, Acta Mater. **54**, 841 (2006)
12. D. Isheim, R. Prakash Kolli et al., An atom-probe. Scripta Mat. **55**, 35–40 (2006)
13. T.W. Montemarano, B.P. Sack, J.P. Gudas, M.G. Vassilaros, H.H. Vanderveldt, J. Ship Prod. **22**, 145–162 (1986)
14. Z.W. Zhang, C.T. Liu, X.-L. Wang et al., Phy. Rev. B **84**, 174114 (2011)
15. M.D. Mulholland, D.N. Seidman, Nanoscale co-precipitation and mechanical properties of a high-strength low-carbon steel. Acta Mater. **59**, 1881–1897 (2011)

16. G.C. Hwang, S. Lee et al., Effect of direct quenching on microstructure and mechanical properties of copper-bearing high-strength alloy steels. Mater. Sci. Eng. A **A252**, 256–268 (1998)
17. L. Ming, Q. Wang et al., A remarkable role of niobium precipitation in refining microstructure and improving toughness of A QT-treated 20CrMo47NbV steel with ultrahigh strength. Mater. Sci. Eng. A **613**, 240–249 (2014)
18. C. Borgia, T. Scharowsky, A. Furrer et al., A combinatorial study on the influence of elemental composition and heat treatment on the phase composition, microstructure and mechanical properties of Ni-W alloy thin films. Acta Mater. **59**, 386–399 (2011)
19. J. Kong, C. Xie, Effect of molybdenum on continuous cooling bainite transformation of low-carbon microalloyed steel. Mater. Des. **27**, 1169–1173 (2006)
20. H.J. Jun, J.S. Kang, D.H. Seo et al., Effect of deformation and boron on microstructure and continuous cooling transformation in low carbon HSLA steels. Mater. Sci. Eng. A **422**, 157–162 (2006)
21. S.K. Putatunda, C. Martis, J. Boileau, Influence of austempering temperature on the mechanical properties of a low carbon low alloy steel. Mater. Sci Eng. A **528**, 5053–5059 (2011)
22. S.W. Thompson, D.J. Colvin, G. Krauss, Metall. Mater. Trans. A **27A**, 1996–1557
23. J.S. Kang, J.-B. Seol, C.G. Park, Three-dimensional characterization of bainitic microstructures in low-carbon high-strength low-alloy steel studied by electron backscatter diffraction. Mater. Charact. **79**, 110–121 (2013)

Effects of Microstructure on the Strain Rate Sensitivity of Advanced Steels

Rakan Alturk, Steven Mates, Zeren Xu and Fadi Abu-Farha

Abstract The dependence of the strain rate sensitivity of advanced ~1 GPa tensile strength steels on the phases present in their microstructures was studied by testing different steels at 0.005 and 500 s^{-1}. The high strain rate tests were performed using a Kolsky bar setup, while the quasi-static tests were performed using a universal testing machine. The two main steels of interest were the Ferrite-Martensite DP980 and the Ferrite-Martensite-Austenite QP980; the latter being a transformation induced plasticity (TRIP) assisted steel. For comparison, ferritic CR5 mild steel and austenitic stainless steel 201 were also tested under the same conditions. Though the differences in the steel chemistries were not taken into account, the results obtained here suggest a strong relationship between the phase-content of the steel and its response to the changes in the loading rate. The relationships between the observed mechanical behavior and the phases present in the microstructure are discussed.

Keywords High rate deformation · Microstructure · AHSSs · Kolsky bar · Rate sensitivity

Introduction

With the ever-increasing push for lightweighting in the automotive sector, Advanced High Strength Steels (AHSSs) continue to infiltrate the industry due to their unique properties and merits [1]. The uniqueness of AHSSs can be primarily

---

R. Alturk (✉) · Z. Xu · F. Abu-Farha (✉)
Clemson University – International Center for Automotive Research (CU-ICAR), Greenville, SC, USA
e-mail: ralturk@clemson.edu

F. Abu-Farha
e-mail: FADI@clemson.edu

S. Mates
National Institute for Standards and Technology, Gaithersburg, MD, USA

The Minerals, Metals & Materials Society, *TMS 2017 146th Annual Meeting & Exhibition Supplemental Proceedings*, The Minerals, Metals & Materials Series,
DOI 10.1007/978-3-319-51493-2_24

attributed to their complex multi-phase microstructures, as compared to conventional single-phase steels, which enable achieving particular blends of strength and ductility. While strength drives the lightweighting potential of the material, ductility defines the practical limits of its stamping. However, beyond forming materials into the various automotive components, materials need to pass specific tests and meet certain criteria to be used in vehicles; some of the most critical ones are crashworthiness-related [2]. Finite element (FE) analyses are typically performed to simulate a crash event; the accuracy of these simulations rely heavily on how well the constitutive equations describe the material behavior.

Automotive materials undergo loading at different strain rates during their lifetime. From quasi-static (<0.1 s^{-1}) to intermediate (around 1 s^{-1}) during stamping, to high rates (>100 s^{-1}) during a crash event. It is well-established in the literature that the deformation behavior and the properties of steels are highly influenced by strain rate. Thus, the governing equations used in the various analyses should be able to incorporate rate effects on the behavior of the material.

Strain rate effects on the mechanical properties of steels have been studied by many researchers over the past decade; examples include these references [3–17]. However, the majority of these studies did not focus on the microstructure and the role it plays in dictating the strain rate sensitivity of the steel grade [3–14]. Few studies considered the microstructure and its correlation to rate sensitivity, particularly in advanced steel grades. Hwang et al. [15] studied the effects of ferrite grain size on the strain rate sensitivity in DP steels. Oliver et al. [16] tested DP and TRIP steels at 0.001 and 200 s^{-1} to compare the changes in the microstructure at different strain rates. Tarigopula et al. [3] studied the effect of strain rate on the mechanical properties and strain localization of DP800. Li et al. [17] studied the effects of strain rate on deformation-induced transformation of austenite in quenched and partitioned steels, up to strain rates of 0.1 s^{-1}.

This work investigates the response of selected advanced steels with complex multiphase microstructures to external loading at rates close to those encountered during a crash event. The response is compared to that at quasi-static rates to evaluate the rate sensitivity of each material. Moreover, the differences in their responses are correlated to the types of constituent phases present in their microstructures. This would be the first step to develop models that can incorporate rate effects in each of the various phases within the microstructure of advanced high strength steels. The main two materials of interest in this study are ~1 GPa tensile strength steel grades: DP980 and QP980. Nevertheless, the deformation of two single-phase steels, ferritic CR5 mild steel and austenitic stainless steel 201, are also investigated to provide some basis for comparison, and establish the rate dependency of the individual phases.

Experiments

Materials and Initial Microstructures

The initial microstructures of the four materials were examined by optical microscopy and scanning electron microscopy; the obtained micrographs are shown in Fig. 1. The CR5, a very ductile grade of mild steel, shows a typical ferrite microstructure with a heterogeneous distribution of small and large grains (~5–25 μm equivalent grain size). The SS 201 is an austenitic stainless steel, the microstructure of which exhibits similar heterogeneous distribution of grains, though larger than the CR5. The micrograph of DP980, a dual phase steel, clearly shows the distinction between ferrite (light) and martensite (dark) in the microstructure. The area fractions of the two phases (both are very fine) were measured to be ~30% ferrite and ~70% martensite. The QP980, on the other hand, is mainly composed of ~40% ferrite, ~50% martensite and ~10% retained austenite. The ferrite and martensite grains are fine (similar to the DP980), while the austenite islands are particularly small (~0.5–1 μm); an example is indicated in the micrograph in Fig. 1d. As noted, phase distinction in this material is very difficult; thus the reported phase fractions were obtained from both optical microscopy and EBSD maps (not presented here).

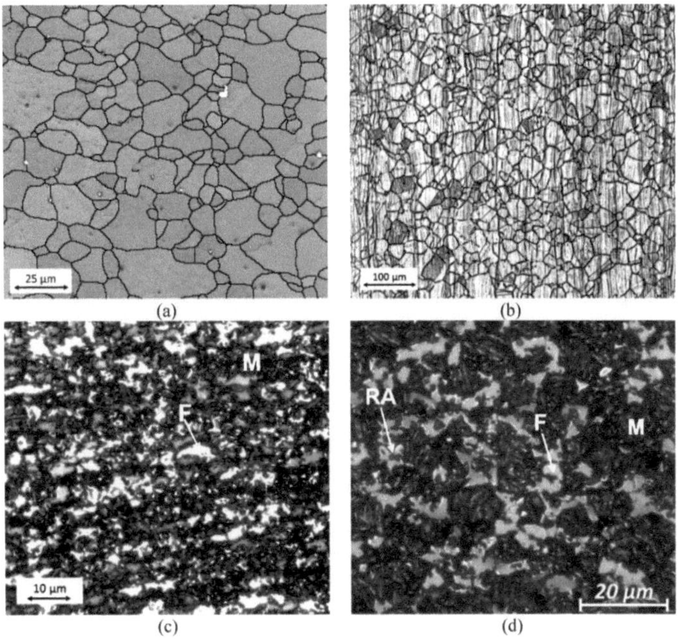

Fig. 1 Micrographs showing the microstructure of the as-received material for **a** CR5, **b** SS 201, **c** DP980 and **d** QP980

Experimental Setups and Procedure

The four materials were tested in tension under both quasi-static and high rate conditions using the setups shown in Fig. 2. The quasi-static tension tests were performed on a universal electromechanical load frame, at a constant crosshead speed of 18 mm/min. A standard ASTM E8 specimen geometry was used, and the corresponding strain rate of deformation was ~0.005 s^{-1}. For accurate strain measurements, a 3D digital image correlation (DIC) system was used to monitor material deformation during testing. High strain rate testing, on the other hand, was performed at the National Institute for Standards and Technology (NIST) using the Kolsky bar setup shown in Fig. 2b. Details on the testing can be found elsewhere [18]. For force measurements, 1000 Ω strain gauges were bonded to the transmission bar. For strain measurements, two DIC high speed cameras were used to capture images of the specimens during testing at a frame rate of 75,000 frames/s, at a resolution of 192 × 336 pixels. The actual strain rate obtained during Kolsky bar testing depends on the pressure, the incident bar length and the specimen geometry. Testing reported in this work was carried out at a particular set of conditions to achieve the desired strain rate; the pressure was set to ~30 psi, the incident bar length was chosen to be 1 m, and with a 10 mm gage length specimen, a strain rate of ~500 s^{-1} was achieved (measured by DIC). After testing, all DIC images, for both quasi-static and high rate testing, were processed in a consistent fashion using the same commercial DIC software to obtain the full-field history of strains during testing.

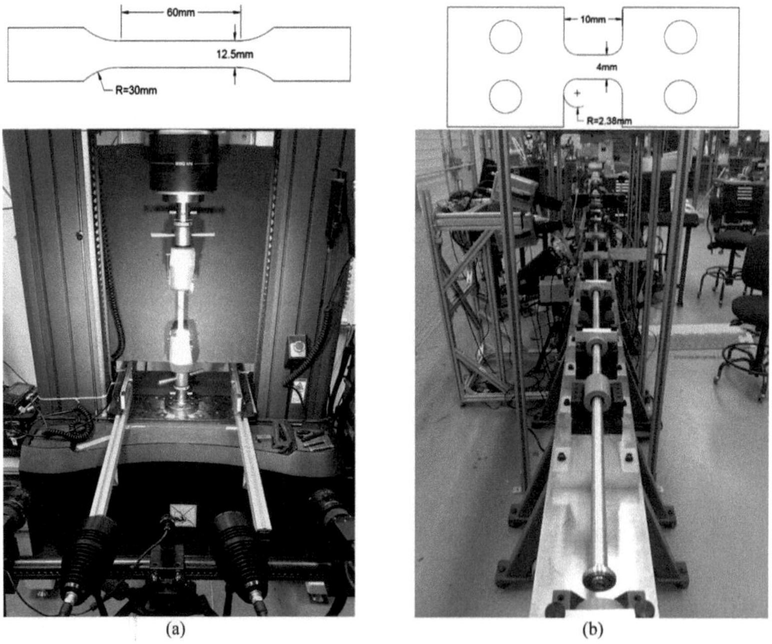

Fig. 2 Experimental setups and test specimens used for **a** quasi-static testing and **b** high rate testing (Kolsky)

All test specimens were prepared by waterjet cutting to obtain a good edge quality. While quasi-static testing is standardized (following the ASTM-E8), there is no clear standard for the specimen geometry to be used at high rates in Kolsky bar testing. Also, it is challenging to achieve stress equilibrium in the gauge area while keeping a uniaxial stress state as there is a tradeoff between stress uniformity and the desired maximum strain rate that is also influenced by the strain hardening behavior of the material. The geometry used here is based on previous optimization that was done in a separate study.

Results

Flow Behavior

Engineering stress-strain curves obtained at low and high strain rates are shown for all four materials in Fig. 3. Part (a) shows the clear distinction in behavior for CR5 at 0.005 and 500 s^{-1}. A significant increase in the yield strength and the ultimate

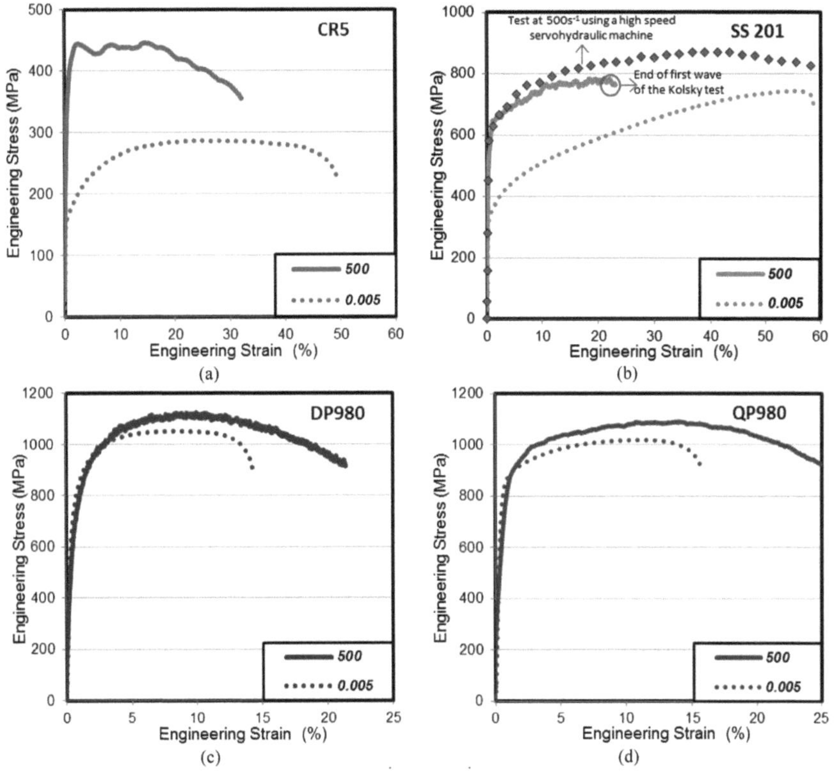

Fig. 3 Engineering stress/strain curves at 0.005 and 500 s^{-1} for **a** CR5, **b** SS 201, **c** DP980 and **d** QP980

strength is noted for the higher strain rate, accompanied by a reduction in the uniform and total elongation of the material; this is indicative of strong strain rate sensitivity. The shape of the stress-strain curve is also different at high strain rate; the material exhibits lower strain hardening at 500 s^{-1}. This is a typical behavior of mild steels and several researchers have reported the same behavior in previous studies [7, 11, 19]. Positive strain rate sensitivity is also observed in the austenitic stainless steel 201, with a significant increase in the yield strength. Important to note here is that, for the SS 201, fracture could not be achieved after the first pulse during Kolsky bar testing (due to the extreme ductility of the material), so the dynamic stress/strain curve shown in Fig. 3b is incomplete. It is therefore not possible to make a clear conclusion in regards to the tensile strength, uniform and total elongation of the material at ~500 s^{-1}. Nevertheless, testing was performed on this material in another study using a high speed servo-hydraulic machine, the results of which are included in Fig. 3b. The plot shows an increase in yield strength, a slight increase in the tensile strength and a drop in the uniform ductility; total elongation is similar to that at quasi-static conditions.

A different behavior is observed in the multiphase steels; the DP980 and QP980 (Fig. 3c, d); overall, there is a small difference between the quasi-static results and the high strain rate results. The yield and ultimate strengths change slightly with the increase in strain rate. Total elongation is notably larger at high rate in both cases; the change in the uniform elongation, however, is not as notable, and stays almost the same for the DP980. The stress/strain curves do not show the change in shape observed in CR5 and SS 201; i.e., significant drop in strain hardenability at higher rates. On the contrary, the DP980 in particular seems to strain harden more during high rate deformation, while QP980 does not show any notable change between the two rates.

Strain Rate Sensitivity

To take a closer quantitative look at the strain rate sensitivity, the percentage change in four mechanical properties between 0.005 and 500 s^{-1} was calculated and plotted in Fig. 4. Part (a) shows how strain rate affects the yield strength in the four materials. As expected, the largest increase in the yield strength was observed in CR5, which showed an increase of ~105%, followed by SS 201, which showed an increase of ~79%. On the other hand, the yield strength of DP980 and QP980 only increased by ~2% and ~9%, respectively. The same trend was observed when comparing the change of ultimate tensile strength in Fig. 4b, but the values were different. The difference in the values is due to the difference in the strain hardening behavior of the materials at high rates. As observed in the stress strain curves of CR5, the material doesn't show any significant hardening behavior in the high rate regime. While for SS 201, the material does harden at high rates but significantly less than in the quasi-static rate regime. As a result, the percentage increase in

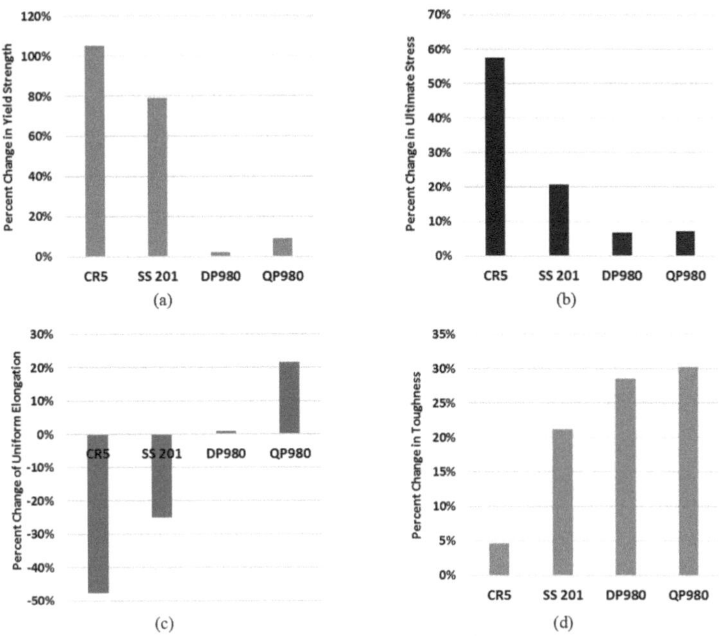

Fig. 4 Percent change between 0.005 and 500 s^{-1} of **a** Yield stress, **b** Ultimate tensile strength, **c** Uniform elongation and **d** toughness

ultimate tensile strength for CR5 was ~57%, and that of SS 201 was ~21%. For DP980 and QP980, strain hardenability was not significantly affected at high rates, confirmed by the similar percentage increase of both the ultimate tensile strength and the yield strength for the two materials (~7%).

Uniform elongation is an indication of material ductility and has a strong influence on the energy absorption capabilities of the materials, both of which are very important properties to be considered when selecting materials to be used on structural members in vehicles. As in the case of tensile strength, the phases in the microstructure affect the response of the material to changes in strain rate. Figure 4c shows the percentage change in uniform elongation for the four materials. The uniform elongation of both CR5 and SS 201 decreased at high rates, agreeing with the classic theory of strain rate sensitivity, which dictates that ductility decreases with the increase in strain rate (for materials with a positive strain rate sensitivity). DP980 showed almost no change in the uniform elongation which agrees with the insignificant strain rate sensitivity observed in terms of the yield and ultimate tensile strengths. A similar behavior would be expected of QP980, as the increase in the strength was not significant; however, its uniform elongation increased by ~21% at the higher rate. This behavior does not conform to the definition of positive nor negative strain rate sensitivity.

Energy absorbed until fracture is critical to the performance of the material during a vehicle crash incident, and this property can be represented by toughness.

Though toughness depends on both strength and ductility, it is difficult to predict its change by simply looking at the changes in strength and ductility curves in Fig. 4a–c; their combined effect is what matters. Figure 4d compares the effect of strain rate on the toughness of the four materials. As noted, the percentage increase of toughness for DP980 and QP980 was the highest (~28% and ~30%, respectively), followed by SS 201 which showed an increase of ~21%, and finally CR5 showed a negligible increase of only ~5%. In summary, it is clear here that the four materials (driven by their different microstructures) responded differently to increasing strain rate, leading to net change in toughness observed in Fig. 4d. Total ductility reduction in CR5 at high rates counter-acted the increase in the strength, leading to a small increase in toughness. Toughness increase in SS 201 was strength-driven, since total ductility remained unchanged, while toughness increase in both DP980 and QP980 was primarily ductility-driven.

Discussion

All the aforementioned results suggest that the phases present in the microstructure of a steel greatly affects its response to increasing loading rates. Based on Orowan's principle [20], strain rate is directly proportional to dislocation velocity and other dislocation parameters. When materials are loaded at high rates, the density and the speed of dislocations are higher; but it is well-known that the dislocation velocity is directly related to the critical resolved shear stress [21], suggesting that higher stress is needed to overcome the critical resolved shear stress to move the dislocations at the speed required for the strain rate. This in turn implies higher yield and flow stresses during deformation at high rates, and hence reduced ductility.

The above coincides with the behavior of CR5 at high rates, where the ultimate tensile strength increased significantly; but when looking at the results of DP980, the strain rate sensitivity dropped at higher rate. To explain this, it is important to look closely at what happens during high rate testing. Under quasi-static conditions, test duration is long giving more time for heat generated during the test to be dissipated, thus a small increase in temperature is observed; it is therefore well established in the literature that isothermal conditions take place during quasi-static testing. This is not the case during high rate testing where the test occurs in a very short period of time; the 500 s^{-1} tests carried out in this study were completed within 0.4 ms. In this short period of time, the generated heat doesn't have enough time to be dissipated and thus the specimen's temperature increases significantly, changing the condition from isothermal to adiabatic. For CR5 mild steel, which consists of soft and ductile ferrite only, the effect of temperature rise softening is not as significant as the work hardening due to the higher loading rate, thus the net effect is significant increase in strength at higher rates. On the other hand, there are two opposing effects taking place in the DP980: work hardening which occurs in the ferrite; and softening due to temperature rise, which occurs primarily in the martensite, a hard and a brittle phase. The latter was reported in the work of Wang et al. 2013 [22], which showed a slight decrease in strength and an appreciable increase in ductility in a Martensitic steel as the result of increase in deformation rate. The combined effect of the aforementioned is the reduced strain rate sensitivity observed in the DP980. In regards to ductility, the net change in uniform elongation is almost zero, since the drop in ductility of ferrite is balanced by the increase in ductility in the martensite. On the other hand, the total elongation of the material increases, since the post-necking elongation portion increases (this is the case with all four steels regardless of the microstructure). Note the post-necking elongation of ~12–20% strains in all four materials at 500 s^{-1}—significantly higher than the quasi-static post-necking strains. This is believed to be primarily driven by the localized temperature increase, shifting the necking from localized to diffused, and thus prolonging the deformation.

For QP980, the small increase in strength at higher loading rate is similar to that observed in DP980, which can be attributed to the similarities in the microstructure in terms of the presence of martensite and ferrite. The slightly lower martensite and slightly higher ferrite volume fractions in QP980 (compared to DP980) results in the slightly higher rate sensitivity (Fig. 4a, b). But in terms of ductility (Fig. 4c), QP980 showed an increase of ~21% in uniform ductility at higher rate, which is significantly higher than that exhibited by DP980. Since this cannot be driven by the ferrite/martensite portion of the microstructure, it is believed to be associated with the austenite present in QP980. However, the effects and exact mechanisms are not clear. While austenite is known to exhibit deformation induced transformation into martensite, causing additional resistance to necking and thus enhanced tensile ductility and strain hardening, it is well established in the literature that the austenite-to-martensite transformation is inhibited at high rates since temperature increase reduces the chemistry driving force and increases the stacking fault energy, which have the combined effect of inhibiting the transformation. The results presented in Talonen et al. 2005 [23] and Lichtenfeld et al. 2006 [24] clearly show that for several austenitic stainless steels. The results obtained in this work also support that; note the drop in uniform ductility in the SS 201 at high rates (Figs. 3b and 4c). So this does not really agree nor explain why QP980 exhibits improved uniform ductility, on the level of ~20%, at high rates. In fact, the effects of high loading rates on the austenite-to-martensite transformation in TRIP-assisted steels (including QP steels) is not fully understood. While some researchers claim that transformation should be inhibited at high rates (based on the results in fully austenitic stainless steels), others indicate the opposite. For instance, Liu et al. [25] showed that at strain rates higher than 100 s^{-1}, QP980 shows increased ductility (which is similar to the results shown here in Fig. 3d). Liu et al. also measured the austenite content in the material after deformation and showed that the transformation is actually faster at higher rates. In summary, there are some uncertainties in this area that require further investigation to help understand the transformation kinetics dependence on strain rate (in austenite), and thereafter the impacts of this on the response of TRIP-assisted steels (such as the QP980).

Summary and Conclusions

The effects of the phase content on the strain rate sensitivity in two ~1GPa tensile strength steels, Ferrite-Martensite DP980 and Ferrite-Martensite-Austenite QP980, were investigated. Ferritic CR5 mild steel and austenitic stainless steel 201 were also studied for comparison. The materials were deformed at 0.005 and 500 s^{-1}, and their mechanical response was examined. DP980 showed almost no rate sensitivity; slight increase in strength and drop in uniform ductility at higher deformation rate. This was attributed to the opposite effects of ferrite and martensite balancing out. Ferrite is known to have a positive strain rate sensitivity, which was confirmed by the results obtained here for CR5. Martensite is known to have a negative strain rate sensitivity that is primarily driven by high temperature softening at higher rates. QP980 was also shown to exhibit negligible rate sensitivity in regards to strength; however, QP980 showed notable enhanced uniform ductility, which cannot be explained based on the known responses of the individual phases. Even austenite, the highly ductile phase, is known to exhibit a drop in uniform ductility at high rates due to the inhibition of austenite-to-martensite transformation, which was also confirmed by the results obtained here for the SS 201. Further investigation of austenite-to-martensite transformation kinetics and its rate dependence is ongoing to help explain the response of TRIP-assisted steels (such as QP980) at higher rates. Toughness in both DP980 and QP980 was shown to improve significantly at rates close to crash conditions, which compared favorably to the conventional CR5 mild steel that showed no increase in toughness. It is important to note here that the differences in behaviors of these steels are not merely driven by their microstructures and phases contents; other parameters like chemical composition play important roles. The ferrite in CR5, for instance, is not exactly the same as the ferrite in DP980 or QP980. The comparisons made in this study on the basis of phases were done to extract general trends to help explain the responses of steels with two or more phases in their microstructures.

Acknowledgements The authors greatly acknowledge the support of the Department of Energy, for funding this work. The support of General Motors, ThyssenKrupp Steel and AK Steel for supplying the materials used in this work is also acknowledged.

This material is based upon work supported by the Department of Energy under Cooperative Agreement Number DE-EE0005976, with United States Automotive Materials Partnership LLC (USAMP). This report was prepared as an account of work sponsored by an agency of the United States Government. Neither the United States Government nor any agency thereof, nor any of their employees, makes any warranty, express or implied, or assumes any legal liability or responsibility for the accuracy, completeness, or usefulness of any information, apparatus, product, or process disclosed, or represents that its use would not infringe privately owned rights. Reference herein to any specific commercial product, process, or service by trade name, trademark, manufacturer, or otherwise does not necessarily constitute or imply its endorsement, recommendation, or favoring by the United States Government or any agency thereof. The views and opinions of authors expressed herein do not necessarily state or reflect those of the United States Government or any agency thereof.

References

1. S. Keeler, M. Kimchi, Advanced high-strength steels application guidelines V5. WorldAutoSteel (2015)
2. S. Curtze, V.T. Kuokkala, M. Hokka, P. Peura, Deformation behavior of TRIP and DP steels in tension at different temperatures over a wide range of strain rates. Mater. Sci. Eng. A **507** (1), 124–131 (2009)
3. V. Tarigopula, O.S. Hopperstad, M. Langseth, A.H. Clausen, F. Hild, A study of localisation in dual-phase high-strength steels under dynamic loading using digital image correlation and FE analysis. Int. J. Solids Struct. **45**(2), 601–619 (2008)
4. H. Huh, S.B. Kim, J.H. Song, J.H. Lim, Dynamic tensile characteristics of TRIP-type and DP-type steel sheets for an auto-body. Int. J. Mech. Sci. **50**(5), 918–931 (2008)
5. H. Huh, J.H. Lim, S.H. Park, High speed tensile test of steel sheets for the stress-strain curve at the intermediate strain rate. Int. J. Automot. Technol. **10**(2), 195–204 (2009)
6. S. Curtze, V.T. Kuokkala, M. Hokka, P. Peura, Deformation behavior of TRIP and DP steels in tension at different temperatures over a wide range of strain rates. Mater. Sci. Eng. A **507** (1), 124–131 (2009)
7. H. Huh, H.J. Lee, J.H. Song, Dynamic hardening equation of the auto-body steel sheet with the variation of temperature. Int. J. Automot. Technol. **13**(1), 43–60 (2012)
8. S. Xu, D. Ruan, J.H. Beynon, Y. Rong, Dynamic tensile behaviour of TWIP steel under intermediate strain rate loading. Mater. Sci. Eng. A **573**, 132–140 (2013)
9. J.H. Kim, D. Kim, H.N. Han, F. Barlat, M.G. Lee, Strain rate dependent tensile behavior of advanced high strength steels: experiment and constitutive modeling. Mater. Sci. Eng. A **559**, 222–231 (2013)
10. J. Qin, R. Chen, X. Wen, Y. Lin, M. Liang, F. Lu, Mechanical behaviour of dual-phase high-strength steel under high strain rate tensile loading. Mater. Sci. Eng. A **586**, 62–70 (2013)
11. J. Huh, H. Huh, C.S. Lee, Effect of strain rate on plastic anisotropy of advanced high strength steel sheets. Int. J. Plast. **44**, 23–46 (2013)
12. D.O.N.G. Danyang, L.I.U. Yang, W.A.N.G. Lei, S.U. Liangjin, Effect of strain rate on dynamic deformation behavior of DP780 steel. Acta Metall. Sin. **49**(2), 159–166 (2013)
13. X. Yang, L.G. Hector Jr., J. Wang, A combined theoretical/experimental approach for reducing ringing artifacts in low dynamic testing with servo-hydraulic load frames. Exp. Mech. **54**(5), 775–789 (2014)
14. S. Li, Y. Kang, G. Zhu, S. Kuang, Effects of strain rates on mechanical properties and fracture mechanism of DP780 dual phase steel. J. Mater. Eng. Perform. **24**(6), 2426–2434 (2015)
15. B.C. Hwang, T.Y. Cao, S.Y. Shin, S.H. Kim, S.H. Lee, S.J. Kim, Effects of ferrite grain size and martensite volume fraction on dynamic deformation behaviour of 0 · 15C–2 · 0Mn–0 · 2Si dual phase steels. Mater. Sci. Technol. **21**(8), 967–975 (2005)
16. S. Oliver, T.B. Jones, G. Fourlaris, Dual phase versus TRIP strip steels: microstructural changes as a consequence of quasi-static and dynamic tensile testing. Mater. Charact. **58**(4), 390–400 (2007)
17. S. Li, D. Zou, C. Xia, J. He, Effect of strain rate on deformation-induced martensitic transformation of quenching and partitioning steels. Steel Res. Int. (2015)
18. S. Mates, F. Abu-Farha, Dynamic tensile behavior of a quenched and partitioned high strength steel using a Kolsky bar, in *Proceedings of the 2015 SEM Annual Conference and Exhibition (SEM 2015)*, Costa Mesa, CA, 08–11 June 2015 (2015)
19. P. Verleysen, J. Peirs, J. Van Slycken, K. Faes, L. Duchene, Effect of strain rate on the forming behaviour of sheet metals. J. Mater. Process. Technol. **211**(8), 1457–1464 (2011)
20. K. Qin, L.M. Yang, S.S. Hu, Interpretation of strain rate effect of metals, in *Dynamic Behavior of Materials*, vol. 1, pp. 21–27. Springer International Publishing (2014)
21. J.J. Gilman, W.G. Johnston, *Dislocations and Mechanical Properties of Crystals*, ed. by J. Fisher, W. Johnston, R. Thomson, T. Vreeland, vol. 116 (1957)

22. W. Wang, M. Li, C. He, X. Wei, D. Wang, H. Du, Experimental study on high strain rate behavior of high strength 600–1000 MPa dual phase steels and 1200 MPa fully martensitic steels. Mater. Des. **47**, 510–521 (2013)
23. J.A. Lichtenfeld, C.J. Van Tyne, M.C. Mataya, Effect of strain rate on stress-strain behavior of alloy 309 and 304L austenitic stainless steel. Metall. Mater. Trans. A **37**(1), 147–161 (2006)
24. J. Talonen, H. Hänninen, P. Nenonen, G. Pape, Effect of strain rate on the strain-induced $\gamma \rightarrow \alpha'$-martensite transformation and mechanical properties of austenitic stainless steels. Metall. Mater. Trans. A **36**(2), 421–432 (2005)
25. C. Liu,L. Wang, Y. Liu, Effects of strain rate on tensile deformation behavior of quenching and partitioning steel. Mater. Sci. Forum **749**, 401–406 (2013)

Part VI
Advanced Materials in Dental and Orthopedic Applications

Characterization of Chitin Synthesized from Snail Shell

S.O. Adeosun, O.P. Gbenebor, E.I. Akpan and S.A. Olaleye

Abstract Thermogravimetric analysis was used to analyze chitin thermal stability present in unprocessed and processed snail shell samples using different concentrations of HCl and NaOH. The result shows that 42% of $CaCO_3$ in the snail shell decomposes between 699–779 °C while decomposition of chitin (2.11%) starts at 275 °C and ends at 335 °C. De-mineralization and de-proteinization of the snail shell using 1.5 M HCl and 0.4 M NaOH eliminates $CaCO_3$ formation and sample shows existence of 60% chitin decomposing between 346 and 412 °C. The XRD of virgin snail shell showing the main diffraction face intensity for $CaCO_3$ occurring at $2\theta = 29.5^o$ with other $CaCO_3$ peaks occur at $2\theta =$ 30.2°, 39.3°, 41.0°, 44.2°, 45.9°, 48.9°, 51.5°, 53.1° and 55.6°. Chitin present occurs at reflection $2\theta = 36.3^o$ with weak intensity and the XRD shows no evidence of $CaCO_3$ but two chitin crystalline peaks at 20.6° and 26.3° with 78% lowest crystallinity index.

Keywords Snail shell · Chitin · Synthesis · HCL · NaOH · Extraction parameters · Characterization

---

S.O. Adeosun (✉) · O.P. Gbenebor
Department of Metallurgical and Materials Engineering, University of Lagos, Lagos, Nigeria
e-mail: samsonoluropo@yahoo.com

O.P. Gbenebor
e-mail: gbeneborphilips@yahoo.co.uk

E.I. Akpan
Institut fur Verbundwerkstoffe GmbH, 67663 Kaiserslautern, Germany
e-mail: emma_eia@yahoo.com

S.A. Olaleye
Department of Mechanical Engineering, University of Lagos, Lagos, Nigeria
e-mail: solaleye@unilag.edu.ng

The Minerals, Metals & Materials Society, *TMS 2017 146th Annual Meeting & Exhibition Supplemental Proceedings*, The Minerals, Metals & Materials Series, DOI 10.1007/978-3-319-51493-2_25

Introduction

Chitin consists of linear chains of (1, 4)-linked 2-acetylamido-2-deoxy-b-D-glucose [1–5] and is the second most abundant natural occurring polysaccharides found essentially in shells of crustaceans. Chitin offers advantages for today's industrial applications as it is generally found in marine food product wastes such as shrimp, snail, periwinkle and crab. It is biological, biodegradable and renewable, and pose no threat to environmental conservation. Chitin is mostly extracted from crustaceans by acid treatment (demineralization) followed by alkaline extraction to solubilize proteins (Deproteinization). A decolorization step is often added to remove leftover pigments and obtain a colorless product. Rinaudo [3] has shown that these treatments are mostly specialized to chitin source because of the difference in the ultrastructure of the initial materials.

Chitin and its derivatives are well known for applications in biomedical and food industries, pharmaceutical, agriculture, textiles, cosmetics and water treatment plants, paper production, textile finishes, photographic products, cements, heavy metal chelating agents, membranes, hollow fibers, and waste removal [6–9]. They are specifically used in the biomedical industries for wound healing, as drug delivery agents, anti-cholesterolemic agents, blood anticoagulants, anti-tumor agents, immunoadjuvants, tissue engineering scaffolds, wound dressings, separation membranes, antibacterial coatings, stent coatings, and sensors [7, 8, 10–14]. Other interesting uses of chitin and it's derivative includes use as excipient and drug carrier in film, gel or powder, hydroxyapatite chitin–chitosan composite bone-filling material with a self-hardening property for guided tissue regeneration in treatment of periodontal bony defects [3, 15]. The use of chitin and its derivatives for these applications is owing to their polyelectrolite properties, the presence of reactive functional groups, gelforming ability, high adsorption capacity, biodegradability, biocompatibility and non-toxicity [6].

However, chitin is limited in its bio-polymeric applications because it is highly insoluble in water and in some common organic solvents with low chemical reactivity. It dissolves only in solvents such as N, N-dimethylacetamide (DMAc)-LiCl, hexafluoroacetone or hexafluoro-2-propanol. It is soluble in aqueous acidic medium due to the presence of amino groups [16]. The insolubility of chitin is due to the presence of high amount of acetyl groups. Deacetylation of chitin yields chitosan that is soluble under acidic conditions. Chitosan is ideally a fully N-deacetylated chitin. However, total deacetylation of chitin has not been realized and when the degree of deacetylation of chitin reaches ~50%, it becomes soluble in aqueous acidic media and is then called chitosan. The solubility of chitosan in water and organic solvents, has made it useful as hydrogels, films and fibres for packaging and medical applications [3, 4]. However, chitosan possess poor thermal stability and mechanical properties owing to low crystallinity. It is not useful as load bearing material in bio-medical applications due to its poor mechanical properties [17]. Therefore, to be useful in applications such as scaffolds, internal fixations, sutures and drug delivery systems, there is need for improvement in its mechanical and thermal properties [4, 18, 19].

It has been shown that the amount of N-acetyl groups in processed chitin is dependent on the processing variables (temperature, time, concentration of acid and base) during demineralization and deprotenization. Solubility, physicochemical characteristics, biodegradability, immunological activity and solution properties of chitin is basically determined by the degree of N-acetylation (the ratio of 2-acetamido-2-deoxy-d-glucopyranose to 2-amino-2-deoxy-d-glucopyranose structural units) while the mechanical properties is tied to crystallinity [20–22]. On the other hand, sorption ability of chitin is inversely proportional to the degree of crystallinity. Thus, controlling the process parameters will lead to control of properties. However, the effect of process variables on the properties of chitin and its derivative has not been given wide attention by researchers. In this study, the effect of acid and alkali concentrations on thermal stability and crystallinity of chitin extracted from snail shell is examined. The experiment was designed with increasing concentrations of acid during demineralization and increasing concentrations of alkali during deproteinization and the effect of this on thermal stability and crystallinity of the extracted chitin studied.

Materials

Chitin Extraction

Shells of snails were scraped free of loose tissues, washed in water, dried and ground to a 250 μm particle size in a ball mill. Demineralization was carried out at room temperature (32 °C) by soaking 100 g of ground samples in 1.5, 1.7 and 1.9 M HCl. For each concentration, the process was repeated several times until evolution of gaseous effluent ceased and the number of baths (in terms of acid volume) used was noted. The demineralized samples were washed with distilled water to neutral (pH7.0), filtered and dried in the oven at 70 °C for 4 h to constant weights. Deproteinization was carried out on each sample by heating mineral—devoid samples in concentrations of 0.4, 0.8 and 1.2 M NaOH solutions in a beaker at 100 °C for 1 h. At the end of this period, samples were filtered and soaked in fresh sets of alkali solutions (0.4, 0.8 and 1.2 M) for 18 h at room temperature (32 °C) for effective protein removal. The samples were washed with distilled water to pH 7.0, filtered and oven dried at 70 °C. Depigmentation and bleaching was carried out by soaking extracted chitin in 1 M H_2O_2 for 24 h at 32 °C. The chitin was then washed in distilled water and dried for 4 h in an oven at 70 °C before characterizations.

Thermogravimetric Analysis (TGA)

In TGA, the change in mass of a sample is usually measured as a function of temperature or time under an inert (nitrogen) or oxidative (air) atmosphere. TGA was used to quantitatively determine the constituent of minerals and organic compound (including the biopolymer) in the samples. Analysis of samples was carried out on TGA Q500 instrument in Soochow University, Suzhou, China. 2 mg of samples was heated to 750 °C at 10°/minute heating rate. In this test, the temperature for the onset of thermal decomposition (T_{onset}), the temperature at which decomposition rate was rapid (T_{max}), chitin content and yield were deduced from the thermograms.

X-Ray Diffraction (XRD)

The XRD of the samples was conducted using PANanalytical X' Pert PRO MPD X-ray diffraction system PW3040/60 machine in Soochow University, Suzhou, China. Samples were exposed to a monochromatic Cu Kα radiation (k = 1.5406 Å), operating at 40 kV and 40 mA. The samples were registered in a zero background sample holder to avoid external background interferences. The diffractograms were registered in the range of 7° to 90° (2θ.)n a step scan mode of 0.026261 at a counting time of 17.34 s per step. In order to calculate the crystallinity index (CrI), a method which has been employed previously for chitin was used and the expression is given in Eq. 1.

$$\mathrm{CrI}\ (\%) = [A_c/(A_c + Aa)]\ \mathrm{x}\ 100 \tag{1}$$

where A_c and A_a represent the intensities of the crystalline and amorphous regions respectively.

Scanning Electron Microscopy (SEM) with Energy Dispersive X-ray Analysis (EDS)

An ASPEX 3020 model variable pressure SEM from Soochow University, Suzhou, China operated with an electron intensity beam 15 kV and equipped with Noran-Voyager energy dispersive spectroscope was used to observe the morphological features of all samples. The samples to be observed under the SEM were mounted on a conductive carbon imprint left by the adhesive tape prepared by placing the samples on the circular holder and coated for 5 min to enable it conduct electricity. For EDS analysis, samples were analyzed at an accelerating voltage of 15 kV.

Results and Discussion

Effect of Acid/Alkali Concentrations Treatments on Thermal Properties of Extracted Chitin

Activation energy is one of the most consistent methods of determining the stability of organic compounds. Apparent activation energies for the thermal degradation of synthesized compounds were determined from the TG curves as shown in Fig. 1. The activation energy typically ranges between 25 and 35 kJ/mol except for samples treated with 1.7 M acid/0.4 M alkali (94 kJ/mol) and 1.7 M acid/0.8 M alkali (57 kJ/mol). Activation energies between 25 and 35 kJ/mol are typical activation energies for chitosan that have been reported by other researchers [23, 24]. On the other hand, the higher activation energy (94 or 57 kJ/mol) is typical of chitin that is in agreement with the work of Nam et al. [25]. The reason for low activation energy of these samples is attributed to the presence of high content of chitosan in the synthesized samples. Higher activation energy for thermal degradation of chitin is an indication that the acetylated side-chains increased the thermal stability of the molecular main chains to which they are linked. Figure 2 shows the effect of concentrations on the values of activation energy. It is observed that low alkaline concentration favours higher activation energy at average acid concentration.

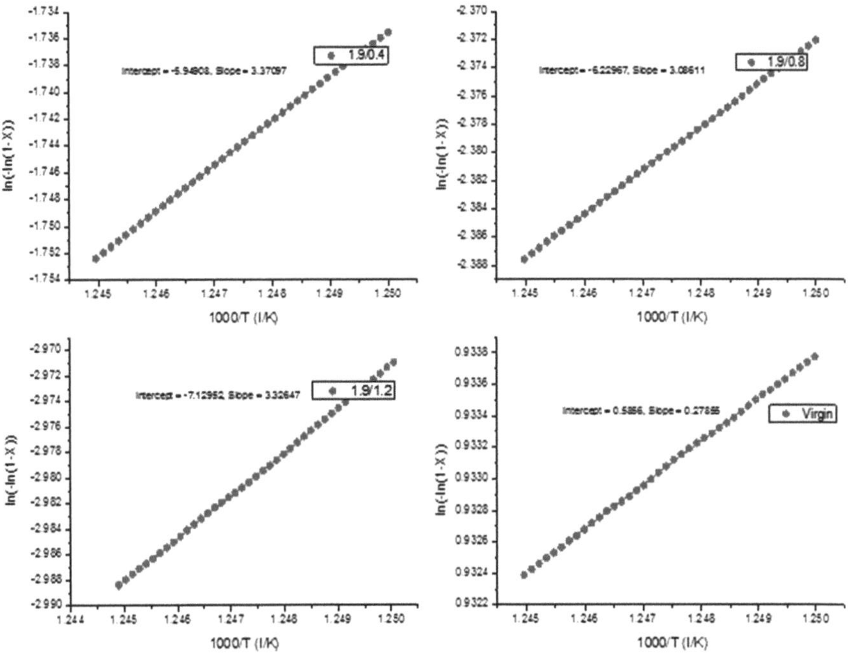

Fig. 1 Activation energy of extracted chitin (legend shows concentration of acid/alkali in M)

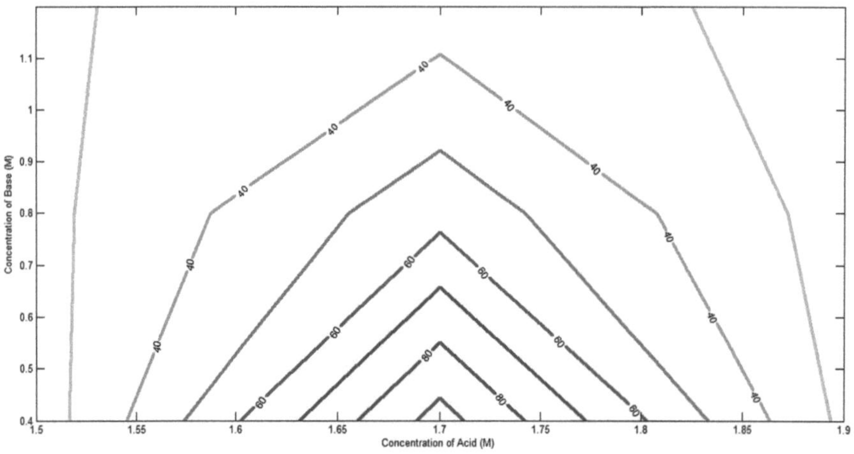

Fig. 2 Effect of acid/alkali concentrations treatments on activation energy of extracted chitin

This is true as increase in alkaline concentration during deprotenization has been found to increase the formation of less thermally stable chitosan compared to chitin. The line of steepest ascent is vertically downward in the region of average acid concentration (see Fig. 2).

Figure 3 shows the TGA and its corresponding DTG thermograms of the treated samples. Generally all the thermograms show two decomposition steps with a wide exothermic peak. The first occurs between 50–110 °C, and is attributed to water evaporation. The second occurs at 250–400 °C and is attributed to degradation of the polysaccharide structure of the molecule, including the dehydration of polysaccharide rings with the polymerization and decomposition of chitin acetylated and deacetylated units. The virgin unprocessed snail shell sample shows second step degradation at 550–850 °C, which is attributed to the degradation of calcium carbonate. Thus, the maximum degradation temperature exhibited treated samples shows the influence and impact of the acid/alkali concentration treatments. Temperature is found to increase with increases in alkaline concentration during deprotenization but is independent of acid concentration during demineralization. The temperature increases from 396 to 398 °C with NaOH concentration increment from 0.4 to 1.2 M at fixed acid concentration (1.5 M). This trend is repeated for acid concentrations of 1.7 and 1.9 M. Figure 4 shows that the line of steepest ascent is in the direction of increasing alkali concentration (see arrows 1, 2 and 3). The highest change in maximum degradation temperature is in samples demineralized with 1.9 M HCl. It is evident as in Fig. 3 that the maximum degradation temperature is strongly dependent on NaOH concentrations during deprotenization. This attributed to the decreases in the crystalline order that occurred after deacetylation of the side chain of chitin, leading to decrease in thermal stability lower than sample with higher amount of acetyl groups. This thermal stability relationship with deacetylation has also been observed by [24, 26]. These results suggest a higher thermal stability for

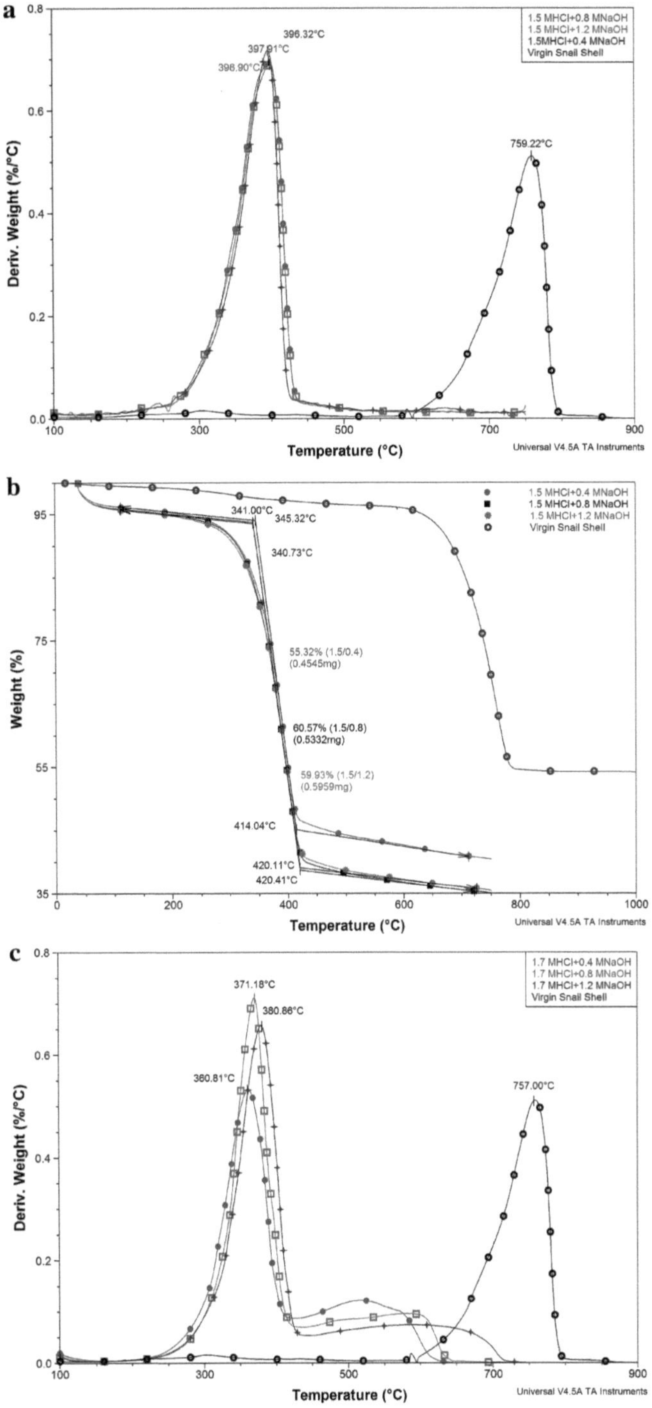

Fig. 3 TGA and DTG curves of extracted chitin

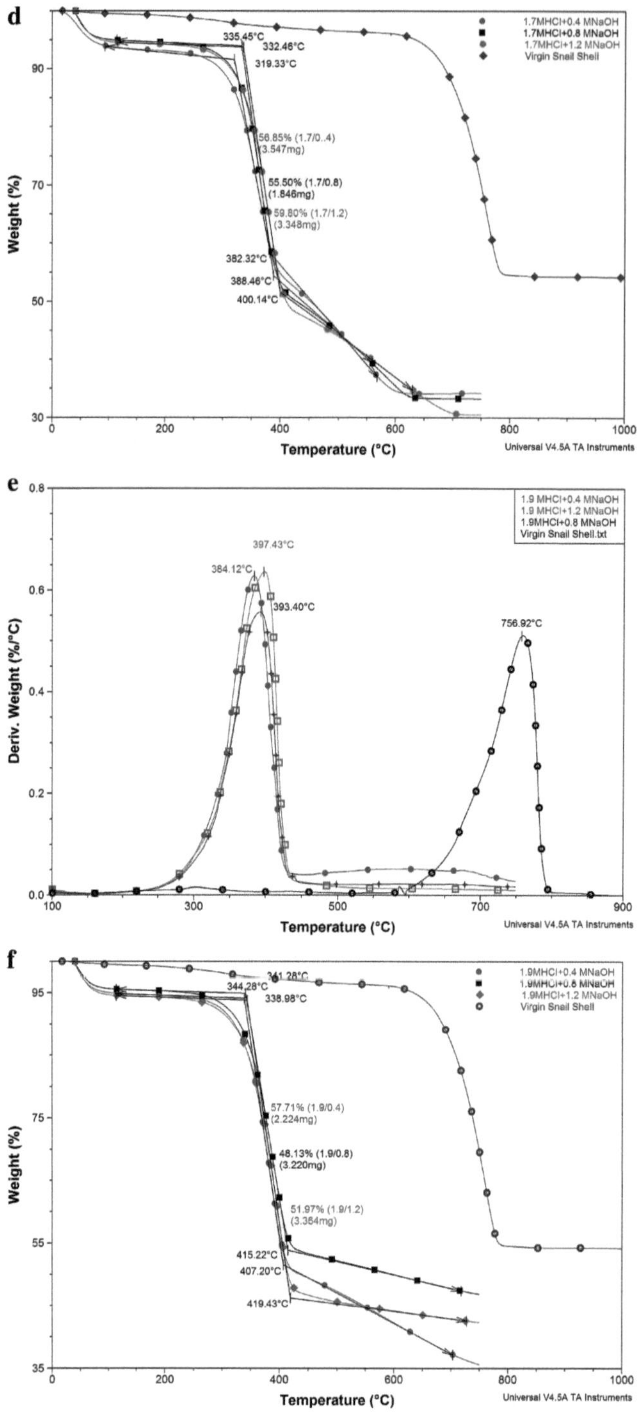

Fig. 3 (continued)

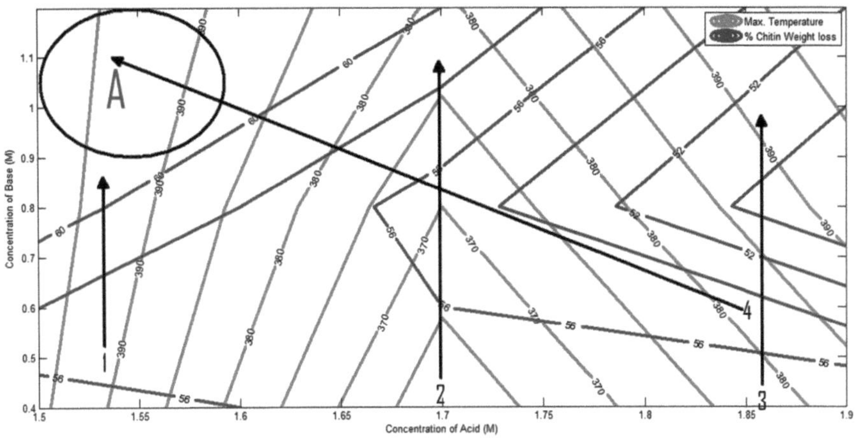

Fig. 4 Contour plot of maximum temperature and yield of extracted chitin

samples treated with higher concentrations of NaOH and HCl. Report from previous studies show a maximum degradation temperatures between 350 and 390 °C for chitin synthesized from crab, krill, shrimp, crayfish, and insects [27–30]. However, this research shows that a good combination of the NaOH and HCl concentrations in the extraction of chitin can enhance its thermal stability. The synthesized chitins from this study displayed weight loss in two steps (see Fig. 3a, c and e).

The weight loss in the first step results from the evaporation of water, and the second is due to the deterioration of polysaccharide molecules. Kaya et al. [31], Juárez-de la Rosa et al. [27], Wang et al. [32], Gonil et a. [29] and Jang et al. [30] also observed weight losses in chitins isolated from bryozoa crab, krill, shrimp, crayfish, and insects in two major steps between 30–650 °C. The weight loss of chitin in this study ranges between 48 and 61% and shows a strong dependence on the choice of acid and alkaline concentrations combination. Figure 3 shows that the line of steepest ascent is in the reverse diagonal towards increase in alkaline concentration (see arrow 4 in Fig. 4). Thus, decreasing HCl concentration while increasing NaOH concentration will improve chitin yield. Examining Fig. 4 a combination of high weight loss and high maximum degradation temperature can be attained within the region of high alkaline and low acid concentration (See region 'A' in Fig. 4).

Effect of concentrations of acid/alkali treatments **on chitin crystallinity**

XRD analysis was applied to detect the crystallinity and crystal size of the isolated chitin. Figure 5 shows the XRD patterns of samples. Generally all treated samples possess three prominent peaks at 19.01°, 20.55°, 23.65°, 29.32° except for the virgin snail sample that shows an extra peak at 30.2°. The peaks at 19.01°, 20.55°, 23.65°, 29.32° are attributed to the presence of chitin/chitosan in the sample. The peak at 30.2° is attributed to the presence of $CaCO_3$. The XRD confirms that the treated samples contain a combination of α- chitin and chitosan as have been reported [30, 32, 33]. From the Figure the peak at 30.2 ° that is attributed

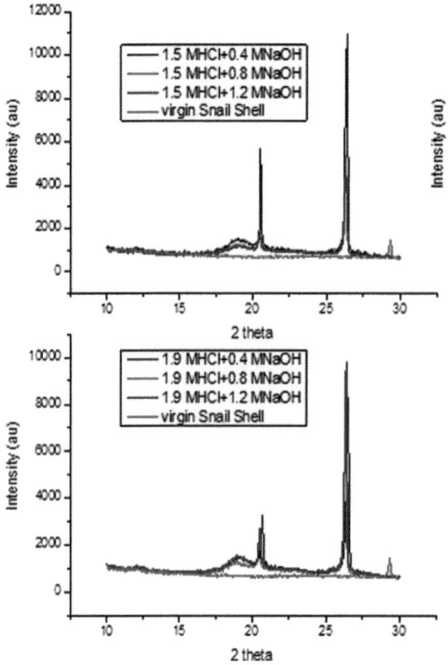

Fig. 5 XRD patterns showing effect of concentrations of acid/alkali treatments on chitin structure

to the presence of $CaCO_3$, was completely removed from all treated samples and this confirms the effectiveness of the demineralization process in the removal of the mineral elements. It is also noted that all treated samples possess very high crystalline peaks showing structural changes brought about by the treatment method. To show the effect of treatment on crystallinity a contour plot is presented in Fig. 6. Thus, higher crystallinity is achieved with the use of an average concentration of NaOH and a lower concentration of HCl and the line of steepest is in reverse horizontal direction (see arrow in Fig. 6).

However, extreme low crystallinity can be attained with a high concentration of both acid and base. A high crystallinity is an indication that an ordered supermolecular structure is achieved in the extracted chitin sample. A moderate crystallinity will result with the use of a moderate concentration of acid and base. A low crystallinity indicates complete conversion to chitosan while a moderate crystallinity indicates the presence of chitosan and chitin, as evident from XRD peaks distribution pattern [34–36]. An increase in the degree of crystallinity of chitin is associated with increase in density of structured area, tensile strength, and average elongation at rupture [36]. However, a high crystallinity of the original chitin brings about low reactivity with high C-2 amino group content of the N-acetylated [37]. These results indicates that crystallinity index of synthesized samples is dependent

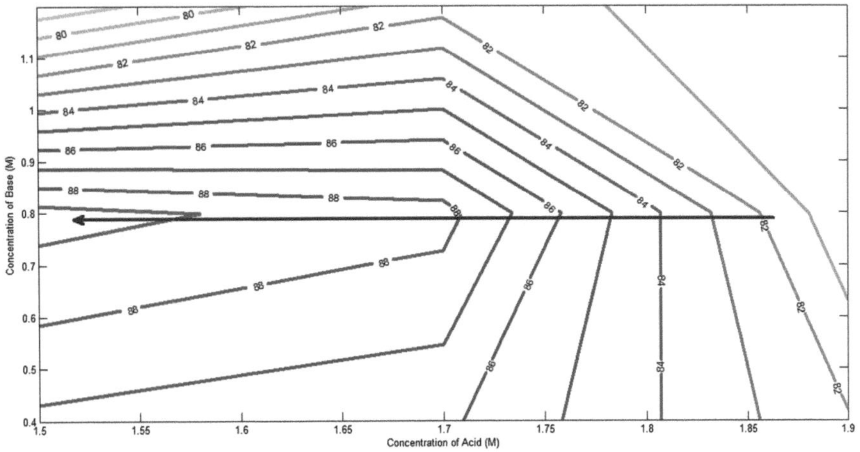

Fig. 6 Contour plot showing the effect of treatment on the crystallinity of chitin samples

on the acid/alkali concentrations. It is worthy of note that other researchers have reported crystallinity values 64, 66.6, 59.86, and 56.94% for chitins extracted from brown shrimp, pink shrimp, crab, and crayfish respectively [38] while in this study the lowest crystallinity index of chitin extracted from snail shell is 78%.

Conclusion

Chitin was extracted from snail shells with varying concentrations of alkali and acidic for the deproteination and demineralisation processes. It was found that thermal stability and crystallinity of the extracted chitin is dependent on the concentration of acid and alkali used for the extraction. Contour plots show that increasing alkali concentration irrespective of the acid concentration leads to increase in maximum degradation temperature of chitin while increasing alkali media concentration and decreasing acid concentration favours both yield and maximum degradation temperature of chitin. It was also found that higher crystallinity is achieved with the use of moderate alkali concentration and low acid concentration.

Acknowledgements The research group acknowledge with grateful thanks the University of Lagos, Nigeria, for providing the platform for the TETFUND RESEARCH GRANT (CRC NO. 2015/19) used for this research. We equally recognise the assistance provided by staff of Metallurgical Laboratory, University of Lagos.

References

1. T. Kean, M. Thanou, Chitin and chitosan: sources, production and medical applications, in *Renewable Resources for Functional Polymers and Biomaterials*, ed. by P.A. Williams (RSC Publishing, 2011), pp. 292–318
2. R.A.A. Muzzarelli, Chitin and chitosan hydrogel, in *Handbook of Hydrocolloids*, eds. by G. O. Phillips, P.A. Williams (Oxford, UK, Woodhead Publishing Ltd, 2009), pp. 849–876
3. M. Rinaudo, Chitin and chitosan: properties and applications. Prog. Polym. Sci. **31**, 603–632 (2006)
4. M. Rinaudo, Main properties and current applications of some polysaccharides as biomaterials. Polym. Int. **57**, 397–430 (2008)
5. K.M. Varum, O. Smidsrod, Chitosans, in *Food Polysaccharides and their Applications*, eds. by M. Stephen, G.O. Phillips, P.A. Williams (Boca Raton Fl, USA, CRC Taylor and Francis, Publishers, 2006), pp. 497–520
6. H. Sawssen, G. Olfa, Y. Islem, J. Kemel, N. Moncef, Chitin extraction from crab shells by Bacillus bacteria. Biological activities of fermented crab supernatants. Int. J. Biol. Macromol. **79**, 167–173 (2015)
7. R.A.A. Muzzarelli, *Naturally Chelating Polymers* (Pergamon Press, New York, 1973)
8. R.A.A. Muzzarelli, *Chitin* (Pergamon Press, New York, 1977)
9. J.-F. Revol, R.H. Marchessault, In vitro chiral nematic ordering of chitin crystallites. Int. J. Biol. Macromol. **15**, 329–350 (1993)
10. Y. Oshima et al., Clinical application of chitin non-woven fabric as wound dressing. Eur. J. Plast. Surg. **10**, 66–69 (1987)
11. K. Watanabe et al., Chem.6-O-Carboxymethyl-chitin (CM-chitin) as a drug carrier. Pharm. Bull. **38**, 506–509 (1990)
12. D.N. Hon, *Polysaccharides in Medicinal Applications* (Marcel Dekker, New York, 1996), pp. 631–649
13. R.A.A. Muzzarelli et al., Chitin nanofibrils/chitosan glycolate composites as wound medicaments. Carbohydr. Polym. **70**, 274–284 (2007)
14. J.P. Zikakis, *Chitin, Chitosan, and Related Enzymes* (Academic Press, Orlando, FL, 1984)
15. M. Ito, Y. Matahira, K. Sakai (eds.), *The Application of Chitin Chitosan to Bone Filling Materials*, vol. 4 (Kichin, Kitosan Kenkyu, Publication Nippon Kichin, Kitosan Gakkai, 1998), pp. 142–143)
16. M. Mahlous et al., Gamma irradiation-aided chitin/chitosan extraction from prawn shells. Nucl. Instrum. Methods Phys. Res. B **265**, 414–417 (2007)
17. M.H. Mohammed, P.A. Williams, O. Tverezovskaya, Extraction of chitin from prawn shells and conversion to low molecular mass chitosan. Food Hydrocolloids **31**, 166–171 (2013)
18. E. Khor, Chitin: a biomaterial inwaiting. Curr. Opin. Solid State Mater. Sci. **6**, 313–317 (2002)
19. A. Yang, R. Wu, Mechanical properties and interfacial interaction of a novel bioabsorbable chitin fiber reinforced poly (-caprolactone) composite. J. Mater. Sci. Lett. **20**, 977–979 (2001)
20. R.J. Nordtveit, K.M. Varum, O. Smidsrod, Degradation of partially N-acetylated chitosans with hen egg white and human lysozyme. Carbohyd. Polym. **29**, 163–167 (1996)
21. N. Zydowicz (ed.), *Advances in Chitin Science*, vol. 1, (Lyon, 1996), 262
22. G. Peluso et al., Chitosan-mediated stimulation of macrophage function. Biomaterials **15**, 1215–1220 (1994)
23. L.S. Guinesi, E.T.G. Cavalheiro, The use of DSC curves to determine the acetylation degree of chitin/chitosan samples. Thermochim. Acta **444**, 128–133 (2006)
24. F.S. Kittur et al., Characterization of chitin, chitosan and their carboxymethyl derivatives by differential scanning calorimetry. Carbohydr. Polym. **49**, 185–193 (2002)
25. Y.S. Nam et al., Effect of the degree of deacetylation on the thermal decomposition of chitin and chitosan nanofibers. Carbohydr. Polym. **80**, 291–295 (2010)

26. T. Wanjuna, W. Cunxin, C. Donghu, Kinetic studies on the pyrolysis of chitin and chitosan. Polym. Degrad. Stab. **87**, 389–394 (2005)
27. B.A. Juárez-de la Rosa et al., Effects of thermal treatments on the structure of two black coral species chitinous exoskeleton. J. Mater. Sci. **47**, 990–998 (2012)
28. Y. Wang et al., Crystalline structure and thermal property characterization of chitin from Antarctic krill (Euphausia superba). Carbohydr. Polym. **92**, 90–97 (2013)
29. P. Gonil, W. Sajomsang, Applications of magnetic resonance spectroscopy to chitin from insect cuticles. Int. J. Biol. Macromol. **51**, 514–522 (2012)
30. M.-K. Jang et al., Physicochemical characterization of a-chitin, b-chitin and c-chitin separated from natural resources. J. Polym. Sci., Part A: Polym. Chem. **42**, 3423–3432 (2004)
31. M. Kaya et al., First chitin extraction from Plumatella repens (Bryozoa) with comparison to chitins of insect and fungal origin. Int. J. Biol. Macromol. **79**, 126–132 (2015)
32. M.-T. Yen, J.-L. Mau, Physico-chemical properties of chitin from shiitake stipes and crab shells. Ann. Tainan Woman's Coll. Arts Technol. **23**, 229–240 (2004)
33. M.-T. Yen, J.-L. Mau, Selected physical properties of chitin prepared from shiitake stipes. LWT—Food Sci. Technol. **40**, 558–563 (2007)
34. Y. Liu et al., Absorption behaviors and structure changes of chitin in alkali solution. Carbohydr. Polym. **72**, 235–239 (2008)
35. J. Marszalek, L. Szosland, M. Mucha, Immiscible polymer blends containing dibutyrylchitin as environmentally friendly materials. Mol. Cryst. Liq. Cryst. **354**, 427–433 (2000)
36. G.A. Vikhoreva, I.N. Gorbacheva, L.S. Gal'braikh, Synthesis and properties of water-soluble derivatives of chitin: a review. Fiber Chem. 31, 573–8493 (1999)
37. C.K.S. Pillai et al., Chitin and chitosan polymers: chemistry, solubility and fiber formation. Prog. Polym. Sci. **34**, 641–678 (2009)
38. E.S. Abdou et al., Improved antimicrobial activity of polypropylene and cotton nonwoven fabrics by surface treatment and modification with chitosan. J. Appl. Polym. Sci. **108**, 2290–2296 (2008)

Titanium-Magnesium Composite for Dental Implants (BIACOM)

Martin Balog, Mateja Snajdar, Peter Krizik, Zdravko Schauperl, Zlatko Stanec and Amir Catic

Abstract We report on the study on the titanium-magnesium (Ti–Mg) bioactive metal-metal composite utilized for a fabrication of dental implants. The biomedical Ti-12vol.%Mg composite, named BIACOM, is manufactured using a cost effective approach, where a mixture of elemental Ti and Mg powders is extruded at low temperature to sound profiles. Microstructure of composite comprises filaments of biodegradable Mg component, which are arrayed along extrusion direction and are homogenously distributed within permanent, bioinert Ti matrix. Compared to Ti Grade 4, the reference material used for dental implants, the properties of as-extruded composite include significantly reduced Young's elastic modulus (92.1 GPa) and low density (4.12 g.cm^{-3}), while the mechanical strength of Ti Grade 4 is maintained. Dynamic testing of dental implants fabricated from as-extruded composite, realized to follow the ISO 14801 standard for endosseous dental implants, confirms fatigue performance of BIACOM implants equal to the one of the refer-

M. Balog (✉) · P. Krizik
Institute of Materials and Machine Mechanics, Slovak Academy of Sciences, Dubravska Cesta 9, 84513 Bratislava, Slovakia
e-mail: martin.balog@savba.sk

P. Krizik
e-mail: peter.krizik@savba.sk

M. Snajdar · Z. Schauperl
Faculty of Mechanical Engineering and Naval Architecture, University of Zagreb, Ivana Lucica 5, 10000 Zagreb, Croatia
e-mail: mateja.snajdar@fsb.hr

Z. Schauperl
e-mail: zdravko.schauperl@fsb.hr

Z. Stanec
Private Dental Practice, Ljudevita Gaja 62a, 10430 Samobor, Croatia
e-mail: zlatko.mec@gmail.com

A. Catic
School of Dental Medicine, University of Zagreb, Gunduliceva 5, 10000 Zagreb, Croatia
e-mail: catic@sfzg.hr

The Minerals, Metals & Materials Society, *TMS 2017 146th Annual Meeting & Exhibition Supplemental Proceedings*, The Minerals, Metals & Materials Series, DOI 10.1007/978-3-319-51493-2_26

ence material. Exposure of as-extruded composite samples to Hank's solution, realized in order to simulate behavior in human body over the time after implantation, yields gradual dilution of Mg from composites surface and volume. Corroded Mg leaves at prior Mg filament sites pores within Ti matrix, which remains intact. This provides further decrease of Young's modulus and enhances macro and micro roughness at implants surface. As a result, BIACOM shows improved mechanical compatibility (i.e., reduction of stress-shielding) and better osseointegration potential.

Keywords Powder · Composite · Implant · Titanium · Magnesium · Dental

Introduction

Because of their high strength, good fracture toughness and damage tolerance metallic biomaterials are preferentially applied in load-bearing application over ceramic and polymeric biomaterials [1]. Among the other metallic biomaterials, such as cobalt-chromium (Co-Cr), stainless steels, nickel-titanium (Ni-Ti) alloys, etc., commercial purity (CP) titanium (Ti) and Ti alloys have been preferred and widely employed as implant (and prosthodontic) materials in orthopedics and dentistry [2]. This is due their excellent biocompatibility and corrosion resistance, low density and good mechanical properties, in addition to Young's elastic modulus (E) of Ti and Ti alloys (E = 105 − 120 GPa) being similar to that of bone (<40 GPa), when compared to other biocompatible metallic counterparts [3]. Still, elasticity mismatch remains critical issue, which may result in peri-implant bone atrophy during functional and parafunctional loading and possible overloading in patients with temporomandibular dysfunctions, caused by stress-shielding effect over the surrounding bone [4, 5]. Previous attempts to reduce E of Ti were aimed at obtaining low modulus β-type Ti alloys by addition of alloying elements such as Zr, Nb, Ta and Mo (TNTZ) [6]. Although all these elements have shown to have a non-toxic effect on human body their high price and melting points, much higher than that of Ti, presented major issues for high volume production [7]. Significant reduction of E is possible also by introduction of residual porosity to Ti and Ti alloy structures [8, 9]. In addition to reduced E, porous Ti and Ti alloys implants benefit from their improved ingrowth of new tissue and bone tissue attachment [10]. However, residual porosity leads also to substantial deterioration of other mechanical properties [11]. Thus porous Ti and Ti alloy structural implants are predetermined mostly for less intense or non-load-bearing applications under non fatigue conditions, such as scaffold structures, etc. [12].

As far as production concerns, metallic Ti, despite its high content in the earth's crust, is considered to be rare and expensive material [13]. The reason is that the conventional metallurgical methods are ineffective in case of Ti highly reactive at elevated temperatures [14], and problematic and ineffective machining, which results in a great loss of expensive material [15]. For these reasons there is growing interest in powder metallurgy (PM) as a cost-effective approach of direct production of complex Ti parts [16]. Moreover, PM approach brings advantage of improved chemical and microstructural homogeneity as well as possibility to produce various composite structures [17]. PM Ti parts may be fabricated by a wide range of techniques, where press-and-sinter approach is technically most simple and economically most attractive approach capable of production of near-to-shape parts. However, limiting fact is that the sintering has to be realized under relatively high vacuum by batch type process, which restricts effectiveness and reproducibility, in addition to relatively high residual porosity, which predetermines applications under non fatigue conditions [18].

Unlike Ti, magnesium (Mg) presents a rather novel material in the field of biomedical implants. Since it is a natural occurring element in human body, in recent years Mg and Mg alloys have been in focus of many studies concerning biodegradable implants, which are finally replaced by natural tissue [19]. Besides its excellent biocompatibility and biodegradation potential, the mechanical properties similar to that of bone (e.g., E = 45 GPa), Mg is relatively affordable and low density metal. Upon exposure to environment rich with chloride ions Mg corrodes intensively, which leads to fast implant degradation thus not leaving enough time for surrounding tissue to heal [20]. Fast corrosion rates of Mg in body environment yields excess of corrosion byproducts, which have toxic effect on surrounding tissue and can cause necrosis [21]. Thus previous studies on Mg as implant material mostly investigated corrosion rates when exposed to bodily fluids and means of its control [22]. Some other studies addressed utilization of Mg or Mg alloys as surface coatings applied onto a surface of Ti and Ti alloy to improve implants bioactivity [23]. All above-mentioned issues, concerning both Ti and Mg biocompatible materials, create great need for new type of biomaterial which presents improvement to current concepts.

Motivated by these factors, the objective of this study is to present a new type of biomedical metal-metal composite, named BIACOM (bio active composite metal), fabricated by cost effective PM warm extrusion technique. Hybrid BIACOM structure comprises bio-inert Ti matrix, which remains permanent after implantation and provides mechanical performance of implant, and biodegradable Mg component, unmixable element in Ti. Firstly, present Mg lowers E compared to conventional Ti and Ti alloys, which should reduce stress shielding phenomenon. Secondly, Mg provides basis for selective biodegradation of implant by creating pores on surface and volume of implant, where there is exposure to corrosive medium, while Ti matrix stays intact. Diluted Mg is replaced by new bone formation after implantation, by which it should improve osseointegration, it further lowers E and improves mechanical compatibility with bone. Unlike few other pioneering works on Ti–Mg composites [8, 11, 24], reduction of E of BIACOM does not deteriorate markedly other mechanical properties. Thus, BIACOM implants are aimed for intense load-bearing application, such as dental implants, often subjected to rigorous fatigue loading and is to be compared directly to Ti Grade 4, the reference material of dental implants. This study introduces concept of

BIACOM, wherein microstructure and mechanical properties of BIACOM are characterized in as-fabricated condition and after exposure to body fluid—Hank's solution (HS).

Experimental

Affordable hydrogenation–dehydrogenation (HDH) process commercial purity CP Ti 99.4 wt% powder with the median particle size (d_{50}) of 84.5 µm (<150 µm fraction) and gas atomized Mg 99.8 wt% powder with d_{50} = 30 µm powder were used in this study (Fig. 1). Particle size distribution of both powders was determined by FRITSCH Analysette 22 laser diffraction system. A mixture of Ti + 12 vol.% Mg powders was homogenized for 30 min in Turbula shaker. Powder mixture was cold compacted by cold isostatic pressing (CIP) at 150 MPa. CIP blanks were compressed by uniaxial vacuum pressing (VP) at 420 °C and 425 MPa. VP blanks were consolidated by direct extrusion (DE) at 475 °C and the reduction ratio of 16:1 to a final Ti + 12 vol.% Mg composite profile with the diameter of 7.5 mm.

The density of as-extruded composite was measured using the Archimedes' principle. Microstructure of as-extruded composite was characterized using scanning electron microscope (SEM, JEOL 7500 machine) equipped with energy dispersive X-ray spectrometer (EDS, Oxford). Tensile specimens with the gauge diameter of 3 mm and the length of 21 mm were machined along the longitudinal direction of extruded composite. Eight tensile bars were tested using a Zwick Roell 1474 machine at a strain rate of ~$5 \times 10^{-4} s^{-1}$. E of as–extruded composite was measured by dynamic mechanical analysis (DMA, TAQ800 machine) using $4 \times 1 \times 55$ mm^3 bars and the three-point bending method. DMA specimens were extracted from as-extruded composite profile longitudinally to the extrusion

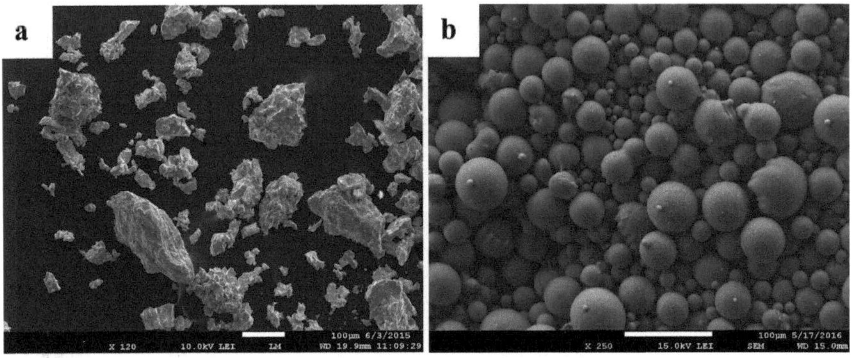

Fig. 1 SEM images of HDH Ti (**a**) and atomized Mg (**b**) powders

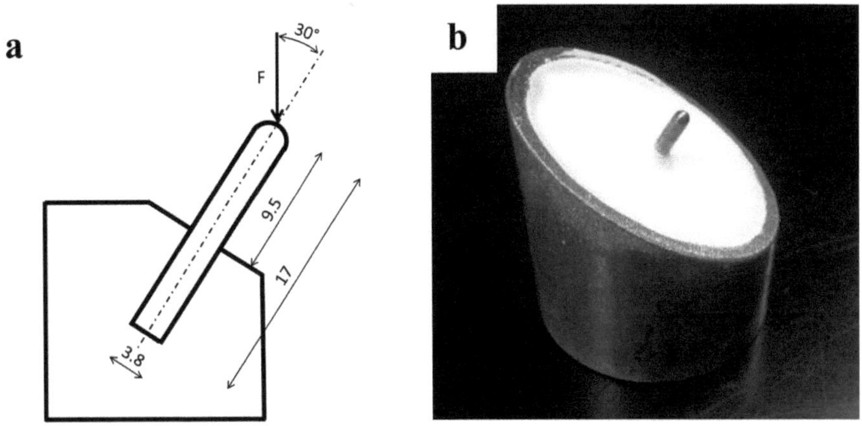

Fig. 2 Fatigue testing set up: schematics (**a**); the implant embedded in a supporting PMMA resin structure (**b**)

direction by wire electrical discharge machining (WEDM) and then dry polished. The dimensions of the DMA specimens were measured with an accuracy of 0.005 mm. DMA testing of as-extruded composite was conducted at room temperature with a frequency of 1 Hz, a preload force of 0.5 N, and a sinusoidal force having amplitude of 3.5 N for 4 min. The E was measured using two samples, and each sample was measured four times on each side. The error of the E measurements was <1%. Fatigue testing of mini-implant shaped specimens machined from as-extruded composite was conducted to follow the ISO 14801 standard for endosseous dental implants [25]. Implants were mounted in a PMMA holder at an angle of 60° with the horizontal in an experimental setup shown in Fig. 2. Dry environment testing was conducted using Walter + bai AG, LFV 50–HH servo hydraulic testing machine, equipped with DIGIWIN 2000-EDC120 digital control system. Loading was set to vary following a sinusoidal pattern with the frequency of 15 Hz and was applied vertically through the center of the hemisphere at the end of implant. Testing was conducted for 5×10^6 cycles, with the minimum force set to 10% of the maximum force applied.

DMA specimens fabricated from as-extruded composite were immersed to HS kept at 37 °C. After 1 day dilution in HS the specimens were subsequently immersed to 10 $g.l^{-1}$ CrO_3 solution for 1 min to clean the corrosion product and then ultrasonically cleaned in deionized water for another minute. DMA testing and SEM observation were performed on diluted specimens. 3D surface analyze of diluted samples was realized by Alicona InfiniteFocus machine. The same procedure was conducted after additional 2 and 3 days immersion in Hank's solution, resulting in an overall immersion time of 6 days.

Results and Discussion

Microstructure of as-Extruded Ti + 12 Vol.% Mg

Used PM consolidation approach results in sound Ti + 12 vol.% Mg composite profile with a low density of 4.12 ± 0.02 g.cm^{-3}. The density of as-extruded composite is ~9% lower compared to CP Ti Grade 4. Further decrease of composites density is expected due to depletion of Mg. The apparent density is 98.8 ± 0.44% of the theoretical density. The porosity of ~1.2% is fully acceptable assuming that composite is manufactured via PM approach and that porosity will further escalate later due to Mg dilution upon implantation. Microstructure of as-extruded composite in transversal direction is characterized by Mg phase embedded in Ti matrix (Fig. 3a). Mg component is well dispersed in Ti matrix and no distinct Mg conglomerates are observed (Fig. 3c, e). Mg powder particles are deformed severely during warm extrusion by Ti powder particles of significantly higher strength and deformation resistance. As a result, microstructure of the composite in transversal direction shows Mg filaments of large aspect ratio elongated and arrayed into extrusion direction (Fig. 3b, d and f). Break-through pressure determined during warm DE is only ~720 MPa. The pressure limit of industrial extrusion presses with larger diameter containers is typically at ~850 MPa. Thus, a future up-scaling of applied consolidation process is feasible. The coefficient of thermal expansion (CTE) of Mg is approx. three fold higher than the one of Ti. Rather large CTE mismatch could lead to disintegration and formation of porosity at Ti–Mg interfaces during cooling down after extrusion due to pronounced thermal shrinkage of Mg phase embedded in Ti matrix. Nevertheless, detailed microstructural characterization reveals that Ti–Mg interfaces are generally free of porosity and voids (Fig. 4). As confirmed by EDS line analyses, Ti–Mg interfaces are not enriched for O (Fig. 4). This points out that intense formation of TiO_2 and MgO layers on surface of Ti and Mg powders is avoided during consolidation process [26]. Is to be mention that machining of implants from extruded BIACOM profiles has to be realized under condition where no or non-corrosive coolant agents are used. Formation of surface porosity on BIACOM implants prior implantation is undesirable because of bacteria may likely be trapped within surface pores.

The Mechanical Properties of as-Extruded Ti + 12 Vol.% Mg

The tensile tests of as-extruded composite confirm very good reproducibility of the mechanical properties, as demonstrated by eight stress-strain curves (Fig. 5). The average values of obtained ultimate tensile strength (UTS), 0.2% strain offset yield stress ($YS_{0.2}$) and elongation (ε) are summarized in Table 1. Expectedly, composite possesses lower strengths, when compared to Mg free PM Ti fabricated using the

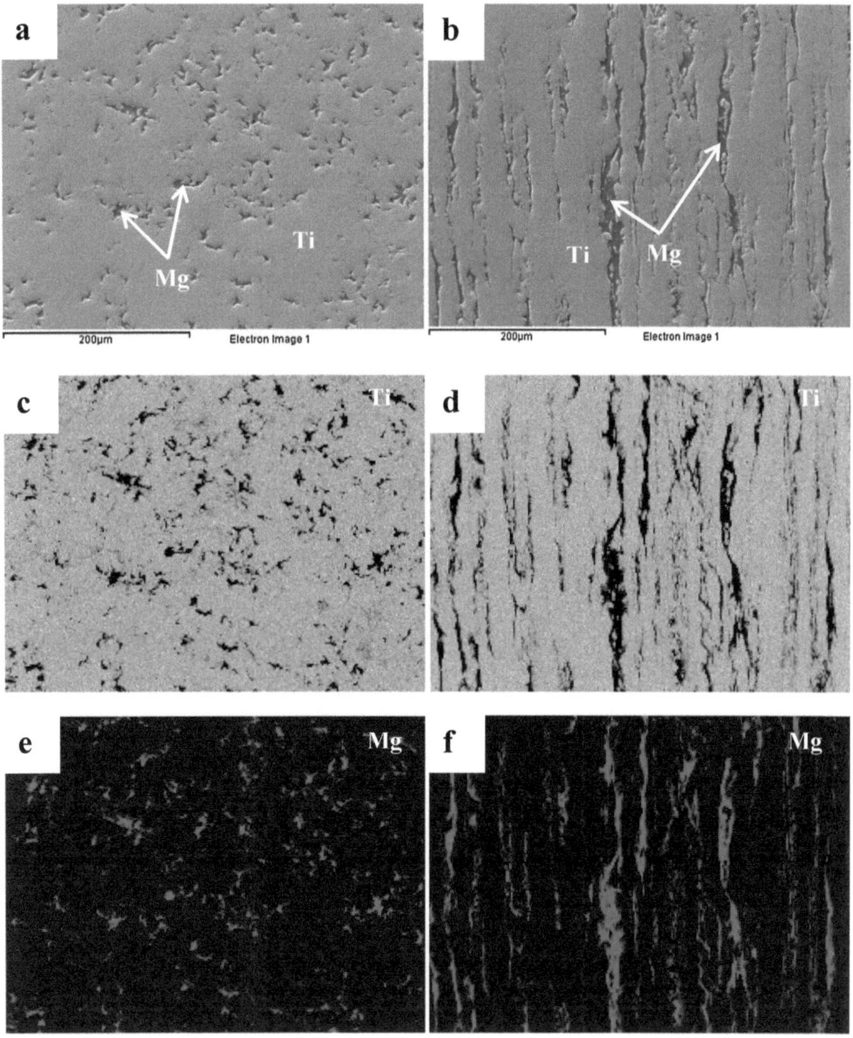

Fig. 3 SEM micrographs of as-extruded Ti + 12 vol.% Mg composite in **a** transversal and **b** longitudinal direction. The representative Ti (**c**, **d**) and (**e**, **f**) Mg elemental maps obtained by EDS analyses

same procedure [26], Table 1. This implies that Mg phase in composite does not contribute effectively to a load transfer during plastic deformation and composite strength is governed solely by Ti matrix. Despite of clean Ti–Mg interfaces, free of any voids and pores (Fig. 4), no metallurgical bonding between unmixable Ti and Mg components is established. Nevertheless, composite reaches reasonably good strengths (UTS, $YS_{0.2}$), which exceed the tabular strengths for Ti Grade 4 PM90

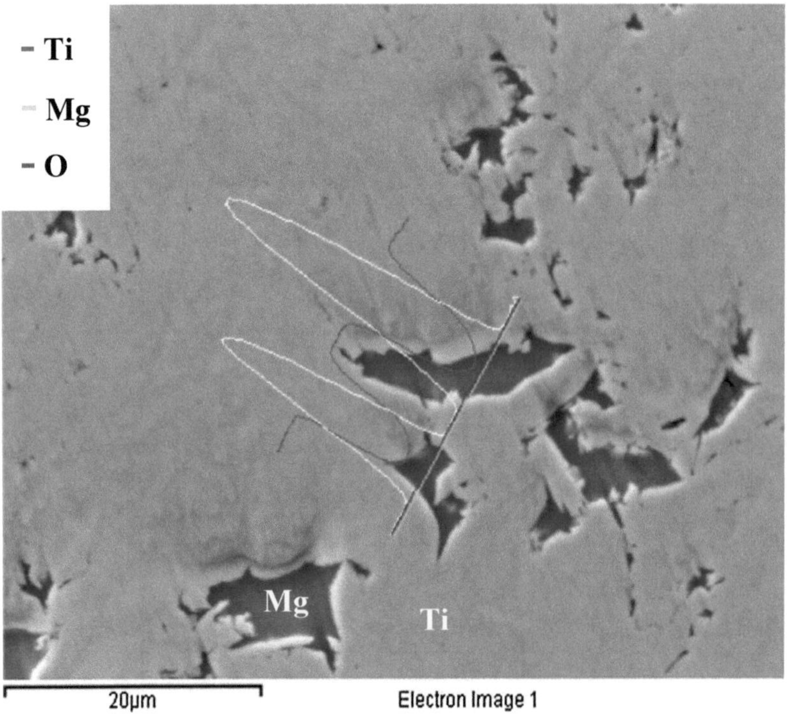

Fig. 4 EDS line analysis performed across of Ti–Mg interfaces in as-extruded Ti + 12 vol.% Mg composite shown in transversal direction

(Table 1) and are comparable to the minimum values for cast CP Ti Grade 4 as defined by the ASTM B988 and B265 standards, respectively.

As-extruded composite has E of 92.1 GPa (Table 2). This represents valuable ~12% decrease compared to CP Ti Grade 4 with E = 105 GPa, by which reduction of stress-shield phenomenon for BIACOM implants may be expected. It needs to be emphasized, that composite with lowered E possesses acceptably high strength, necessary for endosseous implants used in intense load-bearing application. Following a simple rule of mixture and assuming that Mg component with E_{Mg} = 45 GPa effectively contributes to load transfer in elastic region, calculated E for Ti + 12 vol.% Mg is 97.8 GPa (E_{Ti} = 105 GPa is used for calculation). A large discrepancy between the calculated and measured E (97.8 and 92.1 GPa, respectively) points out again to inferior metallurgical bonding at Ti and Mg interfaces. If only Ti matrix with E_{Ti} = 105 GPa would contribute to transfer of elastic load, the calculated E value equals to 92.4 GPa, what was in good agreement with measured E (92.1 GPa). This suggests that contribution of Mg component to load transfer in both elastic and plastic region is very limited.

Compared to cast and PM CP Ti Grade 4 material studied composite shows rather low ε = 1.4% (Table 1). Lower ε is due to the presence of Mg phase and

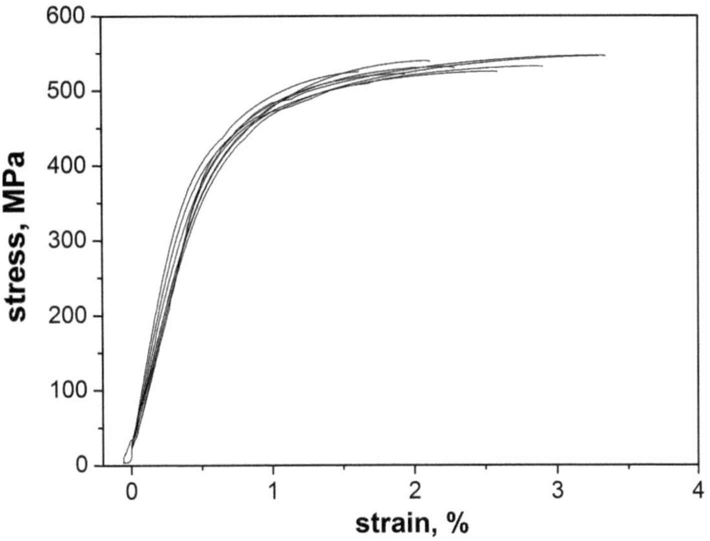

Fig. 5 The tensile stress-strain curves of as-extruded Ti + 12 vol.% Mg composite

Table 1 Ultimate tensile strength (UTS), 0.2% strain offset yield stress ($YS_{0.2}$) and elongation (ε) of as-extruded Ti + 12 vol.% Mg composite obtained from tensile tests including standard deviations. Data for PM Ti fabricated using the same procedure [26], for PM fabricated Ti Grade 4 PM90 according to ASTM B265 and minimum requirements for the reference cast CP Ti Grade 4 material according to ASTM B265 are included too

Material	References	UTS (MPa)	$YS_{0.2}$ (MPa)	ε (%)
PM Ti + 12 vol.% Mg	This study	529 ± 7	450 ± 12	1.4 ± 0.3
PM Ti	[26]	701 ± 4	512 ± 11	13 ± 2
Ti Grade 4 PM90	ASTM B988	495	435	16
CP Ti Grade 4 minimum	ASTM B265	550	483	15

Table 2 Young's modulus (E) of Ti + 12 vol.% Mg composite in as-extruded state and after dilution in Hank's solution for 1, 3 and 6 days

Sample state	E (GPa)
As-extruded	92.1 ± 0.6
Immersion time 1 day	91.8 ± 0.6
Immersion time 3 days	91.4 ± 0.9
Immersion time 6 days	91.4 ± 0.6

inferior bonding at Ti–Mg interfaces. Mg filaments act as stress concentrators and reduce ε, eventually. Though it needs to be emphasized, that mechanical properties (incl. ε) obtained in tension represent more valuable information than data from often used compression tests [8, 11]. Furthermore, composite reaches ε with an excellent reproducibility ($\varepsilon = 1.4 \pm 0.3\%$), what renders any discussion on composites brittleness. Furthermore, neither there is an actual reason nor any relevant

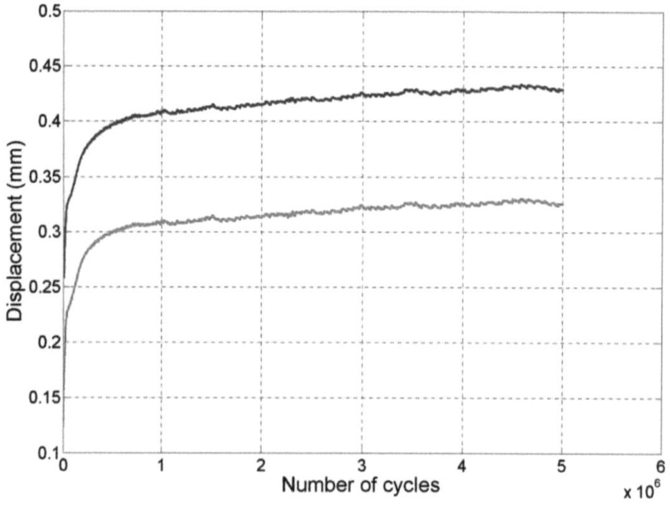

Fig. 6 The maximum (*blue curve*) and minimum (*red curve*) displacements determined during fatigue testing of Ti + 12 vol.% Mg composite implants under at sinusoidal 30–300 N loading

standard specifies ε being higher than 1.4% for already fabricated dental implant. Thus, relatively low ε does not limit utilization of BIACOM composite as material used for fabrication of endosseous dental implants.

The dental implants must comply with ISO14801 standard for fatigue testing. The ISO 14801 standard is designed to reproduce the unfavorable conditions that must be successfully endured by an endosseous dental implant [25]. It is widely believed that PM products suffer from poor fatigue life performance. In addition, a presence of Mg may negatively contribute to fatigue performance of composite. Nevertheless, dynamic testing of composite samples shows that under proscribed rigorous conditions three samples endured 5×10^6 cycles at sinusoidal 30–300 N loading (Fig. 6). No micro-cracks are detected during subsequent surface analysis of the tested samples at the clamping point. The fatigue limits obtained for BIACOM specimens are at the level of the fatigue limits of ~300 N reported for CP Ti Grade 3 implants of the same diameter [27]. Thus, in spite of present porosity and presence of Mg phase, dental implants fabricated of studied composite have capability of withstanding required fatigue limits.

Dilution of Ti + 12 Vol.% Mg in Hank'S Solution

To get insight into mechanical property changes due to dilution in conditions similar to body environment, E of as-extruded composite as a function of dilution time in HS is pursued. Table 2 summarizes composites E diluted for 1, 3 and 6 days. E decreases slightly after 1 day immersion from 92.1 to 91.8 GPa and then

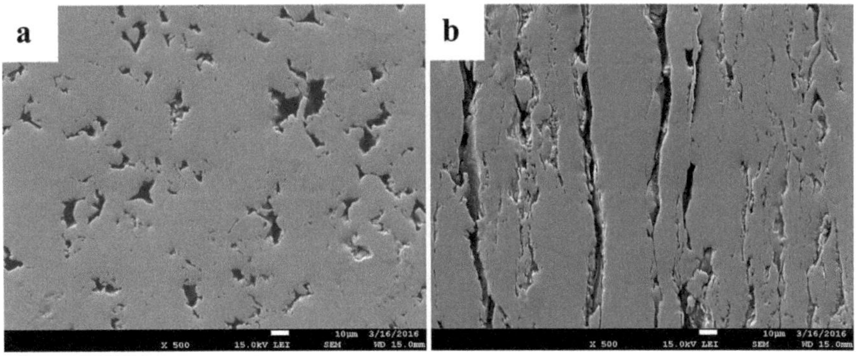

Fig. 7 SEM micrographs of Ti + 12 vol.% Mg composite diluted in Hank's solution for 1 day in **a** transversal and **b** longitudinal direction

it tends to drop gradually down to 91.4 GPa after 3 days immersion. Increase of immersion time to 6 days does not lead to further change and E remains constant. It needs to be point out that slight variations of E are at limit of used DMA analyze. Nevertheless, SEM of immersed samples reveals obvious depletion of Mg areas from surface and volume of composite already after one day exposure to Hank's solution (Fig. 7). Corroded Mg leaves pores at prior Mg filaments sites, while bioinert Ti matrix remains intact and provides mechanical performance. Thus, subtle but repeatedly determined decrease of composites E after immersion is undoubtedly attributed to corrosion of Mg. Subtle decrease of E = 0.7 GPa after 3 days corrosion in HS is attributed to limited contribution of Mg to transfer of load in elastic region as discussed for as-extruded composite. Theoretically, if all Mg contributed effectively to E of BIACOM with 12 vol.% Mg, i.e., contribution of Mg is 5.4 GPa, decrease of 0.7 GPa would represent overall dilution of ~13 vol.% Mg present in composite. Moreover, taking in account a circumstance, that dilution of Mg from a surface got saturated at a certain depth, it is obvious that changes in E due to immersion depend on actual cross-section of DMA specimen. For thinner DMA specimen E would decrease more profoundly. This would be a case of crucial thin wall thread areas of BIACOM implants. In any case, it is apparent that E of composite would decrease further after implantation and thus mechanical compatibility between implant and bone structure will improve further. Moreover, Mg diluted from a surface of BIACOM provides bases for enhanced macro and micro porosity (Fig. 8). As a result, bonding strength due to ingrowth of new tissue and bone tissue attachment is expected to improve as corroded Mg is replaced by new bone formation after implantation and it proceeds towards implants volume.

Nevertheless, further systematic in vitro and in vivo testing of BIACOM, which is part of future scope to be published soon (Fig. 9), is needed mainly to address critical corrosion rate of Mg component in body environment and subsequent osseointegration process of implanted BIACOM.

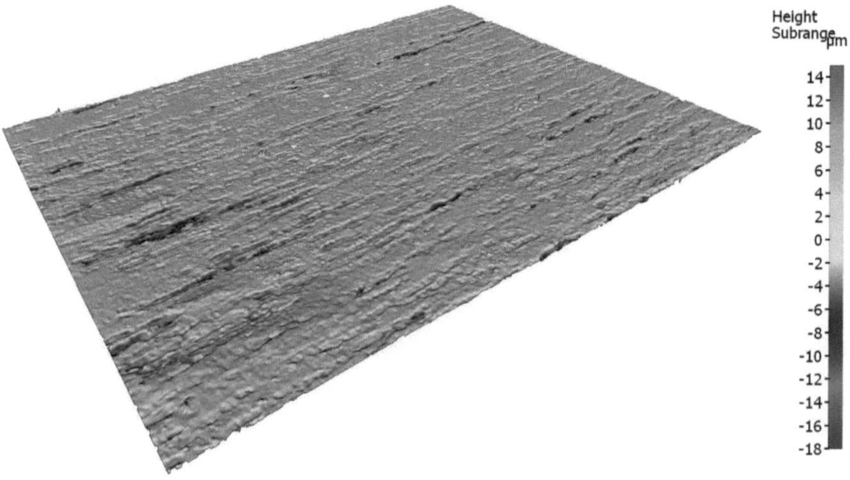

Fig. 8 3D surface analysis of Ti + 12 vol.% Mg composite diluted in Hank's solution for 6 days in longitudinal direction

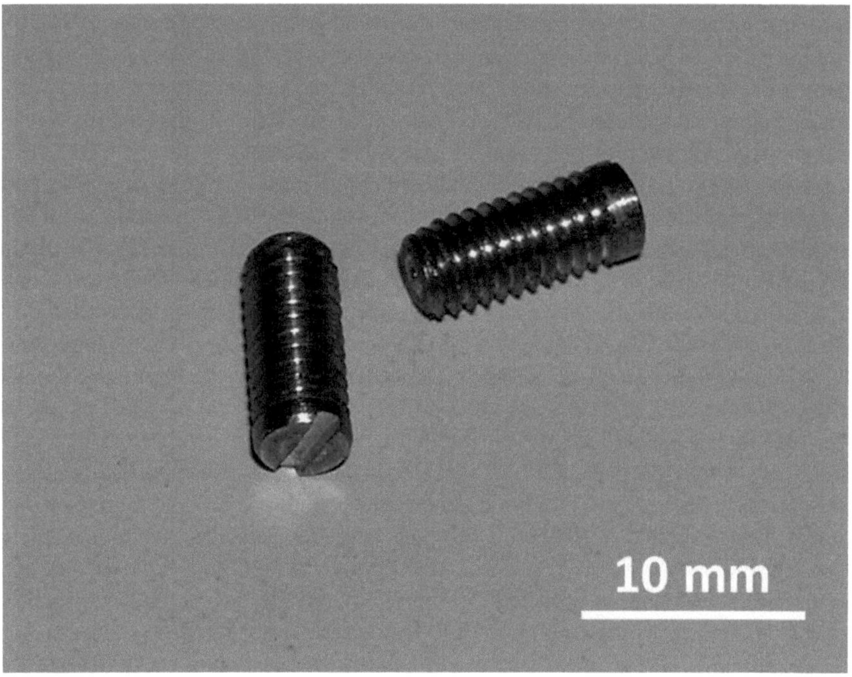

Fig. 9 Ti + 12 vol.% Mg composite implants ready for in vivo studies

Conclusions

In this study a new type of Ti-12vol.%Mg bimetallic composite material, manufactured through a cost effective approach of the powder metallurgical warm extrusion of elemental Ti and Mg powders, was presented. Microstructure of the composite consisted of Mg filaments arrayed along extrusion direction and homogenously distributed within Ti matrix. The as-extruded composite showed reduced Young's elastic modulus of 92.1 GPa and low density of 4.12 g.cm^{-3}, while the mechanical strength and fatigue performance of Ti Grade 4, the reference material used for dental implants, was maintained. Upon testing in Hank's solution biodegradable Mg component gradually and selectively diluted and it formed pores at prior sites within Ti matrix, which stayed intact. Corrosion of Mg provided further decrease of Young's modulus and enhanced macro and micro surface roughness. Based on performed mechanical and corrosion testing it is anticipated that the bioactive composite can be used as a material to manufacture biomedical implants with improved mechanical compatibility and osseointegration potential for intense load-bearing applications.

Acknowledgements Financial support from the ITMS 26240220088, the SRDA APVV-0556-12, the VEGA 2/0065/16 and 2/0158/16, and the SAS—TUBITAK 2014/5 projects, and MESS No. 065-0650446-0435 University support "Research of new ceramic materials and technologies in dental prosthodontics" is gratefully acknowledged. The authors thank Lubomir Orovcik of IMSAS for help with SEM characterization and Sanja Solic of CTS for 3D surface analyze.

References

1. B. Aksakal, M. Gavgali, B. Dikici, J. Mater. Eng. Perform. **19**(6), 894–899 (2010)
2. K. Ida, Y. Tani, S. Tsutsumi, T. Togaya, T. Nambu, K. Suese, T. Kawazoe, M. Nakamura, H. Wada, Dent. Mater. J. **4**, 191–185 (1985)
3. M. Long, H.J. Rack, Biomaterials **19**, 1621–1639 (1998)
4. J.M. Anderson, A. Rodriguez, D.T. Chang, Elsevier Semin. Immunol. **20**, 86–100 (2008)
5. P. Moshayedi, G. Ng, J.C. Kwok, G.S. Yeo GS, C.E. Bryant, J.W. Fawcett, K. Franze, J. Guck, Biomaterials **35,** 3919–3925 (1985)
6. M.T. Mohammed, Z.A. Khan, A.N. Siddiquee, Int. J. Chem. Nucl. Mater. Metall. Eng. **8**, 772–777 (2014)
7. J.I. Qazi, H.J. Rack, Adv. Eng. Mater. **7**, 993–998 (2005)
8. Z. Esen, B. Dikici, O. Duygulu, A.F. Dericioglu, Mater. Sci. Eng. A **573**, 119–126 (2013)
9. Z. Esen, S. Bor, Mater. Sci. Eng. A **528**, 3200–3209 (2011)
10. C.S.Y. Jee, N. Ozguven, Z.X. Guo, J.R.G. Evans, Metall. Mater. Trans. B **31**, 1345–1352 (2000)
11. G. Jiang, C. Wang, Q. Li, J. Dong, G. He, Mater. Sci. Eng. C **47**, 142–149 (2015)
12. B.V. Krishna, S. Bose, A. Bandyopadhyay, Acta. Biomater. **3**, 997–1006 (2007)
13. M. Qian, Powder Metal. **46**, 29–44 (2010)
14. W. Kroll, Z. Metallkunde **29**, 189–192 (1937)
15. J.E. Barnes, W. Peter, C.A. Blue, Mater. Sci. Forum **618–619**, 165–168 (2009)
16. M. Qian, F.H. Froes, *Titanium Powder Metallurgy: Science, Technology and Applications* (Elsevier Science, 2015) ISBN: 9780128009109

17. C. Poletti, M. Balog, T. Schubert, V. Liedtke, C. Edtmaier, Compos. Sci. Tech. **68**, 2171–2177 (2008)
18. P.J. Anderson, V.M. Svoyatkovsky, F.H. Froes, Y. Mahajan, D. Eylon, *Modern Developments in Powder Metallurgy, Metal Powder Industries Federation* (Princeton NJ, 1981), pp. 537–549
19. M.P. Staiger, A.M. Pietak, J. Huadmai, G. Dias, Biomaterials **27**, 1728–1734 (2006)
20. R. Zeng, W. Dietzel, F. Witte, N. Hort, C. Blawert, Adv. Mater. 03–14 (2008)
21. N.T. Kirkland, Corros. Eng Sci. Tech. **47**, 322–328 (2012)
22. G. Song, Corros. Sci. **49**, 1696–1701 (2007)
23. O. Duygulu, A.A. Kaya, G. Oktay, F.C. Sahin, Mater. Sci. Forum **546–549**, 417–420 (2007)
24. G. Jiang, Q. Li, C. Wang, J. Dong, G. He, Mater. Desi. **67**, 354–359 (2015)
25. EN ISO 14801:2003, *Fatigue Test for Endosseous Dental Implants* (2003)
26. M. Balog, J. Viskic, P. Krizik, Z. Schauperl, M. Snajdar, Z. Stanec, A. Catic, Key Eng. Mater. **704**, 351–359 (2016)
27. P. Sevilla, C. Sandino, M. Arciniegas, J. Martínez-Gomis, M. Peraire, F.J. Gil, Mater. Sci. Eng. C **30**, 14–19 (2010)

Part VII
Alloys and Compounds for Thermoelectric and Solar Cell Applications V

Thermoelectric Behaviour of Polyvinyl Acetate/CNT Composites

Hussein Badr, Mostafa Amr Youssef, Hebatullah Sayed Abd Elsalam, Mirna Abd Elrahman Mohamed, Mohamed Gamal El-kholy, Iman S. El-Mahallawi and Ahmed A. Abdel-Rehim

Abstract In this work an organic polymer/carbon nanotubes composite is introduced for low cost, versatile, nontoxic, light weight and high performance thermoelectric materials. Organic polymers have low thermal conductivity which is desirable for thermoelectric field. Composite films of polyvinyle Acetate and multi wall carbon nanotubes (MWCNT) are demonstrated as a new candidate material for thermoelectric application. MWCNT wt. percentage ranging from 1 to 70 wt% were manufactured and tested for thermopower and electrical conductivity. Dioctyl sodium sulfosuccinate is used as the wetting agent in ratios of 2, 25 and 100% of CNT weight content. A study of the effect of the dispersantto CNT ratio on the electrical conductivity and thermopower of the composite was also carried out. The composite thermoelectric properties were measured as a function of CNT concentration at room temperature.

Keywords Thermoelectric materials • Organic polymers • Carbon nanotubes • Power factor • Dispersants

Introduction

Harvesting electricity from waste heat by means of thermoelectric devices has been gaining scientific and technological interest over the past decades. A thermoelectric device is a solid state device that directly converts the heat flow into current flow and vice versa the current flow to temperature difference without moving parts or

H. Badr · M.A. Youssef · H.S.A. Elsalam · M.A.E. Mohamed · M.G. El-kholy · I.S. El-Mahallawi (✉)
Faculty of Engineering, Cairo University, Cairo 12613, Egypt
e-mail: saiman@eng1.cu.edu.eg; ielmahallawi@bue.edu.eg

A.A. Abdel-Rehim
Faculty of Engineering, The British University in Egypt, El Shorouk, Egypt

A.A. Abdel-Rehim
Sabbatical leave from Benha University, Shoubra, Egypt

The Minerals, Metals & Materials Society, *TMS 2017 146th Annual Meeting & Exhibition Supplemental Proceedings*, The Minerals, Metals & Materials Series,
DOI 10.1007/978-3-319-51493-2_27

working fluids [1]. This material capability makes thermoelectrics a challenging issue for thermoelectricity generation and refrigeration. It is very effective in generating low power and compact devices [2]. The same thermoelectric device is also used for cooling systems particularly in applications that require robustness and silence [3].

The material's ability to act as a thermoelectric device is determined by the material physical parameters: thermal conductivity, electrical conductivity and Seebeck coefficient. Seebeck coefficient is defined as the voltage induced by temperature difference between the device two ends [4]. The dimensionless thermoelectric figure of merit (ZT) is a common measure of a material's energy conversion efficiency [5]:

$$ZT = \frac{S^2\sigma}{K}T$$

where S (in V/K) is Seebeck coefficient, σ(in S/m) is electrical conductivity K (in W/mK) is thermal conductivity, and T (in K) is the absolute temperature, However, in the absence of precise data on the thermal conductivity, the performance is assessed with the TE power factor (PF) [5]:

$$PF = S^2\sigma$$

where S (in V/K) is Seebeck coefficient, σ(in S/m) is the electrical conductivity.

The challenge to develop thermoelectric materials with superior performance is to tailor the interconnected thermoelectric physical parameters: electrical and thermal conductivity and Seebeck coefficient. Semiconductors are the most suitable class of materials for thermoelectric applications as they have best combination of physical properties to achieve the largest figure of merit. Until now, Bi_2Te_3 alloy and its derivatives are the most efficient thermoelectric materials with a relatively large ZT ~ 1 and PF ~ 1000 $\mu W/mK^2$ [6]. Although the tellurides group showed promising results, the research is directed to the organic materials because of the problems associated with inorganics: high density, high cost, brittleness, poor processability, toxicity and finally rareness of elements [2, 7]. Organic polymers possess interesting properties of light weight, low cost, versatility, ease of fabrication [7] and low thermal conductivity (~0.1 – 0.5 W/m K) which is desirable for thermoelectric applications [6]. Thus using polymer based composites in replacement of Bi_2Te_3 with the same performance can increase the power density six or seven times due the large difference of densities between Bi_2Te_3 (~7.86 gm/cm^3) and polymers (~1.2 gm/cm^3) [3]. Synthetic conducting polymers (with doping) and/or electrically conductive nanofiller materials are used for fabricating thermoelectric composites to overcome the disadvantage property of low electrical conductivity of polymers and thus obtaining high value of dimensionless figure of merit. Both carbon nanotubes and silicon nanowires are used as nanofiller materials due to their unique features of electrical conductivity but carbon nanotubes are widely used due to the problems associated with silicon nanowire fabrication.

Yu et al. prepared a composite of 20 wt% carbon nanotubes in a matrix of polyvinyle acetate (PVAc) and obtained an electrical conductivity of 4800 S/m and Seebeck coefficient around ~45 μV/K, the figure of merit was calculated to be ~0.006 [3]. Replacing the carbon nanotubes mixture (metallic and semiconducting double and triple CNT) to single walled carbon nanotubes and increasing the percentage from 20 to 60 wt% raised the electrical conductivity to 1.35×10^5 S/m but lowered the Seebeck coefficient to ~15 μV/K due to increase of the number of conductive paths [2]. Marioarty et al. made another thermoelectric composite of 12 wt% carbon nanotubes in (PVAc) using a semiconducting stabilizer: sodium deoxycholate (DOC) or meso-tetra (4-carboxyphenyl) porphine (TCPP). This composite exhibited an electrical conductivity of 1474 and 7108 S/m for multiwall carbon nanotubes (MWCNT) and double wall carbon nanotubes (DWCNT). The Seebeck coefficient of DWCNT composite was also examined to show 71 and 78 μV/K with DOC and TCPP [5]. These groups of experiments showed that carbon nanotubes in all forms are promising for thermoelectric generation with some optimization to the performance and cost wise.

Experimental Work

Materials Used

Poly vinyle acetate (PVAc) is the polymer used in this study. The polymer was used in the emulsion form that is 42 wt% solids in water as the composite matrix starting material. MWCNTs, with a diameter range from 17–35 nm, was used as the electrically conductive nano filler material. The MWCNTs were fabricated by chemical vapor deposition method (CVD) multi Quartz tubes. MWCNTs contain amorphous carbon and iron impurities, which are added as a catalyst during the CNT manufacturing. MWCNTs were used in as received state, without any chemical modification. An anionic dioctyl sodium sulfosuccinate wetting agent was added to the aqueous suspension as a surfactant to stabilize the nanotubes in water during composite preparation. Dioctyl phthalate (AOT) was the plasticizer used to overcome macro cracking that occurs during polymer composite dehydration.

MWCNTs/PVAc Composite Preparation

MWCNTs: dioctyl sodium sulfosuccinate (AOT) with the weight percentage of 1: 0.02 were balanced and distilled water was added and mixed with a magnetic stirrer for 10 min at 300 rpm. The PVAc emulsion was added to the suspension with a predetermined weight percentage and then stirred for 5 h at 1000 rpm. The plasticizer was finally added to the suspension with a weight percentage of 0.5–2:

1 MWCNTs and stirred for 2 h at 1000 rpm. The suspension was cast onto an aluminum substrate (purity ~99.5%) using film applicator for thick film product (thickness ~1 mm). The cast was left for drying suspensions under ambient conditions for 2 days to ensure that all moisture content was removed prior to working. Samples with different MWCNTs weight percentage (10, 20, 30, 50, 60 and 70) were prepared to study their thermoelectric behavior.

MWCNTs/PVAs Composite Characterization

In order to study the thermoelectric behavior of the fabricated composite, Seebeck coefficient and electrical conductivity measurements were recorded to get the power factor which is an indication for the energy conversion efficiency. Samples with standard dimensions of 5 $\times$ 20 mm^2 in surface area and thickness of ~0.7 mm were cut to be characterized. For Seebeck coefficient measurements, a heater was attached to one end of the sample and the other end was kept at room temperature to apply a temperature difference. The hot and cold temperature was measured at both ends using two thermocouples and the potential difference was measured with a multimeter. Seebeck coefficient was obtained from the voltage—temperature curve from the relation ($S = \Delta V/\Delta T$) where ΔV is the thermoelectric potential difference and ΔT is the temperature difference. The electrical resistance was measured with a multimeter then converted to electrical conductivity by multiplying a geometric factor. Also the electrical conductivity variation with temperature increase was recorded.

Results and Discussion

Seebeck Coefficient and Power Factor

Figure 1a shows the enhancement that occurs to the PVAc based composite as a function of the MWCNT weight percentage. Pure PVAc is an insulating organic polymer with electrical conductivity in the order of 10^{-13} S/m [8], while carbon nanotube is a high conducting nano material with electrical conductivity in the range 10 to 10^7 S/m depending on the chirality [9]. The electrical conductivity of the fabricated composite is shown to increase linearly with the MWCNT content increase to reach 0.25, 98 and 172 S/m at 10, 50 and 70 wt% at 90 °C. This large enhancement occurred as the CNT contents increased, which is attributed to the increase in charge carrier concentrations and to the creation of a highly conducting CNT network, which assists the charge carrier mobility.

Seebeck coefficient, which is the material's ability to convert the thermal energy into electrical energy, was recorded to fully determine the composite power factor. Seebeck coefficient was found to increase from 3.2 to 7.5 μV/K (at $\Delta T = 60$) as the

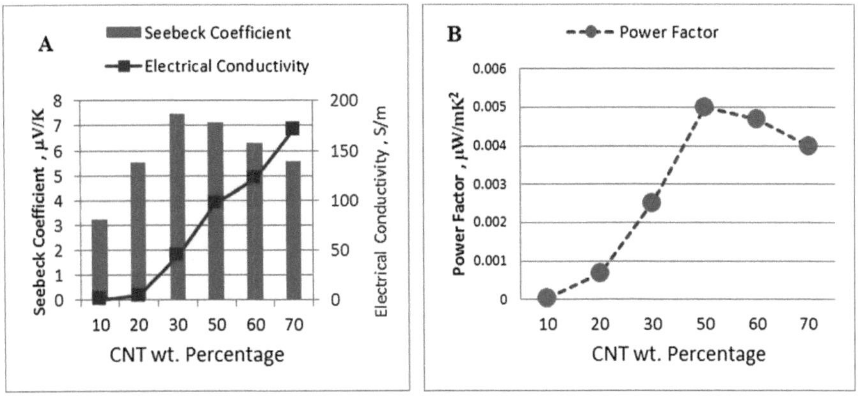

Fig. 1 Thermoelectric properties of composites containing 10, 20, 30, 50, 60 and 70% MWCNT. **a** Electrical conductivity (*red line*) and Seebeck coefficient (*blue columns*) as a function of MWCNT wt%. **b** Calculated power factor as a function of MWCNT wt% (color in online)

CNT content increase from 10 to 30 wt%. 30 wt% CNT resulted the peak value of Seebeck coefficient, where Seebeck coefficient started to decrease slightly to reach 5.6 at 70 wt% CNT (at $\Delta T = 60$), afterwards. After 50 wt% CNT, though Seebeck coefficient values continued to decrease, the charge carrier concentrations increased due to the large enhancement in carriers mobility and the composite electrical conductivity, accordingly. The composite's Seebeck coefficients showed positive values, which indicate that, the dominant charge carriers are holes. The majority of holes is referred to the method of CNT fabrication in oxygen rich atmosphere that results in oxygen doped CNT [10].

The power factor was calculated (as shown in Fig. 1b) for all samples from the relation $PF = S^2.\sigma$ where S is the Seebeck coefficient and σ is the composite electrical conductivity. The composite's power factor is found to increase as the CNT content increase and the composite containing 50 wt% CNT has the largest power factor of 0.005 $\mu W/mK^2$ (at $\Delta T = 60$). After that the power factor decreases as the Seebeck coefficient decreases (Seebeck coefficient is represented by a squared value in the power factor formula).

The Effect of the Dispersant Content on the Thermoelectric Properties of the Fabricated Composite

The first stage in this work aimed at finding the optimum value of CNT addition to get the largest power factor ("Seebeck Coefficient and Power Factor"). Hence, the following stage aimed at investigating the effect of the dispersant content on the thermoelectric properties of the fabricated composite. An anionic surfactant dioctyle sodium sulfosuccinate (AOT) is the dispersant used through this study.

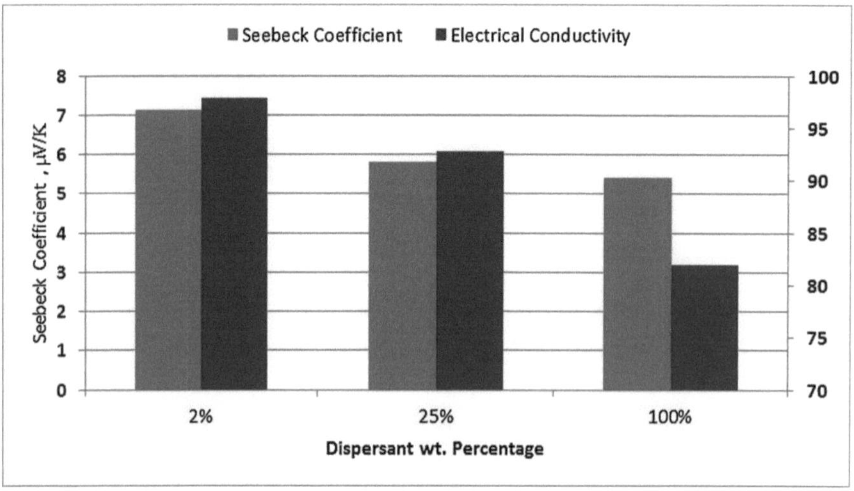

Fig. 2 The composite Seebeck coefficient and electrical conductivity as a function of dispersant content

Daniel L. Akins et al. reported that AOT is used for debungling the CNTs and stabilizing them through the polymer emulsion without affecting the CNT electronic properties [11]. It is shown from Fig. 2 that Seebeck coefficient decreases from 7.13 to 5.4 µV/K as the dispersant content increases from 2 to 100 wt. percentage. Also the electrical conductivity decrease from 98 to 82 S/m at the same dispersant content. The addition of excess AOT would result in a measurable decrease in the composite's Seebeck coefficient and electrical conductivity. That is attributed to the formation of a dispersant thick layer that covers the CNTs and prevents the tube-to-tube connection and, accordingly, impedes the carriers mobility and current flow.

Power Measurements

In order to go through the engineering approach in product design, all studied material properties were transferred to measurable values of output power. The maximum safe working temperature of the fabricated composite is 110 °C to avoid any thermal degradation or sample deformation. The output voltage and current of the fabricated composite was measured as a function of temperature.

As shown from Fig. 3a, The samples containing 30, 50 and 60 wt% CNT show almost the same output voltage of 0.02 mV at 40 °C, the voltage increases to 0.57 mV at 110 °C and the sample containing 70% CNT show a maximum voltage of 0.41 mV at 110 °C. This is not the case with the current measurement as the composite containing 50% CNT shows the largest output current of 0.7 µAmpere at 40 °C, and the output current increases to 8 µAmpere at 110 °C (as shown in Fig. 3b). On the other hand the composite containing 70 wt% CNT shows the

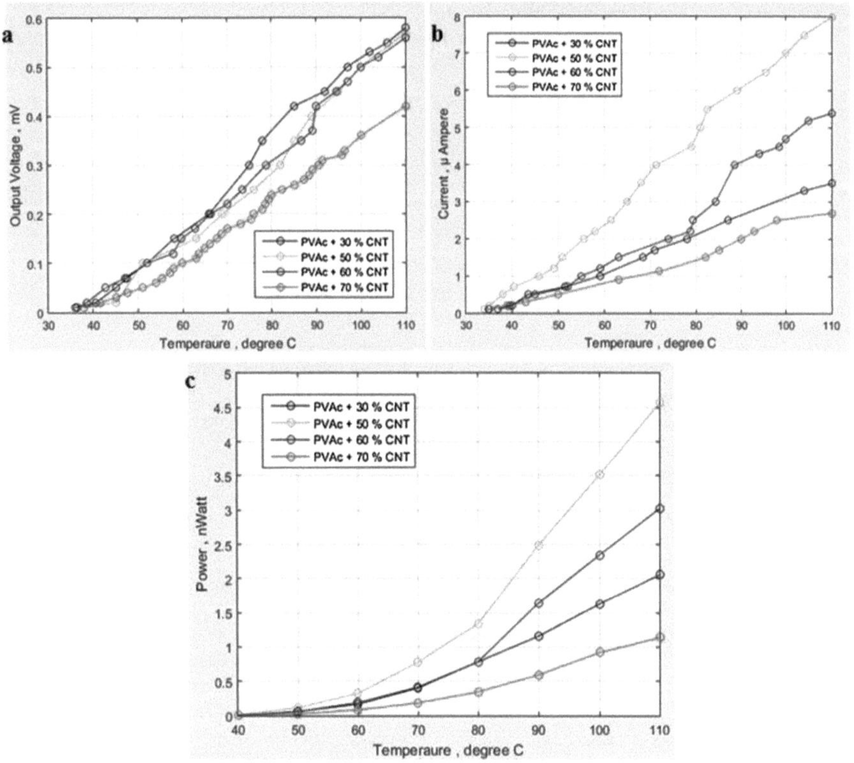

Fig. 3 The output **a** voltage, **b** current, and **c** power of the fabricated composite as a function of the hot side temperature in °C

lowest output current of 0.28 at 110 °C. Consequently this composite showed the lowest output power of 1.2 nWatt at 110 °C while the composite containing 50 wt% CNT showed the largest output power of 4.5 nWatt at 110 °C (as shown in Fig. 3c). These values are comparable to the values reported by Corey A. Hewitt et al., where an output power of 137 nWatt was achieved for 72 connected samples of multi-layered MWCNT/polyvinylidene fluoride composite at $\Delta T = 50$ °C [12].

Conclusion

In summary, the following conclusions may be drawn from this work:

(1) Polyvinyle Acetate based composites containing different weight percentage of MWCNT up to 70 wt% were fabricated, and demonstrated that the optimum value of MWCNT addition for thermoelectric performance is 50 wt%.

(2) Though the highest Seebeck coefficient was 7.5 μV/K (at ΔT = 60) at 30 wt% CNT and the highest electrical conductivity was achieved from the composite containing 70 wt% CNT, the largest power factor was 0.005 at 50 wt% CNT.
(3) The maximum output power of 4.5 nWatt at 110 °C was achieved from a 5 mm thick × 20 mm^2, which proves promising future for this type of thermoelectric material.
(4) 2 wt% dispersant is the optimum value of dispersant content and the addition of excess dispersant to improve the CNT dispersion has a negative effect on the composite thermoelectric properties, as it reduces Seebeck coefficient and the electrical conductivity and hence lowers the output power.

Acknowledgements The authors wish to thank the Non-ferrous Material Extraction and Processing Lab in the Department of Metallurgical Engineering for hosting this work. The authors also wish to thank the Egyptian Academy of Science and ITEDA for sponsoring this work, within their Students' Support Programme.

References

1. C. Meng, C. Liu, S. Fan, A promising approach to enhanced thermoelectric properties using carbon nanotube networks. Adv. Mater. **22**, 535–539 (2010)
2. C. Yu, K. Choi, L. Yin, J.C. Grunlan, Light-weight flexible carbon nanotube based organic composites with large thermoelectric power factors. ACS Nano **5**, 7885–7892 (2011)
3. C. Yu, Y.S. Kim, D. Kim, J.C. Grunlan, Thermoelectric behavior of segregated-network polymer nanocomposites. Nano Lett. **8**(12), 4428–4432 (2008)
4. D.M. Rowe, *Thermoelectrics Handbook—Macro to Nano* (CRC Press, 2005), pp. **1**–2
5. G.P. Moriarty, J.N. Wheeler, C. Yu, J.C. Grunlan, Increasing the thermoelectric power factor of polymer composites using a semiconducting stabilizer for carbon nanotubes. Carbon **50**, 885–895 (2012)
6. C. Bounioux, Pablo Dıaz-Chao, M. Campoy-Quiles, M.S. Martın-Gonzalez, A.R. Goni, R. Yerushalmi-Rozene, C. Muller, Thermoelectric composites of poly(3-hexylthiophene) and carbon nanotubes with a large power factor. Energy Environ. Sci. **6**, 918–925 (2013)
7. Q. Yao, L. Chen, W. Zhang, S. Liufu, X. Chen, Enhanced thermoelectric performance of single-walled carbon nanotubes/polyaniline hybrid nanocomposites. ACS Nano **4**, 2445–2451 (2010)
8. G. Wang, Z. Tan, X. Liu, S. Chawda, J.-S. Koo, V. Samuilov, M. Dudley, *Conducting MWNT/poly(vinyl acetate) Composite Nanofibres by Electrospinning* (IOP Publishing Ltd, 2006), pp. 5829–5835
9. T.W. Ebbesen, H.J. Lezec, H. Hiura, J.W. Bennett, H.F. Ghaemi, T. Thio, Electrical conductivity of individual carbon nanotubes. Nature **382**, 54–56 (1996)
10. P.G. Collins et al., Extreme oxygen sensitivity of electronic properties of carbon nanotubes. Science **287**, 1801–1804 (2000)
11. S.C. Wang, H. Yang, S. Banerjee, I.P. Herman, D.L. Akins, *AOT Dispersed Single-Walled Carbon Nanotubes for Transistor Device Application* (Elsevier B.V., 2007), pp. 843–845
12. C.A. Hewitt, A.B. Kaiser, S. Roth, M. Craps, R. Czerw, D.L. Carroll, Multilayered carbon nanotube/polymer composite based thermoelectric fabrics. Nano Lett. **12**(3), 1307–1310 (2012)

Part VIII
Applications of Solidification Fundamentals

Microstructure Characteristics of A356 Nanocomposites Manufactured via Ultrasonic Cavitation Processing Under Controlled Solidification Conditions

Yang Xuan and Laurentiu Nastac

Abstract It has been recently demonstrated that microstructure as well as mechanical properties of metal-matrix-nano-composites can be highly improved by using ultrasonic cavitation. The main reason is ultrasonic cavitation processing can help to spread and break the clusters of nanoparticles in the melt during the fabrication process. The present investigation attempted to evaluate how different solidification conditions effect the microstructure of nanocomposites and how nanoparticles influence the evolution of microstructure during solidification. An A356 alloy and Al_2O_3 nanoparticles are used as the matrix and the reinforcements, respectively. Ultrasonic cavitation was applied during the induction melting to assist in the dispersion of the nanoparticles. Ultrasonic cavitation was also applied during solidification for different cooling conditions. Experimental results indicated that ultrasonic cavitation processing and Al_2O_3 nanoparticles play an important role for microstructure refinement. In addition, Al_2O_3 nanoparticles can refine the eutectic Si.

Keywords Ultrasonic cavitation during solidification · Nanocomposites · Microstructure · Controlled solidification conditions

Introduction

Since Al matrix composites can offer desirable properties at low price like high specific strength, high wear resistance, and controllable expansion coefficient, etc., they have been treated as suitable materials in automobile, aerospace and military applications [1–9]. However, normally micro-size particles have been used to

Y. Xuan (✉) · L. Nastac
Department of Metallurgical and Materials Engineering, The University of Alabama, Box 870202, Tuscaloosa, AL 35487, USA
e-mail: yxuan@crimson.ua.edu

L. Nastac
e-mail: lnastac@eng.ua.edu

The Minerals, Metals & Materials Society, *TMS 2017 146th Annual Meeting & Exhibition Supplemental Proceedings*, The Minerals, Metals & Materials Series,
DOI 10.1007/978-3-319-51493-2_28

strengthen the matrix, which could significantly jeopardize the ductility of these composites. It has been recently proven that by using nano-size ceramic particles as reinforcement to fabricate metal-matrix-nano-composites (MMNCs) can markedly improve matrix properties, while maintaining good ductility and high temperature creep resistance. The most important part of the fabrication of MMNCs is to achieve uniform dispersion of the nano-size particles [10–12].

It has been also shown that the Ultrasonic Stirring Technology (UST) can be used during the alloy solidification process to refine the matrix microstructure and to change it's the grain morphology from dendritic to globular/non-dendritic [13–18]. Zhang et al. [19] indicated that not only the matrix, but also the precipitate phase can be refined via ultrasonic cavitation processing during solidification.

Ultrasonic cavitation processing is indeed a promising way to fabricate MMNCs since it will induce nonlinear effects like cavitation and acoustic streaming into the melt, which are very useful to spread and break the nanoparticle clusters in the melt [20–23]. However, there are only few studies that focused on the ultrasonic cavitation effects on the distribution of nanoparticles during the solidification process and how the nanoparticles distribution influence the microstructure of the matrix or of the precipitate phase.

The present study uses A356 alloy as matrix and Al_2O_3 as reinforcement nano-particles to fabricate MMNCs, then to evaluate how different solidification conditions effect the microstructure of this nanocomposite and how nanoparticles influence the evolution of microstructure during solidification.

Experimental Approach

In the present work, commercial aluminum alloy A356 was used as the metallic matrix. Jia and Nastac [20] indicated that one of the optimum type of nano-size particles is Al_2O_3, and the optimum amount of those particles are 1.0 wt%. Thus, in this study, Al_2O_3 (spherical shape, average diameter is about 20 nm) was used as reinforcement particles. The ultrasonic equipment parameters are as follows: maximum power is 2.4 kW; frequency is18 kHz, the diameter of the Nb ultrasonic probe is 40 mm and the probe amplitude is 20 μ. The sketch of the UST system is shown in Fig. 1.

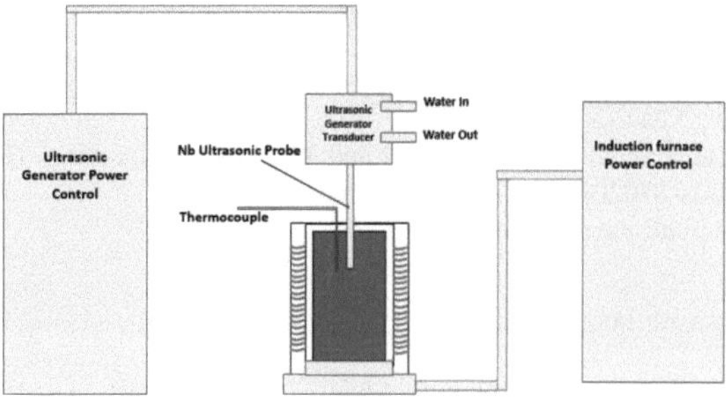

Fig. 1 Sketch of the ultrasonic cavitation processing system

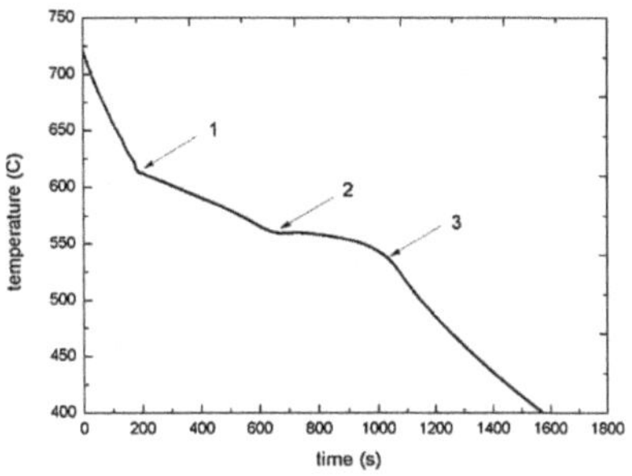

Fig. 2 Temperature—time curve of present alloy

An induction furnace by Inductotherm was used to melt the alloy. When the melt temperature reached about 750 °C, the ultrasonic probe was inserted to about 50 mm below the melt surface to perform ultrasound stirring. 1.0 wt% Al_2O_3 was added into the cavitation area during a 15 min time-frame. K-type thermocouples inserted 50 mm beneath the melt surface were used to evaluate the temperature during the experiment. Temperature-time curve of furnace cooling melt has been plotted to identify the phase transformation temperature, as shown in Fig. 2. The liquidus temperature of this alloy is about 614 °C (inflexion 1) Inflexion 2 indicates the start of the eutectic phase at about 560 °C. Inflexion 3 shows the end of solidification at about 538 °C. In order to eliminate probe temperature effect in the experiment, the ultrasonic probe stayed in the melt until melt temperature decrease to about 600 °C. Ultrasonic cavitation processing was controlled by turning on/off the ultrasonic generator power.

Three types of experiments were designed: non-treatment UST processing, above-liquidus UST processing and below-liquidus UST processing. In the non-treatment process, after 15 min ultrasonic cavitation processing at 750 °C, the furnace power was shut down to let the melt cooled inside the furnace. The second approach, above-liquidus processing, ultrasonic cavitation was applied into the molten alloy over a range of temperatures from 650 to 620 °C when melt cooled inside the furnace. The third experiment was similar with second onr but the ultrasonic cavitation processing temperature range was changed to 614–600 °C. The microstructures of the specimens based on these experiments were then characterized in details via OM.

Results and Discussion

Figure 3 shows the comparison of obtained microstructures of the nanocomposite without the use of ultrasonic cavitation processing during solidification, with ultrasonic cavitation processing at 650 to 620 °C during solidification, and with ultrasonic cavitation processing at 614 to 600 °C during solidification. From Fig. 3a it can be seen that when the ultrasonic cavitation processing during solidification was not used, the microstructure was dendritic in shape with a large grain size. By

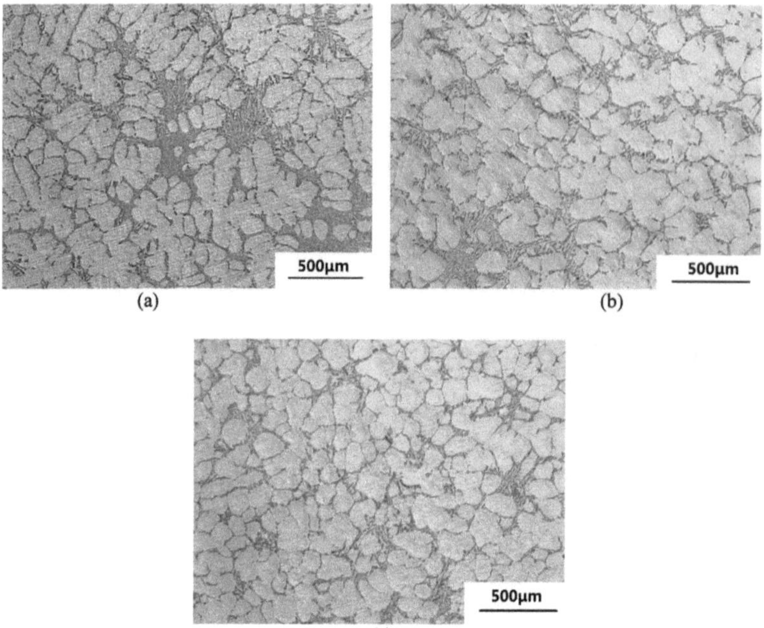

Fig. 3 Optical micrographs of cast samples at 50X **a** no ultrasonic cavitation processing during solidification, **b** applied ultrasonic cavitation over a range of temperatures from 650 to 620 °C during solidification, **c** applied ultrasonic cavitation over a range of temperatures from 614 to 600 °C during solidification

applying the ultrasonic cavitation during solidification, regardless of the temperature range, the microstructure is refined significantly and the grain morphology changed from dendritic to globular, as shown in Fig. 3b, c. By comparing the microstructures from Fig. 3a, b and c it can be concluded that for A356 with 1.0 wt % Al_2O_3 nanoparticles the ultrasonic cavitation processing plays a key role to refine the microstructure during solidification. There are two hypotheses about ultrasonic cavitation refinement during solidification [14–17]. One is ultrasound will increase the nucleation potential of the melt. This is because when applying ultrasonic cavitation, the temperature and pressure characteristics of the melt will change periodically at high frequencies, which can then locally increase the number of nuclei. In addition, strong convection produced by ultrasonic cavitation can promote diffusion of solute to increase nucleation, too. Other hypothesis is ultrasonic cavitation can produce shock waves, which can induce the breakage of the dendrites tips to refine the microstructure. In the present work, the alloys obtained by applying the ultrasonic cavitation at 650 to 620 °C range, which is higher than the liquidus temperature (614 °C), still got fine globular grain structure. Since there are few dendritic grains grown at 650 to 620 °C range, this may suggest that in the given experiments the dominant mechanism for grain refinement is likely that the increase in nucleation potential was not because of dendrite fragmentation.

Jian et al. [14] indicated that for A356, the microstructure refinement can be observed only when the ultrasonic cavitation is applied until to lower than liquidus temperature (614 °C) during solidification. In addition, it has been proven that for Al-based alloys without nanoparticles, there is no grain refinement when the ultrasonic cavitation was applied only at the range that is higher than the liquidus temperature [16, 19, 24]. In the present results, as Fig. 3b shown, the use of ultrasonic cavitation at 650 to 620 °C (6 °C above liquidus temperature) still refined the microstructure significantly, which means that Al_2O_3 nanoparticles affect the microstructure evolution to refine the microstructure in the solidification process. More studies are required to determine the mechanism of the effects of nanoparticles on the microstructure refinement during solidification.

Figure 4 shows the eutectic silicon shape of the alloys produced under different solidification conditions. The arrows indicate the eutectic Si shape in different locations. One is the eutectic Si phase that been refined with short length (shorter than 10 μm). The reason of refinement is silicon atoms will aggregate around Al_2O_3 nanoparticles (20 nm in present experiment) in melt, thus forming some precursors. The agglomerations and attachment of these precursors and the individual silicon atoms will cause nucleation and growth of silicon particles [25]. The results show

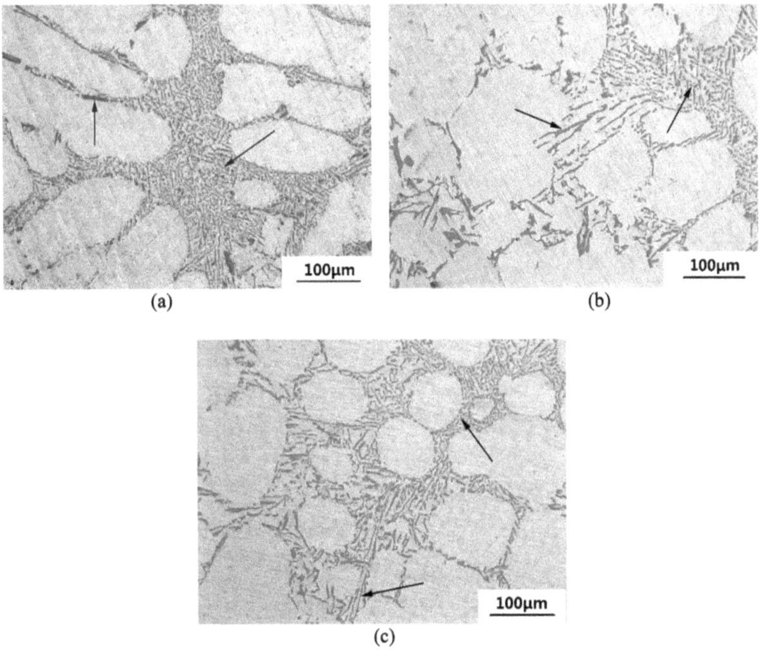

Fig. 4 Eutectic Si phase refinement at 200X **a** no ultrasonic cavitation during solidification, **b** applied ultrasonic cavitation over a range of temperature from 650 to 620 °C during solidification, **c** applied ultrasonic cavitation over a range of temperature from 614 to 600 °C during solidification

that area with well refined eutectic Si phase can be observed regardless of the solidification conditions. This indicate that ultrasonic cavitation processing during solidification may have significant effect on eutectic Si refinement. On the other side, the refinement was not complete, coarse eutectic Si (longer than 100 μm) still can be found in the alloy, as shown by the arrows in Fig. 4. However, the reason why the partial refinement still exists requires more study.

Conclusion

A microstructure refinement and a morphological transition from dendritic grain to globular grain were obtained when the ultrasonic cavitation was applied to an A356 alloy reinforced with Al_2O_3 nanoparticles during solidification. Ultrasonic cavitation processing during solidification plays an important role to refine the microstructure by increasing the heterogeneous nucleation in the melt and by enhancing the dendrite fragmentation. A356 with Al_2O_3 nanoparticles can have globular/non-dendritic grain when ultrasonic cavitation is applied above the liquidus temperature, but the reason of the refinement needs more study. In addition, the addition of Al_2O_3 nanoparticles and not the use of ultrasonic cavitation during solidification can refine the eutectic Si during solidification.

Acknowledgements Authors wish to express thanks to the China Scholarship Council (CSC) for the financial support.

References

1. B. Su, H.G. Yan, G. Chen, J.L. Shi, J.H. Chen, P.L. Zeng, Study on the preparation of SiCp/Al-20Si-3Cu functionally graded material using spray deposition. Mater. Sci. Eng. A **527**, 6660–6665 (2010)
2. S.M. Seyed Reihani, Processing of squeeze cast Al6061-30vol% SiC composites and their characterization. Mater. Des. **27**, 216–222 (2006)
3. D. Bozic, B. Dimcic, O. Dimcic, J. Stasic, V. Rajkovic, Influence of SiC particles distribution on mechanical properties and fracture of DRA alloys. Mater. Des. **31**, 134–141 (2010)
4. M.P. De Cicco, X. Li, L.-S. Turng, Semi-solid casting (SSC) of zinc alloy nanocomposites. J. Mater. Process. Technol. **209**(18–19), 5881–5885 (2009)
5. C.S. Goh, J. Wei, L.C. Lee, M. Gupta, Simultaneous enhancement in strength and ductility by reinforcing magnesium with carbon nanotubes. Mater. Sci. Eng. A **423**(1), 153–156 (2006)
6. A.B. Elshalakany, T.A. Osman, A. Khattab, B. Azzam, M. Zaki, Microstructure and mechanical properites of MWCNTs reinforced A356 aluminum alloys cast nanocomposites fabricated by using a combination of rheocasting and squeeze casting techniques. J. Nanomaterials **2014**, 1–14 (2014)
7. X. Liu, S. Jia, L. Nastac, Ultrasonic cavitation-assisted molten metal processing of cast A356-nanocomposites. Int. J. Metalcast. (AFS) **8**(3), 51–58 (2014)

8. S. Jia et al., Ultrasonic processing of 6061-based nanocomposites for high performance applications, in *Advances in the Science and Engineering of Casting Solidification: an MPMD Symposium Honoring,* eds. by D.M. Stefanescu, L. Nastac (Wiley, 2015), pp. 31–36
9. S. Jia, D. Zhang, Y. Xuan, L. Nastac, An experimental and modeling investigation of aluminum-based alloys and nanocomposites processed by ultrasonic cavitation processing. Appl. Acoust. **103**, 226–231 (2016)
10. K. Akio, O. Atsushi, K. Toshiro, T. Hiroyuki, Fabrication process of metal matrix composite with nano-size SiC particle produced by vortex method **49**(4), 149–154 (1999)
11. S.L.U. Filho, R. Rodriguez, J.C. Earthman, E.J. Lavernia, Synthesis of diamond reinforced Al-Mg nanocrystalline composite powder using ball milling. Mater. Sci. Forum **416–418**, 213–218 (2003)
12. J.R. Groza, Sintering of nanocrystalline powders. Int. J. Powder Metalurgy. **35**(7), 59–66 (1999)
13. X. Jian, T.T. Meek, Q. Han, Refinement of eutectic silicon phase of aluminum A356 alloy using high-intensity ultrasonic vibration. Scripta Mater. **54**, 893–896 (2006)
14. X. Jian, H. Xu, T.T. Meek, Q. Han, Effect of power ultraound on solidification of aluminum A356 alloy. Mater. Lett. **59**, 190–193 (2005)
15. X. Liu, Y. Osawa, S. Takamori, T. Mukai, Grain refinement of AZ91 alloy by introducing ultrasonic vibration during solidification. Mater. Lett. **62**, 2972–2875 (2008)
16. G. Wang, M.S. Dargusch, D.H. StJohn, Effect of ultrasonic treatment on the as-cast grain structure during the solidification of an Al-2Cu alloy, in *Paper presented at the 9th Pacific Rim International Conference on Advanced: Materials and Processing (PRICM9)*, Kyoto, Japan, August 1 to 5, 2016, pp. 344–347
17. G. Wang, M.S. Dargusch, M. Qian, D.G. Eskin, D.H. StJohn, The role of ultrasonic treatment in refining the as-cast grain structure during the solidification of an Al-2Cu alloy. J. Cryst. Growth **408**, 119–124 (2014)
18. M. Qian, A. Ramirez, A. Das, Ultrasonic refinement of magnesium by cavitation: clarifying the role of wall crystals. J. Cryst. Growth **311**, 3708–3715 (2009)
19. L. Zhang, D.G. Eskin, A. Miroux, L. Katgerman, *Formation of Microstructure in Al-Si Alloys Under Ultrasonic Melt Treatment*, (Light Metals 2012, ed. Carlos E. Suarez, TMS, March 11–15, 2012, Orlando, FL), pp. 999–1004
20. S. Jia, L. Nastac, The influence of ultrasonic stirring on the solidification microstructure and mechanical properties of A356 alloy. Chem. Mater. Eng. **1**(3), 69–73 (2013)
21. D. Zhang, L. Nastac, Numerical modeling of the dispersion of ceramic nanoparticles during ultrasonic processing of aluminum-based nanocomposites. J. Mater. Res. Technol. **3**(4), 296–302 (2014)
22. D. Zhang, S. Jia, P. Allison, L. Nastac, Numerical modeling of the dispersion of ceramic nanoparticles during ultrasonic processing of 6061-based nanoparticles, in *Proceedings of the "Advanced Metal Casting Technologies" Symposium, TMS*, October 12–16, 2014, Pittsburgh, PA, pp. 695–702
23. S. Jia, D. Zhang, L. Nastac, Experimental and numerical analysis of the 6061-based nanocomposites fabricated via ultrasonic processing. J. Mater. Eng. Perform. **24**(6), 2225–2233 (2015)
24. J.I. Youn, Y.J. Kim, Nucleation enhancement of Al alloys by high intensity ultrasound. Japan. J. Appl. Phys. **48**, 07GM14 (2009)
25. Y. Zhang, H. Zheng, Y. Liu, L. Shi, R. Xu, X. Tian, Cluster-assisted nucleation of silicon phase in hypoeutectic Al-Si alloy with further inoculation. Acta Mater. **70**, 162–173 (2014)

Part IX
Biological Materials Science

Development of Sponge Structure and Casting Conditions for Absorbable Magnesium Bone Implants

Stefan Julmi, Christian Klose, Ann-Kathrin Krüger, Peter Wriggers and Hans Jürgen Maier

Abstract In the case of bone defects, there are two different methods to close such defects. One option is to use bone autografts, but therefore the bone graft has to be cut off from the same person's hip. In this case the patient has to undergo an additional surgery, which bears complications, like causing inflammations. Absorbable, open-pored implants minimize these risks. Synthetic bone implants are typically made of ceramics, bioglass or polymers. In this study, magnesium alloys were investigated as absorbable porous bone substitute materials in which the bone can grow into. The main advantages are the design flexibility to produce individual implants by investment casting and mechanical properties similar to the bone. In order to adapt the degradation behavior to the bone's ingrowth behavior, the implant material has to be alloyed and coated. Moreover, to meet the mechanical requirements, finite element simulations of the sponge structure were used during the design phase of the structures and compression tests were conducted for experimental validation.

Keywords Sponge structure · Absorbable magnesium · Bone implants

Introduction

Natural bone is reconstructed continuously to repair defects and prevent ageing of the bone. Therefore the bone cells, i.e. osteoclasts, remove the old bone before the osteoblasts can rebuild the structure [1]. If a bone defect exceeds a critical size, the body's own healing mechanisms cannot close this defect. In this case the defect has to be closed by surgery. A widely used technique is the use of allogenic or

S. Julmi (✉) · C. Klose · H.J. Maier
Leibniz Universität Hannover, Institut für Werkstoffkunde (Materials Science), An der Universität 2, Garbsen, 30823, Germany
e-mail: julmi@iw.uni-hannover.de

A.-K. Krüger · P. Wriggers
Leibniz Universität Hannover, Institute of Continuum Mechanics, Appelstraße 11, Hannover, 30167, Germany

The Minerals, Metals & Materials Society, *TMS 2017 146th Annual Meeting & Exhibition Supplemental Proceedings*, The Minerals, Metals & Materials Series, DOI 10.1007/978-3-319-51493-2_29

autologous bone grafts. When using the body's own bone, the patient has to undergo a second surgery, which can cause inflammation and pain [2]. Allogenic bone grafts can lead to infections or transmission of diseases [3].

To avoid these problems, synthetic materials can be used in bone substitute implants. Traditional implant materials for this application can be categorized in ceramics, bioglass, polymers and metals. Ceramics and polymers can be synthetic materials like zirconia (ZrO_2) and polymethylmethacrylate (PMMA) or natural materials like calcium phosphates (CaP) and collagen [4, 5].

Absorbable polymers do not have the necessary strength and can cause foreign body reactions. In contrast, ceramics show good biocompatibility and high strength, but the Young's modulus is higher than that of the bone, which may result in stress shielding. Additionally, they have low ductility and, as a result, might fail quickly under shifting loads.

Magnesium (Mg) alloys, which will be investigated in this study, have an elastic modulus near that of the human bone and show higher ductility than ceramics as well as good biocompatibility and a sufficient yield strength [6].

To adapt the mechanical properties to the requirements, the material can be alloyed. The degradation behavior can be adapted by alloying as well as by coating the material. Previous studies demonstrated a reduced corrosion rate of the Mg alloy LAE442 in comparison with other Mg alloys [7]. Furthermore, an increase of ductility is observed by alloying with lithium [8]. Moreover, alloying Mg with Lanthanum results in strengthening and a good corrosion resistance [9, 10].

In a related study, implants were produced by infiltration casting [11], but with this method, a reproducible, defined pore structure could not be achieved. In contrast, with the use of an investment casting process individual parts with defined geometry and a high design flexibility can be produced. The basis of the present approach is the generative production process called rapid prototyping (RP). This technology is now used not only in biomedical engineering, but also in other industries like aerospace. As it is the case in patient-individual implants, RP is predestined for the production of complex parts with a small production quantity. Another advantage is the possibility to simulate the mechanical behavior of the actual implant geometry under stress and the mold filling during the casting process to adjust the pore structure. The aim of this study was to obtain a structure that can be casted and has the mechanical properties needed for the application as a bone substitute material.

Materials and Methods

Magnesium Alloys

The Mg alloys LAE442 (4 wt% Li, 4 wt% Al, 2 wt% rare earth mischmetal, balance Mg) and La2 (2 wt% La, balance Mg) were used. The alloy LAE442 was produced as proposed by Seitz et al. [12] and the La2 alloy was produced according to Weizbauer et al. [13].

Simulation-Assisted Design of the Sponge Structure

There are three requirements to the implants. They must have a pore size between 150–500 μm for optimized bone ingrowth [4], they should carry the patient's load during the ingrowth time of the bone and the desired geometry has to be castable depending on the alloy, i.e. the mold has to be filled completely by the melt. Three different implant geometries were designed. The implants will be tested in a future in vivo study with rabbits and therefore the cylindrical implants had a diameter of 4 mm and a height of 5 mm. Using the CAD models, the stress distribution was simulated with a compressive load of 60 N, as it is the case in the loaded rabbit tibia [14]. For the simulation the magnesium alloy LAE442 was used with a yield strength of 148 MPa, a compressive yield point of 150 MPa and a compressive strength of 420 MPa [12]. The simulation was implemented in Abaqus by using a network of tetrahedrons and the cylindrical shaped bodies were loaded along the z-axis.

Casting Simulation and Investment Casting

The first step of the investment casting process was the implant's design using the CAD tool SolidWorks. The CAD model was then cut into layers with a thickness of 0.0127 mm. The sliced shape was then produced by a fused deposition modelling (FDM) method using a T612 3D printer (Solidscape, Inc., Merrimack, USA), with which the wax gating system was created. This casting tree was embedded in a special mold material for magnesium investment casting. In this case, Gilcast AM (BK Giulini GmbH, Ludwigshafen, Germany) was used. The curing process was carried out according to the manufacturer's instruction [15].

After curing, the mold was kept at 300 °C. The casting process was performed in a laboratory induction furnace MC50 (Indutherm GmbH, Walzbachtal-Wössingen, Germany) using a ceramic crucible with graphite lining. At first the furnace went through a preheating process to evaporate liquids and then the alloy was placed in the crucible. Before casting, the chamber underwent two cycles of evacuation and flushing with argon inert gas (Ar). Afterwards, the Ar pressure was set to 1 bar in order to prevent oxidation. To cast the alloy, the furnace was heated up to 740 °C with 40% of the heating capacity. Once the melting temperature was reached, the whole casting chamber was tilted in order to let the melt flow into the mold. After the casting process, the chamber stayed in the tilted position and a pressure of 2 bar of argon gas was applied in the casting chamber to promote the mold filling.

After demolding, the form filling was analyzed using a μCT80 computer tomograph (SCANCO Medical AG, Brüttisellen, Switzerland) with a nominal spatial resolution of 10 μm, an exposure time of 600 ms, a tube voltage of 70 kVp and a tube current of 114 μA. In the present study, a structure with a pore size of 400 μm based on the geometry of the structure with 500 μm was employed.

Compressive Test

To investigate the mechanical behavior under load, the implants were tested according to the standard DIN 50134 using a Z010 testing machine (Zwick GmbH & Co. KG, Ulm, Germany). The load velocity was adjusted to 10^{-2} s^{-1}. The ratio of length to diameter deviates from the standard of 1.5–2. The ratio used was chosen according to necessary size for the rabbits (length: 5 mm, diameter: 4 mm).

Results

Maximum Pore Size: 500 μm

The samples tested in this study are shown in Fig. 1. The models in Fig. 1 A and B are having a similar web thickness, while the vertical elements in model A are continuous along the z-axis and in model B the vertical elements are shifted. In model C the pores are more spherical with a narrow size distribution. Here, the struts are connected at right angles.

Mold Filling of the Structures in Fig. 1 with the Alloy LAE442

Figure 2 shows casted samples representative for the three different geometries. The two structures A and B with similar web thicknesses are completely filled with no cast defects or unfilled struts. By contrast, the molds of the samples in Fig. 1c were not all completely filled in the casting process and large parts of the samples show casting defects. In Fig. 2 the difference between the good and the defective samples can be seen clearly. Typical casting defects were not completely filled samples, broken bars or filled pores.

Mechanical Simulations

Due to the bad mold filling of the shape in Fig. 1c, only the mechanical simulations of the ones in Fig. 1a, b are compared in Fig. 3. If the shape has continuous elements in z-direction, the effective stress on the struts is lower. This allows for a better stress transmission along the z-axis. But this is only the case for a z-directed force.

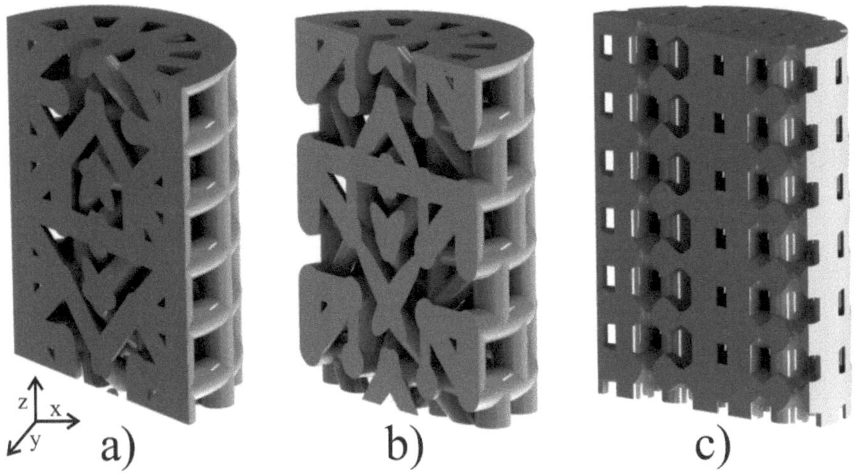

Fig. 1 Cross section of three different shapes designed with CAD

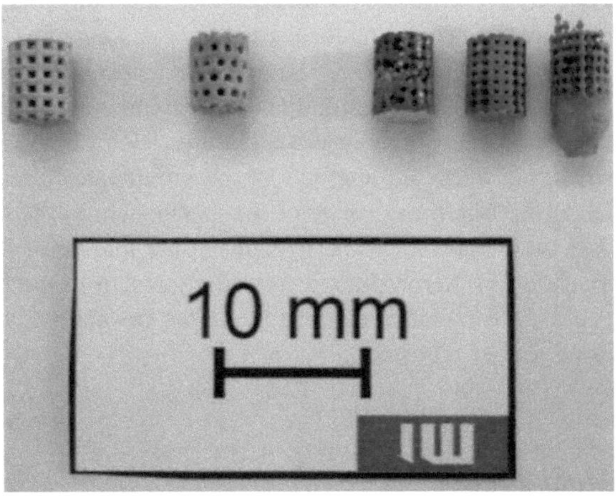

Fig. 2 Casted shapes of the previously mentioned structures. A is the on the *left*, followed by B and the three structures on the *right* are implants of the structure C

Due to the complete mold filling and the better behavior under stress, the shape with the continuous elements was slightly changed as it is shown in Fig. 4. In this final structure the struts have a thickness of 0.4 and 0.5 mm with a theoretical porosity of 41.35%. In Fig. 4 the CAD model is compared with the corresponding cast sample, recorded by μCT. The overall structure of the cast shape corresponds with the designed shape for both alloys. All of the struts are filled, however the thickness varies and the surface is not as flat as in the CAD constructed file.

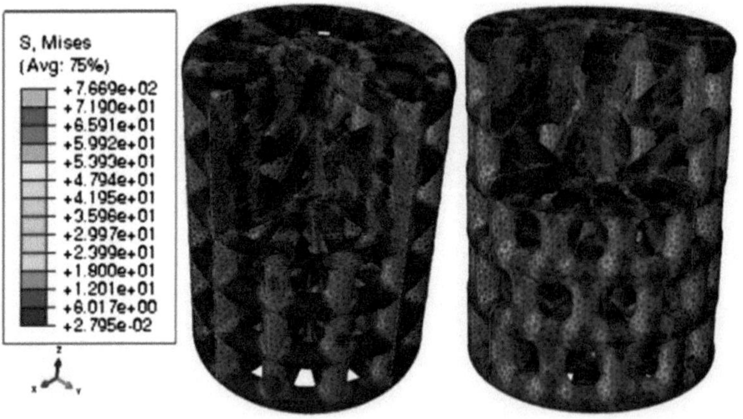

Fig. 3 Stress distribution in MPa of the structures a and b in Fig. 1 under the load of 60 N

Maximum Pore Size: 400 μm

Figure 5 shows a second shape, which was constructed on the basis of the previous experiments. The difference is that this structure has smaller elements and pores. In the model all the struts are having a thickness of 0.4 mm and the porosity is slightly lower (37.7%). As a result, the effective horizontal cross section area is larger. In the middle of the sample are irregularities. Here, it seems as if there are lots of small struts, but these are artefacts of the volume calculation by the CT. For this calculation the brightness range of the magnesium in the scans has to be set. But in this case there was still some gypsum in the middle of the shape and there is no hard transition in the brightness from the gypsum to the pores. As a result the gypsum couldn't be excluded completely for the calculation and also the solving process has to be further adjusted.

Mold Filling

As expected, the casting of finer structures is more challenging because the pores, which are filled with gypsum during the casting process, are smaller and therefore the gypsum is less stable. Additionally, the elements are smaller and therefore the melt solidifies earlier, which results in a not completely filled mold. Figure 5 shows that the structure can be filled completely, but the gating system has to be adapted in order to make the melt flow into the cavities with a smaller pressure thus preventing damage of the mold. Compared with the designed shape, the cast structure has the same derivations as in case of the structure with bigger pores. All the elements are filled, but there is a rougher surface.

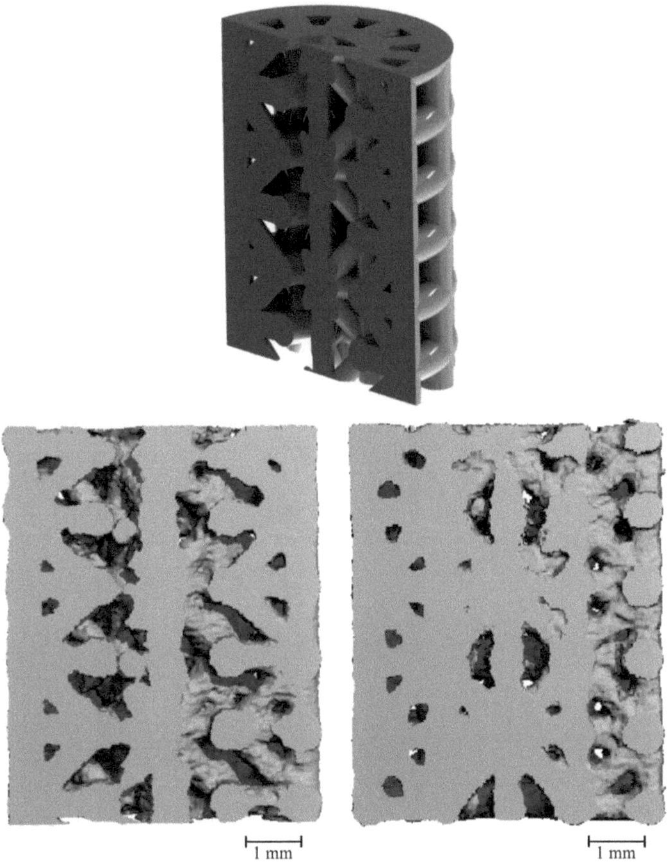

Fig. 4 CAD file and casted foam of the final shape for the implants with LAE442 on the *left* and La2 on the *right*

Mechanical Simulation

The stress distributions of the structures with maximum pore sizes of 500 μm is in the range of 1.13–1.61 MPa. At the hubs the stress is locally higher and reaches up to 21.5 MPa. In the structure with a maximum pore size of 400 μm, the stress on the vertical elements between the horizontal elements with 5.4–10.8 MPa is significantly smaller. Only where the vertical elements cross the horizontal ones, the stress is in the range of 10.8–16.1 MPa. The lower stress is due to the larger effective cross section area (Fig. 6).

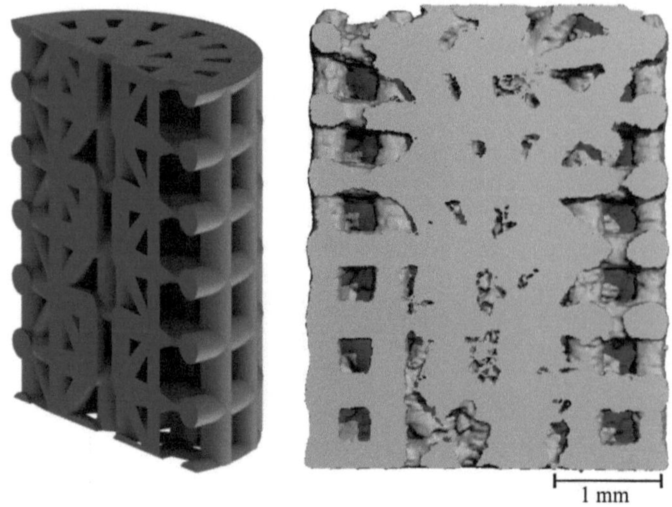

Fig. 5 CAD file and cast sample of the final shape with maximum pore size of 400 μm

Fig. 6 Stress distribution of the structures in Figs. 3a and 5 under the load of 60 N in MPa

Compressive Tests

As shown in the force-strain-curves in Fig. 7 and the corresponding values in Table 1, with 341 N and 383 N, respectively, the compression yield point F_{dF} in both structures is well above the expected load of 60 N in the rabbit tibia, in which the implants should be inserted. After the elastic limit there is a long region of plastic deformation until the structure collapses. Depending on the structure, the implants have compressive strength of 1608 or 1921 N, with a corresponding deformation of 21.4 or 33.2%.

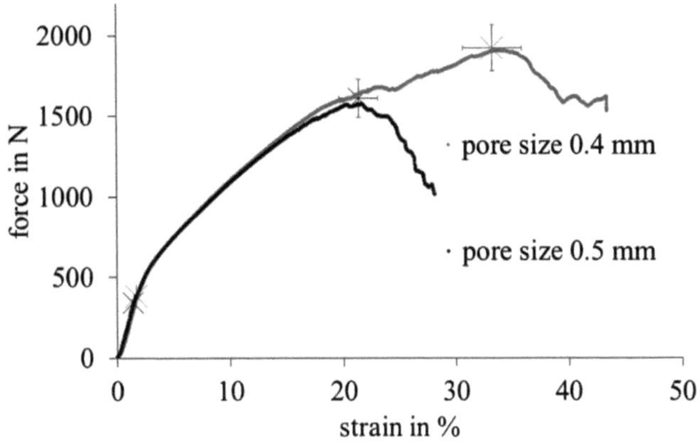

Fig. 7 Exemplary force-strain-diagram of one sample with larger pores (max. 0.5 mm) and one with smaller pores (max. 0.4 mm)

Table 1 Results of the compression test

Maximum pore size in mm	Compressive strength F_{max} in N	Strain at F_{max} in %	Compression yield point F_{dF} in N	Strain at F_{dF} in %
0.5	1608	21.4	341	1.45
0.4	1921	33.2	383	1.72

Discussion

The mold filling experiment has shown that the structure with the homogenous element thickness was filled completely, while several casting defects appeared in the other structures. The problem here is that the thickness of the elements is significantly smaller than 0.4–0.5 mm in some areas. Here, the melt can solidify faster, hence the elements can be blocked. In Fig. 2 an implant is shown in which some pores are filled. This suggests that the melt has broken some of the gypsum struts during the casting.

The structures with a complete mold filling and without defects are simulated for their stress distribution under a load of 60 N. This investigation has shown, that the effective stress is smaller, if the elements in z-direction (direction of the force in the experiments) are not tilted after every horizontal strut. If the struts are shifted, the stress lines are accumulated at the hubs, which causes higher stress peaks. Compared with the results of Seitz et al. [12], the simulated stress is lower than the experimentally determined yield point of 150 MPa. However, in the present study, the shapes were produced by investment casting whereas Seitz et al. used extruded materials. The production method has a large influence on the microstructure and accordingly the mechanical properties.

Based on the compression tests, the assumption made after the simulation, that the structures made of LAE442 are stable enough can be confirmed. The magnesium sponges in Figs. 4 and 5 have enough strength for the application as an implant in a rabbit tibia. Here the yield point is in both cases much larger than the required 60 N.

If the magnesium alloys LANd442, Nd2, ANd42 and LNd42 corrode, there is a significant volume loss and a loss in the three-point-bending strength [16]. It can be expected, that other mechanical properties are decreasing as well, because the corrosion leads to more defects and less material. The load will therefore be carried by a smaller cross-section, which results in a higher stress. Therefore, it is beneficial, that there is a buffer in the compression yield point.

Still there are other aspects, which have an influence on the corrosion, like saturation effects of the corrosion products [17].

Finally the bone ingrowth behavior, which depends on the pore size, is very important and degradation of the implant should be tailored accordingly. Therefore additional in vivo experiments are needed to find out which structure is stable long enough such that the bone can stabilize the defect.

The investigation of the casted structures with the CT showed that the whole structure could be casted close to the constructed shape. There is also no big difference between the two casted alloys. The LAE442 as well as the La2 alloy could fill the structure shown in Fig. 4. Furthermore the alloy LAE442 can fill the structure with the smaller pores and struts. The only differences are the rough surface and the different sizes in the struts. This is due to different process related reasons. The parts were first shaped layer by layer by FDM. This always results in a rougher surface in the z-direction, which cannot be compensated by surface treatments. Furthermore, the gypsum causes a rough surface depending on the size of the particles. The sponge-like structure however, has no serious defects or missing elements and can therefore be produced by investment casting.

Conclusion

In this study, a manufacturing process to produce sponge-like structures by investment casting was investigated and different implant geometries and alloys were simulated and casted. It has been demonstrated that the constructed shapes could be casted successfully by investment casting. Additionally, the implants appear to have a sufficient compressive strength and a pore structure that is necessary for an optimized bone ingrowth behavior.

Acknowledgements This research is sponsored by the German Research Foundation (DFG) within the project "Interfacial effects and ingrowing behavior of magnesium-based foams as bioresorbable bone substitute material" (grant no. MA 1175/52-1).

References

1. S.C. Manolagas, Birth and death of bone cells: basic regulatory mechanisms and implications for the pathogenesis and treatment of osteoporosis. Endocr.Soc. **21**(2), 115–137 (2000)
2. P.C. Missiuna et al., Anatomically safe and minimally invasive transcrestal technique for procurement of autogenous cancellous bone graft from the mid-iliac crest. Can. J. Surg. **54**(5), 327–332 (2011)
3. A.J. Salgado, O.P. Coutinho, R.L. Reis, Bone tissue engineering: state of the art and future trends. Macromol. Biosci. **4**, 743–765 (2004)
4. Bone regeneration and repair, *Biology and Clinical Applications* (Humana Press Inc, Totowa, NJ, 2005)
5. T.V. Thamaraiselvi, S. Rajeswari, Biological evaluation of bioceramic materials: a review. Trends Biomater. Artif. Organs **18**(1), 9–17 (2004)
6. X. Gu, Y. Zheng, A review on magnesium alloys as biodegradable materials. Frontiers Mater. Sci. China **4**(2), 111–115 (2010)
7. F. Witte et al., In vivo corrosion of four magnesium alloys and the associated bone response. Biomaterials **26**, 3557–3563 (2005)
8. H. Haferkamp et al., Develompent, processing and applications range of magnesium lithium alloys. Mater. Sci. Forum **350–351**, 31–42 (2000)
9. T.L. Chia et al., The effect of alloy composition on the microstructure and tensile properties of binary Mg-rare earth alloys. Intermetallics **17**, 481–490 (2009)
10. N. Birbilis et al., On the corrosion of binary magnesium-rare earth alloys. Corros. Sci. **51**, 683–689 (2009)
11. F. Bach et al., Magnesium sponges as a bioabsorbable material: attributes and challanges. Int. J. Mater. Res. **98**, 609–612 (2007)
12. J. Seitz et al., The effect of different sterilization methods on the mechanical strength of magnesium based implant materials. Adv. Eng. Mater. **13**, 1146–1151 (2011)
13. A. Weizbauer et al., Novel magnesium alloy Mg–2La caused no cytotoxic effects on cells in physiological conditions. Mater. Sci. Eng. **C 41**, 267–273 (2014)
14. J. Reifenrath et al., Axial forces and bending moments in the loaded rabbit tibia in vivo. Acta Vet. Scand. **54**(21), 1–7 (2012)
15. Giullini, http://bk-giulini-pcg.com/Deutsch/Gips/Giesserei_Modellbau/Giesserei.htm. Accessed 27 June 2016
16. J. Seitz et al., Comparison of the corrosion behavior of coated and uncoated magnesium alloys in an in vitro corrosion environment. Adv. Biomater. **13**, B1–B11 (2011)
17. X. Gu et al., In vitro corrosion and biocompatibility of binary magnesium alloys. Biomaterials **30**, 484–498 (2009)

Osteoporosis and Fatigue Fracture Prevention by Analysis of Bone Microdamage

Gerardo Presbítero, David Gutiérrez and David Taylor

Abstract We analyzed the conditions under which microcracks, generated by fatigue, affect the fracture properties of bones; this has clinical relevance to stress fractures and osteoporosis. A novel theoretical model was developed to describe microcrack behaviour, using probabilistic analysis and the concept of a characteristic length. In this way we identified effects of aging and gamma radiation sterilisation, which weaken bone and cause accelerated development of microcracks. This work will help in the development of better predictive models to understand and prevent stress fractures and osteoporosis-related fragility fractures.

Keywords Osteoporosis • Weibull distribution • Fatigue fractures • Gamma radiation • Bone microdamage

---

G. Presbítero (✉)
Faculty of Engineering, National Autonomous University of Mexico,
University Centre for Advanced Technology (PUNTA-UNAM), 66629 Monterrey,
Mexico
e-mail: presbitg@unam.mx

D. Gutiérrez
Center for Research and Advanced Studies (Cinvestav), 66600 Monterrey,
Mexico
e-mail: dgtz@ieee.org

D. Taylor
Trinity Centre for Bioengineering, Trinity College Dublin, College Green,
Dublin 2, Ireland
e-mail: dtaylor@tcd.ie

The Minerals, Metals & Materials Society, *TMS 2017 146th Annual Meeting & Exhibition Supplemental Proceedings*, The Minerals, Metals & Materials Series,
DOI 10.1007/978-3-319-51493-2_30

Introduction

In this work it is presented mechanical tests performed within the facilities of Trinity College Dublin, Ireland, and Cinvestav Monterrey, Mexico, in which we were able to describe with use of statistical analysis—Weibull distribution, how microdamage generates and grows under fatigue stresses towards the fracture of bones. Behaviour of microdamage in bone generated by fatigue stresses has been extensively studied: accumulation of microdamage in bone affects in the development of fatigue fractures, and is highly associated with the process of aging and osteoporosis. When damage accumulation is too high for the repair process, it elevates considerably the level of microdamage in bone [1, 2]. Accumulation of microdamage causes decrease in the resistance to fracture or fracture toughness [3], and fracture of bone by fatigue is caused by "impact or growth of a dominant crack that grows in conformation to the progress of fatigue loading" [4]. These types of microcrack developments have been studied by several researchers, in studies done in vivo [5] as well as in vitro [6, 7] just to mention a few of them. Young bone develops fatigue damage in a very different way than does old bone, some studies in this theme have been published by [8–10]. Regarding emission of gamma radiation in human bones with the objective of sterilization, [11] remarked an explanation to microcrack behaviour under these conditions, concluding that gamma radiation sterilization reduces the amount of overall damage, suggesting that there may be an association between the damage process in bone tissue and the collagen structure, nevertheless, it is also suggested about the details of these associations to be "still obscure and remain to be investigated". We performed fatigue tests developed in bovine and human bones, in which analysis of microcracking behaviour with the use of fluorescence techniques under epifluorescence microscopy, in a similar way as the analyses done by [12–14], we were able to describe microdamage development as the intrinsic mechanical properties of the bones were reduced. Following our previous work [15] in which it is defined how the values of Weibull distribution parameters differ and vary in a systematic way with number of cycles during in vitro testing. Two-parameter Weibull distribution analysis was used to explain the differences of the conditions in which microcracks grow towards the fracture of bones in relation to the decrease in its mechanical properties. The mechanical properties were altered by use of gamma radiation, submitted for sterilization purposes, as well as due to aging (usage of bones of 32 and 55 years old). These differences were used in the development of a new theoretical model in which volume (V), number of cycles to failure (N_f), applied stress (σ) and Weibull constants are applied in one single equation.

Materials and Methods

Compression Tests and Radiation Treatment

Several mechanical tests submitted in human and bovine bones have been developed by our group to study corresponding microdamage developments. Specifically, fatigue compression tests were developed in cortical samples of mid diaphysis regions of bovine femurs, tested at a frequency of 3 Hz with a stress of 70 MPa, cycling between 11 and 81 MPa. This stress range was applied until reaching fracture or one million cycles. For the mentioned tests, cylindrical specimens of 35 mm of length and diameter of 7 mm were obtained. The specimens were then placed in a VS 450 CNC machine lathe to waist the centre regions. The CNC program turned the centre region of the specimen to a gauge diameter of 5 mm and a gauge length of 7 mm. In a similar manner, compression tests were submitted in cortical samples of mid diaphysis regions of 32 and 55 years old human femurs, donated by the Bone and Tissue Bank of the Autonomous University of Nuevo Leon, Mexico. The mid-diaphysis sections of human femurs were dissected, submitted under lyophilisation procedure and kept at room temperature. The human bones were submitted under Cobalt-60 radiation source for approximately 2–3 h at the National Institute of Nuclear Research (ININ), Mexico, and separated afterwards in three differing groups: control, emission of gamma radiation of range 15–25 kGy, and emission of gamma radiation of range and 26–30 kGy, following segmentation by placing in a plaster mounting to obtain samples of dimensions $3 \times 3 \times 4.5$ mm for each group with the use of a CNC machine. In the case of the human femurs, the samples were tested at a frequency of 3 Hz under a stress of 94 MPa.

Microdamage Analysis

The microdamage observation procedure included the application of a dye sequence as specified by [16]. This consists in the application of different fluorochromes to look for microfractures while increasing the number of cycles under a fatigue test under compressive stresses. The specimens were initially labelled with one of the fluorochromes to label damage in vivo. The specimen was first allowed to thaw and placed overnight in the vial of 0.0005 M Alizarin Complexone, in a vacuum desiccator. This ensured the penetration of the stain in the specimen. For the case of the bovine bones, having stained with Alizarin Complexone, the sample was subjected under the compressive fatigue forces staining with Xylenol Orange, as the experiment progressed the Xylenol was removed from the bath, and the bath and the specimens were rinsed with distilled water prior to refilling the bath with the subsequent dye in the sequence (Calcein). For the case of the human bones, calcein was used subsequently to the use of alizarin complexone. Calcein labelled

Table 1 Summary of data sources

Reference	Bone type	Stress (MPa)	Number of cycles	α	a_0 (µm)	a_{max} (µm) (P = 0.9)
Gellasch and Kalscheur [17]	Canine radius	In vivo	In vivo	3.2	69.7	90.1
Burr [18]	Canine femur	2700 $\mu\varepsilon$	426,000	3.1	78.5	102.8
Khan [19]	Ovine rib	In vivo	In vivo	4.1	95.0	116.7
O'Brien et al. [13]; Presbitero et al. [15]*	Bovine tibia	In vivo	In vivo	3.0	88.9	116.9
Presbitero et al. [15]*	Bovine tibia	70 MPa	50,000	2.4	113.1	159.9
O'Brien et al. [13]	Bovine tibia	80 MPa	50,000	2.2	108.7	158.5
Presbitero et al. [15]*	Bovine tibia	70 MPa	Million cycles non-fractured	2.3	113.8	163.5
Presbitero et al. [15]*	Bovine tibia	70 MPa	Fracture av.159,700	2.3	126.0	181.1
O'Brien et al. [13]	Bovine tibia	80 MPa	Fracture av. 88,000	2.0	129.4	198.5

*The reported results were obtained by the group Fergal J. O'Brien, T.Clive Lee and David Taylor and Gerardo Presbitero.

microcracks towards the fracture of the specimens, or the conclusion of the tests (after reaching one million cycles). Microcracking was observed using fluorescence microscopy, at $\times 125$ of magnification, and measured in terms of mean number of cracks occurring per section, crack numerical density (number of cracks occurring per mm^2), crack surface density (µm of crack length occurring per mm^2), and mean crack length (µm).

Statistical Analysis

In our previous work [15], we introduced a statistical method in which it is possible to accurately describe microcrack developments by the two-parameter Weibull distribution through the following equation:

$$P = 1 - e^{-(a/a_0)^{\alpha}} \qquad (1)$$

which relates the cumulative probability P with crack length.

By linearization of the Weibull distribution, it is possible to obtain the parameters a_0 and α, being in (1) the characteristic length and variability constants, respectively. Additionally, it was determined the parameter $a_{max}(P = 0.9)$, which is the value of crack length which corresponds to a P value of 0.9, i.e. the length below which 90% of all cracks are found (measure of the largest crack present). Table 1 gives the corresponding data for several types of bones, compared to our study [15].[1]

Theoretical Model

A new theoretical model involving applied stress, volume and number of cycles to failure in one same equation, is developed for the obtaining of predictions involving the effect of microcracks, by their generation as well as their growth, to provoke fracture of bones. This theoretical model involves the following deduction:

A sample of bone, of volume (*V*), contains a number of microcracks, *n*, giving a microcrack density of *n/V*. Weibull distribution (Eq. 1) can be expressed in a relationship that depends on the stress range (σ), number of cycles (*N*), crack density (n/mm^3), characteristic crack length value (a_0), and type of bone (animal species). Given *n* cracks in a volume *V*, the length of the longest crack, a_{max}, will be some function of *n*:

$$a_{max} = f(n) \quad \text{or alternatively,} \quad n = g(a_{max})$$

The numerical density of cracks # is the number of cracks per unit volume, so # = *n/V*.

At the same time numerical density is a power law relationship with number of cycles (*N*):

$$\#^k = \beta N \tag{2}$$

β will be a constant working as a function of σ, for which the model will follow a power-law behaviour. Equation (1) considers the probability *P* of one crack to be equal to or less than a specified length, the probability of *n* cracks to be less than some length *a* is:

$$\varphi = \left[1 - e^{-\left(\frac{a}{a_0}\right)^{\alpha}}\right]^n \tag{3}$$

Applying corresponding substitutions among these relationships, it is obtained the following relationship:

$$N_f = \frac{1}{\beta}\left(\frac{g(a_{\max})}{V}\right)^k = \frac{1}{\beta}\left\{\frac{\ln \varphi}{V}\left[\ln\left[1 - e^{-(a_{\max}/a_0)^{\alpha}}\right]\right]^{-1}\right\}^k \tag{4}$$

---

[1]The reported results were obtained by the group Fergal J. O'Brien, T.Clive Lee and David Taylor and Gerardo Presbitero.

This relates φ for n microcracks, in a relationship involving volume (V), number of cycles to failure (N_f) and maximum crack length (a_{max}) in one single equation.

In order to establish involvement of stress, stress intensity range (K) is applied with the use of (a_{max}):

$$\Delta K = F\Delta\sigma\sqrt{\pi a} \tag{5}$$

Failure occurs when this reaches some critical value K_{max} (and assuming $F = 1$). Under further substitutions among these relationships, it is obtained the following relationships:

$$N_f = \frac{1}{\beta}\left[\frac{\ln \varphi}{V \ln \Phi}\right]^k \tag{6}$$

$$V = \frac{\ln \varphi}{(N_f\beta)^{1/k}}[\ln \Phi]^{-1} \tag{7}$$

Being:

$$\Phi = 1 - e^{-\left(\Delta K_{\max}^2 / a_0\pi\Delta\sigma^2\right)^\alpha}$$

In this way, relationships involving volume, number of cycles to failure and applied stress are obtained, in which the use of constants β and K_{max} will have influence on the predictions with respect to experimental data. Weibull constants (variability and characteristic length) are used for bones at the point of fracture.

Results

For the case of gamma radiation emission in human samples, using unpaired t-tests with a significance level of 0.05 (MINITAB® Release 14.1), significant differences ($p < 0.05$) were found taking into account mean number of cycles to fracture between differing ages (control groups), as well as under the submission of gamma radiation. Furthermore, there was a drastic fatigue-life reduction after emission of radiation among the control groups. Regarding microdamage distributions, significant differences were found as well ($p < 0.05$): for every increase in gamma radiation range, bone age, and fatigue life, there was considerable increase in the development of microcracks.

With use of the microcracks data, Weibull distributions were obtained and the characteristic length and variability constants calculated. These results are presented in Table 2. In addition, it is presented in Table 2 the value of crack length which corresponds to a P value of 0.9, considered as a measure of the largest crack present, which we refer as a_{max}.

Table 2 Weibull constants of microdamage developments

		α	a_0	$a_{max}(P = 0.9)$
Presbitero et al. [15]; Young bovine	In vivo	3.0	88.9	116.9
70 MPa*	Fracture	2.3	126.0	181.1
Presbitero et al. [15]; Young bovine	In vivo	3.0	88.9	116.9
80 MPa*	Fracture	2.0	129.4	198.5
Control 32 yrs	In vivo	3.83	83.66	104.01
Human 94 MPa	Fracture	3.06	136.99	179.91
15–25 kGy 32 yrs	In vivo	3.45	76.37	97.26
Human 94 MPa	Fracture	2.16	177.39	260.99
26–30 kGy 32 yrs	In vivo	3.06	108.94	143.07
Human 94 MPa	Fracture	2.00	259.51	393.73
Control 55 yrs	In vivo	2.39	159.72	226.43
Human 94 MPa	Fracture	2.16	205.37	302.07
15–25 kGy 55 yrs	In vivo	2.01	134.38	203.59
Human 94 MPa	Fracture	2.12	288.72	428.23
26–30 kGy 55 yrs	In vivo	2.53	123.02	171.06
Human 94 MPa	Fracture	2.00	352.14	535.13

*The reported results were obtained by the group Fergal J. O'Brien, T.Clive Lee and David Taylor and Gerardo Presbitero.

As it can be observed from the table, α decrease, while a_0 and $a_{max}(P = 0.9)$ increase as the mechanical properties of the bones are reduced. After obtaining the Weibull distributions data, the theoretical model can be employed with the use of the following information:

It was calculated the values of the constants k and β with use of our experimental numerical densities results and Eq. (2); the obtained values were 0.4 and 2.6×10^{-6}, respectively. Using a K_{max} value of 2.2 $\mathrm{MPa(m)}^{1/2}$, which is the approximate stress intensity range developed in human cortical bones [20–22], we substituted the corresponding Weibull distribution constants for the case of bones cycled until fracture. Additionally, for the case of plotting relationships of Stress versus Number of cycles to failure, we determined three constant volumes (10, 140 and 3000 mm^3). All these values were substituted in Eq. (6), in order to obtain a relationship such as:

$$N_f = \frac{1}{2.6 \times 10^{-6}} \left[\frac{\ln 0.9}{V \ln \Phi} \right]^{0.4}; \ \Phi = 1 - e^{-\left(\Delta K_{max}^2 / 129.4 \pi \Delta \sigma^2\right)^2}$$

The final σ versus N_f relationship is shown in Fig. 1, in which the experimental data correlate these predictions; lower experimental values of volume tend to correlate the model plotted for 10 mm^3, whereas higher experimental values of volume correlate with the model when plotting at 3000 mm^3. Experimental data acquired from [23–28].

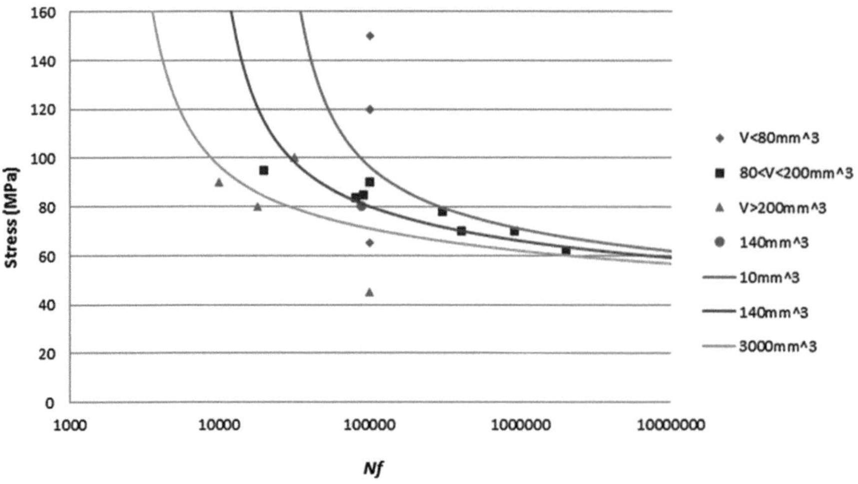

Fig. 1 Stress versus Nf predictions of the model, for three different volumes. Each model prediction correlates the corresponding experimental data

Similarly, σ versus V, and N_f versus V plots can be obtained correlating with experimental data. Finally, if Weibull distribution parameters are substituted as the mechanical properties are reduced, it is possible to obtain relationships such as the ones shown in Fig. 2:

For the case of each of the relationships of Fig. 2a, b, it is clearly observed that as the characteristic lengths increase due to the decrease of the mechanical properties of the bones, it makes possible to predict the behavior of bones under fatigue towards the fracture depending on their mechanical conditions.

Discussion

Attempts have been done to find ways of describing mathematically the way in which microdamage evolves in bones. In our work, with the usage of microdamage experimental data, we were able to describe with two-parameter Weibull distribution analysis the way microcracks grow in cortical bones under fatigue stresses towards the development of fractures in vivo conditions. We firstly worked in the development of fatigue tests in samples taken from the cortical zones of bovine and human femurs, which were at the same time labelled with chelating agents until the fracture. After the fracture of the samples, thin slices of approximately 100–150 μm were cut and observed through fluorescence microscopy. These observations allowed to perform measurements of the way microcracks grew from in vivo conditions to the fracture. Several types of measurements were developed, including number of cracks occurring per section, crack numerical density (number

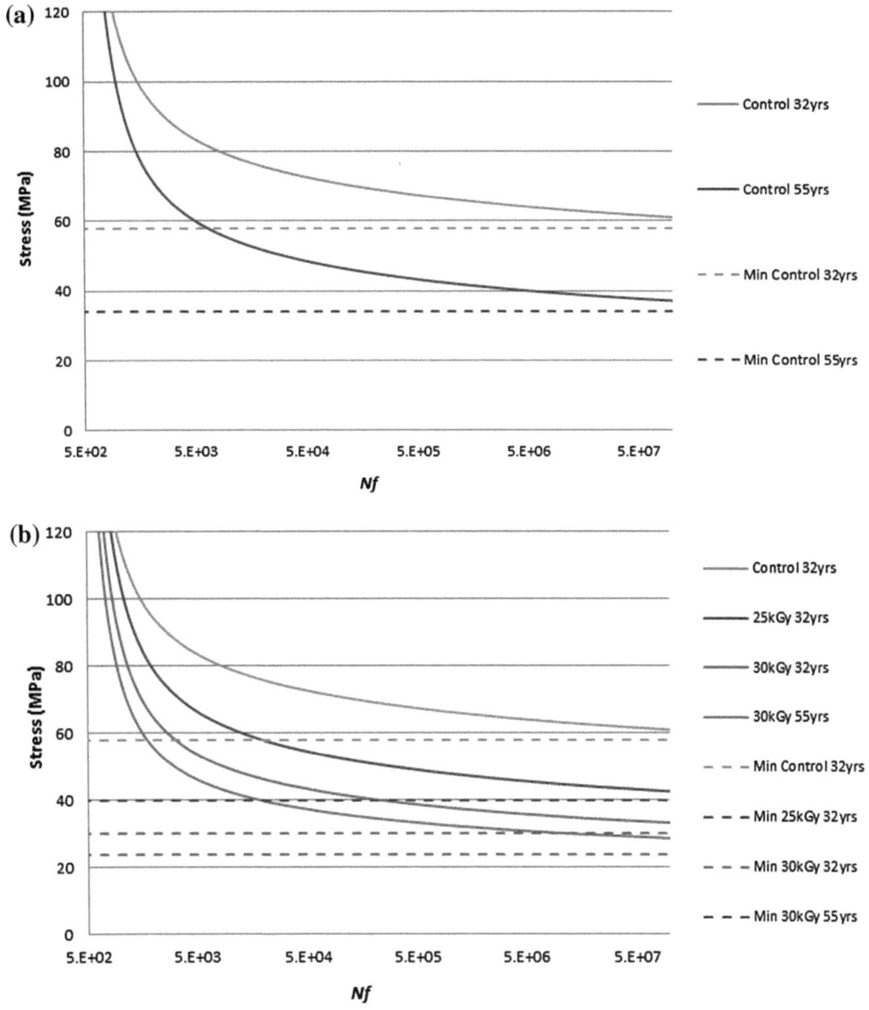

Fig. 2 **a** Reduction of stresses with increase of characteristic lengths (a_0), as mechanical properties of bones are reduced due to aging. **b** Reduction of stresses with increase of characteristic lengths (a_0), as mechanical properties of bones are reduced due to gamma ray emission

of cracks occurring per mm^2), crack surface density (μm of crack length occurring per mm^2), and mean crack length (μm), in which using unpaired t-tests with a significance level of 0.05 to look for significant differences (MINITAB® Release 14.1), we found significant differences ($p < 0.05$) between differing ages and for every increase in gamma radiation range (15–25 and 25–30 kGy). These results allowed to deduce through Weibull distribution analysis, the behaviour of micro-cracks at failure instances as the mechanical conditions of the bones decreased.

Extensive number of studies exist regarding bones mechanical behaviour due to age and emission of gamma radiation, just some of these were the ones published by [29–35]; in the case of our studies we aimed to study the microdamage behaviour with use of the two-way Weibull distribution. With the results obtained in Table 2, it was observed that the variabilities (α) tended to decrease, and characteristic lengths (a_0) as well as $a_{max}(P = 0.9)$ tended to increase at fracture instances when the mechanical properties of the bones decreased (increasing age and gamma radiation intensities).

In addition, a new theoretical model was introduced, in which important factors such as applied stress, sample volume and number of cycles to failure were integrated in one equation. Other factors were included as well in the model, such as taking into account that numerical density is a power law relationship with number of cycles (N), and by including the behavior of the two-parameter Weibull distribution parameters obtained from the statistical analysis performed to microcrack lengths of bones fatigued until fracture.

This way of describing the mechanical behavior of bones under submitted fatigue stresses has never been reported, and it is presented as a novel method that helps to predict the behavior of bones towards their fracture with several components included in one same relationship. Following the relationships obtained in Fig. 2, it can be observed that as the mechanical properties of bones decrease, characteristic lengths increase, giving an explanation for which it is extremely difficult to fracture healthy human bones under stresses considered under extreme in vivo conditions i.e., $\geq$40 MPa (stress applied in military regimes and during extreme stair climbing or other rigorous movements [36]). It has to be mentioned that other variables such as bone mineral density (BMD) have to be taken into account, as under certain conditions bones of athletes and army members do fracture. Hence, when decreasing the mechanical properties due to age or gamma radiation, Weibull characteristic lengths increase, reaching domains in which it may be possible to predict instances when bones with these conditions will fracture under fatigue stresses comparable to the ones produced during normal running (approx. 20–30 MPa [37]). We are aware that further experimental data is essential for the approval of the presented model (case of Fig. 2), and future work will be based on these aspects. Even so, this work has increased our understanding of microdamage develop in bone and will allow in the near future to refine predictions towards the prevention of fatigue fractures of bones of adults suffering from osteoporosis, athletes, and members of the army, and will be useful for the development of new clinical approaches.

References

1. H.M. Frost, Some ABC's of skeletal pathophysiology: 5. Calcif. Tissue Int. **49**, 229–231 (1991)
2. D.B. Burr et al., Bone microdamage and skeletal fragility in osteoporotic and stress fractures. J. Bone Miner. Res. **12**, 6–15 (1997)

3. B.D. Agarwal, L.J. Broutman (eds.), *Analysis and Performance of Composite Materials* (Wiley, New York, 1980), pp. 287–362
4. P. Zioupos, C. Kaffy, Tissue heterogeneity, composite architecture and fractal dimension effects in the fracture of ageing human bone. Int. J. Fract. **139**, 407–424 (2006)
5. H.M. Frost, Presence of microscopic cracks in vivo in bone. Henry Ford Hosp. Med. Bull. **8**, 25–35 (1960)
6. T.M. Wright, W.C. Hayes, Tensile testing of bone over a side range of strain rates: effects of strain rate, microstructure and density. Med. Biol. Eng. **14**, 671–680 (1976)
7. D.R. Carter, W.C. Hayes, Compact Bone Fatigue Damage: A Microscopic Examination. Clin. Orthop. Relat. Res. **127**, 265–274 (1977)
8. R. Wank, K.J. Jepsen, Aged rat bone fatigues in a fundamentally different manner from yonder adult bone. 47 annual meeting, Orthopaedic Research Society (2001)
9. T. Diab et al., Age-dependent fatigue behavior of human cortical bone. Eur. J. Morphol. **45** (1/2), 53–59 (2005)
10. T.L. Norman, T.M. Little, Y.N. Yeni, Age-related changes in porosity and mineralization and in-service damage accumulation. J. Biomech. **41**, 2868–2873 (2008)
11. O. Akkus, Fracture resistance of gamma radiation sterilized cortical bone allografts. J. Orthop. Res. **19**, 927–934 (2001)
12. T.C. Lee et al., Sequential labelling of microdamage in bone using chelating agents. J. Orthop. Res. **18**, 322–325 (2000)
13. F.J. O'Brien, D. Taylor, T.C. Lee, Microcrack accumulation at different intervals during fatigue testing of compact bone. J. Biomech. **36**(7), 973–980 (2003)
14. F.J. O' Brien, D. Taylor, T.C. Lee, The effect of bone microstructure on the initiation and growth of microcracks. J. Orthop. Res. **23**(2), 475–480 (2005)
15. G. Presbitero, F.J. O'Brien, T.C. Lee, D. Taylor, Distribution of microcrack lengths in bone in vivo and in vitro. J. Theor. Biol. **7**(304), 164–171 (2012)
16. F.J. O'Brien, D. Tayor, T.C. Lee, An improved labelling technique for monitoring microcrack growth in compact bone. J. Biomech. **35**(4), 523–526 (2002)
17. K.L. Gellasch, V.L. Kalscheur, Fatigue microdamage in the radial predilection site for osteosarcoma in dogs. Am. J. Vet. Res. **63**(6), 896–899 (2002)
18. D.B. Burr, Does microdamage accumulation affect the mechanical properties of bone? J. Biomech. **31**(4), 337–345 (1998)
19. M.F. Khan, *Microdamage and Remodelling in an Ovine Model* (NUI, Royal College of Surgeons in Ireland, 2010)
20. D. Vashishth, J.C. Behiri, W. Bonfield, Crack growth resistance in cortical bone: concept of microcrack toughening. J. Biomech. **30**(8), 763–769 (1997)
21. P. Lucksanasombool, W.A. Higgs, Fracture toughness of bovine bone: influence of orientation and storage media. Biomaterials **22**, 3127–3132 (2001)
22. R.K. Nalla et al., Aspects of in vitro fatigue in human cortical bone: time and cycle dependent crack growth. Biomaterials **26**(14), 2183–2195 (2005)
23. D.R. Carter, W.E. Caler, Cycle-dependent and time-dependent bone fracture with repeated loading. J. Biomech. Eng. **105**, 166–170 (1983)
24. P. Zioupos, X.T. Wang, J.D. Currey, Experimental and theoretical quantification of the development of damage in fatigue tests of bone and antler. J. Biomech. **29**(8), 989–1002 (1996)
25. D. Taylor, Fatigue of bone and bones: an analysis based on stressed volume. J. Orthop. Res. **16**, 163–169 (1998)
26. D. Taylor et al., Compression data on bovine bone confirms that a "stressed volume" principle explains the variability of fatigue strength results. J. Biomech. **32**(11), 1199–1203 (1999)
27. D. Taylor, J.G. Hazenberg, T.C. Lee, The cellular transducer in damage-stimulated bone remodelling: a theoretical investigation using fracture mechanics. J. Theor. Biol. **225**(1), 65–75 (2003)
28. J.J. Kruzic, R.O. Ritchie, Fatigue of mineralized tissues: cortical bone and dentin. J. Mech. Behav. Biomed. Mater. (2007)

29. M.B. Schaffler, K. Choi, C. Milgrom, Aging and matrix microdamage accumulation in human compact bone. Bone **17**(6), 521–525 (1995)
30. J.D. Currey et al., Effects of ionizing radiation on the mechanical properties of human bone. J. Orthop. Res. **15**(1), 111–117 (1997)
31. E.J. Mitchell et al., The effect of gamma radiation sterilization on the fatigue crack propagation resistance of human cortical bone. J. Bone Joint Surg. Am. **86-A**(12), 2648–2657 (2004)
32. O. Akkus, R.M. Belaney, Sterilization by gamma radiation impairs the tensile fatigue life of cortical bone by two orders of magnitude. J. Orthop. Res. **23**, 1054–1058 (2005)
33. Z. Zhou et al., Mechanical strength of cortical allografts with gamma radiation versus ethylene oxide sterilization. Acta Orthop. Belg. **77**(5), 670–675 (2011)
34. A. Kaminski et al., Effect of gamma irradiation on mechanical properties of human cortical bone: influence of different processing methods. Cell Tissue Bank **13**, 363–374 (2012)
35. A. Islam et al., Gamma radiation sterilization reduces the high-cycle fatigue life of allograft bone. Clin. Orthop. Relat. Res. **4**(3), 827–835 (2016)
36. U. Andreaus, M. Colloca, A. Toscano, Mechanical behavior of a prosthetized human femur: a comparative analysis between walking and stair climbing by using the finite element model. Biophys. Bioeng. Lett. **1**, 1–15 (2008)
37. D.B. Burr, C. Milgrom, D. Fyhrie, M. Forwood, M. Nyska, A. Finestone, S. Hoshaw, E. Saiag, A. Simkin, In vivo measurement of human tibial strains during vigorous activity. Bone **18**(5), 405–410 (1996)

Bone Remodeling Under Tooth Loading

Kangning Su, Li Yuan and Jing Du

Abstract The stability and success rate of orthopedic and dental implants are affected by their surrounding bone quality. Bone adapts to mechanical loads through remodeling activities to achieve new equilibrium in strain/stress state. The object of this study is to develop a numerical algorithm to simulate bone remodeling activities under mechanical loading. Finite element method is used to calculate the strain/stress distribution in the alveolar bone under tooth loading. The bone density remains unchanged near the equilibrium point of the mechanical stimulus; under greater or smaller mechanical stimulus, it increases or decreases. Iterations are performed to simulate the evolution of bone density. Effects of model geometry and adjacent teeth are studied. Effects of various applied loads and boundary conditions are compared. Simulation results are validated using computed tomography (CT) data of human mandibles. The implications of the results on patient-specific treatment and the insights for clinical techniques are also discussed.

Keywords Bone remodeling · Finite element · Dentistry

Introduction

Teeth are attached to jaw bone through the periodontal ligament (PDL). Bone is a living tissue. It has the ability to adapt to the change of mechanical stimulus [1]. Generally, the increase and decrease of mechanical stimulus promote bone

---

K. Su · J. Du (✉)
Department of Mechanical and Nuclear Engineering,
Pennsylvania State University, State College, USA
e-mail: jingdu@psu.edu

K. Su
e-mail: conysu666@gmail.com

L. Yuan
Shenzhen People's Hospital, 2nd Clinical Medical College
of Jinan University, Guangzhou, China

The Minerals, Metals & Materials Society, *TMS 2017 146th Annual Meeting & Exhibition Supplemental Proceedings*, The Minerals, Metals & Materials Series,
DOI 10.1007/978-3-319-51493-2_31

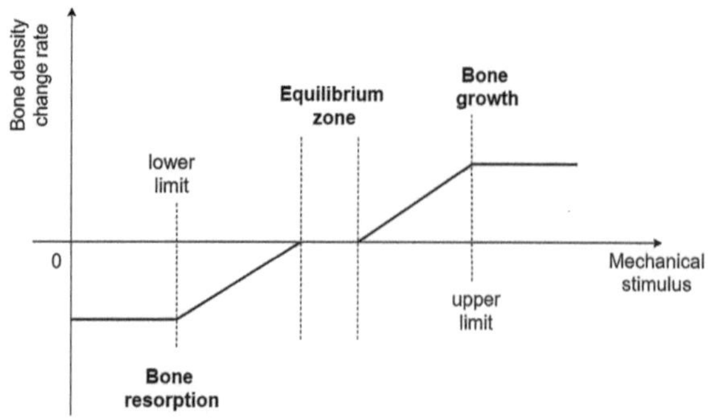

Fig. 1 Bone density change rate against mechanical stimulus

formation and resorption activities, respectively. The resorption and formation processes of bone are called bone remodeling. It has been studied using various animal experiments [2, 3] and numerical simulations, such as finite element method [4, 5].

In numerical simulations, bone remodeling is usually represented by the change of bone density. A commonly used rule [6] for bone density change is presented in Fig. 1. The mechanical stimulus can be mechanical strain [7], daily stress [8], strain energy density [9] or micro-damage [7]. Inside the equilibrium zone, the desired value of mechanical stimulus retains the bone density. When the applied mechanical stimulus is less than the desired value, bone density decreases due to bone disuse. In contrast, when the applied mechanical stimulus is greater than equilibrium, bone overuse causes the increase of bone density.

In the simplified model in Fig. 1, the rate of density change is linearly proportional to the difference between the applied mechanical stimulus and the equilibrium zone. When the mechanical stimulus is beyond the upper or lower thresholds, the bone density changes at a constant rate. In other works [5], parabolic relation between density change rate and mechanical stimulus were used. In such cases, when the applied mechanical stimulus was much greater than equilibrium, it caused overload bone resorption.

In this study, the trabecular bone density is assumed to be uniform initially. Finite element method is used to calculate the strain energy density in teeth and their surrounding tissues under normal chewing and biting forces. The bone density change was calculated using the bone remodeling algorithm. Iterations are carried out to simulate the adaptation of bone density over multiple time steps. Several model geometries and boundary conditions are used to study the effects of adjacent teeth. Computed tomography (CT) data of human mandibles are used to validate the simulation results. The implications of the results on patient-specific treatment and the insights for clinical techniques are also discussed.

Bone Remodeling Simulations

Finite Element Model

The finite element simulations were carried out using the Abaqus FEA software package (Dassault Systemes Simulia Corporation, Providence, RI) to calculate the strain energy density in the tooth-bone structures under normal chewing force. 2-dimensional (2D) models were built to simplify the geometry of natural human teeth in the jaw bone, as shown in Fig. 2. Figure 2a shows lateral incisor and its surrounding soft and hard tissues. Figure 2b shows central and lateral incisors, canine and their surrounding tissues. The dimensions of the mandible bone and the geometry of the lateral incisor were the same in both models. They each consisted of enamel, dentin, pulp, cementum, periodontal ligament (PDL), gingiva, cortical bone and trabecular bone. The geometry and dimensions of the models were based on the average anatomy of natural human teeth [10]. The thickness of cortical bone layer was 2 mm.

4-node linear quadrilateral element was used in the mesh. An identical load was applied on each tooth to simulate the normal chewing and biting force [11]. It consisted of a lateral force of 10 N to the right (mesial direction) and a vertical force of 100 N. The bottom of the model was fixed. Two kinds of boundary conditions were used for the left and right boundaries of the model. In one case, the two sides were constrained to move only in the vertical direction. In the other case, they were set to be free.

The material properties of the teeth and their surrounding tissues used in the simulations are listed in Table 1 [12]. All the materials were assumed to be linear elastic and isotropic.

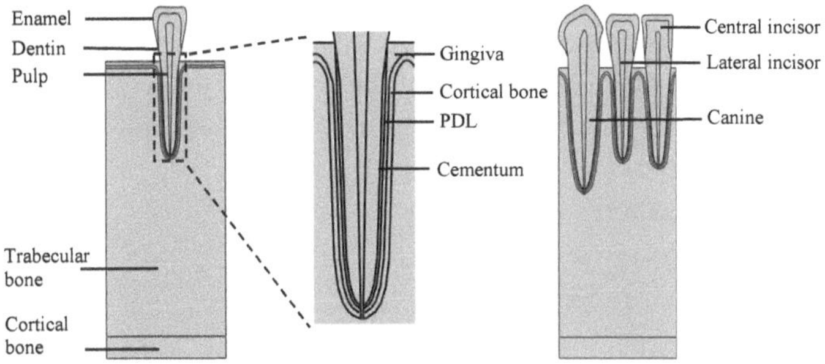

Fig. 2 Finite element model of **a** lateral incisor and **b** central and lateral incisors, canine and their surrounding soft and hard tissues

Table 1 Material properties used in the finite element simulations

Materials	Enamel	Dentin	Pulp	Cementum	Periodontal ligament	Gingiva	Cortical bone	Cancellous bone
Young's modulus (GPa)	79.6	18.6	0.15	13.7	0.2	0.2	13.7	1.5
Poisson's ratio	0.3	0.31	0.49	0.3	0.45	0.45	0.3	0.33

Bone Density and Young's Modulus

Carter and Hayes [13] suggested an empirical relationship between trabecular bone density and its Young's modulus:

$$E = C\rho^3 \tag{1}$$

where E is the Young's modulus in GPa, C is a constant, which in this case is 3740, and ρ is the apparent density in g/cm^3. Equation (1) was used to calculate the bone density in the initial condition and to change the Young's modulus of trabecular bone in the following steps.

Bone Remodeling Algorithm

The flow chart of the bone remodeling algorithm is presented in Fig. 3. The algorithm was implemented using Python scripts to set up and run finite element simulations in Abaqus. The density and Young's modulus in the trabecular bone were assumed to be uniform in the initial conditions. The strain energy density in the tooth-bone structure was calculated using the model presented in "Finite Element Model". The bone density change was calculated using the remodeling rules [14] given by

$$\text{Bone loss at constant rate: } \Delta\mathrm{\rho} = -0.05 \cdot \mathrm{B} \cdot \Delta\mathrm{t}, \text{ if} \frac{\mathrm{U}}{\mathrm{\rho}} < \Xi_l \tag{2}$$

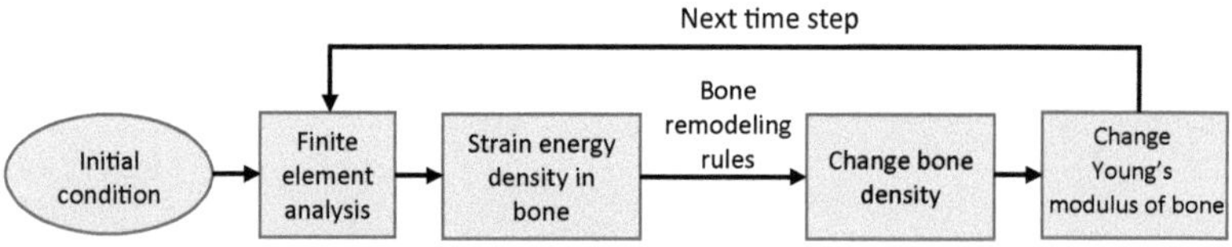

Fig. 3 Flow chart of the bone remodeling algorithm

$$\text{Bone loss: } \Delta\rho = B\left(\frac{U}{\rho} - (1-\delta)\Xi\right) \cdot \Delta t, \text{ if } 0.05 \cdot B + (1-\delta)\Xi < \frac{U}{\rho} < (1-\delta)\Xi \tag{3}$$

$$\text{Bone equilibrium: } \Delta\rho = 0, \text{ if} (1-\delta)\Xi < \frac{U}{\rho} < (1+\delta)\Xi \tag{4}$$

$$\text{Bone growth: } \Delta\rho = B\left(\frac{U}{\rho} - (1+\delta)\Xi\right) \cdot \Delta t, \text{ if} (1+\delta)\Xi < \frac{U}{\rho} < 0.05 \cdot B + (1+\delta)\Xi \tag{5}$$

$$\text{Bone growth at constant rate: } \Delta\rho = 0.05 \cdot B \cdot \Delta t, \text{ if } \frac{U}{\rho} > \Xi_u \tag{6}$$

where U is the strain energy density, Ξ is the reference strain energy density per unit bone mass, Ξ_u and Ξ_l are the upper and lower limit of strain energy density shown in Fig. 1. δ is the range factor for the equilibrium zone, B is the remodeling constant, and t is time. In this case, Ξ is 0.0036 J/(g cm^3), δ is 10% and B is 10 (time unit · g/cm^5). A small time increment was used to assure convergence, thus Δt was set to be 0.2 s. In natural teeth, the trabecular bone is less dense than cortical bone. Therefore, ρ is set to be less or equal to 2.0 g/cm^3. Also, it is set to be greater than 0.1 g/cm^3. [15] The Young's modulus of each element was then calculated using Eq. (1).

Results and Discussion

Strain Energy Density and Bone Density

For the 3-tooth model, simulation results at the 5th, 10th, 15th and 25th time steps are presented in Figs. 4 and 5, respectively, for constrained and free boundary conditions. It can be observed that the strain energy density initially concentrated near the apical region of the teeth. As time goes by, strain energy density dissipated into other regions of the structure and resulted in a more uniform distribution. Moreover, the overall strain energy density in trabecular bone reduced over time.

Bone density was set to be uniform in trabecular bone initially. During the bone remodeling processes, bone density gradually changed under load. The bone density increased between tooth roots, where strain energy density was greater. It also increased in the right side of the model (mesial side), towards the direction of the applied load.

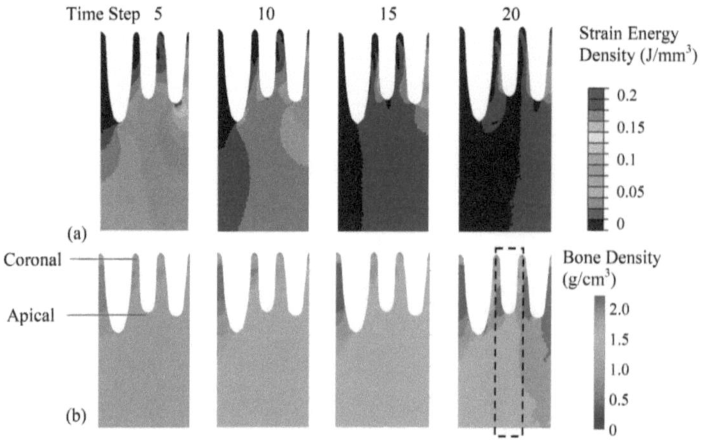

Fig. 4 **a** Strain energy density and **b** bone density in the 3-tooth model with constrained boundary conditions

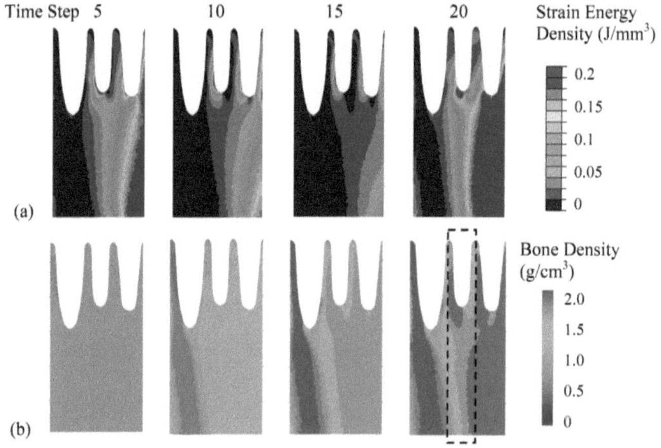

Fig. 5 **a** Strain energy density and **b** bone density in the 3-tooth model with free boundary conditions

Boundary Conditions

The results from models using constrained or free boundary conditions were similar in the vicinity of the lateral incisor. For the two boundary conditions, the bone density intensity increased between lateral incisor and its adjacent teeth during the remodeling process. Generally, the bone density value was lower in the coronal region and higher in the apical region. Below the lateral incisor, the bone density distribution is more uniform when the boundary was constrained (Fig. 4b).

When the boundary was set free, the variation in the bone density below the lateral incisor is more severe (Fig. 5b).

The actual teeth movement is constrained by their adjacent teeth and jaw bone. Therefore, the constrained boundary condition may be a better representation of the actual situation. Furthermore, the boundaries should be set far away from the region of interest. In other words, the results in the middle of the model (near the lateral incisor) are more reliable than those close to the left and right boundaries of the model.

Adjacent Teeth

Figure 6 shows the results obtained from the 1-tooth model with constrained boundary conditions. Strain energy density was higher towards the direction of applied load. The overall variation of the strain energy density decreased during bone remodeling process. It increased below the tooth in the direction of the applied load. The bone density also increased in the similar regions below and to the right of the tooth. In contrast, bone density to the left of the tooth drastically decreased. When adjacent teeth were missing, with fewer constraints, the density variation in the left (distal) and right (mesial) sides of the tooth was more severe.

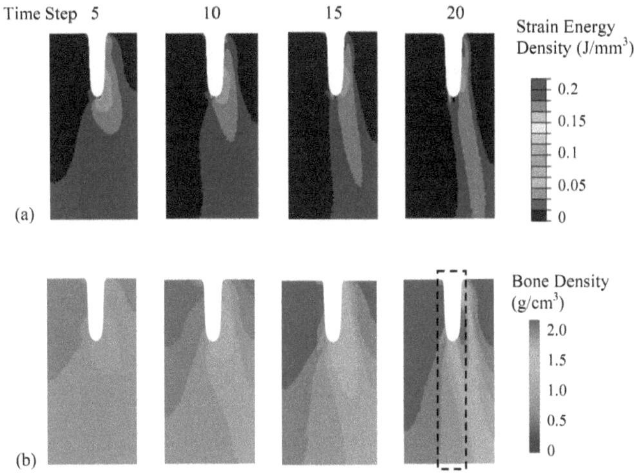

Fig. 6 **a** Strain energy density and **b** bone density in the 1-tooth model with constrained boundary conditions

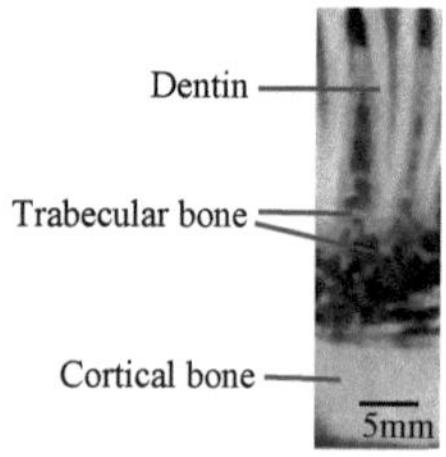

Fig. 7 A slice of dental cone beam computed tomography (CT) images

Computed Tomography (CT) Scans

A virtual slice of the dental computed tomography (CT) images of a human jaw is presented in Fig. 7. The greyscale of the image represents the density of the bone. As shown in Fig. 7, the density of cortical bone is much higher than trabecular bone. The trabecular bone around the lateral incisor is relatively denser than those below the tooth. There is not significant variation in the bone density beneath the tooth. These are in agreement with the bone density distribution obtained from 3-tooth model with constrained boundary conditions.

Implications

The current results suggest that there is a relationship between the applied load and the bone density distribution in jaw bone. Human teeth have the natural tendency to move towards the front and center (mesial drift) [16]. Moreover, jaw bone level decreases rapidly in the first 6 months to 2 years following tooth extraction. [17] Bone remodeling simulations can provide insights to these dental phenomena. However, there is a need to develop more accurate bone remodeling models.

There are several limitations in the current model. For example, the initial bone density was set to be uniform. Also, other stimulus for bone remodeling, such as biochemical signals, were ignored in the model. Further work is also needed to assess the actual geometry of teeth in jaw bone and the anisotropic and viscoelastic mechanical properties of the tissues.

Conclusion

This paper presents the results of the numerical simulations of jaw bone remodeling processes during normal chewing and biting activities. Bone increased at the regions where the mechanical stimulus (strain energy density) was greater, and *vice versa*. The variations in mechanical stimulus in the whole structure reduced over time. The overall level of the mechanical stimulus also reduced. With constrained boundaries far from the region of interest, results obtained from 3-tooth model were in agreement with the dental CT data. Without constrained boundaries, the results were not comparable to CT data in the regions below the tooth. Without the presence of the adjacent teeth, bone density in the left (distal) side of the tooth decreased drastically in the 1-tooth model.

Acknowledgements This study is supported by the Department of Mechanical and Nuclear Engineering and Institute for Cyber Science (ICS) at the Pennsylvania State University. The authors are grateful to Dr. Reuben Kraft for useful technical discussions.

References

1. J. Wolff, *The Law of Bone Remodeling* (Springer, New York, 1986)
2. L.P. Maia, D.M. Reino, A.B. Novaes, V.A. Muglia, M. Taba, M.F. de Morais Grisi, S.L.S. de Souza, D.B. Palioto, Influence of periodontal biotype on buccal bone remodeling after tooth extraction using the flapless approach with a xenograft: a histomorphometric and fluorescence study in small dogs. Clin. Implant Dent. Relat. Res. **17**, e221–e235 (2015)
3. B.A. Gultekin, P. Gultekin, S. Yalcin, Application of finite element analysis in implant dentistry. Finite Elem. Anal. New Trends Dev. 21–54 (2012)
4. C. Bourauel, D. Vollmer, A. Jäger, Application of bone remodeling theories in the simulation of orthodontic tooth movements. J. Orofac. Orthop. / Fortschritte der Kieferorthopädie **61**(4), 266 (2000)
5. J. Li, H. Li, L. Shi, A.S.L. Fok, C. Ucer, H. Devlin, K. Horner, N. Silikas, A mathematical model for simulating the bone remodeling process under mechanical stimulus. Dent. Mater. **23**(9), 1073–1078 (2007)
6. J. Martínez-Reina, J. Ojeda, J. Mayo, On the use of bone remodelling models to estimate the density distribution of bones. Uniqueness of the solution. PLoS One **11**(2), e0148603 (2016)
7. B.P. McNamara, D. Taylor, P.J. Prendergast, Computer prediction of adaptive bone remodelling around noncemented femoral prostheses: the relationship between damage-based and strain-based algorithms. Med. Eng. Phys. **19**(5), 454–463 (1997)
8. D.J. Adams, A.A. Spirt, T.D. Brown, S.P. Fritton, C.T. Rubin, R.A. Brand, Testing the daily stress stimulus theory of bone adaptation with natural and experimentally controlled strain histories. J. Biomech. **30**(7), 671–678 (1997)
9. A. Mellal, H.W.A. Wiskott, J. Botsis, S.S. Scherrer, U.C. Belser, Stimulating effect of implant loading on surrounding bone. Comparison of three numerical models and validation by in vivo data. Clin. Oral Implants Res. **15**(2), 239–248 (2004)
10. D.L. Bird, D.S. Debbie S, *Modern Dental Assisting* (Elsevier Health Sciences, 2013)
11. T.M.G.J. van Eijden, J.H. Koolstra, P. Brugman, W.A. Weijs, A feedback method to determine the three-dimensional bite-force capabilities of the human masticatory system. J. Dent. Res. **67**(2), 450–454 (1988)
12. M. Su, H. Chang, Y. Chiang, J. Cheng, L.-J. Fuh, C.-Y. Wang, C.-P. Lin, Modeling viscoelastic behavior of periodontal ligament with nonlinear finite element analysis. J. Dent. Sci. **8**(2), 121–128 (2013)
13. D.R. Carter, W.C. Hayes, The compressive behavior of bone as a two-phase porous structure. J. Bone Joint Surg. Am. **59**(7), 954–962 (1977)
14. D. Lin, Q. Li, W. Li, P. Rungsiyakull, M. Swain, Bone resorption induced by dental implants with ceramics crowns **45**, 1–7 (2009)

15. Z. Lian, H. Guan, S. Ivanovski, Y.-C. Loo, N. Johnson, H. Zhang, Finite element simulation of bone remodelling in the human mandible surrounding dental implant. Acta Mech. **217**(3–4), 335–345 (2011)
16. A. Nanci, A.R. Ten Cate, *Ten Cate's Oral Histology: Development, Structure, and Function* (Elsevier, 2013), p. 379
17. D.A. Atwood, Bone loss of edentulous alveolar ridges. J. Periodontol. **50**(4s), 11–21 (1979)

Investigating Biochemical Constituents of *Cymbopogon citratus* Leaf: Prospects on Total Corrosion of Concrete Steel-Reinforcement in Acidic-Sulphate Medium

Joshua Olusegun Okeniyi, Elizabeth Toyin Okeniyi, Olubanke Olujoke Ogunlana, Taiwo Felicia Owoeye and Oluseyi Ebenezer Ogunlana

Abstract In this paper, the biochemical constituents of *Cymbopogon citratus* leaf were investigated by atomic absorption spectroscopy (AAS), Fourier transform infrared spectroscopy (FT-IR) and phytochemical screening analyses for assessing its steel-rebar corrosion-protection prospects. AAS results showed that the leaf contained Fe = 4,641.025 μg/g, Mn = 849.5069 μg/g, Cu = 171.045 μg/g, Pb = 13.2938 μg/g, Ni = 11.5187 μg/g, Cd = 4.9310 μg/g, but Cr = 0.0 μg/g. FT-IR indicates the leaf-extract contained S-, N-, O-containing heteroatoms and aromatic compounds, which are rich in π-electrons and that are known to inhibit steel-rebar corrosion. Phytochemical screening results showed that *Cymbopogon citratus* leaf-extract contains tannins, phlobatannins, saponins, glycosides, flavonoids, steroids and terpenoids. Macrocell corrosion tests, as per ASTM G109-99a, showed that the plant reduced total corrosion in 0.5 M H_2SO_4-immersed steel-reinforced

J.O. Okeniyi (✉)
Mechanical Engineering Department, Covenant University, Ota 112001, Nigeria
e-mail: joshua.okeniyi@covenantuniversity.edu.ng

E.T. Okeniyi (✉)
Petroleum Engineering Department, Covenant University, Ota, Nigeria
e-mail: lizzoken@yahoo.com

O.O. Ogunlana
Biochemistry Programme, Department of Biological Sciences, Covenant University, Ota, Nigeria

T.F. Owoeye
Chemistry Department, Covenant University, Ota, Nigeria

O.E. Ogunlana
Biochemistry Programme, Department of Biological Sciences, Crawford University, Igbesa, Nigeria

The Minerals, Metals & Materials Society, *TMS 2017 146th Annual Meeting & Exhibition Supplemental Proceedings*, The Minerals, Metals & Materials Series,
DOI 10.1007/978-3-319-51493-2_32

concrete samples. These indicated positive prospects of *Cymbopogon citratus* as an eco-friendly inhibitor of steel-reinforcement corrosion in concretes designed for acidic sulphate service-environment.

Keywords *Cymbopogon citratus* • Eco-friendly natural-plant • Steel-reinforcement in concrete • Total-corrosion modelling • Inorganic and organic bio-constituents • Phytochemical analyses • Acidic-sulphate medium

Introduction

Use of corrosion inhibitors has been identified in studies [1–6] as a less costly and easily applicable method in comparisons to the other known methods for mitigating corrosion risks in steel-reinforced concrete. In spite of this, problems ensuing from the toxicity of traditional inhibitors [7–10] are making researchers turn to extracts from natural plant sources. Reasons for this include the consideration that plant extract abounds with biologically safe organic constituents that are also known to inhibit reinforcing steel corrosion in aggressive environments [4, 8, 11–13]. Chemicals and phyto-constituents from natural plants have the potentials of being made up of eco-friendly electron donors that also inhibit the corrosion process [7, 11, 14]. Unlike these, synthetic chemical inhibitors could be lethal to living organisms, by disrupting biochemical process and inducing malfunctioning or damage of internal organs. However, usage of extract from plant as corrosion inhibitor requires that the bio-constituents from the natural plant need to be characterized for understanding the biochemical ligands that are produced by the plant [11, 15]. Also, investigations of mitigating effectiveness on metallic corrosion by such plant extract needs to be done in specified medium of corrosive test-system [15–17].

Cymbopogon citratus, lemongrass, is a perennial, aromatic plant of tropical area habitat in Africa, Asia, Central and South America. In these regions, fresh and dried leaves of the plant are used for medicinal purposes and as ingredients in food [18–21]. According to literature, leaf-extract of *Cymbopogon citratus* is not toxic [18, 22], rather infusions, decoctions and essential oils from the plant exhibit health benefits including positive bioactivities of antioxidant, anti-inflammatory, anti-carcinogenic, antihypertensive, anti-diabetic, anti-mutagenic and antimicrobial properties [22–24]. While studies had deliberated on the chemical constituents of this natural plant, indication from literature include the fact that such constituents could vary according to the geographical origin of the plant such that needs for bio-constituents of different extract are still required [18, 21]. These constitute motivation for this study, especially on the basis that there is dearth of study in which characterization by the use of Fourier transform infrared, FT-IR, spectroscopy techniques has been detailed for *Cymbopogon citratus*. Also, the phytochemical and inorganic characterization of the plant from our study location could find comparison with that from other regions even as no study has detailed the composition of *Cymbopogon citratus* with respect to its prospect on corrosion-protection of steel-reinforcement in concrete. Therefore, this study deliberates on the biochemical characterization of the inorganic and organic constituents of *Cymbopogon citratus* leaf and the test on prospects of the plant on the corrosion protection of concrete steel-reinforcement in aggressive medium.

Experimental

Leaves of *Cymbopogon citratus* (*C. citratus*) *Poaceae* were collected fresh at Idiroko Road, Canaanland, Ota, Ogun State, Nigeria. The leaves were identified at the Forestry Herbarium Ibadan (FHI), Nigeria and a sample was deposited there with the voucher FHI. No. 109497. The leaves of *C. citratus* for this study were dried under cover in a well aerated room maintained at 20 °C by a 5 ton air-conditioning system.

Study of inorganic (heavy metal) constituent using the AAS (Atomic Absorption Spectrometer) technique followed the procedure that had been detailed in studies [16, 25]. This was initiated by ashing some dried and pulverised leaves of *C. citratus* in a muffle furnace maintained for 2 h at 500 °C. The ash content obtained by this = 6.3989%. From the ashed material, 0.25 g was soaked overnight in 5 ml of 1:1 nitric-perchloric acid mixture. This was then heated at 150 °C for 1 h in a reflux condenser before raising the temperature gradually to 235 °C by heating and the heating then maintained for another 2 h for the occurrence of dense fume. The test-material was then removed from the heating block, cooled to 100 °C before the addition of 1:1 HCl, and which was followed by heating to white fumes until the resultant solution obtained was colourless. The colourless solution was then poured into 100 ml flask and washed with water for five times, at which the washing was added to the flask each of the time. The volume was then made up to the 100 ml mark of the flask with water. Then, the obtained solution was filtered and the filtrate from this was taken to the Hollow Cathode Lamps equipped Model S Series, AA Spectrometer from Thermo Electron Corporation®.

Two methods were employed for the characterization of organic constituent, which was initiated by obtaining leaf-extract of *C. citratus*, using procedure that had been detailed in studies [13, 26]. The first method, the FT-IR analyses, include smearing of *C. citratus* leaf-extract between a pair of mounting discs that was then mounted onto the FT-IR instrument of the Perkin-Elmer FT-IR System, Spectrum BX. Rendering of the FT-IR spectra obtained from this approach to the Euclidean Search of Fluka library by the Perkin-Elmer® instrument [16] was also done in this study.

The second method for organic constituent characterization employs phytochemical screening methods using the procedures prescribed in [27] and detailed in [16]. For this, 2 g of *C. citratus* leaf-extract was added to 20 ml of distilled water for making test sample that was then subjected to the phytochemical test-procedures detailed in [16] for the tests for tannins, phlobatannins, saponins, flavonoids, steroids, terpenoids, glycosides and alkaloids.

Anticorrosion testing employs electrochemical measurements of macrocell current technique [28] as per ASTM G109-99a [29]. For this, steel-reinforced concrete samples into, which had been admixed different concentrations of *C. citratus* leaf-extract, were used for the corrosion tests. Admixture concentration design, in steel-reinforced concrete specimens, includes 0 g/L (i.e. blank or control sample), 1.667 g/L *C. citratus* and 6.667 g/L *C. citratus* leaf-extract, where the g/L were of concrete mixing water. Each 100 × 100 × 200 mm steel-reinforced concrete sample, having ∅12 mm reinforcing steel embedment, was formulated and cast as described in studies [13, 30]. Each of the steel-reinforced concrete samples was immersed in plastic bowl containing 0.5 M H_2SO_4 test-medium. From these experimental setups, macrocell corrosion current was measured versus $Cu/CuSO_4$ (CSE) reference electrode, from Tinker and Rasor®, using zero resistance ammeter (ZRA), from Corrosion Service® [28, 31, 32]. The measurements were taken first at five days interval for 40 days and thereafter at seven days interval for the next 7 weeks. These constitute 16 data-point measurements, from each steel-reinforced concrete sample, in 89 days experimental period for the corrosion test-investigation. From the macrocell current measurements, total corrosion, *TC* in coulomb (C), was modeled using the equation [25, 28, 29]:

$$TC_k = TC_{k-1} + [(t_k - t_{k-1}) \times (I_k + I_{k-1})/2]; \quad k = 2, 3, \ldots, 16 \tag{1}$$

where: $t \equiv$ time (s) between the previous, (k–1)th, and the present (kth) measurement of macrocell corrosion current; $I_k \equiv$ macrocell corrosion current (A) at time t_k.

Results and Discussion

Results of the AAS analyses for heavy metals are as shown in Table 1. This showed that the tested plant was rich in iron micronutrient and low in other heavy metals, which indicates that the growth environment of the studied *C. citratus* was not toxic with other heavy metals that could have resulted in the deficiency of Fe [16, 33].

Results of the organic constituent characterisation of *C. citratus* are presented in Fig. 1, for the FT-IR spectra, and in Table 2, for the phytochemical screening analyses. Figure 1 showed that the extract exhibited overlaps of many organic heteroatoms. Notable from the figure include the free O–H stretching vibration of hydroxyl group absorbed at 3731.85 cm^{-1}, a high frequency that, according to [34],

Table 1 AAS Test-results of inorganic constituents from *C. citratus*

Inorganic constituent	Iron (Fe)	Manganese (Mn)	Copper (Cu)	Lead (Pb)	Nickel (Ni)	Cadmium (Cd)	Chromium (Cr)
Concentration (μg/g)	4641.025	849.5069	171.045	13.2938	11.5187	4.9310	–

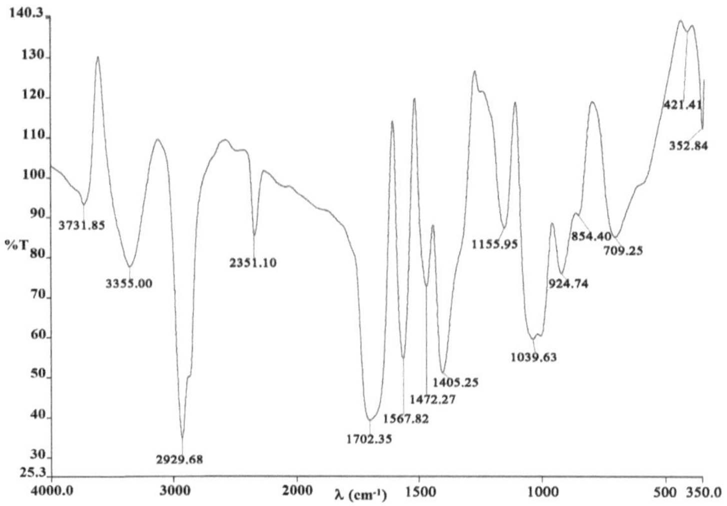

Fig. 1 FT-IR spectra of the leaf extract of *C. citratus*

suggests characterization of phenol antioxidants class of compounds that could be found in food. This exhibits fingerprint of phenol (a hydroxyl group bonded to an aromatic hydrocarbon group) or tertiary alcohol O–H in-plane bending vibration at 1405.25 cm^{-1}. Another fingerprint for this group includes the C–O stretching bands at 1155.95 cm^{-1}, suggesting tertiary alcohol, and at 1039.63 cm^{-1}. The 3355.00 cm^{-1} exhibited potencies of overlaps of N–H stretch of the secondary amine group, having fingerprint of NH_2 scissoring at 1567.82 cm^{-1} and N–H wagging at 854.40 cm^{-1} and 709.25 cm^{-1}, and/or of carboxylic acid due to the additional O–H stretch at 2929.68 cm^{-1} and C=O at 1702.35 cm^{-1}. This carboxylic acid group suggestion followed from the overlapping of the group's fingerprints in the spectra including the in-plane bending of C–O–H at 1405.25 cm^{-1} and the O–H out-of-plane bending at 924.74 cm^{-1}. In the same light, the absorption at 709.25 cm^{-1} could also include C–Cl stretch of chloro halogenated organic compounds, just as the 1039.63 cm^{-1} could also represent weakened frequency, due to electron extraction, of the C=S thiocarbonyl while the absorption at 2351.10 cm^{-1} could be weakened frequency of the S–H thiols.

The rendering of the obtained spectra from Fig. 1 to the Euclidean Search of Fluka library by the Perkin-Elmer® instrument gave hit-list that suggests constitution of biochemical compounds in *C. citratus* leaf-extract. These biochemical compounds are detailed in 3-D optimised structure of the compounds in Fig. 2. These organic biochemical compounds indicated two N-, four S- and nine O-containing as well as eight aromatic compounds as ligands that could be constituted in *C. citratus* leaf-extract.

Phytochemical screening analyses of *C. citratus* leaf-extract showed that it contains seven out of the eight phyto-constituents tested; only alkaloids was not identified as been present in the extract. These phytochemical test-results find

Table 2 Phytochemical constituent of *C. citratus**

Tannins	Phlobatannins	Saponins	Glycosides	Flavonoids	Steroids	Terpenoids	Alkaloids
+	+	+	+	+	+	+	−

*Present ≡ +; Absent ≡ −

agreements with that from [18] wherein it was confirmed that *C. citratus* contained tannins, saponins, flavonoids and steroids. That study did not test for phlobatannins, glycosides and terpenoids although alkaloids was found to be present in *C. citratus* in that study but not observed by the phytochemical test in this study.

It is worth noting that plant extract with FT-IR spectra and phytochemical test-results finding similarity with that obtained in this study had been employed in other studies for inhibiting steel corrosion in aggressive media [13, 25, 26, 35, 36]. Among these, the plant extract in [25, 26, 35] inhibited steel corrosion in 0.5 M H_2SO_4. Also, the report in [35] showed that ligands with aromatic, lone pairs and π electron donors, such as were found in this study, exhibited affinity for carbon steel. In that study in [35] for instance, the ligands exhibiting corrosion protection were characterized from the protective film on the metal. These considerations make *C. citratus* leaf-extract suspect of positive prospect, also, for concrete steel reinforcement corrosion risk reduction in sulphuric acid medium.

The trends of electrochemical test-results of macrocell corrosion current measurements from steel-reinforced concrete samples admixed with different concentrations of *C. citratus* in the 89-day experimental period are presented in Fig. 3. The plots in Fig. 3 shows that the blank steel-reinforced concrete sample, with 0 g/L *C. citratus* leaf-extract, exhibited high macrocell current. The steel-reinforced concrete sample with 3.333 g/L *C. citratus* exhibited corrosion current that tends to increase towards and thereafter undulating about the corrosion current obtained from the blank sample. This suggests dynamism of corrosion attacks on the steel-reinforcement and the consistent mitigation of the metallic corrosion by the *C. citratus* admixture. The steel-reinforced concrete admixed with 6.667 g/L *C. citratus* not only exhibited lower macrocell corrosion current than the blank sample throughout the experimental period but also, a reducing trend of corrosion current from the 68th day through the 89th day. Thus, that the macrocell corrosion current from the 3.333 g/L *C. citratus* steel-reinforced concrete sample tended towards that from the blank sample indicates insufficiency of the 3.333 g/L *C. citratus* concentration at effectively mitigating the acidic corrosion attack. Therefore, the increment of the *C. citratus* concentration to 6.667 g/L constitutes the attainment of sufficient or suitable concentration of *C. citratus* admixture for the effective mitigation of corrosion risk in the tested H_2SO_4-immersed steel-reinforced concrete samples. This suitability of concentration for effective corrosion mitigation is a major criterion for indicating acceptability of a substance as corrosion inhibitor [37–39].

The analysis of the corrosion current measurements for each steel-reinforced concrete sample, as per ASTM G109-99a [29], through Eq. (1), gives the total corrosion model that plotted for each sample in Fig. 4. From this figure, it could be

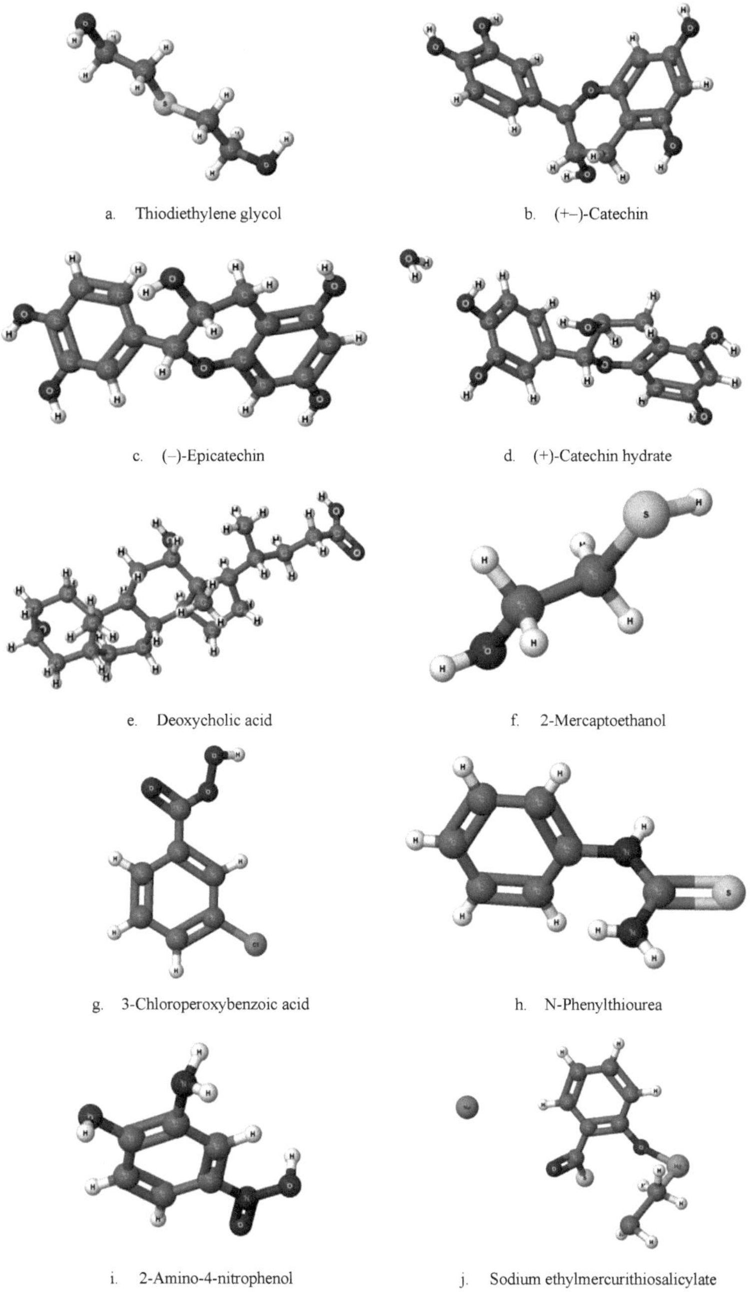

Fig. 2 Optimised structures of compounds from the Euclidean search hit-list on FT-IR spectra of *C. citratus* leaf-extract

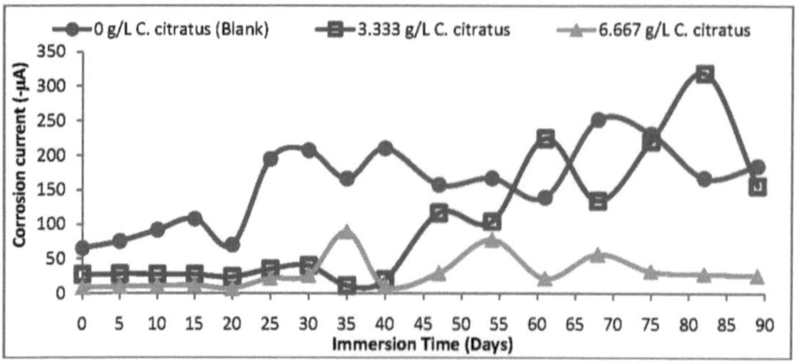

Fig. 3 Trend of corrosion current measurements from steel-reinforced concrete with *C. citratus* leaf-extract admixture concentrations in 0.5 M H_2SO_4 medium

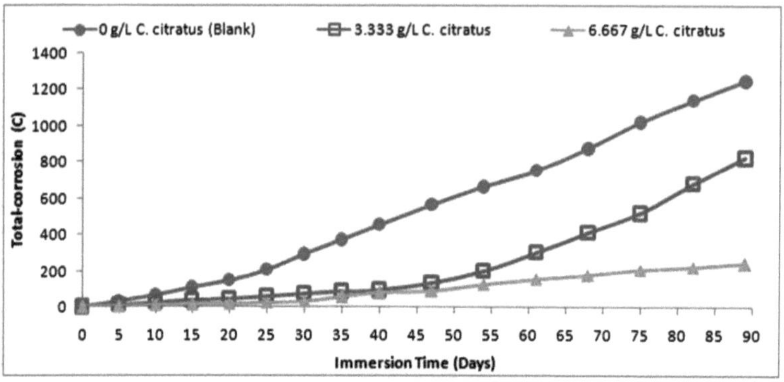

Fig. 4 Total corrosion effects of *C. citratus* leaf-extract admixture on 0.5 M H_2SO_4-immersed steel-reinforced concrete

noted that the undulations of corrosion current that could be observed in Fig. 3 are converted into linear forms of test-response by the total corrosion model in Fig. 4. This linear form of corrosion test-result clearly shows the consistently increasing total corrosion from the blank sample, indicating the aggressiveness of the sulphuric acid medium on concrete steel-reinforcement in the absence of corrosion inhibitor. The figure also showed the tending of the total corrosion from the 3.333 g/L admixed sample towards that from the blank sample and the highly mitigated total corrosion from the steel-reinforced concrete with 6.667 g/L *C. citratus* admixture.

Therefore, it could be deduced from this study that *C. citratus* leaf-extract admixture in concrete reduced total corrosion of the embedded steel-reinforcement in the 0.5 M H_2SO_4-immersed concrete, relative to the blank sample without admixture. These bare implications of positive prospects on the suitability of *C. citratus* leaf-extract as an environmentally-friendly admixture in concrete for the corrosion protection of concrete steel-reinforcement in concrete designed for corrosive service-environment.

Conclusions

In this study, it was established that:

- The leaf of *C. citratus* contains non-toxic levels of inorganic (heavy metal) micronutrients, through results from AAS experiment;
- Leaf-extract from *C. citratus* is constituted of S-, N-, O-containing heteroatoms and aromatic compounds that are rich in lone pairs and π-electron that are known to exhibit coordinate affinity with steel metal used for concrete reinforcement, through results from characterization experiment of FT-IR;
- Phyto-constituents that could enhance steel-rebar corrosion-protection in aggressive environment are contained in *C. citratus* leaf-extract, through phytochemical screening test-experiments;
- Concentrations of *C. citratus* leaf-extract admixture reduced total corrosion in 0.5 M H_2SO_4-immersed steel-reinforced concrete samples, relative to blank sample without admixture in the same medium, through corrosion test-experimental employing measurement of macrocell current technique.
- From these results of biochemical constituent characterisation and corrosion-protection ability, it could be concluded from the study that *C. citratus* exhibited positive prospects for eco-friendly corrosion mitigation on steel-reinforcement in concrete designed for the acidic sulphate environment.

References

1. D.M. Bastidas, M. Criado, S. Fajardo, A.L. Iglesia, J.M. Bastidas, Corrosion inhibition mechanism of phosphates for early-age reinforced mortar in the presence of chlorides. Cement Concr. Compos. **61**, 1–6 (2015)
2. J.O. Okeniyi, I.O. Oladele, O.M. Omoniyi, C.A. Loto, A.P.I. Popoola, Inhibition and compressive-strength performance of $Na_2Cr_2O_7$ and $C_{10}H_{14}N_2Na_2O_8 \cdot 2H_2O$ in steel-reinforced concrete in corrosive environments. Can. J. Civ. Eng. **42**, 408–416 (2015)
3. F.-L. Fei, J. Hu, J.-X. Wei, Q.-J. Yu, Z.-S. Chen, Corrosion performance of steel reinforcement in simulated concrete pore solutions in the presence of imidazoline quaternary ammonium salt corrosion inhibitor. Constr. Build. Mater. **70**, 43–53 (2014)
4. J.O. Okeniyi, C.A. Loto, A.P.I. Popoola, Corrosion inhibition performance of *Rhizophora mangle* L bark-extract on concrete steel-reinforcement in industrial/microbial simulating-environment. Int. J. Electrochem. Sci. **9**, 4205–4216 (2014)
5. J.O. Okeniyi, I.J. Ambrose, I.O. Oladele, C.A. Loto, A.P.I. Popoola, Electrochemical performance of sodium dichromate partial replacement models by triethanolamine admixtures on steel-rebar corrosion in concretes. Int. J. Electrochem. Sci. **8**, 10758–10771 (2013)
6. Y. Tang, G. Zhang, Y. Zuo, The inhibition effects of several inhibitors on rebar in acidified concrete pore solution. Constr. Build. Mater. **28**, 327–332 (2012)
7. J.O. Okeniyi, O.A. Omotosho, O.O. Ajayi, C.A. Loto, Effect of potassium-chromate and sodium-nitrite on concrete steel-rebar degradation in sulphate and saline media. Constr. Build. Mater. **50**, 448–456 (2014)

8. S.A. Asipita, M. Ismail, M.Z. Abd Majid, Z.A. Majid, C. Abdullahb, J. Mirza, Green *Bambusa arundinacea* leaves extract as a sustainable corrosion inhibitor in steel reinforced concrete. J. Clean. Prod. **67**, 139–146 (2014)
9. J.O. Okeniyi, O.A. Omotosho, O. Ajayi, O.O. James, C.A. Loto, Modelling the performance of sodium nitrite and aniline as inhibitors in the corrosion of steel-reinforced concrete. Asian J. Appl. Sci. **5**, 132–143 (2012)
10. O.A. Omotosho, J.O. Okeniyi, O.O. Ajayi, C.A. Loto, Effect of synergies of $K_2Cr_2O_7$, K_2CrO_4, $NaNO_2$ and aniline inhibitors on the corrosion potential response of steel reinforced concrete in saline medium. Int. J. Environ. Sci. **2**, 2346–2359 (2012)
11. M. Ismail, P.B. Raja, A.A. Salawu, Developing deeper understanding of green inhibitors for corrosion of reinforcing steel in concrete, in *Handbook of Research on Recent Developments in Materials Science and Corrosion Engineering Education*, ed. by H. Lim (IGI Global, Hershey, PA, 2015), pp. 118–146
12. J.O. Okeniyi, C.A. Loto, A.P.I. Popoola, *Rhizophora mangle* L. effects on steel-reinforced concrete in 0.5 M H_2SO_4: implications for corrosion-degradation of wind-energy structures in industrial environments. Energy Procedia **50**, 429–436 (2014)
13. J.O. Okeniyi, C.A. Loto, A.P.I. Popoola, Electrochemical performance of *Anthocleista djalonensis* on steel-reinforcement corrosion in concrete immersed in saline/marine simulating-environment. Trans. Indian Inst. Met. **67**, 959–969 (2014)
14. A.K. Satapathy, G. Gunasekaran, S.C. Sahoo, K. Amit, P.V. Rodrigues, Corrosion inhibition by *Justicia gendarussa* plant extract in hydrochloric acid solution. Corros. Sci. **51**, 2848–2856 (2009)
15. J.O. Okeniyi, C.A. Loto, A.P.I. Popoola, "Anticorrosion performance of *Anthocleista djalonensis* on steel-reinforced concrete in a sulphuric-acid medium. HKIE Trans. (2016). doi:10.1080/1023697X.2016.1201437
16. J.O. Okeniyi, O.O. Ogunlana, O.E. Ogunlana, T.F. Owoeye, E.T. Okeniyi, Biochemical characterisation of the leaf of *Morinda lucida*: prospects for environmentally-friendly steel-rebar corrosion-protection in aggressive medium, in *TMS2015 Supplemental Proceedings* (Wiley, Hoboken, NJ, USA, 2015), pp. 635–644. doi:10.1002/9781119093466.ch78
17. J.O. Okeniyi, C.A. Loto, A.P.I. Popoola, *Morinda lucida* effects on steel-reinforced concrete in 3.5% NaCl: Implications for corrosion-protection of wind-energy structures in saline/marine environments. Energy Procedia **50**, 421–428 (2014)
18. C.E. Ekpenyong, E.E. Akpan, N.E. Daniel, Phytochemical constituents, therapeutic applications and toxicological profile of *Cymbopogon citratus* Stapf (DC) leaf extract. J. Pharm. Phytochem. **3**, 133–141 (2014)
19. K. Manvitha, B. Bidya, Review on pharmacological activity of *Cymbopogon citratus*. Int. J. Herbal Med. **1**, 5–7 (2014)
20. V. Francisco, A. Figueirinha, G. Costa, J. Liberal, M.C. Lopes, C. García-Rodríguez, C.F.G. C. Geraldes, M.T. Cruz, M.T. Batista, Chemical characterization and anti-inflammatory activity of luteolin glycosides isolated from lemongrass. J. Func. Foods **10**, 436–443 (2014)
21. G. Shah, R. Shri, V. Panchal, N. Sharma, B. Singh, A.S. Mann, Scientific basis for the therapeutic use of *Cymbopogon citratus*, stapf (Lemon grass). J. Adv. Pharm. Technol. Res. **2**, 3–8 (2011)
22. S.K. Olorunnisola, H.T. Asiyanbi, A.M. Hammed, S. Simsek, Biological properties of lemongrass: an overview. Int. Food Res. J. **21**, 455–462 (2014)
23. S.R. Kanatt, S.P. Chawla, A. Sharma, Antioxidant and radio-protective activities of lemon grass and star anise extracts. Food Biosci. **6**, 24–30 (2014)
24. V. Francisco, G. Costa, A. Figueirinha, C. Marques, P. Pereira, B.M. Neves, M.C. Lopes, C. García-Rodríguez, M.T. Cruz, M.T. Batista, Anti-inflammatory activity of *Cymbopogon citratus* leaves infusion via proteasome and nuclear factor-κB pathway inhibition: contribution of chlorogenic acid. J. Ethnopharmacol. **148**, 126–134 (2013)
25. J.O. Okeniyi, O.A. Omotosho, O.O. Ogunlana, E.T. Okeniyi, T.F. Owoeye, A.S. Ogbiye, E. O. Ogunlana, Investigating prospects of *Phyllanthus muellerianus* as eco-friendly/sustainable

material for reducing concrete steel-reinforcement corrosion in industrial/microbial environment. Energy Procedia **74**, 1274–1281 (2015)
26. J.O. Okeniyi, C.A. Loto, A.P.I. Popoola, Electrochemical performance of *Phyllanthus muellerianus* on the corrosion of concrete steel-reinforcement in industrial/microbial simulating-environment. Portugaliae Electrochimica Acta **32**, 199–211 (2014)
27. H.O. Edeoga, D.E. Okwu, B.O. Mbaebie, Phytochemical constituents of some Nigerian medicinal plants. Afr. J. Biotechnol. **4**, 685–688 (2005)
28. J.O. Okeniyi, C.A. Loto, A.P.I. Popoola, Modelling *Rhizophora mangle* L bark-extract effects on concrete steel-rebar in 0.5 M H_2SO_4: implications on concentration for effective corrosion-inhibition, in *TMS2015 Supplemental Proceedings* (Wiley, Hoboken, NJ, USA, 2015), pp. 751–758. doi:10.1002/9781119093466.ch92
29. ASTM G109-99a, *Standard Test Method for Determining the Effects of Chemical Admixtures on the Corrosion of Embedded Steel Reinforcement in Concrete Exposed to Chloride Environments* (ASTM International, West Conshohocken, PA, 2005)
30. J.O. Okeniyi, O.A. Omotosho, C.A. Loto, A.P.I. Popoola, Corrosion rate and noise resistance correlation from $NaNO_2$-admixed steel-reinforced concrete. Asian J. Sci. Res. **8**, 454–465 (2015)
31. G.E. Abdelaziz, A.M.K. Abdelalim, Y.A. Fawzy, Evaluation of the short and long-term efficiencies of electro-chemical chloride extraction. Cem. Concr. Res. **39**, 727–732 (2009)
32. W.J. McCarter, Ø. Vennesland, Sensor systems for use in reinforced concrete structures. Constr. Build. Mater. **18**, 351–358 (2004)
33. United States Environmental Protection Agency (USEPA), *Ecological Soil Screening Levels for Iron* (USEPA, Washington, DC, 2003)
34. J. Coates, Interpretation of infrared spectra, a practical approach, in *Encyclopedia of Analytical Chemistry*, ed. by R.A. Meyers (Wiley, Chichester, 2000), pp. 10815–10837
35. P. Mourya, S. Banerjee, M.M. Singh, Corrosion inhibition of mild steel in acidic solution by *Tagetes erecta* (Marigold flower) extract as a green inhibitor. Corros. Sci. **85**, 352–363 (2014)
36. M. Sangeetha, S. Rajendran, J. Sathiyabama, A. Krishnaveni, P. Shanthy, N. Manimaran, B. Shyamaladevi, Corrosion inhibition by an aqueous extract of *Phyllanthus amarus*. Portugaliae Electrochimica Acta **29**, 429–444 (2011)
37. J.O. Okeniyi, O.M. Omoniyi, S.O. Okpala, C.A. Loto, A.P.I. Popoola, Effect of ethylenediaminetetraacetic disodium dihydrate and sodium nitrite admixtures on steel-rebar corrosion in concrete. Eur. J. Environ. Civil Eng. **17**, 398–416 (2013)
38. T.A. Söylev, M.G. Richardson, Corrosion inhibitors for steel in concrete: State-of-the-art report. Constr. Build. Mater. **22**, 609–622 (2008)
39. A.M. Vaysburd, P.H. Emmons, Corrosion inhibitors and other protective systems in concrete repair: concepts or misconcepts. Cement Concr. Compos. **26**, 255–263 (2004)

Two-Step Sintering Effects on the Microstructure and Mechanical Properties of Forsterite Scaffolds

Fariborz Tavangarian, Lindsay Childs, Guoqiang Li, Dakota Wooten and Bryant Cornwell

Abstract In this study, we evaluate the effects of two-step sintering on the microstructure, compressive strength and porosity of forsterite scaffolds for tissue engineering application. Forsterite slurry was prepared and pre-cut foams were immersed in the slurry for 1 h. The saturated foams were then annealed at various times and temperatures using the two-step sintering method. The samples were characterized by X-ray diffraction (XRD), scanning electron microscopy (SEM) and transmission electron microscopy (TEM) techniques. The results showed that two-step sintering method can prevent the grain growth which occurs during the sintering process of scaffolds at high temperatures and subsequently provides higher compressive strength compared to conventional sintering method. The first step of the sintering process, which occurs at a higher temperature, is to provide initial binding through sintering and localized melting between the contact surfaces of adjacent grains and the second step at a lower temperature is to strengthen and complete the sintering process.

Keywords Compressive strength · Scaffold · Two-step sintering

F. Tavangarian (✉)
Mechanical Engineering Program, School of Science, Engineering and Technology, Penn State Harrisburg, Middletown, PA 17057, USA
e-mail: f_tavangarian@yahoo.com

L. Childs · D. Wooten · B. Cornwell
Department of Engineering and Technology Management, Morehead State University, Morehead, KY 40351, USA

G. Li
Department of Mechanical and Industrial Engineering, Louisiana State University, Baton Rouge, LA 70803, USA

The Minerals, Metals & Materials Society, *TMS 2017 146th Annual Meeting & Exhibition Supplemental Proceedings*, The Minerals, Metals & Materials Series,
DOI 10.1007/978-3-319-51493-2_33

Introduction

In recent years, scaffolds have attracted the attention of many scientists as structural components in bone defects. Their porosity, mechanical properties, bioactivity and biodegradability are the most important factors that determine their usefulness in various applications. For example, in many maxillofacial surgeries, there is no need to have scaffolds with excellent mechanical properties since in those applications the implant is not under load. However, in load bearing applications, scaffolds are expected to tolerate enough pressure without breaking or failure in the defect location. In these applications the strength of the scaffolds should be compatible with the strength of the surrounding tissue. If the strength of the scaffold is less than that of surrounding tissue, it is possible that during the healing process the implant (scaffold) fails and subsequently there is a need for a second surgery and/or prolonged healing time. In the other situation, mechanical strength mismatch between implant and the surrounding bone is the main reason of stress shielding, surrounding bone resorption and implant loosening [1, 2]. Furthermore, this situation causes an extensive micro-motion between implants and the surrounding bone tissue, which prevents bone formation. This situation inhibits the osseointegration of the scaffold [3]. Thus, developing new implants with compatible mechanical strength and elastic modulus to that of surrounding bone is required. If we can provide scaffolds with appropriate strength and proper elastic modulus, biological fixation can happen through bone ingrowth and finally faster healing period is expected.

Forsterite (Mg_2SiO_4) ceramic is a new bioceramic with good biocompatibility [4]. It shows better compressive strength and fracture toughness compared to hydroxyapatite [5]. On the other hand, since the melting point of forsterite is around 1890 °C, its sintering temperature is too high which inversely affects its mechanical properties due to the grain growth phenomena which occurs during the sintering process. Therefore, two-step sintering method seems a very useful technique to overcome this problem.

Highly porous scaffolds with open structure are required to treat bone defects to ensure bone oxygenation and angiogenesis [6, 7]. Macroporosity with pore size in the range of 200–800 μm is needed to allow the migration of nutrient supplies and removal of waste materials. On the other hand, microporosity with pore size less than 10 μm is necessary for capillary ingrowth and cell - matrix interactions [8–10]. It should be noticed that increasing the pore size decreases the mechanical properties of the scaffold and hence a balance between the pore size and mechanical properties are required for load bearing applications. In this study, forsterite scaffolds were prepared by two-step sintering method to prevent grain growth and subsequent reduction in the pore size. The mechanical properties and microstructure of the forsterite scaffolds were characterized and the optimum sample synthesis parameters were determined. The current method can be used to fabricate scaffolds with desired porosity and acceptable mechanical properties for load bearing applications.

Experimental Procedure

Forsterite powder was synthesized according to the procedures described in [11]. A slurry was prepared with adding ethanol to the forsterite powder and then pre-cut sponges were soaked in the slurry for 1 h. Then the sponges were dried at 100 °C and subsequently were heat treated. Heat treatment of the samples was carried out by two-step sintering (TSS) process. In TSS method two temperatures were selected to holding the specimens. The first temperature was T1 which was required for the initial attachment and formation of necking between adjacent grains. The second temperature was called T2 which was required for completion of sintering process and improving the strength of the scaffolds. T1 was selected as 1500 °C and T2 was selected as 1000 °C. Three different cycles were selected in this study. In Cycle 1, abbreviated as C1, the samples heated to T1 with the heating rate of 10 °C/min, then the samples were hold at that temperature for 60 min and then the samples were cooled down to 1000 °C with the cooling rate of 30 °C/min, remain at 1000 °C for 1000 min and then cooled down to room temperature with the cooling rate of 10 °C/min. The samples prepared by cycle 1 is called S1. In the second cycle, which was called C2, the holding time at 1500 °C was 5 min and the rest of the cycle was similar to C1. These samples were called S2. In the third cycle, called C3, the samples were remained at 1000 °C for 0 min and the rest of the cycle was similar to C1. These samples were called S3. Samples S4, S5, and S6 was similar to samples S1, S2, and S3, respectively, however in their case T1 was set to 1400 °C. Please see Table 1 for further information.

A Philips X'PERT MPD diffractometer with Cu Kα radiation ($\lambda = 0.154056$ nm) was used for X-ray diffractometry (XRD) analysis to confirm the formation of forsterite after heat treatment. XRD patterns were recorded in the 2θ range of 20–80° (a step size 0f 0.04° and a time per step of 1 s). The morphology of the scaffolds was observed by scanning electron microscopy (SEM) using a s (Quanta3D FEG, FEI Company, USA) at an acceleration voltage of 5 kV. The transmission electron

Table 1 Designation, compressive strength and porosity of the forsterite scaffolds

Designation	Heat treatment cycles				CS (MPa)	CM	Porosity (%)
	T1 (°C)	HT at T1 (min)	T2 (°C)	HT at T2 (min)			
S1	1500	60	1000	1000	0.19 ± 0.02	0.78 ± 0.05	78
S2	1500	5	1000	1000	0.36 ± 0.06	1.36 ± 0.02	82
S3	1500	60	1000	0	0.17 ± 0.04	0.88 ± 0.06	81
S4	1400	60	1000	1000	0.21 ± 0.03	0.68 ± 0.06	81
S5	1400	5	1000	1000	0.26 ± 0.04	0.92 ± 0.03	84
S6	1400	60	1000	0	0.22 ± 0.02	0.76 ± 0.02	82

HT Holding time; *CS* Compressive strength; *CM* Compressive modulus microscopy (TEM; JEM-1400) technique was utilized to characterize the morphology and the structure of the prepared samples. Forsterite powder particle size was measured by UTHSCSA Image Tool 3 software.

Results and Discussions

The XRD pattern of the forsterite sample after heat treatment by C1 is shown in Fig. 1. The XRD peaks correspond to the characteristic peaks of forsterite structures (XRD JCPDS data file No. 34-0189). Figure 2 shows the TEM of the forsterite sample after heat treatment by C1. As can be seen in Fig. 2 the crystallite size of the forsterite sample is still less than 100 nm even after sintering at 1500 °C for 60 min. This can be due to the extremely small crystallite size of the forsterite powder after preparation according to [11] and before heat treatment based on C1.

The compressive strength, compressive modulus and porosity of the specimens after various heat treatment cycle are shown in Table 1. As can be seen, the compressive strength of the samples was in the range of the compressive strength of spongy bone in human body which is around 0.2–4 MPa [12–14].

As can be seen in this table, the compressive strength of S2 is the highest through this experiment. In comparison to S1 sample, it is clear that holding the samples at 1500 °C for 60 min resulted in a lower compressive strength due to the detrimental effects of grain growth on the mechanical properties of S1. Furthermore, with comparing sample S2 and S3, it can be concluded that holding the samples at 1000 °C for 1000 min plays a critical role in the strengthening of the specimens. In the case of S3, since the sample is not hold at 1000 °C, the sintering process is not completed and therefore it did not have enough strength.

For those specimens sintered at 1400 °C almost similar results were observed. Compared to those samples sintered at 1500 °C, samples sintered at 1400 °C showed lower compressive strength. Based on the obtained results it can be

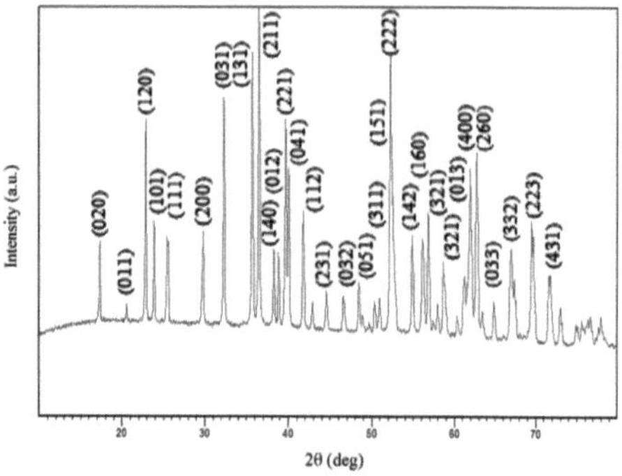

Fig. 1 XRD pattern of forsterite after heat treatment by cycle 1

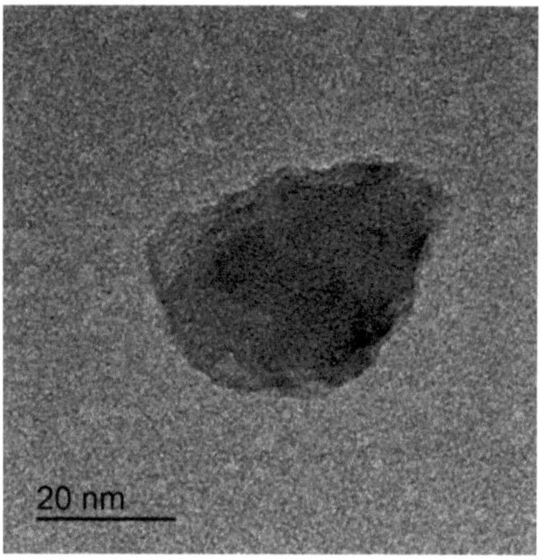

Fig. 2 TEM micrograph of forsterite after heat treatment by cycle 1

concluded that initial sintering temperature is crucially important for strengthening the scaffolds. With increasing the sintering temperature to 1500 °C the porosity of the specimens decreased. Higher porosity was observed in those specimens sintered at lower time and for less holding time at higher temperatures.

Figure 3 shows the macroporous network and the microstructure of the S2 scaffold. As can be seen in Fig. 3, the structure is completely porous with the pore size of about 100 μm. It should be mentioned that very small amount of enstatite ($MgSiO_3$) can exist in the structure. At around 1557 °C, enstatite is dissociated into forsterite and a SiO_2-rich liquid [15]. SiO_2 may form a glassy phase which caused a better connection between the forsterite particles and formation of a denser structure which consequently increased the compressive strength of the prepared scaffolds. This can be another reason for the high compressive strength of the samples sintered at 1500 °C compared to those sintered at 1400 °C.

In order to produce an ideal scaffold for biomedical application, porosity and compressive strength should be similar to that of damaged bone. In the case that the strength of the scaffold is less than that of injured bone, under load bearing condition the scaffold may fail under pressure. In this situation there is a need for second surgery to replace the implant (scaffold). If the strength of the scaffold is more than the injured site, it may cause stress shielding which cause the loosening of the scaffold, possible bleeding and finally result in a clinical failure. On the other hand, it is demonstrated that nanostructure forsterite is bioactive and biodegradable which can release magnesium and silicon in the surrounding system [16]. The nanostructure forsterite scaffold (S2 specimen) developed in this study had appropriate porosity and mechanical properties simultaneously, and hence can be a great candidate as an implant for bone defects to promote three-dimensional bone cells ingrowth and subsequently accelerate the healing process.

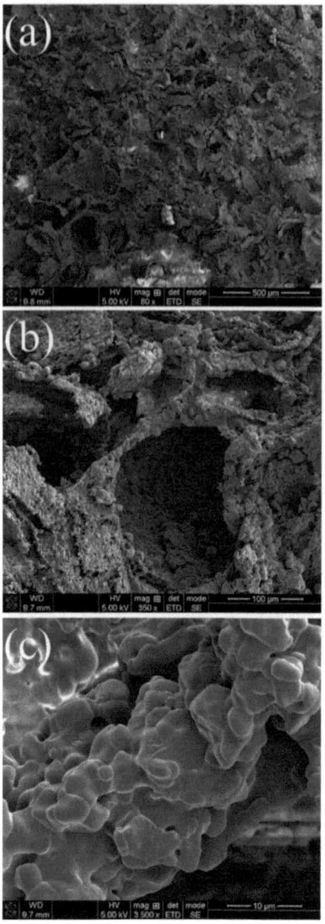

Fig. 3 SEM micrographs of S2 at various magnifications: **a** 80x, **b** 350x, and **c** 3500x. As can be seen in section (**b**), pores with the size of about 100 μm formed in the scaffolds. On the other hand, part **c** shows that grains are sintered completely and attached to each other. In many locations, necking can be observed

Conclusion

In this study nanostructure forsterite scaffold with appropriate mechanical properties and porosity were developed for load bearing applications. The compressive strength, elastic modulus, and porosity of S2 was 0.36 MPa, 1.36 MPa, and 82%, respectively. Two-step sintering method was utilized to prevent grain growth during the sintering process and also to provide enough energy to develop strong bonds between the forsterite grains. The produced forsterite scaffolds had a nanostructure which is an important parameter in the case of forsterite ceramic because micron-sized forsterite has not shown bioactivity and biodegradability in vitro. The results obtained in this study can open a new horizon for developing scaffolds with enough compressive strength and appropriate porosity for load bearing applications.

References

1. W.C. Head, D.J. Bauk, R.H. Emerson, Titanium as the material of choice for cementless femoral components in total hip arthroplasty. Clin. Orthop. Relat. Res. **311**, 85–90 (1955)
2. B.V. Krishna, S. Bose, A. Bandyopadhyay, Low stiffness porous Ti structures for load-bearing implants. Acta Biomater. **3**, 997–1006 (2007)
3. J. Duyck et al., The influence of micro-motion on the tissue differentiation around immediately loaded cylindrical turned titanium implants. Arch. Oral Biol. **51**, 1–9 (2006)
4. F. Tavangarian, R. Emadi, Improving degradation rate and apatite formation ability of nanostructure forsterite. Ceram. Int. **37**, 2275–2280 (2011)
5. S. Ni, L. Chou, Preparation and characterization of forsterite (Mg_2SiO_4) bioceramics. Ceram. Int. **33**, 83–88 (2007)
6. M. Vallet-Reg, Revisiting ceramics for medical application. Dalton Trans. **44**, 5211–5220 (2006)
7. V. Karageorgiou, D. Kaplan, Porosity of 3D biomaterial scaffolds and osteogenesis. Biomaterials **26**, 5474–5491 (2005)
8. A.J. Salgado, O.P. Coutinho, R.L. Reis, Bone tissue engineering: state of the art and future trends. Macromol. Biosci. **4**, 743–765 (2004)
9. M. Beheshti et al., Reduction of nanowire agglomeration via an intermediate membrane in nanowires preparation for nanosensors application, in *ASME 2015 International Mechanical Engineering Congress and Exposition* (American Society of Mechanical Engineers, 2015)
10. M. Beheshti et al., Micron level placement of nanowires via real time observation under optical microscope on a desired nanochannel for nanosensors application, in *ASME 2016 International Mechanical Engineering Congress and Exposition* (American Society of Mechanical Engineers, 2016)
11. F. Tavangarian, R. Emadi, Synthesis of pure nanocrystalline magnesium silicate powder. Ceram.-Silik. **54**, 122–127 (2010)
12. S.I.R. Esfahani, F. Tavangarian, R. Emadi, Nanostructured bioactive glass coating on porous hydroxyapatite scaffold for strength enhancement. Mater. Lett. **62**, 3428–3430 (2008)
13. R. Emadi et al., Nanostructured forsterite coating strengthens porous hydroxyapatite for bone tissue engineering. J. Am. Ceram. Soc. **93**, 2679–2683 (2010)
14. A. Douy, Aqueous synthesis of forsterite (Mg_2SiO_4) and enstatite ($MgSiO_3$). J. Sol-Gel. Sci. Technol. **24**, 221–228 (2002)
15. F. Tavangarian, R. Emadi, Enhancement mechanism of nanostructure forsterite formation rate by mechanical activation and ammonium chloride. NANO **6**, 131–138 (2011)
16. F. Tavangarian, R. Emadi, Nanostructure effects on the bioactivity of forsterite bioceramic. Mater. Lett. **65**, 740–743 (2011)

Part X
Bulk Metallic Glasses XIV

Material Behavior in Micro Milling of Zirconium Based Bulk Metallic Glass

Boyuan Xie, Maroju N. Kumar, David P. Yan and Xiaoliang Jin

Abstract Bulk metallic glasses (BMGs) are a group of metallic materials with amorphous microstructure. BMGs have been increasingly used in various applications such as optical molds, sport equipment and biomedical components. The machining mechanism for the BMGs is distinct from that of the crystalline metal alloys due to the differences in their microstructures and mechanical properties. This paper presents experimental investigations on the material behavior in micro milling of zirconium based bulk metallic glass (Zr-based BMG) . Micro milling of the BMG with high spindle speed up to 105,000 rpm was conducted. Surface morphology, microstructural evolution and crystallization of the machined surface were characterized by optical and scanning electron microscopy, X-Ray diffraction and energy dispersive X-Ray spectroscopy. The results detailed the responses of the Zr-BMG to the milling conditions which enables the machining conditions to be optimized to achieve an improved surface quality and machining efficiency.

Keywords Bulk metallic glass • Micro milling • Crystallization

Introduction

Amorphous solids are defined as materials in which the atoms and the molecules are not arranged in a regular lattice pattern. Commonly used amorphous materials include silicate glasses, BMGs, and plastics. BMGs are a group of newly developed metallic materials with amorphous microstructure produced by rapid cooling of

B. Xie · M.N. Kumar · X. Jin (✉)
School of Mechanical and Aerospace Engineering, Oklahoma State University, 218 Engineering North, Stillwater, OK 74078, USA
e-mail: xiaoliang.jin@okstate.edu

D.P. Yan
College of Science and Technology, University of Wisconsin-Green Bay, 2420 Nicolet Drive, Green Bay, WI 54311, USA

The Minerals, Metals & Materials Society, *TMS 2017 146th Annual Meeting & Exhibition Supplemental Proceedings*, The Minerals, Metals & Materials Series, DOI 10.1007/978-3-319-51493-2_34

melted metal alloys [1]. Compared to their crystalline counterparts, BMGs have superior mechanical properties, including higher hardness, improved corrosion resistance, and no grain boundaries on surface. BMGs have been increasingly used in various industrial areas such as optics, aerospace, sports, and biomedical applications. Geometric accuracy and surface quality are critical factors which influence the functionality of BMG components. Precision machining techniques such as micro turning, micro milling, grinding and polishing are used to fabricate the components with desired quality [2–5]. Compared to conventional crystalline metals, the fundamental mechanisms of chip formation and surface generation in machining BMGs is different due to their unique microstructure and mechanical properties [6]. Furthermore, crystallization of BMGs microstructure may occur as a result of the stress and temperature variations in the material deformation zone, depending on the machining conditions [7]. Therefore, it is critical to investigate the machinability of BMG in order to improve the process efficiency and the components quality.

Bakkal et al. performed turning of Zr-based BMG at various cutting speeds, tool rake angles and coating properties [6]. It was found that the occurrence of light emission in the turning process was related to the melting of BMG chips. High cutting speed significantly reduced the forces due to thermal softening effect. Surface roughness in turning BMG was better than those of Al6061 and SS304. Fast tool wear and edge chipping were found at short machining time. X-Ray diffraction (XRD) results showed the crystallization of machined chips at high cutting speed, while no crystallization occurred on the machined surface. XRD peaks matched the pattern of monoclinic ZrO_2, confirming that the oxidation was associated with the light emission and high flash temperature during the turning process [7]. The chip morphology showed the presence of shear bands, void formation, and enhanced viscous flow at higher cutting speed [8]. Crystalline grain, residual amorphous region, and phase separation were observed in the amorphous–crystalline transition region of machined chips by TEM examinations [9]. Jiang and Dai [10] performed the turning experiments on Zr-based BMG, and concluded that the formation of lamellar chips was due to the repeated shear-band formation in the primary shear zone (PSZ). A coupled thermomechanical orthogonal cutting model considering force, free volume and energy balance in the PSZ was developed to quantitatively characterize the lamellar chip formation. Bakkal et al. performed drilling on Zr-based BMG with HSS and WC–Co tools [11]. It was observed that holes with good surface finish could be produced when the spindle speed was below the critical value causing chip light emission. However, large size burr formations at the edge of the holes were identified, and a new type of chip with long tangled ribbon shape was observed. Thrust force, torque, and tool wear were determined for drilling of BMG with different tool materials and cutting conditions [12]. An experimental study on grinding of Zr-based BMG was conducted by Bakkal et al. [13]. It was found that grinding forces reduced with the increasing of cutting speed as specific grinding energy increased. Machined surface and chip morphology showed that material removal and surface formation of the BMG were mainly due to the ductile chip formation and ploughing as well as brittle fracture of some particles from the edge of the tracks. Conventional Al_2O_3 wheel was not suitable for grinding of the BMG in dry conditions due to extensive light emission. Bakkal and Nakşiler [14] further conducted investigations on end milling of Zr-based BMG. The effects of feedrate on the cutting force, burr formation, and surface roughness were determined with constant spindle speed of 5000 rpm. It was found that BMG required less cutting force compared to SS304, and higher forces than Al6061 under the same cutting conditions. The rollover burr formations were observed along the milling slot. Furthermore burr formations were characterized in the spiral and slot milling paths [15].

The results in the literature demonstrate that mechanical machining technology is able to achieve desirable geometric accuracy and surface quality for BMG components if the machining conditions are properly selected. The machinability and performance are considerably different for turning, milling, drilling and grinding processes. Regarding to milling process specifically, there has been no investigation on the surface microstructure, and the effect of coolant which is critical in machining BMG has not been reported. In this study, the surface microstructure and potential surface crystallization in milling of BMG is characterized by scanning electron microscopy (SEM), XRD, and energy dispersive X-Ray spectroscopy (EDS). The critical values of spindle speed and feedrate which correspond to the onset of surface crystallization are identified. This paper provides new findings in the machinability of micro milling of BMG, and the results can be used to determine the cutting conditions in order to improve the surface quality and the process efficiency.

Experiment Setup and Procedure

Experiment Setup

The milling experiments on Zr-based BMG were conducted with spindle speed from 10,000 to 105,000 rpm. Two micro-milling machines with different speed range capacities were used, including a commercial MDA precision micro milling machine (Fig. 1a) with speed in the range of 10,000–60,000 rpm, and an in-house developed high-speed micro milling machine at 60,000–160,000 rpm (Fig. 1b). A two-flute AlTiN coated end mill (MICROCUT) with 3.175 mm diameter and 30° helix angle was used. The workpiece is Zr-based BMG (model: LM105) provided by Liquidmetal Technologies Inc. The composition and the mechanical properties of the BMG workpiece are listed in Tables 1 and 2 respectively. The dimension of the workpiece is 21 mm × 14 mm × 2 mm.

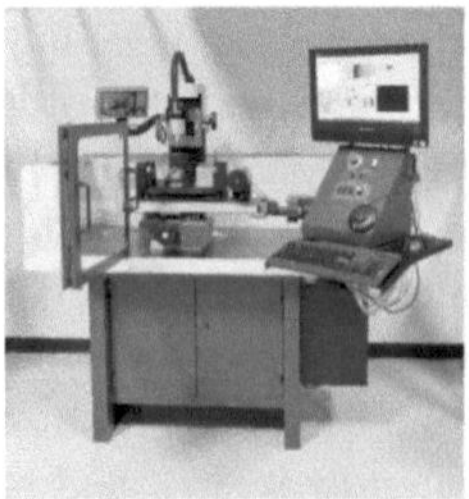
(a)

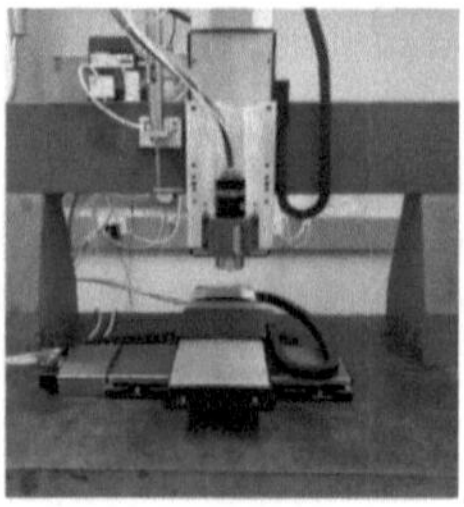
(b)

Fig. 1 Milling machines **a** MDA precision micro-machining center, **b** high speed micro milling machine

Table 1 Compositions of LM105 BMG [16]

Constituent elements	Zr	Ti	Cu	Ni	Al
Weight (%)	65.7	3.3	15.6	11.7	13.7

Table 2 Mechanical Properties of LM105 BMG [16]

Density (g/cm^3)	Hardness (HRC)	Yield strength (MPa)	Young's modulus (GPa)	Fracture toughness (MPa • $m^{1/2}$)	Poisson's ratio	Glass transition temperature (°C)	Crystallization temperature (°C)
6.57	53	1524	92.7	~40	0.38	399	468

Experiment Procedure

Side milling of the BMG workpiece was conducted with the axial depth of cut of 2 mm. A small radial depth of cut 0.45 mm was selected due to the fast tool wear rate and edge chipping at a larger depth. The feedrate value ranges from 0.5 μm/flute to 10 μm/flute. Higher radial depth of cut or feedrate leads to complete tool wear in one cutting path. Hitachi S-4800 scan electron microscope and Zeiss Observer.Z1 m metalloscope are used to determine the morphology and microstructure of the machined surface. The cross section of the surface is finely polished for the examination purpose. Bruker AXS D8 Discover X-ray diffractometer was used to investigate the surface crystallization due to the milling process. EDS was used to determine the constituent elements of the machined chip and surface. At each spindle speed, different values of feedrate were tested in order to determine the milling condition which corresponds to the onset of surface crystallization.

Light Emission and Machined Surface Property in Milling of Zr-Based BMG

Light Emission

Light emission was observed in the micro milling of Zr-based BMG. The occurrence and strength of the light emission depend on the machining conditions. Table 3 shows the cutting parameters and the results of light emission strength which is classified into three levels, explained below:

(1) No light emission is generated during the whole milling process, shown as Fig. 2a, and defined as level 1;
(2) There is no light emission at the beginning of the milling process, however, the intermittent light appears in the later stage of the process due to the heat accumulation in the material removal region, shown in Fig. 2b and defined as level 2;
(3) Strong light emission starts to develop after about 3–5 s of the milling process, shown as Fig. 2c. Associated with the light generation, workpiece material is melted and deposited on the machined surface. This light emission strength is defined as level 3.

The results in Table 3 show that the strength of light emission increases with feedrate at constant spindle speed, and the increase of spindle speed enhances the generation of light emission. This is due to a higher stress and temperature when

Table 3 Classifications of the light emission levels in the different experiments

Test number	Spindle speed (rpm)	Feedrate (μm/flute)	Light emission level
1	50,000	0.5	Level 1
2	50,000	3	Level 2
3	50,000	4.5	Level 3
4	60,000	0.5	Level 1
5	60,000	1	Level 2
6	60,000	3	Level 3

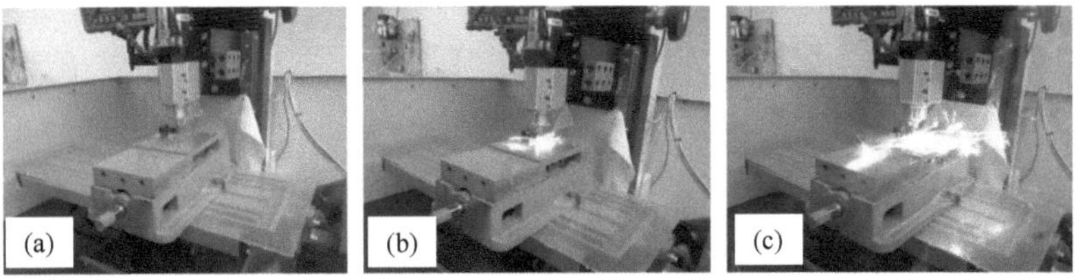

Fig. 2 Photos of the light emission during milling of LM105 BMG used to classify the emission levels: **a** level 1, **b** level 2, **c** level 3

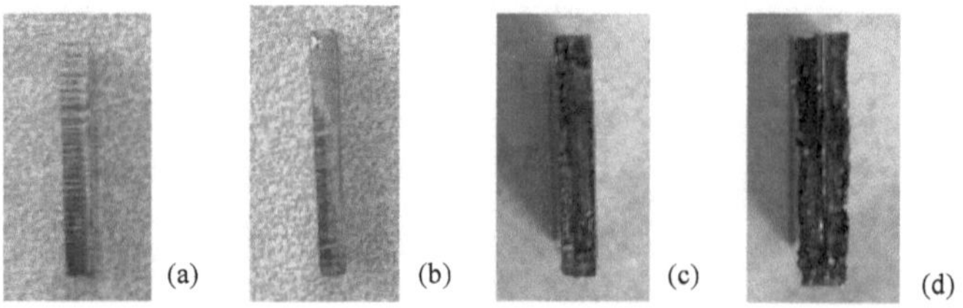

Fig. 3 Photos of the machined surface: **a** test number 2, **b** test number 5, **c** test number 3, **d** test number 6

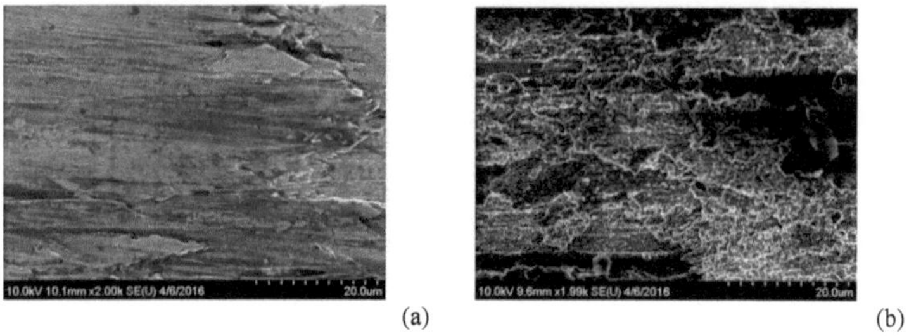

Fig. 4 Micrographs of the machined surface investigated by SEM (2000x magnification). **a** at the beginning of machining with no light emission; **b** with extensive light emission and surface melting

increasing the spindle speed and feedrate. Figure 3 shows the pictures of machined surface corresponding to the test numbers listed in Table 3. Figure 3a and b are based on the process with level-2 light emission. It was found that there is no significant change of the machined surface morphology between the regions with and without light emission. On the other hand, for Fig. 3c and d which correspond to the level-3 light emission, the machined surface involved with the melting of workpiece material turned into black color, and the surface roughness is higher compared with the surface obtained in levels 1 and 2. Therefore, it was proved that the melting of the surface material is detrimental to the surface quality in the milling process, while light emission itself is not directly related to the surface morphology.

Characterization of Machined Surface by SEM and Metalloscopy

The machined surface in the micro milling of Zr-based BMG was investigated by SEM and metalloscopy. Figure 4 shows the SEM images (2000x magnification) of machined surface under the test number 3 (Table 3) with the level 3 of light emission. Figure 4a shows the machined surface at the beginning stage of the milling process when light emission was not generated. Figure 4b shows the

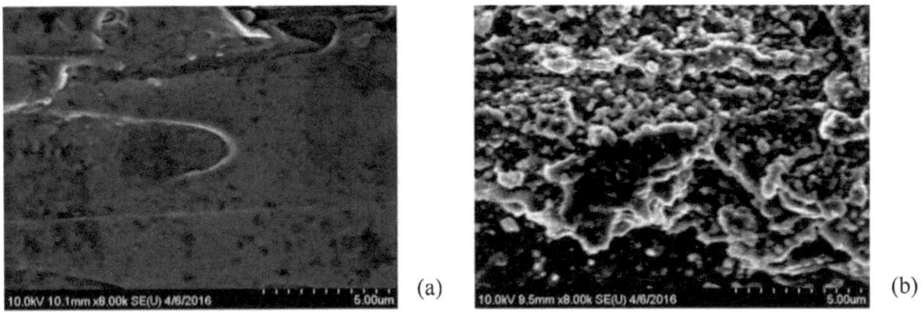

Fig. 5 Micrographs of the machined surface investigated by SEM (8000x magnification). **a** at the beginning of machining with no light emission; **b** with extensive light emission and surface melting

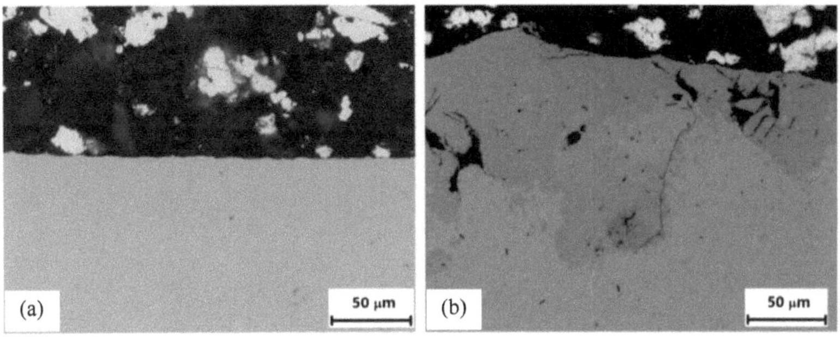

Fig. 6 Micrographs of the cross section of the machined surface: **a** test number 5, level 2 light emission; **b** test number 6, level 4 light emission

machined surface when extensive light emission was developed and melted material was deposited on the surface. Figure 5 shows the SEM results at the same location as Fig. 4, and with higher magnification (8000x). It was found that the surface melting causes uneven surface morphology compared to the location with no light emission region, and results in the solidified grain clusters at the size of 1-3 μm.

The cross section of the machined surface was further examined by the metalloscope in order to investigate the microstructure property. The workpiece was cut into half and the cross section was finely polished for the examination. Figure 6a and b show the cross section of the machined surface at the light emission levels of 2 and 3 when the spindle speed is 60,000 rpm, corresponding to the test number 5 and 6 given in the Table 3. The surface cracks and voids are visible in the Fig. 6b. When the feedrate increases, the temperature is high which results in the melting of the material. The voids formation is due to the dynamic solidification process, which may cause degradation of the mechanical properties of the machined components.

Crystallization of Workpiece Material

Crystallization of original amorphous microstructure of the workpiece in the milling process was examined through XRD. Figure 7 shows the XRD results of the surface for the original and the machined surface. The machining parameters labeled as (b), (c) and (d) correspond to the process with the light emission levels of 1, 2 and 3 respectively. It can be observed from the diffraction patterns that the process with the light emission levels of 1 and 2 resulted in the similar broad peaks covering same theta value as the original surface, representing an amorphous microstructure. However, the diffraction pattern corresponding to the level 3 of light emission show strong characteristic peaks, proving that the crystallization of the surface microstructure occurred. Therefore, it can be concluded that the crystallization of the original amorphous microstructure of the workpiece surface is caused by the surface melting and solidification, which are associated with the level 3 light emission in the milling process.

Figure 8 shows that the crystallization occurred on the both chip and machined surface at the level 3 light emission, with clear sharp peaks in the XRD results. The experimental XRD curves were compared with the standard material diffraction

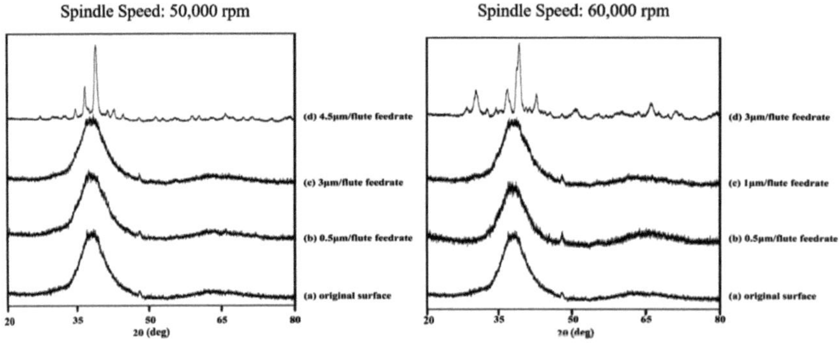

Fig. 7 XRD diffraction patterns showing the comparison of the original and machined surfaces

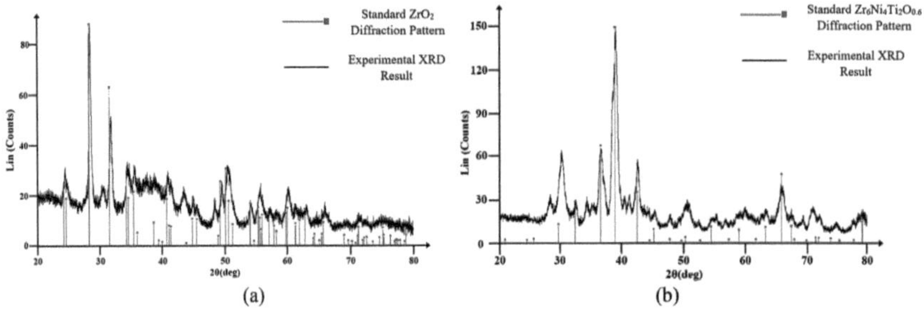

Fig. 8 XRD diffraction patterns **a** the chip and **b** the machined surface (spindle speed: 60,000 rpm, feedrate: 3.0 μm/flute)

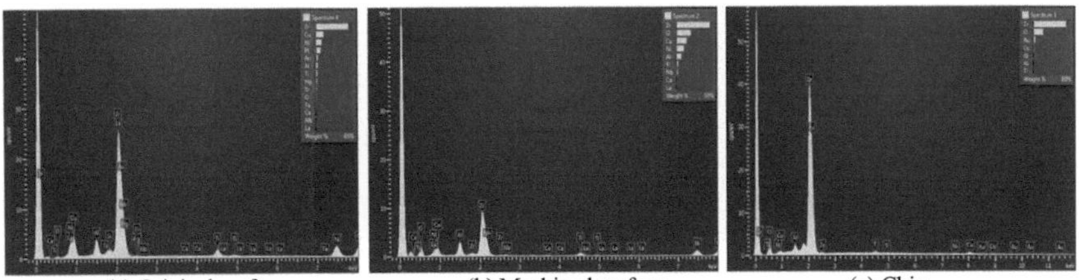

Fig. 9 EDS patterns showing the comparison of the original surface, machined surface and chip (corresponding to the machining condition with the Level 3 light emission). **a** Original surface **b** machined surface **c** chip

patterns. It is shown that the crystalline phase in the chip is monoclinic ZrO_2, while the major peaks in the pattern from the machined surface correspond to the crystalline phase of $Zr_6Ni_4Ti_2O_{0.6}$. The phase difference between the chip and the machined surface may be due to the fact that the chip carried away the most of the heat during the machining and was exposed to the oxygen (from air) directly compared to the machined surface. These results agreed that the crystallization is associated with the oxidation in machining Zr-BMG. EDS examination was further conducted to confirm the oxidation on the crystallized chip and the machined surface. Figure 9 shows the EDS spectra for the surfaces (original and machined) and chips at the level 3 light emission. In Fig. 9a, no oxygen presents in the spectrum of the original surface, while the analysis of Fig. 9b shows that the content of oxygen on the machined surface is more than 10% by weight than the original one. Similarly, Fig. 9c shows the presence of the oxygen on the chip. These results demonstrate that the oxidation exists on the machined surface and the chip at the cutting condition corresponding to the level 3 light emission.

Experimental results proved that at a constant spindle speed, higher feedrate intends to cause crystallization of the original amorphous workpiece materials due to a larger material removal load. A critical feedrate value exists which corresponds to the onset of the surface crystallization. When feedrate exceeds the critical value, the material melting occurs in the milling process, and the solidified material forms a dark region on the machined surface, leading to the surface crystallization of the workpiece. Therefore, there is a need to determine the cutting conditions which can achieve the highest material removal rate without causing any crystallization. In this study, a series of experiments were performed to determine the cutting conditions corresponding to the onset of the surface crystallization. The spindle speed varied from 10,000 to 60,000 rpm, with increment step of 5,000 rpm. At each spindle speed, feedrate gradually increased from 0.5 to 10 μm/flute with 0.5 μm/flute increment, until the surface crystallization was observed. The tool edge geometry after each cut was examined under the microscope to avoid excessive tool wear. Figure 10 shows the identified crystallization transition curve which represents the cutting condition corresponding to the onset of the surface crystallization. The critical feedrate value decreased asymptotically when the spindle speed increased. Below 30,000 rpm, there was no crystallization even when the feedrate increased

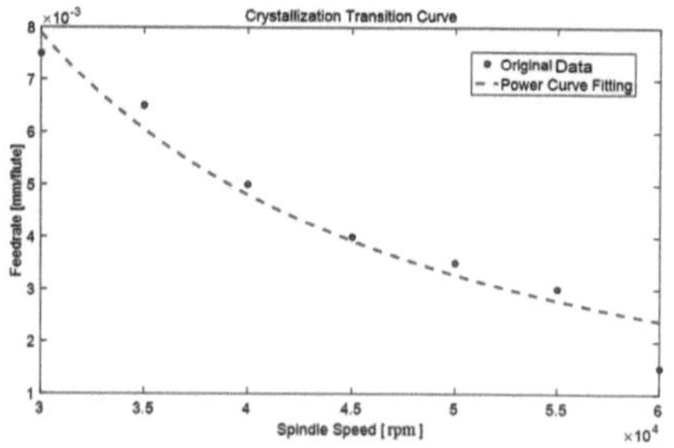

Fig. 10 Identified crystallization transition curve for LM105 BMG

up to 10 μm/flute. Based on the identified results, the critical cutting parameters were curve fitted by a power function, expressed as:

$$f = 407200n^{-1.723} \quad (1)$$

where f is the critical feedrate with the unit of mm/flute, and n is the spindle speed with the unit of rev/min. The cutting condition in the region below the transition curve represents the process without the surface crystallization, and vice versa.

Conclusion

This paper investigated the machinability of Zr-based BMG material in micro milling process. The effect of cutting conditions on the light emission was investigated and classified into 3 levels. The surface morphology and crystallization were also characterized through the SEM, metalloscopy, XRD and EDS. The constituent elements of the machined chip and surface were determined and compared. The critical values of the spindle speed and feedrate corresponding to the onset of the surface crystallization were identified. The conclusions based on the results are listed below:

(1) The surface morphology is modified due to the melting and solidification of the workpiece material in the milling process. As a result, surface cracks and voids are generated, which are detrimental to the mechanical properties of the machined components.
(2) The surface melting causes the crystallization of the material on the both chip and machined surface. However, the remaining constituent elements between the chip and the surface are different. A crystallization transition curve for the machined surface exists which represents the onset of the surface crystallization. The results provide the cutting conditions to achieve the maximum efficiency during machining without causing the occurrence of the crystallization.

Acknowledgements This work is supported by the National Science Foundation (NSF) CAREER award CMMI—1553815.

References

1. C. Suryanarayana, A. Inoue, *Bulk Metallic Glasses* (CRC Press, 2010)
2. J. Chae, S.S. Park, T. Freiheit, Investigation of micro-cutting operations. Int. J. Mach. Tools Manuf **46**(3), 313–332 (2006)
3. M. Rahman, H.S. Lim, K.S. Neo, A.S. Kumar, Y.S. Wong, X.P. Li, Tool-based nanofinishing and micromachining. J. Mater. Process. Technol. **185**(1), 2–16 (2007)
4. Y. Altintas, X. Jin, Mechanics of micro-milling with round edge tools. CIRP Ann. Manuf. Technol. **60**(1), 77–80 (2011)
5. X. Jin, Y. Altintas, Chatter stability model of micro-milling with process damping. J. Manuf. Sci. Eng. **135**(3), 031011 (2013)
6. M. Bakkal, A.J. Shih, R.O. Scattergood, Chip formation, cutting forces, and tool wear in turning of Zr-based bulk metallic glass. Int. J. Mach. Tools Manuf **44**(9), 915–925 (2004)
7. M. Bakkal, C.T. Liu, T.R. Watkins, R.O. Scattergood, A.J. Shih, Oxidation and crystallization of Zr-based bulk metallic glass due to machining. Intermetallics **12**(2), 195–204 (2004)
8. M. Bakkal, A.J. Shih, R.O. Scattergood, C.T. Liu, Machining of a Zr–Ti–Al–Cu–Ni metallic glass. Scripta Mater. **50**(5), 583–588 (2004)
9. M. Bakkal, Electron microscopy of bulk metallic glass machining chips. J. Non-Cryst. Solids **355**(45), 2220–2223 (2009)
10. M.Q. Jiang, L.H. Dai, Formation mechanism of lamellar chips during machining of bulk metallic glass. Acta Mater. **57**(9), 2730–2738 (2009)
11. M. Bakkal, A.J. Shih, S.B. McSpadden, C.T. Liu, R.O. Scattergood, Light emission, chip morphology, and burr formation in drilling the bulk metallic glass. Int. J. Mach. Tools Manuf **45**(7), 741–752 (2005)
12. M. Bakkal, A.J. Shih, S.B. McSpadden, R.O. Scattergood, Thrust force, torque, and tool wear in drilling the bulk metallic glass. Int. J. Mach. Tools Manuf. **45**(7), 863–872 (2005)
13. M. Bakkal, E. Serbest, I. Karipçin, A.T. Kuzu, U. Karagüzel, B. Derin, An experimental study on grinding of Zr-based bulk metallic glass. Adv. Manuf. **3**(4), 282–291 (2015)
14. M. Bakkal, V. Nakşiler, Cutting mechanics of bulk metallic glass materials on meso-end milling. Mater. Manuf. Process. **24**(12), 1249–1255 (2009)
15. M. Bakkal, V. Nakşiler, B. Derin, Machinability of bulk metallic glass materials on milling and drilling, in *Advanced Materials Research*, vol. 83 (Trans Tech Publications, 2009), pp. 335–341
16. Liquidmetal Material Properties. Liquidmetal Technologies Inc

Part XI
Computational Thermodynamics and Kinetics

Comparison of the Phase-Field Models to Predict the Recrystallization Kinetics

Julia Kundin

Abstract Continuous models for the prediction of the grain growth and recrystallization kinetics can be roughly classified into two methods. One method is multi-phase-field approaches. The second method is the phase-field model of Kobayashi, Warren and Carter (KWC) with two or three order parameters that make the model more efficient and more promising for the description of polycrystals. The present work proposes an extension to the KWC model to incorporate the recrystallization process by means of an additional order parameter for recrystallized grains. The results of the simulation of the grain growth and recrystallization process obtained by this extended KWC model are compared to existing results in the literature obtained by the original KWC phase-field model. The procedure of the parameter estimation for a simple isotropic case (no orientation dependence of the surface energy and kinetic parameters) is suggested. The ability of the model to describe the grain boundaries with low misorientation and to incorporate the grain boundary diffusion with the impurity segregation is discussed.

Keywords Grain growth · Phase-field modeling

Introduction

During the past two decades, the phase field approach has been employed to simulate grain growth phenomena [1–4]. Fan and Chen [1], Chen and Yang [2] first suggested a multi-phase-field algorithm for the grain growth based on a Ginzburg-Landau-type polynomial potential, which allows suppressing the ghost phases on the grain boundaries. The Fan and Chen's algorithm was used by Suwa et al. [5–7] and Moelans et al. [8, 9] for the simulation of the grain growth and the migration of recrystallization boundaries.

J. Kundin (✉)
Department of Engineering Science, University Bayreuth, 95440 Bayreuth, Germany
e-mail: juliafkg@web.de

The Minerals, Metals & Materials Society, *TMS 2017 146th Annual Meeting & Exhibition Supplemental Proceedings*, The Minerals, Metals & Materials Series,
DOI 10.1007/978-3-319-51493-2_35

Kim et al. [10] developed an efficient computation scheme for the phase-field simulation of ideal grain growth based on a multi-phase-field approach of Steinbach and Pezzolla [11]. This model has several merits in modeling grain growth in comparison to the models which follow the Fan and Chen algorithm. First, the force balances at triple junctions are maintained for both the isotropic and anisotropic grain boundary energies. Second, the anisotropies in interface energy and mobility as functions of misorientation and inclination can be easily implemented. Third, a double obstacle potential is adopted, which results in a clear-cut boundary width without smearing phase field into grains. Details of the model were described in [12]. This effective phase-field simulation of grain growth allows an unlimited number of orientation variables and high computational efficiency in two and three dimensions.

A principally different approach to the phase-field modeling of grain growth was suggested by Warren et al. [13–15]. They introduced a frame-invariant phase-field model for solidification of many arbitrarily oriented grains using only three continuous fields: a solid phase order parameter, a coarse-grained measure of the degree of crystalline order, and the crystalline orientation. In comparison to the previously described multi-phase approaches, the model of Warren, Kobayashi and Carter (KWC model) is numerically significantly more efficient. Moreover, the model allows for two modes of grain boundary dynamics at once: rotation and motion, and allows to include thermodynamic variables which control the state of the system. However, the model still shows difficulties by taking into account the high anisotropy of the physical properties (e.g. surface energies).

In the last years, the phase-field models were applied to the simulation of the recrystallization process. The multi-phase-field model [10] was combined with crystal plasticity theory and FEM by Takaki et al. [16–18] to simulate the static and dynamic recrystallization. The misorientation between recrystallized and deformed grains can be straightforwardly incorporated into the multi-phase-field model as it was shown by Takaki et al. [16] and Miyoshi et al. [19, 20]. Again by Takaki et al. [21, 22] the KWC model was applied where only two order parameters for the solid phase and for the crystalline orientation were used. However, the misorientation between recrystallized and deformed grains is still not incorporated into the KWC model despite its efficiency.

In the present work we suggest an extension of the KWC model of the grain growth to describe the recrystallization kinetics. In order to distinguish the recrystallized and deformed grains and to incorporate the misorientation between them, we suggest to use an additional order parameter which will be responsible for the recrystallized grains. We incorporate this order parameter into the model and demonstrate the simulation results for a polycrystalline structure in comparison to the previous works [21, 22]. Our purpose is to demonstrate an efficient method for the recrystallization process which can describe the misorientation between any grains with the precision comparable to more time consuming multi-phase-field models.

Model Description

We consider the free energy density of an alloy system following the KWC model, which includes the order parameter ϕ associated with the local crystalline order and the crystallographic orientation field θ. Then, to describe the recrystallization, we introduce the order parameter ξ, which is responsible for the measurement of the recrystallized grains. $\xi = 1$ indicates a pure recrystallized grain and $\xi = 0$ indicates the initial deformed grains. Thus, this parameter indicates the grains with reduced defect density. The difference of the defect density is the driving force of the recrystallization. The total free energy density is

$$\mathcal{F} = \int dV \left[f(\phi) - \mu_0 p(\phi) + \frac{\alpha^2}{2}\Gamma^2(|\nabla\phi|, \theta) + sg(\phi)|\nabla\theta| + \frac{\varepsilon^2}{2}h(\phi)|\nabla\theta|^2 + \right.$$
$$\left. s_\xi g(\phi)|\nabla\xi| + \frac{{\varepsilon_\xi}^2}{2}h(\phi)|\nabla\xi|^2 + q(\xi) - \mu_\xi m(\xi) n(\phi) \right], \qquad (1)$$

where s and ε are coupling parameters between ϕ and $|\nabla\theta|$, s_ξ and ε_ξ are coupling parameters between ϕ and $|\nabla\xi|$, μ_0 is the chemical driving force of solidification, μ_ξ is the driving force of recrystallization, while f, q, p, g, h, m, and n are monotonic functions of ϕ and ξ. In the following, we choose $f(\phi) = a^2\,\phi^2(1-\phi)^2$, $q(\xi) = a^2\,\xi^2(1-\xi)^2$, $p(\phi) = g(\phi) = h(\phi) = n(\phi) = \phi^2(3-2\phi)$, and $m(\xi) = \xi^2(3-2\xi)$.

The kinetic equations for the order parameters ϕ, θ, and ξ are

$$\tau_\phi \frac{\partial\phi}{\partial t} = \alpha^2\nabla^2\phi - \frac{\partial f(\phi)}{\partial\phi} + \mu_0\frac{\partial p(\phi)}{\partial\phi} - \frac{\partial g(\phi)}{\partial\phi}s|\nabla\theta| - \frac{\partial h(\phi)}{\partial\phi}\frac{\varepsilon^2}{2}|\nabla\theta|^2 -$$
$$\frac{\partial g(\phi)}{\partial\phi}s_\xi|\nabla\xi| - \frac{\partial h(\phi)}{\partial\phi}\frac{{\varepsilon_\xi}^2}{2}|\nabla\xi|^2 + \mu_\xi\, m(\xi)\frac{\partial n(\phi)}{\partial\phi}, \qquad (2)$$

$$\tau_\theta h(\phi)\frac{\partial\theta}{\partial t} = \nabla\cdot\left[h(\phi)\varepsilon^2\nabla\theta + g(\phi)s\frac{\nabla\theta}{|\nabla\theta|}\right] \qquad (3)$$

and

$$\tau_\xi h(\phi)\frac{\partial\xi}{\partial t} = \nabla\cdot\left[h(\phi)\varepsilon_\xi^2\nabla\xi + g(\phi)s_\xi\frac{\nabla\xi}{|\nabla\xi|}\right] - \frac{\partial q(\xi)}{\partial\xi} + \mu_\xi\,\frac{\partial m(\xi)}{\partial\xi}n(\phi), \qquad (4)$$

Note that τ_θ is responsible for the rotation of grains, i.e., the larger τ_θ, the slower the rotation. From the experiment it is known that low angle boundaries migrate slower than high angle boundaries. Furthermore, a segregation on the boundaries reduces the mobility [23]. These phenomena can be taken into account by the phenomenological dependencies of the dynamic parameters τ_ϕ and τ_ξ on the misorientation between the grains $\Delta\theta$ and on the concentration c. The misorientation can be calculated in each grid point by $\Delta\theta = |\nabla\theta|d$, where d is an average boundary width.

The dependency of the boundary mobility during the recrystallization on the misorientation is known in the literature as $\tau_\phi(\Delta\theta)^{-1} = M_0\left[1 - \exp\left(-5\left(\Delta\theta/\Delta\theta_m\right)^4\right)\right]$ [16, 24].

The segregation of impurities can be included by diffusion equation $\partial c/\partial t = \nabla\left[D\nabla c\right]$. Here, the effective diffusion coefficient is defined using the bulk and grain boundary diffusion coefficients, i.e., $D = D^{\text{bulk}}\phi + D^{\text{gb}}(1-\phi)$.

The model parameters used in the simulations below are:
$\{\alpha^2, s, \varepsilon^2, a, s_\xi, \varepsilon_\xi^2\, \tau_\phi, \tau_\xi\, \tau_\theta\} = \{0.56, 0.9, 0.32, 0.5, 0.8, 0.32, 1, 1, 1\}$.

Estimation of the Stored Energy

The driving force of the recrystallization is the stored energy in the deformed grains. The stored energy is estimated as a function of the dislocation density difference $E_{\text{store}} = 0.5(\rho - \rho_0)\mu b^2$, where ρ_0 and ρ are the dislocation densities in deformed and recrystallized grains respectively, and μ is the elastic shear modules. The stored energy is calculated following [21] as a function of the smoothed dislocation density. We assume $E_{\text{store}} = 0$ in recrystallized grains and in liquid phase, and $E_{\text{store}} \neq 0$ in deformed grains. In the numerical simulation, the smoothing is done by averaging the order parameter ξ around a grid point i in a circle C:

$$\langle E_{\text{store}}\rangle = E^0_{\text{store}}(1 - \langle\xi\rangle), \tag{5}$$

where

$$\langle\xi\rangle = \frac{\sum_{j\in C}\xi_j\omega_j}{\sum_{j\in C}\omega_j} \tag{6}$$

with $\omega = 1/|r_i - r_j|$ being the inverse distance between grid points i and j as shown in Fig. 1. The required number of grid points inside a considered circle C is defined from the numerical reason. We have chosen the radius of the circle $R = 20$ and $E^0_{\text{store}} = 0.084$.

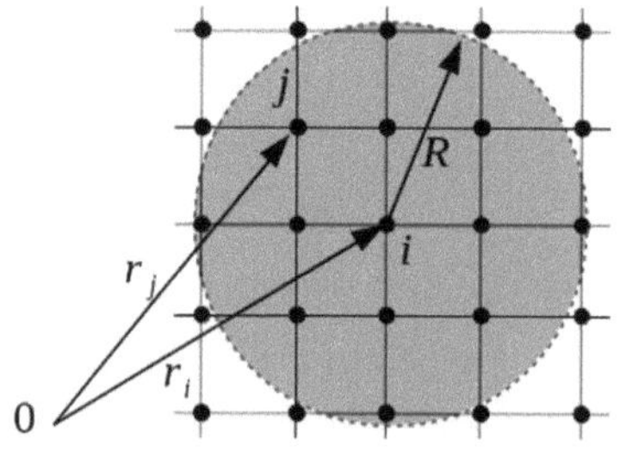

Fig. 1 Smoothing area inside a circle with radius R used for the calculation of the stored energy

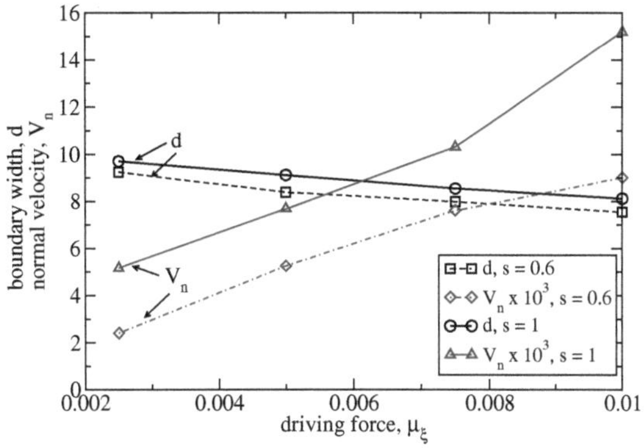

Fig. 2 The dimensionless boundary width, d, and the normal growth velocity, V_n, estimated numerically during the steady-state recrystallization process as functions of the recrystallization driving force, μ_ξ, for various parameters s

Numerical Results for the Steady-State Moving-Boundary

Here we examine the extended KWC model with the model functions $p = g = h = n = \phi^2(3 - 2\phi)$ in all evolution equations. The application of a uniform function of higher order in the full model including the recrystallization order parameter should serve for better stability and precision of the simulation.

The numerical test is carried out for a bicrystal with two grains in the simulation domain of size 300×900. One grain is the recrystallized grain with the order parameter $\xi = 1$, the second grain is the initial deformed grain. During the recrystallization, the boundary moves in the direction of the initial grain. At the same time, the boundary width changes until it reaches a steady-state value, which was estimated. The plot in Fig. 2 shows how the steady-state boundary width and the normal velocity of the moving boundary depend on the recrystallization driving force. Two parameters $s = 0.6$ and $s = 1.0$ are used. s is responsible for the grain boundary formation due to the misorientation gradients. The boundary width decreases with increasing driving force and increases with increasing parameter s, while the steady-state growth velocity increases with increasing driving force in conformity with theory [15]. Furthermore, the velocity is faster for the larger s.

Nucleation and Growth of a Recrystallized Grain

The numerical results of the growth of a recrystallized grain on the boundary between two deformed grains are shown in Figs. 3, 4 and 5. Figure 3 shows the microstructure (order parameters) at $t = 3600$ for the test with $\tau_\theta = 1$. In Fig. 4 the profiles of the order parameters ϕ, θ, and ξ are plotted without rotation ($\tau_\theta = 50$) and with rotation ($\tau_\theta = 1$). The comparison of the boundary width and the growth kinetics is presented in Fig. 5. From these results can be deduced that the growth velocity of recrystallized grain is faster and the boundary width is larger for the test

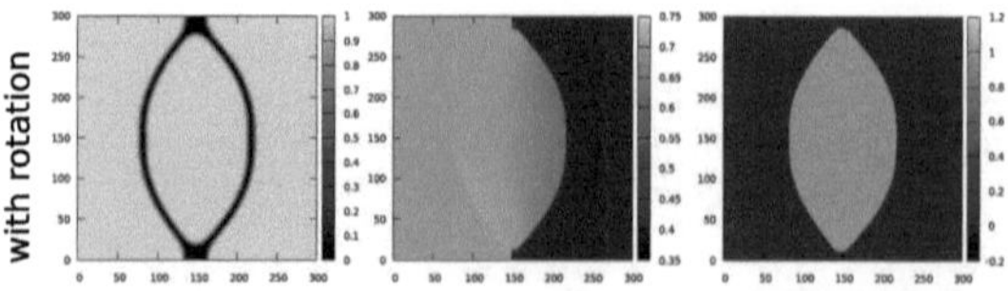

Fig. 3 Order parameters ϕ (*column 1*), θ (*column 2*), and ξ (*column 3*) during the recrystallization with grain rotation at $t = 3600$

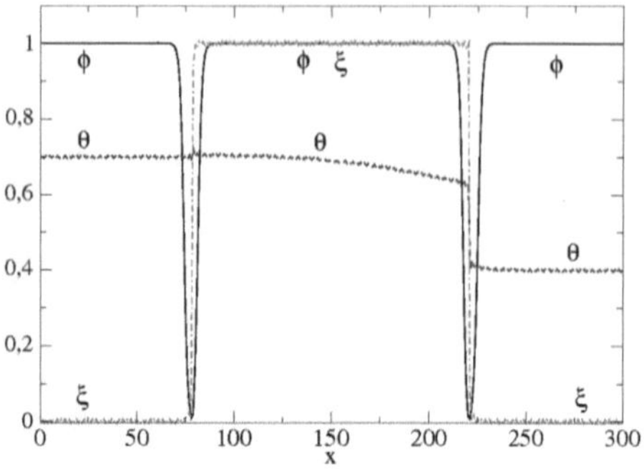

Fig. 4 Profiles of the order parameters ϕ, θ, and ξ (from *left* to *right*) during the recrystallization for the test with grain rotation at $t = 3600$

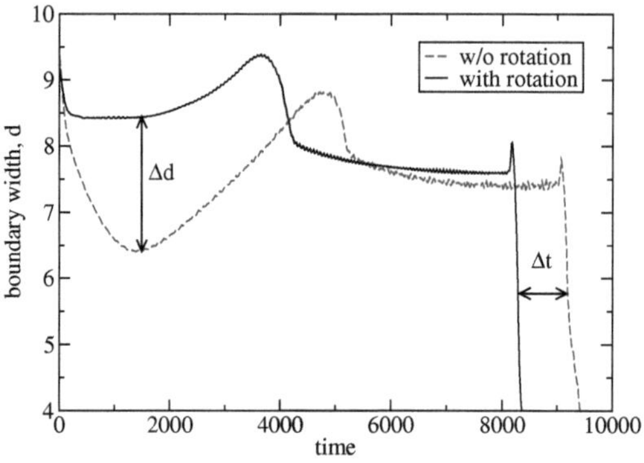

Fig. 5 Comparison of the boundary width and the growth kinetics for the tests with and without grain rotation

with rotation (small τ_θ). This behavior shows that the growth velocity depends on the misorientation between the grains and the rotation, and slows down for larger misorientation without rotation. Note that we do not take into account the mobility dependence on the misorientation in this test.

Numerical Results for Multigrain Systems

As it was demonstrated above for a small system, the growth kinetics of a recrystallized grain depends on the mobility of the grain to rotate. Moreover, the growth velocity strongly correlates with the driving force of the recrystallization and the boundary width. To investigate these phenomena in large systems, we have performed series of simulations with 3 grains in the simulation domain of size 300 ×

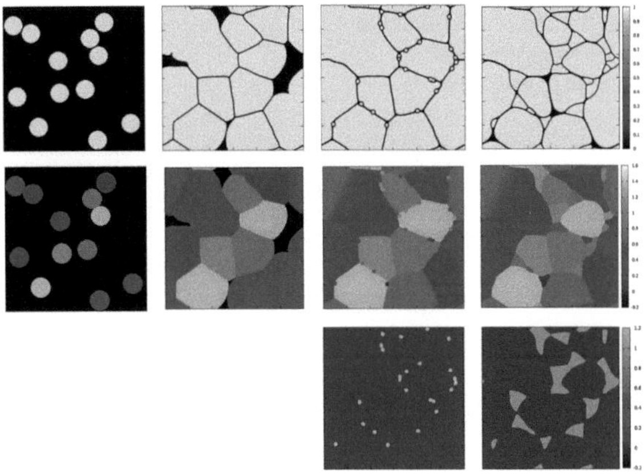

Fig. 6 Order parameters ϕ (*row 1*), θ (*row 2*), and ξ (*row 3*) during the recrystallization in the large system of size 1200 with grain rotation at $t = 80, 1400, 2080, 3340$ from *left* to *right*

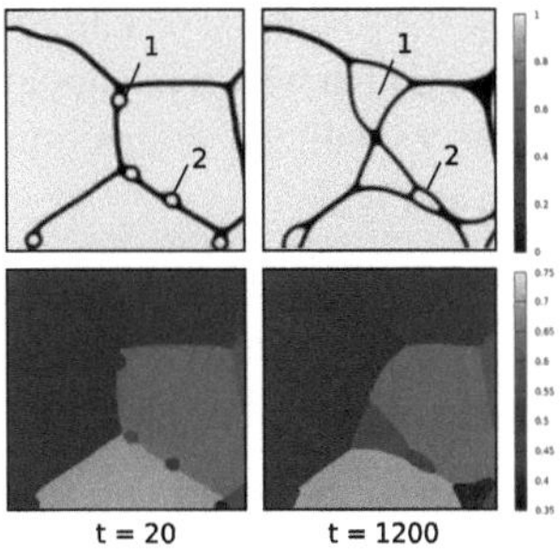

Fig. 7 The illustration of the recrystallized grain growth with various positions on the deformed grain boundaries: grain 1 near the triple junction and grain 2 on the pure boundary between two grains

300 Δx and with 12 grains in the simulation domain of size $1200 \times 1200\,\Delta x$. We start with small nuclei of the solid phase in the liquid phase and let them grow until full solidification. Then, we assume that the grains are deformed and insert small nuclei of the recrystallized phase randomly on the grain boundaries. We have chosen 6 nuclei in the simulation domain of size 300 and 40 nuclei in the simulation domain of size 1200. In Fig. 6, the microstructure evolution is illustrated as the evolution of three order parameters: ϕ (upper row), θ (middle row), and ξ (bottom row). The simulation time increases from left to right as $t = 80, 1400, 2080, 3340$. At $t = 1360$ the nuclei of the recrystallized grains of size 9 Δx are inserted.

Figure 7 illustrates the recrystallization dynamic of two grains with various positions on the deformed grain boundaries. Grain 1 which lies near the triple junction grows faster than grain 2 which lies on the pure boundary between two grains. This phenomena indicates the effect of grain boundaries including triple junctions on the recrystallization kinetic.

The evolution of the recrystallized volume fraction f_ξ with time is shown in Fig. 8 for both system sizes. The average growth velocity df_ξ/dt for the volume fraction interval from 0 to 0.5 can be estimated as 0.00275 and 0.00116 for the small and large systems respectively. The ratio between the growth velocities is 2.37. This corresponds to the ratio between the densities of nuclei $n_{300} : n_{1200} = 2.38$,

Fig. 8 The evolution of the recrystallized volume fraction f_ξ with time

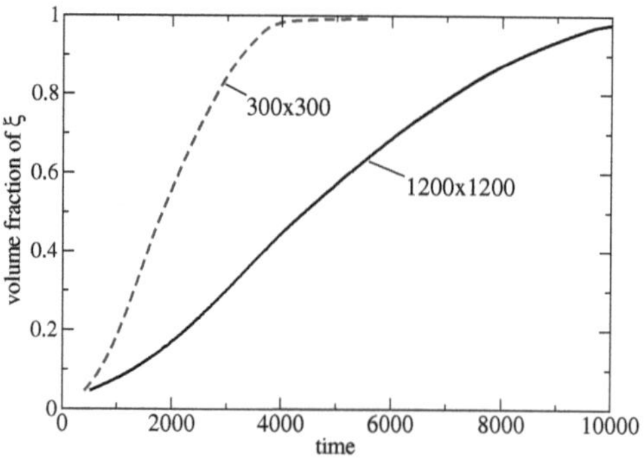

where $n_{300} = 6/300^2 = 6.6 \times 10^{-5}$ nuclei/Δx^2 is the density in the small system and $n_{1200} = 6/1200^2 = 2.7 \times 10^{-5}$ nuclei/Δx^2 is the density in the large system. An initial period in which the nuclei start to grow, a stable linear period, and a period where the recrystallized grains come into contact and the growth slows down can also be seen in Fig. 8.

Comparison with the KWC Model with Two Order Parameters

In the standard KWC model, two order parameters are used for the recrystallization simulations in Refs. [21, 22]. There are the phase-field order parameter ϕ, which represents the recrystallized grains, and the order parameter θ, which represents the crystallographic orientation. In this model, the specific features of the grain boundaries between initial deformed grains and their effect on the recrystallization process are not taken into account. Therefore, the resulting morphology of recrystallized grains does not depend in general on the initial configuration of grain boundaries. On the contrary, in the extended model proposed above, all boundaries are considered and the interaction between the recrystallized grains and the initial grain boundaries can be studied including the movement of the initial grain boundaries during the coarsening and the rotation of grains. The misorientation and the dislocation density in the deformed grains as well as on the deformed grain boundaries can be readily taken into account.

Conclusion

In the present study, an extended phase-field model based on the KWC model of the grain growth is applied, which takes into account the growth of the recrystallized grains along with the coarsening of initial grains in the deformed matrix. We suggested and tested an order parameter ξ being responsible for the recrystallized region and their evolution equation which incorporates the interaction between ξ and the solid phase order parameter ϕ. The numerical study by means of the full set of evolution equations shows the effects of the misorientation, rotation, and initial grain structure on the recrystallization kinetics.

References

1. D. Fan, L.Q. Chen, Acta Mater. **45**, 611–622 (1997)
2. L.Q. Chen, W. Yang, Phys. Rev. B **50**, 15752–15756 (1994)
3. C.E. Krill, L.Q. Chen, Acta Mater. **50**, 3057–3073 (2002)
4. A. Kazaryan, Y. Wang, S.A. Dregia, B.R. Patton, Acta Mater. **50**, 2491–2502 (2002)
5. Y. Suwa, Y. Saito, Mater. Trans. **44**, 2245–2251 (2003)
6. Y. Suwa, Y. Saito, H. Onodera, Comput. Mater. Sci. **40**, 40–50 (2007)
7. Y. Suwa, Y. Saito, H. Onodera, Mater. Trans. **49**, 704–709 (2008)
8. N. Moelans, B. Blanpain, P. Wollants, Phys. Rev. Lett. **101**, 025502 (2008)
9. N. Moelans, A. Codfrey, Y. Zhang, D.J. Jensen, Phys. Rev. B **88**, 054103 (2013)
10. S.G. Kim, D.I. Kim, W.T. Kim, Y.B. Park, Phys. Rev. E **74**, 061605 (2006)
11. I. Steinbach, F. Pezzolla, Phys. D **134**, 385–393 (1999)
12. S.G. Kim, W.T. Kim, T. Suzuki, M. Ode, J. Crys, Growth **261**, 135–158 (2004)
13. J.A. Warren, W.C. Carter, R. Kobayashi, Phys. A **261**, 159–166 (1998)
14. R. Kobayashi, J.A. Warren, W.C. Carter, Phys. D **140**, 141–150 (2000)
15. J.A. Warren, R. Kobayashi, A.E. Lobkovsky, W.C. Carter, Acta Mater. **51**, 6035–6058 (2003)
16. T. Takaki, A. Yamanaka, Y. Tomita, *APCOM07 in Conjunction with EPMESC XI, December 3–6, Kyoto, Japan* (2007)
17. T. Takaki, T. Hirouchi, Y. Hisakuni, A. Yamanaka, Y. Tomita, Mater. Trans. **49**(11), 2559–2565 (2008)
18. T. Takaki, A. Yamanaka, Y. Tomita, Adv. Struct. Mater. **64**, 441–459 (2015)
19. E. Miyoshi, T. Takaki, Comput. Mater. Sci. **112**(A) 44–51 (2016)
20. E. Miyoshi, T. Takaki, Comput. Mater. Sci. **120**, 77–83 (2016)
21. T. Takaki, A. Yamanaka, Y. Higa, Y. Tomita, J. Comput. Aided Mater. Des. **14**, 75–84 (2007)
22. T. Takaki, A. Yamanaka, Y. Tomita, Mater. Sci. Forum 558–559, 1195–1200 (2007)
23. R.D. Doherty et al., Mater. Sci. Eng. A **238**, 219–274 (1997)
24. F.J. Humphreys, Acta Mater. **45**, 4231–4240 (1997)

RETRACTED CHAPTER: Surface Reaction and Transport in Oxides Formed on FeCrAl Alloys in High Temperature Nitridation Environments

Christine Geers, Vedad Babic, Lars-Gunnar Johansson and Itai Panas

Abstract An extended experimental study on corrosion of commercial FeCrAl alloys in predominantly N_2 atmosphere at high temperatures raises fundamental questions concerning nitrogen permeation through oxide scale components, each commonly perceived to constitute a robust barrier. State-of-the-art microscopy combined with DFT calculations is employed to unravel the underlying mechanisms for how chromia nodules support nitrogen dissociation and subsequently act windows for internal nitridation. A similar analysis addressed the role of yttria nodules shown to exhibit apparent expulsion of nitrogen, while favoring rapid alumina growth underneath the nodules. In this case, as for the slowly growing bare alpha alumina scale, the origin for the protection towards nitrogen permeation was identified as the inability to satisfy the necessary conditions for N_2 dissociation.

Keywords Oxide transport properties · Surface reaction · DFT

Introduction

The resilience of high-temperature alloys to withstand a corrosive atmosphere relies on the buildup of a slow-growing oxide scale thus controlling the rate of a process which eventually leads to the consumption of said alloy. Besides the composition of the corrosive atmosphere, rate of chemical transformation is controlled by temperature. Therefore, for applicability it is necessary choose high temperature alloys

The original version of the chapter was revised: The Minerals, Metals & Materials Society, editor of the publication *TMS 2017 146th Annual Meeting & Exhibition Supplemental Proceedings*, is retracting the book chapter, "Surface Reaction and Transport in Oxides Formed on FeCrAl Alloys in High Temperature Nitridation Environments" (DOI 10.1007/978-3-319-51493-2_36). The corresponding author requested this retraction because an image was used for which the authors do not hold copyright. The authors could not obtain permission to use the image. The erratum to the chapter is available at 10.1007/978-3-319-51493-2_78

C. Geers (✉) · V. Babic · L.-G. Johansson · I. Panas
Energy and Materials, Chalmers University of Technology, Gothenburg, Sweden
e-mail: geersc@chalmers.se

The Minerals, Metals & Materials Society, *TMS 2017 146th Annual Meeting & Exhibition Supplemental Proceedings*, The Minerals, Metals & Materials Series,
DOI 10.1007/978-3-319-51493-2_36

to match the specific corrosive atmosphere including the crucial growth kinetics of the effective barrier oxide.

An unexposed alloy is characterized by the simultaneous presence of each constituent at the material surface. In the case of FeCrAl(RE) alloys aluminum is commonly the least noble metal in the matrix, thus acting oxygen sink. Five weight percent of aluminum have proven to be limiting in terms of form- and weldability. It is the growth of alpha alumina which commonly ensures formation of a protective oxide scale, the purpose of which is to offer a kinetic barrier between the corrosive species and the underlying metal [1]. Protection is provided by the low electronic and ionic conductivities of α-alumina whereby decelerated corrosion rate is achieved [2].

In this study, two oxidants are offered, oxygen and nitrogen. Despite the preferential oxidation of aluminum also the other alloying elements form competitive oxides and nitrides. This, in conjunction with the kinetic control being essential for the resulting corrosion resistance, is why the very initiation of a corrosion process is so crucial to the resulting performance of the alloy over time. Here, we describe how simultaneous presence of three different oxides—alumina, chromia and yttria—impact the performance of a FeCrAl(RE) alloy differently in a high temperature environment dominated by nitrogen and with only traces of oxygen. The impact of the competition for oxygen among the alloying elements is not easy to foresee because it crucially sustains transient compounds supporting molecular nitrogen dissociation as well as acting windows for nitrogen permeation, a consequence of which is catastrophic nitride precipitation within the alloy.

Experimental and Computational Methods

The nitridation experiments where performed at 900 °C in a N_2-5%H_2 gas mixture with approximately 16 ppm oxygen as impurity in the nitrogen gas. Equilibrating the oxygen impurity with hydrogen provides around 35 ppm steam as oxidizing species to the gas stream and an oxygen partial pressure of 10^{-23} bar. This amount renders yttria, alumina and even chromia thermodynamically stable. The gas mixture was streamed in a horizontal tube furnace with a flow of 6.3 cm/min with a sufficient purging time before and after the heating to avoid oxygen levels above the impurity limit. 15 × 15 × 2 mm coupons consisting of the high temperature creep resistant alloy Kanthal APMT™, see chemical composition in Table 1, was placed in the center of the heat constant zone for 1 week. After the experiments the specimens underwent thorough investigations focusing on surface, scale and internal corrosion species by

Table 1 Chemical composition of alloy Kanthal APMT

wt%	C	Si	Mn	Mo	Cr	Al	Fe	Other
Nom.				3.0	21.0	5.0	Balance	Y, N
Min.	–	–	–		20.5			
Max.	0.08	0.7	0.4		23.5	–		

using surface SEM (scanning electron microscopy) studies, when indicated coupled with FIB (focused ion beam) and cross sectioning, either manually via epoxy embedding, cutting and polishing or via the BIB (broad ion beam) technique.

DFT calculations employed the CASTEP code [3] within the Materials Studio 6.1 suite [4]. The Purdue-Burke-Ernzerhof functional within the generalized gradient approximation was used with a 500.0 eV plane-wave cut-off energy in conjunction with on-the-fly computed pseudopotentials. Convergence criteria for the geometry optimization calculations included 10^{-3} Å (displacement) and 10^{-5} eV/atom (energy). A Monkhorst-Pack k-point grid with 0.04 $Å^{-1}$ k-points separation was employed [5]. Spin-polarized calculations were performed throughout.

Results and Discussion

The surface of FeCrAl alloy Kanthal APMT™ showed defects in the α-alumina scale after 1 week of exposure in the high temperature nitriding condition. A large defect area is shown in the SEM top view image of Fig. 1. The defect appears brighter in the backscatter electron image since chromium is the major oxide former in this area instead of aluminum. Some even brighter particles are in the vicinity which contain so called reactive elements in this case yttrium as main component. Besides the preferential alumina, chromium and yttrium in form of their oxides are secondary oxide species in contact with the gas phase. The α-Al_2O_3 scale is the major oxide component, covering most of the alloy surface after exposure to the corrosive atmosphere, and underneath a very thin and homogeneous scale, the alloy is intact with no sign of internal corrosion. Where secondary oxide species are present, the consequences can easily be seen by inspecting the cross section images in Fig. 1. Consider the image in the middle of Fig. 1, taken from an alloy section where chromia has been found on the surface instead of alumina. Below the chromia, it reveals a massive nitridation attack. On the other hand, no nitridation attack can be found underneath a reactive element particle but rather enhanced oxide growth is observed instead as highlighted in the FIB cross section on the right of Fig. 1.

Thus, three totally different corrosion morphologies and chemistries are present at these positions on the same alloy material exposed to the same environment. In order to explain the processes leading to these disparate behaviors, each oxide

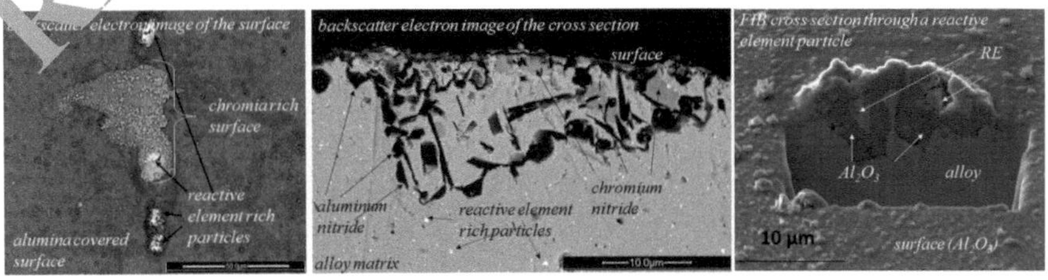

Fig. 1 SEM images of the surface and cross sections of a Kanthal APMT™ specimen after 1 week exposure

species was subject to analysis. Consequences of bringing each of oxide species in contact to the sandwiching gas and alloy environments were explored by means of 1st principles DFT calculations.

Chromia

The internal precipitates underneath the chromia covered surface as seen in the image in the middle of Fig. 1 comprise aluminum nitrides (dark) and chromium nitrides (medium grey), telling of rapid ingress of nitrogen into the alloy. Consequently the chromia on the surface is not able to protect the interior in spite of the fact that it is thermodynamically stable under the given oxygen activity, i.e. 10^{-23} atmospheres exceeding the dissociation pressure of chromia by a factor of 100. The transparency of chromia for nitrogen is arrived at by considering the reduction of chromia by aluminum in the underlying alloy acting reducing agent. Thus, reduction of chromia by aluminum acting oxygen sink is able to cause transient oxygen vacancies which act dissociative surface sites for N_2 as well as hosts for the dissociation products, nitrogen from the gas environment acting substitutional oxidation agent:

$$Cr_2O_3(s) + \frac{y}{2}N_2(g) + yAl(s) \rightarrow Cr_2N_yO_{3-\frac{3y}{2}}(V_O^{**})_{\frac{y}{2}}(s) + \frac{y}{2}Al_2O_3(s) \qquad (1)$$

Validation of the above understanding hypothesis was provided by DFT calculations followed each step along the nitrogen ingress via chromia and the summarizing reaction named in Eq. 1 is exothermic. The reaction path was elucidated in detail in a previous publication [6]. It included conditional ability of a chromia slab model to support molecular N_2 dissociation, as well as diffusion into the chromia lattice by creating transient chromium-oxynitride species with intermediate stabilities.

Alumina

Alumina on the other hand does not exhibit the possibility to dissociate nitrogen on the surface. DFT calculations tell that even the unrealistic 100% coordinatively unsaturated Al sites terminated surface of an alumina slab model, is unable to dissociate N_2.

Yttria

Yttria, being even more stable than alumina thermodynamically, does not support nitrogen dissociation on its surface either. However, the interface between yttria and alumina has proven to offer the ability to sustain a constant oxygen activity along its circumference, leading to isotropic growth of alumina on the yttria particle. Even though only ppms of water are present in the nitriding environment a rapid hydroxylation process along the reactive element particle is implied. The high concentration of nitrogen in the gas environment expelled from the aluminum reservoir due to inability of yttria surfaces as well as yttria/alumina interface to dissociate N_2.

Conclusion

High temperature corrosion processes in environments causing degradation such as internal nitridation cannot be fully understood by relying sole on thermodynamic equilibrium data of oxide scales in contact with the gas phase. While the corresponding Ellingham-Richardson diagram offers robust boundary conditions for the feasibility of the coexisting candidate oxides resulting from the exposure to a certain environment, cautious investigation are required to unravel the time evolutions decisive for capturing the resulting corrosion processes. Indeed, competition and cooperation among alloying elements as well in their transient reaction products in response to the corrosive atmosphere may determine the resilience of the alloy. This understanding was demonstrated for the FeCrAl alloy Kanthal APMT™ under nitriding conditions with limited access to oxygen, for three of oxides coexisting at the alloy surface in presence. Thermodynamics dictate that alumina, chromia and yttria are stable under the given conditions which was confirmed by the experimental findings. However, the surface reaction and transport properties regarding oxygen and nitrogen are different for each one:

- Alumina exhibits slow transport of oxygen while nitrogen is expelled due to inability to support of N_2 dissociation at the oxide/gas interface.
- Chromia possesses redox properties allowing the transport of oxygen and nitrogen by metastable intermediates even through the bulk oxide. Being nobler than aluminum, transient reduced chromia surface sites are able to adsorb and as well as dissociate N_2;
- Yttria displays rapid transport of oxygen at the alumina/yttria interface. However, similar to alumina its inability to support of N_2 dissociation renders any nitrogen transport in yttria bulk irrelevant.

What eventually drives transport of nitrogen through the oxide scales and into the metal matrix is the ability of nitrogen to form stable aluminum nitrides, analogous to internal oxidation. As elucidated by DFT slab calculations, the crucial discriminating factor leaving yttria and alumina protective towards nitrogen permeation, in contrast to chromia, is the inability of the two former to display sufficiently reducing surface sites to dissociate molecular nitrogen.

Acknowledgements This work was carried out within the Swedish High Temperature Corrosion Centre (HTC) and was supported by Sandvik Heating Technology AB.

References

1. S. Chevalier, in *Developments in High Temperature Corrosion and Protection of Materials* (Woodhead Publishing, 2008), pp 290–328; B.A. Pint, in *Shreir's Corrosion*, ed. by J.A. Richardson, Editor-in-Chief: Tony (Elsevier, Oxford, 2010), pp. 606–645
2. A.H. Heuer, T. Nakagawa, M.Z. Azar, D.B. Hovis, J.L. Smialek, B. Gleeson, N.D.M. Hine, H. Guhl, H.S. Lee, P. Tangney, W.M.C. Foulkes, M.W. Finnis, Acta Mater. **61**, 66[illegible]–6[illegible] (2013)
3. M.D. Segall, J.D. Lindan Philip, M.J. Probert, C.J. Pickard, P.J. Hasnip, S.J. Clark, M.C. Payne. J. Phys. Cond. Matter **14**, 2717 (2002)
4. Materials Studio 6.0, Accelry Inc
5. H.J. Monkhorst, J.D. Pack, Phys. Rev. B **13**, 5188–5192 (1976)
6. C. Geers, V. Babic, N. Mortazavi, M. Halvarsson, B. Jönsson, L.-G. Johansson, I. Panas, J.-E. Svensson, in *Oxid Met* 2017, p. accepted

Atomic-Scale Modeling of Fe-Al-Mn-C Alloy Using Pair Models and Monte-Carlo Calculations

Jérôme Dequeker, Alexandre Legris, Rémy Besson and Ludovic Thuinet

Abstract The Fe-Al-Mn-C system is widely studied for automotive applications due to its good mechanical properties and relatively low density. Our work is devoted to the atomic-scale modeling of this system and to start with, we focused on the well-documented Fe-Al binary system. More precisely, we tested the capability to reproduce its phase diagram combining ab initio calculations and cluster expansion methods. Several models were built using different input atomic configurations: pure iron, a substitutional aluminium atom diluted in iron, pairs of substitutional aluminium atoms located at different neighbour shells and complementary structures (B2, B32 and $D0_3$). Long-range order parameters (occupation of sublattices) were defined to analyse the equilibrium configurations generated by Monte-Carlo runs in the semi-grand canonical ensemble. Phase diagrams were plotted for each model and compare well with experimental ones.

Keywords Ab initio · Monte-Carlo · Cluster expansion · Phase diagram · Automotive application

Introduction

Automotive weight reduction leads to the study of third generation duplex steels [1, 2] containing iron, aluminium, manganese and carbon. To reach the aimed mechanical properties (ultimate tensile strength above 1 GPa, uniform elongation larger than 15%) for low densities (5–10% lighter than common steels), duplex Fe-Al-Mn-C alloys are foreseen. This work is interested in fundamental aspects related to these alloys and seeks for atomistic models to describe the thermodynamics of Fe-Al on a Body Centered Cubic (BCC) lattice. Since chemically more

J. Dequeker (✉) · A. Legris · R. Besson · L. Thuinet
Université Lille, CNRS, INRA, ENSCL, UMR 8207—UMET,
Unité Matériaux et Transformations, 59000 Lille, France
e-mail: jerome.dequeker@ed.univ-lille1.fr

The Minerals, Metals & Materials Society, *TMS 2017 146th Annual Meeting & Exhibition Supplemental Proceedings*, The Minerals, Metals & Materials Series,
DOI 10.1007/978-3-319-51493-2_37

complex quaternary (quinary if vacancies are included) systems are foreseen, we choose to work with simple lattice hamiltonians built using pair interactions fitted on DFT energy calculations. Different models are tested depending on the range of interactions and on the input ab initio database. Phonon effects are taken into account within the framework of the harmonic approximation and phase diagrams are estimated using Monte-Carlo (MC) simulations in the semi-grand canonical ensemble. Our results obtained using "direct" methods are compared to previous diagrams obtained either by direct [3] or "inverse" Cluster Variation Method (CVM) and MC methods [4–6].

Method

The system under study consists of a BCC lattice where each site is occupied by an iron, aluminium or manganese atom, although only Fe-Al results are discussed in this paper. The system energy is calculated using a cluster expansion based on an Ising-like model containing only point and pair interactions and is given by

$$E = J_\phi + \sum_{i=Al,Mn} J_1^i \sum_n p_n^i + \sum_q \sum_{i,j} J_2^{q,ij} \sum_{\{n,m<n\}_q} p_n^i p_m^j$$

where i and j are non-iron chemical species variables (i.e. Al or Mn) and n and m are lattice sites. p_n^i is a chemical species indicator: it is equal to 1 only if site n is occupied by an i atom (Al or Mn) and 0 otherwise. The first energetic coefficient J_ϕ corresponds to the energy of pure iron. J_1^i is related to the substitution energy of an iron atom by another species $i \in \{Al, Mn\}$ and $J_2^{q,ij}$ is a pair coefficient for two substitutional atoms qth neighbours of each other. Neighbouring shells are represented on Fig. 1.

The various pair interactions $J_2^{q,ij}$ coefficients are fitted on an energy database built from ab initio calculations performed with the Vienna ab initio Package (VASP) [7–10] using the PAW-PBE [11–13] potential and a cut-off energy $E_{cut} = 560$ eV. For compounds (B2, B32, $D0_3$), 2 × 2 × 2 unit cells (16 atoms) were used. The Brillouin zone was sampled with a 24 × 24 × 24 Monkhorst-Pack mesh for the unitary cubic cell. Some ab initio calculations were performed on

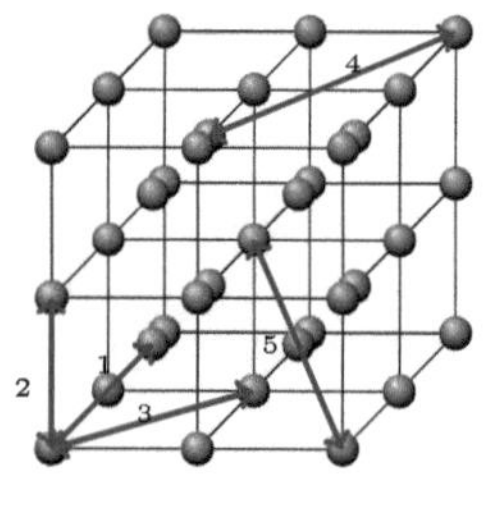

Fig. 1 Illustration of various neighbouring shells in the BCC lattice

higher cell sizes up to 3 × 3 × 3 unit cells (54 atoms). All the DFT calculations included magnetic effects. VASP calculations were run with different initial magnetic moments and configurations with lowest final energy were retained.

To take into account vibrational effects, for each structure in the input database, the phonon spectrum ω_j was calculated at 0 K at the Gamma point, and the phonon free energy was determined in the framework of the harmonic approximation using [14]:

$$F_0 = k_B T \sum_j \ln[2\sinh \frac{\hbar\omega_j}{2k_B T}],$$

where ω_j are the eigen frequencies.

All the structures were found to be stable (positive eigen frequencies) except FeAl-B32. Since FeAl-B32 has been experimentally observed [15], this feature has to be investigated further in future works. In order to maintain an homogeneous treatment, FeAl-B32 was given the same vibrational free energy as FeAl-B2.

Several models were developed to describe the system. The FeAl-SS3c1-T600K notation, for instance, indicates that in this model interactions up to third substitutional-substitutional neighbours are considered and complementary structures in set c1 (B2, B32 and D03) were included in the ab initio input database. The $J(600\,K)$ interactions were fitted from energies containing the vibrational contribution to each structure calculated at 600 K. In near future, $J(T)$ will be calculated using the same methodology.

MC runs (3 million equilibration steps followed by 9 million equilibrium steps) were performed on systems containing up to 3,456 atomic sites (12 × 12 × 12 cells) for temperatures ranging from 400 to 1100 K using a chemical potential step $\delta\mu_{Al}$ of 0.001 eV/at ($\delta\mu_{Al} = \mu_{Al} - \mu_{Fe}$, with $\mu_{Fe} = J_\phi/2$). This generates equilibrium configurations from which order parameters are calculated and allow to build up phase diagrams. To this end, the system is subdivided in four interpenetrating Faced Centered Cubic (FCC) sublattices α, β, γ and δ (cf Fig. 2) and their average occupations (p_α, p_β, p_γ and p_δ) are calculated. Long-range order parameters are then determined [16, 17] and are associated to various ordered states (cf. Table 1).

A typical analysis of the MC data is illustrated in Figs. 3 and 4 for the binary FeAl-SS2c1 model at 500 K. In Fig. 3, the order parameters p_i are plotted as a function of $\delta\mu_{Al}$ and allow us to determine the state of order, while Fig. 4 illustrates the concentration of aluminium versus chemical potential dependence.

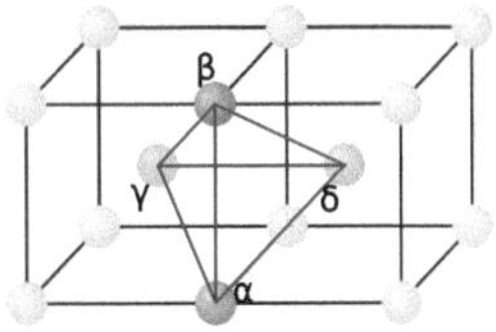

Fig. 2 Four FCC sublattices in BCC structures

Table 1 Correspondence between compounds and FCC sublattice occupations

Compounds	Occupation
A2	$\alpha=\beta=\gamma=\delta$
B2	$\alpha=\delta\neq\beta=\gamma$
B32	$\alpha=\beta\neq\delta=\gamma$
$D0_3$	$\alpha\neq\beta\neq\gamma=\delta$

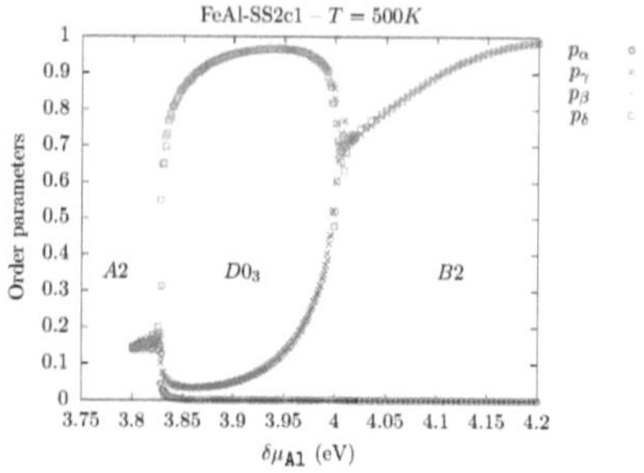

Fig. 3 Long range order parameters as a function of chemical potential difference ($\delta\mu_{Al}$)

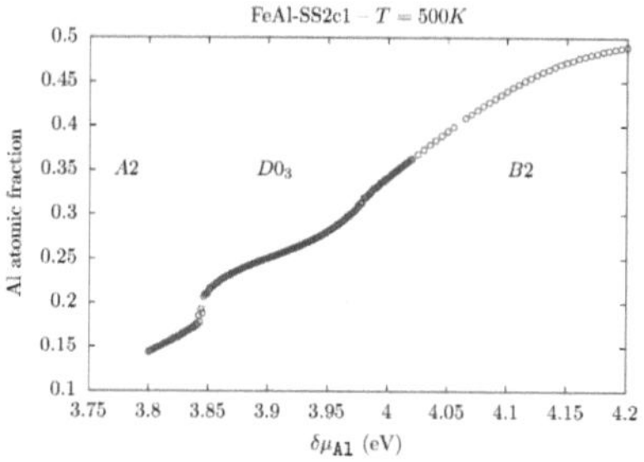

Fig. 4 Al atomic fraction as a function of chemical potential difference ($\delta\mu_{Al}$)

Table 2 DFT calculations included in the database to fit the energy coefficients for each binary Fe-Al model

	Pure iron	1Al	2Al 1-2nn	2Al 3nn	2Al 4nn	2Al 5nn	B2-B32-$D0_3$
FeAl-SS2	✓	✓	✓				
FeAl-SS3	✓	✓	✓	✓			
FeAl-SS2c1	✓	✓	✓				✓
FeAl-SS3c1	✓	✓	✓	✓			✓
FeAl-SS4c1	✓	✓	✓	✓	✓		✓
FeAl-SS5c1	✓	✓	✓	✓	✓	✓	✓
FeAl-SS2c1-T600K	✓	✓	✓				✓
FeAl-SS3c1-T600K	✓	✓	✓	✓			✓

Results

For all proposed binary Fe-Al models, the configurations included in the input database are given in Table 2, and the numerical values of the energy coefficients are reported in Table 3, the results (from [4]) obtained using inverse methods

Table 3 Numerical values of the energy coefficients (eV) for binary Fe-Al models fitted on DFT calculations or determined from experiments using inverse methods [4] (last column)

J (eV)	FeAl-SS2	FeAl-SS3	FeAl-SS2c1	FeAl-SS3c1	FeAl-SS4c1	FeAl-SS5c1	FeAl-SS2c1-T600 K	FeAl-SS3c1-T600 K	Inverse methods [4]
J_{ϕ}	−16.6117	−16.6117	−16.6096	−16.6112	−16.6105	−16.6100	−16.8023	−16.8019	–
J_1^{Al}	3.8389	3.8389	3.7601	3.7955	3.7477	3.7150	3.7831	3.7717	–
$J_2^{1,AlAl}$	0.0363	0.0363	0.1966	0.1958	0.2808	0.2931	0.2056	0.2058	0.0215
$J_2^{2,AlAl}$	0.0506	0.0506	0.0462	0.0457	0.0459	0.2536	0.0522	0.0524	0.0053
$J_2^{3,AlAl}$		−0.0853		−0.0054	0.0023	0.0077		0.0018	−0.0199
$J_2^{4,AlAl}$					−0.0282	−0.0322			0
$J_2^{5,AlAl}$						−0.1557			−0.0008

based on diffuse scattering experiments are also reported. Note that there is no attenuation of the pair contribution for 5th neighbor in our calculations, and that in inverse methods the 4th neighbor interactions were neglected because J^4 is not included in the octahedron or cube approximations of the CVM.

Ground State Analysis

In Fig. 5 are presented the lattice ground states as function of J^1 and J^2, in an analysis restricted to 2nd neighbors [18]. From this schematics, we conclude that the presence of complementary structures is needed in order to obtain the correct ground state at 0 K for the models going up to 2nd neighborhood. These results are confirmed by the calculated phase diagrams (cf. Figs. 6 and 8).

From Ref. [18] it is also possible to perform the same ground state analysis for the 3rd neighbour models. Here again, it can be concluded that complementary structures are needed to obtain the correct ground state properties.

Phase Diagrams

Following the ground state analysis, phase diagrams have been determined from MC calculations and are represented on Figs. 6 and 7 for FeAl-SS2 and FeAl-SS3

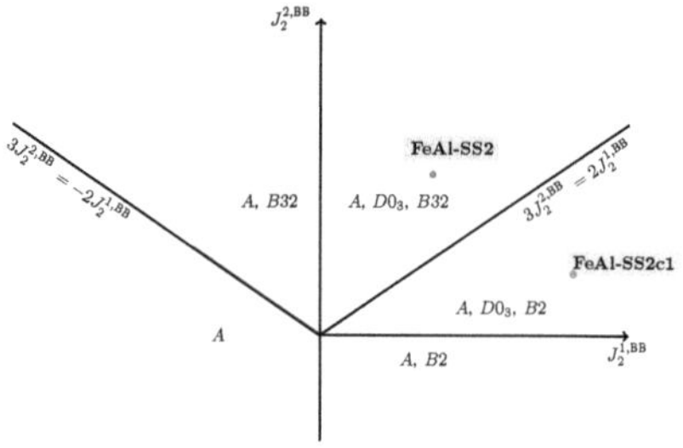

Fig. 5 Ground states of BCC Fe-Al as a function of the pair interaction coefficients

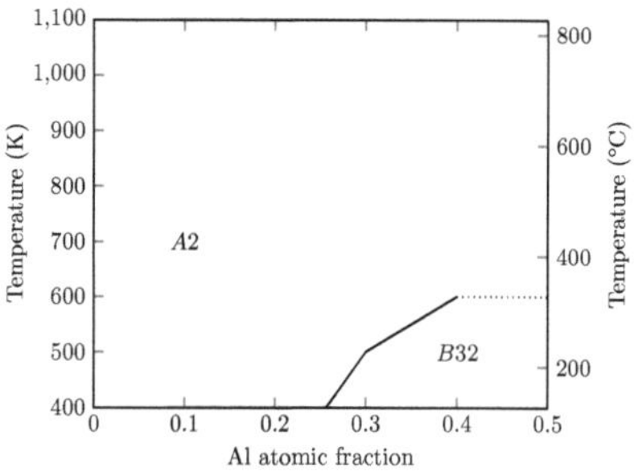

Fig. 6 FeAl-SS2 calculated phase diagram

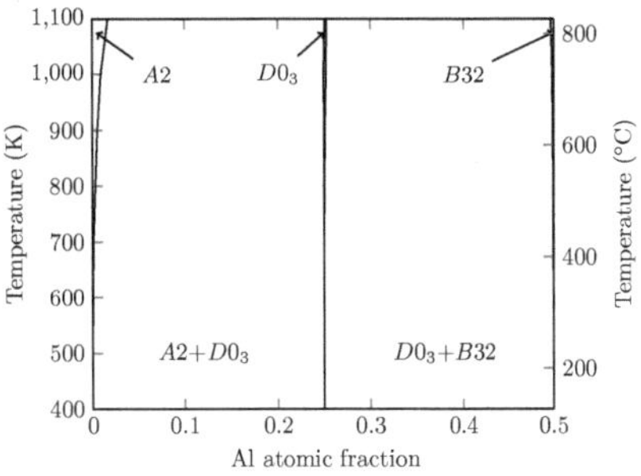

Fig. 7 FeAl-SS3 calculated phase diagram

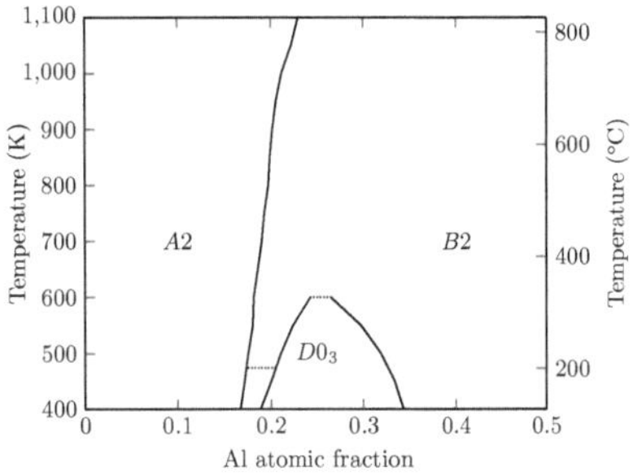

Fig. 8 FeAl-SS2c1 calculated phase diagram

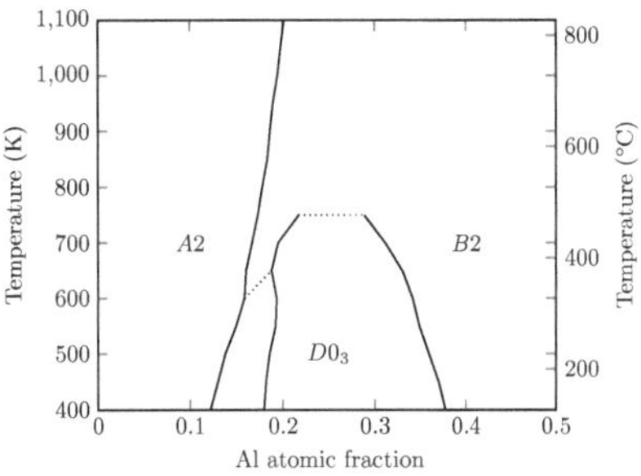

Fig. 9 FeAl-SS3c1 calculated phase diagram

models. The $D0_3$ compound is not predicted by the SS2 model and the deviation from stoichiometry in $D0_3$ allowed by SS3 is not large enough compared to experimental data. In both cases B32 is predicted whereas B2 appears in the experimental phase diagram. All these remarks are in agreement with the ground state analysis of Fig. 5.

When complementary structures ($D0_3$, B2 and B32) are added to the DFT input database, the obtained phase diagram (Figs. 8 and 9) are closer to the experimental one. We also can notice the presence of a two-phase domain between A2 and $D0_3$. The main difference between the phase diagrams given by these two models is the transition temperature between $D0_3$ and B2. The B2 phase appears at a much higher temperature with SS3c1 than with SS2c1. The experimental transition temperature at Fe_3Al stoichiometry is around 820 K [19], which agrees well with the prediction of SS3c1.

Including phonons in the input ab initio database modifies the energy parameters of the SS2c1 and SS3c1 models (see Table 3). The most remarkable difference is the sign change of $J_2^{3,AlAl}$. The general shape of the phase diagram is not changed but modifications of the transition temperature between $D0_3$ and B2 are observed: the inclusion of phonons increases the transition temperature for the SS2c1 model by 100 K, whereas the opposite is obtained for the SS3c1 model with a decrease of the transition temperature by 150 K.

Discussion

All the 2nd neighbour models are in qualitative agreement with previous theoretical works concerning both the ground states [18] and phase diagram [20] calculations. Adding complementary structures in the DFT calculations database is needed to predict the presence of B2 instead of B32 for FeAl, and to obtain a two-phase A2 + $D0_3$ domain. Moreover, increasing the range of the pair interactions to the 3rd neighbour shell leads to a better transition temperature between $D0_3$ and B2, at least when phonon effects are not included.

The ab initio-based pair energy models proposed in this work are thus able to reproduce the major features of the iron-rich side of the Fe-Al phase diagram. However, a main discrepancy between our results and the experimental Fe-Al phase diagram concerns the existence, around 600 °C and above the A2 + D03 field, of a narrow A2 + B2 two-phase domain, absent from the present simulations. From various simulation works, the origin of this feature is still under debate. To obtain it, some authors [6] pointed out the necessity to include explicitly magnetic degrees of freedom in the energy description of the alloy, whereas others [5] were able to reproduce it without including magnetism, simply by considering purely "chemical" interactions such as those used in the present work, the requirement being then that the fourth-neighbour interactions have to be included in order to predict A2 + B2 domain. Fourth neighbour pair interactions are often neglected because they are not present in the BCC cube or in the irregular octahedron approximation of the CVM [4, 5]. Increasing the pair interaction range beyond the 3rd interaction range is still under investigation. This issue clearly illustrates that further fundamental work is necessary to deepen our understanding of the physical mechanisms underlying order and magnetism in Fe-Al alloys.

The direct method developed allows us to describe with a qualitative agreement the binary Fe-Al system on the Fe-rich side. This methodology is currently being extended to include manganese and carbon.

Acknowledgements The authors are indebted to the Agence Nationale de la Recherche project MeMnAl-Steels for its financial support under contract ANR-RNMP-0002.

References

1. I. Zuazo, B. Hallstedt, B. Lindahl, M. Selleby, M. Soler, A. Etienne, A. Perlade, D. Hasenpouth, V. Massardier-Jourdan, S. Cazottes, X. Kleber, Low-density steels: complex metallurgy for automotive applications. JOM **66**(9), 1747 (2014)
2. M.J. Yao, P. Dey, J.-B. Seol, P. Choi, M. Herbig, R.K.W. Marceau, T. Hickel, J. Neugebauer, D. Raabe, Combined atom probe tomography and density functional theory investigation of the Al off-stoichiometry of κ-carbides in an austenitic Fe-Mn-Al-C low density steel. Acta Mater. **106**, 229 (2016)
3. P.G. Gonzales-Ormeno, H.M. Petrilli, C.G. Schon, Ab initio calculation of the bcc Fe-Al phase diagram including magnetic interactions. Scripta Mater. **54**, 1271 (2006)
4. V. Pierron-Bohnes, S. Lefebvre, M. Bessiere, A. Finel, Short range order in a single crystal of Fe-19.5 at.% Al in the ferromagnetic range measured through X-ray diffuse scattering. Acta Metall. Mater. **38**(12), 2701 (1990)
5. C. Bichara, G. Inden, Monte Carlo calculation of the phase diagram of BCC Fe-Al alloys. Scr. Metall. Mater. **25**(11), 2607 (1991)
6. F. Schmid, K. Binder, Modelling order-disorder and magnetic transitions in iron-aluminium alloys. J. Phys.: Condens. Matter **4**(13), 3569 (1992)
7. G. Kresse, J. Hafner, *Ab initio* molecular dynamics for liquid metals. Physical Review B **47** (1), 558 (1993)
8. G. Kresse, J. Hafner, Ab initio molecular-dynamics simulation of the liquid-metal–amorphous-semiconductor transition in germanium. Phys. Rev. B **49**(20), 14251 (1994)
9. G. Kresse, J. Furthmüller, Efficiency of ab-initio total energy calculations for metals and semiconductors using a plane-wave basis set. Comput. Mater. Sci. **6**(1), 15 (1996)
10. G. Kresse, J. Furthmüller, Efficient iterative schemes for ab initio total-energy calculations using a plane-wave basis set. Phys. Rev. B **54**(16), 11169 (1996)
11. P.E. Blöchl, Projector augmented-wave method. Phys. Rev. B **50**(24), 17953 (1994)
12. G. Kresse, D. Joubert, From ultrasoft pseudopotentials to the projector augmented-wave method. Phys. Rev. B **59**(3), 1758 (1999)
13. J.P. Perdew, K. Burke, M. Ernzerhof, Generalized gradient approximation made simple. Phys. Rev. Lett. **77**(18), 3865 (1996)
14. O. Madelung, *Introduction to Solid-State Theory* (Springer, Berlin, 1996)
15. M. Becker, W. Schweika, B32 order in iron rich Fe-Al alloys. Scr. Mater. **35**(11), 1259 (1996)
16. L. Anthony, B. Fultz, Kinetics of B2, DO3, and B32 ordering: results from pair approximation calculations and Monte Carlo simulations. J. Mater. Res. **9**(2) (1994)
17. B.B. Lindahl, B.P. Burton, M. Selleby, Ordering in ternary BCC alloys applied to the Al-Fe-Mn system. CALPHAD **51**, 211 (2015)
18. A. Finel, Contribution à l'étude des effets d'ordre dans le cadre du modèle d'Ising: états de base et diagrammes de phase. PhD thesis, 1987
19. F. Stein, M. Palm, Re-determination of transition temperatures in the Fe-Al system by differential thermal analysis. IJMR **98**(7), 580 (2007)
20. H. Ackermann, G. Inden, R. Kikuchi, Tetrahedron approximation of the cluster variation method for BCC alloys. Acta Metall. **37**(1), 1 (1989)

Control Technique Study of Non-metallic Inclusions in Low Carbon Steel by Rare Earth Final Deoxidization

Bowen Peng, Fangjie Li, Shaobo Zheng and Huigai Li

Abstract Thermodynamic calculations were carried out to analyze the formation mechanism of inclusion in low carbon steel by rare earth final deoxidization. The deoxidization and alloying process of steelmaking was stimulated in laboratory. The inclusions in as-cast samples were extracted with the method of non-aqueous solution electrolysis. Scanning electron microscope (SEM) with energy dispersive spectrometer (EDS) was used to analyze the chemical composition, three-dimensional morphology and size distribution of the inclusions. The calculation result showed that the Al–Mg-Ce-O complex inclusions could form when the content of rare earth increases to a proper value, and it has lower melting point than magnesium aluminate spinel, which is harmful to the property of steel. The experimental result showed that, the three-dimensional morphology of Al–Mg-Ce-O complex inclusions is spherical, and the mean diameter of that is about 5 μm. Furthermore, the formation mechanism and the control technique of complex inclusion were discussed.

Keywords Thermodynamic calculations · Formation mechanism · Control technique

Introduction

Negative effects of inclusions on properties of low carbon steel depend on their composition, shapes and size distributions [1, 2]. Hard, large and angular inclusions should be avoided because they could not be deformed together with steel matrix

B. Peng (✉) · F. Li · S. Zheng · H. Li (✉)
State Key Laboratory of Advanced Special Steel, Shanghai Key Laboratory of Advanced Ferrometallurgy, School of Materials Science and Engineering, Shanghai University, Shanghai 200072, China
e-mail: pengbw1993@163.com

H. Li
e-mail: lihuigai@shu.edu.cn

The Minerals, Metals & Materials Society, *TMS 2017 146th Annual Meeting & Exhibition Supplemental Proceedings*, The Minerals, Metals & Materials Series,
DOI 10.1007/978-3-319-51493-2_38

during hot rolling and could easily result in stress concentrations at steel/inclusion interface [3, 4]. On the contrary, inclusions with low temperature have better deformability [5] and tend to be spherical in morphology, and they are much easier to be removed from steel. Therefore, the method of forming better inclusions of relative lower melting points, smaller in size and spherical in morphology was significant and it was researched and practiced by many scientists. The control technique of non-metallic inclusions by rare earth final deoxidization was proposed in this paper, because the inclusions containing rare earth (Ce) are of lower melting points than spinel inclusions, which is beneficial to control the size and sharp of inclusions.

As we all know, China has substantial deposits of rare earth, and rare earth have strong deoxidizing capacity. For example, Ce is an important kind of rare earth, and its deoxidizing capacity is much stronger than aluminum and similar to magnesium as it is shown in the oxygen potential diagram. Besides, the inclusions containing Ce element are relative flexible in general, therefore Ce final deoxidization is supposed to be a great control technique of non-metallic inclusions.

The inclusions composition can be estimated from the thermodynamic calculation as a function of steel composition and temperature, provided that the equilibrium between steel and inclusion could be established. The experiments in this paper are designed based on the results of thermodynamic calculation, the control technique is supposed to be proven in both theory and practice.

Thermodynamic Calculation and Analysis

Some thermodynamic calculations were carried out to evaluate formation mechanism of inclusion in low carbon steel by Ce final deoxidization, the nominal compositions of steel were given in Table 1. Formation of $MgO \cdot Al_2O_3$ and Al_2O_3-MgO-Ce_2O_3 system inclusions were calculated. As $MgO \cdot Al_2O_3$ could be easily formed in low carbon steel although [Mg] and [Al] content are of quite low level, $Al_2O_3/MgO \cdot Al_2O_3/MgO$ stability diagram phase was calculated. Considering that $MgO \cdot Al_2O_3$ was not stable when [Ce] entered molten steel, stability phase diagram of $MgO/MgO \cdot Al_2O_3/CeAlO_3$/Al–Mg-Ce-O was also calculated.

Equilibrium constants used in the present study were given in Table 2 and interaction coefficients used in calculation were listed in Table 3.

Activities of components in molten steel could be expressed by Eqs. (1) and (2):

Table 1 Nominal chemical compositions of steel (mass%)

C	Si	Mn	O	S	P	Ce	Al	Mg
0.001	0.25	0.35	0.001	<0.001	<0.001	0.01	0.28	<0.001

Table 2 Equilibrium constants used in the thermodynamic calculation

Equation	Log K	Ref
$Al_2O_3(s) = 2[Al] + 3[O]$	20.57–64000/T	[6]
$MgO(s) = [Mg] + [O]$	−4.28–4700/T	[7]
$Ce_2O_3(s) = 2[Ce] + 3[O]$	20.2–81100/T	[8]
$MgAl_2O_4(s) = Al_2O_3(s) + MgO(s)$	−0.32–980/T	[9]
$[Ce] + Al_2O_3(s) = CeAlO_3(s) + [Al]$	12.9–22168/T	[10]

Table 3 Interaction coefficients used in thermodynamic calculations at 1873 K [2]

i	j						
	C	Si	Mn	P	S	Al	O
Al	0.091	0.0056	0.012	0.033	0.030	0.045	−6.6
O	−0.436	−0.131	−0.021	0.07	−0.133	−3.9	−0.2
Mg	−0.15	−0.09	–	–	−1.38	−0.12	–
Ce	−0.077	−0.092	–	–	–	–	−5.03

$$\lg f_{[i]} = \sum e_i^j [\text{mass}\%j] \quad (1)$$

$$a_{[i]} = f_i [\text{mass}\%i] \quad (2)$$

In Eq. (1), e_i^j is the interaction coefficient of element j to i; while $a_{[i]}$ is the activity of constituent [i] in molten steel.

Calculation of $Al_2O_3/MgO \cdot Al_2O_3/MgO$ stability diagram was carried out as the follows:

Calculation of $MgO/MgO \cdot Al_2O_3$ boundary

$$4(MgO) + 2[Al] = (MgO \cdot Al_2O_3) + 3[Mg] \quad (3)$$

$$\text{Log K} = -33.09 + 50880/T \quad (4)$$

$$K = \frac{a_{[Mg]}^3 \cdot a_{MgO \cdot Al_2O_3}}{a_{[Al]}^2 a_{MgO}^4} \quad (5)$$

Calculation of $MgO \cdot Al_2O_3/Al_2O_3$ boundary

$$4(Al_2O_3) + 3[Mg] = 3(MgO \cdot Al_2O_3) + 2[Al] \quad (6)$$

$$\text{Log K} = 34.37 - 46950/T \quad (7)$$

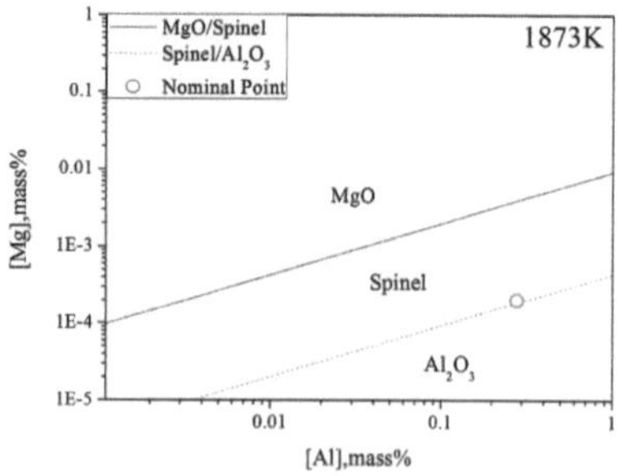

Fig. 1 Al_2O_3/$MgO \cdot Al_2O_3$/MgO stable diagram

$$K = \frac{a^2_{[Al]} \cdot a^3_{MgO \cdot Al_2O_3}}{a^4_{[Al_2O_3]} a^3_{[Mg]}} \tag{8}$$

When MgO/$MgO \cdot Al_2O_3$ boundary was calculated, activities of MgO and $MgO \cdot Al_2O_3$ were 0.8 and 0.99 respectively. And activities of $MgO \cdot Al_2O_3$ and Al_2O_3 were 0.47 and 1 respectively when $MgO \cdot Al_2O_3$/Al_2O_3 boundary was calculated. Dissolved oxygen content was taken as 0.001% when activity coefficients of aluminum and magnesium were estimated.

Calculated phase stability diagram for Al_2O_3/$MgO \cdot Al_2O_3$/MgO is shown as Fig. 1. As it can be seen, the nominal content of dissolved [Al] and [Mg] in molten steel are mainly in the formation region near or rightly on the boundary of spinel/Al_2O_3. It predicts the observation of $MgO \cdot Al_2O_3$ and Al_2O_3 inclusions in experiment. Spinel system inclusions could be formed in steel as chemical reaction products.

However, alumina inclusions are unstable though quite low level of [Ce] exists in molten steel and would be prone to be changed into $CeAlO_3$ or Al–Mg-Ce-O inclusions, whose molting point would be lower than spinels. Therefore, MgO/$MgO \cdot Al_2O_3$/$CeAlO_3$/Ce–Al-Mg-O stable phase diagram was calculated.

MgO/$CeAlO_3$ boundary was calculated by Eq. (9)

$$4CeAlO_3(s) + 3[Mg] = 3[MgO \cdot Al_2O_3] + 4[Ce] + 2[Al] \tag{9}$$

$$\mathrm{Log}\,K = -17 + 41712/T \tag{10}$$

$$K = \frac{a^3_{MgO \cdot Al_2O_3} \cdot a^4_{[Ce]} \cdot a^2_{[Al]}}{a^4_{CeO \cdot Al_2O_3} a^3_{[Mg]}} \tag{11}$$

Activity of $CeAlO_3$ and $MgO \cdot Al_2O_3$ are taken as 0.1 and 1 respectively during calculation. Dissolved oxygen content was taken as 0.001% when activity coefficients of aluminum and magnesium were estimated. The calculated stability phase of MgO, spinel, $CeAlO_3$ and Ce–Al-Mg-O was indicated in Fig. 2. Data point indicated by circle in the figure was nominal point in present study. As it can be

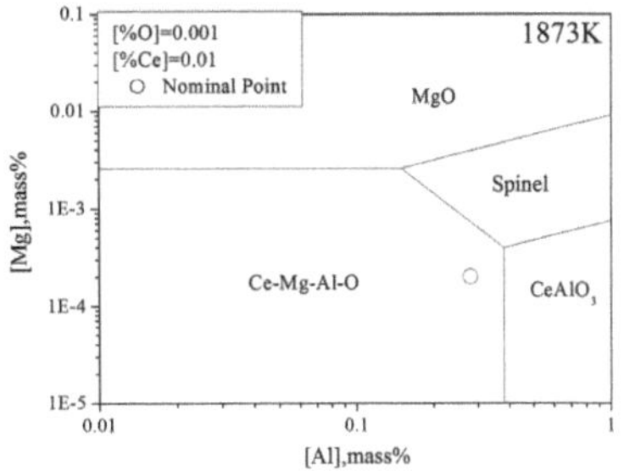

Fig. 2 Phase stabililty diagram of inclusion composition considering Ce in steel

seen, Al_2O_3, MgO and $MgO \cdot Al_2O_3$ inclusions in this study were not stable even when Ce content in steel was 0.01%, Al–Mg-Ce-O phase and $CeAlO_3$ are more possible to be the final stable phases, both of which are flexible.

Illustration on Formation Mechanism of Inclusions During Reaction

Although the final stable phases of inclusions had been proved by thermodynamic calculations above, further understanding on formation mechanism of inclusions was still necessary. As a result, a vivid description on evolution of inclusions during reaction process by Ce final deoxidization would be helpful.

The evolution of inclusions was illustrated by Fig. 3 and described as following:

(1) During the process of Al alloying, Al_2O_3 clusters were quickly and largely formed in molten steel in the environment of high dissolved Al, as indicated in Fig. 3a. Meanwhile MgO would be reduced from slag which was illustrated in reaction (13) with dissolved Al and formed some of spinel gradually by reaction (14), as indicated in Fig. 3b.

$$2[Al] + 3[O] = (Al_2O_3) \quad (12)$$

$$2[Al] + 3(MgO) = 3[Mg] + (Al_2O_3)_{inclusion} \quad (13)$$

$$2[Al] + 4(MgO) = (MgO \cdot Al_2O_3)_{inclusion} + 3[Mg] \quad (14)$$

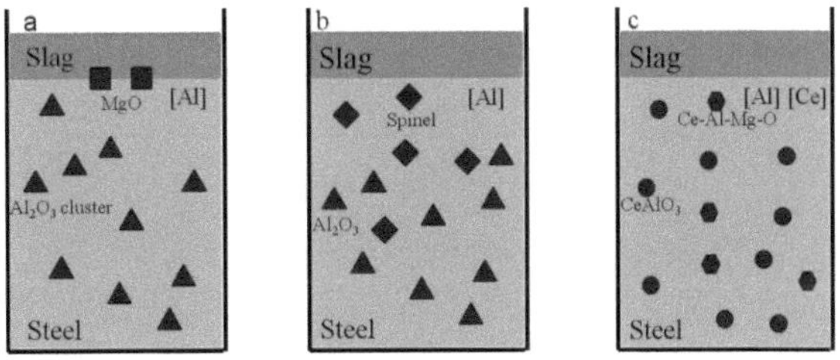

Fig. 3 Illustration on formation mechanism of inclusions during reaction

(2) During Ce final deoxidization, the formed massive Al_2O_3 and some of $MgO \cdot Al_2O_3$ were not stable when Ce entered steel and would be transferred into $CeAlO_3$ and Al–Mg-Ce-O by reaction (15) and (16), which surrounded the original the cores. As a result, the $CeAlO_3$ and Al–Mg-Ce-O inclusions formed, as indicated in Fig. 3c.

$$[Ce] + (Al_2O_3)_{inclusion} = (CeAlO_3)_{inclusion} + [Al] \quad (15)$$

$$m[Ce] + n(MgO \cdot Al_2O_3)inclusion = (xCe_2O_3 \cdot yMgO \cdot zAl_2O_3)_{inclusion} \quad (16)$$

Experiments

Deoxidization experiments were carried out in a vertical electric resistance furnace with $MoSi_2$ heating bars and Al_2O_3 reaction tube to smelt a kind of steel which conforms to the nominal composition. 250 g steel materials (chemical compositions were given in Table 4), Mn and 25 g slag materials (chemical compositions were given in Table 5) were charged into a Magnesia crucible. Then, the magnesia crucible was set in the even temperature zone. Slag was the mixture of reagent grade CaO, SiO_2, Al_2O_3, MgO, FeO, MnO, and CaO was dehydrated and pulverized before mixed. Two ends of the reaction tube were sealed with silicon rubber plugs during experiment. Before heating, the reaction tube was evacuated by flowing Ar gas flow (1.5 L/min) for 10 min, the Ar gas flow was reduced to 1.0 L/min during heating. When temperature in even temperature zone reached 1873 K, the Si–Fe alloys and Al were charged into the melt separately, the Ce was added into the melts to deoxidize eventually. Several samples were picked up when the melts were held for 20 min. After that, crucible with the melt was picked out and cooled in air.

Steel samples were prepared for inclusions analysis. The inclusions in as-cast samples were extracted with the method of non-aqueous solution electrolysis (The electrolysis device was given in Fig. 4). Scanning electron microscope (SEM) with energy dispersive spectrometer (EDS) was used to analyze the chemical composition, three-dimensional morphology and size distribution of the inclusions.

Table 4 Chemical compositions of steel material (mass%)

C	Si	Mn	P	S	O	N
0.08	0.004	0.045	0.007	0.004	0.0168	0.0013

Table 5 Chemical compositions of slag material (mass%)

CaO	SiO_2	Al_2O_3	MgO	FeO	MnO	R
41	34	12	8	1	2	1.2

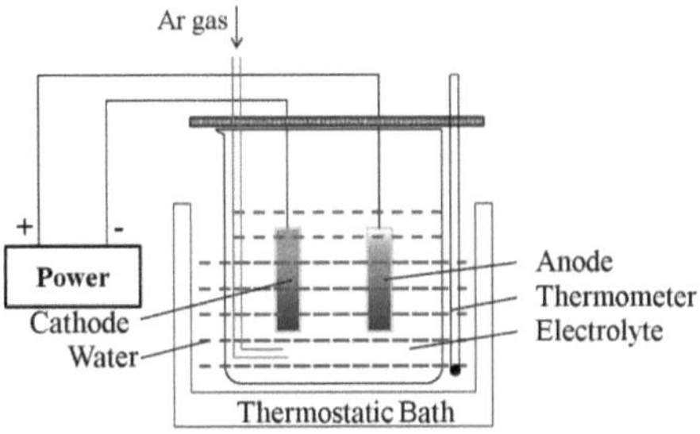

Fig. 4 Electrolysis devices

Results and Discussion

More than 30 images were taken at a magnification of ×3000 by SEM, inclusions larger than 1 μm were counted. The statistical results including the chemical composition, three-dimensional morphology and size distribution were shown in Table 6. The frequency comparison between spinel system and Ce system was given in Fig. 5.

There are mainly five types of detected inclusions, totally three size ranges are classified. Obviously, the Ce system inclusions, such as Al–Mg-Ce-O and $CeAlO_3$, are both spherical in morphology. The mean diameter of that is about 5 μm or even less, which is considered to be beneficial structure for the steel. However, the inclusions, such as spinels and Al_2O_3, are both very large and sharp, besides they are not as flexible as the Ce system inclusions in deformability, so the spinel system inclusions were considered to be harmful to all kinds of steel, because they may dramatically affect the strength and others properties of steel. The three-dimensional

Table 6 Chemical composition, three-dimensional morphology and size distribution of inclusions

Inclusion		Size		
		1–5 μm	5–10 μm	>10 μm
Spinel system	MgO		□	□
	Al_2O_3	△	△△△	△△△△
	$MgO \cdot Al_2O_3$		◇◇◇	◇◇◇◇
Ce system	$CeAlO_3$	○○○○	○○	
	Al–Mg-Ce-O	◎◎◎◎	◎◎	

□: MgO; △: Al_2O_3; ◇: $MgO \cdot Al_2O_3$; ○: $CeAlO_3$; ◎: Al–Mg-Ce-O
The numbers of different shape indicate its relative proportion

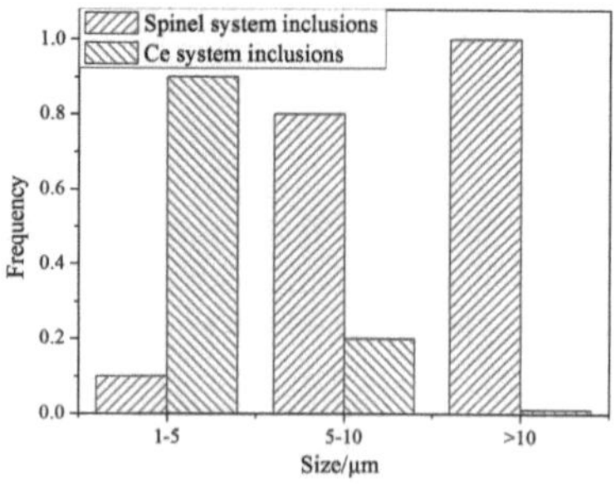

Fig. 5 Frequency comparison between spinel system and Ce system inclusions

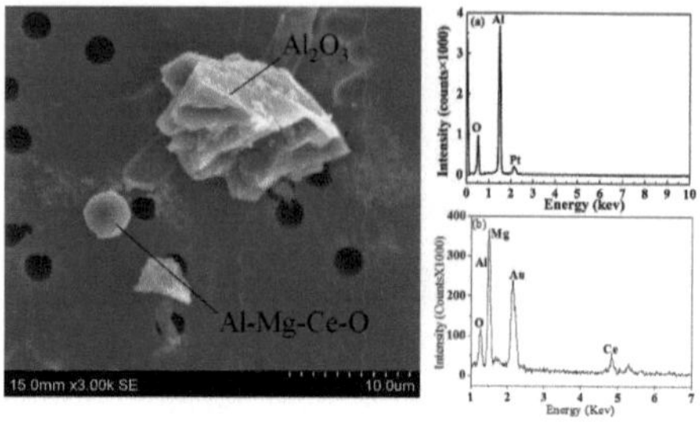

Fig. 6 Typical three-dimensional morphology of spinel system and Ce system inclusions

morphology comparison between the typical Ce system inclusion and spinel system inclusion was shown in the Fig. 6.

The difference between the morphology and size distribution of the two types of inclusions could be explained from the standpoint of the thermodynamic calculation and mechanism analysis. Ce is more active to react with oxygen than Mg and Al. It's easy to capture the oxygen from spinel system inclusion, so large numbers of Al–Mg-Ce-O system inclusions would form after Ce was charged in the melt. The melting point of Ce system inclusions, such as Al–Mg-Ce-O, is lower than that of the spinel system inclusions, such as Al_2O_3, so the former inclusions are spherical rather than sharp as spinels. Meanwhile, the lower melting point property of the Ce system inclusions provides the better chance to be seized by the slag once they grown up, which results in the smaller size of Al–Mg-Ce-O system inclusions in sample than that of spinel system inclusions. If suitable string was put into the melt during the experiment, more inclusions might float into the slag, which is much better for the steel.

The experiment results maintained that the control technique of complex inclusion by rare earth final deoxidization is reasonable and viable, 0.01% dissolved Ce could transfer most of spinel into complex inclusion with lower melting point. The size distribution and morphology of the Ce-system inclusions are also desirable for steel properties. This is just a beginning of the inclusion control technique. Further researches and practices are necessary.

Conclusion

Thermodynamic calculations were carried out to analyze the formation mechanism of inclusion in low carbon steel by rare earth final deoxidization and smelting experiments were practiced in laboratory to coordinate the thermodynamic calculations. Based on results of both calculations and experiments, the following conclusions are drawn:

(1) Al–Mg-Ce-O system inclusions were more stable than spinel system inclusions in Ce final deoxidized steel.
(2) Al–Mg-Ce-O system inclusions are spherical in morphology and have smaller size distribution than spinel system inclusions
(3) The control technique of complex inclusion by rare earth final deoxidization is reasonable and viable.

Acknowledgements This work was supported by 973 Project (No. 2011CB012902). Support by the Instrumental Analysis and Research Center of Shanghai University are gratefully acknowledged.

References

1. C. Bertrand, J. Molinero, S. Landa, R. Elvira, M. Wild, G. Barthold, H. Schifferl, Metallurgy of plastic inclusions to improve fatigue life of engineering steels. Ironmaking Steelmaking **30** (2), 165–169 (2003)
2. H. Suito, R. Inoue, Thermodynamics on control of inclusions composition in ultraclean steels. ISIJ Int. **36**(5), 528–536 (1996)
3. Z.G. Yang, S.X. Li, J.M. Zhang, J.F. Zhang, G.Y. Li, Z.B. Li, Y.Q. Weng, The fatigue behaviors of zero-inclusion and commercial 42CrMo steels in the super-long fatigue life regime. Acta Mater. **52**(18), 5235–5241 (2004)
4. M.D. Chapetti, T. Tagawa, T. Miyata, Ultra-long cycle fatigue of high-strength carbon steels part II: estimation of fatigue limit for failure from internal inclusions. Mater. Sci. Eng., A **356** (1), 236–244 (2003)
5. J.I. Takamura, S. Mizoguchi, Metallurgy of oxides in steels. I. Roles of oxides in steels performance, in *IISC. The Sixth International Iron and Steel Congress,* vol. 1, Oct 1990, pp. 591–597
6. Recommended Values of Equilibrium Constants for the Reactions in Steelmaking, Japan Society for the Promotion of Science, 19th Committee, Tokyo (1984), 255
7. H. Itoh, M. Hino, S. Ban-Ya, Deoxidation equilibrium of magnesium in liquid iron. Tetsu-to-Hagane (J. Iron Steel Inst. Jpn.) **83**(10), 623–628 (1997) (Japan)
8. J.X. Chen, Common chart data for steel making, 550–551 (1984)
9. J.H. Park, H. Todoroki, Control of MgO Al_2O_3 spinel inclusions in stainless steels. ISIJ Int. **50**(10), 1333–1346 (2010)
10. C.R. Li, H. Yang, H. Wen, Thermodynamic analysis of modification treatment of rare earth elements on inclusions in hard wire steel. Hot Working Technol. **39**(8), 12–14 (2010)

Effect of Cooling Rate on Phase Transformation and Microstructure Evolution in a Large Size Forged Ingot of Medium Carbon Low Alloy Steel

Emna Ben Fredj, Hadi Ghasemi Nanesa, Davood Shahriari, Jean-Benoit Morin and Mohammad Jahazi

Abstract In the present study, the influence of cooling rate on the kinetics of phase transformation is investigated in large size forged ingots made of medium carbon low alloy steel. In particular, the volume fraction of martensite after quenching is determined. Dilatometry tests were conducted on Gleeble® 3800 thermomechanical simulator to determine the transformation temperatures and identify the phases after different heat treatment cycles. The heat treatment tests were carried out at 900 °C with a heating rate of 5 °C/s, followed by holding for 1800 s after which the samples were cooled at 3 and 15 °C/s. Optical and electron microscopic analyses, X-Ray Diffraction and micro-hardness measurements were utilized to identify and quantify the volume fraction of martensite and retained austenite for each condition. The calculated volume fraction of martensite using mathematical analysis was compared with those calculated by Koistinen-Marburger equation.

Keywords Dilatometry · Phase transformation · Martensite · Koistinen-Marburger equation

E.B. Fredj (✉) · H.G. Nanesa · D. Shahriari · M. Jahazi
Department of Mechanical Engineering, École de Technologie Supérieure, 1100 rue Notre-Dame Ouest, Montréal, QC H3C 1K3, Canada
e-mail: emna.ben-fredj.1@ens.etsmtl.ca; benfredj.emna@gmail.com

H.G. Nanesa
e-mail: hadi.ghasemi-nanesa.1@ens.etsmtl.ca

D. Shahriari
e-mail: Davood.Shahriari@etsmtl.ca

M. Jahazi
e-mail: Mohammad.Jahazi@etsmtl.ca

J.-B. Morin
FINKL STEEL – SOREL, 100 McCarthy Street, St-Joseph-de-Sorel, QC J3R 3M8, Canada
e-mail: jbmorin@finkl.com

The Minerals, Metals & Materials Society, *TMS 2017 146th Annual Meeting & Exhibition Supplemental Proceedings*, The Minerals, Metals & Materials Series, DOI 10.1007/978-3-319-51493-2_39

Introduction

The utilization of ever large components made by plastic injection moulding is continuously increasing in the transport industry [1]. This issue requires very large steel blocks that are employed as moulds for the plastic injection moulding process. The alloys which are essential for these moulds, have to possess superior mechanical properties as well as good machinability, wear and corrosion resistance [2, 3]. Specifically, homogeneous microstructure and uniform hardness are crucial requirements needed to be considered in the final product [4, 5]. Ingot casting followed by open die forging, quenching and tempering are the main manufacturing steps used to produce the moulds. Among these stages, the heat treatment cycle (quench and temper) is critical as it largely determines the final mechanical properties [6]. However, in the case of large size forged ingots, the large temperature gradients between the surface and the core, lead to important microstructural discrepancies through the working volume of the forged piece. The volume fraction of phases, grain size [7, 8] and precipitations [9] are the main microstructural elements affected by the temperature gradient. Such variations may generate internal stresses, distortions and cracks [6]. The focus of this study will be on the martensitic transformation occurring mainly on the surface region of large size forged blocks during hardening treatment. The kinetics of martensitic transformation will be studied using a numerical method to determine the volume fraction of martensite and retained austenite after quenching.

It is well-known that the transformation of austenite to martensite is associated with a volume change that produces distortion [6, 10, 11]. Many investigations have been reported on martensitic transformation in different steels and models have been proposed predicting the volume fraction of martensite [12–15]. Koistinen and Marburger (K-M) [12] have proposed an empirical equation for the volume fraction of martensite:

$$f_m = 1 - exp\{-b(M_s - T)\} \tag{1}$$

where f_m is the volume fraction of martensite, M_s is the start temperature for martensitic transformation, T is a temperature below M_s and b represents the rate of martensitic transformation varying with the chemical composition of the steel [12]. Koistinen [12] considered a b value of 0.011 for the kinetics of martensitic transformation in carbon steels with different carbon contents and quenching temperatures (24 °C, −24 °C and −79 °C). Seok-Jae [14] evaluated different models for studying the kinetics of athermal martensite transformation and showed that the parameter b depends on chemical composition [16]. Based on Eq. (1), In order to determine f_m at any temperature below M_s, b and M_s must be known.

Yamamoto [17] developed a model to determine these two parameters based on dilatation curve obtained from dilatometry tests. This model is based on the sum of three variations [17, 18]:

- Shrinkage of retained austenite as the temperature decreases, caused by the expansion coefficient.
- Shrinkage of newly formed martensite as temperature decreases.
- Expansion due to austenite-martensite transformation caused by the difference in volume between austenite and martensite.

Thus, Yamamoto's model is described by the following main equations:

$$d_{i+1} = d_i - \left(f_m \alpha_m + f_\gamma \alpha_\gamma\right) \Delta T \tag{2}$$

$$E_{i+1} = d_{i+1} + \varepsilon_{\gamma-m} f_m \tag{3}$$

where i is the ith experimental point, d indicates the shrinkage induced by the expansion coefficient, E is the total dilatation at ith temperature, α_m is the expansion coefficient of martensite, α_γ is the expansion coefficient of austenite, $\varepsilon_{\gamma-m}$ is the strain due to the lattice volume change between austenite and martensite, f_m is the volume fraction of martensite determined by K-M equation, f_γ is the volume fraction of austenite and ΔT is the change of temperature between any two temperatures.

α_m and α_γ can be directly obtained from the dilatometry curves. However, it is not feasible to obtain unique values for b, M_s and $\varepsilon_{\gamma-m}$. This indicates the fact that various f_m may exist for a given material quenched under similar conditions.

Qiu et al. [19] modified the above equation as follows:

$$E = E_0 + \left(\varepsilon_{\gamma-m} - \alpha_m M_s - \frac{(\alpha_\gamma - \alpha_m)}{b}\right) + \alpha_m T + \left(\frac{(\alpha_\gamma - \alpha_m)}{b} - \varepsilon_{\gamma-m}\right) e^{-b(M_s - T)} \tag{4}$$

where E is the total dilatation and E_0 is the dilatation at M_s Temperature. Using the above equation, b, M_s and $\varepsilon_{\gamma-m}$ can be adjusted until accurate fitting of experimental dilatation results.

In the present work, using the modified Yamamoto equation, the volume fraction of martensite and retained austenite will be quantified in a medium carbon low alloy steel. Dilatometry experiments are used for two cooling rates, simulating the cooling conditions in the surface region of a large size forged ingot. Microstructure evolution is studied and the volume fraction of phases are quantified.

Materials and Experimental Procedures

The composition of the medium carbon low-alloy steel used in this study is shown in Table 1.

Table 1 The chemical compositions (*wt%*) of the investigated steel

C	Mn	Si	Ni	Cr	Mo	Cu
0.35	0.99	0.41	0.5	1.86	0.53	0.16

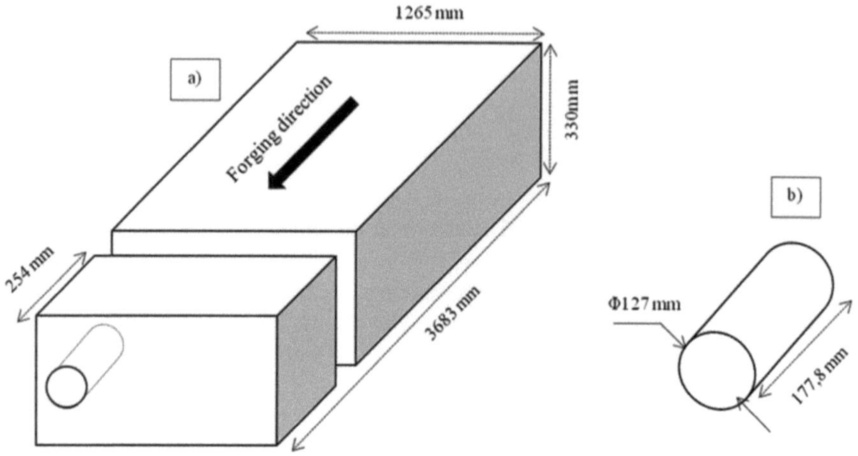

Fig. 1 **a** The position and orientation of the cut block from the forged ingot, **b** Dimensions of the cylindrical block

The ingot (40 tonnes) was forged by open die forging at high temperature (1260 °C) and then air cooled. Parallelepiped block (254 × 330 × 1265 mm) was cut from the ingot. Small bars were then extracted near the surface region. Figure 1 illustrates the position and orientation of each part. The circular block was machined to 5 mm in diameter and 84 mm in length for dilatometry tests.

The ISO-Q dilatometry module of the Gleeble® 3800 thermomechanical simulator was used to carry out the tests. In the Gleeble system, the change in diameter was measured using the C-gauge transducer. To control the temperature, type K thermocouples were welded to the surface of each sample and the temperature variation was controlled within 1 °C. As shown in Fig. 2, the dilatometry samples were heated to 900 °C, at a heating rate of 5 °C/s and then were held for 1800 s. They were then cooled to room temperature using cooling rates of 3 and 15 °C/s. After testing, samples were cut and polished following standard metallographic procedures. Then, they were etched using the following etchant: NaOH (40 g) + H_2O (60 g) + $NaNO_3$ (15 g) [20]. The microstructures of the samples were observed using the confocal microscope Olympus LEXT4100 and for detailed analysis, the scanning electron microscope Hitachi SU-8230 was utilized.

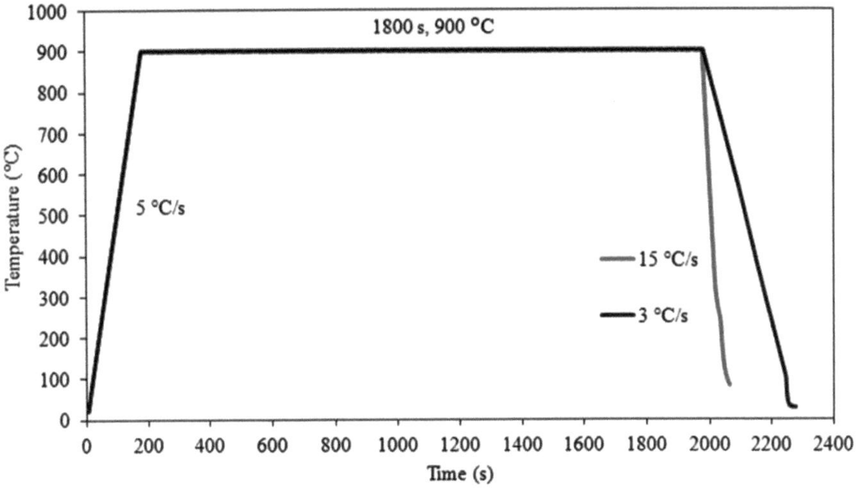

Fig. 2 Heat treatment cycle of dilatometer experiment

X-Ray Diffraction was used to identify and quantify the volume fraction of phases (martensite and retained austenite), on X'Pert3 MRD PANalytical using Cu Kα radiation with a step size of 0.0835. Microhardness tests were conducted according to ASTM E384–11 standard to determine the hardness of the phases.

Results and Discussion

The dilatation curves for the studied steel are shown in Fig. 3 for cooling rates of 3 and 15 °C/s. Both conditions correspond to the formation of martensite. The transformation of martensite occurred at 334 °C and 339 °C for 3 and 15 °C/s, respectively. The linear zone represents the contraction of austenite during the continuous cooling. There is no transformation in this zone.

The second zone corresponds to the expansion of austenite to martensite. It is characterized by a deviation of the thermal expansion linearity. Finally, the third zone contributes to the end of martensitic transformation and is characterized by the contraction of the samples. The curves in Fig. 4 show the comparison between the experimental results and those obtained by K-M equation. Using the least square method [19] the five variables, E0, M_s, α_m, α_γ and $\varepsilon_{\gamma-m}$ in Eq. (4) were determined from the experimental dilatometry tests. The values of these constant for both cooling rates are given in Table 2.

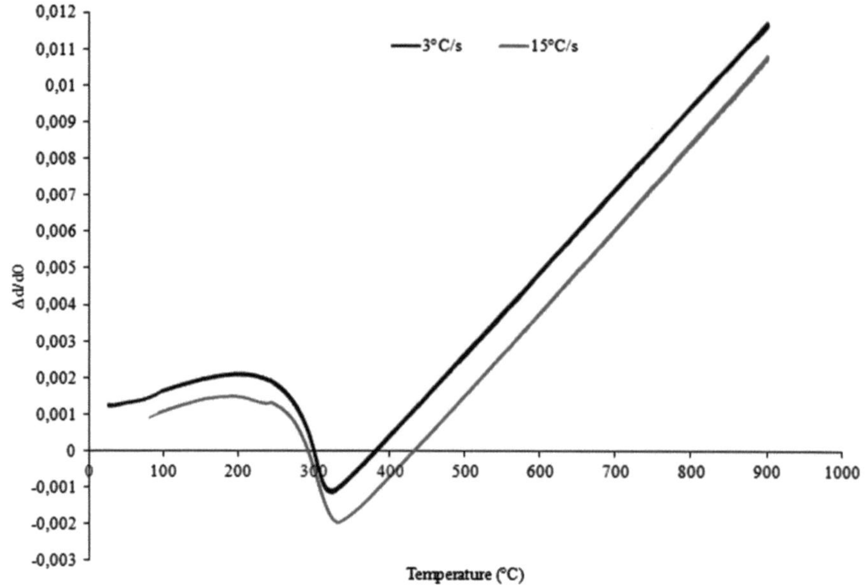

Fig. 3 Dilatometry curves showing change in diameter of samples during cooling at 3 and 15 °C/s

The martensitic start temperature for 3 °C/s as obtained by experiments is 334 °C and predicted to be 314 °C as predicted by the model. For 15 °C/s, M_s is found to be 339 °C experimentally and calculated to be 317 °C.

It can be seen that the constant *b* for 3 °C/s is less than the one for 15 °C/s. It has been reported by several researchers that the constant *b* depends on the chemical composition of the steel [14]. However, the results obtained in the present work indicate that *b* is also a function of the cooling rate; in accordance with those previously obtained by Ghasemi Nanesa et al. [21] who also reported different *b* values as a function of cooling rate in a high carbon tool steel.

Microstructure Examination

Figure 5a, b show the microstructure of the samples after the heat treatment cycle. The presence of a martensitic microstructure can be clearly observed in both conditions. Carbides are not observed in these microstructures as shown in the SEM images presented in Fig. 6a, b. Microhardness results confirmed the presence of martensite with a hardness of 736 HV for 3 °C/s and 729 HV for 15 °C/s.

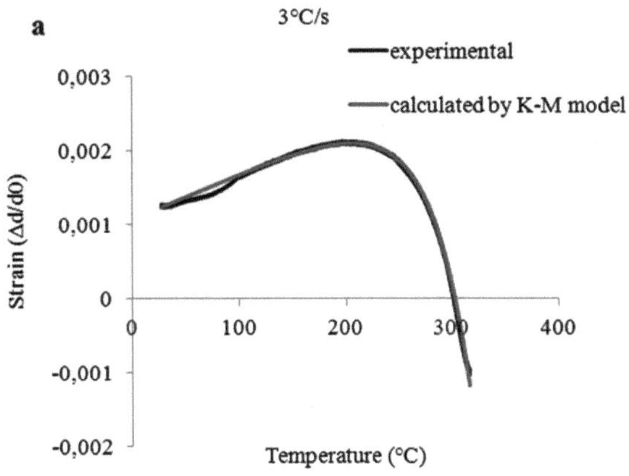

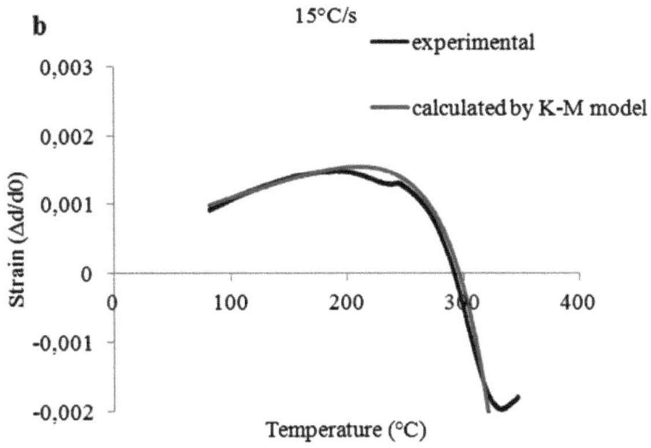

Fig. 4 Comparison of dilatation curves between experimental and calculated by K-M model **a** 3 °C/s, **b** 15 °C/s

Table 2 Data employed for fitting experimental curves

Cooling rate (°C/s)	b (°C^{-1})	E_0	α_m	$\varepsilon_{\gamma-m}$	α_γ	M_s *exp* (°C)	M_s calculated by model (°C)
3	0.026	−0.00099	6 E-06	0.00452	2,0125 E-05	334	314
15	0.027	−0.00158	6 E-06	0.00450	2,0000 E-05	339	317

XRD results did not show the presence of retained austenite in the microstructure (see Fig. 7) indicating that the microstructure of the samples is 100% martensite after cooling to room temperature. The volume fraction of martensite calculated by the K-M model is represented in Fig. 8. The best proposed models which described the martensitic transformation for both cooling rates, 3 and 15 °C/s, can be written as:

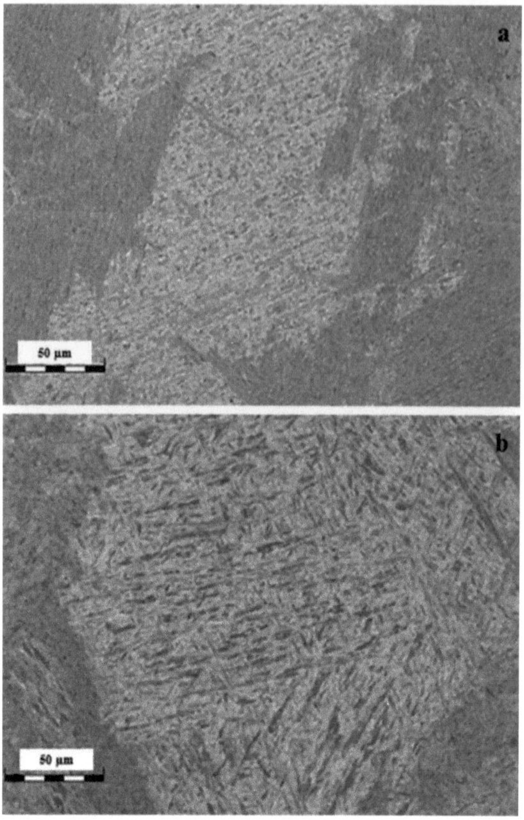

Fig. 5 Optical images of studied steel at various cooling rates: **a** 3 °C/s, **b** 15 °C/s

$$f_{m(3°C/s)} = 1 - exp\{-0.026(314 - T)\} \quad (5)$$

$$f_{m(15°C/s)} = 1 - exp\{-0.027(317 - T)\} \quad (6)$$

The proposed models (Eqs. 5–6) predict 99.94 and 99.96% of martensite for 3 and 15 °C/s, respectively. Thus, the fraction of the retained austenite will be 0.06 and 0.04%. The low volume fraction of retained austenite for both cooling rates, calculated by the K-M model, could be related to calculation errors when compared with the XRD results.

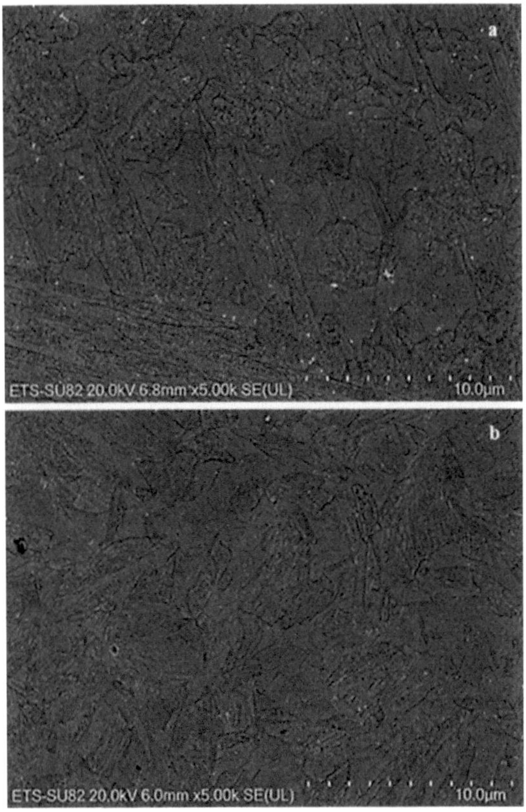

Fig. 6 SEM images of studied steel at cooling rates of: **a** 3 °C/s, **b** 15 °C/s. The microstructures show the presence of fully martensitic structure

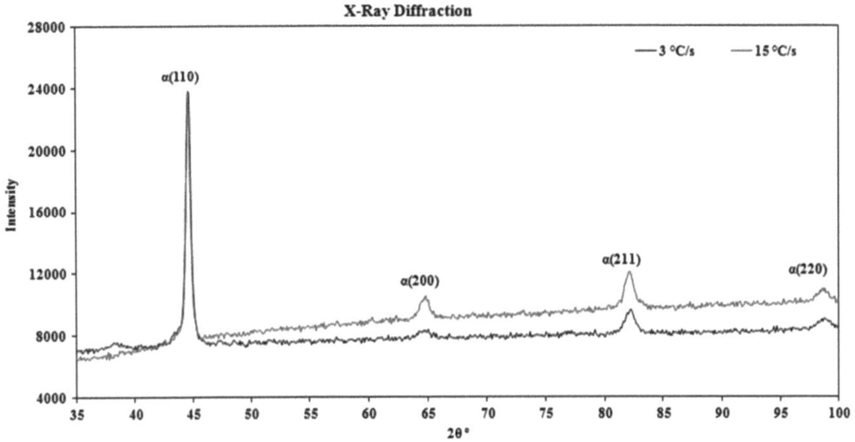

Fig. 7 X-Ray Diffraction for 3 and 15 °C/s

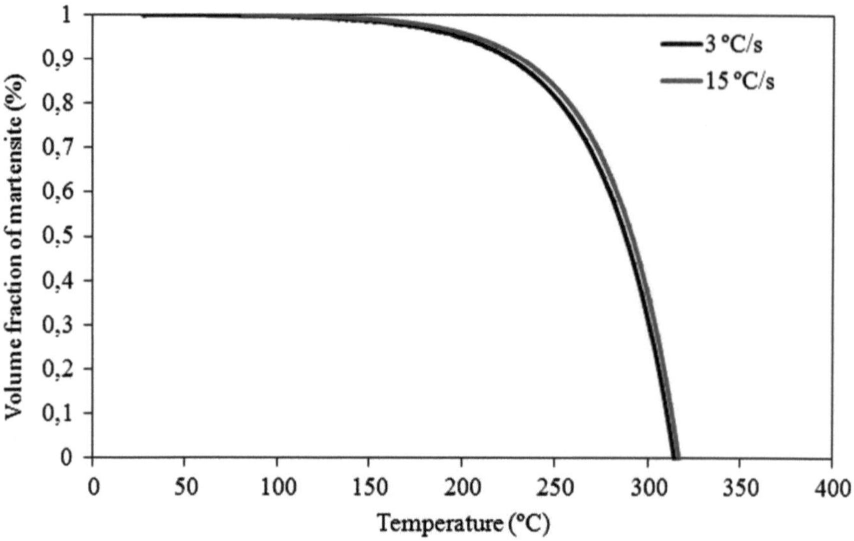

Fig. 8 Volume fraction of martensite for 3 and 15 °C/s

Conclusions

The effect of cooling rate on martensitic transformation of a medium carbon low alloy steel was investigated in this study and the following conclusions can be drawn:

1. Kinetics study using dilatometry on Gleeble® 3800 showed only the presence of martensite for 3 and 15 °C/s cooling rates.
2. The comparison between experimental results with those calculated by K-M model did not show any difference on the volume fraction of martensite.
3. The best fits of the dilatation curves allowed determination of the constant *b* for both conditions. This value is 0.026 for 3 °C/s and 0.027 for 15 °C/s, indicating that *b* increases slightly with increasing the cooling rate.

Acknowledgements The authors gratefully acknowledge the financial support of FRQNT Grant number FR08218 and MITACS Grant number IT03151. The authors also appreciate the collaboration of Finkl Steel Group (Sorel Forge), for the technical support in this study.

References

1. Injection Molded Plastic Market Analysis By Raw Material (Polypropylene, ABS, HDPE, Polystyrene), By Application (Packaging, Consumables & Electronics, Automotive, Building & Construction, Medical) And Segment Forecasts To 2022 (2016), http://www.grandviewresearch.com/industry-analysis/injection-molded-plastics-market
2. D.D. Wegman, E.A. Wanner, W.P. Rehrer, S. Widge, Heat, corrosion, and wear resistant steel alloy and article, Google Patents (1990)
3. B. Mordike, T. Ebert, Magnesium: properties—applications—potential. Mater. Sci. Eng. A **302**(1), 37–45 (2001)
4. F. Beaudet, C. Blais, H. Lehuy, B. Voyzelle, G. L'espérance, J.-P. Masse, M. Krishnadev, Improvement of hardenability and static mechanical properties of P20 + 0.5 Ni mold steel through additions of vanadium and boron. ISIJ Int. **52**(3), 424–433 (2012)
5. K. Chadha, D. Shahriari, M. Jahazi, Constitutive modelling of ingot breakdown process of low alloy steels. La Metallurgia Italiana—Int. J. Ital. Assoc. Metall. **4**, 5–12 (2016)
6. L.-A. Steels, Defects and Distortion in Heat-Treated Parts (1991)
7. S. Serajzadeh, Modelling of temperature history and phase transformations during cooling of steel. J. Mater. Process. Technol. **146**(3), 311–317 (2004)
8. S. Sista, Z. Yang, T. DebRoy, Three-dimensional Monte Carlo simulation of grain growth in the heat-affected zone of a 2.25 Cr-1Mo steel weld. Metall. Mater. Trans. B **31**(3), 529–536 (2000)
9. C. Maidorn, D. Blind, Solidification and segregation in heavy forging ingots. Nucl. Eng. Des. **84**(2), 285–296 (1985)
10. C. Cayron, One-step model of the face-centred-cubic to body-centred-cubic martensitic transformation. Acta Crystallogr. Sect. A: Found. Crystallogr. **69**(5), 498–509 (2013)
11. D. Deng, FEM prediction of welding residual stress and distortion in carbon steel considering phase transformation effects. Mater. Des. **30**(2), 359–366 (2009)
12. D. Koistinen, R. Marburger, A general equation prescribing the extent of the austenite-martensite transformation in pure iron-carbon alloys and plain carbon steels. Acta Metall. **7**(1), 59–60 (1959)
13. F. Huyan, P. Hedström, A. Borgenstam, Modelling of the fraction of martensite in low-alloy steels. Mater. Today: Proc. **2**, S561–S564 (2015)
14. S.J. Lee, Evaluation of empirical kinetics models of athermal martensite transformation in plain carbon and low alloy steels, in *Advanced Materials Research* (Trans Tech Publications, 2013), pp. 39–44
15. H.K. Bhadeshia, Multiple, simultaneous, martensitic transformations: implications on transformation texture intensities, in *Materials Science Forum* (Trans Tech Publications, 2013), pp. 9–13
16. S. Van Bohemen, J. Sietsma, Effect of composition on kinetics of athermal martensite formation in plain carbon steels. Mater. Sci. Technol. **25**(8), 1009–1012 (2009)
17. J. Yamamoto, S. Meguro, Y. Muramatsu, N. Hayakawa, K. Hiraoka, Analysis of residual stress reduction by using Low transformation-temperature materials. QJ JWS **25**, 560–568 (2007)
18. J. Yamamoto, S. Meguro, Y. Muramatsu, N. Hayakawa, K. Hiraoka, Analysis of martensite transformation behaviour in welded joints of low transformation-temperature materials. Weld. Int. **23**(6), 411–421 (2009)
19. H. Qiu, J. Qi, F. Yin, K. Hiraoka, Determination of parameters for fitting the dilatation curve of austenite-martensite transformation in Cr-Ni steels. ISIJ Int. **49**(1), 146–148 (2009)
20. M. Goune, O. Bouaziz, S. Allain, K. Zhu, M. Takahashi, Kinetics of bainite transformation in heterogeneous microstructures. Mater. Lett. **67**(1), 187–189 (2012)
21. H.G. Nanesa, M. Jahazi, R. Naraghi, Martensitic transformation in AISI D2 tool steel during continuous cooling to 173 K. J. Mater. Sci. **50**(17), 5758–5768 (2015)

Formation and Control of CaS Inclusion in Gear Steel 20MnCr5

Jie Xu, Jianxun Fu, Yanxin Wu and Xu Li

Abstract The modification of $(CaO)m(Al_2O_3)n$ inclusion and the fact of CaS inclusion formation in gear steel 20MnCr5 after calcium treatment have been studied by thermodynamic equilibrium calculation. Based on the thermodynamic equilibrium, when the liquid steel containing 0.035% [Al], the mass fraction of sulfur is lower than 0.00628% and the ratio of w(Ca)/w(Al) is higher than 0.483, the lower melting point $12(CaO)\cdot 7(Al_2O_3)$ inclusion will form in the molten steel. Calculated results show that with calcium treatment of liquid steel the stable CaS forms easily and the aluminum deoxidizing product Al_2O_3 is easily modified to low-melting calcium aluminate $12(CaO)\cdot 7(Al_2O_3)$. In order to completely modify the Al_2O_3 to low-melting calcium aluminate $(CaO)\cdot(Al_2O_3)$ and $12(CaO)\cdot 7(Al_2O_3)$, the calcium treatment of liquid steel should carry out in conditions of low [S] and ≥ 1600 °C, and the sulfur wire feeding operation after soft argon stirring in refining process, it is available to obviously decrease the occurrence of nozzle blocking.

Keywords 20MnCr5 · Calcium treatment · Thermodynamics · Nozzle clogging

Introduction

Calcium treatment of the oxide and sulfide inclusions to improve the quality of steel has been widely used in the steelmaking process. As we all know, calcium treatment can modify the solid Al_2O_3 inclusions to liquid calcium-aluminate to avoid

J. Xu · J. Fu (✉) · Y. Wu · X. Li
State Key Laboratory of Advanced Special Steel, Shanghai Key Laboratory of Advanced Ferrometallurgy, School of Materials Science and Engineering, Shanghai University, Shanghai, China
e-mail: fujianxun@shu.edu.cn

J. Xu
e-mail: xiabaxujie@163.com

The Minerals, Metals & Materials Society, *TMS 2017 146th Annual Meeting & Exhibition Supplemental Proceedings*, The Minerals, Metals & Materials Series, DOI 10.1007/978-3-319-51493-2_40

nozzle clogging of aluminum killed steel in continuous casting, and reduce the harm of Al_2O_3 inclusions to the steel matrix. When molten steel containing low-level of sulfur, calcium treatment can also improve the machinability of steel. But for the higher sulfur content steel, if the control of calcium treatment process is not proper, it can not well modify the oxide and sulfide inclusions, and generate a higher melting point CaS inclusions on the contrary, thus resulting in nozzle clogging and making casting difficult.

Some studies show that complex inclusions that generated by calcium sulfide and oxide inclusions are formed in the production process [1]. There are also other studies show that complex inclusions are formed in the solidification process [2]. And some studies believe that the sulfur reacts with calcium-aluminate will generate CaS [3].

MnCr series gear steel is still one of the leading steel of gear steel in China, and sulfur content in the 20MnCr5 is higher in order to improve the cutting property. Because of its high market demand, the production have a larger proportion in the gear steel. To improve the mechanical properties and surface quality of this steel, and to avoid nozzle clogging in the casting process, it usually feed calcium at end of refining period to modified the oxide inclusions after the addition of calcium and control the sulfide formation in the steel. However, this kind of steel is often encountered nozzle clogging problem in the process of continuous casting. In this paper, the thermodynamic calculation of sulfur-containing aluminum killed steel is specific to 20MnCr5 gear steel, and the effects of other elements in steel on the activity of [Al], [S] and [Ca] are considered. The corresponding [Al]-[S], [Al]-[Ca] and [Ca]-[S] diagrams were calculated to determine the inclusion composition of molten steel after calcium treatment, in order to improve the refining process and avoid the clogging of the nozzle.

Thermodynamic Calculation and Analysis

Calculation Basis

20MnCr5 gear steel is one kind of steel that to increase sulfur content in steel to improve the mechanical properties, the chemical composition of the national standard is shown in Table 1. The thermodynamic calculation of this paper is based on the composition of the elements in Table 2. The interaction coefficients of [Al], [S] and [Ca] in liquid steel are shown in Table 3, and the activity of oxide inclusions in $(CaO)m(Al_2O_3)n$ are shown in Table 4.

Table 1 Chemical composition of 20MnCr5 gear steel

Steel	C	Si	Mn	P	S	Cr	Al	[O]
20MnCr5	0.17–0.23	<0.12	1.1–1.5	<0.035	0.020–0.035	1.0–1.3	0.020–0.055	<20

Table 2 The composition of liquid steel used in this paper

Steel	C	Si	Mn	P	S	Cr	Al	[O]	Ca
20MnCr5	0.20	0.10	1.2	0.015	0.025	1.15	0.035	20	0.0010

Table 3 Interaction coefficients of element at 1873 K

e_i^j	C	Si	Mn	P	S	Cr	Al	O	Ca
S	0.11	0.063	−0.026	0.29	−0.028	−0.011	0.035	−0.27	−100
Al	0.091	0.0056	0.012	0.05	0.03	0.0025	0.45	−6.698	−0.047
O	−0.45	−0.131	−0.021	0.07	−0.133	−0.04	−3.9	−0.2003	0
Ca	−0.34	−0.097	0	0	−125	0	−0.072	−678	−0.002

Table 4 Activity of CaO and Al2O3 in $(CaO)m(Al_2O_3)n$ melt at 1873 K

Items	$3(CaO)\cdot(Al_2O_3)$		12C7A		CA	
	a_{CaO}	a_{Al2O3}	a_{CaO}	a_{Al2O3}	a_{CaO}	a_{Al2O3}
Ye (KTH Calculation model) [4]	1	0.0065	0.53	0.027	0.085	0.30
Rein/Chipman (Experimental measurements) [5]	1	0.0050	0.34	0.064	0.110	0.30
Fujisawa (Experimental measurements) [6]	1	0.0100	0.53	0.027	0.050	0.61
Korousic (Calculation model) [7]	1	0.0050	0.36	0.038	0.050	0.50
Sun (model + measuremets) [8]	1	0.0038	0.36	0.053	0.074	0.43
Wei (Bjorkvall model) [9]	1	0.0041	0.37	0.036	0.110	0.18
Average value	1	0.0057	0.42	0.041	0.080	0.39

The Formation and Control of Calcium Aluminate

According to the liquidus temperature and solidus temperature of tested steel were calculated, respectively 1785.1 K and 1711.1 K. For the 20MnCr5 gear steel, there are the following reactions in the liquid steel after the calcium treatment.

$$2[Al] + 3[O] = Al_2O_3(s) \quad \Delta G^\theta = -1225417 + 393.8T\,J/mol \tag{1}$$

$$[Ca] + [O] = CaO(s) \quad \Delta G^\theta = -629998 + 144.75T\,J/mol \tag{2}$$

Formula (1) subtract formula (2):

$$2[Al] + 3CaO(s) = Al_2O_3(s) + 3[Ca] \quad \Delta G^\theta = 664577 - 40.45T\ J/mol \tag{3}$$

$$\lg f_i = \sum_j e_i^j[j] \tag{4}$$

The activity coefficients of each element in liquid steel are calculated from the data of Tables 2 and 3. The activity coefficients of the elements are shown in Table 5.

According to the formula (3) and the Henry interaction coefficients in Table 5 at 1873 K temperature, it can be obtained that the effect of [Als]-[Ca] balance and w(Ca)/w(Al) to the modification of inclusion, as shown in Figs. 1, 2 and 3.

According to Fig. 1, Fig. 2 shows that with the increase of [Als] content in steel, the content of calcium to formation the low melting point inclusions increase, but the w(Ca)/w(Al) ratio reduce. Under the condition of the actual w(Als) of the tested steel is 0.0035%, when the concentration of w(Ca) is 0.0015%, the Al_2O_3 inclusion can be modified to $(CaO) \cdot (Al_2O_3)$. To completely modify the high melting point

Table 5 Henry activity coefficient f_i(1873 K)

Elements	Al	Ca	S
f_i	1.096	2.75E-05	0.774

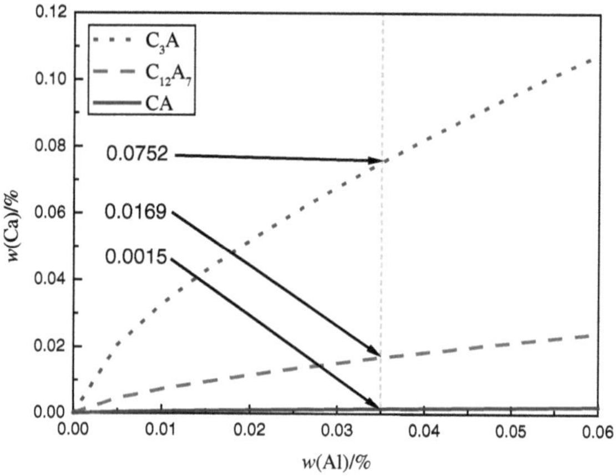

Fig. 1 [Als]-[Ca] equilibrium diagram of different inclusions under 1873 K

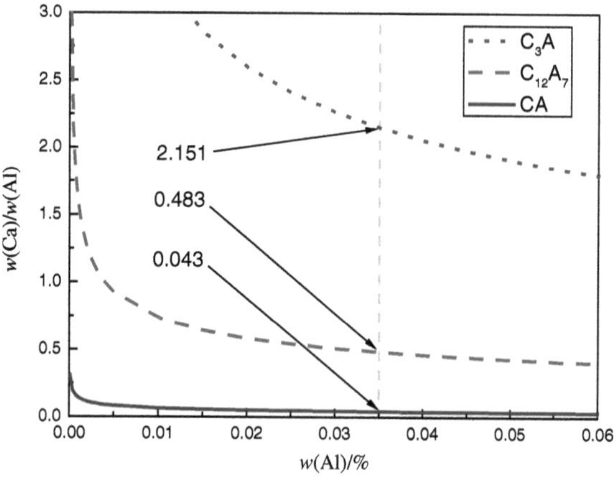

Fig. 2 Relationship between w(Ca)/w(Al) and inclusion composition in liquid steel

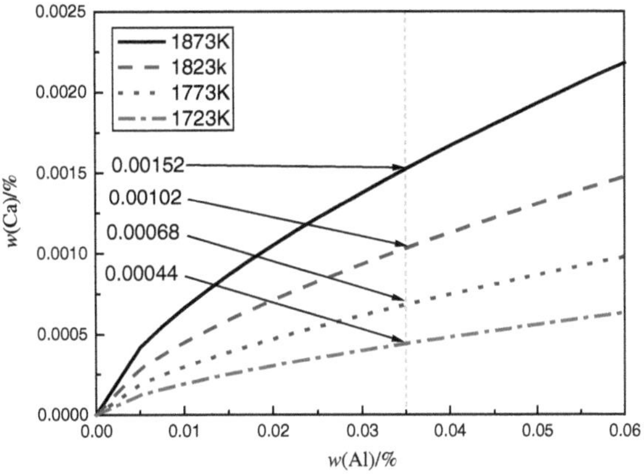

Fig. 3 The [Al]-[Ca] equilibrium relationships of the formation of (CaO)·(Al2O3) at different temperatures

Al_2O_3 inclusion to the low melting point inclusions $12(CaO)\cdot 7(Al_2O_3)$, it should be controlled in the steel w(Ca)/w(Al) > 0.483, or w(Ca) > 0.0169%, but it is difficult to achieve the actual production. When w(Als) is 0.0035%, the scope of the calcium required to make the liquid inclusion is 0.0015%–0.0169%. Therefore, the molten steel refining temperature should be greater than 1873 K, and it should be ensured enough time to float, then the inclusion can be removed.

As it can be seen from Fig. 3, in the case of a certain [Als] content, the calcium content that modify the Al_2O_3 inclusion completely to $(CaO)\cdot(Al_2O_3)$ with the reduction of temperature gradually decrease.

Formation and Control of CaS

The melting point of CaS is 2500 °C, it is a solid inclusions in liquid steel, it will cause the nozzle clogging in the continuous casting process. The reaction of [Ca] and [S] in 20MnCr5 steel generates CaS according to the following equation:

$$[Ca] + [S] = CaS(s) \quad \Delta G^{\theta} = -542531 + 124.15T\ J/mol \tag{5}$$

The modification process of Al_2O_3 inclusions can be simply understood as a process of [O] and [S] striving for [Ca], in which [O] is usually determined by [Al], so the degree of Al_2O_3 modification can be directly estimated by the following formula:

$$3CaO + 2[Al] + 3[S] = 3CaS(s) + Al_2O_3(s) \quad \Delta G^{\theta} = -963016 + 332T\ J/mol \tag{6}$$

The solubility of CaS in calcium aluminate in the steelmaking temperature is very small, the highest content can reach up to 5%. When chemical reactions of [Ca] and (Al_2O_3) occur, even if a very small amount of [Ca] is involved in the desulfurization reaction, the generated CaS is very easy to reach saturation. So this article in the thermodynamic calculation, take $a_{CaS} = 1$. The formula (5) can be obtained in Fig. 4.

As it can be seen from Fig. 4, when the steel liquid w(S) is 0.025%, the dissolved calcium need to reach 0.00425% to precipitate CaS inclusion in steelmaking temperatures. So it does not generate CaS inclusions over 1873 K. With the decrease of temperature, the content of dissolved calcium is gradually reduced, so the CaS inclusion is more easy to precipitate during the solidification of molten steel. Because the activity of calcium or dissolved calcium content in the steel is not easy to measure, it can also be based on the [Als] and sulfur content in the steel to determine the precipitation of CaS inclusion. According to the reaction formula (6), the effects of [Als]-[S] equilibrium and temperature on [Als]-[S] equilibrium in steel with different inclusions can be obtained (vide Figs. 5 and 6).

As it can be seen from Fig. 5, when steel w (Als) = 0.035% at a temperature of 1873 K, the w(S) will be required to achieve 0.00628% to form CaS on the surface

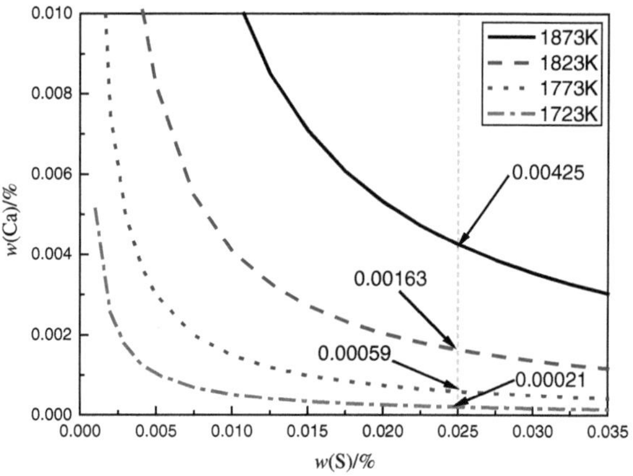

Fig. 4 The equilibrium relationship of [Ca]-[S] to form CaS at different temperatures

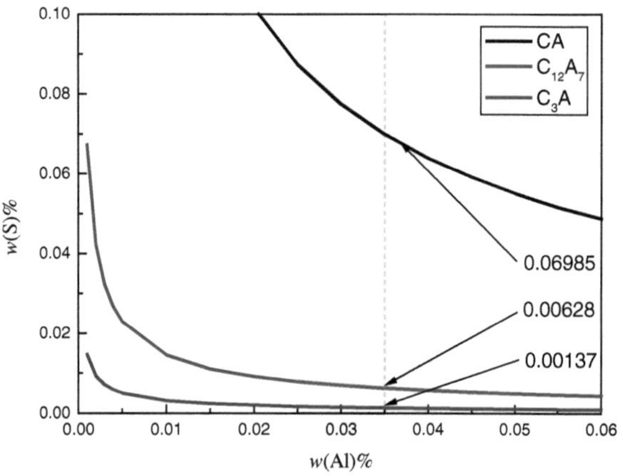

Fig. 5 [Als]-[S] equilibrium relationships of different inclusions at 1873 K temperature

Fig. 6 [Als]-[S] equilibrium relationship of 12(CaO) · 7 (Al2O3) at different temperatures

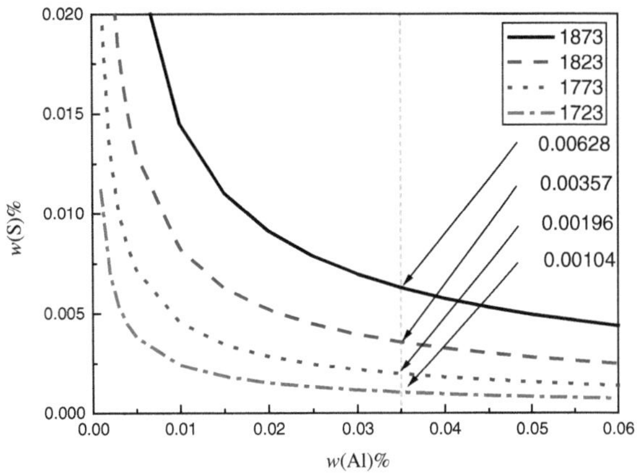

of $12(CaO) \cdot 7(Al_2O_3)$, on the surface of $(CaO) \cdot (Al_2O_3)$ and $3CaO) \cdot (Al_2O_3)$ to precipitate CaS, the w(S) will be required to be respectively 0.06985% and 0.00137%. Therefore, from the point of view of [Als]-[S] equilibrium, it is proved that the CaS is not likely to precipitate on the surface of $12(CaO) \cdot 7(Al_2O_3)$ inclusions when the temperature is higher than 1873 K. In addition, the higher the content of sulfur, the narrower the range of low melting point inclusions. This is due to calcium treatment at higher sulfur content in steel, the dissolution of calcium firstly react with sulfur to produce CaS inclusions wrapped around the Al_2O_3 periphery, hindering the process of modification. At 1873 K temperature, when the w (Als) = 0.035%, it is difficult to modify Al_2O_3 inclusion to $12(CaO) \cdot 7(Al_2O_3)$ inclusions, but easily to become $(CaO) \cdot (Al_2O_3)$ as long as the w(S) > 0.00628%. The melting point of $(CaO) \cdot (Al_2O_3)$ is 1605 degrees Celsius, if the molten steel temperature is above this value, $(CaO) \cdot (Al_2O_3)$ inclusion is liquid that can be gathered in molten steel and then floating up. As it can be seen from Fig. 6, if the calcium aluminate inclusions have not been fully floating removal in molten steel at high temperature, the range of low melting point $12(CaO) \cdot 7(Al_2O_3)$ inclusion becomes narrower with the decrease of temperature during the solidification process of molten steel. The $(CaO) \cdot (Al_2O_3)$ inclusion will be formed in the form of solid inclusions that will cause the nozzle clogging during continuous casting and that is similar to the brittle inclusions which can reduce the quality of steel.

Recommendation of Process Improvement

In summary, when the sulfur content in liquid steel is high (>0.00628%), it is difficult to modify Al_2O_3 inclusion to $12(CaO) \cdot 7(Al_2O_3)$ inclusions. The calcium treatment can not only achieve the expected effect, but bring the problem to the continuous casting process in the high sulfur steel, because the CaS and not fully modified calcium aluminate ($(CaO) \cdot (Al_2O_3)$) will plug the nozzle.

For 20MnCr5 gear steel [S] is about 0.025%, it is difficult to achieve the expected goal if the content of [Al] and [S] is high in steel for calcium treatment according to the theoretical calculation. Therefore, calcium treatment should be carried out when the sulfur content is low in liquid steel. Under the condition that the refining slag desulphurization ability is better and the deoxidation of liquid steel is sufficient, sulfur in molten steel is easily reduced to below 0.01%, or even lower. In this case calcium treatment can easily modify the Al_2O_3 inclusions to the liquid calcium aluminate. Inclusions will aggregate and float after the full soft stirring while the molten steel eliminate residual calcium. The amount of CaS generated by the reaction should be much less than that feeding sulfur line before the calcium treatment process. At the same time, the CaS will generate in liquid calcium aluminate as the core of precipitation, and CaS can also be melted into liquid calcium aluminate, thus it will greatly reduce the degree of nozzle clogging, and then feeding sulfur line to make the content of sulfur in molten steel adjusted to the required level.

Conclusion

(1) From thermodynamic analysis: To completely modify the high melting point Al_2O_3 inclusion to the low melting point inclusions $12(CaO)\cdot 7(Al_2O_3)$, it should be controlled in the steel w(Ca)/w(Al) > 0.483, or w(Ca) > 0.0169%, but it is difficult to achieve the actual production.

(2) CaS inclusion can not spontaneously form under the condition of this chemical composition of liquid steel at steelmaking temperature; Before the calcium treatment, the w(S) should be controlled under 0.00628% so it can maximize the effect on the modification of the Al_2O_3 inclusion.

(3) When the acid soluble aluminum content is 0.035%, the range of the calcium that required to modify the Al_2O_3 inclusion is 0.0015%–0.0169%. The calcium treatment can not only achieve the expected effect, but bring the problem to the continuous casting process in the high sulfur steel.

(4) In the refining process, calcium treatment is carried out after aluminium-deoxidizing at the temperature of molten steel is greater than 1873 K. In this case calcium treatment can easily modify the Al_2O_3 inclusions to the liquid calcium aluminate. Then inclusions will aggregate and float after the full soft stirring. This will greatly reduce the probability of nozzle clogging.

Acknowledgements This work is supported by the Foundation of State Key Laboratory of Development and Application Technology of Automotive Steels (Baosteel Group), National Natural Science Foundation of China (Nos. 51671124, 51474142) and the Program for Professor of Special Appointment (Eastern Scholar) at Shanghai Institutions of Higher Learning (2012).

References

1. S.R. Story, T.J. Piccone, R.J. Fruehan et al., Inclusion analysis to predict casting behavior. Iron Steel Technol. **1**(9), 163–169 (2004)
2. L. Holappa, M. Hämäläinen, M. Liukkonen et al., Thermodynamic examination of inclusion modification and precipitation from calcium treatment to solidified steel. Ironmaking Steelmaking **30**(2), 111–115 (2013)
3. S.K. Choudhary, A. Ghosh, Thermodynamic evaluation of formation of oxide-sulfide duplex inclusions in steel. ISIJ Int. **48**(11), 1552–1559 (2008)
4. G. Ye, P. Jonsson, T. Lund, Thermodynamics and kinetics of the modification of Al2O3 inclusions. ISIJ Int. **36**(Suppl), S105–S108 (1996)
5. B. Song, Q. Han, Equilibrium of calcium vapor with liquid iron and the interaction of third elements. Metall. Mater. Trans. B **29**(2), 415–420 (1998)
6. T. Fujisawa, Inclusion modification in Al-killed steel, in *Proceedings of the Sixth International Iron and Steel Congress* (1990), pp. 201–205
7. V. Presern, B. Korousic, J.W. Hastie, Thermodynamic conditions for inclusions modification in calcium treated steel. Steel Res. **62**(7), 289–295 (1991)
8. Z.Q. Sun, Y.C. Che, M.F. Jiang et al., Equilibrium diagram for Fe-Al-Ca-O-S system. J. Iron Steel Res. (2005)
9. J. Wei, G. Yan, Z. Tian et al., Improvement of CSP castability in low carbon Al-killed steel. J. Univ. Sci. Technol. Beijing (2005)

Investigations on the Mechanical Deformation of Amorphous Alloy Nanowires Using Phase-Field Modeling and Thermodynamics Avalanche Models

Guangping Zheng

Abstract The relations between the mechanical properties of amorphous alloy nanowires and their intrinsic deformation defects are elucidated. In particular, the size effects on the deformation mechanisms which are caused by the density of deformation defects in the interiors of nanowires are quantitatively determined by phase-field modeling. The simulation provides deep insights into the emergence of ductility in amorphous alloy nanowires. The fractures of the amorphous alloy nanowires are found to be well characterized by an avalanche model with some features of self-organized criticality.

Keywords Amorphous alloys • Size effects • Phase-field modeling • Fracture • Avalanche model

Introduction

Although amorphous alloy nanowires have been well studied by experiments and simulations [1–5], the size effects on mechanical properties especially on their tensile ductility have not been well characterized and fully understood. The plastic deformation of amorphous alloys in the microscopic scale is essentially associated with deformation defects created under the accumulation of local shear stresses. These local mechanistic events can be divided into homogenous deformation and inhomogeneous deformation, depending upon the temperatures, applied strain rates and compositions of the glassy alloys [6]. In order to understand the size effects on the ductility of amorphous alloy nanowires, the origins of their ductility, i.e., either the homogenous deformation or the multiple events of inhomogeneous deformation, have to be determined.

G. Zheng (✉)
Department of Mechanical Engineering, Hong Kong Polytechnic University, Kowloon, Hong Kong, China
e-mail: mmzheng@polyu.edu.hk

The Minerals, Metals & Materials Society, *TMS 2017 146th Annual Meeting & Exhibition Supplemental Proceedings*, The Minerals, Metals & Materials Series, DOI 10.1007/978-3-319-51493-2_41

In this work, size effects on the mechanical behaviors of amorphous alloys under tensile deformation are investigated by phase-field modeling. The emergence of ductility in amorphous alloy nanowires is elucidated in microscopic and mesoscopic scales (10–800 nm). The relations between the mechanical properties and the density of deformation defects in the nanowire interiors and the diameters of nanowires are quantitatively determined.

Modeling of the Plastic Deformation of Amorphous Alloy Nanowires

Shear localization or shear band formation is generally believed to be directly related with strain softening. When strain increases, the shear modulus of glassy solid would be reduced, resulting in the localized deformation in higher strain rates due to the decrease of viscosity of the glassy solid. Possible mechanisms for this phenomenon, including the local production of free volumes due to atomic dilatations, local evolution of structural disorder due to the operations of shear transformation zones (STZ), redistribution of internal stresses associated with STZ operations, and local heating, have been elucidated by experiments. These mechanisms have all considered that the strain softening is accompanied by local structural changes of the amorphous alloys. Therefore the free-energy functional of glassy alloy under external loading can be expressed in terms of a local free-energy density f(ε_{ij}, ρ) that is a function of the defect field $\rho(r)$ at the local position $\boldsymbol{r}$ only, the energy for deviations from spatial uniformity of defect distribution, and the kinetic energy associated with the local displacement field $\boldsymbol{u(r)}$, as follows [7]:

$$F = \int \{f(\rho, \varepsilon_{ij}) + \frac{\rho_0}{2}[\dot{\vec{u}}]^2 + \frac{\kappa}{2}|\nabla\rho|^2\}dV, \tag{1}$$

where ρ is the normalized local density of deformation defects and ρ_0 is the mass density of the glassy alloy; $\boldsymbol{\varepsilon}_{ij}$ is the local strain tensor; κ is the 'interfacial' energy between the regions with different densities of deformation defects. When a crack appears, it becomes the surface energy. The local free-energy density of a glassy alloy can be described as,

$$f(\rho, \varepsilon_{ij}) = e[\varepsilon_{ij}] + \frac{a_0}{2}\rho^2 + \frac{b_0}{3}\rho^3 + \frac{c_0}{4}\rho^4 + (\frac{a_1}{2}\rho^2 + \frac{b_1}{3}\rho^3)\{e[\varepsilon_{ij}] - e_0\}. \tag{2}$$

The coefficients a_0, b_0, c_0 depend on the energy landscape of deformation defects, independent of $\boldsymbol{\varepsilon}_{ij}$. e is the local strain energy and e_0 is the strain energy at the elastic limit when the applied stress σ is close to the fracture strength σ_F. Based on Eq. (2), one could define an energy barrier between the un-activated ($\rho = 0$) and activated ($\rho \sim 1$) states of deformation defects in the un-deformed glassy alloys.

When ρ increases to a critical value $\rho_s = 0.8$ under plastic deformation, deformation defects are activated.

Based on the Ginzburg-Landau formulism, the shear band propagation and evolution are simulated by solving the governing equations of $\rho(\mathbf{r})$ and $\mathbf{u}$ = [u v w], as follows [7]

$$\rho_0 \frac{\partial^2 \vec{u}}{\partial t^2} = \mu \nabla \cdot [(1 + \frac{a_1}{2}\rho^2 + \frac{b_1}{3}\rho^3)\nabla \vec{u}], \quad (3)$$

and

$$\tau_\rho \frac{\partial \rho}{\partial t} = \kappa \nabla^2 \rho - (a_0\rho + b_0\rho^2 + c_0\rho^3) - \rho(a_1 + b_1\rho)\{e[\varepsilon_{ij}] - e_0\}, \quad (4)$$

$$\varepsilon_{ij} = \frac{1}{2}(\frac{\partial u_i}{\partial x_j} + \frac{\partial u_j}{\partial x_i}), \quad (5)$$

where i, j = 1, 2, 3 and τ_ρ is the characteristic time for the activation of deformation defect, μ is the shear modulus of a bulk sample. The Vitrelloy-1 ($Zr_{41}Ti_{14}Ni_{10}$-$Cu_{12.5}Be_{22.5}$) is chosen as the material in the modeling. The definitions of coefficients a_0, b_0 and c_0, a_1 and b_1 can be found in Ref. [7].

Equations (3)–(5) are solved numerically in the amorphous alloy nanowire systems with a fixed height $L = 800$ nm under uniaxial tensions applied along the z-direction. The diameters of the nanowires are $D = 40 \sim 400$ nm. Tetrahedrons are used to discrete space and the backward differentiation scheme is employed to discrete and integrate time in the numerical solutions of Eqs. (3)–(5). The uniaxial tension is applied on the top surface, and the displacement constraint in z-direction only is imposed on the bottom surface, and other surfaces are kept free. Before mechanical deformation, the intrinsic defects (defined as the points with $\rho(\boldsymbol{r}) = 1.0$) are randomly distributed in the interior of nanowires with a volume fraction of N_0 and on sample surfaces with a content of n_0, except for the top and bottom surfaces. The regions without intrinsic defects are at an ideal glassy state with $\rho(\boldsymbol{r}) = 0$.

Results and Discussion

Plastic Deformations of Amorphous Alloy Nanowires Under Tensions

Figure 1 shows the plastic deformations in nanowires with different diameters, where the densities of deformation defects in the nanowire interiors ($N_0 = 9.8\%$) and side surfaces are independent on the diameters. In amorphous alloy nanowires with a diameter of D = 400 nm, the regions with significant plastic strains are localized. As shown in Fig. 1f, a shear band can be identified as the dash surface

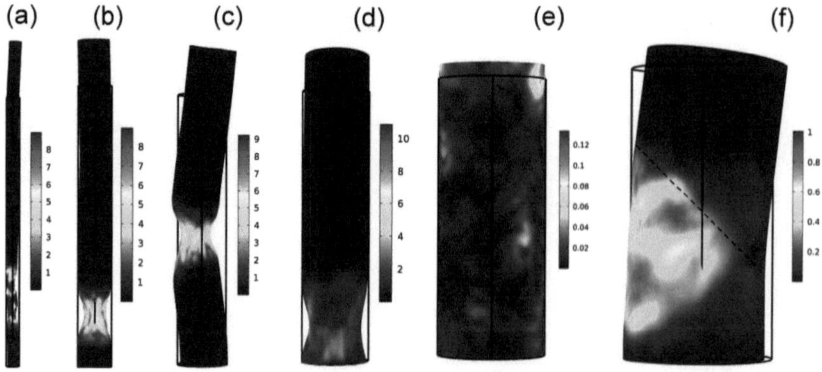

Fig. 1 The deformation plots of amorphous alloy nanowires (volume plots of displacement vectors **u**) under plastic deformations: **a** D = 40, **b** D = 100 nm, **c** D = 150 nm, **d** D = 200 nm, **e** D = 320 nm, and **f** D = 400 nm; The *dash line* represents the surface where the shear band propagates. The *color bars* show the values for equivalent plastic strains. The *solid frames* represent the un-deformed nanowires

rotating at ~43° from the cross-sectional plane, suggesting that the simulation results are consistent with those in bulk amorphous alloys observed experimentally. In the amorphous alloy nanowires with a diameter of D = 320 nm, significant shear localization still exists but it occurs at the edges of the top surface. While the plastic deformation in the nanowire interiors seems to be less heterogeneous compared to that with D = 400 nm, as shown in Fig. 1e. When the nanowire diameter is lower than 200 nm, necking instead of shear banding can be observed in the nanowire interior (Fig. 1a–d). Remarkably, homogeneous plastic deformations are much evident in amorphous alloy nanowires with D = 40 nm and D = 100 nm, as shown in Fig. 1a, b, respectively. Moreover, their plastic strains are much larger than those in nanowires with D $\geq$ 320 nm, signaling significant ductility of nanowires with D $\leq$ 200 nm.

The tensile curves of the amorphous alloy nanowires with D = 100, 150, 200, 320 and 400 nm at a temperature T = 300 K are shown in Fig. 2. The fracture strength σ_F, which is defined as the critical tensile stress for the propagation of shear bands across the nanowire's side surfaces, can be found to significantly increase with decreasing diameter. As shown in Fig. 2, the tensile curves for nanowires with D = 100, 150, 200 nm have distinct features of stress overshoots, suggesting the deformation mechanisms in nanowires with d $\leq$ 200 nm could be much different with those in nanowires with D = 320 and 400 nm.

In the stress-strain curves for the bulk amorphous alloys, the stress overshot is an indicator of homogeneous plastic deformation at high temperatures which are close to the glass transition temperature. At T = 300 K, the homogeneous plastic deformation in amorphous alloy nanowires with D $\leq$ 200 nm implies that the size effects on the deformation mechanisms could be related with the imperfections on the nanowire surfaces. During the plastic deformation processes in nanowires, instead of

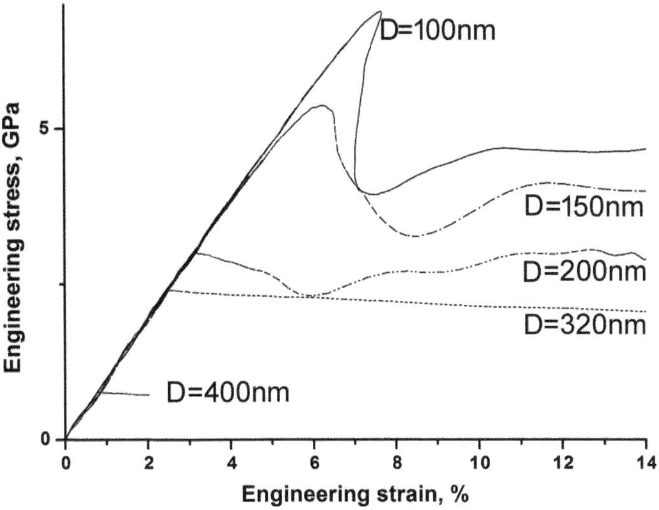

Fig. 2 Tensile curves of amorphous alloy nanowires with different diameters

a balanced condition of defect creation and annihilation, the easily activated surface imperfections on the nanowire surface result in temporarily stronger defect creation than the defect annihilation process. Thus the strain softening and hardening occur alternatingly and several peaks in the stress-strain curve exist.

It should be noted that no strain hardening process can be observed to occur in nanowires with D $\geq$ 320 nm, as shown in Fig. 2. Therefore, although the elongations to failure of nanowires are all significantly larger than those of their bulk amorphous alloy counterparts, only the ductility of nanowire with D $\leq$ 200 nm is meaningful.

We further investigate the effects of density of defects in the nanowire interiors (N_0) on the ductility of amorphous alloy nanowires. As shown in Fig. 3a, N_0 does

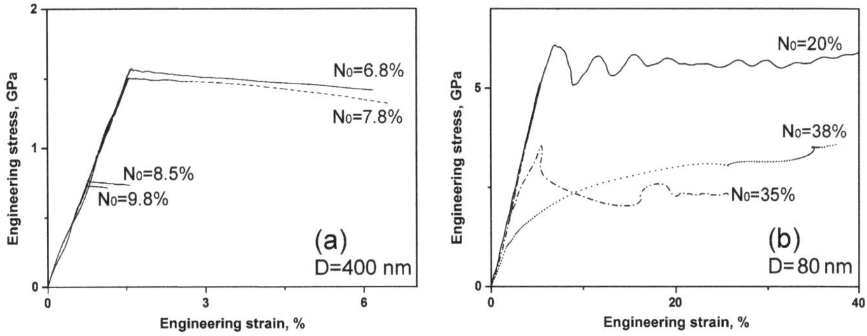

Fig. 3 Tensile curves of amorphous alloy nanowires with different densities of defects in their interiors. **a** D = 400 nm; **b** D = 80 nm

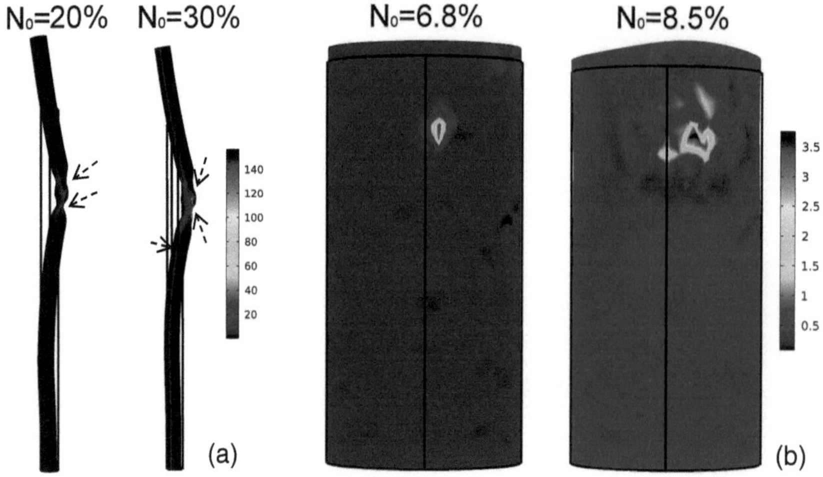

Fig. 4 Plastic deformation of nanowires with different defect densities N_0 in their interiors. **a** D = 40 nm; The *arrows* indicate the regions of necking in the nanowires. **b** D = 400 nm. The *color bars* show the values for equivalent plastic strains. The *solid frames* represent the un-deformed nanowires

not change the meaningful ductility of amorphous alloy nanowires with D = 400 nm. The reasons are that the increase in N_0 does not lead to the change in the deformation mechanism of heterogeneous plastic deformation in the nanowires, as shown in Fig. 4b. Similar effects of N_0 on the ductility of nanowire with D = 320 nm are observed. On the contrary, the ductility of nanowire with D $\leq$ 200 nm is found to increase with increasing N_0, as shown in Fig. 3b, using the nanowire with D = 80 nm as an example. The simulation results well demonstrate that the homogeneous plastic deformation is responsible for the ductility in amorphous alloy nanowires with D $\leq$ 200 nm, which could be further enhanced by increasing the density of defects in the nanowire interiors. Using the nanowire with D = 40 nm as an example, the increase of ductility with increasing N_0 could be closely associated with the events of necking, which occur more frequently along the nanowire with higher N_0, as shown in Fig. 4a.

Avalanche Model of Fracture of Amorphous Alloy Nanowires

As shown in Fig. 3, the fracture strength σ_F is extremely sensitive to the density of defects N_0 in the nanowire interiors. As shown in Fig. 5a, the fracture strength decreases with increasing N_0, and at a critical value (N_c) of N_0 it drops to almost zero (<0.01 GPa). The effect of N_0 on the fracture strength can be further elucidated by the scaling relations among $N_c(D)$, $\sigma_F(D)$ and D, since previous theoretical and experimental investigations have shown that the fracture events in disordered

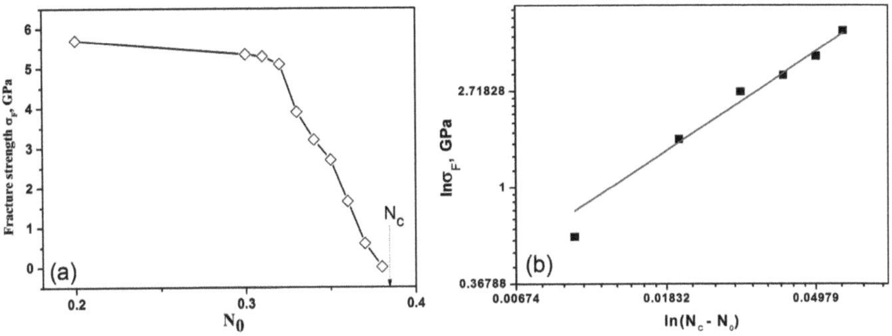

Fig. 5 **a** The relation between the fracture strength and the density of defects in the nanowire interiors. D = 80 nm. **b** Log-log plots of the scaling relation Eq. (8)

materials can be well characterized by avalanche models with characteristics of self-organized criticality [8, 9]. Here we apply the avalanche model to describe the fracture of the amorphous alloy nanowires where their interior deformation defects dominate such process. We define the variables that control the fracture process of amorphous alloy nanowire as $R \equiv [N_c - N_0]/N_c$ and $h \equiv \sigma_F - \sigma$.

Under the applied stress σ, there is a correlation length ξ(R, h) which characterizes the size of activated deformation defects on the nanowire interiors with an intrinsic defect density N_0. The scaling for ξ(R, h) is generally given by

$$\xi \sim R^{-\nu}\Phi(h/R^{\beta\delta}), \tag{6}$$

3where Φ is a scaling function independent of the system size *D*. When R = 0 or h = 0, ξ diverges. β and ν are scaling exponents defined by the finite-size scaling relation for the maximum size ΔL of the deformation defects at h = 0 (or $\sigma = \sigma_F$): $\Delta L = D^{-\frac{\beta}{\nu}}\psi(D^{\frac{1}{\nu}}R)$, where Ψ is the scaling function. δ is defined by the scaling relation between σ_F and the plastic strain ε_f which is related with ΔL when R → 0, as follows: $\sigma_F \sim \varepsilon_f^\delta \sim (\frac{\Delta L}{L})^\delta$. Therefore the scaling between ΔL and σ_F can be written as $\Delta L \sim (\sigma_F)^{\frac{1}{\delta}}$, and

$$\sigma_F = D^{-\frac{\beta\delta}{\nu}}\psi(D^{\frac{1}{\nu}}R). \tag{7}$$

Based on Eq. (7), the relation between the fracture strength and the critical intrinsic interior defect density N_0 of nanowire can be described as

$$\sigma_F \sim (N_c - N_0)^{\beta\delta}. \tag{8}$$

Figure 5b shows such scaling relation in the amorphous alloy nanowire with a diameter of D = 80 nm. The exponents in Eq. (8) is found to be βδ = 1.06 ± 0.06.

Conclusions

The size effects on the mechanical behaviors of amorphous alloys under tensile deformations are investigated by phase-field modeling. Homogeneous plastic deformations are demonstrated to be responsible for the ductility of amorphous alloy nanowires with D $\leq$ 200 nm. Heterogeneous plastic deformation leads to shear banding in amorphous alloy nanowires with D $\geq$ 320 nm, which exhibit no strain hardening during the tensile deformation. The increases of intrinsic deformation defects in the interior of nanowires with D $\leq$ 200 nm are effective in enhancing the ductility of nanowires. Whereas the intrinsic deformation defects in the interior of nanowires with D $\geq$ 320 nm do not result in meaningful ductility of nanowires. The fractures of the amorphous alloy nanowires are found to be well characterized by an avalanche model with some features of self-organized criticality, and the relation between the fracture strength and the intrinsic deformation defects in the interior of nanowires is determined.

References

1. G. Ravichandran, A. Molinari, Analysis of shear banding in metallic glasses under bending. Acta Mater. **53**, 4087–4095 (2005)
2. Z. Zhang, H. Zhang, Effect of aspect ratio on the compressive deformation and fracture behaviour of Zr-based bulk metallic glass. Philos. Mag. Lett. **85**, 513–521 (2005)
3. X.L. Wu, Y.Z. Guo, Q. Wei, W.H. Wang, Prevalence of shear banding in compression of $Zr_{41}Ti_{14}Cu_{12.5}Ni_{10}Be_{22.5}$ pillars as small as 150 nm in diameter. Acta Mater. **57**, 3562–3571 (2009)
4. F. Delogu, Molecular dynamic study of size effects in the compression of metallic glass nanowires. Phys. Rev. B **79**, 184109 (2009)
5. C.Q. Chen, Y.T. Pei, J.T.M. De Hosson, Effects of size on the mechanical response of metallic glasses investigated through in situ TEM bending and compression experiments. Acta Mater. **58**, 189–200 (2010)
6. J. Lu, G. Ravichandran, Pressure-dependent flow behavior of $Zr_{41.2}Ti_{13.8}Cu_{12.5}Ni_{10}Be_{22.5}$ bulk metallic glass. J. Mater. Res. **18**, 2039–2049 (2003)
7. G.P. Zheng, Application of phase-field modelling to deformation of metallic glasses. Curr. Opin. Solid State Mater. Sci. **15**, 116–124 (2011)
8. E. Vives, J. Goicoechea, J. Ortin, A. Planes, Universality in models for disorder-induced phase transitions. Phys. Rev. E **52**, R5–R8 (1995)
9. Z. Halasz, F. Kun, Slip avalanches in a fiber bundle model. Europhys. Lett. **89**, 26008–26011 (2010)

Thermodynamic Modeling of Al-Fe-Cr Ternary System

Shusen Wang, Zhu Li, Ziwei Qin, Shihua Wang, Xionggang Lu and Chonghe Li

Abstract The Al-Fe-Cr ternary system is one of the most important systems to developing aluminum alloys. Up to now, although several reviews about this ternary system have been available, however no thermodynamic assessment was published. In this paper, previous work of the Al-Fe-Cr ternary system and its related binary systems are reviewed. Then, Al-Cr binary system was re-assessed in order to better describe the phases $Al_8(Cr, Fe)_5$ in ternary system. Based on the re-assessed Al-Cr binary system and Al-Fe, Fe-Cr binary systems assessed before, as well as the experimental data reported in the literature, the Al-Fe-Cr ternary system was assessed by means of CALPHAD method. Isothermal sections of 1433, 1348, 1273, 1173, 1073 and 973 K in Al-rich corner and some invariant reactions are calculated. It is shown that the present calculations are in good agreement with the most of the experimental results.

Keywords Ternary alloy system · Al-Fe-Cr · Calphad method · Optimization

Introduction

The phase diagram of the Al-Fe-Cr ternary system is fundamental to developing the novel structural and functional alloys containing these three elements. Sikka et al. [1, 2] reported that the addition of 2–5 mass% Cr the ductility of Fe_3Al-based alloy

S. Wang · Z. Li · Z. Qin · S. Wang · X. Lu · C. Li (✉)
State Key Laboratory of Advanced Special Steel, Shanghai Key Laboratory of Advanced Ferrometallurgy, School of Materials Science and Engineering, Shanghai University, Shanghai 200072, China
e-mail: chli@staff.shu.edu.cn

S. Wang
e-mail: 982707836@qq.com

Z. Qin · X. Lu · C. Li
Shanghai Special Casting engineering technology research center, Shanghai 201605, China

The Minerals, Metals & Materials Society, *TMS 2017 146th Annual Meeting & Exhibition Supplemental Proceedings*, The Minerals, Metals & Materials Series, DOI 10.1007/978-3-319-51493-2_42

can be improved to 20%, which should be acceptable for many practical applications. In addition, Ferritic-structured Fe-Cr-Al alloys exhibits good oxidation resistant and can be used as accident tolerant light water reactor fuel cladding [3]. What's more, the Al-Fe-Cr ternary system is also one of the sub ternary systems of the Al-Ti-Fe-Cr quaternary system, which is the basis system for many nanostructured alloys and low cost titanium alloys that have been developed [4–8].

Taillandier et al. [9] determined the liquidus surface of the ternary system by thermal analysis in 1932, and now several sets of experimental data of this ternary system are available and some assessments and reviews have been reported. However, no full thermodynamic description has been published yet.

In this study, the available thermodynamic data for binary systems Al-Fe and Fe-Cr are reviewed [10, 11]. The Al-Cr binary system was re-assessed in order to better describe the phases $Al_8(Cr, Fe)_5$ in ternary system. Also, the thermodynamic parameters of the Ti-Fe-Cr system are optimized and ternary interaction parameters of every phase (Four ternary phases O1, H, E and D3 are included) are presented. Finally, three sectional phase diagrams from 973 to 1433 K and the invariant reaction temperatures are calculated.

Binary Systems

The Al-Fe system has been assessed for many times [10, 12–15]. Compared with others' assessments, Sundman published a new description of the Al–Fe system which can describe disordered A2 and the B2, D03 and B32 ordering phases. Thus, Sundman's thermodynamic parameters for the Al-Fe system are adopted in this work.

The Fe-Cr binary system has been assessed by Andersson and Sunderman [16]. Then Lee [17] revised the thermodynamic description for the purpose of getting a better agreement between the calculated and experimental data concerning the BCC_A2/liquid tie-lines and the liquidus temperatures in the Fe-rich region of three ternary systems Fe-Cr-C, Fe-Cr-Ni and Fe-Cr-Mn. Although now several versions of re-assessed data are available, which are compatible with our previous calculations, we decided to select the thermodynamic parameters from SGTE for pure Fe and the sublattice model $(A)_8(B)_4(A,B)_{18}$ for sigma phase, and in this work the old version of the Fe-Cr description [16, 17] is accepted.

Several thermodynamic descriptions of the Al-Cr system are available [18–21]. In this work, the Al-Cr binary system was re-assessed by making some modifications on Chen et al.' work [19]. The model of phase Al_8Cr_5 was changed from $(Al, Cr)_{17}(Al,Cr)_8$ to $(Al,Cr)_8(Al,Cr)_5$, which should be the same with Al_8Fe_5 phase in Al-Fe binary system [10] for those two phases in binary systems were treated as one phase $Al_8(Cr, Fe)_5$ in ternary system. Also, the parameters were optimized in order to be better describe the binary system.

Table 1 The optimized thermodynamic parameters of Al-Fe-Cr system

Phase	Parameter (in J/mol)	
Liquid	${}^{0}L^{Liquid}_{Al,Cr} = -36800 + 1.8 \times T$	${}^{1}L^{Liquid}_{Al,Cr} = -14800 + 6.3 \times T$
(Al, Fe, Cr)	${}^{0}L^{Liquid}_{Al,Cr,Fe} = -303566.95 + 195 \times T$	
BCC_A2	${}^{0}L^{BCC\_A2}_{Al,Cr} = -58452.5 + 8.6 \times T$	${}^{0}L^{BCC\_A2}_{Al,Cr,Fe} = -121786.55 + 71.60 \times T$
(Al, Fe, Cr)		
FCC_A1	${}^{0}L^{FCC\_A1}_{Al,Cr} = -30946 + 8.30177 \times T$	
(Al, Fe, Cr)		
Al_2Fe	${}^{0}G^{Al2Fe}_{Al:Cr} = -10000 + 2 \times {}^{0}G^{FCC\_A1}_{Al} + {}^{0}G^{BCC\_A2}_{Cr}$	
$(Al)_2(Cr,Fe)$	${}^{0}L^{Al2Fe}_{Al:Cr,Fe} = 55545.66 - 66.72 \times T$	
Al_5Fe_2	${}^{0}G^{Al5Fe2}_{Al:Cr} = -10000 + 5 \times {}^{0}G^{FCC\_A1}_{Al} + 2 \times {}^{0}G^{BCC\_A2}_{Cr}$	
$(Al)_5(Cr,Fe)_2$	${}^{0}L^{Al5Fe2}_{Al:Cr,Fe} = -100000$	
$Al_{13}Fe_4$	${}^{0}G^{Al13Fe4}_{Al:Cr:Al} = -8900 + 0.765 \times {}^{0}G^{FCC\_Al}_{Al} + 0.235 \times {}^{0}G^{BCC\_A2}_{Cr}$	
$(Al)_{0.6275}(Fe,Cr)_{0.235}(Al,Va)_{0.1375}$	${}^{0}G^{Al13Fe4}_{Al:Cr:Va} = -12000 + 0.6275 \times {}^{0}G^{FCC\_A1}_{Al} + 0.235 \times {}^{0}G^{BCC\_A2}_{Cr}$	
	${}^{0}L^{Al13Fe4}_{Al:Fe,Cr:Va} = 0.13 \times T^2 - 330.98 \times T + 205868.77$	
$Al_7Cr(\theta)$	${}^{0}G^{Al7Cr}_{Al:Cr} = -103570 + 24.8 \times T + 7 \times {}^{0}G^{FCC\_A1}_{Al} + {}^{0}G^{BCC\_A2}_{Cr}$	
$(Al)_7(Al,Cr,Fe)$	${}^{0}G^{Al7Cr}_{Al:Al} = 30000 + 8 \times {}^{0}G^{FCC\_A1}_{Al}$	
	${}^{0}L^{Al7Cr}_{Al:Al,Cr} = -28000 + 13.5 \times T$	${}^{0}L^{Al7Cr}_{Al:Cr,Fe} = -480000$
$Al_{11}Cr_2(\eta)$	${}^{0}G^{Al11Cr2}_{Al:Cr} = -147600.9 + 11 \times {}^{0}G^{FCC\_A1}_{Al} + 2 \times G^{BCC\_A2}_{Cr}$	
$(Al,Fe)_{11}(Al,Fe,Cr)_2$	${}^{0}G^{Al11Cr2}_{Al:Al} = 60000 + 13 \times {}^{0}G^{FCC\_A1}_{Al}$	${}^{0}L^{Al11Cr2}_{Al:Al,Cr} = 172602.3 - 190.4713 \times T$
	${}^{0}G^{Al11Cr2}_{Fe:Al} = 60000 + 11 \times {}^{0}G^{BCC\_A2}_{Fe} + 2 \times {}^{0}G^{FCC\_A1}_{Al}$	
	${}^{0}G^{Al11Cr2}_{Al:Fe} = 40000 + 11 \times {}^{0}G^{BCC\_A2}_{Al} + 2 \times {}^{0}G^{BCC\_A2}_{Fe}$	
	${}^{0}G^{Al11Cr2}_{Fe:Cr} = 20000 + 11 \times {}^{0}G^{BCC\_A2}_{Fe} + 2 \times {}^{0}G^{BCC\_A2}_{Cr}$	

(continued)

Table 1 (continued)

Phase	Parameter (in J/mol)	
	${}^{0}G^{Al11Cr2}_{Fe:Fe} = 20000 + 13 \times {}^{0}G^{BCC\_A2}_{Fe}$	${}^{0}L^{Al11Cr2}_{Al:Cr,Fe} = -447300 + 100 \times T$
Al4Cr(μ)	${}^{0}G^{Al4Cr}_{Al:Cr} = 9.5 \times T - 79800 + 4 \times {}^{0}G^{FCC\_A1}_{Al} + {}^{0}G^{BCC\_A2}_{Cr}$	
$(Al)_4(Al,Cr,Fe)$	${}^{0}G^{Al4Cr}_{Al:Al} = 60500 + 5 \times {}^{0}G^{FCC\_A1}_{Al}$	${}^{0}L^{Al4Cr}_{Al:Al,Cr} = 44.003 \times T - 103000$
	${}^{0}G^{Al4Cr}_{Al:Fe} = 270000 + 4 \times {}^{0}G^{FCC\_A1}_{Al} + {}^{0}G^{BCC\_A2}_{Fe}$	
	${}^{0}L^{Al4Cr}_{Al:Cr,Fe} = -150000$	${}^{0}L^{Al4Cr}_{Al:Al,Fe} = -150000$
$\gamma 2$	${}^{0}G^{\gamma 2}_{Al:Al:Cr} = 18 \times T - 489900 + 17 \times {}^{0}G^{FCC\_A1}_{Al} + 9 \times {}^{0}G^{BCC\_A2}_{Cr}$	
$(Al,Cr)_{12}(Al,Cr,Fe)_5(Al,Cr,Fe)_9$	${}^{0}G^{\gamma 2}_{Cr:Al:Cr} = -18 \times T - 78000 + 21 \times {}^{0}G^{BCC\_A2}_{Cr} + 5 \times {}^{0}G^{FCCF\_A1}_{Al}$	
	${}^{0}G^{\gamma 2}_{Al:Cr:Cr} = -30000 + 12 \times {}^{0}G^{FCC\_A1}_{Al} + 14 \times {}^{0}G^{BCC\_A2}_{Cr}$	
	${}^{0}G^{\gamma 2}_{Cr:Cr:Cr} = 50000 + 26 \times {}^{0}G^{BCC\_A2}_{Cr}$	${}^{0}G^{\gamma 2}_{Al:Al:Al} = 36600 + 26 \times {}^{0}G^{FCC\_A1}_{Al}$
	${}^{0}G^{\gamma 2}_{Al:Cr:Al} = 21 \times {}^{0}G^{FCC\_A1}_{Al} + 5 \times {}^{0}G^{BCC\_A2}_{Cr}$	${}^{0}G^{\gamma 2}_{Cr:Al:Al} = 12 \times {}^{0}G^{BCC\_A2}_{Cr} + 14 \times {}^{0}G^{FCC\_A1}_{Al}$
	${}^{0}G^{\gamma 2}_{Cr:Cr:Al} = 17 \times {}^{0}G^{BCC\_A2}_{Cr} + 9 \times {}^{0}G^{FCC\_A1}_{Al}$	
	${}^{0}G^{\gamma 2}_{Al:Fe:Cr} = -500000 + 12 \times {}^{0}G^{FCC\_A1}_{Al} + 14 \times {}^{0}G^{BCC\_A2}_{Cr}$	
	${}^{0}G^{\gamma 2}_{Al:Fe:Cr} = -1500$	${}^{0}G^{\gamma 2}_{Cr:Fe:Cr} = 100000 + 5 \times {}^{0}G^{BCC\_A2}_{Fe} + 21 \times {}^{0}G^{BCC\_A2}_{Cr}$
	${}^{0}G^{\gamma 2}_{Al:Fe:Al} = 1000000 + 21 \times {}^{0}G^{FCC\_A1}_{Al} + 5 \times {}^{0}G^{BCC\_A2}_{Fe}$	
	${}^{0}G^{\gamma 2}_{Cr:Fe:Al} = -200000 + 9 \times {}^{0}G^{FCC\_A1}_{Al} + 5 \times {}^{0}G^{BCC\_A2}_{Fe} + 12 \times {}^{0}G^{BCC\_A2}_{Cr}$	
	${}^{0}G^{\gamma 2}_{Al:Al:Fe} = -620000 + 17 \times {}^{0}G^{FCC\_A1}_{Al} + 9 \times {}^{0}G^{BCC\_A2}_{Fe}$	
	${}^{0}G^{\gamma 2}_{Al:Cr:Fe} = 300000 + 12 \times {}^{0}G^{FCC\_A1}_{Al} + 5 \times {}^{0}G^{BCC\_A2}_{Cr} + 9 \times {}^{0}G^{BCC\_A2}_{Fe}$	
	${}^{0}G^{\gamma 2}_{Cr:Al:Fe} = -100000 + 17 \times {}^{0}G^{FCC\_A1}_{Al} + 9 \times {}^{0}G^{BCC\_A2}_{Fe}$	
	${}^{0}G^{\gamma 2}_{Cr:Cr:Fe} = 200000 + 17 \times {}^{0}G^{BCC\_A2}_{Cr} + 9 \times {}^{0}G^{BCC\_A2}_{Fe}$	
	${}^{0}G^{\gamma 2}_{Cr:Fe:Fe} = 12 \times {}^{0}G^{BCC\_A2}_{Cr} + 14 \times {}^{0}G^{BCC\_A2}_{Fe}$	
	${}^{0}G^{\gamma 2}_{Al:Fe:Fe} = 810000 + 12 \times {}^{0}G^{FCC\_A1}_{Al} + 14 \times {}^{0}G^{BCC\_A2}_{Fe}$	

(continued)

Table 1 (continued)

Phase	Parameter (in J/mol)	
	${}^{0}L^{\gamma 2}_{Al,Cr:Al:Cr} = {}^{0}L^{\gamma 2}_{Al,Cr:Al:Al} = -59507.1 - 6.62835 \times T$	${}^{0}L^{\gamma 2}_{Al,Cr:Al:Fe} = -1155625 + 625 \times T$
$\gamma 1$	${}^{0}G^{\gamma 1}_{Al:Cr} = -13 \times T - 216220 + 8 \times {}^{0}G^{FCC\_A1}_{Al} + 5 \times {}^{0}G^{BCC\_A2}_{Cr}$	
$(Al,Cr,Fe)_8(Al,Cr,Fe)_5$	${}^{0}G^{\gamma 1}_{Cr:Al} = 390114.8 + 5 \times {}^{0}G^{FCC\_A1}_{Al} + 8 \times {}^{0}G^{BCC\_A2}_{Cr}$	
	${}^{0}L^{\gamma 1}_{Al:Al,Cr} = -83500 - 6.4091 \times T$	${}^{1}L^{\gamma 1}_{Al:Al,Cr} = -49000 + 60 \times T$
	${}^{0}L^{\gamma 1}_{Cr:Al,Cr} = -59000 - 230 \times T$	${}^{1}L^{\gamma 1}_{Cr:Al,Cr} = -129000 + 60 \times T$
	${}^{0}L^{\gamma 1}_{Al,Cr:Al} = -45244.4 - 9.9705 \times T$	${}^{1}L^{\gamma 1}_{Al,Cr:Al} = 14176.682 + 17.55163 \times T$
	${}^{0}L^{\gamma 1}_{Al,Cr:Cr} = -21944.4 + 27.9705 \times T$	${}^{1}L^{\gamma 1}_{Al,Cr:Cr} = 33576.682 + 9.55163 \times T$
	${}^{0}L^{\gamma 1}_{Cr:Al,Fe} = {}^{0}L^{\gamma 1}_{Al,Cr:Fe} = 1722028 - 1411 \times T$	
$AlCr_2$	${}^{0}G^{AlCr2}_{Al:Cr} = -45950 + {}^{0}G^{FCC\_A1}_{Al} + 2 \times {}^{0}G^{BCC\_A2}_{Cr}$	
$(Al,Cr)(Al,Cr)_2$	${}^{0}G^{AlCr2}_{Cr:Al} = 75279.2 + 2 \times {}^{0}G^{FCC\_A1}_{Al:Al} + {}^{0}G^{BCC\_A2}_{Cr}$	
	${}^{0}G^{AlCr2}_{Al:Al} = 15000 + 3 \times {}^{0}G^{FCC\_A1}_{Al}$	${}^{0}G^{AlCr2}_{Cr:Cr} = 15000 + {}^{0}G^{BCC\_A2}_{Cr}$
	${}^{0}G^{AlCr2}_{Al:Al,Cr} = {}^{0}G^{AlCr2}_{Cr:Al,Cr} = -31502.4 + 5.7033 \times T$	
	${}^{0}G^{AlCr2}_{Al,Cr:Al} = {}^{0}G^{AlCr2}_{Al,Cr:Cr} = -11282.4 - 6.37 \times T$	
O1	${}^{0}G^{O1}_{Al:Al} = {}^{0}G^{FCC\_A1}_{Al}$	${}^{0}G^{O1}_{Al:Cr} = 3000 + 0.76 \times {}^{0}G^{FCC\_A1}_{Al} + 0.24 \times {}^{0}G^{BCC\_A2}_{Cr}$
$(Al)_{0.72}(Al,Cr,Fe)_{0.28}$	${}^{0}L^{O1}_{Al:Cr,Fe} = -90000$	${}^{1}G^{O1}_{Al:Cr,Fe} = -15000$
	${}^{0}L^{O1}_{Al:Al,Fe} = 1000$	${}^{0}L^{O1}_{Al:Al,Cr} = -6000$
	${}^{0}L^{O1}_{Al:Al,Cr,Fe} = 8.49975 \times T^2 - 21640 \times T + 13634100$	
D3	${}^{0}G^{D3}_{Al:Al} = 500 + {}^{0}G^{FCC\_A1}_{Al}$	${}^{0}G^{D3}_{Al:Cr} = 2000 + 0.72 \times {}^{0}G^{FCC\_A1}_{Al} + 0.28 \times {}^{0}G^{BCC\_A2}_{Cr}$
$(Al)_{0.72}(Al,Cr,Fe)_{0.28}$	${}^{0}G^{D3}_{Al:Fe} = 2000 + 0.72 \times {}^{0}G^{FCC\_A1}_{Al} + 0.28 \times {}^{0}G^{BCC\_A2}_{Fe}$	
	${}^{0}L^{D3}_{Al:Cr,Fe} = 0.27127 \times T^2 - 691.54716 \times T + 354134.40862$	

(continued)

Table 1 (continued)

Phase	Parameter (in J/mol)	
H	${}^{0}G^{H}_{Al:Cr:Al} = 1.22 \times {}^{0}G^{FCC\_A1}_{Al}$	${}^{0}G^{H}_{Al:Cr:Cr} = {}^{0}G^{FCC\_A1}_{Al} + 0.31 \times {}^{0}G^{BCC\_A2}_{Cr}$
$(Al)(Cr)_{0.09}(Al,Cr,Fe)_{0.22}$	${}^{0}G^{H}_{Al:Cr:Fe} = {}^{0}G^{FCC\_A1}_{Al} + 0.09 \times {}^{0}G^{BCC\_A2}_{Cr} + 0.22 \times {}^{0}G^{BCC\_A2}_{Fe}$	
	${}^{0}L^{H}_{Al:Cr:Cr,Fe} = -95390.9 + 33.3 \times T$	${}^{0}L^{H}_{Al:Cr:Al,Fe} = -41460 + 20 \times T$
	${}^{0}L^{H}_{Al:Cr:Al,Cr,Fe} = -298000$	
E	${}^{0}G^{E}_{Al:Cr:Fe} = -13290.5 - 1.5 \times T + 0.77 \times {}^{0}G^{FCC\_A1}_{Al} + 0.21 \times {}^{0}G^{BCC\_A2}_{Cr} + 0.03 \times {}^{0}G^{BCC\_A2}_{Fe}$	
$(Al)_{0.77}(Cr)_{0.21}(Fe)_{0.02}$		

Ternary System

There are four ternary phases in Al-Fe-Cr ternary system, O1, H, E and D3. Pavlyuchkov et al. [20] reported that the O1 phase was different with Al_4Cr phase for the reason that O1 phase was base-centered orthorhombic structure while the later one was hexagonal structure. Later the structure of O1 phase was proved to be the similar with the Al4Cr phase by Czochralski method [21]. Pavlyuchkov et al. demonstrated that the phase structure of D3 was the same as $Al_{70}Pd_{13}Mn_{17}$ using X-ray analysis method [22]. The atomic site occupation of H ternary phase was studied by Mo et al. with high resolution electron microscope and it was reported that there were 31 TM sites (TM: Cr or Fe) and 100 Al sites in one asymmetric unit [23]. The E phase has a primitive orthorhombic structure and a small compositional region around $Al_{77}Cr_{12}Fe_2$ [24].

The Al-Fe-Cr ternary system has been reviewed for several times [24–30]. Ghosh et al. [29] summarized several isothermal sections, because in his work four ternary phases O1, H, E and D3 were not considered the phase relations are not reproduced in this work. Pavlyuchkov et al. [30] determined the isothermal sections of 973, 1073, 1173, 1273, 1348 and 1433 K by experiments. In his work, the above 50 at% Al part were measured according to the experiments and the above four ternary phases were also considered. The invariant reactions in the ternary system are all existed at the Al rich corner, Khoruzha et al. [28] have measured them and determined the temperature of the invariant reactions.

Thermodynamic Modeling and Parameters Optimization

There are three types of thermodynamic models used in this ternary system: the single lattice association model, which is used in ordered solution phases such as Liquid BCC_A2 and FCC_A1; The two sub lattice model is applied in phases like BCC_B2, Al4Cr, $\gamma 1$ and $Al_{13}Fe_4$; The others, the $Al_{13}Fe_4$ phase is used the model of $(Al)_{0.625}(Fe,Cr)_{0.235}(Al,Va)_{0.1357}$, the Al8Cr5_α phase used the model of $(Al,Cr)_{12}(Al,Cr,Fe)_5(Al,Cr,Fe)_9$, the ternary phase O1 and D3 used the model $(Al)_{0.72}(Al, Cr, Fe)_{0.28}$ and the H phase used $(Al)(Cr)_{0.09}(Al, Cr, Fe)_{0.22}$. Finally, the stoichiometric ratio model is applied in the E phase. For any phase m, the Gibbs free energy formula algorithm is:

$$G_m = G_m^{ref} + \Delta_{mix} G_m^{id} + {}^E G \quad (1)$$

$$G_m^{ref} = \sum_{i=Al,Fe,Cr} x_i G_{m,i}^0 \quad (2)$$

$$\Delta_{mix} G_m^{id} = RT \sum_{i=Al,Fe,Cr} x_i \ln x_i \quad (3)$$

G_m^{ref} represents the Gibbs free energy of all the pure components, its extend formula is shown as (2). $\Delta_{mix}G_m^{id}$ represents the contribution of ideal mixed entropy to the Gibbs free energy of m phase,its extend formula is shown as (3). ^{E}G represents the excess Gibbs free energy of m phase, which is determined by the extent of the departure from the ideal solution, ^{E}G is the key part of the optimization of phase diagram,the calculation process of Gibbs free energy and extend formulas are available in [31]. The optimized thermodynamic parameters of Al-Fe-Cr system are shown in Table 1.

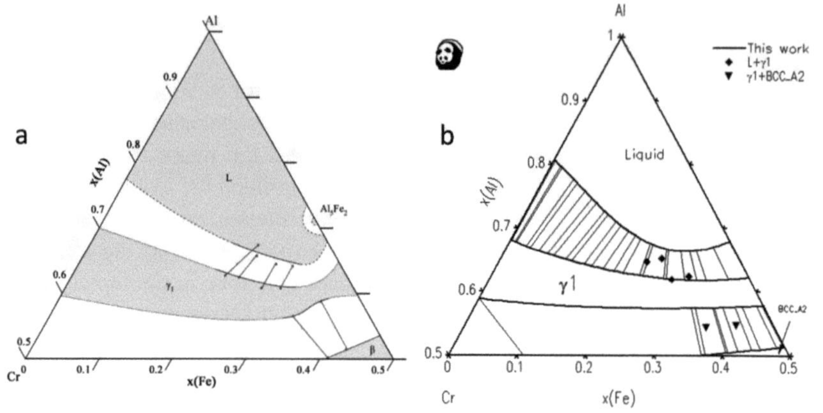

Fig. 1 The calculated isothermal sections at 1433 K (**b**) at the Al corner along with the correspond isothermal sections experimental measured by Pavlyuchkov et al. (**a**) [30]

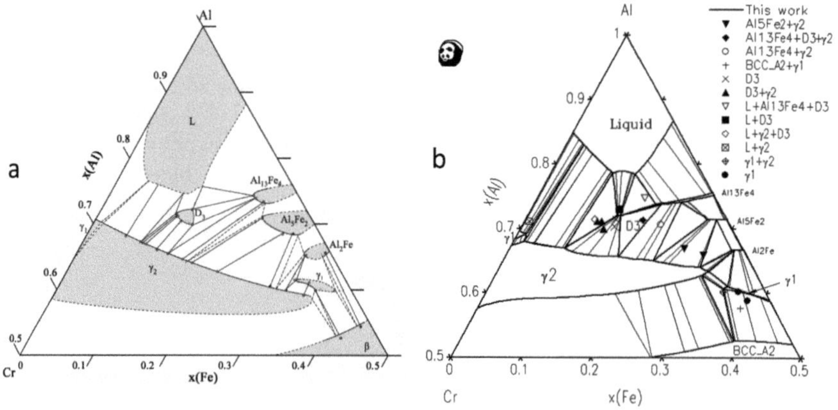

Fig. 2 The calculated isothermal sections at 1348 K (**b**) at the Al corner along with the correspond isothermal sections experimental measured by Pavlyuchkov et al. (**a**) [30]

Results and Discussion

The calculated isothermal sections at 1433, 1348 and 1273 K at the Al corner along with the correspond isothermal sections experimentally measured by Pavlyuchkov et al. [30] are shown in Fig. 1a, b, Fig. 2a, b, and Fig. 3a, b, respectively. It is shown that the calculated phase region shape of the isothermal sections are similar with the ones constructed by Pavlyuchkov et al. [30], What's more, there is no difference between the calculated results and experimental results in the phase equilibrium that presented in the isothermal sections. Also, the experimental points which were used to identify the phase composition were reproduced in Figs. 1b, 2b, and 3b, it demonstrates that the present evaluation is consistent with almost all of the experimental results within the error. Besides, the model of $Al_{11}Cr_2$ (η) used in this work is $(Al,Fe)_{11}(Al,Fe,Cr)_2$, which is based on the binary system model $(Al)_{11}(Al,Cr)_2$. Because according to the model used the max Cr content in theory is

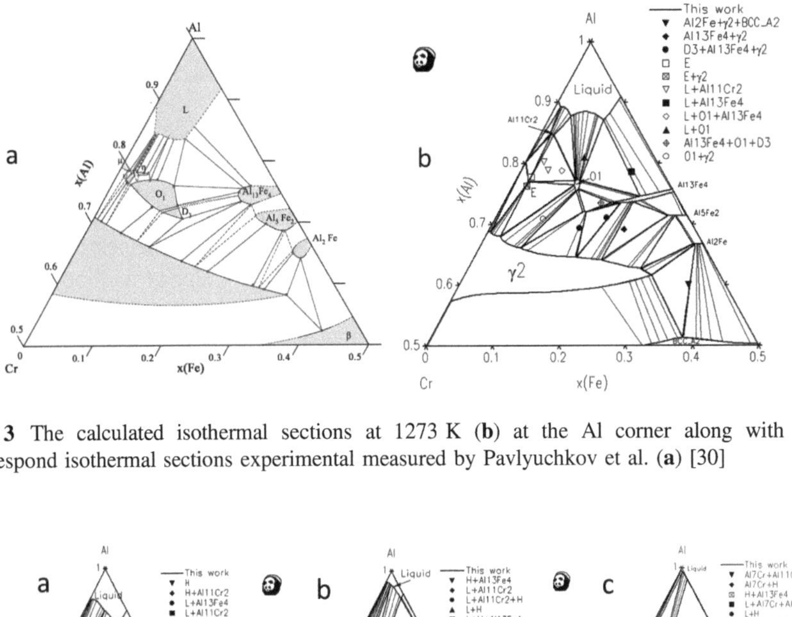

Fig. 3 The calculated isothermal sections at 1273 K (**b**) at the Al corner along with the correspond isothermal sections experimental measured by Pavlyuchkov et al. (**a**) [30]

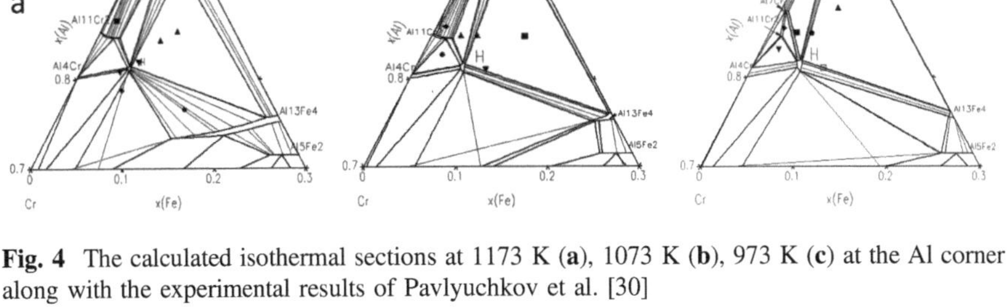

Fig. 4 The calculated isothermal sections at 1173 K (**a**), 1073 K (**b**), 973 K (**c**) at the Al corner along with the experimental results of Pavlyuchkov et al. [30]

Table 2 Calculated and experimental results for the invariant reactions in Al-Fe-Cr system

Reactions	This work			Experimental [28]		
	T(K)	x(Al)	x(Cr)	T(K)	x(Al)	x(Cr)
L + D3 → O1 + $Al_{13}Fe_4$	1316	0.820	0.081	1348	0.748	0.130
L + γ2 → E + O1	1316	0.824	0.123	1318	0.824	0.143
L + γ2 → E + Al_4Cr	1304	0.864	0.130	1308	0.830	0.150
L + E → Al4Cr + $Al_{11}Cr_2$	1279	0.881	0.106	1258	0.845	0.132
L + $Al_{11}Cr_2$ → Al_7Cr + H	1002	0.993	0.005	1023	0.980	0.015

15.4 at% and couldn't be higher, the $Al_{11}Cr_2$ (η) phase optimization has some defects in this work.

The calculated isothermal sections at 1173, 1073 and 973 K at the Al corner along with the experimental results of Pavlyuchkov et al. [30] are shown in Fig. 4. The present calculated results are agree well with the most experimental results. Table 2 lists the temperature of the invariant reactions at the Al-rich corner as well as the values reported by Khoruzha et al. [28], it can be seen that the calculated results are in good agreement with the experimental ones.

Conclusion

The previous investigations for the Al-Fe-Cr ternary system and the thermodynamic descriptions of the sub-binary systems are reviewed. Based on the selected models, the Al-Fe-Cr ternary system is assessed by means of the Calphad method using the ternary experimental data in the literature. The isothermal sections of the Al-rich corner at 1433, 1348, 1273, 1173, 1073 and 973 K, as well as the ternary invariant reactions have been calculated. It is shown that the present thermodynamic assessment is in good agreement with the majority of experimental results. This assessment may provide the theoretical support to designing the alloys of Al-Fe-Cr system.

Acknowledgements The authors thank the CompuTherm LLC, U.S. for the use of Pandat software. Also, Thank the National Basic Research Program of China (973 Program No. 2014CB643403), the National Nature Science Foundation of China (Grant No. 51225401 and 51374142) and the Science and Technology Fund of Scientific Committee of Shanghai (No. 14JC1491400).

References

1. V.K. Sikka et al., Powder processing of Fe_3Al-based iron-aluminide alloys. Mrs Online Proc. Libr. **213**(213), 901–906 (1990)
2. J.R. Knibloe et al., Microstructure and mechanical properties of Fe_3Al alloys with chromium. J. Mater. Sci. **28**(8), 2040–2048 (1993)

3. K.G. Field et al., Deformation behavior of laser welds in high temperature oxidation resistant Fe-Cr-Al alloys for fuel cladding applications. J. Nucl. Mater. **454**(1–3), 352–358 (2014)
4. M. Zawrah et al., Microstructure and hardness of nanostructured Al-Fe-Cr-Ti alloys through mechanical alloying. Mater. Sci. Eng., A **355**(1), 37–49 (2003)
5. L. Shaw et al., Thermal stability of nanostructured $Al_{93}Fe_3Cr_2Ti_2$ alloys prepared via mechanical alloying. Acta Mater. **51**(9), 2647–2663 (2003)
6. L. Shaw et al., Compressive behavior of an extruded nanocrystalline Al-Fe-Cr-Ti alloy. Scripta Mater. **50**(7), 921–925 (2004)
7. B. Gunawarman et al., Mechanical properties and microstructures of low cost β titanium alloys for healthcare applications. Mater. Sci. Eng., C **25**(3), 304–311 (2005)
8. W. Guo et al., Hot compressive behavior of Ti-3.0Al-3.7Cr-2.0Fe-0.1B titanium alloy. Trans. Nonferrous Met. Soc. China **22**(12), 2965–2971 (2012)
9. M.C.H. Taillandier, Contribution to Al-Fe-Cr Alloys, Parts I and II (in French). Rev. Met. **29**, 315–325, 348–356 (1932)
10. B.O. Sundman et al., An assessment of the entire Al-Fe system including D03 ordering. Acta Mater. **57**(10), 2896–2908 (2009)
11. J.O. Andersson et al., Thermodynamic properties of the Cr-Fe system. Calphad **11**(1), 83–92 (1987)
12. A.T. Phan et al., Phase equilibria and thermodynamics of the Fe-Al-C system: critical evaluation, experiment and thermodynamic optimization. Acta Mater. **79**(30), 1–15 (2014)
13. M. Seierstein, *COST 507, Thermochemical Database for Light Metal Alloys*, vol. 1 (European Communities, Luxembourg, 1998), pp. 34–39
14. U.R. Kattner et al., *Phase Diagrams of Binary Iron Alloys*, (Materials Park, OH, American Society for Metals, 1993), pp. 12–28
15. M.H.G. Jacobs et al., Phase behavior and thermodynamic properties in the system Fe-Al. Calphad **33**(1), 170–178 (2009)
16. J.O. Andersson et al., Thermodynamic properties of the Cr-Fe system. Calphad **11**(1), 83–92 (1987)
17. B.J. Lee, Revision of the thermodynamic descriptions of the Fe-Cr and Fe-Ni Liquid phases. Calphad **17**(3), 251–268 (1993)
18. M. Audier et al., Phase equilibria in the Al-Cr system. J. Alloy. Compd. **220**(220), 225–230 (1995)
19. L.Y. Chen et al., Thermodynamic modeling of the Ti–Al–Cr ternary system. J. Alloy. Compd. **509**(5), 1936–1946 (2011)
20. D.V. Pavlyuchkov et al., Complex orthorhombic phase in the Al-Cr-Fe system. Powder Metall. Met. Ceram. **47**(11), 698–701 (2008)
21. B. Bauer et al., Crystal growth of Al-Rich complex metallic phases in the system Al-Cr-Fe using the Czochralski method. Zeitschrift Für Anorganische Chemie **637**(13), 2052–2058 (2011)
22. D. Pavlyuchkov et al., Stable decagonal quasicrystals in the Al-Fe-Cr and Al-Fe-Mn alloy systems. J. Alloy. Compd. **477**(1), L41–L44 (2009)
23. Z.M. Mo et al., Structure of v-$Al_{80.61}Cr_{10.71}Fe_{8.68}$, a giant hexagonal approximant of a quasicrystal determined by a combination of electron microscopy and X-ray diffraction. Acta Crystallogr. A **56**(3), 392–401 (2000)
24. V.G. Khoruzha et al., The Al-Cr-Fe phase diagram. I. Phase equilibria at subsolidus temperatures over composition range 58–100 at.% Al. Powder Metall. Met. Ceram. **50**(1), 83–97 (2011)
25. Y. Liang et al., Thermodynamic modeling of the Al-Cr system. J. Alloy. Compd. **460**(1), 314–319 (2008)
26. B. Hu et al., Thermodynamic reassessment of the Al-Cr-Si system with the refined description of the Al-Cr system. Thermochim. Acta **561**(37), 77–90 (2013)
27. V. Raghavan, Al-Cr-Fe (Aluminum-Chromium-Iron). J. Phase Equilib. Diffus. **33**(1), 55–58 (2012)

28. V.G. Khoruzha et al., The Al-Cr-Fe phase diagram. II. Liquidus surface and phase equilibria for crystallization of 58–100 at.% Al alloys. Powder Metall. Met. Ceram. **50**(3), 217–229 (2011)
29. G. Ghosh et al., *Light Metal Systems, Part 1* (Springer, Berlin, 2004), pp. 320–351
30. D. Pavlyuchkov et al., Al–Cr–Fe phase diagram. Isothermal sections in the region above 50 at % Al. Calphad **45**(24), 194–203 (2014)
31. Z.Y. Xu, L. Li, *Thermodynamics of Phase Diagram (in Chinese)* (Science Press, Beijing, 2009), pp. 161–171

Part XII
Defects and Properties of Cast Metals

A Modeling and Experimental Investigation on the Formation of Acicular Silicon and Sludge in High Pressure Die Casting of a Modified A383 Alloy

Mikko Kärkkäinen, Tao Liu, Laurentiu Nastac, Luke Brewer, Vishweshwar Arvikar and Ilya Levin

Abstract A numerical model is being developed to predict the formation of sludge (e.g., Fe-containing intermetallic compounds) and acicular silicon during high pressure die casting (HPDC) of a modified aluminium A383 alloy. The iron-rich sludge-phase and silicon rich phases with needle-like morphology play a major detrimental role in the mechanical properties, especially the fracture characteristics, of high pressure die cast engine components. The modelling approach consists of using a macroscopic fluid flow, heat transfer and solidification model to predict the cooling rates in the HPDC A383 process. The results are being compared against experimental measurements (e.g., amount, type and size of detrimental phases) in A383 samples extracted from various regions of an HPDC engine block. The approach can be applied to predict the effects of cooling rate on the final phase microstructure.

Keywords High pressure die casting · A383 alloy · Solidification · Microstructure · Modelling

Introduction

Iron-rich intermetallics, especially the β-phase, has been shown to have a deleterious effect on mechanical properties in aluminium casting [1, 2]. Also, the shape and size of silicon rich phases has been shown to have a significant effect on the elongation of aluminum alloys [3, 4]. These phases are of interest when optimising any aluminum casting process.

The objective of this work is to apply a transient modeling approach to get insights into the effects of cooling rate on the solidification structure of HPDC A383

M. Kärkkäinen · T. Liu · L. Nastac (✉) · L. Brewer
Department of Metallurgical and Materials Engineering, The University of Alabama, Box 870202, Tuscaloosa, AL 35487, USA
e-mail: lnastac@eng.ua.edu

V. Arvikar · I. Levin
Nemak Alabama, Sylacauga, USA

The Minerals, Metals & Materials Society, *TMS 2017 146th Annual Meeting & Exhibition Supplemental Proceedings*, The Minerals, Metals & Materials Series, DOI 10.1007/978-3-319-51493-2_43

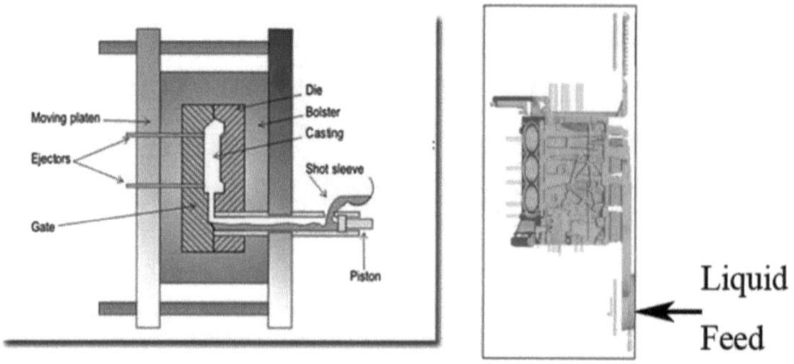

Fig. 1 Diagram of HPDC process and simulation geometry

alloy. The strategy consists of comparing cooling rates simulated by a commercial transient macroscopic model (**NovaFlow & Solid CV** by Novacast (http://www.Novacast.se) with experimentally measured phase fractions, phase sizes and types. The results can assist in achieving better control of the solidification structure and mechanical properties in the final cast product. A schematic of the HPDC process and the simulation is illustrated in Fig. 1.

Simulation Methods

NovaFlow & Solid CV is a commercial mold filling and solidification simulation software developed by NovaCast Systems AB. The software uses a Gauss-Seidel solver with a control volume mesh to solve heat, mass transfer and Navier-Stokes equations as well as free surface fluid flow in 3D via volume of fluid (VOF) method.

The current simulation of the HPDC A383 process used a quasi-equilibrium solidification model with the assumption of no convective heat or mass transfer inside the melt. No segregation was assumed to occur either. A convective cooling heat transfer boundary condition was assumed at the water channel—casting boundary and the mold-air boundary. The main process parameters used in the simulation are summarised in Table 1. The parameters are typical for an HPDC process.

All thermal properties for the A383 alloy and the die material H13 used in the simulation were a function of temperature. A modified phase diagram approach is

Table 1 Process parameters

Initial temperature	630 °C
Filling time	125 ms
Mold thickness	25 mm
Average die temperature	300 °C
Cooling water temperature	20 °C
Air temperature inside the mold	45 °C
Cavity pressure	82.5 MPa

Table 2 Simulated melt composition (wt%)

Al	Si	Cu	Zn	Mn	Ni	Sn	Mg	Fe
83.45	10.5	2.5	1.5	0.5	0.3	0.15	0.1	1.0

used by **NovaFlow &Solid CV** to account for the multicomponent melt chemistry and to model the evolution of the liquid and solidification phases (primary and eutectic) in A383. The geometry of the mesh was 555 × 1080 × 1255 mm, with a cell size of 5 mm. The melt chemistry of the A383 alloy used for the simulation is shown in Table 2.

Experimental Procedure

The A383 cast alloys were made by the High Pressure Die Casting (HPDC) process at **Nemak Alabama**. Samples were cut from the casting locations with different solidification times according to the results of simulation. Samples used for optical microscopy were made by metallographic techniques with diamond suspension in 9, 3, and 1 μm. The microstructure was analysed using a **Nikon Epiphot 200** microscope. The characteristics of sludge and eutectic Si particles, including area and roundness, were measured and quantified using **Photoshop CS6** in conjunction with the optical micrographs. For each sample, at least 3 fields of view at a magnification of 500X were examined from different parts of the sample surface. An energy dispersive x-ray spectroscopy (EDS) analysis of the sludge phase composition was performed using a JEOL 7000 scanning electron microscope (SEM) with an Oxford Instruments EDS detector system (Fig. 2).

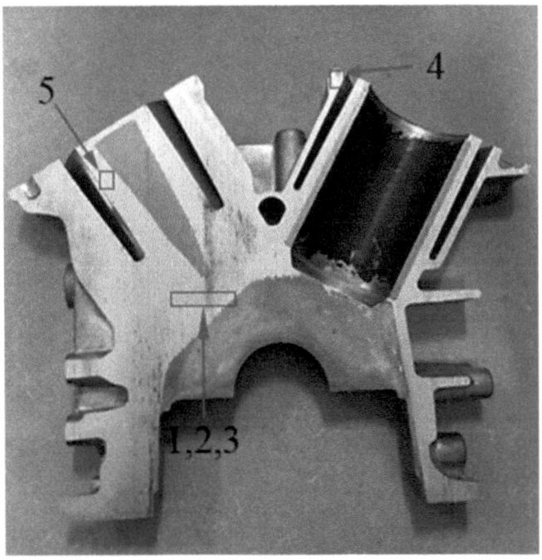

Fig. 2 Location of samples. Samples 1, 2, 3 are the front, rear and center surfaces of the block extracted from the locations shown

Results and Discussion

Figure 3 shows the simulated local solidification time for the cross section used in the measurements. Local solidification time is given by the time required for the point to cool from liquidus to solidus temperature. The figure also has zoomed in view of the location of samples 1, 2 and 3 from a perpendicular axis, plotting the solidification time of these samples. The pictures in Fig. 3 show that solidification time is the shortest at the top of the cylinders, which are the furthest points from feeding.

The local solidification time in these locations is shorter than 1 s. The cylinder walls also rapidly solidify, in the range of 2–10 s. This is explained by the cylinder having a large ratio of area of contact over volume with the mold and an adjacent cast iron "sleeve". The bulkhead, the arc-like geometry below the cylinders, has a solidification time in the range of 30–40 s, which is approximately the same duration as the time the A383 alloy is held inside the mold during the HPDC process. The path along the main feeding vein of the castings has the longest solidification time, 60–80 s. However in practice, the cooling time may be slightly different, since the casting is removed from the mold at approximately 30–45 s after filling and exposed to air cooling, and these steps were not simulated.

The current simulated solidification times do not directly account for "cycling", i.e., the repeated casting of multiple engine blocks in the same mold, which may introduce variation in the mold surface temperature boundary condition. Instead the average temperature of the mold over multiple cycles was used. According to the simulation, the effect of water cooling channels in the casting appears to be highly localised to the inner surface of the channels, and marginal in effect elsewhere.

Figure 4 shows the inverse cooling rate for samples 1, 2 and 3. The cooling rates of the samples coincide with the local solidification time. It can be seen that the cooling rates of the bulkhead outer surfaces, where samples 1 and 3 were taken,

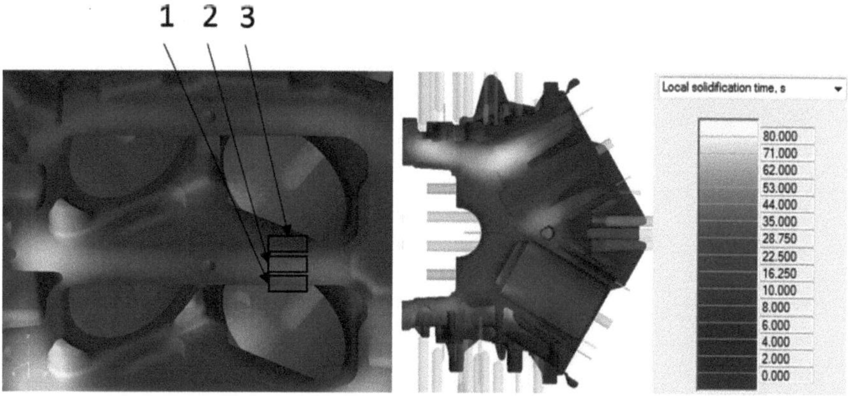

Fig. 3 Local solidification time (in seconds) profile in the A383 casting. Cross-section of casting with close-up of sample 1, 2 and 3 shown on the *left*

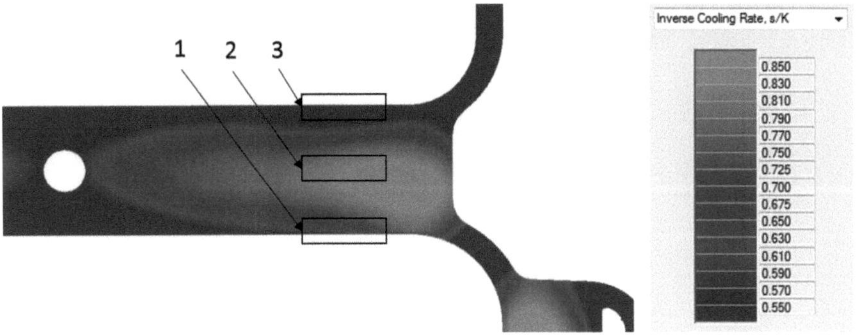

Fig. 4 Inverse cooling rate of the samples 1, 2 and 3, (s/K)

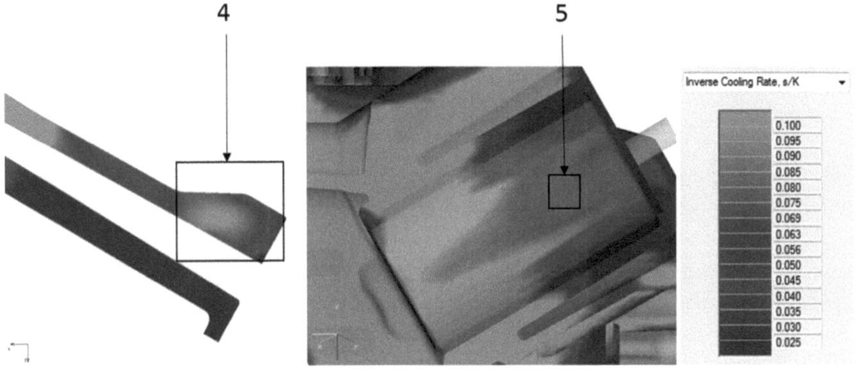

Fig. 5 Inverse cooling rate of the samples 4 and 5, (s/K)

are not symmetrical. Sample 1 has a lower cooling rate than sample 3 due to the flow pattern of the filling. Figure 5 shows the simulated inverse cooling rates of samples 4 and 5. Sample 4, located at the most remote location, the rim circling the right cylinder, actually has a lower cooling rate than the inner cylinder surface where sample 5 was taken. This may be explained by the pattern of flow during filling. The calculated cooling rates for all of the samples are shown in Table 3.

Figure 6 shows micrographs used for the silicon size and roundness analysis. Samples taken close to the metal-mold interface, coinciding with a high cooling rate

Table 3 Cooling rates of the samples

Sample	Cooling rate, K/s
1	1.7
2	1.2
3	2.8
4	10
5	25

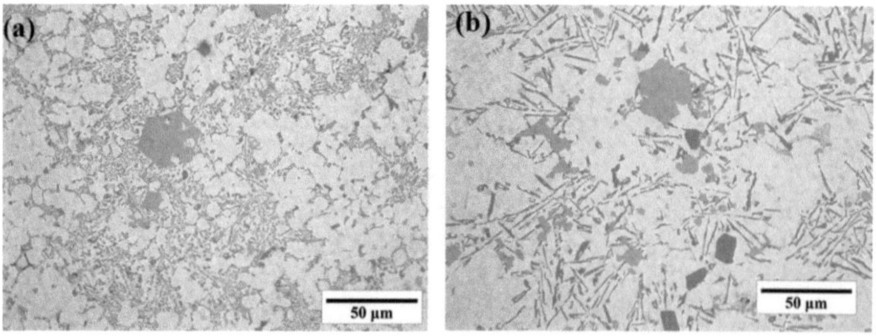

Fig. 6 **a** Microstructure from metal-mold interface. Large plate-like dark grey features are sludge particles. Fine eutectic Al–Si structure in primary aluminum matrix. **b** Microstructure from center of the casting, with acicular silicon, sludge particles, and coarse eutectic grains

Table 4 Sludge composition (wt%)

Al	Si	Cu	Zn	Fe	Mn	Cr	Ni
58.46	11.21	1.85	0.71	18.15	7.14	2.29	0.28

(Fig. 6a), have a fine eutectic structure. The silicon morphology is globular and the area is small. Samples taken from lower cooling rate areas (Fig. 6b) show an acicular or plate-like morphology, and significantly increased size. According to previous experimental work, large elongated needle-like silicon particles are especially harmful for the ductility of the alloy. This figure also shows larger dark grey plate-like features. These are sludge particles, iron-rich intermetallics. In the analysed samples, the sludge content was below 2%. The approximate average composition of the sludge according to EDS-analysis is shown in Table 4.

Figure 7 shows the results of the quantitative analysis of eutectic silicon mean size and roundness. At least 3 micrographs were analysed for each sample and the

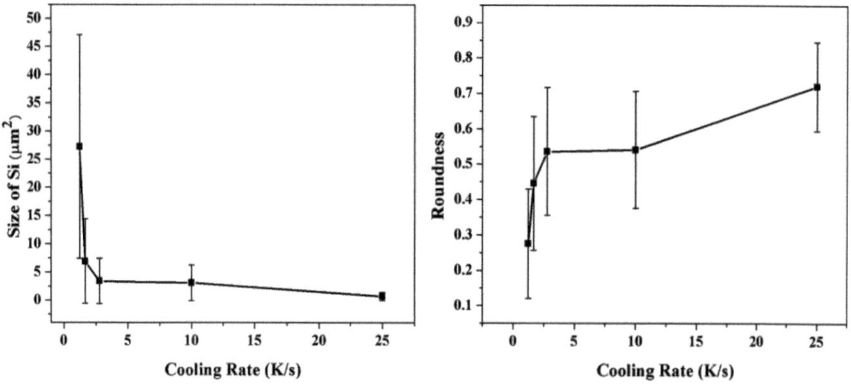

Fig. 7 Cooling rate versus silicon size and roundness

average values are shown in Fig. 6. A roundness value of 1 indicates a spherical particle; while particles with a roundness value approaching zero are needle-like, or acicular, in appearance. There appears to be a logarithmic correlation between the mean particle size and cooling rate. Between the cooling rates 10 and 25 K/s, a major shift occurs from coarse acicular structure to fine globular eutectic, which can be seen in the difference in roundness and size. Another large difference can be observed at cooling rates below 1 K/s, where the mean silicon size rapidly increases and the morphology becomes almost entirely acicular.

Figure 8 illustrates the size of the sludge particles and fraction of the sludge phase in the samples. The results show that the size of the sludge particles increases rapidly at below 2.5 K/s cooling rate, suggesting a different growth rate for the sludge. At cooling rates above 2.5 K/s, no difference is observed in mean sludge size. The fraction of sludge is higher by a factor of 2 in sample 2, taken from the center of the 'bulkhead' of the casting, than the other samples, which were taken close to the metal-mold interface. Again, cooling rate above 2.5 K/s was seen to have a negligible effect of the sludge fraction, as it did with the mean particle size.

There are multiple ways of formation and growth of the sludge phase. Uniformity of the sludge size and fraction would support the conclusion that sludge is formed before the casting, in the holding furnace or in shot sleeve, where it has time to grow uniformly before the HPDC casting process where large thermal gradients exist. However, the large mean size of sludge in the samples taken close to the center of the casting suggests that the sludge particles show major growth and/or coalescence during the solidification interval in the casting, which leads to another conclusion that sludge nucleation and growth can take place during casting as well. Further experiments and simulations are needed to verify the sludge formation timeframe and mechanism.

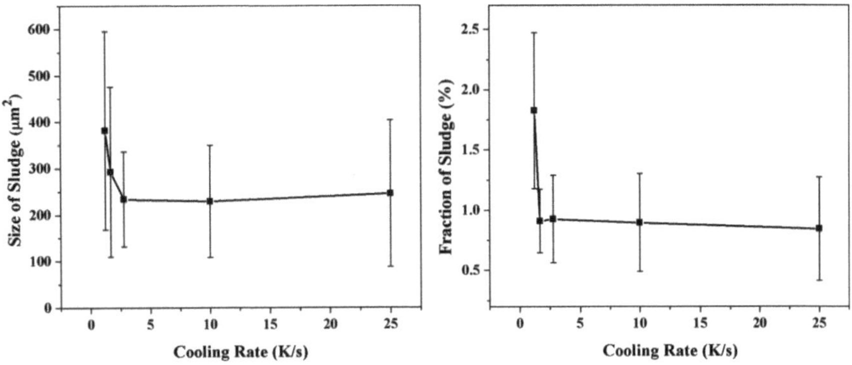

Fig. 8 Mean size and area fraction of sludge particles as a function of cooling rate

Conclusions

The sludge particle size is relatively large at cooling rates below 2.5 K/s. At higher cooling rates, variance is negligible. Fewer sludge particles are found at the metal-mold interface, indicating less trapping of pre-existing sludge formed before the casting. Microstructure modelling and further experiments including differential scanning calorimetry (DSC) are needed to verify the sludge formation mechanism.

The size and roundness of eutectic silicon appear to have a logarithmic correlation with the simulated cooling rate. However, sample 4 is an outlier, and has coarser silicon particle size than the simulated cooling rate indicates. The current results suggest that local cooling rate is a key factor in determining the morphology of the eutectic, which, assuming no defects exist, to a large degree determines the mechanical properties of the casting.

The quasi-equilibrium solidification model appears to be on the right track for predicting solidification time of alloy in HPDC, since the eutectic silicon morphology has a good correlation with the simulated local solidification time. However, more experimental indicators for solidification time and cooling rate are needed for validation.

References

1. S. Ferraro, G. Timelli, Influence of sludge particles on the tensile properties of die-cast secondary aluminum alloys. Metall. Mater. Trans. B **46**, 1022 (2015)
2. G. Timelli, A. Fabrizi, S. Capuzzi, S. Ferraro, The role of Cr additions and Fe-rich compounds on microstructural features and impact toughness of AlSi9Cu3(Fe) diecasting alloys. Mater. Sci. Eng. A **603**, 58–68 (2014)
3. A.M.A. Mohamed, F. Samuel, S.A. Kahtani, Microstructure, tensile properties and fracture behavior of high temperature Al–Si–Mg–Cu cast alloys. Mater. Sci. Eng. A **577**, 64–72 (2013)
4. A.M. Samuel, G.H. Garza-Elizondo, H.W. Doty, F.W. Samuel, Role of modification and melt thermal treatment processes on the microstructure and tensile properties of Al–Si alloys. Mater. Des. **80**, 99–108 (2015)

Hot-Tearing of Multicomponent Al-Cu Alloys Based on Casting Load Measurements in a Constrained Permanent Mold

Adrian S. Sabau, Seyed Mirmiran, Christopher Glaspie, Shimin Li, Diran Apelian, Amit Shyam, J. Allen Haynes and Andres F. Rodriguez

Abstract Hot-tearing is a major casting defect that is often difficult to characterize, especially for multicomponent Al alloys used for cylinder head castings. The susceptibility of multicomponent Al-Cu alloys to hot-tearing during permanent mold casting were investigated using a constrained permanent mold in which the load and displacement were measured. The experimental results for hot tearing susceptibility are compared with those obtained from a hot-tearing criterion based on temperature range evaluated at fraction solids of 0.87 and 0.94. The Cu composition was varied from approximately 5–8 pct. (weight). Casting experiments were conducted without grain refining. The measured load during casting can be used to indicate the severity of hot tearing. However, when small hot-tears are present, the load variation cannot be used to detect and assess hot-tearing susceptibility.

This manuscript has been authored by UT-Battelle, LLC under Contract No. DE-AC05-00OR22725 with the U.S. Department of Energy. The United States Government retains and the publisher, by accepting the article for publication, acknowledges that the United States Government retains a non-exclusive, paid-up, irrevocable, world-wide license to publish or reproduce the published form of this manuscript, or allow others to do so, for United States Government purposes. The Department of Energy will provide public access to these results of federally sponsored research in accordance with the DOE Public Access Plan (http://energy.gov/downloads/doe-public-access-plan).

A.S. Sabau (✉) · A. Shyam · J. Allen Haynes
Oak Ridge National Laboratory, Oak Ridge, TN 37831, USA
e-mail: sabaua@ornl.gov

S. Mirmiran · C. Glaspie
Fiat Chrysler Automobiles North America, LLC, Auburn Hills, MI 48326, USA

S. Li · D. Apelian
Worcester Polytechnic Institute, Worcester, MA 01609, USA

A.F. Rodriguez
Nemak Monterrey, 66000 Garcia, Mexico

The Minerals, Metals & Materials Society, *TMS 2017 146th Annual Meeting & Exhibition Supplemental Proceedings*, The Minerals, Metals & Materials Series,
DOI 10.1007/978-3-319-51493-2_44

Keywords Hot-tearing • Load measurements • Aluminum-copper alloys

Introduction

One of the challenges for the high-volume production of automotive cast metal parts is the reduction of casting defects. Hot tearing is one of the most detrimental solidification casting defects [1]. The appearance of hot-tear defects depends on both the local state of stress and interdendritic feeding. When the interdendritic liquid flow can feed local deformation regions, in which dendrite arms are pulled apart over a significant distance, the space between displaced dendrite arms is filled, and the hot tears are "healed". As the solidification proceeds, the solid fraction increases, area open to interdendritic fluid flow decreases, and liquid feeding of solidification shrinkage and hot tears becomes more difficult. Thus, at high volume fraction of solid, when an opening of the dendritic network by tensile deformation remains unfilled, as liquid feeding is difficult, cracks nucleate and grow, causing a "hot tear" defect [2]. Identifying criteria for hot-tear formation has been the subject of numerous studies [3, 4]. For Al-Cu alloys, Campbell and Clyne [5] found that the peak of the lambda curve for hot tearing susceptibility correlated with the nonequilibrium freezing range. Other criteria are based on the critical values for the strain, accumulated plastic strain, or strain rate [2]. One of the most recent criteria that was developed for solidification cracking in welding was also applied for estimating hot tearing during casting [6]. Most of the studies on hot-tearing of Al alloys were mainly dedicated to the DC casting [7]. Li et al. [8] reviewed hot tearing molds for permanent mold castings.

In this study, an Al-Cu multicomponent alloy system is considered [9]. The Cu composition was varied from approximately 5–8 pct. (weight). The susceptibility of four multicomponent Al-Cu alloys to hot-tearing during permanent mold casting was investigated using a constrained permanent mold in which the load and displacement were measured. The experimental results for hot tearing susceptibilities were compared with those obtained from the hot-tearing criterion presented by [6]. This criterion was based on temperature range evaluated at fraction solids of 0.87 and 0.94.

Solidification Characteristics for Hot Tearing Criteria

The Cu content of the four alloys (DA1, DA2, DA6, and DA7), whose hot-tearing resistance was characterized in this study, was 4.95, 6.6, 7.3, and 8 wt%, respectively [9]. The microstructure model in ProCAST was used to conduct thermodynamic

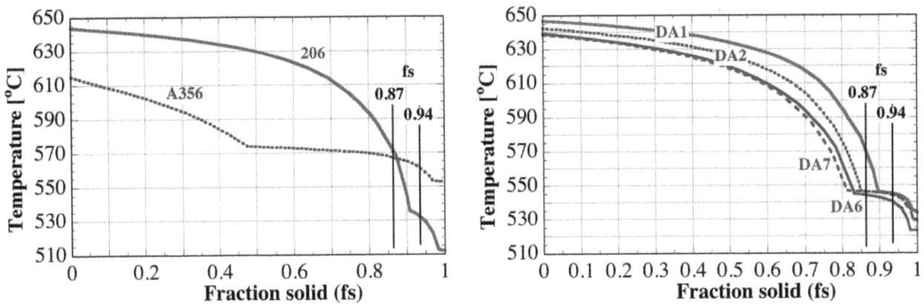

Fig. 1 Calculated solid fraction using the Al material database Computherm and microstructure module in ProCAST for: **a** 206 and A356, and **b** four new alloys considered

simulations for the four alloys considered based on the Scheil model. The variation of the fraction solid (f_s) as a function of temperature (T) is shown in Fig. 1. This fraction solid data can be used in different hot-tearing criteria [2, 6]. For reference, the f_s data on 206 and A356 aluminum alloys was also presented. The solidification characteristics can be used as a first-order estimate of hot-tear resistance. For example, alloys that freeze over a larger temperature range with only a small amount of final eutectic liquid are expected to be less hot tear resistant [5, 10]. The freezing range is almost the same for all DA# considered (from a minimum of 108 °C for DA7 to 116 °C for DA6). Thus, based on freezing range alone, one would expect that DA# family would not be hot tear resistant.

Assessment of Hot-Tearing Resistance Based on Visual Inspection of Castings

The casting experiments to assess the hot-tear resistance were conducted using an instrumented constrained rod mold, which was developed by WPI and CANMET [8] to measure the load and/or temperature during solidification for a restrained casting. The casting consists of two tapered arms that emerge from the bottom of the riser. Three types of castings were made for each alloy, temperature measurement (TM), load measurement (LM), displacement measurement (DM). In the TM castings, the temperature was measured with two thermocouples, whose tips are indicated by arrows in Fig. 2. If the front view of the casting is considered to be on the vertical casting surface in which the thermocouples are placed, then the LHS (left-hand-side) arm will have its end fixed for all casting types and most measurements will be taken from the right-hand-side (RHS) arm. In this section, pictures taken of the casting surface in each arm-riser joint region, where hot tearing is expected to occur, were presented to assess the level of hot tearing resistance.

The pictures of the RHS arm can always be identified by the thermocouple indent mark near the joint transition. Due to the fact that the end of the LHS arm is always fixed, while the RHS arm has a more relaxed geometric constraint at its end,

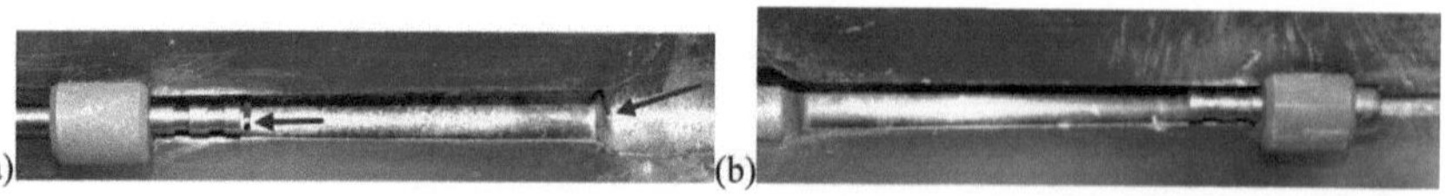

Fig. 2 Picture of mold setup for load measurement type castings showing: **a** mold cavity for the RHS arm with the location of two thermocouples. The thermocouples were intentionally inserted in the mold cavity to their location for the temperature measurement castings, and **b** mold cavity for the LHS arm with the bushing used to fix the arm end

Table 1 Assigned visual hot-tearing indicator (VHTI) for different types of hot-tears

VHTI	Label	Description
0	None	No hot tearing
1	Very short hairline tear (VSH)	Hairline tear near the surface; length less than 1/4 the circumference of the arm
2	Short hairline tear (SH)	Hairline tear near the surface; length less than 1/2 the circumference of the arm
3	Long hairline tear (LH)	Hairline tear near the surface; length approx. the entire circumference of the arm
5	Medium tear (M)	Deep tear that extends over the entire circumference of the arm
8	Severe tear (S)	Wider tear than medium tear that extends over the entire circumference of the arm
10	Complete tear (C)	Partial or complete separation of the arm

hot tearing defects are expected to appear more pronounced on the LHS arm than on the RHS arm. The castings were visually inspected and the severity of cracking was ranked according to the classifications considered in Table 1. For reference, Kamga et al., [11] used four hot tear classifications, which correspond to SH, LH, S, and C in Table 1. For the sake of simplicity, images for representative LM castings are presented in Figs. 3, 4, 5, and 6 for the DA1, DA2, DA6, and DA7 alloys. The Figs. 3 and 4 indicate that severe hot tearing is seen for DA1 and DA2 castings. The hot-tearing defects were found to be limited to very thin hairline cracks for the DA6 and DA7 castings. The summary of the visual inspection ranking of hot-tearing severity is shown in Table 2.

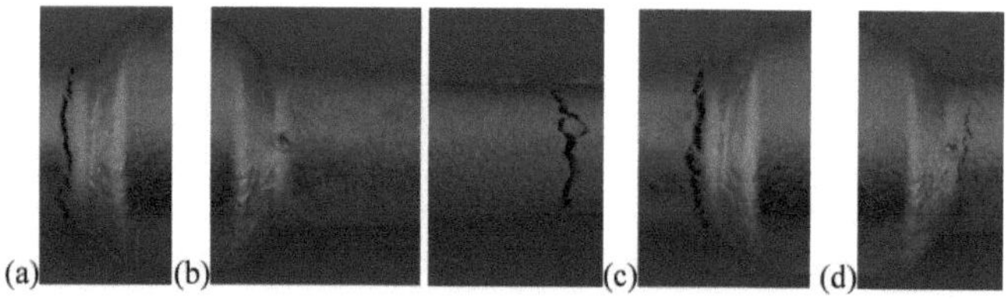

Fig. 3 Pictures of the DA1 alloy load measurement castings, showing hot tearing in: **a** L4 left arm, **b** L4 right arm, **c** L5 left arm, and **d** L5 right arm

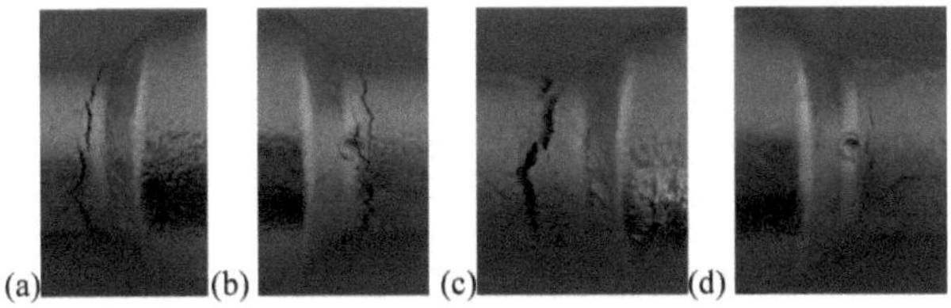

Fig. 4 Pictures of the DA2 alloy load measurement castings, showing hot tearing in: **a** L3 left arm, **b** L3 right arm, **c** L4 left arm, and **d** L4 right arm

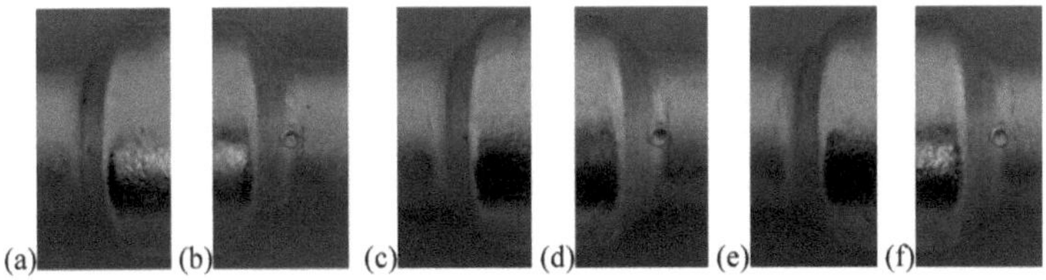

Fig. 5 Pictures of the DA6 alloy load measurement castings, showing: **a** L1 left arm, **b** L1 right arm, **c** L2 left arm, **d** L2 right arm, **e** L3 left arm, **f** L3 right arm

Fig. 6 Pictures of the DA7 alloy load measurement castings, showing: **a** L1 left arm, **b** L1 right arm, **c** L2 left arm, and **d** L2 right arm

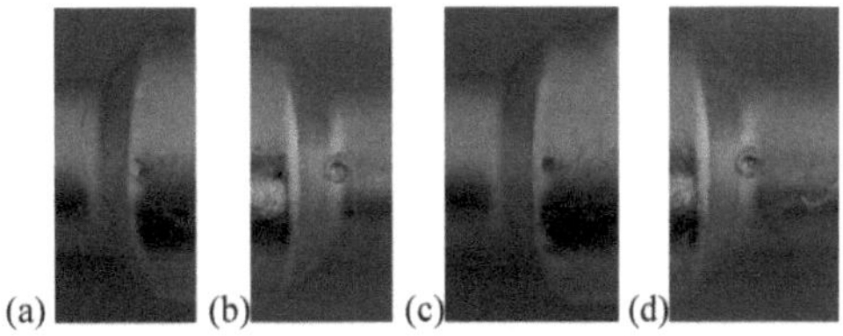

Table 2 Summary of hot-tearing defects in LM castings based on visual inspection

Alloy	No.	LHS arm	VHTI LHS arm	RHS arm	VHTI RHS arm
DA1	L4	S/C (9)	9.5	C (10) arm SH (2)	6.5
	L5	C (10)		LH (3)	
DA2	L3	M (5)	7.5	LH (3)	2.5
	L4	C (10)		SH (2)	
DA6	L1	None (0)	1	VSH (1)	2
	L2	None (0)		SH (2)	
	L3	LH (3)		LH (3)	
DA7	L2	VSH (1)	0.5	None (0)	0.5
	L3	None (0)		VSH (1)	

Load Measurement During Casting Solidification

The load measurements during casting in the WPI mold can be used to obtain quantitative data on the hot-tearing defects, as it was shown for A356 and 206 alloys [8] by correlating the load variation with the solidification behavior. Specifically, Li et al. [8] correlated sudden changes in the load rate with onset of hot-tearing and crack propagation. The "zero" time origin was selected for both the TM and LM castings to be that instant at which molten metal enters the right casting arm, i.e., as evidenced by the initial rise in the measured temperature on the riser side. The raw data for measured load, i.e., unsmoothed by any data post processing algorithm, is shown in Fig. 7 as a function of time during casting solidification for the four new alloys considered. For each LM case, the load onset time (i.e., cross-over from the initial negative values to positive values) is marked on each figure. As shown in Fig. 7, the lowest and highest load levels were measured for DA1 and DA7, respectively. DA1, DA2, and DA6 were found to exhibit a wide range load variation while the load measurement for DA7 showed very good reproducibility.

The variation in the load rate was found to be very important by Li et al. [8] for the WPI castings. The 7-point moving-average data for load rate is shown in Fig. 8 as a function of time during casting solidification for the four new alloys considered. Li et al. [8] indicated that a "V" like signature of the load rate variation was found to occur for hot-tearing, the left-top point on the "V" indicating the "crack onset" and the end of the "V" indicating the end of the "crack propagation". This

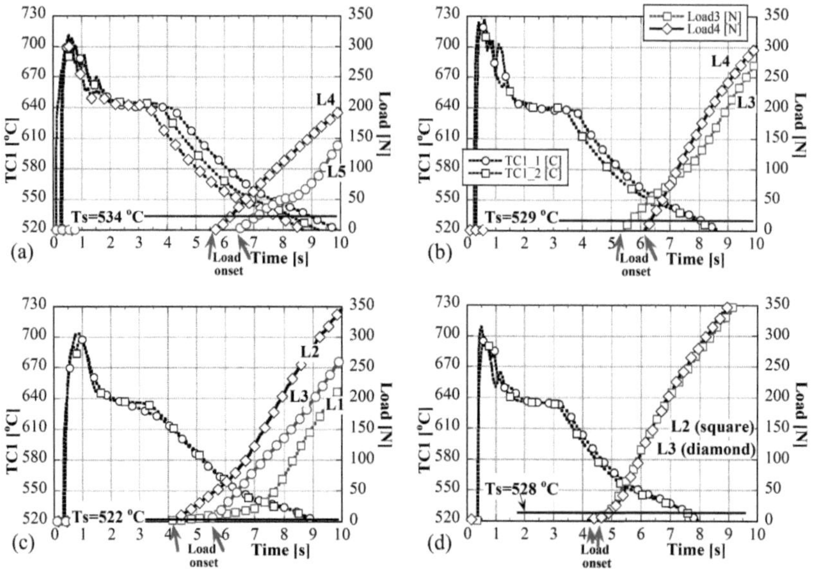

Fig. 7 Measured load during casting for alloys: **a** DA1, **b** DA2, **c** DA6, and **d** DA7

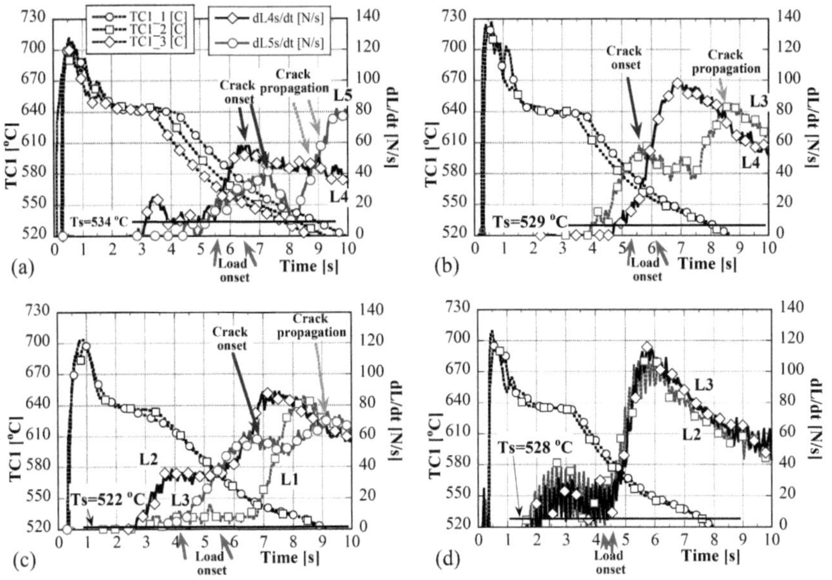

Fig. 8 Calculated load rate during casting solidification for alloys: **a** DA1, **b** DA2, **c** DA6, and **d** DA7

"V" signature was identified for each of the LM data presented in Fig. 8. The typical "V" signature for hot-tearing can be easily observed for DA1 castings (Fig. 8a). For DA2, only the L3 casting shows the hot-tearing signature (Fig. 8b). The L4-DA2 casting does not exhibit the "V" hot-tearing signature, indicating that for the short hairline tear (SH) (Fig. 4d for the right arm), the load measurement is not sensitive enough to identify this type of very small hot-tearing. This is also evidenced for the L1-DA6 and L2-DA6 load data, i.e., a "V" shape in the load variation could not be noticed to characterize the Very short hairline tear (VSH) and Short hairline tear (SH) seen in Figs. 5b–d, respectively. On other hand, the Long hairline tear (LH) seen in Fig. 5f, which is actually extended over the entire arm circumference unlike the SH tear seen in Figs. 5b–d, exhibits the typical "V" shape signature in the load variation (Fig. 8c). The load rate exhibited the most consistent variation for the DA7 castings (Fig. 8d).

Discussions and Conclusions

The susceptibility of multicomponent Al-Cu alloys to hot-tearing during permanent mold casting was investigated using the WPI constrained permanent mold in which the load and displacement were measured. First, it has to be mentioned that the alloys with 7.3 and 8 wt. Cu were found to be the most hot-tear resistant. Second, the data for the hot-tearing indicator introduced by Kou [6], which is based on

Table 3 Comparison between hot-tearing indicators based on thermodynamic data and experimental data

Alloy	DA1	DA2	DA6	DA7
Cu wt. [%]	4.95	6.6	7.3	8
KHTI – $\Delta T/\Delta\sqrt{f}$ between $f_s = 0.87$ and $f_s = 0.94$	620	61	126	66
Expected hot-tear resistance based on KHTI	Very low	High	High	High
VHTI (scale 0–10)	8	5	1.5	0.5
Hot-tear resistance based on VHTI	Very low	Low	Low-High	High

variation of $T(\sqrt{f})$ between $f_s = 0.87$ and $f_s = 0.94$, is compared directly with the hot-tearing indicator evaluated from the visual inspection of the WPI castings (Table 2). It has to be noted that it is more convenient to use Kou's hot-tearing indicator (KHTI) rather than conduct casting experiments, as KHTI can be obtained solely from thermodynamic simulations. For the data shown in Table 3, the VHTI was calculated by averaging the VHTI for the left arm and right arm (Table 2). The values of the Kou's hot-tearing indicator (KHTI) based on temperature slope in the $T(\sqrt{f})$ curve indicate that DA2, DA6, and DA7 are expected to be hot-tearing resistant alloys while the VHTI data indicate that DA7 and, to some extent DA6, are hot-tear resistant alloys. The ranking of the hot-tear resistance of the least hot-tear resistant alloy, DA1, and the most hot-tear resistant alloy, DA7, are in good agreement between the KHTI and the experimental VHTI. However, the DA2 alloy, which experimentally is shown to have a poor hot-tear resistance, the KHTI predicts a very high hot-tear resistance. This comparison between experimental and computational data is very important, as provides more information that can be used to assess the range of applicability of the KHTI criterion to accurately predict hot-tearing resistance of Al alloys.

Table 4 VHTI for the right casting arm (Table 2), load at solidus and load rate average

Alloy	LM No.	VHTI for RHS arm	Load at solidus [N]	Load rate average [N/s]
DA1	L4	2	141	45.7
	L5	3	63	28.4
DA2	L3	3	166	46
	L4	2	125	75
DA6	L1	1	153	31.5
	L2	2	283	59.6
	L3	3	196	54.7
DA7	L2	0	246	76.6
	L3	1	252	79

Third, in an attempt to correlate the hot-tearing with the quantitative variables related to load and load rate values, the VHTI for the right casting arm (Table 2) are shown in Table 4 together with load at solidus point and load rate average. It has to be mentioned that when small hot-tears are present, the load variation cannot be used to detect and assess hot-tearing susceptibility. Although it is not a strong correlation, it can be seen from the data shown in Table 4 that the higher the load at solidus temperature, the lower the hot-tearing indicator. Excluding the L1-DA6 data on average of the load rate, it was found that there is a relatively strong correlation in the average of the load rate and the hot-tearing indicator, i.e., the higher the average of the load rate the lower the hot-tearing indicator. The data also show that at higher VHTI values larger crack growth times were observed. Thus, load at solidus point and load rate average over the freezing range, which were obtained from the measured load data, can be used to indicate the hot tearing resistance of the Al-Cu multicomponent alloys.

Acknowledgements This work was performed under a Cooperative Research and Development Agreement (CRADA) with the Nemak Inc., and Fiat Chrysler Automobiles (FCA) for the project "High Performance Cast Aluminum Alloys for Next Generation Passenger Vehicle Engines." Research sponsored by the U. S. Department of Energy, Office of Energy Efficiency and Renewable Energy, Vehicle Technologies Office, as part of the Propulsion Materials Program under contract DE-AC05-00OR22725 with UT-Battelle, LLC.

References

1. J. Campbell, *Castings*. (Butterworth and Heineman, Ltd, 1991)
2. M. Rappaz, J.-M. Drezet, M. Gremaud, A new hot-tearing criterion. Metall. and Materials Trans. **30A**, 449–455 (1999)
3. U. Feurer, Giesserei Forsch. **2**, 75–80 (1976)
4. T.W. Clyne, G.J. Davies, The Influence of Composition on Solidification Cracking in Binary Alloy Systems. Brit. Foundryman. **74**, 65–73 (1981)
5. J. Campbell, T.W. Clyne, Cast Metals **3**(4), 224–226 (1991)
6. S. Kou, A criterion for cracking during solidification. Acta Mater. **88**, 366–374 (2015)
7. C. Davidson, D. Viano, L. Lu, D. John St, Observation of crack initiation during hot tearing. Int. J. Cast Met. Res. **19**, 59–65 (2006)
8. S. Li, K. Sadayappan, D. Apelian, Characterization of hot tearing in al cast alloys: methodology and procedures. Int. J. Cast Metals Res. **24**(2), 88–95(8) (2011)
9. A. Shyam, J.A. Haynes, P.J. Maziasz, A.S. Sabau, D. Shin, Y. Yamamoto, C.R. Glaspie, J.A. Gonzalez-Villarreal, S. Mirmiran, A.F. Rodriguez-Jasso, S. Roy, J. Talamantes-Silva, L. Zhang, Aluminum Alloy Compositions and Methods of Making and Using the Same. Non-Provisional Application Filed; U.S. Patent Application No. 15/160,926; ORNL Ref. No. 3569.0, Filled on May 20, 2016
10. R.A. Rosenberg, M.C. Flemings, H.F. Taylor, Nonferrous binary alloys hot tearing. AFS Trans **58**, 518–528 (1960)
11. H.K. Kamga, D. Larouche, M. Bournane, A. Rahem, Hot tearing of aluminum–copper B206 alloys with iron and silicon additions. Mat. Sci. and Eng. A **527**, 7413–7423 (2010)

Effect of Molybdenum Content, Pouring Temperature and Cooling Rate on the Casting Defects of High Chromium White Cast Iron

Izudin Dugic

Abstract High chromium white cast are commonly used in application requiring excellent abrasion resistance, as central parts for pumps. The specifications and requirements applied for the white cast iron components are among the most stringent used within the iron foundry branch. One of the biggest problems for the production of these components is hot tearing. Irrespective of the name, this phenomenon represents the formation of an irreversible failure (crack) in the still semisolid casting. This paper aims to investigate the effect of molybdenum content, pouring temperature and cooling rate on the casting defect hot tearing. The procedure to achieve this was to study one casting component, impeller, in a production scale. The experiments showed that molybdenum content and pouring temperature had an important influence on casting defects. It was also observed that the solidification rate has a strong effect on the hot cracking.

Keywords White cast iron • Molybdenum • Pouring temperature • Cooling rate • Hot tearing and casting

Introduction

High chromium white cast iron is considered as one of the most useful wear resistance materials and their usage are widely spread in industry as example central parts for pumps. Industrial pumps in general are especially designed to meet the highest conveying requirements for the use within the industry. These pumps are capable of handling different materials; lubricating, abrasive, chemically neutral, or corrosive product plus fibrous matter and matter containing solids with great ease. In the mining industry, high solids content and large particles can dramatically increase wear costs and have a negative impact on the pump's operation [1, 2].

I. Dugic (✉)
Faculty of Technology, Department of Mechanical Engineering,
Linnaeus University, 35195 Vaxjo, Sweden
e-mail: izudin.dugic@lnu.se

The Minerals, Metals & Materials Society, *TMS 2017 146th Annual Meeting & Exhibition Supplemental Proceedings*, The Minerals, Metals & Materials Series,
DOI 10.1007/978-3-319-51493-2_45

The European Standard (EN 12513:2011) defines the grades of abrasion resistant white cast irons [3]. It specifies the grades in terms of chemical composition and hardness. The types of abrasion resistant white cast irons covered by this standard are:

(a) unalloyed or low alloy cast irons (EN-GJN-HV350);
(b) nickel-chromium cast irons covering two general types:

- 4% Ni 2% Cr cast irons (EN-GJN-HV550);
- 5% Ni 9% Cr cast irons (EN-GJN-HV600);

(c) high chromium cast irons covering four ranges of chromium content:

- 11% < Cr ≤ 14% (EN-GJN-HV600(XCr11));
- 14% < Cr ≤ 18% (EN-GJN-HV600(XCr14));
- 18% < Cr ≤ 23% (EN-GJN-HV600(XCr18));
- 23% < Cr ≤ 28% (EN-GJN-HV600(XCr23));

and for each chromium content range, three ranges of carbon content.

An very important class of these materials is high chromium white cast iron, EN-GJN-HV600(XCr23). Their exceptional wear resistance is the result of their high carbide content, which forms along with austenite during solidification as a proeutectic or eutectic phase depending on alloy composition. This is particularly depending upon carbon and chromium content. High chromium white irons are typically cast as hypoeutectic alloys having a primary phase of dendritic austenite with the eutectic made up of austenite and M_7C_3 carbides [2, 4].

Table 1 presents a typical chemical composition used for the production of as-cast EN-GJN-HV600(XCr23) castings [3].

The Vickers hardness test shall be carried out in accordance with EN ISO 6507 and for the EN-GJN-HV600(XCr23) should be min 600 HV (or 53 HRC, Rockwell C) [3].

The wear resistance and mechanical properties of high chromium white cast iron mainly depend on type, size, number, morphology of hard carbides and the matrix structure (γ or α). The high chromium cast iron having a structure consisting of complex carbides in matrix which, in the hardened condition, is predominantly martensitic but which can also contain some austenite or other transformation products of austenite. Maratray and Usseglio-Nanot [5] showed that the carbon and

Table 1 Chemical composition of EN-GJN-HV600(XCr23) high chromium cast iron

Chemical composition in % (mass fraction)								
C	Si max	Mn	P max	S max	Cr max	Ni max	Mo max	Cu max
>1.8–2.4 >2.4–3.2 >3.2–3.6	1.0	0.5–1.5	0.08	0.08	23.0–28.0	2.0	3.0	1.2

chromium contents control the volume fraction of eutectic carbides according to the Eq. (1):

$$\%E.C. = 12.33(\%C) + 0.55(\%Cr) - 15.2 \quad (1)$$

where, %E.C. is the volume percent of eutectic carbides, %C is the weight percent carbon, and %Cr is the weight percent chromium.

According to Albertin and Sinatora [6], the hardness and wear resistance of high chromium white cast iron is possible to improve with the increase of EC.

The required structure and hardness of abrasion resistant cast irons are obtained by the selection of a suitable chemical composition and the manufacturing method. The effect of different alloying elements on mechanical properties of high chromium white cast iron have been discussed by many authors and lots of works have been done in the foundries [7–11].

The specifications and requirements applied for the white cast iron components are among the most stringent used within the iron foundry branch. From the data given in literature [4–11] it follows that there are still various problems that occur during manufacture of this cast material and have not been properly solved until the present day. One of the biggest problems for the production of these components is hot tearing or hot cracking, or hot shortness.

Due to the influence of various parameters on the quality and mechanical properties of high chromium white cast iron, the aim of the current work is to highlight the influence of molybdenum content, pouring temperature and cooling rate on the casting defects of high chromium white cast iron.

Experimental

Alloys and Materials

The XY Foundry produces several different impellers for pumps. The European standard EN-GJN-HV600(XCr23) has the status of the XY Foundry standard in relevant cast impeller for pumps as shown in Table 1. The hardness test shall be carried out in accordance with EN ISO 6507 and should be 55 ± 5 HRC.

In the experimental works a typical "Impeller" was selected, illustrated in Fig. 1, with the diameter of 260 mm and weight of 7.3 kg and tested in production scale.

In order to investigate the influence of molybdenum content and pouring temperature on the casting defects, the castings were made in shell moulding using resin coated sand. The sand contains iron oxide to reduce the thermal shock. Average grain size of the sand was 0.18 mm and grain concentration 78.46%. Loss on ignition was 4.85% and shell forming time 15 s. A three series of castings experiments were carried out. The chemical composition was determined by the light emission spectrometer—ARL 3460. The chemical compositions of the base melt for each casting and pouring temperature is shown in Table 2.

Fig. 1 The casting component "Impeller"

Table 2 The chemical composition of the base melts

Experiment	Element in wt %								%E. C	Ratio %Cr/%C	Pouring temp °C
	C	Si	S	P	Mn	Ni	Cr	Mo			
I	2.54	0.78	0.04	0.02	0.55	0.31	24.82	0.97	29.77	9.77	1535
II	2.51	0.77	0.03	0.02	0.65	0.41	24.56	1.48	29.25	9.78	1545
III-1 III-2 III-3	2.47	0.79	0.03	0.02	0.57	0.23	25.06	1.98	29.04	10.15	1560 1570 1580
IV	2.93	1.60	0.03	0.03	0.28	0.26	25.52	0.25	34.96	8.71	1540
V	2.91	1.59	0.03	0.03	0.29	0.27	25.50	0.22	34.71	8.76	1542
VI	2.90	1.60	0.03	0.03	0.30	0.27	25.51	0.23	34.59	8.80	1538

For each of the different chemical composition of the melt, 45 shells were moulded and cast. In the third experiment (experiment III), for each of the three different pouring temperature (III-1 1560 °C, III-2 1570 °C and III-3 1580 °C) 15 shell were moulded and cast. This yield a total 135 castings, thereby providing a sound statistical basis for evaluation of the effect of molybdenum content and pouring temperature on the casting defects.

Different cooling rate after pouring were tested using silica sand and a water based phenolic resin system cured by the use of CO_2 gas by Laempe machine LFB25. One "adapter" is manufactured to use same pattern which is used for the shell moulding. The silica sand with particle size (MK) of 0.27 mm (AFS 57) were mixed with a resin content of 2.3% based on the weight of the sand, see Figs. 2 and 3.

Fig. 2 Adapter with the pattern

Fig. 3 The castings lower mould

Table 3 The high of the mould used in the experiments

Name	Experiment								
	A	B	C	D	E	F	G	H	I
Lower mould, cm	11	16	11	16	20	16	20	11	20
Upper mould, cm	11	11	16	16	16	20	20	20	11

Each mould had length of 508 mm and width of 455 mm but the height of the lower and upper mould were various, see Table 3. The reason to have different hight of the lower and upper mould was to have different cooling rate.

For each different height of the mould 25 castings were moulded. To have a better emission of gases from the mould, two air directions are drilled in each feeder with a drill of 15 mm in diameter. The gluing of the lower and upper mould was made with wood glue.

A three series (IV, V and VI) of castings experiments were carried out. The chemical composition were determined by the light emission spectrometer—ARL 3460 and are shown in Table 2. The Melt IV was used to cast the experiment A, B, C. The Melt V was used to cast the experiment D, E and F. The Melt VI was used to cast the experiment G, H and I. All the casting experiments yield a total 225 castings, thereby providing a sound statistical basis for evaluation of the effect of different cooling rate on the casting defects.

Results and Discussions

Castings Defects

After sand blasting all castings were investigated first by an ocular inspection, and then with the Liquid penetrant testing in accordance with EN 1371-1:2011. The results are given in Table 4. The hot tearing often occurs in same locations on this casting. These locations are shown in Fig. 4.

Mechanical Testing

The hardness tests were performed on each casting component in according to SS-EN ISO 6508:2006. In Table 5 the average value for all measurements are shown.

Table 4 Statistical listing of casting defects in % found on the castings

Casting defect	Experiment													
	I	II	III			IV			V			VI		
			III-1	III-2	III-3	A	B	C	D	E	F	G	H	I
Hot tearing	4	0	40	73	80	16	12	12	4	8	8	4	8	12
Veining	0	0	0	0	0	0	0	0	4	8	12	16	12	8
Cold shut	9	56	0	0	0	0	0	0	0	0	0	0	0	0

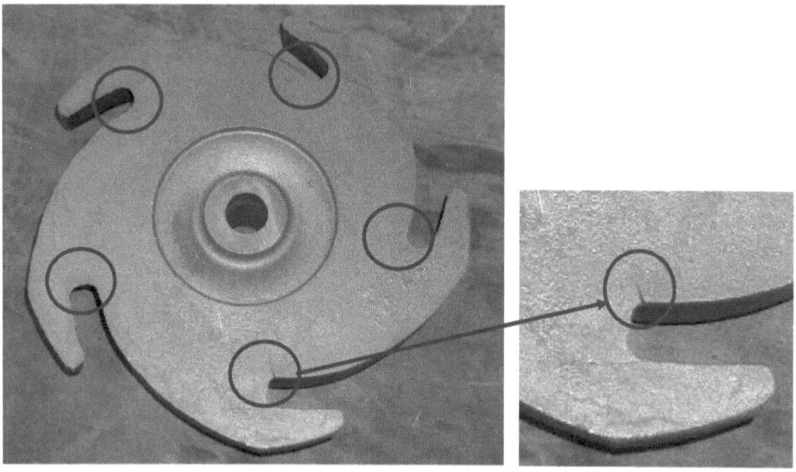

Fig. 4 The casting component "Impeller". The *circles* show the areas where the hot tearing usually occur

Table 5 The average value of hardness measurements in Rockwel, HRC

Condition	Experiment													
	I	II	III			IV			V			VI		
			III-1	III-2	III-3	A	B	C	D	E	F	G	H	I
Untreated	52	53	54	53	54	52	51	51	50	49	48	45	46	46

Casting Simulation

A series of simulations were performed with the casting simulation program MagmaSoft® using the add-on module MagmaIron®, especially developed for cast iron simulation. The "Impeller" with its gating system was modeled for the simulation. In the simulation, the mould filling sequence as well as the solidification sequence was considered. Figure 5 shown 3-D result from the simulation, showing strain ration when 50% of the mould is solidified. Strain rate is elevated at radius (the arrows show the areas). The material has not solidified there yet and there is a risk of hot cracking.

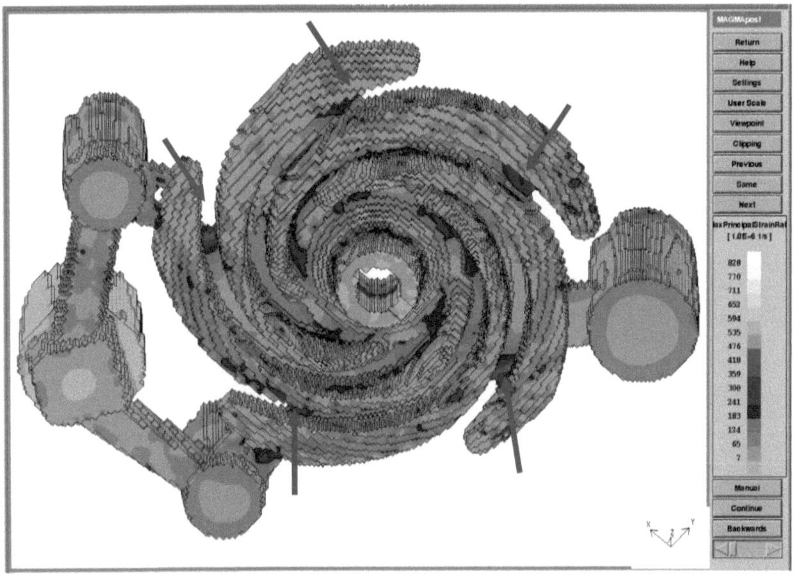

Fig. 5 3-D result from the simulation showing Strain ration when 50% of the mould is solidified

Conclusions

Molybdenum content, pouring temperature and cooling rate appeared to be key process parameters for eliminating casting defect hot tearing. With increasing of the pouring temperature hot tearing increase. When we use the lower pouring temperature, the other casting defect occurs, the cold shut.

In the experiments with different cooling rate, the worst cases of hot tearing were obtained with experiment A. The experiments show that the best results regarding hot tearing have been obtained in experiments D and G, only 4% for both two experiments. However, in experiment G with the highest lower and upper mould is obtained a large tendency to other castings defect, veining.

By simulation it is possible to detect the areas where hot tearing ale likely to be formed.

Acknowledgements The author would like to thank Linnaeus University, Faculty of Technology, Department of Mechanical Engineering, Växjö, Sweden.

References

1. T.W. Chenje, D.J. Simbi, E. Navara, Relationship between microstructure, hardness, impact toughness and wear performance of selected grinding media for mineral ore milling operations. Mater. Des. **25**, 11 (2006)
2. A. Studnicki, J. Kilarski, M. Przybył, J. Suchoń, D. Bartocha, Wear resistance of chromium cast iron—research and application. J. Achiev. Mater. Manuf. Eng. **16**(1–2) (2006)
3. CEN, European Committee for Standardization, Ref. No. EN 12513:2011
4. L. Baotong, L. Jingli, C. Stefano, Corrosion and wear resistance of chrome white irons–a correlation to their composition and microstructure. Metall. Mater. Trans. A **37**, 3029 (2006)
5. F. Maratray, R. Usseglio-Nanot, Factors affecting the structure of chromium and chromium-molybdenum white irons. Climax Molybdenum S.A. (1970)
6. E. Albertin, A. Sinatora, Effect of carbide fraction and matrix microstructure on the wear of cast iron balls tested in a laboratory ball mill. Wear **250**(1–12) 492–501. 13th International Conference on Wear of Materials (2001)
7. Khaled M. Ibrahim, Adel A. Nofal, Effect of titanium addition on structure and properties of the as-cast high-cr-mo white iron. China Foundry **7**(4), 438–446 (2010)
8. C. Scandian, C. Boher, J.D.B. de Mello, F. Rézaï-Aria, Effect of molybdenum and chromium contents in sliding wear of high-chromium white cast iron: the relationship between microstructure and wear. Wear. **267**(1–4) 401–408. 17th International Conference on Wear of Materials (2009)
9. J.W. Choi, S.K. Chang, Effects of molybdenum and copper additions on microstructure of high chromium cast iron rolls. ISIJ International **32**(11) (1992)
10. C.R. Loper, H.K. Baik, Influence of molybdenum and titanium on the microstructures of Fe-C-Cr-Nb white cast irons. AFS Trans. **97**, 1001–1008 (1989)
11. P. Dupin, J.M. Schissler, Influence of additions of silicon, molybdenum, vanadium, and tungsten upon the structural evolution of the as-cast state of a high-chromium cast iron (20% Cr, 2.6%C). AFS Trans. **92**, 355–360 (1984)

Effect of Various Aluminum Content on the Formation of Inclusion

Yan Luo, Lifeng Zhang, Wen Yang and Ping Shen

Abstract More attention is paid to high aluminum steel which attains hardness, abrasive resistance and corrosion resistance. However, the development of subsurface nitrides and oxides in a nitrogen-rich atmosphere during smelting process and solidification makes a significant effect on the properties of high aluminum steel. It is important to understand how theses nitrides and oxides form and their consequences for the quality of steel products. In the current study, the effect of various aluminum content on the formation of inclusions in a nitrogen-rich atmosphere was investigated using vacuum induction furnace and high-temperature contact angle device. The growth and evolution of the precipitation in a subsurface oxygen-depleted region were evaluated by means of optical microscope and scanning electron microscope. The depth of the internal oxide and nitride zones were different in the samples containing various aluminum content. Increasing aluminum content to a given value has an influence on modifying the surface oxide scale composition and increasing its continuity, which gradually decreased the internal precipitation zones with increasing aluminum content. For the sample with higher aluminum content, a thick continuous oxide layer formed and completely prevented nitrogen penetration into the bulk of the samples.

Keywords High aluminum steel · Inclusion · Evolution

Introduction

More attention is paid to high aluminum steel for their hardness, abrasive resistance and corrosion resistance. However, high aluminum (Al) content in steel possibly lead to the formation of nitrides at high temperature in rich nitrogen environments because aluminum nitrides (AlN) can be easily formed in liquid steel at such high

Y. Luo · L. Zhang (✉) · W. Yang · P. Shen
School of Ecological and Metallurgical Engineering, University of Science and Technology Beijing, Beijing 100083, China
e-mail: zhanglifeng@ustb.edu.cn

The Minerals, Metals & Materials Society, *TMS 2017 146th Annual Meeting & Exhibition Supplemental Proceedings*, The Minerals, Metals & Materials Series, DOI 10.1007/978-3-319-51493-2_46

Al content up to 1 mass% [1–3]. Nitrogen from the air has an effect on continuous casting. The gaseous species can react with dissolved Al in steel to form an oxide layer on the surface and oxides and nitride precipitation in the subsurface. Oxide layers on the surface are first formed because the free energy of formation of alumina is far lower than that of AlN. Oxide films, if dense and continuous, are known to severely limit the penetration of adsorbed nitrogen is low in the oxide. The outer oxide films can be modified die surface during the next process. But oxides and nitride precipitation in the subsurface are likely to fall off and embroil the molten steel. These oxides and nitride precipitates are harmful to both the quality of products and the castability of steels. Considerable investigations have reported that oxide inclusions deteriorate the corrosion and crack [4–6]. Holappa et al. [7] reported there was a bad machining property for conventional steel grades without high oxide inclusion cleanliness. Wegrzyn et al. [8] reported the role of oxide inclusion sites on the transformation of austenite to acicular ferrite in steel weld metal deposits and their toughness. Although AlN are beneficial in restricting grain growth, they can cause embrittlement and induce cracking phenomena such as rock candy fracture during continuous casting and hot rolling [9, 10], the similar results reported by Wilson and Gladman [11]. Vedani et al. [12] reported AlN can affect grain boundary cohesion to promote the development of crack, and is regarded as a detrimental phase for hot ductility of steels. Therefore, these aluminum-rich steel grades placed great challenges to steelmaking and casting processes.

The current work is to study the conditions for the formation of nitrides in the steel with different content of dissolved aluminum.

Experimental Procedure

Approximately 400 g of electrolytic iron and various weight of high purity Al particles (99.999 wt%) were re-melted to obtain different Al samples in a vacuum induction furnace. The compositions of steel ingot were shown in Table 1.

The metal samples were polished before experiments to remove any pre-existing oxides and to provide an adequate surface for observing in the high-temperature contact angle device. The partial pressure of oxygen in the nitrogen gas used in these experiments was measured with a doped zirconia-based oxygen sensor to be about 1×10^{-14} ppm. Each sample was then loaded in cylindrical plates of

Table 1 Composition of steel ingots

[Al]s (wt%)	[Mg] (ppm)	T.O. (ppm)	[N] (ppm)
6.1	<5	11.0	5.9
4.13	5	2.3	23.0
2.42	4	6.8	4.3
1.09	5	9.2	29.8
0.50	5	5.9	15.8

alumina substrate with a diameter of 30 mm and a height of 3 mm, placed in a sealed, and aerated the furnace for 30 min before experiments to remove the air of the furnace.

The heating regime was designed to re-melt the metal samples. The temperature was ramped from room temperature to 1200 °C over the course of 120 min, and then slowly ramped to 1600 °C at the speed of 5 °C/min. The sample was required to be held at 1600 °C for 120 min. In the present experiments, the samples were rapidly cooled around 10 °C/min in a nitrogen atmosphere.

0.5 and 1 wt% Al were expected to undergo a phase transformation from α-Fe to γ-Fe during the heating. Samples with 2, 4 and 8 wt% Al remained α-Fe throughout the present temperature range.

After heat treatment, the composition and morphology of the sub-surface scale layers and precipitates on the surface and cross-section of each sample were determined using scanning electron microscopy with energy dispersive X-ray spectroscopy (SEM-EDS). ASPEX was used to detect the behavior of precipitates in the cross-section of each sample.

Results and Discussion

Figure 1 shows the surface morphology of each sample with various Al content. Samples with 6, 4 and 2 wt% Al was not melted due to the formation of oxide film on the surface, while 1 wt% aluminum sample was partially melted and 0.5 wt% aluminum sample was fully melted at the given temperature under a rich nitrogen atmosphere. The composition of oxide film was iron oxide, alumina, hercynite and iron-aluminum spinel phase, according to Bott's results [13]. Their morphologies varied with different aluminum content. For high aluminum samples, the shape of oxide film was a large irregular plate, yet that of low aluminum samples was like-leaf and small plate.

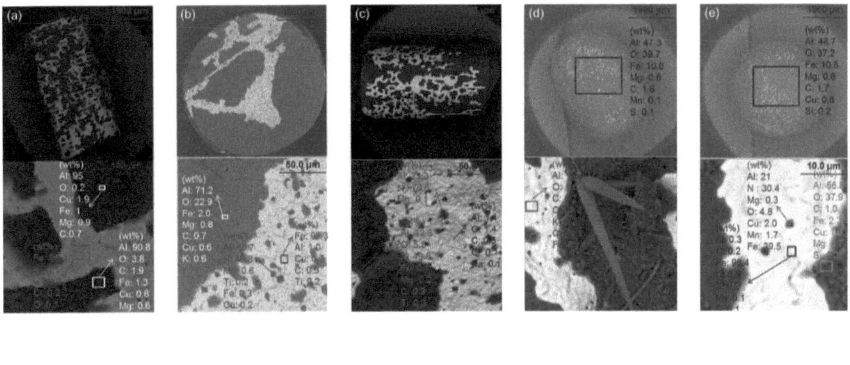

Fig. 1 Surface morphology of samples (**a** 6 wt%, **b** 4 wt%, **c** 2 wt%, **d** 1 wt%, **e** 0.5 wt%)

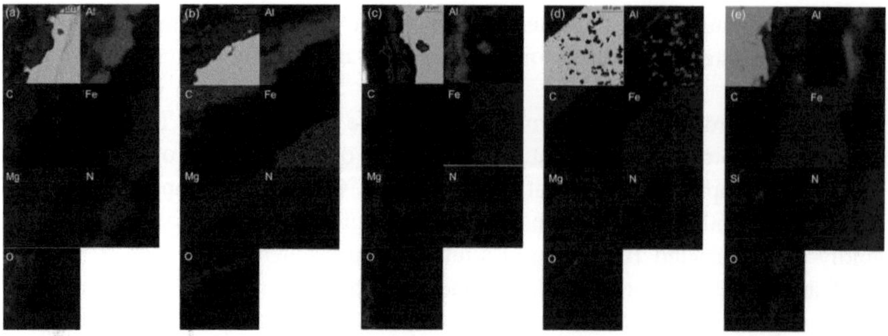

Fig. 2 Mapping of cross-section samples (**a** 6 wt%, **b** 4 wt%, **c** 2 wt%, **d** 1 wt%, **e** 0.5 wt%)

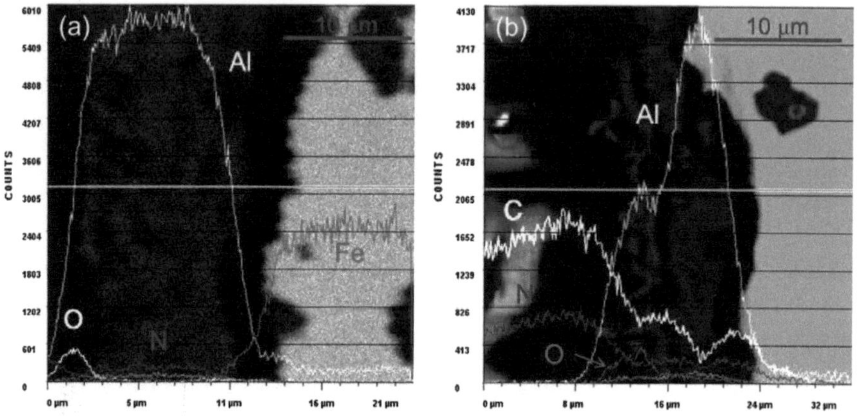

Fig. 3 Line scanning of cross-section of samples (**a** 4 wt%, **b** 2 wt%)

The mapping results of cross-section of samples with various aluminum content is shown in Fig. 2, where the black layer was resin. There existed two reaction layers on the surface region of iron, on the top surface was oxide and on the sub-surface was nitride, as revealed by Figs. 2d and 4b. SEM-EDS scans of the oxide film showed a discontinuous scale composed of aluminum and magnesium. Large aluminum nitride particles were found in the subsurface of each sample. 1 wt% aluminum sample exhibited distinct layers of precipitates. It was a good agreement with Bott's report [13].

Figure 3 shows SEM line scanning results of oxide film of each sample. The oxide film from the surface to the center was mainly alumina, oxynitride, and few nitride, respectively. The average thickness of oxide layer of each sample reduced as aluminum content decreasing. The thickest was up to 8 µm at 6 wt% aluminum sample, and the slimmest was 2 µm at 0.5 wt% aluminum sample. It is indicated that high aluminum sample cannot be melt for the formation of a higher melting alumina and the reaction product between dissolved aluminum in steel and alumina hardly going through thick oxide layer.

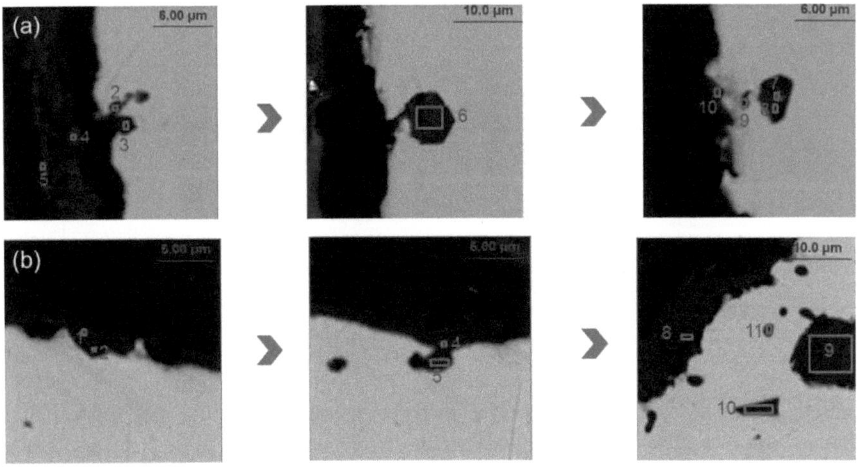

Fig. 4 Evolution of the formation of oxides and precipitates (**a** 2 wt%, **b** 1 wt%)

Fig. 5 Evolution of nitride precipitates in steel

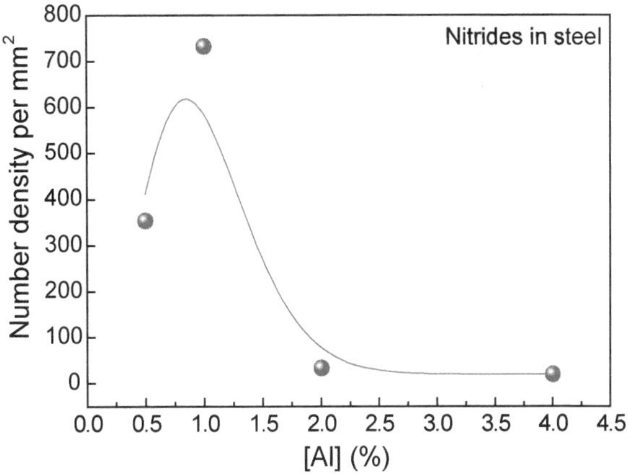

Figure 4 shows the evolution of the formation of oxides and precipitates of samples with various aluminum content. In the present work, the nitrogen activity under the present condition was sufficiently high to allow for the precipitation of nitrides. However, the oxygen potential in the initial studies was sufficiently high to become unfavorable for selective oxidation of aluminum. Therefore, aluminum-based oxides were first formed on the surface since the free energy of formation of alumina is far lower than that of aluminum nitride. Subsequently, the formation of nitrides can happen by a diffusion model that assumes gas accessed to the metal surface and the top oxide scale was permeable.

Figure 5 shows the evolution of nitride precipitates in steel using ASPEX. Al content had a large influence on the distribution of nitrides in steel. The number density of nitrides in steel first increased and then reduced with Al decreasing. It is

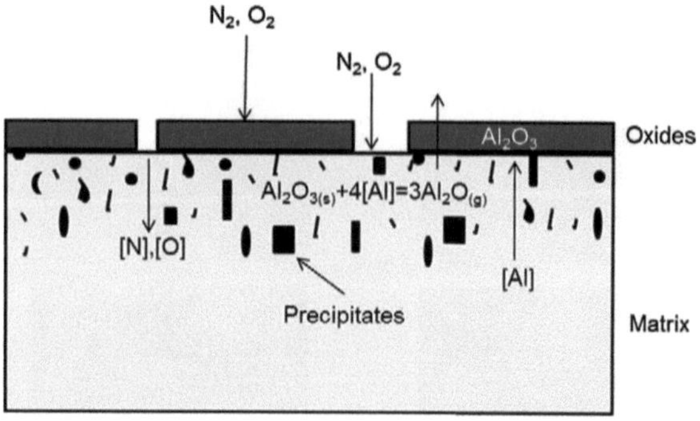

Fig. 6 Schematics of inclusion evolution at a rich nitrogen atmosphere

indicated that the thickness of the oxide layer in high Al sample was thicker than that in low one. The permeability of oxide layer ceased at a critical Al content at which the outer scale was sufficiently compact and nitrides were absent.

Based on the above experimental results, Fig. 6 shows the schematics of inclusion evolution at a rich nitrogen atmosphere. The more thermodynamically stable alumina will first form on the surface. Subsequently, nitrides will appear only at locations where the oxygen activity is sufficiently low and the local nitrogen activity sufficiently high, and will extend to a depth. In the meanwhile, the oxide layer is destroyed owing to the increased dissolution of alumina and to the formation of gaseous suboxides (Al_2O), which can go through the slim oxide layer to accelerate the formation of nitrides.

Conclusions

The development of nitrides and oxides in samples with different Al contents under a nitrogen-rich atmosphere was investigated. The results were as following.

(1) Oxide layer was first formed under the nitrogen-rich atmosphere, and was composed of iron oxide, alumina, hercynite and iron-aluminum spinel phase.
(2) Nitride was formed due to the sufficiently high nitrogen activity and low oxygen activity in the precipitation zone.
(3) The morphology of oxide layer on the sample surface was affected by Al content. Higher Al content lead to the formation of more compact alumina layer on the surface, resulting in smaller amount of nitride formed in the sub-surface region.

Acknowledgements The authors are grateful for support from the National Science Foundation China (Grant No. 51274034, No. 51334002, No. 51604023, No. 51504020, and No. 51404019), Beijing Key Laboratory of Green Recycling and Extraction of Metals (GREM), the Laboratory of Green Process Metallurgy and Modeling (GPM^2) and the High Quality steel Consortium (HQSC).

References

1. O. Grässel, L. Krüger, G. Frommeyer, L.W. Meyer, High strength Fe–Mn–(Al, Si) TRIP/TWIP steels development-properties-application [J]. Int. J. Plast **16**(10), 1391–1409 (2000)
2. W. Kim, J. Kang, C. Park, J. Lee, J. Pak, Thermodynamics of aluminum, nitrogen and AlN formation in liquid iron [J]. ISIJ Int. **47**(7), 945–954 (2007)
3. *The Japan Society for the Promotion of Science*. Steelmaking data sourcebook revised edition [M]. Gordon and Breach Science Publishers (1984)
4. H. Todoroki, S. Inada, Recent innovation and prospect in production technology of specialty steels with high cleanliness [J]. Iron Steel Institute Jpn **8**(8), 575–580 (2003)
5. J.H. Park, H. Todoroki, Control of MgO Al_2O_3 spinel inclusions in stainless steels [J]. ISIJ Int. **50**(10), 1333–1346 (2010)
6. T. Gladman, Development in inclusions control and their effects on steel properties [J]. Ironmaking Steelmaking **19**(6), 457–463 (1992)
7. L. Holappa, M. Hämäläinen, M. Liukkonen, M. Lind, Thermodynamic examination of inclusion modification and precipitation from calcium treatment to solidified steel [J]. Ironmaking Steelmaking **30**(2), 111–115 (2003)
8. T. Wegrzyn, J. Miroslawski, A.P. Silva, D.G. Pinto, M. Miros, Oxide inclusions in steel welds of car body [A]. Mater. Sci. Forum [C] **636**, 585–591 (2010). Trans Tech Publications
9. M. Leger, B. Guillaume, Microfractographic diagnosis of the embrittlement of steel castings by aluminum [J]. Int. Cast Met. J. **5**(3), 40–46 (1980)
10. E.T. Turkdogan, Causes and effects of nitride and carbonitride precipitation during continuous casting [J]. Iron Steelmaker **16**(5), 61–75 (1989)
11. F.G. Wilson, T. Gladman, Aluminium nitride in steel [J]. Int. Mater. Rev. **33**(5), 221–286 (1988)
12. M. Vedani, D. Dellasega, A. Mannuccii, Characterization of grain-boundary precipitates after hot-ductility tests of microalloyed steels [J]. ISIJ Int. **49**(3), 446–452 (2009)
13. J.H. Bott, H. Yin, J. Zhu, S. Sridhar, AlN formation in Fe-Al alloys in N_2-O_2 atmospheres [J]. Mater. Corros. **65**(3), 296–304 (2014)

Effect of Segregated Alloying Elements on the High Strength Steel Properties: Application to the Large Size Ingot Casting Simulation

Chunping Zhang, Davood Shahriari, Abdelhalim Loucif, Mohammad Jahazi, Louis-Philippe Lapierre-Boire and Rami Tremblay

Abstract Macrosegregation is one of the most significant defects which exert a determining effect on the properties of heavy ingots. The objective of this work is to study the influence of segregated solute elements on the physical and mechanical properties of a medium carbon high strength steel during large ingot casting process. The solidification process of a 20 Metric Tons (MT) ingot is simulated using Thercast® FEM code. Different segregation levels of solute elements are picked up from a section along the longitudinal axis in the top of the ingot. Input steel data, including physical and mechanical properties, are determined by means of Thermo-Calc®, JMatPro and a material database. Casting parameters are selected according to actual industrial operational conditions used for casting of large size ingots. Thermic and thermomechanic simulations are employed for calculating the solidification time. Results are analyzed in the framework of diffusion controlled solidification theory and the influence of alloying elements.

C. Zhang (✉) · D. Shahriari · A. Loucif · M. Jahazi
École de Technologie Supérieur, 1100 Rue Notre-Dame Ouest, Montréal, QUÉBEC H3C 1K3, Canada
e-mail: chunping.zhang.1@ens.etsmtl.ca

D. Shahriari
e-mail: Davood.Shahriari@etsmtl.ca

A. Loucif
e-mail: abdelhalim.loucif.1@ens.etsmtl.ca

M. Jahazi
e-mail: Mohammad.Jahazi@etsmtl.ca

L.-P. Lapierre-Boire · R. Tremblay
Finkl-Steel-Sorel, 100 Rue McCarthy, Saint-Joseph-de-Sorel, QUÉBEC J3R 3M8, Canada
e-mail: lplapierre@finkl.com

R. Tremblay
e-mail: rtremblay@finkl.com

The Minerals, Metals & Materials Society, *TMS 2017 146th Annual Meeting & Exhibition Supplemental Proceedings*, The Minerals, Metals & Materials Series,
DOI 10.1007/978-3-319-51493-2_47

Keywords Medium-carbon steel • Large size ingot • Segregated alloying elements • Casting simulation

Introduction

For the production of large size mono-block forgings of high quality multi-component special steels, such as the shaft rotor for steam turbine and the pressure vessels for nuclear reactor, ingot casting is the only method. Macrosegregation, as a compositional heterogeneity, is one of the most significant defects occurring during the solidification process. It influences the thermophysical properties, affects the solidification behavior of casting, deteriorates the microstructure and mechanical properties, proves difficult to remove by subsequent thermomechanical treatment, and has haunted the manufacturers over decades.

Macrosegregation can be controlled by optimization of casting metal composition, dimensions of the ingot and casting/solidification processes. Thus the impact of ingot size [1], mold temperature [2], pouring superheat [3], pouring velocity [4], cooling rate [5], and steel composition [6] on the severity of macrosegregation has been investigated extensively. In order to free from the timely and costly 'trial and error' approach, finite element method (FEM) has been used for the simulation of the casting and solidification process. Numerical methods, however, generally use the input data of the fixed initial nominal composition. In reality, the solidification behavior is not only the nominal composition and processing parameters dependent but also varies with individual segregation domains. This study, therefore, was initiated to explore the effect of different segregation levels of solute elements on the material physical and mechanical properties, the solidification and cooling behavior as well as the solidification time so as to approximate the actual solidification process in large size castings. It will be of experimental and industrial significance for the macrosegregation research, the production process design and optimization of ingot manufacturing efficiency.

For this, three-dimensional thermic and thermomechanic simulations of the solidification process of a 20 MT steel ingot were performed using Thercast® 8.2 version FEM code [7]. Casting parameters were selected according to the actual industrial operational conditions. All the needed data are seldom available in the literature, especially for the steels like the medium carbon low-alloyed multicomponent steel in study. It is known that even small variations in the composition may have a significant effect on the material physical and mechanical properties. Thus, the input steel data, including physical and mechanical properties, were determined by means of Thermo-Calc® and JMatPro. Modeling results were analyzed in the framework of diffusion controlled solidification theory and the influence of segregated solute alloying elements. It should be mentioned that mesh size optimization for computation cost minimization and sensitivity analysis of the results have been performed and its selection was based on the analysis results.

Materials and Methods

A cylindrical steel ingot was cast in a cast-iron mold. The steel melt composition is listed in Table 1. Molten steel was poured from the mold bottom at about 1570 °C and a layer of exothermic powder was over laid on the melt top. A hot-top, made of cast iron and lined with refractory materials inside, was placed on the top of the steel ingot. After solidification and stripping out, the hot-top with 30 cm's head of the ingot was transversely cut off for the study purpose. A plate with about 2.5 cm in thickness was sliced on one side of the axial plane, as shown in Fig. 1a. The slice was then sectioned into over 200 samples and faces in the centerline plane along the longitudinal axis of all the samples were subject to grinding, and then chemical mapping using the Thermo Scientific ARL™ 4460 mass spectrometer. The chemical composition of specimen was obtained by averaging out 3 random spectrometer measurements. Segregation maps of different elements in the longitudinal section of the hot-top and 30 cm's head of ingot were then reconstructed and macrosegregation patterns were identified.

Typical thermophysical data including the temperature dependence of the density, the specific heat capacity, and the phase transformation temperatures of selected segregated solute levels were determined by means of the computational thermodynamics program Thermo-Calc® classic version R with the TCS Fe-containing slag database [8]. Thermomechanical behaviors, such as yield stress, Young's modulus, strain hardening exponent and strain-rate sensibility coefficient and their variations as a function of temperature were developed with the help of JMatPro software, version 9.0 [9]. The simulations of solidification were performed in the finite element code Thercast® by using an Arbitrary Lagrangian-Eulerian (ALE) formulation for computing the thermal convection in liquid pool and mushy zone, and Lagrangian for calculating the deformation in solid regions. The method is applied both to mold filling simulation, where it can provide accurate free surface description, and to the study of thermo-mechanical phenomena occurring in the subsequent cooling down of cast parts: prediction of distortions and stresses.

Table 1 Nominal chemical composition of the investigated steel and the four positions in the upper region of the ingot (wt%)

	C	Mn	P	S	Si	Ni	Cr	Mo	Cu
Nominal	0.35	0.82	0.007	0.002	0.4	0.15	1.79	0.46	0.13
1	0.33	0.82	0.01	0	0.39	0.17	1.83	0.39	0.13
2	0.6	0.94	0.024	0.002	0.5	0.18	2.1	0.52	0.15
3	0.37	0.86	0.013	0.001	0.4	0.17	1.92	0.44	0.14
4	0.44	0.86	0.015	0.001	0.42	0.17	1.89	0.43	0.14

Results and Discussion

Solute distribution along the centerline longitudinal axis of the ingot presents sort of symmetry which confirms the formation of type A macrosegregation. The concentrations of all the elements increase continuously from the ingot lower bordering part to the center upper region, attesting that the liquid-solid interface has set off from the mold wall and globally advanced inwards and upwards to the last region to solidify located at the top center due to the efficient thermal protection of the hot-top. Concentrations of most elements in both the hot-top and the head of the ingot are higher than the nominal one, which indicates that positive segregation has taken place in these regions. As seen in Fig. 2b, the concentration profiles of carbon and chromium along the lines at the top and the bottom of the section in study display all the above-mentioned characteristics. Four representative segregation levels, corresponding to the highest and lowest segregation severity along the ingot centerline and the line next to the mold, were picked up for the analysis. Their

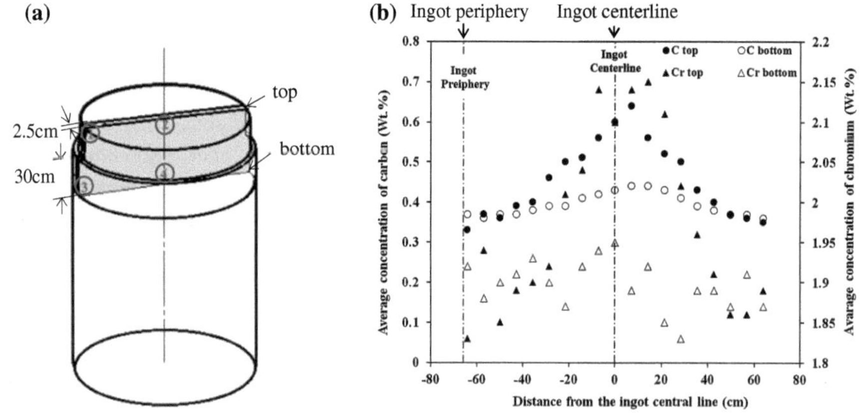

Fig. 1 **a** Cutting diagram of the ingot for chemical analysis and the four analyzed positions, **b** Concentration profiles of carbon and chromium along the lines at the top and bottom of the sectioned slice

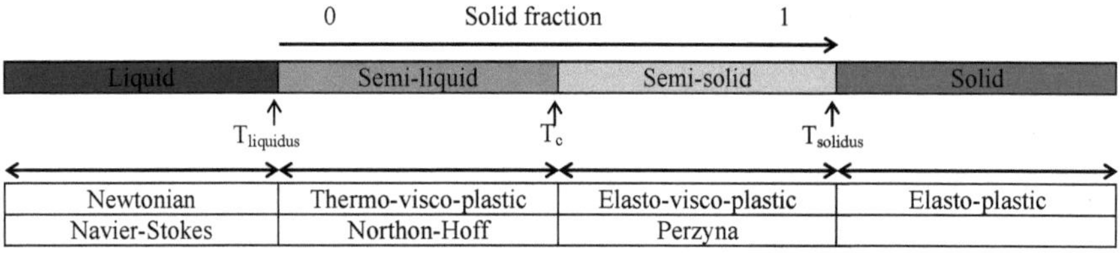

Fig. 2 Material behaviors depending on its condition in the process of solidification [7]

chemical compositions are provided in Table 1 also in order to facilitate the comparison with the nominal composition. Their positions in the section are labeled in Fig. 1a. The numbers given to the four positions are kept consistent hereinafter. The segregation severity level of solute elements in the four studied sites can be sequenced in descending order as 2 > 4 > 3 > 1.

Solidification Model Description

Considering continuum medium as illustrated in Fig. 2, different behaviors of the metal are distinguished by the critical temperature T_c, above which a thermo-viscoplastic behavior is considered and below which a thermo-elasto-viscoplastic one [7]. In order to take into account the complex behavior of solidifying alloys, a hybrid constitutive mold was used. In the liquid state, the metal was treated as a Newtonian fluid and the Navier-Stokes equation with temperature-dependent terms were employed. When the metal begins to solidify, the alloy in the mushy state was modeled as a non-Newtonian fluid obeying a Thermo-viscoplastic Norton-Hoff law. In the calculation of the deformation in solid regions using Lagrangian formulation, the computational grid was allowed to move with the material: this was essential to treat the air gap between mold and the casting. The liquid contraction and the solidification shrinkage were taken into consideration when the ALE scheme was applied to compute the thermal convection in liquid pool and mushy zone so as to prevent the mesh from degenerating and allow tracking the free surfaces.

3D linear tetrahedral elements were used for the discretization of the part and mold components. The unilateral contact condition (loss of contact, that is separation, is accepted) was applied to the boundary between mold and casting. In the resolution stage, a coupled computation of the stress and temperature fields was performed so that equations were solved iteratively using a prescribed time step. Due to the symmetry conditions of the casting system, only one quarter of the 20 MT ingot was modelled, the configuration of which is displayed in Fig. 3. The mold was considered as non-deformable. The finite element mesh of the

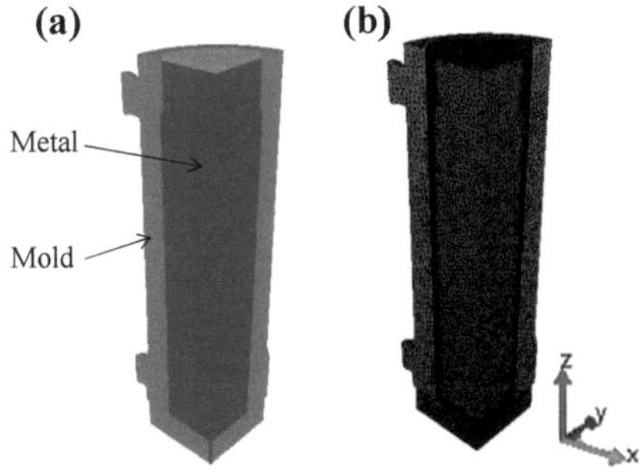

Fig. 3 **a** Geometry of the cast part, **b** Finite element mesh of the casting and of the mold

solidification system consisted of 1,190,000 elements which were selected based on several mesh refinement studies.

Effect of Segregation Levels on Physical Properties

Different segregation levels result in the change of the kinetics of phase transformations, and then the difference in thermophysical and mechanical properties related to the phases formed. Figures 4a and b show the evolution of solid fraction and density with the temperature for the four selected segregated compositions, respectively. As mentioned above, the segregation severity for the four representative positions in descending order is 2 > 4 > 3 > 1. It can be seen from Fig. 4 that the evolution of physical properties is in reverse order to the segregation severity level. Lower segregation results in the progressive augmentation of phase transformation temperature, including the solidus, liquidus and the transformation point from delta ferrite (δ) to austenite (γ). In the highest segregation level, the position 2, no δ - γ transformation took place. In addition, lower segregation levels induce smaller temperature differences between the solidus and liquidus, corresponding to slightly higher element liquid/solid partition coefficient. The observed increased density in lower segregation level may be attributed to more alloying elements dissolved. Such tendency in the density evolution agrees well with the experimental analysis results which indicated low concentration of small porosities in the ingot's center upper region.

The difference in segregation levels, in the meantime, brings about the discrepancy in elastic and plastic behaviors, as observed in Fig. 5a and b. The evolution of the liquid viscosity displays the same decreasing trend with the increase of the segregation severity. The reported deterioration effect of higher content of alloying elements on toughness and plasticity is in accordance with the calculation results [10]. The reduction of density with the increased severity level of segregation may also be the reason. The solute diffusion coefficient in liquid is discovered, based on Stokes-Einstein equation [11], inversely proportional to solvent

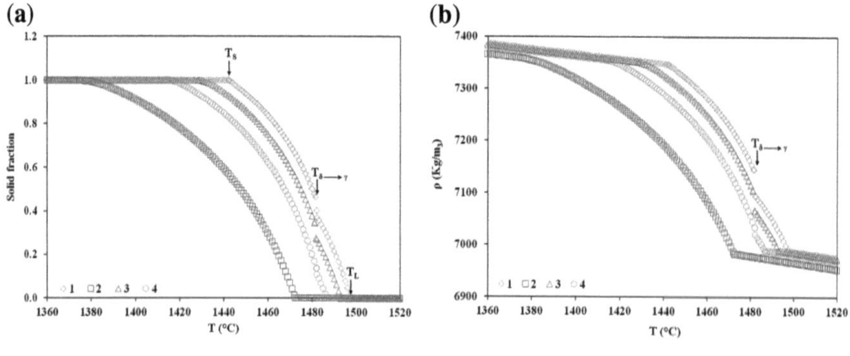

Fig. 4 Temperature dependence of physical properties for different segregation levels

viscosity. Thus, the diffusion coefficients in liquid in the four positions can be sequenced as D2 > D4 > D3 > D1, in agreement with the segregation severity descending order. All these variations in properties due to different segregation levels affect the heat transfer, convection and radiation modes, as well as stress and deformation domain during mold filling, cooling down and the progressive solidification of the ingot.

Effect of Segregation Levels on Solidification Time and Temperature Gradient

The knowledge of solidification time of a large size ingot in the industry is of quite importance in that shorter solidification time will not allow achieving solid state at the center of the ingot and longer time will result in increased segregation levels and porosities. Therefore, solidification time is taken as the simulation result to discuss. Solidification times needed for the four analyzed positions in the casting are different and vary with the segregation severity level. Specifically, solidification time at the center is longer than the one in the mold side, as presented in Fig. 6. It should

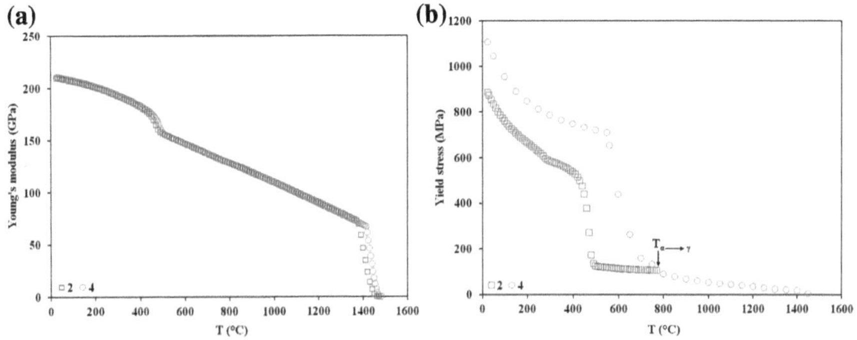

Fig. 5 Evolution of mechanical properties with segregation severity levels

Fig. 6 Plot of solidificaiton time as a function of weight percentage of carbon

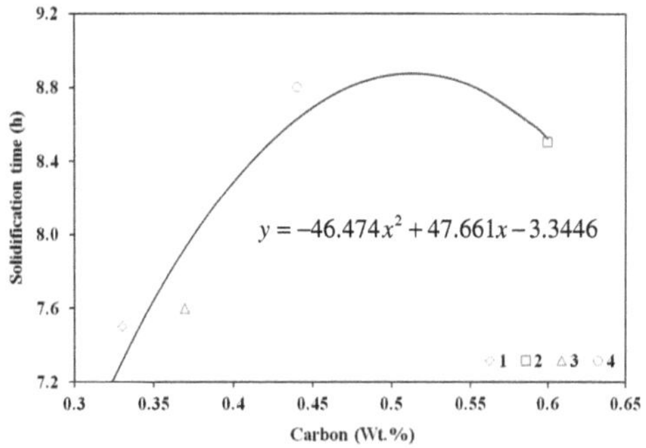

be noted that for presentation purposes, the simulation was stopped for a maximum temperature of 1000 °C to make sure that solid state is achieved at the center of the large size ingot. The observed evolution of solidification time with the percent carbon was approximated with a second degree polynomial function and is illustrated in Fig. 6, where *y* is the solidification time and *x* carbon weight composition. As indicated in the figure, the solidification time attains its peak value at 0.513 wt% carbon, corresponding to the composition located in the interface between the ingot and the hot-top, which probably concerns the pipe formation and the time for its creation.

Figures 7 and 8 illustrate respectively the thermic and thermomechanic simulations of the cooling of the cast part with the composition of the position 4 until the solidus temperature, 1414 °C, is reached. It can be seen that the solidification time

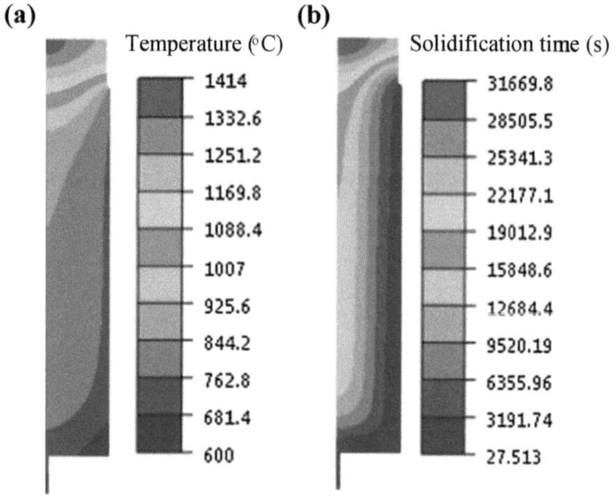

Fig. 7 Thermic simulation: **a** Temperature distribution, **b** Solidification time after 100% solidification of position 4

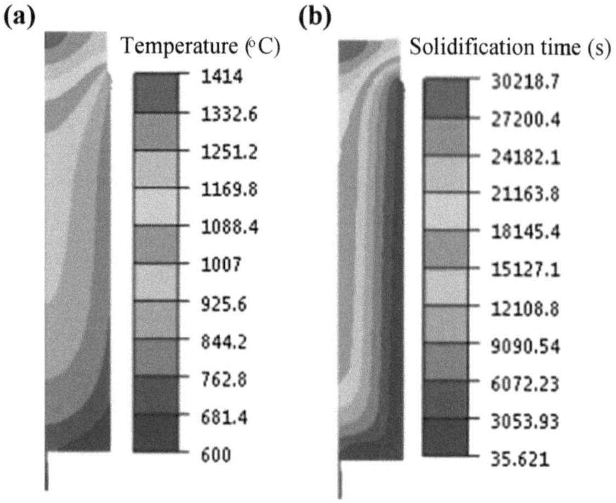

Fig. 8 Thermomecanic simulation: **a** Temperature distribution, **b** Solidification time after 100% solidification of position 4

decreases around 25 min, passing from about 8 h 48 min to 8 h 23 min or so when the influence of the solid deformation is taken into account. According to the simulation results, although the last region to solidify is always located at the top center of the hot-top, solid mechanical deformation gives rise to the alteration of thermal gradient, particularly in the interior of the ingot. The free surface can be seen almost flat in Fig. 8 because of an efficient thermal protection which minimizes the thermal gradient in the hot-top.

Conclusion

This research presented a study on the effect of different segregation levels on the solidification behavior of a 20 MT steel ingot. Comparison analyses of the physical and mechanical properties of representative segregation levels were performed by means of softwares like Thermo-Calc® and JMatPro. Solidification time of varied segregation severities was calculated through thermic and thermomechanic simulations of the mold filling and cooling behavior of steel ingot using Thercast® FEM code. The following main conclusions can be made from this study:

1. The evolution of physical properties and elastic/plastic behaviors is in reverse order to the segregation severity level due to different solute diffusion coefficient in liquid and higher content of alloying elements.
2. By increasing the segregation severity level, solidification time increases. Solid mechanical deformation gives rise to the alteration of thermal gradient and the decrease of solidification time.

Acknowledgements The financial support from the Natural Sciences and Engineering Research Council (NSERC) of Canada in the form of a Collaborative Research and Development Grant (CRDG) under number 470174 is gratefully acknowledged. Finkl Steel-Sorel Co. for providing the material and Transvalor Americas Corp. for permission and technical support on Thercast software are greatly appreciated.

References

1. D.R. Liu et al., Modeling of macrosegregation in steel ingot: influence of mold shape and melt superheat. Kovove Mater. **19**, 1–11 (2011)
2. Y. Zhang et al., Numerical simulation and optimization in solidification of zinc alloy. Adv. Mater. Res. **287–290**, 2902–2905 (2011)
3. T. Pikkarainen et al., Effect of superheat on macrostructure and macrosegregation in continuous cast low-alloy steel slabs, in *4th International Conference on Advances in Solidification Processes, IOP Conf. Series: Materials Science and Engineering*, 117 (2016). 012064
4. J. Zeng, W. Chen, Effect of casting speed on solidification structure and central macrosegregation during continuous casting of high-carbon rectangular billet. La Metallurgia Italiana **7**(8), 51–58 (2015)

5. P.G. Xu, F. Yin, K. Nagai, Solidification cooling rate and as-cast textures of low-carbon steel strips. Mater. Sci. Eng. A **441**, 157–166 (2006)
6. D. Shahriari et al., 3D Numerical simulation of solidification of large size ingots of high strength steel, *SteelSim* 2015. in *6th International Conference on Modelling and Simulation of Metallurgical Processes in Steelmaking*
7. Thercast User Manual, (France: Transvalor, 2012)
8. ThermoCalc Software AB, (Stockholm Technology Park, 2006)
9. JMatPro User's Guide (Sente Software Ltd., 2005)
10. J. He, 21 Chemical Elements and Effects on Steel Mechanical Properties. (Report of OTAI SpecialSteel Co. Ltd., 2015)
11. C. Miller, The Stokes-Einstein law for diffusion in solution. Proc R. Soc. Lond. A., Contain. Pap. Math. Phys. Charact, 106(740) 724–749 (1924)

Thermal-Mechanical Model of Depression Formation in Steel Continuous Casting

Matthew L.S. Zappulla and Brian G. Thomas

Abstract Many serious defects in the continuous casting of steel, including surface cracks and depressions, are related to thermal-mechanical behavior during solidification in the mold. A finite-element model has been developed to simulate the temperature, stress, and shape of the steel shell near the strand surface, as it moves down the mold at the casting speed in a state of generalized plane strain. The thermal model simulates transient heat transfer in the solidifying steel and between the shell and the mold wall. It is coupled with a stress model that features temperature-, composition-, and phase-dependent thermal-elastic-visco-plastic constitutive behavior of the steel. Depressions are predicted to form when the shell is subjected to either excessive compression or tension, but the shapes differ. Predicted depression shapes are compared with previous plant observations. The model reveals new insights into the mechanisms of surface depressions and longitudinal cracks in this process.

Keywords Steel · Solidification · Finite element · Depression · Crack · Modeling

M.L.S. Zappulla · B.G. Thomas (✉)
Formerly with Department of Mechanical Science and Engineering,
University of Illinois at Urbana-Champaign, Champaign, USA
e-mail: bgthomas@illinois.edu

M.L.S. Zappulla
e-mail: zappull1@illinois.edu

M.L.S. Zappulla · B.G. Thomas
Department of Mechanical Engineering, Colorado School of Mines,
1610 Illinois St., Golden, CO 80401, USA

The Minerals, Metals & Materials Society, *TMS 2017 146th Annual Meeting & Exhibition Supplemental Proceedings*, The Minerals, Metals & Materials Series,
DOI 10.1007/978-3-319-51493-2_48

Introduction

In continuous casting, initial solidification behavior within the mold is of the utmost importance. Defects or inconsistencies that appear within the first meter or less of the casting directly affect surface quality in substantial ways. In an ideal situation, adequate machine and narrow face tapering of a caster ensures uniform heat transfer between the mold and steel strand. However, in practice this can be difficult to implement for a variety of reasons. Insufficient taper may lead to reduced or non-uniform heat transfer between the strand and the mold; this can result in narrow face bulging, corner rotation, and shell thinning [1]. Excessive taper causes accelerated mold wear, and possibly even buckling of the shell [1]. Both of these taper issues can cause gutter, longitudinal cracks, and even breakouts [1].

Cracks are one of the most costly defects in continuous casting, often resulting in scarfing, downgrading, or in the worst case scenario, rejection or failure of final products. They arise due to excessive tensile stress together with metallurgical embrittlement. Although the general causes are known, determining the detailed fundamental steps of formation of particular defects and the corresponding best solution(s) remains very difficult. By investigating the effect of different temperature and mechanical histories using computational models, this work aims to identify specific caster situations and conditions that can lead to the formation of different types of surface cracks and the distinguishing depressions that accompany them.

There has been significant previous work to develop damage criteria for hot-tear crack formation [2–5], hot ductility [6–9], strain to failure [10], and sensitivity to cracks [11, 12]. Recent work has examined solidification cracks [12], focusing on micro-scale phenomena. Previous thermal-mechanical modeling suggested that cracks during thin-slab casting with a funnel mold were due to bending around the funnel curves while the shell was still weak [13].

Much recent literature is concerned with the micro-scale phenomena and compositional effects related to cracking [12, 14–16]. Previous work with a macro-scale thermal-mechanical model suggested how off-corner gutter depressions form [1]; however, work on longitudinal and transverse cracks and their associated depressions is rare. More work is needed to understand the fundamental mechanism(s) of depression formation and the steps that can be taken when troubleshooting a caster to solve a particular surface-defect problem. Quantitative answers are needed to questions such as: When does a crack form without a depression? When does a depression form without a crack? When does a sub-surface crack form and propagate (or not) to the surface? What plant practices lead to a given type of crack and shape of depression?

Cracks and depressions are clearly related to each other. Figure 1 shows a few examples of cracks and depressions from different steel plants and previous literature. The wide range of different shapes of the surface defects in these images suggests different formation mechanisms. For example, when a crack opens at the base of a depression it changes the characteristic shape of the defect. Even labeling a given defect is difficult, because words such as gouge, depression, void, crack,

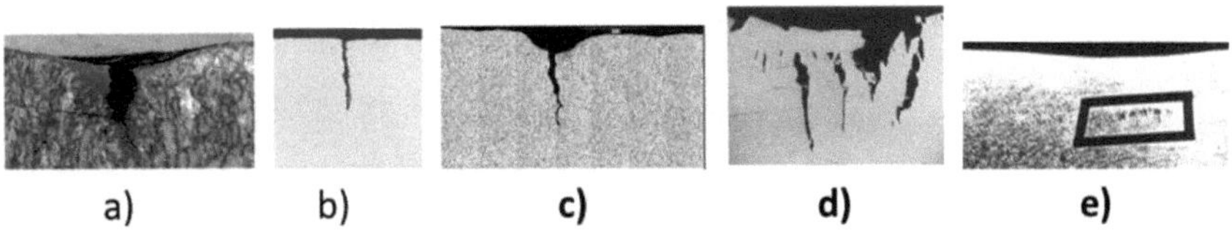

Fig. 1 Observed types of surface defect shapes and openings: **a** depression with high aspect ratio opening, **b** surface crack with no depression [17], **c** depression with narrow opening above crack, **d** depression with multiple sub-surface cracks [17], **e** depression with only subsurface cracks [1]

opened crack, etc. depend on the formation mechanism, which is generally not known and often involves several different stages.

Model Description

The steel solidification process was simulated in this work using a computational thermal-mechanical model [18, 19] of a 2-D slice in a state of generalized plane strain, to predict the displacement, strain, and stress histories of the shell. The governing equations are solved using the finite-element method in ABAQUS/Standard (implicit) [20]. The transient analysis was performed in two steps consisting of a heat transfer analysis, followed by a mechanical analysis. The simulation domain shown in Fig. 2 is a representative portion of the solidifying steel located somewhere along the broad face away from the corner region, where a local surface defect might form. For simplicity, the depression center can modeled as a symmetry plane, and the other edge with a generalized plane strain condition and subjected to fixed displacement.

Heat flux leaving the domain bottom in the region of good (uniform 100%) contact with the mold is given as follows [18, 19], and is applied using user subroutine DFLUX in ABAQUS [20].

$$\dot{q}_{standard} = 6.36(t + 1.032)^{-0.5} \tag{1}$$

The applied heat flux ramps down linearly from 100% at 5 mm to 50% at the depression center, from 1 s to 3 s, as shown in Fig. 3. This is followed by a further linear decrease to 20% at crack center, over a widening region that increases linearly from 5 mm at 3 s to 20 mm wide at 10 s.

The material properties for this study are for a typical low carbon steel grade (0.045 wt%C) used in previous work [18, 19]. Composition and phase transition temperatures are given in Tables 1 and 2.

Temperature- and carbon-dependent properties: enthalpy [21], thermal conductivity [22], density [23, 24], thermal linear expansion [21], and Young's modulus [25] for this steel grade are based on the phase fractions given in Fig. 4. Details of the property data can be found in Sect. 2.7 of [19]. The elastic visco-plastic

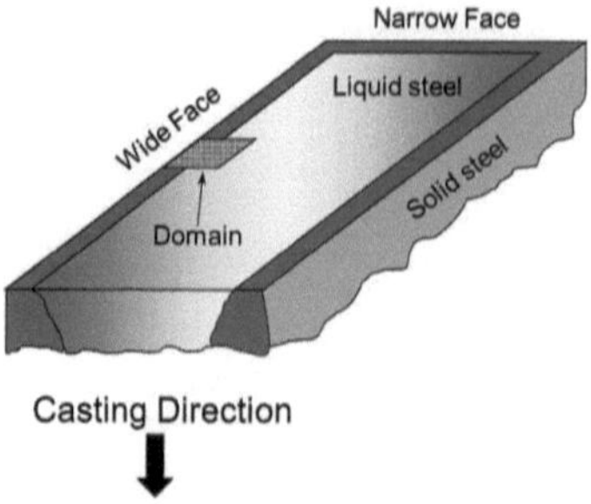

Fig. 2 Solidification of a continuous cast slab (cutaway) showing model domain

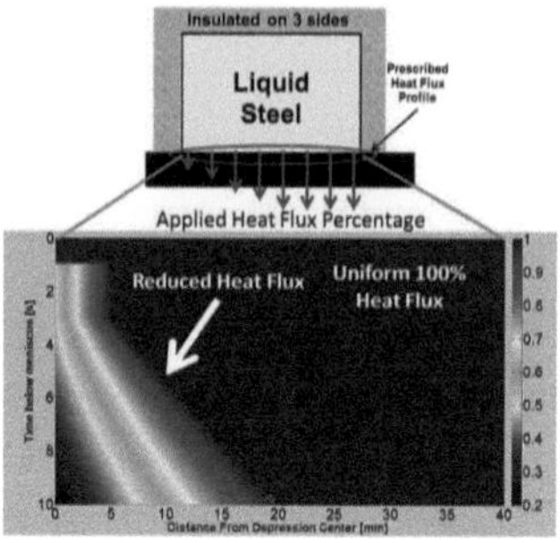

Fig. 3 Thermal boundary conditions and reduced heat transfer applied on domain

Table 1 Low carbon steel composition (wt%C)

Steel	C	Mn	Si	S	P	Cr	Ni	Al	Ti	Cu
Low carbon	0.045	0.20	0.015	0.01	0.01	0.01	0.01	0.04	0.05	0.01

Table 2 Low carbon steel phase transition temperatures (°C)

Steel	Pour	Liquidus	Solidus	Mushy Range	δ start	δ end	γ start
Low carbon	1532	1528	1505	23	1505	1385	1418

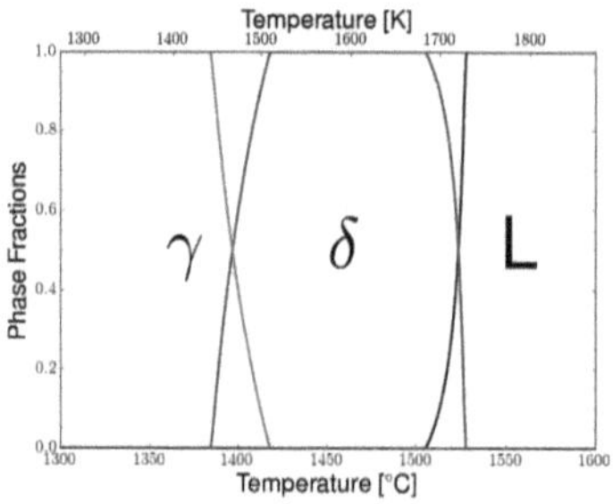

Fig. 4 Phase fractions for the low carbon steel grade examined [19]

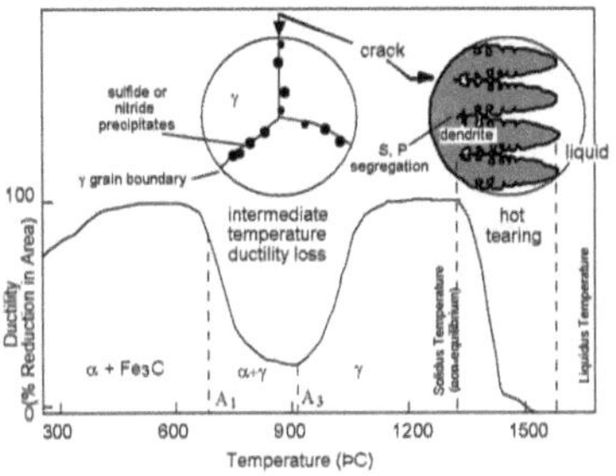

Fig. 5 Low ductility/high embrittlement regions of interest [8]

constitutive models used in this work include both strain-rate- dependent plasticity and time-dependent creep for each steel phase. The δ-ferrite is modeled with the Zhu Modified Power law [26], the austenite phase uses model III from Kozlowski [27], and the liquid phase is a low-strength elastic-plastic material, as described elsewhere [26, 28].

To evaluate the likelihood of crack formation, a damage criterion must be chosen. Most steels exhibit two low ductility troughs, one near the solidus temperature, and a second "intermediate-temperature" drop just above the A_3 temperature, as shown in Fig. 5. Near the solidus temperature, the interdendritic liquid has a strong effect on the ductility of the steel as it is only "mostly" solid. This region (solid fraction between 90–99%) is used to identify "weak areas" susceptible to hot tear cracks. Specifically we are interested when this region experiences high inelastic strain while in a state of tensile stress. Future work will explore more quantitative damage criteria.

Results and Discussion

As observed in the literature, a drop in surface heat flux causes a rise in shell temperature [29]. With uniform heat transfer, the entire surface experiences the same temperature history, given by the lowest line in Fig. 6. When heat flux drops beneath the depression, that part of the surface reheats by several hundred degrees, as observed in other lines in Fig. 6. It should be noted that the austenite layer growth is severely slowed, so that layer becomes much thinner beneath the depression.

Typical temperature contour results at 10 s are shown in the center frame of Fig. 7. The left and right frames show temperature contours in vertical slices through the shell at both the depression center and the region of uniform heat flux. Following these results vertically downwards visualizes the time histories of the temperature profile at these locations. The reduction of the austenite layer beneath the depression is readily apparent as the austenite layer at the depression center is 55% thinner than at the far field. The thickness of solid shell is reduced by only ~1.5 mm (17%).

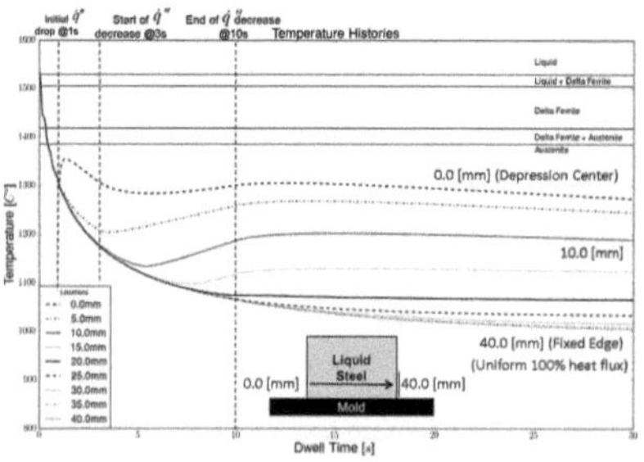

Fig. 6 Temperature histories across the domain surface

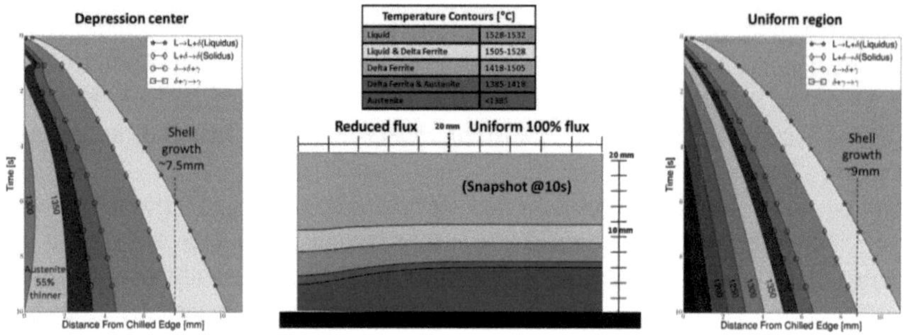

Fig. 7 Reduced heat transfer temperature contours, shell growth, and phase regions

Results for the ideal case of free shrinkage, except for friction at the shell/mold interface, are shown in Fig. 8a. With reduced heat transfer, the shell thins, and when stress concentrates locally, a slight U-shaped depression appears in the domain even without an externally applied force. The tension by friction at the mold interface in combination with the shrinkage of the rest of the shell causes the domain to neck and behave like a tension test specimen.

Results in Fig. 8b show the effect of a continuous applied tension (pull) on the domain edge for domains with different widths. These pull cases can occur in any situation where tension is induced on the shell by issues occurring somewhere outside the small domain; common reasons for this tension could be friction associated with mold flux infiltration problems or bulging due to inadequate taper. Increasing the applied tension results in a much deeper, but not necessarily wider, depression. Wider domains exhibit deeper depressions, for the same conditions.

Figure 9 shows a pull case where the shell was allowed to grow to 3 mm before the tension was applied to the domain. This is scaled along-side a depression from the work of Brimacombe [29] and shows a reasonable qualitative match. The necking formation mechanism is believed to be the predominant creator of depressions, due to the shape of the resulting U-shape depression which is commonly seen in practice.

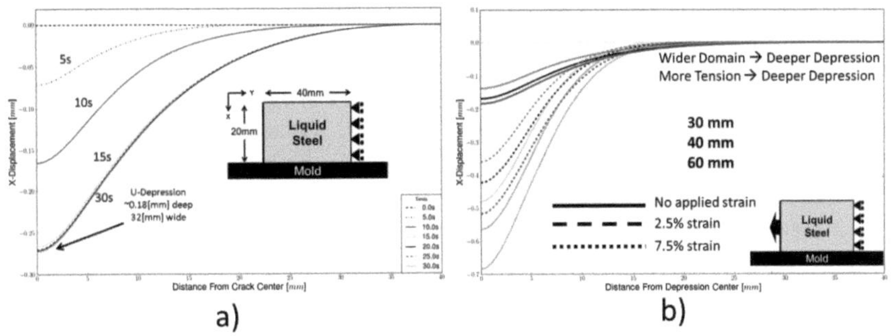

Fig. 8 **a** Formation history of a U-shaped depression (ideal shrinkage), **b** depression shape for different pull cases (10 s), showing effect of strain extent and domain size

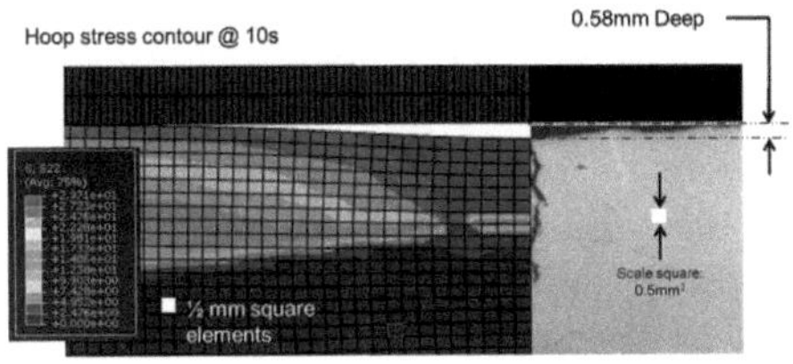

Fig. 9 Simulated necking depression shape (3 mm shell growth, 7.5% applied tensile strain) compared to literature (contours of stress perpendicular to solidification front)

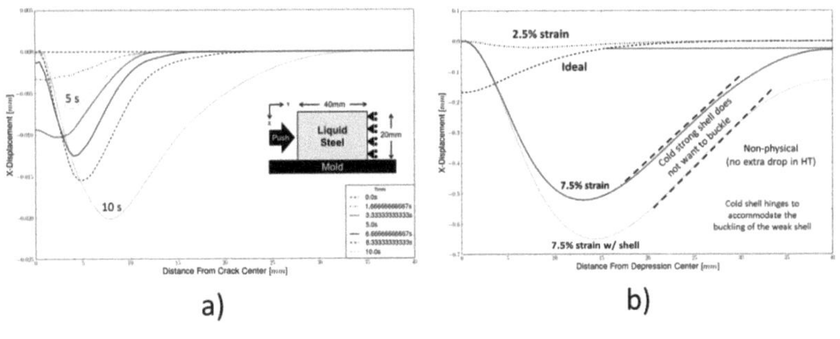

Fig. 10 **a** Formation history of a W-shaped depression, **b** depression shapes for different push cases

Figure 10 shows results for an applied compression (push) case. This situation could arise from an overtapered mold, where the narrow face mold may apply compression on the wide face shell in excess of the desired shrinkage of the strand. When the domain has compression applied to it, the surface of the shell buckles into a W-shaped depression. In the same way that increased tension results in a deeper U-shaped depression, increased compression causes a more severe W-shaped depression. Low levels of compression (e.g., 2.5% case) produce very shallow depressions.

Figure 11 shows hoop stress contours for the severe push case with 7.5% applied strain. As the figure indicates, subsurface tension appears at the base of the W shape, right at the solidification front; this suggests the possibility of a hot tear at this location. W-shaped depressions are uncommon in plant practice; however, Fig. 12 shows a billet that appears to exhibit a W-depression.

Figure 12 is taken from the literature [30] and shows a billet removed from a caster. The caster experienced lubrication issues and became jammed in the mold: removal was accomplished by increasing the drive rolls until the billet finally came free. This resulted in what appears to be a W-shaped depression near the bottom of the photograph (see enlarged inset), caused by a buckling scenario due to the misfit between the solidifying shell and excessive taper of the mold. One possible reason that W-shaped depressions are uncommon in practice is suggested in the photograph. The generation of tension at the solidification front may cause a subsurface crack at the base of either of the two symmetrical depressions. Such a crack would relieve the stresses and might enable the W-depression to evolve into a single depression centered at the subsurface crack. As this crack could form at several places on either side of the W, the final shape of the single surface depression would be expected to wander, as observed in the top portion of Fig. 12. As the current model does not allow for cracks to form, this tension cannot be relieved in the same way that it may in the real process. Another possible reason for the rarity of

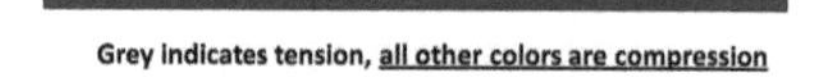

Fig. 11 Push case 7.5% strain hoop stress contours at 10 s

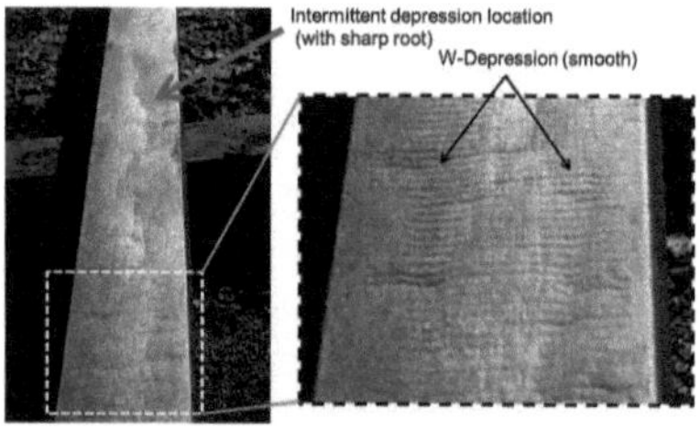

Fig. 12 Billet exhibiting signs of buckling depression mechanism [30, 31]

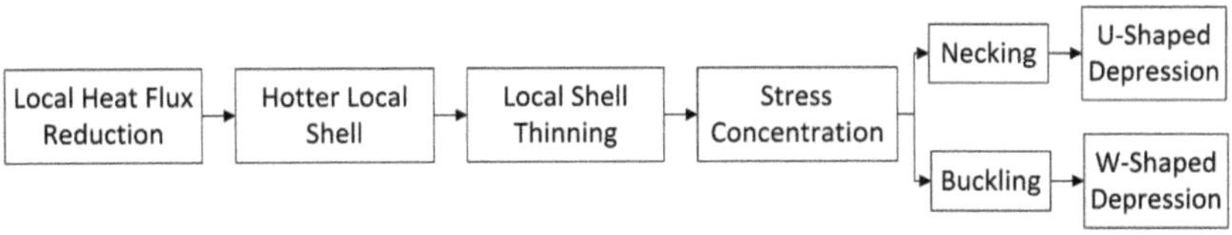

Fig. 13 Depression formation process

W-shaped depressions in practice is that the degree of compression required to buckle the shell in an appreciable way is quite high. Overtaper of only a few percent strain may cause a shallow W but it may be smoothed out in the drive/support rolls sub mold.

Conclusions

Results of this work suggest that there are two different mechanisms for depression formation in continuous casting, one with applied tension and one with applied compression. Additionally, these two mechanisms have quite different characteristic shapes and behaviors (see Fig. 13).

This work demonstrates that non-uniform heat transfer during initial solidification can lead to a depression even with ideal shrinkage (i.e., a perfectly tapered caster). This shows the importance of proper mold conditions regarding mold flux and mold surface upkeep. This work also suggests that the frictional interaction of the steel shell with the mold/flux in combination with shrinkage of the shell and imperfect mold taper can apply tension that could lead to depressions.

Future Work

Future work on this project is especially interested in collecting and characterizing cracks and depressions from operating casters. In addition, simulations will investigate these scenarios and the potential for air gap formation with a fully coupled model that includes detailed physics of the gap. Eventually, a full ¼ symmetry domain of a slab mold to include corner effects is desired as well as an implementation of a quantitative damage or crack criterion. Finally, surface defect modeling to explore grade effects especially into the peritectic range is desired.

Acknowledgements The authors wish to thank the members of the Continuous Casting Center at the Colorado School of Mines for their financial support, as well as the National Science Foundation, grant CMMI-13-00907 Blue Waters, and the National Center for Supercomputing Applications at The University of Illinois at Urbana-Champaign which is supported by the National Science Foundation (awards OCI-0725070 and ACI-1238993) and the state of Illinois.

References

1. B.G. Thomas, A. Moitra, and R. McDavid, Iron Steelmak. ISS Trans. **23**, 143 (1996)
2. M. Rappaz, J. Drezet, M. Gremaud, Metall. Mater. Trans. A **30A**, 449 (1999)
3. Y.M. Won, T. Yeo, D.J. Seol, K.H. Oh, Metall. Mater. Trans. B **31**, 779 (2000)
4. T.W. Clyne, G.J. Davies, Br. Foundrym. **74**, 65 (1981)
5. D.G. Eskin, L. Katgerman, Prog. Mater Sci. **49**, 629 (2004)
6. H.G. Suzuki, S. Nishimura, S. Yamaguchi, Trans. Iron Steel Inst. Jpn. **22**, 48 (1982)
7. P.J. Wray, Metall. Mater. Trans. A **15**, 2059 (1984)
8. B.G. Thomas, J.K. Brimacombe, I.V. Samarasekera, Trans. Iron Steel Soc. **7**, 7 (1986)
9. W.T. Lankford, Metall. Trans. **3**, 1331 (1972)
10. C.H. Yu, M. Suzuki, H. Shibata, T. Emi, ISIJ Int. **36**, S159 (1996)
11. B. Bottger, M. Apel, B. Santillana, D. Eskin, Metall. Mater. Trans. A **44**, 3765 (2013)
12. M.B. Santillana, Thermo-mechanical properties and cracking during solidification of thin slab cast steel. Ph.D. Thesis, Delft University of Technology, 2013
13. L.C. Hibbeler, B.G. Thomas, M.B. Santillana, A. Hamoen, A. Kamperman, in *European Continuous Casting Conference Associazione Italiana di Metallurgia, Riccione* (Italy, 2008)
14. S. Moon, The Peritectic phase transition and continuous casting practice, Ph.D. Thesis, University of Wollongong, 2015
15. M.M. Wolf, K. Kurz, Metall. Mater. Trans. B **12**, 85 (1981)
16. J. Konishi, M. Militzer, I.V. Samarasekera, J.K. Brimacombe, Metall. Mater. Trans. B **33B**, 413 (2002)
17. J.K. Brimacombe, F. Weinberg, E.B. Hawbolt, Metall. Trans. B **10**, 279 (1979)
18. M.L.S. Zappulla, L.C. Hibbeler, B.G. Thomas, Metall. Mater. Trans. A (2016 in review)
19. M.L.S. Zappulla, Grade effects on thermal-mechanical behavior during the initial solidification of continuously cast steels. M.S Thesis, The University of Illinois at Urbana-Champaign, 2016
20. ABAQUS 6.13 Theory Manual (DS SIMULIA Corp, Providence, RI, 2013)
21. K. Harste, Investigation of the shrinkage and the origin of mechanical tension during the solidification and successive cooling of cylindrical bars of fe-c alloys, Ph.D. Thesis, Technischen Universität Clausthal, 1989
22. K. Harste, K. Schwerdtfeger, ISIJ Int. **43**, 1011 (2003)
23. A. Jablonka, K. Harste, K. Schwerdtfeger, Steel Res. **62**, 24 (1991)
24. I. Jimbo, A.W. Cramb, Metall. Mater. Trans. B **24**, 5 (1993)
25. H. Mizukami, A. Yamanaka, T. Watanabe, ISIJ Int. **42**, 375 (2002)
26. C. Li, B.G. Thomas, Metall. Mater. Trans. B **35**, 1151 (2004)
27. P.F. Kozlowski, B.G. Thomas, J.A. Azzi, H. Wang, Metall. Mater. Trans. A **23A**, 903 (1992)
28. S. Koric, B.G. Thomas, J. Mater. Process. Technol. **197**, 408 (2008)
29. J.K. Brimacombe, F. Weinberg, E.B. Hawbolt, Metall. Mater. Trans. B **10**, 279 (1979)
30. C.A.M. Pinheiro, I.V. Samarasekera, J.K. Brimacombe, B. Howes, O. Gussias, Ironmak. Steelmak. **27**, 144 (2000)
31. I.V. Samarasekera, Private Communication (2015)

Study for the Initiation Locations of Longitudinal Surface Cracks on Beam Blank in the Mould of Continuous Casting

Wei Chen, Xingwang Yu, Ying Chen and Baoxiang Wang

Abstract During beam blank's continuous casting production, lots of longitudinal cracks easily appear on the surface. In order to study the initiation locations of longitudinal surface cracks in the mould, a coupled thermo–elasto–plastic finite element model was established to compute the temperature and stress on the software ABAQUS. A user subroutine DFLUX was used to simulate the heat transfer process and thermal field. The stress field under the influence of thermal stress, surface friction, ferrostatic pressure was analyzed by the mean of contact algorithm and restart scheme. According to the results, it can be inferred that the forming locations of longitudinal surface cracks easily appear on the surface of the fillet, the web center, the flange tip and the flange center. Moreover, initiation locations of longitudinal surface cracks should be: on fillet at a distance of 200 mm to the meniscus, and the other three sections at a distance of 180 mm to the meniscus.

Keywords Beam blank · Continuous casting · Surface cracks · Initiation location

Introduction

As a new kind of strand, the beam blank has become the best to produce H-Beam for the advantages of low comprehensive energy consumption, fewer processes, higher rolling yield and so on [1]. Since the first production line was built in 1998 at Maanshan Iron & Steel Corporation in our country, a few companies such as Laiwu Iron and Steel Co, Baotou Iron and Steel Company, Jinxi Iron and Steel Company, etc. also began to produce beam blank [2].

W. Chen (✉) · X. Yu · Y. Chen · B. Wang
College of Metallurgy & Energy, North China University of Science and Technology, Tangshan, China
e-mail: hblgdxzzb@163.com

The Minerals, Metals & Materials Society, *TMS 2017 146th Annual Meeting & Exhibition Supplemental Proceedings*, The Minerals, Metals & Materials Series,
DOI 10.1007/978-3-319-51493-2_49

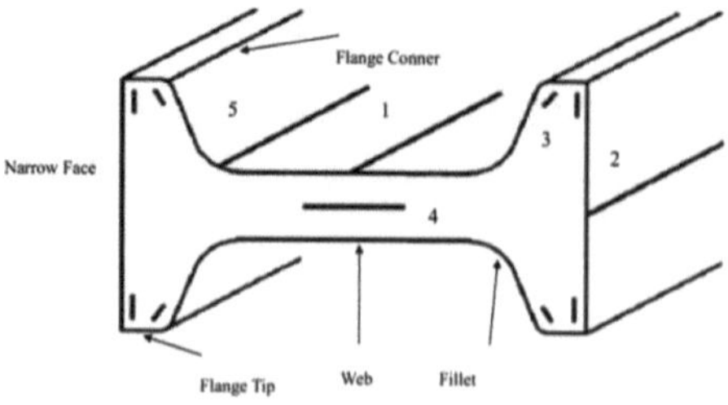

Fig. 1 Transverse shape and typical cracks of beam blank

The beam blank's irregular section, which make the stress state in the blank more complex, results in more defects such as surface cracks, slag inclusion, scratch and so on. These defects would cause the high probability of waste products. According to statistics, the cracks in beam blank have a proportion of 50%, moreover, longitudinal surface cracks are the kind to more easily occurring [3, 4]. The shape and typical cracks of beam blank are shown as Fig. 1.

To solve these problems, many researchers had done lots of works in recent years. For example, Ming [5] described the process characteristics of large section beam blank and did research on the reason of cracks on the web, put forward some methods to control them. Ke-feng [6] developed experiments with Metalloscope and SEM to study the reason why the cracks on the flange of hot rolled H-shaped steel would occur, and put forward methods improving the quality of production. Sakuraya [7] studied the influence of the mould powder and vibration to surface longitudinal cracks. Stress of brittle fracture was obtained by K.H. Kim through the study of Fe-C phase diagram [8].

1. crack at web
2. crack at the center of narrow face
3. internal crack at the top of flange
4. crack at the center of web
5. crack at fillet.

In the field of numerical simulation, Wen [9] established a solidification model of beam- blank in the mould, and designed the taper of the mould. Hai-lun et al. and Wei et al. [10, 11] established two and three dimension solidification model of beam blank, and studied the distribution of temperatures along the section and the gaps between the blank and the mould, then gave out some projects to improve the water seam of the mould.

Our team simulated the three-dimensional flow field of liquid steel in the mould using the method of finite element, and put forward the optimum value of processing parameters to promote the removing of inclusions and improve steel cleanliness [12]. The two-dimensional temperature and stress field were studied during the process of solidification, and a Multi-objective optimization program was set up to optimize the volume of water supply in each cooling zone, which down the defect rate [13–16].

The above works were very useful to improve the quality of beam blank. However, those papers paid more attentions to count cracks' locations, test chemical composition nearby cracks or adjusting process parameters, there was no attention to analyze the process of cracks' initiation and propagation.

This paper is subject to establish a thermal-stress model on the base of software ABAQUS, and analyze the distribution of temperatures and stress on the beam blank using the methods of the contact and restart. At the same time, the thermal stress & friction & ferrostatic pressure are concerned. More attentions are paid to stress along the width and narrow faces.

Mathematical Model

The section of the beam blank is 550 mm × 440 mm × 90 mm. According to the symmetry principle, a 1/4 section model is established. Figure 2 shows the shape and elements of the model.

Heat Transfer Equation

The model is based on the following two-dimensional equation of heat conduction in unsteady state [17]:

$$\rho(T)c(T)\frac{\partial T}{\partial t} = \frac{\partial}{\partial x}(k(T)\frac{\partial T}{\partial x}) + \frac{\partial}{\partial y}(k(T)\frac{\partial T}{\partial y}) + \dot{q} \tag{1}$$

$$\dot{q} = \rho(T)_s L\frac{\partial f_s}{\partial t} \tag{2}$$

Where ρ is density, kg/m^3; T the temperature, °C; t the time, s; c the specific heat, J/(kg · °C); k the conductivity, w/(m · °C); L the latent heat, J/kg.

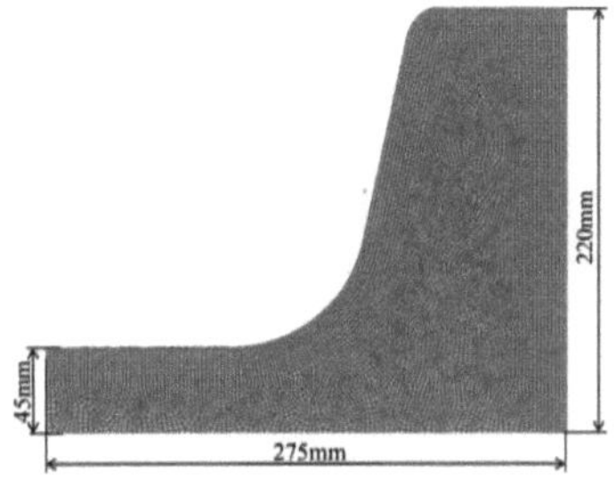

Fig. 2 Elements and geometric model

Thermal Stress Equation

Mechanics equations used the finite element method is [18]:

$$\frac{\partial}{\partial x_j} d\sigma_{ij} + dF_i = 0 \tag{3}$$

$$d\varepsilon_{ij} = \frac{1}{2}\left[\frac{\partial}{\partial x_j} ds_j + \frac{\partial}{\partial x_i} ds_j\right] \tag{4}$$

Where σ_{ij} is stress tensor; dF_i the volume increment; ε_{ij} the strain tensor; ds_i the displacement increment.

The total incremental strain vector has the form of

$$\{d\varepsilon\} = \{d\varepsilon_e\} + \{d\varepsilon_p\} + \{d\varepsilon_T\} \tag{5}$$

Where $d\varepsilon_e$, $d\varepsilon_p$, $d\varepsilon_T$ are incremental elastic strain components, plastic strain components and thermal strain components, respectively.

Solution Conditions

Initial Conditions

The initial condition is: $t = t0$. $t0$ is liquid steel temperature in the tundish.

Boundary Conditions

(1) The symmetric boundary condition is applied to the symmetry plane.
(2) The heat flow in beam blank mould is deduced as [19]:

$$q = a - b\sqrt{t} \tag{6}$$

Where q is surface heat flow density; a is a constant 2,680 here; b is calculated according to the area of the mould copper plate, the temperature and the speed of the water; t the time how long the blank goes through in the mould.

(3) The friction coefficient between the blank and the mould is specified to 0.1.

Table 1 Chemical composition of Q235B (wt%)

Steel	C	Mn	Si	P	S
Q235B	0.14–0.18	0.45–0.55	0.18–0.26	≤0.025	≤0.030

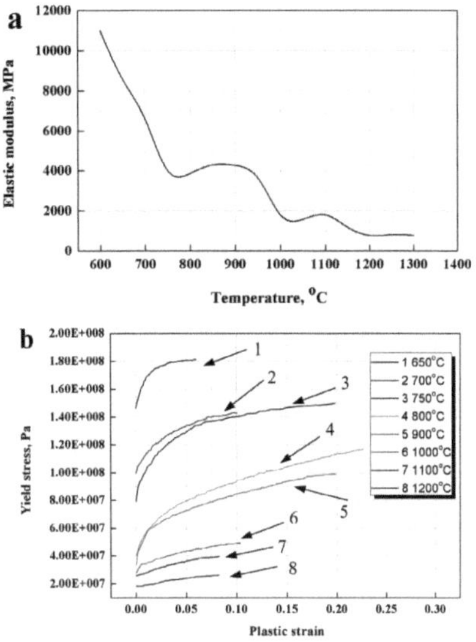

Fig. 3 Elastic modulus **a** and yield stress **b** at different temperature

Production Conditions and Material Property Parameters

The effective length of the mould is 650 mm; the casting speed is 1.0 m/min; the initial temperature is 1,550 °C. The species of steel is Q235B, the composition of steel shows as Table 1. The density of liquid phase is 7,000 kg/m^3, specific heat 840/(kg · °C). The density of solid is 7,400 kg/m^3, specific heat 670/(kg · °C), the latent heat 2.85 × 10^{5}j/kg, the solidus 1,455 °C, the liquidus 1,517 °C, thermal expansion coefficient 1.395 × 10^{-5}/°C. Elastic modulus and yield stress are shown as Fig. 3.

Results

Temperatures in the Beam Blank

The temperatures and the thickness of solidified shell are shown as Fig. 4. According to the picture, the highest and lowest locations of temperature are on the fillet and the flange at the end of the mould. Temperatures of other locations are relative uniform. As for shell thickness, the interior angle of flange is 22 mm, the

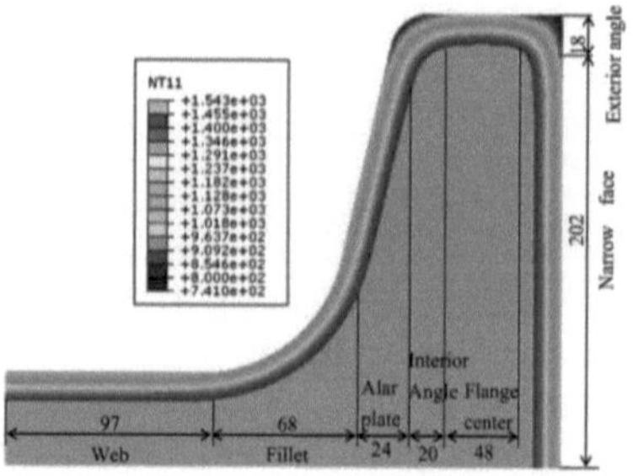

Fig. 4 Temperature and shell thickness contour at the end of the mould

external angle of flange is 28 mm, the thickest location is the fillet as 12 mm. The thickness of solidified shell at other locations is relative uniform.

Temperatures versus time on some of typical locations points are shown as Fig. 5. According to these pictures, as time goes by, the temperature of beam blank on the surface is lower and lower. At the same time, the temperature at the fillet is the highest, and the temperature at the flange is the lowest. The temperature of inner point G decrease a little and the phase is liquid in the whole mould. The temperature is 1,040° Con the fillet, 949 °C on the center of the flange; 813 °C on the interior angle of flange; on other locations of the surface is about 970 °C. The maximum temperature difference is 227 °C.

The simulation results in this paper are close to the previous researches [20].At the same time the temperature and shell thickness are up to the standard of the metallurgical criteria. Therefore, the simulation results is trustworthy.

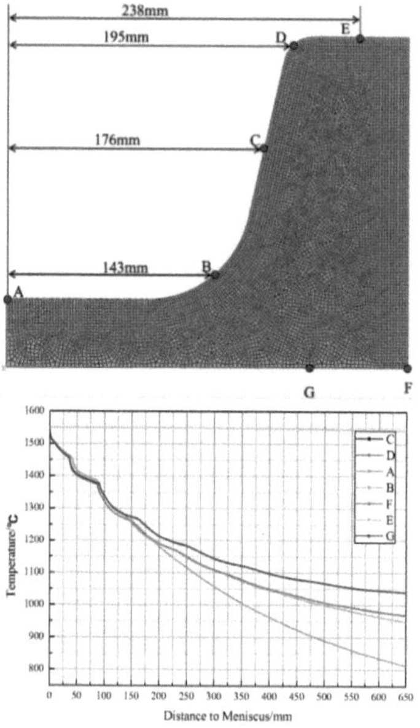

Fig. 5 Temperatures versus time **b** at some typical positions **a** on the beam blank surface

Stress in the Beam Blank

Equivalent Stress

The equivalent stress in the beam blank solidified shell is shown as Fig. 6. Graphs a, b, c and d are the distribution of equivalent stress at distances of 160 mm, 330 mm, 500 mm, 650 mm to meniscus respectively. From these graphs, equivalent stress at the web and flange is the maximum at prophase, fillet the minimum. Furthermore, the stress is not uniform along the transverse section. With continuous casting going on, the location of maximum equivalent stress moves from the surface to the inside. During the process of continuous casting, equivalent stress at the flange exterior angle is maintained at a high level, thus internal crack should easily appear here.

Equivalent stress versus time on the surface is shown as Fig. 7. From the picture, equivalent stress on points A and B reaches maximum of 18Mpa at a distance of 200 mm to meniscus. As continuous casting going on, equivalent stress of these

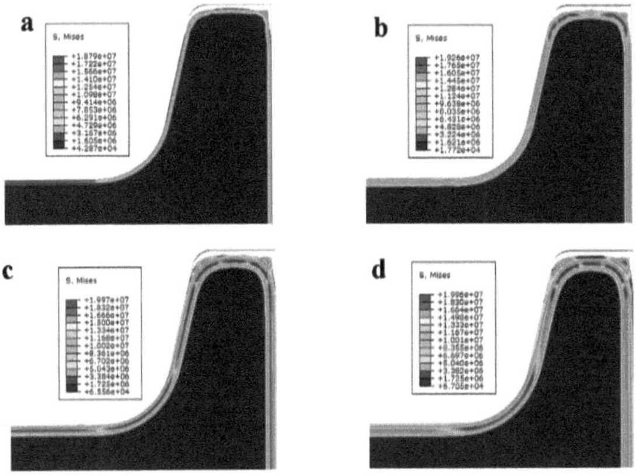

(a)160mm to meniscus; (b)330mm to meniscus; (c)500mm to meniscus; (d)650mm to meniscus

Fig. 6 Mises stress at some typical positions

Fig. 7 Mises stress versus time at some typical points on the beam blank surface

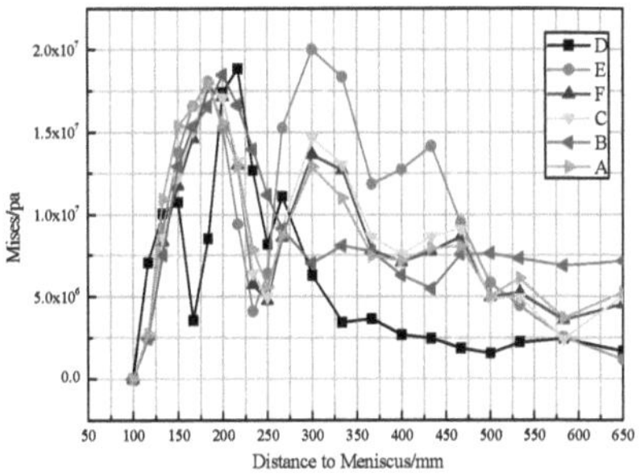

points reaches a minimum numerical at a distance of 240 mm to meniscus. At the end of the mould, the maximum equivalent stress of the beam blank is at the fillet, the minimum is at the center of the flange.

Stress Along Width

Stress versus time along width on these points is shown as Fig. 8. Through this picture, during the process of continuous casting, stress of these points along width goes through from positive to negative then positive. As we all know, the cracks woud not extend under compressive stress, but could healing. However, surface longitudinal cracks should more easily appear at the center of the web, the center of flange, the fillet, because the tensile stress at these positions remains a high level in initial stages in solidification process in the mould. Considering the location the maximum stress appears, it can be concluded that, the height of surface longitudinal cracks easily appearing should probably be: 180 mm to the meniscus at the center of the web and the flange; 200 mm to the meniscus at the fillet. The surface stress at other locations along the width remains at a low level during the process, the cracks should not easily initiate and propagate.

Stress Along Narrow Face

Stress versus time along narrow face on these points is shown as Fig. 9. Through this picture, during the process of continuous casting, stress on these points along narrow face goes through from positive to negative then positive too. Surface longitude cracks should more easily appear at the center of the narrow face and the alar plate, for the tensile stress of these positions remains a high level in initial stages of solidification process in the mould. The height of surface longitudinal cracks easily initiating and propagating should probably be: 180 mm to meniscus at these two locations. The surface stress along narrow face at other locations remains a low level during this process, the cracks should not easily initiate and propagate.

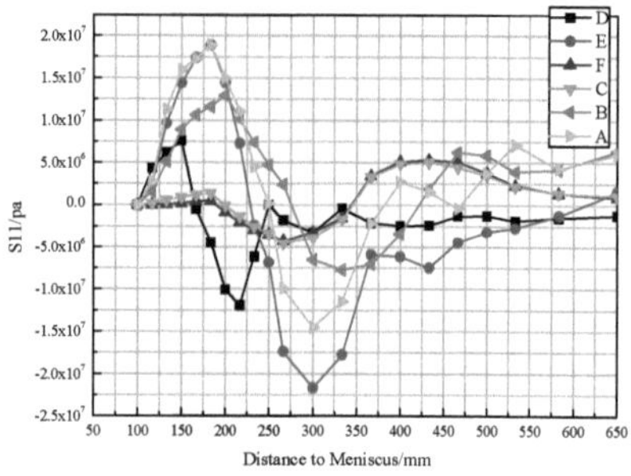

Fig. 8 S11 versus time at some typical points on the beam blank surface

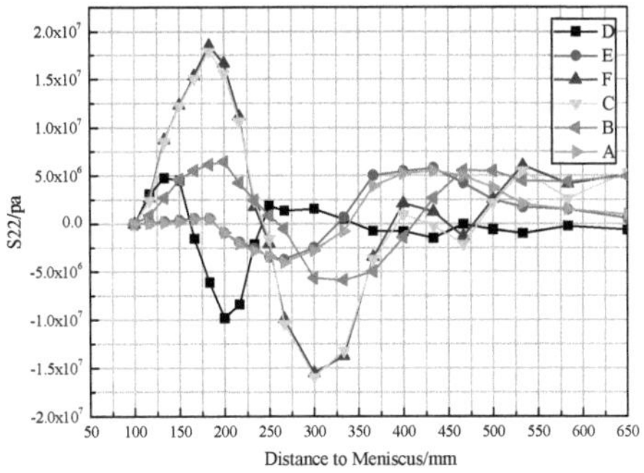

Fig. 9 S22 at some typical points on the beam blank surface

Results

In this paper, a thermal-mechanical coupling model of the beam blank was established in the continuous casting. The temperature, the shell thickness and the stress in the beam blank were analyzed. The main conclusions were as follows:

1. The temperature at the fillet is the highest, flange interior angle the lowest. Temperatures at the web, the narrow face and the width face are relative uniform. The thickness of solidified shell is thinnest at fillet, thickest at flange corner.
2. With continuous casting going on, the maximum equivalent stress moves from the surface to the inside. At the end of the mould, the distribution of equivalent stress is uniform along the section of solidified shell.
3. Stress on the surface goes through from positive to negative then positive, the maximum tensile stress appears at the upper of the mould.
4. The surface longitudinal crack are most easily to occur 200 mm to meniscus at the fillet, 180 mm to meniscus at the web, the flange and the narrow face.

Acknowledgements The authors would like to acknowledge the financial support provided by the Natural Science Foundation of China (51574103).

References

1. L. Wei et al., Thermo-mechanical coupled numerical simulation of mould and strand in beam blank continuous casting. Mater. Sci. Technol. **20**(01), 31–34 (2012)
2. R. Fei et al., Overview and development of beam blank's continuous casting. Foundry Equip. Technol. China: Rev. **11**, 6–7 (2015)
3. L. Jian-hua et al., Genetic analysis of surface cracks in H beam. Iron and Steel **41**(08), 37–40 (2006)
4. W. Xue-jie, C. Kai-ke, On mechanical behaviour at hing temperature for carbon steel. J. Univ. Sci Technol. Beijing **14**(01), 29 (1993)

5. Lv Ming et al., Research and control on web plate crack of the large type beam blank. Iron and Steel **45**(07), 99–102 (2010)
6. W. Ke-feng et al., Analysis of flange crack on hot-rolled H section steel of SS400. Iron and Steel **46**(09), 96–100 (2011)
7. T. Sakuraya, T. Emi, T. Imai, Prevention of the formation of longitudinal facial crack on continuously cast slabs by improving mold powder and mold oscillation. Tetsu- to- Hagane **67**(8), 1220–1228 (1981)
8. K.H. Kim et al., Effect of carbon and sulfur in continuously cast strand on longitudinal surface cracks. ISIJ Int. **36**(3), 6–11 (1996)
9. H. Wen et al., Optimal design of mould taper in continuous casting beam blank. Spec. Cast. Nonferrous Alloys **36**(02), 194–195 (2016)
10. X. Hai-lun, W. Guang-hua, S. Wei, Optimization and design of cooling water channels for copper plate wall of beam blank mold during continuous casting. Steelmaking **28**(06), 65–69 (2012)
11. L. Wei, Y. Bo, X. Yue-xing, Three dimensional simulation of solidification process of beam blank continuous casting. J. Syst. Simul. **24**(02), 301–305 (2012)
12. W. Chen et al., Three-dimensional FEM study of fluid flow in mould for beam blank continuous casting: influence of nozzle structure and parameters on fluid flow. Ironmak. Steelmak. **39**, 560–567 (2012)
13. W. Chen et al., Thermo-mechanical simulation and parameters optimization for beam blank continuous casting. Mater. Sci. Eng. A **499**, 58–63 (2009)
14. W. Chen et al., Numerical simulation of the thermo-mechanical process for beam blank continuous casting. ACTA METALLURGICA SINICA(English letters) **20**(4), 241–250 (2007)
15. W. Chen et al., Optimization of processing parameters for beam blank continuous casting using MOGA combined with FEM. Rev. Adv. Mater. Sci. **33**(2), 337–341 (2013)
16. C. Wei et al., Optimization and control of secondary cooling parameters of continuous casting. Iron and Steel **42**(03), 25–29 (2007)
17. X.-l. Shi, Research on mechanics characteristics of dynamic soft reduction in continuous casting slabs, Ph.D. thesis, Yanshan University, pp. 18–20, 2014
18. J.-c. Ma, Research and application of heat transfer model and thermal stress model in solidification process of billet continuous casting, Ph.D. thesis, Northeastern University, pp. 60–64, 2009
19. L. Juan-juan, B. Yan-ping, Simulation of solidification process of beam blank continuous casting. J. Univ. Sci. Technol. Beijing **27**(02), 174–180 (2005)
20. Z. Li-guang, H. Yi-hua, Mathematical simulation of heat transfer during solidification in mould of beam blank caster. Iron and Steel **43**(11), 49–53 (2008)

The Influence of SEN and Upper Nozzle Design on the Flow Character for the Slab Quality

Yu Yanwen

Abstract The behavior of the liquid steel in the mold will decisively influence the slab quality of a Continuous Casting machine by giving casting conditions, like oscillation parameters and so on. Reasonable flow distribution out of SEN can reduce the impact on the shell formation and the mold level fluctuation and this ensure the proper casting powder melting and uniform heat transfer, promote steel inclusion floating and avoid the involvement of the casting powder into the steel, all this will contribute a better quality of casting slab. A great deal research works had been done on the improvement of nozzle's structure, such as reduce the inner diameter of nozzle in order to minimize the contact areas between steel and the nozzle, to reduce the clogging; Using flat nozzle—slope at the bottom of nozzle, can also reduce clogging, using the Annular shape Nozzle lumen and so on, which can reduce the accumulation of alumina and molten steel impact on wide face of the mold. as well as using the uniform diameter of the Slide gate plates of nozzle, Improve the sealing at the joints to reduce re-oxidation; proper preheating and thermo insulation. All measures mentioned above can contribute to reduce nozzle clogging, but, can not eliminate the SEN clogging completely. This paper introduces a new design of the submerged nozzle to effectively solve the nozzle blockage of heterogeneous adhesion.

Keywords Upper nozzle · Submerged entrary nozzle · Design · Slab quality

Trial in the Site

(See in Fig. 1).

Y. Yanwen (✉)
Baosteel Co., Ltd. Steelmaking Plant, Shanghai, China
e-mail: Yanwenyu88@163.com

The Minerals, Metals & Materials Society, *TMS 2017 146th Annual Meeting & Exhibition Supplemental Proceedings*, The Minerals, Metals & Materials Series,
DOI 10.1007/978-3-319-51493-2_50

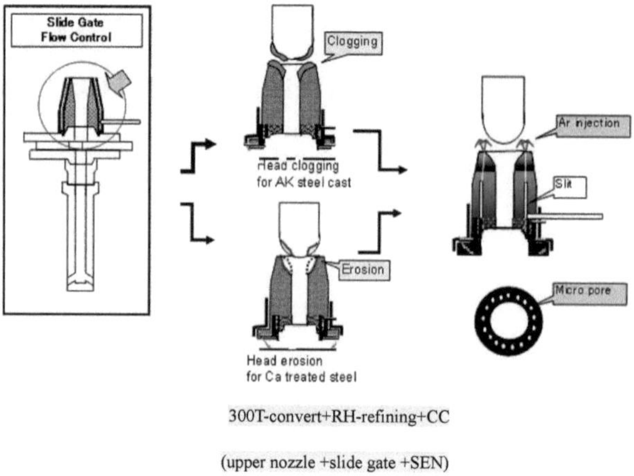

Fig. 1 The schematic drawing about slide gate used in Tundish

The Test on the Different Shapes About Inner Hole Upper Nozzle Anti-clogging

(See in Figs. 2, 3 and 4).

Simulation with Water in the 3 Bottom Shapes of SEN in the Mold

Under the Conditions About the Character in the Mold

		(A)	(B)
① Mold width	mm	1,300	2,000
② Mold thicknes	mm	225	225
③ SEN edge port		15	15
④ immerge depth	mm	260	260
⑤ casting speed	m/min	1.3	1.3
⑥output	T/min	2.7	4.2

The Choice About Bottom Shape of SEN

	current	convent1	convent2
bottom type	mountain shape	parelled shape	concave shape
SEN edge port			

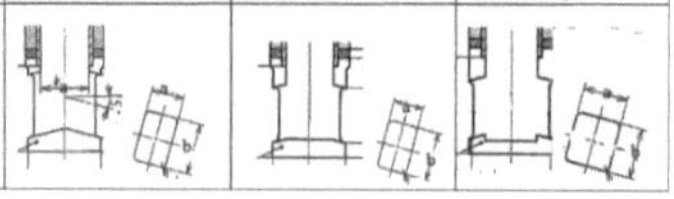

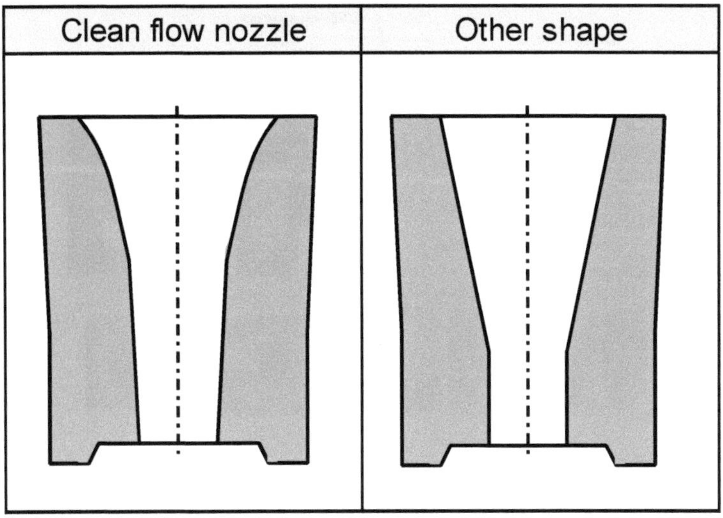

Fig. 2 The different shape of inner hole about upper nozzle schematic drawing

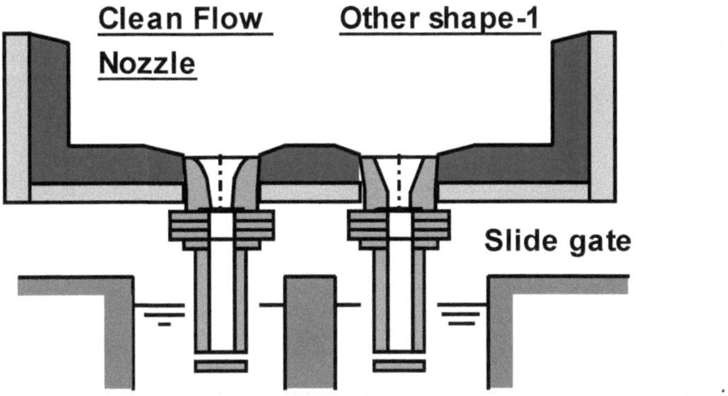

Fig. 3 Two kinds of upper nozzle inner hole shapes in the test used casting the AL-killed steel

The Result About the Water Simulation

(See in Fig. 5)

Investigation About Velocity in the Mold Thickness 2000 Mm Example

(See in Figs. 6 and 7)

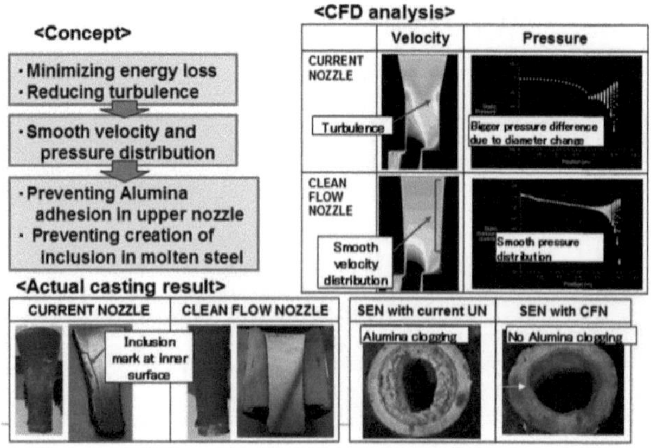

Fig. 4 Clear flow nozzle test

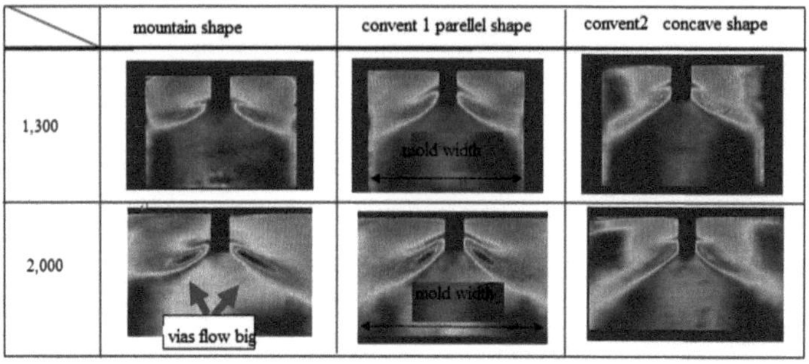

Fig. 5 Flow field in the mold about 3 kinds of shapes for SEN bottom comparison

Clean Flow Nozzle Together with Two Kinds of SEN Casting Experiments

Convent1: the bottom shape of SEN is parallel,
Convent2: the bottom shape of SEN is concave.

Blow insert gas from upper clean flow nozzle at the bottom of tundish when slab is casting. Blow Ar is less than 20L/min for upper clean flow nozzle, the output of molten steel is 3.3 t/min.

Casting ultra-lower carbon Al-Killed steel is more than 400 min, then contrast of two kinds of SEN at the inner hole clogging. Comparison with two kinds of SEN together the clean flow upper nozzle as follow Fig. 8.

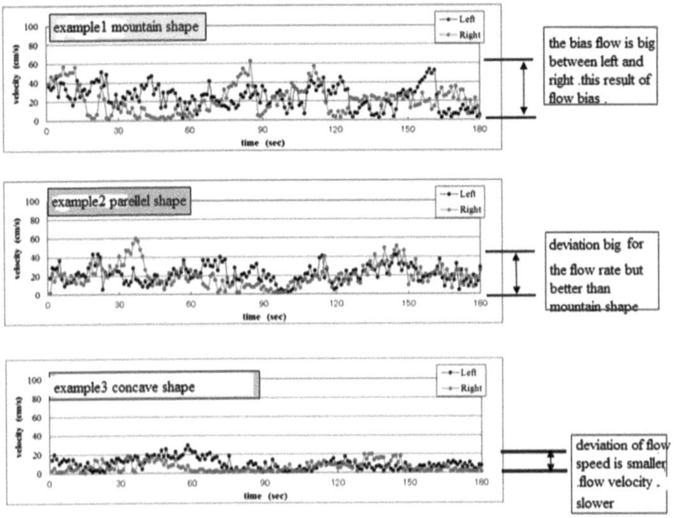

Fig. 6 The velocity between *left* and *right* comparison with 3 kinds bottom **shape of SEN**

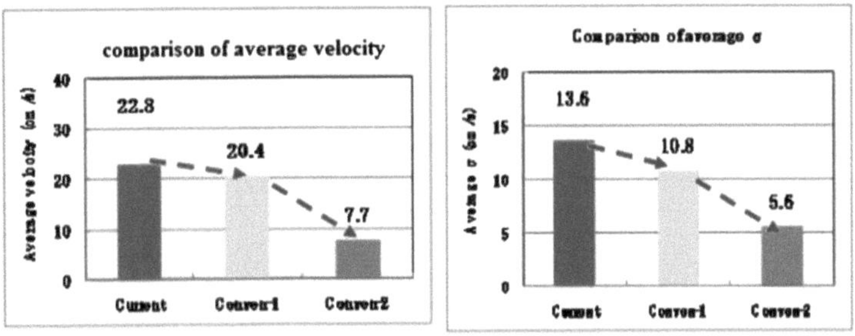

Fig. 7 Comparison of velocity 3 ways

Fig. 8 Comparison with residual clogging

	1ST	2ST
Residual upper nozzle	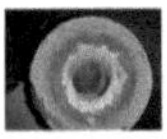	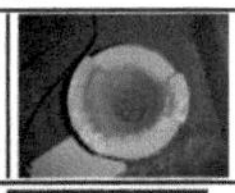
Residual SEN Convent1	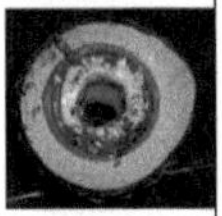	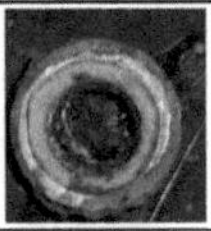
Residual SEN Convent2		
Inclusion rate %	4.6	2.2

Analysis the Mechanism with Clogging

Oxide Sources

Re-oxidation sources come from ladle sand, ladle of tundish slag, mold powder, air aspiration from the joint of LS&CN, refractory, etc. the total oxygen of molten steel increase during casting aspiration air.

The Change of Velocity

① The shape of SEN bottom improved parallel together with clean flow upper nozzle, can control the bias of steel flow. See Fig. 8, adhesion clogging is more than regular shape and improvement effect is not significant
② The shape of SEN bottom improved concave together with clean flow upper nozzle, can be better controlled the bias of steel flow. See Fig. 8, adhesion clogging is less than regular shape and improvement effect is significant.

The Mechanism of Clogging Nozzle

Many of Japanese research think of anti-clogging Al_2O_3 nozzle as follow

$$SiO_2(s) + C(s) \rightarrow SiO(g) + CO(g) \quad (1)$$

$$SiO(g) + Al(L) \rightarrow Al_2O_3(s) + Si(L) \quad (2)$$

$$CO(g) + Al(L) \rightarrow Al_2O_3(s) + C \quad (3)$$

The Whole Process of the Reaction

$$3SiO_2(s) + 3C(s) + 4Al(L) \rightarrow 2Al_2O_3(s) + 3Si + 3C$$

From this analysis, SiO_2 and C are the main facts formatting the Al_2O_3 clogging in conventional refractory.

Conclusions

1. SEN improve the structure design, can be controlled the steel flow in the mold. Especially for the slide gate controlled flow and avoiding the bias flow.

2. Graphite-less and silicon-less or without graphite and silicon material lined in AG tube forms more porous, good insulation, preventing heat loss of Melton steel. Reducing contact surface separate Al_2O_3 out. Slow down the rate between molten steel and refractory reaction.

Bibliography

1. H. Xia et.al., Research on reaction between molten steel and refractors. (China's refractory). pp. 48–56

Effect of the Addition of Ce and Si on Hot Cracking Behavior of SiMn Alloy During the Solidification Process

Zhiqiang Zhou, Zizong Zhu, Yuchuan Ding and Shengnan Zhou

Abstract The effect of the addition of Ce and Si on the hot cracking behavior of SiMn alloy during the solidification process has been studied respectively by scanning electron microscope with energy-dispersive spectroscopy. It was found that a large amount of graphite and SiC particles are scattered in the cracking zones of the original SiMn alloy, which results in hampering of solidification shrinkage of SiMn melt and the stress concentration around the non-metallic inclusions. A small amount addition of Ce had a positive effect on the reduction of hot cracking by grain refining and decreasing the thickness of liquid film. The addition of Si is found to be beneficial on decreasing the hot cracking markedly, which is due to the large reduction of graphite particles and the increasing of eutectic phase in the grain boundary.

Keywords Ce and Si addition · Hot cracking · Liquid film · Eutectic phase

Introduction

SiMn alloy is extensively used as deoxidizer or alloy additive in metallurgical industry. In the past decades, a lot of efforts have been devoted to understand thermodynamics on impurities removal from SiMn alloy, such as phosphorus and sulfur [1–3]. However, very few researchers are interested in issues on solidification and microstructure of SiMn alloy during manufacture process. Hot cracking is one of the major defects of as-cast SiMn alloy during the solidification, which may result in disintegrating into powder during transportation by mechanical collision.

In general, most of the alloys with complex components or high total impurities, as well as small amount of non-equilibrium eutectic alloy, have superior hot cracking tendency. This phenomenon has been studied for several decades, from

Z. Zhou · Z. Zhu (✉) · Y. Ding · S. Zhou
College of Materials Science and Engineering, Chongqing University, Chongqing 400044, People's Republic of China
e-mail: zhuzizong@163.com

The Minerals, Metals & Materials Society, *TMS 2017 146th Annual Meeting & Exhibition Supplemental Proceedings*, The Minerals, Metals & Materials Series,
DOI 10.1007/978-3-319-51493-2_51

both theoretical and experimental standpoints. It's commonly accepted that hot cracking usually originates above the solidus temperature in the semisolid region and propagate along the grain boundary or in the interdendritic liquid film [4, 5]. Researches also show that hot cracking occurs in the final stage of solidification of as-cast alloy and fine grain structures contribute to avoid hot cracking [6]. New hot tearing criterion based on different principles provided by Rappaz [7] and Eskin [8] has limited applicability to casting processes. The criterion is only an indicator of the appearance (or initiation), and not of the propagation of hot tearing. In fact, over the years, many computer models [9–12] have provided a powerful means to predict the occurrence of hot tearing based on Pappza-Drezet-Gremaud (DRG) criterion [7]. The challenge with hot tearing is that it is a complicated defect, even though under exactly the same casting conditions, some casting will tear while others will not. As a result, its prediction and control, is considerably difficult.

Despite amounts of literature on hot cracking, only few efforts have been made on the mechanism of crack formation and propagation in SiMn alloy, as well as approach to eliminate hot cracking. Hot tearing susceptibility in as-casting is strongly affected by the chemical additions of alloys. In this paper, effect of the addition of Ce and Si on the hot cracking behavior of SiMn alloy during the solidification process was studied to solve the issues mentioned above.

Experimental

The SiMn alloy used in this study was commercial alloy characterized by containing kinds of impurity element such as P and C. In order to see the effect of Ce and Si addition on hot cracking behavior during the solidification process, different amounts of Fe-Si-RE alloy and Si-Fe alloy (the chemical composition of the alloys is shown in Table 1) were added respectively to the commercial SiMn alloy in the graphite crucible and heated to 1673 K in a molybdenum silicide ($MoSi_2$) furnace under Ar atmosphere at a flow rate of 200 ml/min. The Argon gas was purified using silica gel also in addition to Mg turnings at 723 K. The melt was stirred manually for completed dissolution of alloy additive and held for 10 min before casting. The melt solidified for several seconds in the air at room temperature. Then the sample surface was addressed by polishing and pickling. The surface microstructure of samples was studied using scanning electron microscopy (SEM) and chemical composition of metal and impurity phase was determined

Table 1 Chemical composition of Si–Mn alloy and Si–Fe alloy wt%

Name	Mn	Si	C	P	S	Fe	Ca	Ce	Ti
Si-Mn	67.49	17.58	1.50	0.95	0.04	12.14	–	–	–
Fe-Si-RE	2.8	40.5	–	–	–	20.9	4.0	28.7	3.1
Si-Fe	0.41	75.05	0.09	0.031	0.01	24.18	–	–	

using energy dispersive X-ray spectroscopy (EDS) (JEOL JSM 6500F). The composition of alloys was determined by the inductively coupled plasma-atomic emission spectroscopy (ICP-OES).

Results and Discussion

Microstructure of Original SiMn Alloy

The typical appearance of the hot cracking feature during the solidification of original SiMn alloy is shown in Fig. 1. The micro cracking zones spread through the whole sample in interdendritic regions and eventually bridging to each other and form reticulations structure. EDS analyses confirmed that a large amount of graphite and SiC particles are scattered in the cracking zones of the original SiMn alloy, which results in hampering of solidification shrinkage of SiMn melt and the stress concentration around the non-metallic inclusions. Meanwhile, the solid-liquid coexistence temperature ranges of SiMn alloy were enlarged due to the solidification segregation of C. It is generally considered that the solidification crack susceptibility of an alloy is strongly depended on the solidification temperature range or mushy zone size, the larger the solidification temperature range, the larger the total cracking length and the greater the tendency for the hot cracking. It is clearly observed that hot cracking forms mainly in the semi-solid mushy zone (Fig. 1b). During the direct-chill (DC) casting of ferroalloys, the name "mushy zone" is usually applied to the total transition region between liquids and solidus temperature during the solidification process. However, in fact, only after the temperature has dropped below the coherency temperature, a real mush is formed. On the microscopic level solidification shrinkage and thermal contraction impose strains and stresses on the solid network in the mushy zone [8]. The deformation behavior of the mushy zone is very critical for the formation of hot cracking.

Enormous effort has been made in recent years to understand the mechanisms and define the criterions of the occurrence in addition to finding out or prediction the hot cracking susceptibility of various alloy systems. In principle, there are two main factors involved in the formation of a hot cracking: deformation of the partially coherent solid by mechanical load development (stress or strain or strain rate) as the restrained thermal contraction of the solidifying metal and lack of interdendritic liquid feeding [7]. In addition, it is important to note that alloy composition, grain size and morphology as well as casting parameters such as mold preheat temperature are some of the significant parameters directly or indirectly affecting the hot cracking susceptibility of casting alloy [13, 14].

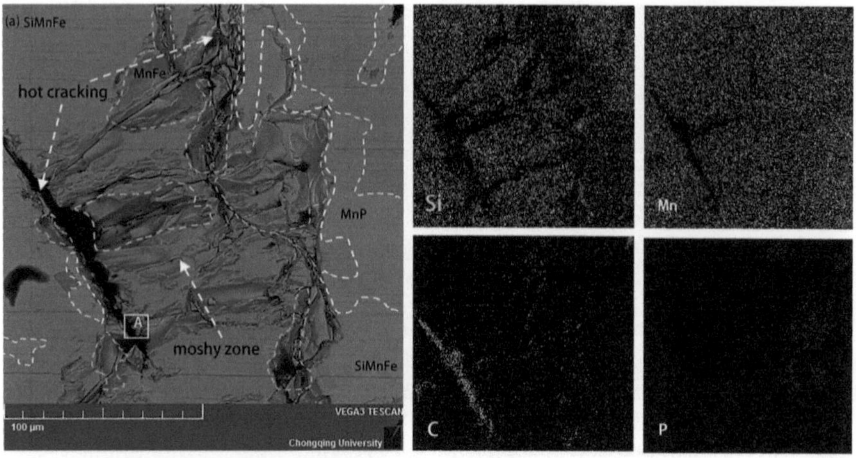

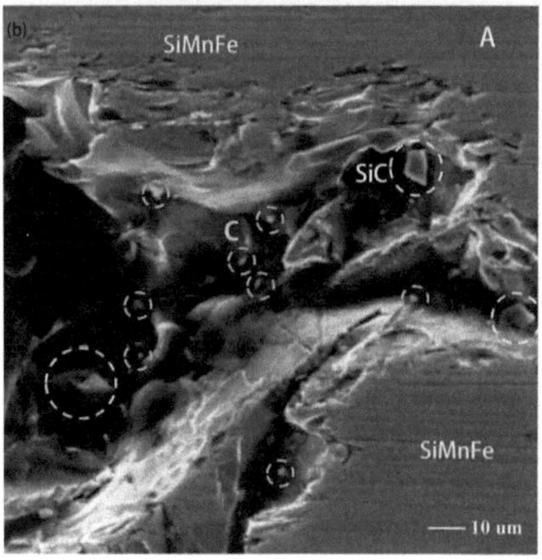

Average composition of main impurity phases in SiMn alloy, wt%.

Phase	C	O	Si	Mn	Color
Graphite	84.63	7.32	1.71	6.34	yellow
SiC	54.24	-	43.93	1.83	white

Fig. 1 **a** The microstructure of original SiMn alloy and mapping of elements distribution, **b** the micrograph of section A in **a** and EDS of impurities phases

Influence of Ce Addition on Hot Cracking Behavior of SiMn Alloy

Hot cracking susceptibility in DC casting of alloy is strongly affected by additions of alloying elements and grain refiners. Grain refinement is one commonly used approach to reduce the hot tear susceptibility of an alloy through changes to the grain size and morphology [15, 16]. Grain size and its morphology are important factors on hot cracking because the liquid metal feeding to the solidification shrinkage and dendrite coherency is determined mainly by them [17].

It can be observed in Fig. 2 that the hot cracking is markedly reduced and it turns to be very micro cracks with the addition of Ce compared with that in Fig. 1. Figure 2 shows the microstructure of SiMn alloy with Ce addition is observed to

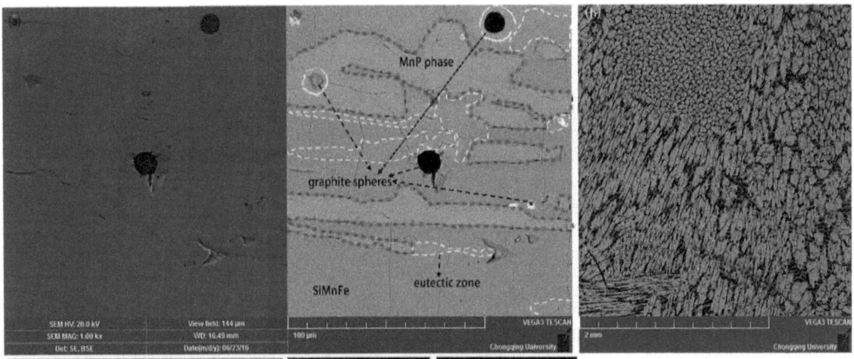

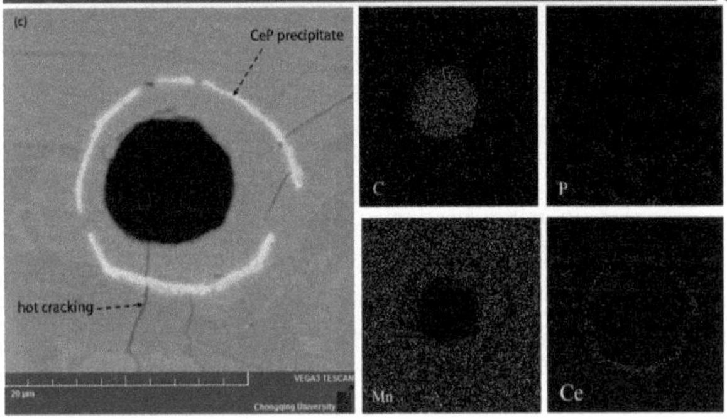

Average composition of CeP precipitate in SiMn alloy

Element	Weight %	Atomic %
C K	10.31	27.20
O K	24.64	48.77
P K	9.94	10.16
Ce L	51.08	11.55
Mn K	4.03	2.32
Totals	100.00	

Fig. 2 **a** and **b** The microstructure of SiMn alloy with 2.5wt% Ce addition before and after pickling, **c** the micrograph and EDS of CeP precipitate

have a finer-equiaxed dendritic microstructures surrounded by plate-like grains. The mechanism of grain refinement on reducing hot cracking is through controlling the grain size or morphology, in particular, delaying the dendrite coherency, and increasing the capillary pressures between grains as they become smaller, as long as the permeability of the mush is not adversely affected by the reduction in grain size [15]. It is known that the tendency of hot cracking is also associated with the properties and thickness of the liquid film. Stress between grains and liquid film will occurs when the solidification shrinkage is hindered at the late stage of solidification. If the shrinkage stress exceeds the tension of the liquid film, the liquid film will rupture and hot cracking forms. However, with Ce addition the grain sizes are refined, which reduces the thickness of the liquid film on the unit area and therefore increases hot cracking resistance of Ce-containing SiMn alloy.

However, it is obviously can be seen that there appeared some spherical particles in SiMn alloy because the graphite is spheroidized with Ce addition. It is interesting that the surface of graphite spheres is covered with a white layer of composition, which is confirmed by EDS analyses as CeP precipitate. The micro cracking mainly initiates from the surface of graphite spheres because of the stress concentration during the solidification shrinkage of the melt. Moreover, CeP is not stable. It can easily react with the moisture in the air when the hot cracking forms and release PH_3 gas, which may result in the further propagation of hot cracking.

Influence of Si Addition on Hot Cracking Behavior of SiMn Alloy

Another technique is through controlling the master alloying elements content by the additions of Si, Mn, Mg, or Fe which can influence the type and amount of intermetallics present during the final stages of solidification when hot tearing occurs [18, 19].

It is found that the impurity carbon content reduces remarkably with the increase of Si addition, which is due to the formation of SiC and the carbide particles float up and dissolve into the slags at slag/metal interface [20]. Hence, the non-metallic inclusions graphite or SiC particles at grain boundaries are eliminated. It is known that pure metal and composition near the eutectic region have high resistance to hot cracking. The intermetallics with a large solidification temperature range is highly susceptible to hot cracking as the alloy takes long time in the vulnerable region of solidification. Therefore, reducing C content is a necessary step to decrease hot cracking. On the other hand, it is

Obviously can be seen that the SiMn alloy with Si addition shows almost no hot cracking compared with that SiMn alloy with Ce addition. As shown in Fig. 3, there appears many banding P-containing eutectic phase of the SiMn alloy with Si addition. Cavities and gaps between grain boundaries also form in the alloys due to thermal contraction or external tension as shown in Fig. 3, which is result from no adequate liquid feeding in the final structure by non-equilibrium eutectics.

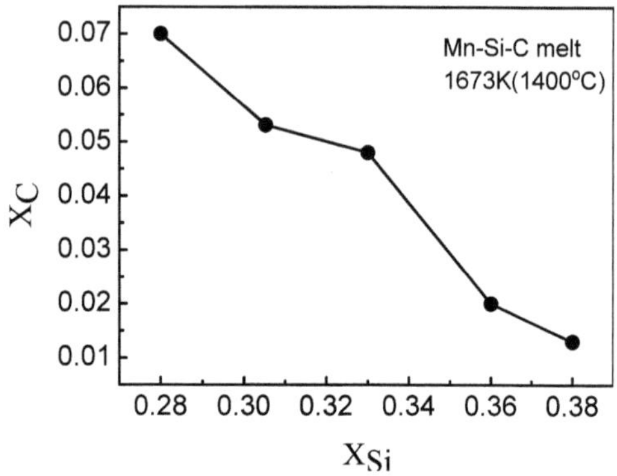

Fig. 3 The relationship between Si and C content in SiMn melt

Conclusions

In this work, the effect of Ce and Si addition on the hot cracking behavior of SiMn alloy during the solidification process has been studied respectively. Based on results obtained in this paper, it was concluded that:

(1) It was found that a large amount of graphite and SiC particles are scattered in the cracking zones of the original SiMn alloy, which results in hampering of solidification shrinkage of SiMn melt and the stress concentration around the non-metallic inclusions.

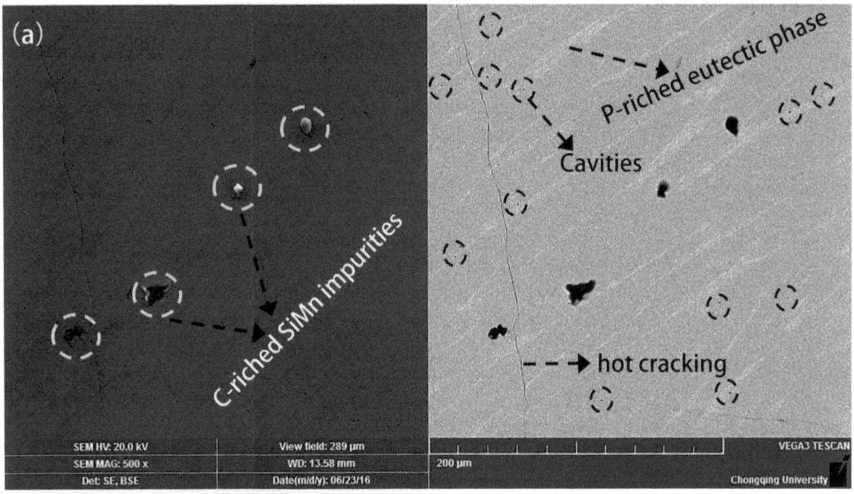

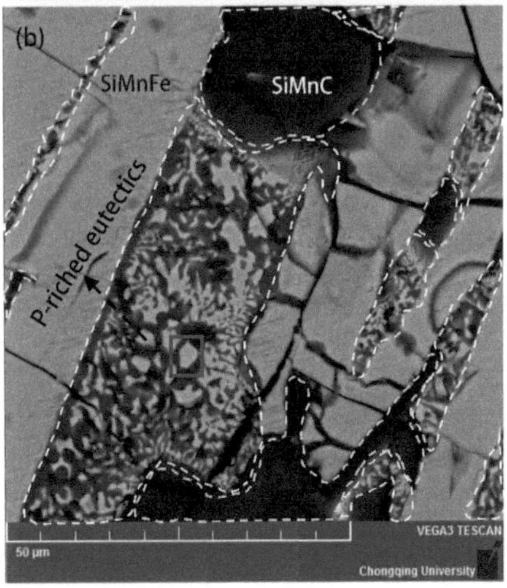

Average composition of P-riched eutectic phase in SiMn alloy

Element	Weight %	Atomic %
C K	8.20	25.38
Si K	12.24	16.20
P K	9.03	10.83
Mn K	58.39	39.50
Fe K	12.13	8.08
Totals	100.00	

Fig. 4 **a** The microstructure of SiMn alloy with Si addition, **b** the microstructure of 25wt%Si-Mn alloy after pickling and EDS of P-riched eutectic phase

(2) A small amount addition of Ce had a positive effect on the reduction of hot cracking by grain refining and decreasing the thickness of liquid film. However, the formation of graphite spheres and CeP precipitate covered on it also more or less lead to micro cracking.
(3) Si addition is found to be most beneficial on decreasing the hot cracking markedly compared with that SiMn alloy with Ce addition, which is due to the large reduction of graphite particles and the increasing of eutectic phase in the grain boundary (Fig. 4).

References

1. S. Oguchi, D.G.C. Robertson, Kinetic model for refining by submerged powder injection: part 1—transitory and permanent-contact reactions. Ironmak. Steelmaking **11**, 262–267 (1984)
2. S. Tabuchi, N. Sano, Thermodynamics of phosphate and phosphide in $CaO\text{-}CaF_2$ melts. Metall. Trans. B **15**(2), 351–356 (1984)
3. J.H. Shin, J.H. Park, Thermodynamics of reducing refining of phosphorus from Si-Mn alloy using $CaO\text{-}CaF_2$ slag. Metall. Mater. Trans. B **43**(6), 1243–1246 (2012)
4. D.G. Eskin, L. Katgerman, Mechanical properties in the semi-solid state and hot tearing of aluminium alloys. Prog. Mater. Sci. **49**(5), 629–711 (2004)
5. M. M'Hamdi, A. Mo, H.G. Fjaer, TearSim: a two-phase model addressing hot tearing formation during aluminum direct chill casting. Metall. Mater. Trans. A **37**(10), 3069–3083 (2006)
6. D. Fabre`guc, A. Deschamps, M. Sue´ry, J.M. Drezet, Non-isothermal tensile tests during solidification of Al–Mg–Si–Cu alloys: mechanical properties in relation to the phenomenon of hot tearing. Acta Mater. **54**(19), 5209–5220 (2006)
7. M. Rappaz, J.M. Drezet, M. Gremaud, A new hot-tearing criterion. Metall. Mater. Trans. A **30** (2), 449–455 (1999)
8. D.G. Eskin, L. Katgerman, A Quest for a New Hot Tearing Criterion. Metall. Mater. Trans. A **38**(7), 1511–1519 (2007)
9. H. Hao, D.M. Maijer, M.A. Wells et al., Modeling the stress-strain behavior and hot tearing during direct chill casting of an AZ31 magnesium billet. Metall. Mater. Trans. A **41**(8), 2067–2077 (2010)
10. M.A. Easton, H. Wang, J. Grandfield et al., Observation and Prediction of the Hot Tear Susceptibility of Ternary Al-Si-Mg Alloys. Metall. Mater. Trans. A **43**(9), 3227–3238 (2012)
11. A.H. Patel, Mathematical modeling of microshrinkage formation during solidification of A356 alloy castings. Dissertations and Theses—Gradworks, 2013, 8(1), 21–27
12. Suyitno, W.H. Kool, L. Katgerman, Integrated approach for prediction of hot tearing. Metall. Mater. Trans. A **40**(10), 2388–2400 (2009)
13. T.W. Clyne, G.J. Davies, The influence of composition on solidification cracking susceptibilityin binary alloy systems. J. Brit. Foundry **74**(4), 65–73 (1981)
14. Z.S. Zhen, N. Hort, Y.D. Huang, O. Utke, N. Petri, K.U. Kainer, Quantitative determination on hot tearing in Mg-Al binary alloys. Int. J. Cast Met. Res. **22**, 331–334 (2009)
15. M.A. Easton et al., An analysis of the effect of grain refinement on the hot tearing of aluminium Alloy. Mater. Forum **28**, 224–229 (2004)
16. M.A. Easton, D.H. StJohn, E. Sweet, Grain refinement and hot tearing of aluminium alloys how to optimise and minimise. Mater. Sci. Forum **630**, 213–221 (2010)
17. A.K. Dahle, L. Arnberg, The rheological properties of solidifying aluminum foundry alloys. Jom J. Miner. Met. Mater. Soc. **48**(3), 34–37 (1996)

18. H.T. Kim, S.W. Nam, S.H. Hwang, Study on the solidification cracking behaviour of high strength aluminum alloy welds: effects of alloying elements and solidification behaviors. J. Mater. Sci. **31**(11), 2859–2864 (1995)
19. L. Sweet et al., Chemical additions to reduce hot tearing in the cast house. TMS Light Met. **510**(1), 1133–1138 (2012)
20. J.H. Park, G.H. Park, Y.E. Lee, Carbide capacity of CaO–SiO2–MnO slag for the production of manganese alloys. ISIJ Int. **50**, 1078–1083 (2010)

Modeling and Predication of Shrinkage Porosity Formation in Steel Ingot

Chaojie Zhang, Yanping Bao, Min Wang and Lechen Zhang

Abstract In order to predict distribution of shrinkage porosity in steel ingot, the feeding process in mushy zone during solidification of steel ingot was analyzed and modeled. $G^2 \cdot L^{-5/3}$ (where G is temperature gradient and L is cooling rate) was put forward as shrinkage porosity criterion. Effects of liquid fraction and specific surface area of dendrites on permeability were both taken into consideration. Integrals with respect to interdendritic feeding flow velocity and pressure drop were evaluated. The derived expression reveals that: the smaller $G^2 \cdot L^{-5/3}$ value is, the higher the critical temperature can be, and the severer shrinkage porosity would occur. Predicted distribution of shrinkage porosity in a 5.5 ton 2Cr13 ingot using criterion $G^2 \cdot L^{-5/3} < 143$ agrees well with experimental sectioning results. Using criterion $G^2 \cdot L^{-5/3}$, an optimization of the 5.5 ton 2Cr13 ingot design was carried out aiming at eliminating shrinkage porosities in the ingot and improving metal yield effectively.

Keywords Shrinkage porosity criterion · Threshold value · Ingot solidification

Introduction

Shrinkage porosity in steel ingot can cause reduction of mechanical properties and decrease metal yield. Method to eliminate these defects by trial and error is time-consuming and expensive. However, with the development of computer technology, method of numerical simulation is becoming widespread in investigation of steel ingot solidification [1–5]. Nowadays, method of combination of

C. Zhang (✉) · Y. Bao (✉) · M. Wang · L. Zhang
State Key Laboratory of Advanced Metallurgy, University of Science and Technology Beijing, Beijing 100083, China
e-mail: zhangchaojie1989@gmail.com

Y. Bao
e-mail: zcj_ustb@163.com

The Minerals, Metals & Materials Society, *TMS 2017 146th Annual Meeting & Exhibition Supplemental Proceedings*, The Minerals, Metals & Materials Series,
DOI 10.1007/978-3-319-51493-2_52

numerical simulation and shrinkage porosity criterion to predict and eliminate these defects is crucial important.

Many researchers have investigated shrinkage porosity in steel ingots and put forward kinds of shrinkage porosity criterions. At the very beginning, shrinkage porosity was predicted through simple methods which rely on solidification patterns from thermal calculation. Then, Niyama et al. [6] put forward the famous criterion, $G/\sqrt{L}$, the temperature gradient divided by the square root of the cooling rate at the end of solidification at each point within a casting, and this criterion is still widely used. Later, Sigworth and Wang [7] developed a simple "geometrical" criterion to describe when shrinkage porosity will occur. Lee et al. [8] proposed a new feeding efficiency parameter to predict the formation of porosity, denoted as $(G \cdot t^{2/3})/V_s$ (where t is local solidification time, and V_s is solidus velocity). Liu and Jia [9] discussed the physical meaning of Niyama criterion. Minakawa et al. [10] reported that $G/R^{1/2}$ was the best parameters which could indicate the onset of porosity, where G and R are the centerline gradient and growth velocity. Casrlson and Beckermann [11] investigated formation of porosity in ingot and put forward a dimensionless Niyama critetion.

However, the derivations of these shrinkage porosity criterions are mainly based on assumptions on permeability in mushy zone to a great extent, and factors such as dendrite arm spacing and specific surface area of dendrites on permeability are not taken into consideration. In this paper, based on mechanism of feeding process in mushy zone, $G^2 \cdot L^{-5/3}$ was obtained as shrinkage porosity criterion through derivation which accounts not only for thermal conditions during casting and material properties, but also for structure of mushy zone such as dendrite arm spacing and specific surface area of dendrites on permeability. The accuracy of criterion $G^2 \cdot L^{-5/3}$ was verified through experimental sectioning of a 5.5 ton 2Cr13 ingot. Finally, an optimization of the 5.5 ton ingot design was carried out to eliminate shrinkage porosities in ingot using criterion $G^2 \cdot L^{-5/3}$.

Materials and Methods

Mechanism of Feeding Process in Mushy Zone

During solidification process of an ingot, dendritic grains grow and consequently lots of small insulated liquid pools form, which can't be fed during the following solidification. As a result, lots of small pores called shrinkage porosities will be remained in the ingot body. Figure 1 schematically illustrates feeding process in mushy zone. There exits feeding flow towards the opposite direction along which dendritic grains grow. With the distance to dendrite root decreases, the velocity of feeding flow decreases due to increase of viscosity. The feeding flow ceases at a certain point where liquid fraction decreases to $f_{L,\ c}$. Consequently shrinkage porosity forms where liquid fraction is smaller than $f_{L,\ c}$.

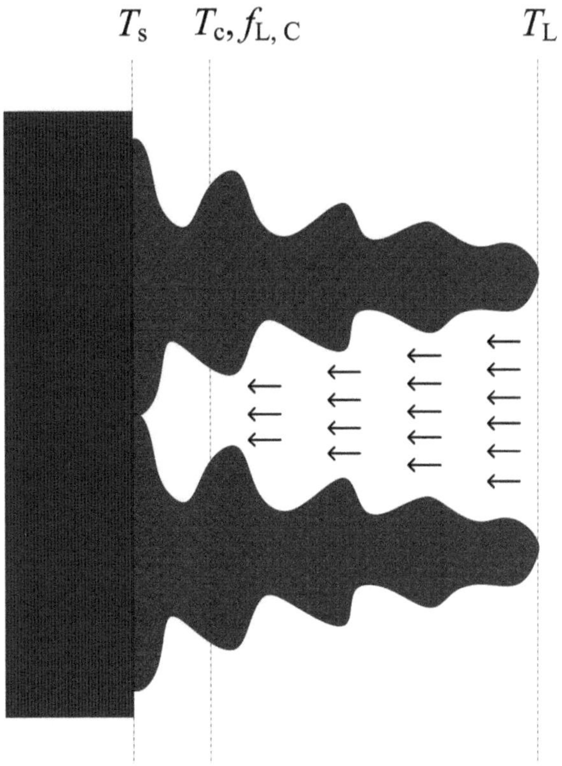

Fig. 1 Schematic of feeding process in mushy zone

Shrinkage Porosity Criterion

Based on the directional solidification process shown in Fig. 1, assumptions are made as followings:

(1) Feeding flow in the mushy zone obeys Darcy's law.
(2) Temperature gradient and cooling rate in the mushy zone are constants.
(3) Feeding flow ceases where liquid fraction reaches $f_{L,\ c}$.

According to Darcy's law, relationship between velocity of feeding flow and pressure gradient in the mushy zone can be expressed as the following equation [6]:

$$\frac{dP}{dx} = -\frac{\mu f_L}{K} v \tag{1}$$

where K is a constant termed the permeability of the mushy zone, P is pressure, μ is viscosity, f_L is liquid volume fraction, v is velocity of feeding flow.

Integrating Eq. (1) over the mushy zone from the critical point to the liquidus yields

$$\Delta P = \int_{x_c}^{x_L} dP = -\int_{x_c}^{x_L} \frac{\mu f_L}{K} v dx \tag{2}$$

Permeability K is mainly determined by porosity and specific surface area of dendritic grains in the mushy zone [12]:

$$K = \frac{\alpha f_L}{S^2} \tag{3}$$

where S is specific surface area, α is material constant.

Assumption was made that dendritic arms in each cross section in the mushy zone were of the same diameter, as shown in Fig. 2.

Denoting d as diameter of dendritic arm, number of dendritic arms in per unit cross section area can be written as

$$n = \frac{1 - f_L}{\pi d^2 / 4} \tag{4}$$

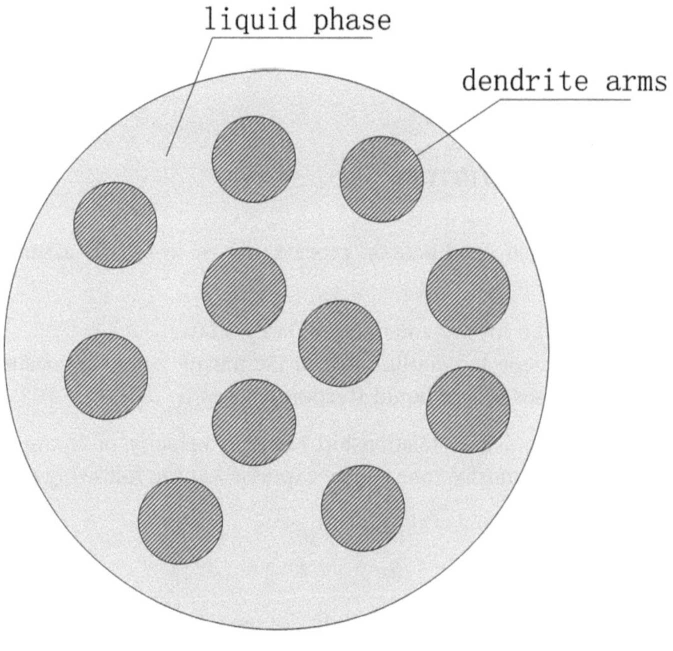

Fig. 2 Schematic of cross section of feeding channel in the mushy zone

Specific surface area of dendritic grains S can be written as

$$S = \frac{n\pi d}{1 - f_L} \tag{5}$$

Substituting Eq. (4) into Eq. (5) yields

$$S = \frac{4}{d} \tag{6}$$

As is known, secondary dendrite arm spacing is mainly depended on cooling rate [13]. Diameter of dendritic arm is closely related to secondary dendrite arm spacing. Therefore diameter of dendritic arm can be determined as

$$d = aL^{-1/3} \tag{7}$$

where L is cooling rate, a is a material constant.

Substituting Eqs. (6) and (7) into Eq. (3) yields

$$K = \frac{\alpha a^2}{16} f_L L^{-2/3} \tag{8}$$

Denoting R as solidification rate and β as total solidification shrinkage rate, $\beta = (\rho_s\text{-}\rho_L)/\rho_L$, velocity of feeding flow v can be expressed as

$$v = \beta R \tag{9}$$

Substituting Eqs. (8) and (9) into Eq. (2) yields

$$\Delta P = \frac{16\mu\beta}{\alpha a^2} R L^{2/3} (x_L - x_c) \tag{10}$$

where

$$x_L - x_c = \frac{T_L - T_c}{G} \tag{11}$$

Using relationship $L = GR$, Eq. (10) becomes

$$\Delta P = \frac{16\mu\beta}{\alpha a^2} (T_L - T_c) G^{-2} L^{5/3} \tag{12}$$

We denote M as a material constant shown as the following equation

$$M = \frac{16\mu\beta}{\alpha a^2} \tag{13}$$

Therefore it can be obtained that

$$T_L - T_c = \frac{\Delta P}{M} G^2 L^{-5/3} \tag{14}$$

As shown in Fig. 1, a larger T_c represents a wider area which can't be fed during solidification, and more serious porosity shrinkages will form there. So, a small value of T_L-T_c indicates serious porosity shrinkages. $\triangle P$ is a constant for a certain casting situation, and M is also a constant for a certain steel grade. According to Eq. (14), a small value of $G^2 \cdot L^{-5/3}$ could lead to a large T_c, and the porosity shrinkages would be serious consequently. Therefore, $G^2 \cdot L^{-5/3}$ can be used as shrinkage porosity criterion.

Some researchers have investigated shrinkage porosity criterion and proposed some different shrinkage porosity criterions [7, 8, 10, 14]. Among which the most widely used is the Niyama criterion that was defined as $G/\sqrt{L}$. However, the Niyama criterion doesn't account for influence of structure of dendritic grain on permeability in the mushy zone. The difference of criterion $G^2 \cdot L^{-5/3}$ between the Niyama criterion in form lies in the L item. This is because the influence of dendritic arm spacing which depends on cooling rate on permeability was taken into consideration during derivation of Eq. (14).

Threshold Value of $G^2 \cdot L^{-5/3}$

In order to predict shrinkage porosity distribution in steel ingot, it's necessary to determine the threshold value of criterion $G^2 \cdot L^{-5/3}$. In this paper, values of criterion $G^2 \cdot L^{-5/3}$ allover steel ingot are calculated by numerical simulation of ingot solidification. Then, the sum of volume fraction of the area with $G^2 \cdot L^{-5/3}$ value below any value can be calculated. The threshold value of $G^2 \cdot L^{-5/3}$ can be obtained with combination of total solidification shrinkage rate and the volume fraction calculated.

A function $F(x)$ shown in Eq. (15) is defined as sum of volume fraction of area whose $G^2 \cdot L^{-5/3}$ value is below x,

$$F(x) = \frac{1}{V_T} \cdot \sum V_i \left\{ \left(G^2 L^{-5/3} \right)_i \leq x \right\} \tag{15}$$

where V_T is the total volume of the ingot, V_i is the volume of cell i, x is a variable which can be any positive value.

With increasing of the variable *x*, the value of function *F(x)* increases. When the value of function *F(x)* is equal to total solidification shrinkage rate, the corresponding value of the variable *x* can be regard as the threshold value of the criterion $G^2 \cdot L^{-5/3}$. Therefore shrinkage porosity criterion can be expressed as the following equation.

$$G^2 L^{-5/3} < x|_{F(x)=\beta} \tag{16}$$

Results and Discussion

Verification of the Shrinkage Porosity Criterion $G^2 \cdot L^{-5/3}$

Melting and casting process of a 5.5 ton 2Cr13 steel ingot was conducted in company ANSTEEL. The ingot was sectioned along its central axis to investigated distribution of shrinkage porosity.

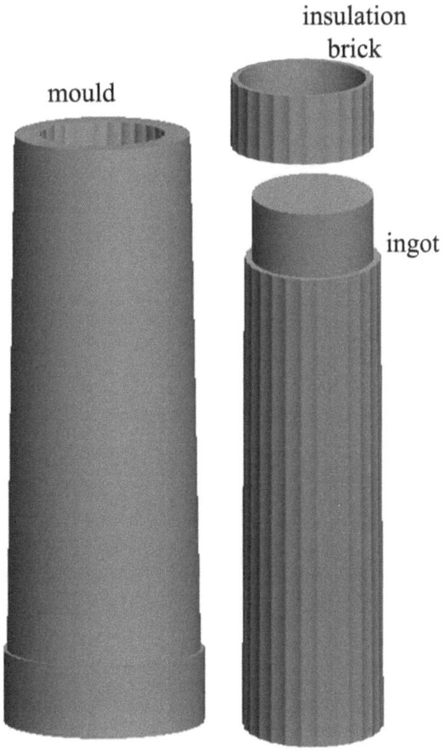

Fig. 3 Geometry of the 5.5 ton steel ingot and the mould

Table 1 Composition of the ingot and the mould

Material	Code	Composition (wt%)					
		C	Mn	Si	P	S	Cr
Melt	2Cr13	0.16–0.25	0–1	0–1	≤0.04	≤0.03	12–14
Mould	Iron	3.2–3.8	0.8–1.2	–	≤0.05	≤0.10	–

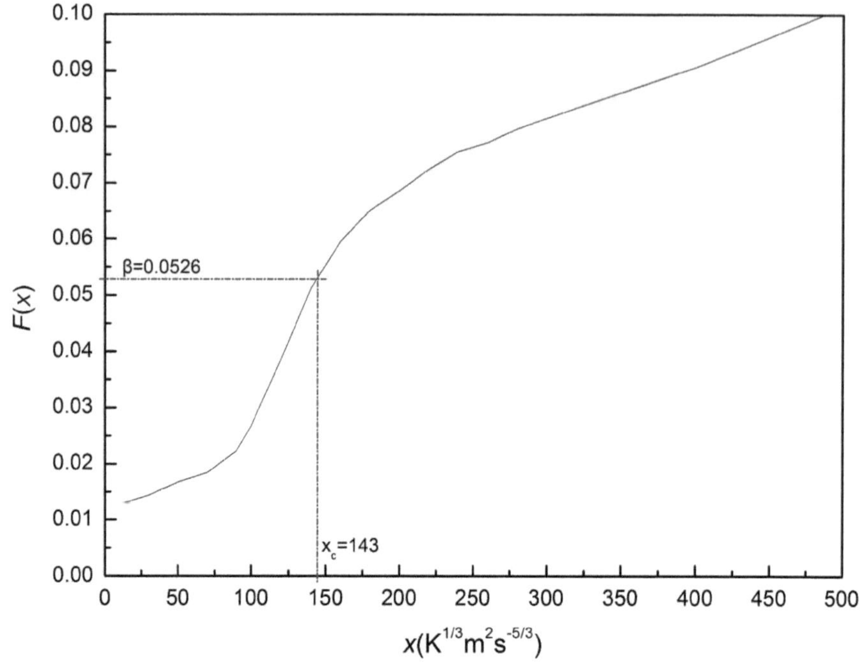

Fig. 4 F(x)-x curve of the 5.5 ton ingot of 2Cr13 steel

Numerical simulation of solidification of the 5.5 ton 2Cr13 steel ingot was conducted. Figure 3 shows the geometry of the 5.5 ton steel ingot and the mould, and Table 1 shows the composition of the ingot and the mould.

Value of $G^2 \cdot L^{-5/3}$ at any region in the ingot was obtained. Then the corresponding *F(x)-x* curve defined in Eq. (15) was calculated as shown in Fig. 4.

The total solidification shrinkage rate of 2Cr13 steel is 0.05263. Therefore the threshold value of $G^2 \cdot L^{-5/3}$ is 143 $K^{1/3}m^2s^{-5/3}$ according to the *F(x)-x* curve in Fig. 4. Namely, $G^2 \cdot L^{-5/3} < 143$ is shrinkage porosity criterion of the 5.5 ton 2Cr13 steel ingot.

Figure 5 shows comparison of the predicted shrinkage porosity using the criterion $G^2 \cdot L^{-5/3} < 143$ with the experimental sectioning. The hot top of the experimental ingot was cut off. It can be seen the predicted distribution of shrinkage porosity is similar to the experimental sectioning results.

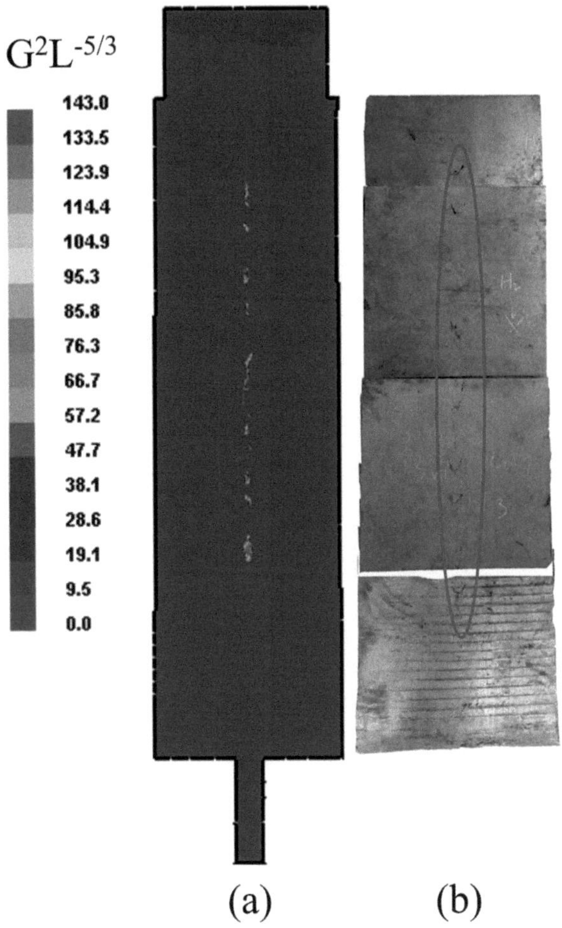

Fig. 5 Comparison of predicted shrinkage porosity with experimental sectioning, **a** predicted, **b** experimental

Table 2 Comparison of shrinkage porosity area ratio between predicted ones and experimental sectioning

Height (mm)	Shrinkage porosity area ratio	
	Experimental (%)	Predicted, $G^2 \cdot L^{-5/3}$ (%)
1776–2220	0.29	0.23
1332–1776	0.64	0.56
888–1332	0.45	0.36
444–888	0.50	0.24
0–444	0.00	0.00

In order to quantize the shrinkage porosity level, shrinkage porosity area ratio was defined as ratio of area of shrinkage porosity in the central plane to the total area of the central plane of the ingot. The area of shrinkage porosity in the central plane was measured by counting number of pixels using digital imaging processing software. Detailed comparison of shrinkage porosity area ratio between predicted ones and experimental sectioning is shown in Table 2. It can be seen that the predicted results are in good agreement with the experimental ones.

Optimization of Design of 5.5 Ton Ingot

Distribution of shrinkage porosity and metal yield of a steel ingot has a strong relationship with design of the ingot, especially the height to diameter ratio (*H*/*D* ratio) and taper [2, 15]. In order to eliminate shrinkage porosities in the 5.5 ton ingot, a series of ingot designs with different *H*/*D* ratios and tapers were carried out.

Table 3 Geometric parameters of the prototype and optimized design of the 5.5 ton steel ingot

Design	Shape	Height/mm	Top diameter/mm	Bottom diameter/mm	Taper (%)	*H*/*D* ratio
Prototype	Polygonal	2520	630	660	1.19	3.4
Optimized	Polygonal	1613	826	786	−2.48	1.7

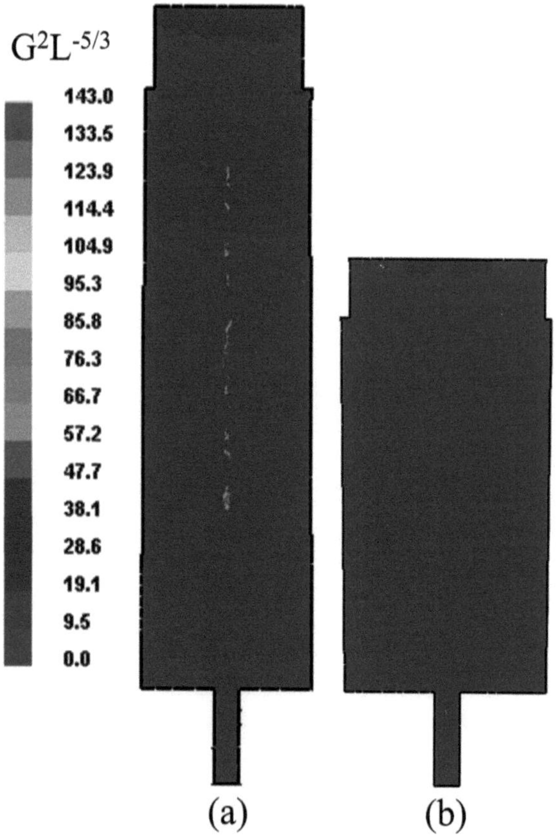

Fig. 6 Distribution of shrinkage porosity of the prototype and optimized 5.5 ton ingot, **a** prototype, **b** optimized

And the shrinkage porosity distribution of each was predicted using the criterion $G^2 \cdot L^{-5/3} < 143$. Finally an optimized design of the 5.5 ton ingot was obtained as shown in Table 3. Figure 6 shows distribution of shrinkage porosity of the prototype ingot and optimized one, respectively. Obviously, shrinkage porosities in ingot are eliminated and metal yield is improved effectively.

Conclusions

A model of the formation of shrinkage porosity during solidification of steel ingot was developed. Feeding process during solidification was investigated theoretically and $G^2 \cdot L^{-5/3}$ was put forward as shrinkage porosity criterion. Verification and application of the shrinkage porosity criterion $G^2 \cdot L^{-5/3}$ was conducted on a 5.5 ton 2Cr13 ingot.

(1) Shrinkage porosity criterion $G^2 \cdot L^{-5/3}$ can be used to predict distribution of shrinkage porosity in steel ingot accurately, which accounts not only for thermal conditions during casting and material properties, but also for structure of mushy zone such as dendrite arm spacing and specific surface area of dendrites on permeability.
(2) Combination of numerical simulation and shrinkage porosity criterion $G^2 \cdot L^{-5/3} < 143$ is a good method to optimize the design of 5.5 ton 2Cr13 ingot to improve the metal yield.
(3) The 5.5 ton 2Cr13 steel ingot with a negative taper of −2.48% and a small *H*/*D* ratio of 1.7 has few shrinkage porosities in ingot body.

Acknowledgements This work was supported by the National Natural Science Foundation of China under Grant No. 51274029. The authors are thankful to ANSTEEL for the support on the field test.

References

1. M. Heidarzadeh, H. Keshmiri, J. Iron. Steel Res. Int. **20**, 78 (2013)
2. A. Kermanpur, M. Eskandari, H. Purmohamad, M.A. Soltani, R. Shateri, Mater Des. **31**, 1096 (2010)
3. N. Lu, Y.L. Jin, S.L. Li, X.G. Ai, X.G. Yuan, China Foundry **10**, 87 (2013)
4. M. Balcar, R. Zelezny, L. Sochor, P. Fila, L. Martinek, Mater Tehnol. **42**, 183 (2008)
5. B. Smetana, M. Zaludova, M. Tkadleckova, J. Dobrovska, S. Zla, K. Gryc, P. Klus, K. Michalek, P. Machovcak, L. Rehackova, J. Therm. Anal. Calorim. **112**, 473 (2013)
6. E. Niyama, T. Uchida, M. Morikawa, S. Saito, Int. Cast Met J. **7**, 52 (1982)
7. G.K. Sigworth, C. Wang, Metall. Trans. B **24,** 349 (1993)
8. Y.W. Lee, E. Chang, C.F. Chuieu, Metall. Mater. Trans. B **21**, 715 (1990)
9. B.Q. Jio, B.C. Liu, Acta Metall. Sin-Engl **9**, 310 (1996)
10. S. Minakawa, I.V. Samarasekera, F. Weinberg, Metall. Mater. Trans. B **16**, 823 (1985)

11. K.D. Carlson, C. Beckermann, Metall. Mater. Trans. A **40A**, 163 (2009)
12. P.C. Carman, Trans-I Chem. E **15**, 150 (1937)
13. M.C. Flemings, Metall. Trans. **5**, 2121 (1974)
14. J. Wang, P. Fu, H. Liu, D. Li, Y. Li, Mater Des. **35**, 446 (2012)
15. K. Tashiro, S. Watanabe, I. Kitagawa, I. Tamura, Trans. Iron Steel Inst. Jpn. **23**, 312 (1983)

Numerical Analysis of Coupled Turbulent Flow and Macroscopic Solidification in a Billet Continuous Casting Mould with Electromagnetic Stirring

H. An, Yanping Bao, M. Wang and L. Zhao

Abstract Mould electromagnetic stirrers (M-EMS) have played an increasingly important role in improving inner quality of special steel billet. A coupled three-dimensional numerical simulation of the electromagnetic field, flow field and temperature distribution in the mold with M-EMS for billet continuous casting was carried out based on the finite element method. The influence of the current and frequency of M-EMS on the flow field and temperature distribution was discussed. According to the simulation results, with the increase of the current, the steady flow became unstable accompanying with the bias flow, and the temperature distribution was unsymmetrical. However, the effect of the frequency on flow field was exactly opposite compared with the current. The significant swirl flow with M-EMS prevented the superheated jet moving downward and weakened the invasion depth of the jet, and thus the location of the hot zone in the mould moved up evidently.

Keywords Billet · Mould electromagnetic stirring · Electromagnetic field · Flow field · Coupled numerical simulation

Introduction

Due to the unique advantages, billet casters are more applied in the production of high value-added wire rod such as spring steel 55SiCrA [1], and M-EMS has played an increasingly important role in controlling the molten steel flow in the mould to improve center segregation and lower porosity in billet casting. M-EMS should promote dissipation of superheat and enhance the columnar-to-equiaxed

H. An · Y. Bao (✉) · M. Wang
State Key Laboratory of Advanced Metallurgy, University of Science and Technology Beijing, Beijing 100083, People's Republic of China
e-mail: hanghangan_ustb@outlook.com

H. An · L. Zhao
School of Metallurgical and Ecological Engineering, University of Science and Technology Beijing, Beijing 100083, People's Republic of China

The Minerals, Metals & Materials Society, *TMS 2017 146th Annual Meeting & Exhibition Supplemental Proceedings*, The Minerals, Metals & Materials Series, DOI 10.1007/978-3-319-51493-2_53

(CET) transition during solidification to reduce internal defects by the way of generating swirling flow in the liquid core of the strand [2, 3]. Because of high temperature of molten steel and mutual coupling of electromagnetic field, flow field and temperature field in the mould with electromagnetic stirring, flow in the mould with EMS are very complicated. So it is very difficult to obtain more useful information in detail on the melt flow with M-EMS in industrial plant trials and laboratory tests [4, 5]. Therefore the numerical simulation on electromagnetic field, flow field and temperature field has been a useful and practicable tool for offering insight into the motion behavior in M-MES. However, very little fundamental work has been reported in this field especially for billet casting.

Fujisaki et al. [6–8] developed a three-dimensional (3D) MHD analysis method for conductive liquid under an alternating magnetic field and studied in detail the magnetic field characteristics in the continuous casting mould with M-EMS. They also analyzed the flow and solidification behaviors of molten steel in the mould with M-EMS. Szekely and Ilegbusi [9] investigated the three-dimensional (3D) velocity field for Newtonian and non-Newtonian melts produced by rotating magnetic field by solving Maxwell's equation for the electromagnetic field. Spitzer et al. [10] derived a two-dimensional analytic solution for the electromagnetic force (EMF) distribution in a round billet mould with M-EMS and numerically studied the flow field in the mould. Ishii et al. [11] numerically analyzed the distribution of the electromagnetic and flow field in horizontal stirring equipment based on FLUENT software. Yu and Zhu [12] analyzed the coupled flow filed, heat transfer and inclusion behavior in a round billet with M-EMS by numerical simulation. A coupled 3D numerical simulation of the electromagnetic and flow fields was developed to study the distribution of the electromagnetic and flow field and the effects of current intensity, frequency and depth of nozzle on flow field in M-EMS in round billet casting based on ANSYS software by Wang et al. [13].

Although the effect of M-EMS on the flow field of molten steel in the mould by different simulation methods was considered and much useful information for practical production was obtained by the above mentioned work, most of research has be focusing on round billet, and paid little attention to explicit 3D flow field in the mould with EMS for billet continuous casting. To get deeper understanding of the character of fluid flow and temperature distribution under the coupled electromagnetic flied, flow flied and heat transfer in the continuous casting of billet with M-EMS, a coupled 3D numerical simulation of the electromagnetic, flow field and temperature distribution was developed in the present work, and the influences of the current intensity and frequency are investigated in detail.

Numerical Models

The numerical analysis for fluid flow and temperature distribution in billet continuous casting with EMS involves three kinds of model, one based on Maxwell's equations solved to calculate the electromagnetic field and the electromagnetic

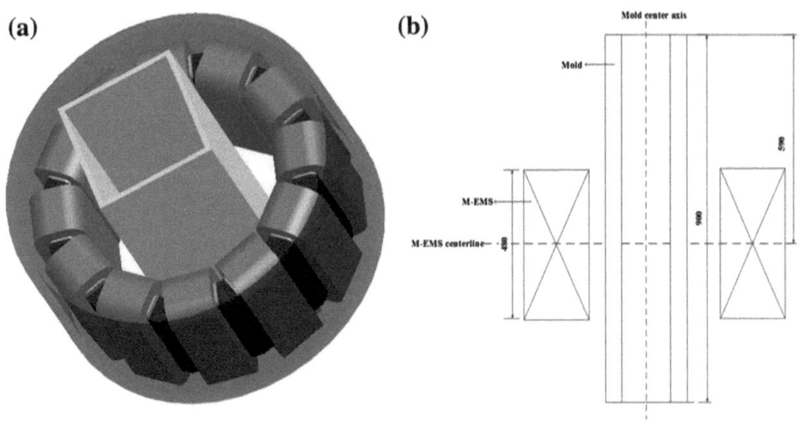

Fig. 1 **a** Schematic illustrations of geometrical model on mould with M-EMS. **b** Schematic illustrations of position of M-EMS

force distribution in the conducting region, another based on Navier–Stokes equation which can compute the flow field driven by the EMF field respectively and the last one based on Energy equation which can be used to calculate the temperature distribution of molten steel in the computational domain. The schematic sketch of the geometrical model of a 3D billet mould with M-EMS is shown in Fig. 1a, which is built by Pro/E soft, and made up of magnetic shielding ring with stainless inner shell, copper winding coils, iron cores and mould cooper. Figure 1b shows installation diagram of M-EMS.

Basic Assumptions

The following assumptions are employed to simplify the mathematical model without losing its meaning:

(i) The molten steel in the mould is assumed to be incompressible Newtonian fluid. The viscosity, the specific heat, and the thermal conductivity are assumed to be constant over temperature.
(ii) The additional magnetic field produced by the molten steel flow in the mould is ignored due to a very small magnetic Reynolds number in the mould with M-EMS. Thus the simulation of electromagnetic field can be decoupled from the calculation of flow-solidification. Mathematical solutions to these problems can be appropriately divided into electromagnetic calculations and fluid flow-solidification calculations.
(iii) The magnetic field generated by the inductive system of M-EMS is considered as quasi-steady state and the displacement current is ignored; the electromagnetic force is obtained in a time averaged form.

Electromagnetic Field Equations

The electromagnetic field distributions in the billet mould including the magnetic flux density and the induced current density are obtained by solving Maxwell's equations, which are given as follows:

$$\begin{cases} \nabla \times H = J \\ \nabla \times E = -\frac{\partial B}{\partial t} \\ \nabla \bullet B = 0 \\ B = \mu H \\ J = \sigma E \end{cases} \tag{1}$$

where H is the magnetic field strength (A m^{-1}); J is the current density (A m^{-2}); t is time (s); E is the electric field strength (V m^{-1}); B is the magnetic flux density (T); μ is the magnetic permeability (H m–1); σ is the electric conductivity.

In the coupled calculation of electromagnetic field and flow field, the time-average induced electromagnetic force can be expressed:

$$F = \frac{1}{2} R_e (J \times B^*) \tag{2}$$

where B* is the complex conjugate, and Re is the real part of $(J \times B^*)$.

Fluid Flow Model Equations

The flow field governing equations include the continuity equation, time averaged Navier–Stokes equation, and two equations k–ε model, which are used to describe the flow field of molten steel in the mould and expressed as follows:

Continuity equation:

$$\frac{\partial(\rho u_i)}{\partial x_i} = 0 \tag{3}$$

Navier–Stokes equations:

$$\frac{\partial(\rho u_i u_j)}{\partial x_j} = -\frac{\partial p}{\partial x_i} + \frac{\partial}{\partial x_j}\left[\mu_{eff}\left(\frac{\partial u_i}{\partial x_j} + \frac{\partial u_j}{\partial x_i}\right)\right] + F_i \tag{4}$$

where ρ, μ, P and g are the density, fluid velocity, pressure and gravitational acceleration respectively.

k–ε model:

$$\frac{\partial(\rho u_i k)}{\partial x_i} = \frac{\partial}{\partial x_i}\left(\frac{\mu_{eff}}{\sigma_k}\frac{\partial k}{\partial x_i}\right) + G - \rho\varepsilon \tag{5}$$

$$\frac{\partial(\rho u_i \varepsilon)}{\partial x_i} = \frac{\partial}{\partial x_i}\left(\frac{\mu_{eff}}{\sigma_k}\frac{\partial \varepsilon}{\partial x_i}\right) + C_1\frac{\varepsilon}{K}G - C_2\frac{\varepsilon^2}{K} \tag{6}$$

where μ_{eff}, k and ε are the effective viscosity, turbulent kinetic energy and its dissipation respectively, σ_k and σ_ε are constant.

The effective viscosity is given as follows:

$$\mu_{eff} = \mu_t + \mu \tag{7}$$

$$\mu_t = \rho C_\mu \frac{k^2}{\varepsilon} \tag{8}$$

where μ_t is the diffusive coefficient of turbulent kinetic energy, μ is viscosity, and C_μ is a constant.

Energy Equation

Energy equation:

$$\frac{\partial}{\partial t}(\rho H) + \frac{\partial}{\partial x_j}(\rho \mu_j H) = \frac{\partial}{\partial x_j}\left[\left(\lambda + C_p\frac{\mu_t}{\sigma_t}\right)\frac{\partial H}{\partial x_j}\right] \tag{9}$$

where λ is the thermal conductivity, σ_t is the turbulent Prandtl number, and $\sigma_t = 1.0$ is assumed.

Boundary Conditions and Solution Method

The electromagnetic field is calculated based on ANSYS Maxell. An air cylinder around the total geometry is used for capturing most of the magnetic induction lines in the surrounding air. Magnetically flux parallel boundary conditions are applied on the external surfaces of the surrounding air cylinder.

Flow field simulation is based on ANSYS-Fluent. The inlet velocity is computed according to the mass conservation between the in-let and outlet based on the casting speed. The values of turbulent kinetic energy and turbulent dissipation rate at the SEN inlet are based on the semi-empirical expression. The free surface is specified as a zero-shear condition. The velocity of the strand wall is specified as casting speed. The outlet guarantees the mass conservation. All walls are set to a non-slip boundary condition, and standard "wall functions" near the wall are used.

The inlet temperature is set as the sum of the liquids temperature and the superheat. The temperature across the inlet plane is fixed to the casting temperature; A fixed temperature equal to the liquid temperature is imposed along all the mould walls; The free surface and the SEN refractory wall are treated as adiabatic.

The numerical modeling was performed by three steps. In the first step, ANSYS Maxell commercial software was used to calculate the 3-D distribution of electromagnetic field in the billet strand by solving Maxwell's equations with finite element method, which can obtain the magnetic field distribution, electromagnetic force field, etc. al. then data files on electromagnetic field were extracted from the post-processor of ANSYS Maxell. In the second step, data files on electromagnetic field in the first step were loaded on the corresponding node as an external body force term in the momentum equation by MHD module of the ANSYS-Fluent software. In the last step, the flow and temperature filed in the mold were computed by solving the turbulent Navier-Stokes equation and energy equation with finite volume method.

The computational domain is meshed by eight-node hexahedral elements, except for the atmosphere region where four node tetrahedral elements are used. In the whole domain, there are about 850000 element grids.

Results and Discussion

In present work, the M-EMS in billet caster from Hang Zhou Iron & Steel Co., Ltd in China is investigated and their production process is analyzed. The process parameters and related physical properties are listed in Table 1.

Experimental Validation for Electromagnetic Field

In order to verify the accuracy of the electromagnetic field models. The magnetic flux density at the center of stirrer in the empty mould was measured using the

Table 1 Simulation conditions in this study

Parameters	Value
Cross-sectional dimension of billet strand/mm^2	150 × 150
Casting speed/m · min^{-1}	2.3
Tundish superheat/k	25
Grade of steel	55SiCrA
Computed length of mould/mm	780
Liquids temperature of molten steel/K	1764

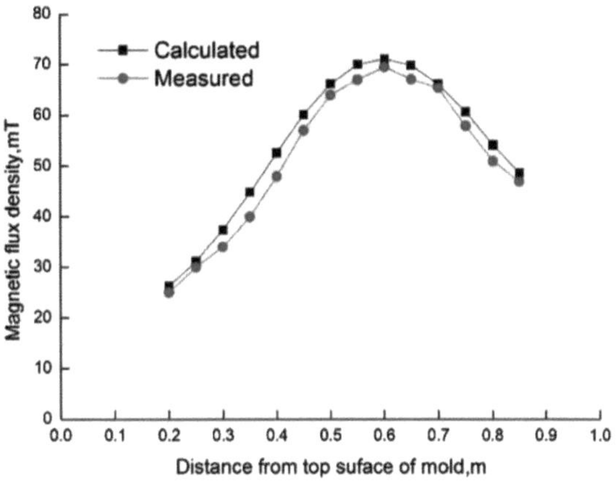

Fig. 2 Comparison of the calculated magnetic flux density with measured datas at the centerline of mould (I = 270 A, f = 3 Hz) along axial direction

KANETEC TM-701 digital teslameter in a real billet caster by adjusting the current intensity and frequency after M-EMS begin operation.

The working principle of this teslameter is based on the Hall Effect, and it consists of a Hall probe and a digital display meter. The calculated results for the magnetic induction are compared with the field test data measured under the condition of the factory, shown in Fig. 2. As shown in Fig. 2, it can be seen that the calculated results are in good agreement with the measured data, which indicates that the developed mathematic model is reasonable for this stirring system. Although there exist some differences between the calculated and measured values, owing to magnetic field leakage and measured or computed error, these deviations are acceptable.

Distribution Characteristic of Magnetic Induction and Electromagnetic Force

The distribution nephogram of magnetic field intensity in the entire mould and on cross section at the center of stirrer are shown both in Fig. 3a, b, which depicts that the magnetic induction is confined mainly in the region surrounded by the inductive coil. As shown in Fig. 3, it can be seen that magnetic field intensity generated by M-MES is unevenly distributed in the entire mould, whose distribution along axial direction is big at both ends but small in the middle, and maximum value appears to be around the center of stirrer mid-plane and in the corner of cross section at the center of stirrer. Figure 4 shows the calculated magnetic flux density at the centerline of mould along axial direction with different current intensities (f = 3 Hz) and frequencies (I = 300 A). The data points of the magnetic flux density are obtained at the center line of the stirrer mid-plane, which also proves the above conclusion.

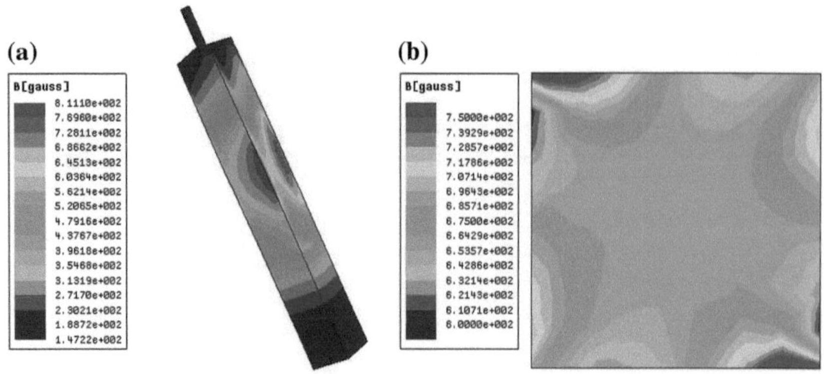

Fig. 3 **a** The distribution nephogram of magnetic field intensity in the entire mould. **b** The distribution nephogram of magnetic field intensity on cross section at the center of stirrer

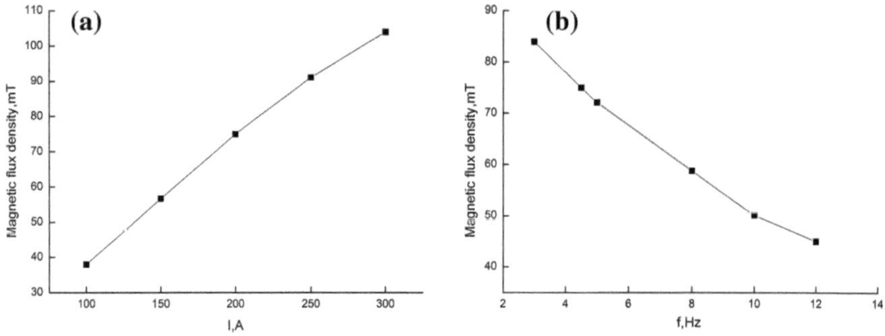

Fig. 4 **a** Magnetic flux density at the center of stirrer mid-plane with different current intensities (f = 3 Hz). **b** Magnetic flux density at the center of stirrer mid-plane with different frequencies (I = 300 A)

The rotary magnetic field across the molten steel will induce the eddy current that interacts with it to generate the electromagnetic force (EMF). The rotary movement driving rotary flow of molten steel could be found. Additionally, the distribution of electromagnetic force is a clockwise rotating action, which is the direction of travelling magnetic field.

Figure 5a, b show the tangential EMF distribution for the different current intensities with the frequency of 2.5 Hz and the different frequencies with a current of 260 A respectively. It is shown that the tangential EMF increases with increasing current intensity or decreasing frequency in the radial directions of the billet. The maximum electromagnetic force occurs at the edge of the billet strand, and the maximum tangential EMF in the radial direction is 2.25 kN m^{-3} at current of 320 A and 1.69 kN m^{-3} with a frequency of 3 Hz respectively.

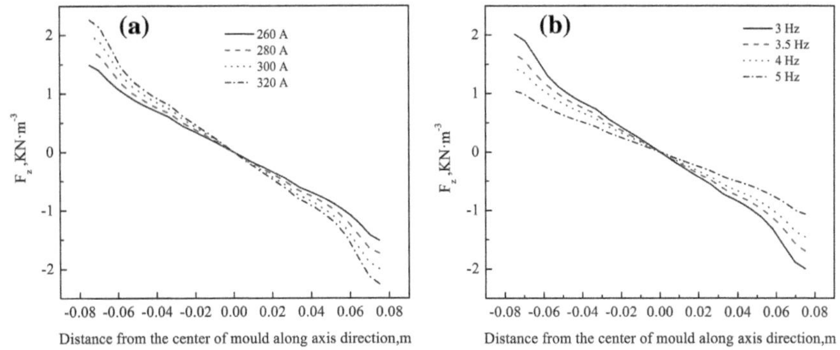

Fig. 5 **a** Distributions of tangential EMF in the radial direction at the center of stirrer with different current intensities (f = 3 Hz). **b** Distributions of tangential EMF in the radial direction at the center of stirrer with different frequencies (I = 300 A)

Flow Field and Temperature Characteristics with M-EMS

Fundamental Characteristics of Flow Field and Temperature Field

In order to analyze the effect of EMS on fluid flow field and temperature distribution, simulated calculation was performed both with and without EMS. Figure 6 show distribution of velocity field along the radial direction at the center of stirrer and 3D streamline in mould, Fig. 7 display temperature distributions at the center symmetry plane. When MES is not applied, the molten steel from SEN flows into the mould, and most of the superheated molten steel directly flows downward, and only a small quantity flows to the free surface of the mould to form a pair of recirculation zones. The recirculating direction of the molten steel is from the center of the mould flowing to the mould wall, the distribution of the streamlines is a radial pattern as a result of molten steel flow downward at the central region and upward near the edge, as shown in Fig. 6a. When M-EMS is applied, the flow field of molten steel in the mould gives rise to an obvious change. The molten steel coming from the SEN forms two pairs of recirculation zones in the mould. Most of the molten steel flows downward from a pair of recirculation zones at the lower region of the mould as well as the upward molten steel from a pair of recirculation zones at the upper region of the mould. The recirculating direction of molten steel is from the mould wall flowing to the center of the mould, the molten steel in the mould has a rotating velocity distribution by the action of the rotating EMF, and the whole rotary action of molten steel appears, as shown in Fig. 6b. Without M-EMS, the superheat of the molten steel in the mould disappears only slowly, the core temperature of the billet is high and the high temperature zone in the billet core extends to the bottom of the mould, as shown in Fig. 7a, and the temperature falls from the billet core to the side of the solidified shell sharply. With M-EMS, it is clearly seen that the super-heated molten steel changes from a vertical downward movement to a

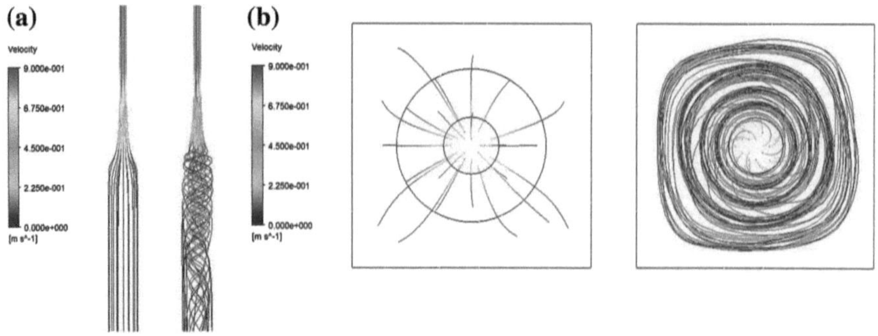

Fig. 6 **a** Distribution of velocity field along radial direction at the center of stirrer and 3D streamline in mould without EMS. **b** Distribution of velocity field along radial direction at the center of stirrer and 3D streamline in mould with EMS (I = 300 A, f = 3 Hz)

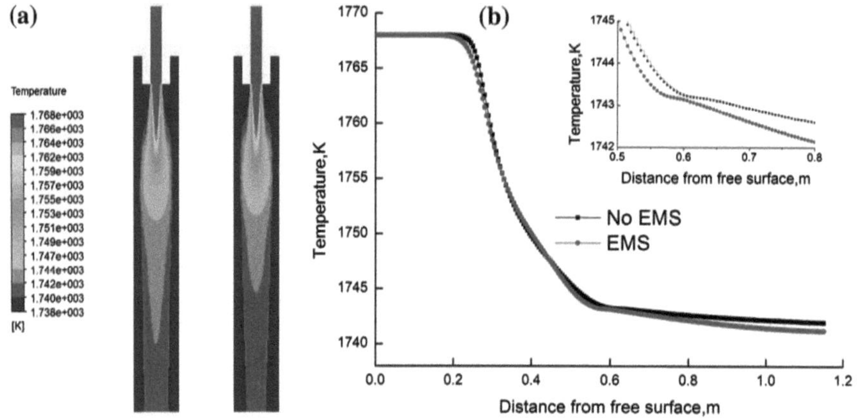

Fig. 7 **a** Temperature nephogram at central symmetry planes of mould without EMS and with EMS (I = 300 A, f = 3 Hz). **b** Distribution of temperature at central line of central symmetry planes of mould without EMS and with EMS (I = 300 A, f = 3 Hz)

horizontal rotation, so that the axial temperature declines quickly and the radial temperature rises. Most of the superheated molten steel remains in the upper region of the mould, as shown in Fig. 7b. The core temperature of the billet reduced dramatically, whereas the temperature at the solidifying front of the solidified shell increases.

Accordingly, the solidification rate and the superheat of the molten steel in the mould are both reduced, which favor for flushing the solidification front, enlarging the area of equiaxed crystal zone and refining the grain. For this reason, the internal quality of the round billet would be enhanced.

The important purpose of this work for the employment of M-EMS in real production aimed at determining the reasonable combination of the current intensity and its frequency. Considering the actual situations of real production in billet

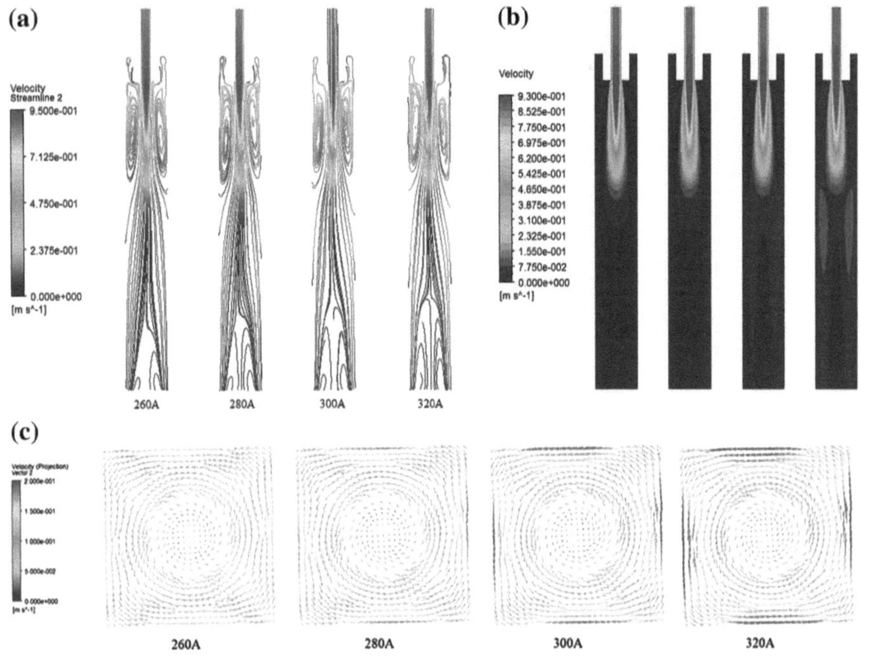

Fig. 8 **a** Streamline map at central symmetry planes of mould for different current intensities (f = 3 Hz). **b** Velocity nephogram at central symmetry planes of mould for different current intensities (f = 3 Hz). **c** Velocity vector of flow field along radial direction at the center of stirrer for different current intensities(f = 3 Hz)

casting, optimal value of the current intensity is of 260, 280, 300 and 320 A and optimal value of the frequency is one of 3, 3.5, 4 and 5 Hz. Therefore, it is necessary to investigate the effect of the current intensity and frequency on the flow field, temperature distribution of molten steel in mould.

Effect of the Current Intensity and Frequency on Flow Field

Figure 8 shows streamline map and velocity nephogram at central symmetry planes of mould, velocity vector of flow field along radial direction at the center of stirrer with the frequency of 3 Hz and the current intensity of 260 A, 280 A, 300 A and 320 A respectively. As shown in Fig. 8, the flow patterns of the molten steel exhibit various distributions at different current intensities. Figure 8a display that there is the upper recirculation zones in the upper region of mould and the rotational flow zone in the lower region of mould with all the current intensity of 260, 280, 300 and 320 A, it has not formed a pair of rounded recirculation zones in the lower region of the mould, with the current intensity of 260 A. It is observer obviously that a pair of rounded recirculation zones in the lower region of the mould appear with the current intensity 320 A from Fig. 8b. The upper recirculation zone appears on either

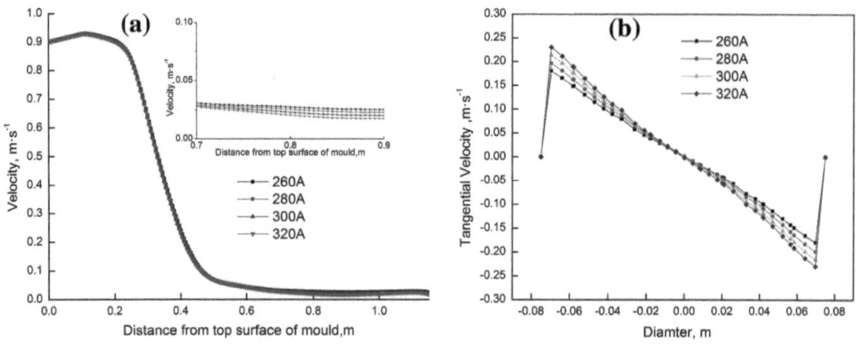

Fig. 9 **a** Distributions of velocity at central line of central symmetry planes of mould for different current intensity (f = 3 Hz). **b** Distributions of tangential velocity in radial direction at the center of stirrer for different current intensity (f = 3 Hz)

side of jet flow formed owing to the interaction of entraining and shearing from molten steel through straight submerged nozzle. When rotating magnetic field induced by MEMS drive the molten steel in the lower region of the mould to generate a rotating movement, the rotational flow zone in the lower region of mould occur. It indicates tangential and upward components of EMF in the mould of billet casting can't bring the rotating molten steel flowing from the wall to the center of the mould when weaker current intensity occur comparatively.

With the current intensity increasing, impacting depth from jet flow decrease, lower rotational flow strengthen, upper circulating region shorten, lower rotational flow zone enlarge and move up.

Figure 9a shows the axial velocity distributions on the central line of central symmetry planes in mould for different current intensity, and Fig. 9b displays the tangential velocity distributions along the radial directions at the center of stirrer for different current intensity. As shown in Fig. 9a, it can be seen that with the increase of current intensity, axial velocity distributions on the central line of central symmetry planes in mould do not change approximately from free surface to the outlet of mould. Axial velocity reaches a maximum at 0.11 m distance below free surface at first and then decrease slowly, there is a sharp turning at 0.44 m distance below free surface where the center of the EMS core is located. With the increase of current intensity, axial velocity decrease by small steps and by a large margin below 0.44 m distance from free surface. As shown Fig. 9b, tangential velocity vector distribution is similar on the whole. Velocity increase first and then decrease from the edge to the center of billet, velocity dropped to nearly zero at the center of billet, The changing tendency of tangential velocity along the radial directions is consistent with that of tangential EMF from Fig. 5. The maximum tangential velocity in the radial direction of the mould increases from 0.053 m s^{-1} with a current of 260 A to 0.076 m s^{-1} with a current of 320 A, and it appears at the location corresponding to the EMS core. The maximum tangential velocity along the radial direction appears at the edge of the billet.

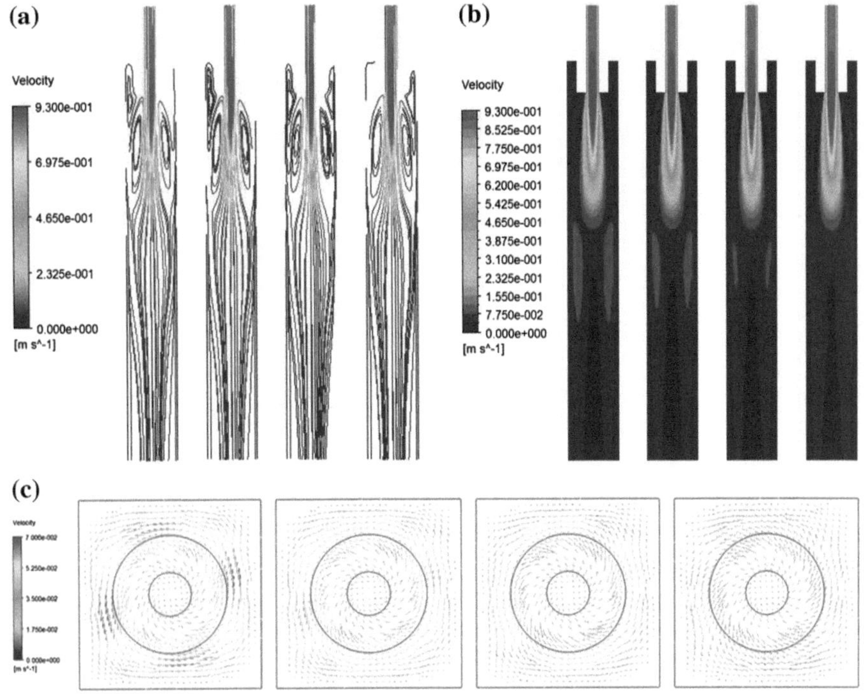

Fig. 10 **a** Streamline map at central symmetry planes of mould for different frequency (I = 300 A). **b** Velocity nephogram at central symmetry planes of mould for different frequency (I = 300 A). **c** Velocity vector of flow field along radial direction at the center of stirrer for different frequency (I = 300 A)

Figure 10 shows streamline map and velocity nephogram at central symmetry planes of mould, velocity vector of flow field along radial direction at the center of stirrer with the current intensity of 300 A and the frequency of 3 Hz, 3.5 Hz, 4 Hz and 5 Hz respectively. It can be seen from Fig. 10 that the recirculating flow of molten steel in the upper regions of mould strengthen, the recirculating velocity in-creases, the center of the vortex of the circulating region gradually moves up with decreasing frequency especially in the case of 3 Hz, It also shows that a pair of recirculation zones in the lower region of the mould appear at the frequency of 3, 3.5 and 4 Hz. With increasing frequency, the recirculation zones get shorter and absolutely vanish with the frequency of 5 Hz.

Figure 11a shows the axial velocity distributions on the central line of central symmetry planes in mould and Fig. 11b displays the tangential velocity distributions along the radial directions at the center of stirrer for different frequency with the current intensity of 300 A. As shown in Fig. 11a, it can be seen that with the increase of frequency, axial velocity distributions on the central line of central symmetry planes in mould do not change approximately from top surface to the outlet of mould. Axial velocity reaches a maximum at 0.11 m distance below free surface at first and then decrease slowly, there is a sharp turning at 0.44 m distance

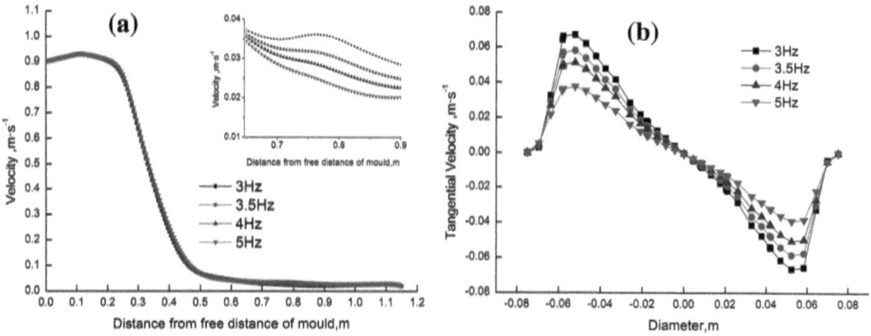

Fig. 11 **a** Distributions of velocity at central line of central symmetry planes of mould for different frequency (I = 300 A). **b** Distributions of tangential velocity in radial direction at the center of stirrer for different frequency (I = 300 A)

below free surface where the center of the EMS core is located. With the increase of the frequency, axial velocity decrease by small steps and by a large margin below 0.44 m distance from free surface.

The changing tendency of tangential velocity with increasing frequency is consistent with the case of EMF. As shown in Fig. 11b, the maximum tangential velocity in the radial direction of the mould decreases from 0.067 m s^{-1} with a frequency of 3 Hz to 0.038 m s^{-1} with a frequency of 5 Hz, and it appears at the location corresponding to the EMS core. The maximum tangential velocity along the radial direction appears at the edge of the billet. It is evident that the effect of decreasing frequency on the flow field of molten steel is more sensitive than that of the increasing current intensity.

Effect of the Current Intensity and Frequency on Temperature Field

Figures 12 and 13 shows temperature nephogram at central symmetry planes of mould and temperature distributions at the central line of central symmetry planes of mould for different current intensity with the frequency of 3 Hz and for different frequency with the current intensity of 300 A respectively.

As shown in Figs. 12a and 13a, the superheat of molten steel and the core temperature of the billet reduced dramatically, the residence time of superheated molten steel increases, and the location of high temperature zone in the billet core also moves up with increasing current or decreasing frequency. With increasing current intensities or decreasing frequency, impacting depth from jet flow de-creases, rotational flow in the lower region of mould strengths, its zone enlarges and move ups, recirculation flow in the lower region of mould is formed. The significant swirl flow with M-EMS prevents the superheated jet moving downward and weakens the invasion depth of the jet flow, and thus the location of the hot zone in the mold moved up evidently. As shown in Fig. 12b, the core temperature in

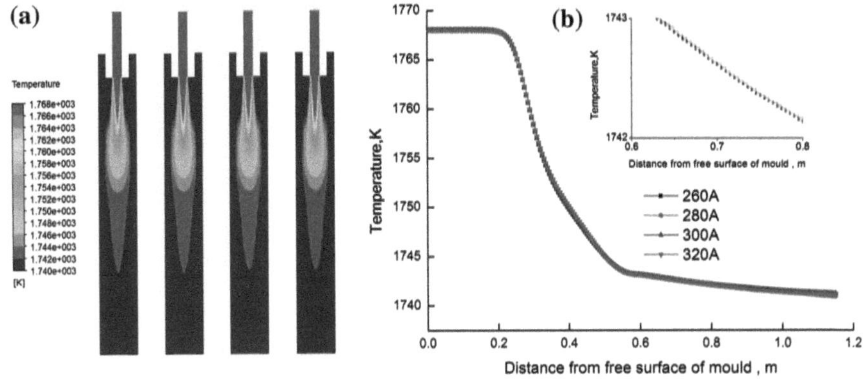

Fig. 12 **a** Temperature nephogram at central symmetry planes of mould for different current intensities (f = 3 Hz). **b** Temperature distributions at central line of central symmetry planes of mould for different current intensities (f = 3 Hz)

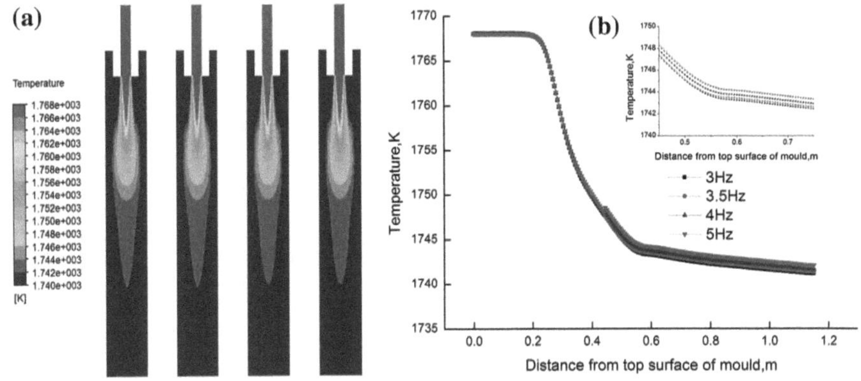

Fig. 13 **a** Temperature nephogram at central symmetry planes of mould for different frequency (I = 300 A). **b** Temperature distributions at central line of central symmetry planes of mould for different frequency (I = 300 A)

mould decreases by a small margin with increasing current intensity. The core temperature at the outlet decreases from 1742.8 K with a current intensity of 260 A and 1742.3 K with the current intensity of 320 A. As shown in Fig. 13b, temperature decreases by a small margin with decreasing frequency. The core temperature at the outlet decreases from 1743 K with a frequency f of 5 Hz and 1742 K with the frequency of 3 Hz, temperature has no obvious change owing to current intensity and frequency with small change.

The temperature in the range of 0.5 m distance below the free surface of the mould has no obvious change with the increasing current intensity or frequency, which is the mainstream zone of molten steel flowing from nozzle into the mould.

The core temperature is rather high, and the high temperature zone of the billet core also extends to the bottom of the mould when the frequency is 5 Hz or the current intensity is 300 A. The core temperature falls and the high temperature zone of the core becomes shorter with increasing current intensity or decreasing frequency. Meanwhile, the position of a given temperature value moves upward continuously with increasing current intensity or decreasing frequency, and it reflects the same results with the change shown in Figs. 12 and 13, namely, the location of hot area moves upward continuously. It is evident decreasing frequency has greater impact on the temperature field of molten steel than that of the increasing current intensity.

Conclusions

In this paper, a coupled 3D numerical simulation of the electro-magnetic field, fluid flow, heat transfer and macroscopic solidification was carried out based on the FEM to investigate the electromagnetic field characteristic and the multiphase flow behavior in the billet mould with a rotary electromagnetic stirrer. The effects of the current intensity and frequency are discussed, and the main conclusions are as follows:

1. There are good consistencies between the predicted magnetic flux density by electromagnetic field models and the measured data. Magnetic field intensity distribution along axial direction is big at both ends but small in the middle, and the data points of the magnetic flux density are obtained at the center of the stirrer mid-plane.
2. The rotating tangential EMF in the radial directions of the billet is parallel to the edge of the billet, which increases with the current intensity and decreases with frequency when the frequency is more than the optimum frequency value, and the maximum occurs at the edge of the billet.
3. With the increase of incurrent intensity, the upper recirculating flow strength and its center vortex move up; In addition, the lower rotational flow strength and a pair of lower recirculation zones is more prone to form. Owing to the increasing current intensity with small change, the core temperature at the outlet of the mould decreases by a small margin, and solidified shell thickness becomes smaller slightly at the outlet of mould but its distribution of becomes more uniform.
4. The effect of the frequency on flow field was exactly opposite compared with the current, and obviously decreasing frequency has greater impact on the electromagnetic field, flow field, temperature distribution and solidification of molten steel in the mould than that of increasing current intensity.

Acknowledgements The authors would like to acknowledge the financial support provided by the Natural Science Foundation of China (No. 51404022) and State Key Laboratory of Advanced Metallurgy Foundation (41614014).

References

1. V. Ludlow, A. Norman ton, A. Anderson, Ironmak. Steelmak. 32(1), 68–69 (2005)
2. A.W. Cramb, The making, shaping and treating of steel of 11th, Casting Volume, Association for Iron Steel Technology, Warren-dale, USA, 287 (2003)
3. M. Barna, M. Javurek, J. Reiter et al., World Iron Steel **5**, 29 (2012)
4. M. Barna, M. Javurek, J. Reiter, J et al. Iron Steel Res. Int. **19**, 884 (2012)
5. P.A. Davidson, F. Boysan, Appl. Sci. Res. **44**, 241 (1987)
6. K. Fujisaki, T. Yamada, S. Satoh, J. Appl. Phys. **89**, 6734–6736 (2001)
7. K. Fujisaki, I.E.E.E. Trans, Ind. Appl. **39**, 3541–3545 (2003)
8. R. Hirayama, K. Fujisaki, T. Yamada, IEEE Trans. Magn. **40**, 2056–2058 (2004)
9. O.J. Ilegbusi, J. Szekely, ISIJ Int. **429**, 462–468 (1989)
10. K. Spitzer, M. Dubke, K. Schwerdtfeger, Metall. Mater. Trans. B: Process Metall. Mater. Process. Sci., **17B**,119–131 (1986)
11. T. Ishii, S. Sazhin, S. Makhlouf, Ironmak. Steelmak. **23**, 67–72 (1996)
12. H.Q. Yu, M.Y. Zhu, Ironmak. Steelmak. **39**(8), 574–575 (2012)
13. B.X. Wang, W. Chen, Y. Chen et al., Ironmak. Steelmak. **42**(1), 63–64 (2015)

Part XIII
Environmentally Assisted Cracking: Theory and Practice

Hydrogen Embrittlement and Hydrogen-Enhanced Strain-Induced Vacancies in α-Iron

Y. Matsumoto, N. Kurihara, H. Suzuki and K. Takai

Abstract Clarifying the states of hydrogen present in iron and steel is important in order to understand hydrogen embrittlement mechanisms and develop materials with high resistance to hydrogen embrittlement. Although it is widely recognized that the fracture strain of iron and steel decreases with increasing amounts of absorbed hydrogen, it is not yet well understood whether hydrogen directly decreases the fracture strain. Therefore, the objective of this study is to clarify the atomic-scale changes in strained α-iron specimens containing hydrogen. Low temperature thermal desorption spectroscopy (L-TDS), which can heat samples from lower temperatures than conventional TDS, was used to identify the peak temperatures and hydrogen states corresponding to various lattice defects in α-iron. The results indicate that new hydrogen trap sites in strained α-iron specimens containing hydrogen are enhanced compared to those without hydrogen. These sites are not dislocations, but hydrogen-enhanced strain-induced vacancies, because they are removed during aging at 30 °C.

Keywords Hydrogen · Pure iron · Vacancy · Thermal desorption spectroscopy

Y. Matsumoto (✉)
Graduate School of Science and Technology, Sophia University, 7-1 Kioi-Cho, Chiyoda-Ku, Tokyo 102-8554, Japan
e-mail: yuya_mattsu@eagle.sophia.ac.jp

N. Kurihara
Graduate School of Science and Technology, Sophia University, currently at DNP Ltd., Tokyo, Japan

H. Suzuki · K. Takai
Faculty of Science and Technology, Department of Engineering and Applied Science, Sophia University, Tokyo, Japan

The Minerals, Metals & Materials Society, *TMS 2017 146th Annual Meeting & Exhibition Supplemental Proceedings*, The Minerals, Metals & Materials Series,
DOI 10.1007/978-3-319-51493-2_54

Introduction

Recently, lattice defects in iron and steel have been studied to better understand hydrogen embrittlement mechanisms and develop materials with high resistance to hydrogen embrittlement. One of the authors reported previously that plastic strain markedly increases the amount of lattice defects in hydrogen-charged specimens of pure iron and Inconel 625. These lattice defects were removed by annealing at 200 °C, so most of the strain-induced lattice defects were not dislocations but vacancies and vacancy clusters [1, 2]. Furthermore, a loss in ductility could not be recovered when a higher density of lattice defects induced by both strain and hydrogen remained in the specimens, even though hydrogen was not present in the fractured specimens.

Therefore, clarification of the hydrogen states present in iron and steel is important to understand hydrogen embrittlement [3]. Recently, thermal desorption spectrometry (TDS) has been widely used to measure the hydrogen content and evaluate hydrogen trapping states in various materials [4]. However, it is difficult to distinguish and identify these trapping sites from the hydrogen present using TDS, because in general heating can only be carried out from room temperature. Fortunately, the authors have developed a low-temperature thermal desorption spectroscopy (L-TDS) method in which the sample can be heated from low temperatures. Unlike existing TDS methods, L-TDS makes it possible to distinguish and identify trapping sites, including various lattice defects such as dislocations, vacancies, and vacancy clusters.

In the present study, the effects of hydrogen on the behavior of vacancies were examined in pure iron using L-TDS.

Experimental

Figure 1 shows the procedure used to investigate the effects of hydrogen on the behavior of vacancies in pure iron. An iron specimen with a purity of 99.98% (6 mass ppm carbon) was used. Tensile testing was used to apply strain to flat specimens (0.5 mm thick, 30 mm gauge length, 10 mm wide). Pure iron was annealed at 900 °C for 2 h in Ar, followed by furnace cooling to eliminate all of the lattice defects. The surfaces of the heat-treated specimens were polished with emery paper to remove any residual oxide film. Hydrogen charging was carried out by cathodic electrolysis using a galvanostat at a current density of 50 A/m^2 for 30 min, which was sufficient for the hydrogen to saturate the specimen. An aqueous solution of H_2SO_4 (pH 2.5) to which 0.09 mass% NH_4SCN was added was used. The temperature of the electrolyte was 30 °C. The specimens were strained by 13.3% at a strain rate of 1.6×10^{-5}/s under these hydrogen charging conditions. These hydrogen charging conditions were as above. Figure 2 shows the strain-displacement curve of the specimens. After straining, the specimens were aged for

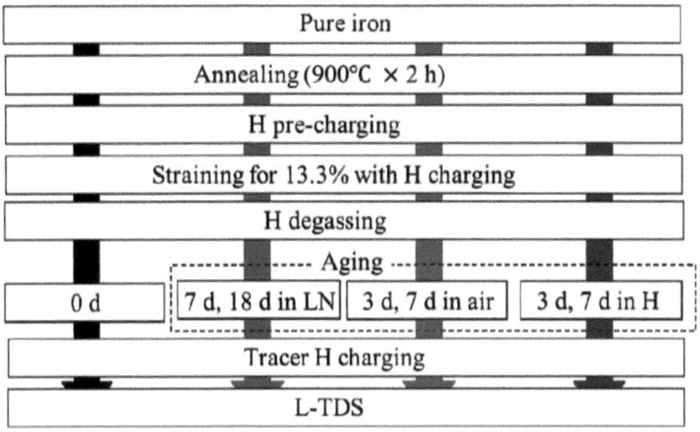

Fig. 1 Procedure for clarifying effects of hydrogen on behavior of vacancies in various aging conditions

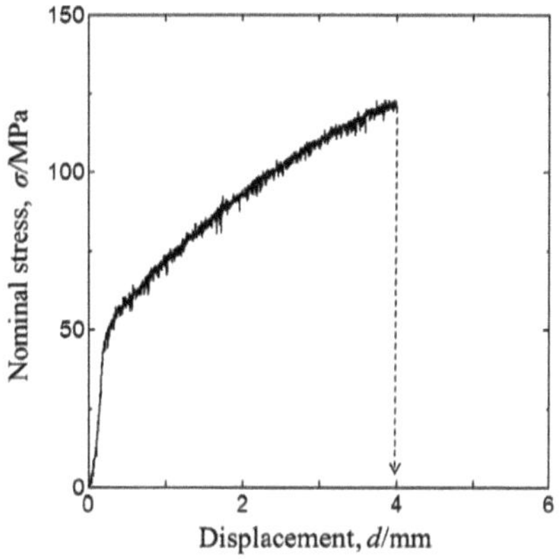

Fig. 2 Stress-displacement *curve* for a pure iron specimen with an applied strain of 13.3%

60 min in air at 30 °C to completely degas the absorbed hydrogen. The deformation layer on the surface was removed by mechanical polishing followed by chemical polishing, which reduced the specimens to a thickness of 0.3 mm. The specimens were then aged for several days in LN (liquid nitrogen) or with or without hydrogen. In the aging with hydrogen, hydrogen charging conditions were carried out by cathodic electrolysis at a current density of 5 A/m^2 in 0.1 N NaOH aqueous solutions. The temperature of the electrolyte was 30 °C. Tracer hydrogen was charged into the specimens by immersion in a 0.01 mass% NH_4SCN solution maintained at 30 °C. After 45 min, the specimens were immediately immersed in LN to prevent the release of hydrogen. The specimens were then mounted on the heating stage of the L-TDS and heated at a rate of 1 °C/min from –200 to 200 °C.

Results and Discussion

Tracer Hydrogen Desorption Spectra of Pure Iron Aged in LN

Figure 3 shows tracer hydrogen desorption spectra of specimens that were aged in LN for 7 and 18 d after straining with hydrogen charging. Desorption of the tracer hydrogen began at approximately 50 °C and ended at approximately 150 °C. Figure 3 shows no significant difference among the specimens aged in LN for 0, 7, and 18 d.

Previous studies have reported that vacancies are formed by straining with hydrogen charging and these vacancies are called hydrogen-enhanced strain-induced vacancies [1, 2]. In this study, tracer hydrogen was trapped only by hydrogen-enhanced strain-induced vacancies because the amount of tracer hydrogen was very small. Therefore, Fig. 3 shows that the tracer hydrogen desorption spectra of hydrogen-enhanced strain-induced vacancies did not change after aging in LN for 7 d or even 18 d. In other words, the hydrogen-enhanced strain-induced vacancies were not removed by 18 d of aging in LN.

Tracer Hydrogen Desorption Spectra of Pure Iron Aged in Air at 30 °C

Figure 4 shows tracer hydrogen desorption spectra of specimens aged in air at 30 °C for 3 and 7 d after straining with hydrogen charging. The amount of tracer hydrogen and the height of the tracer hydrogen peak decreased after 3 d and 7 d of aging in air. Furthermore, the tracer hydrogen desorption from the 3 d and 7 d specimens began at approximately 100 °C, whereas the tracer hydrogen desorption from the 0 d specimen began at approximately 50 °C.

Fig. 3 Tracer hydrogen desorption spectra of specimens aged in LN for 0, 7, and 18 days

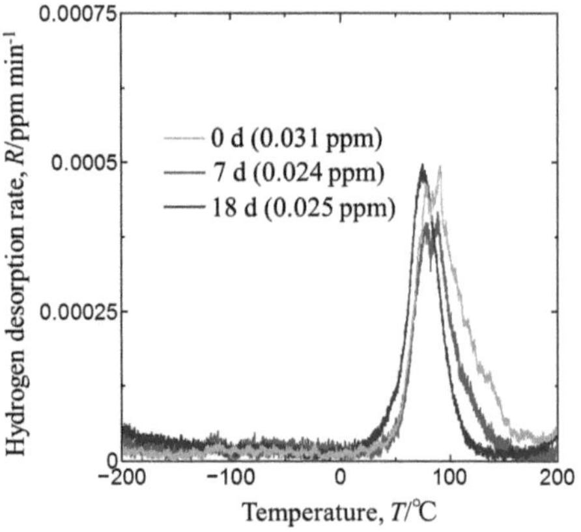

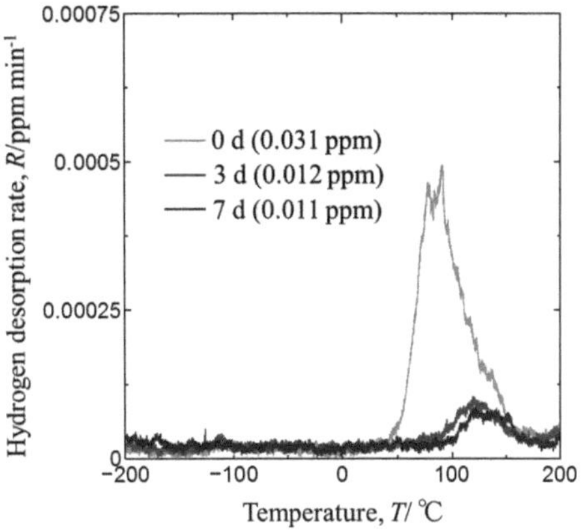

Fig. 4 Tracer hydrogen desorption spectra of specimens aged in air at 30 °C for 0, 3, and 7 days

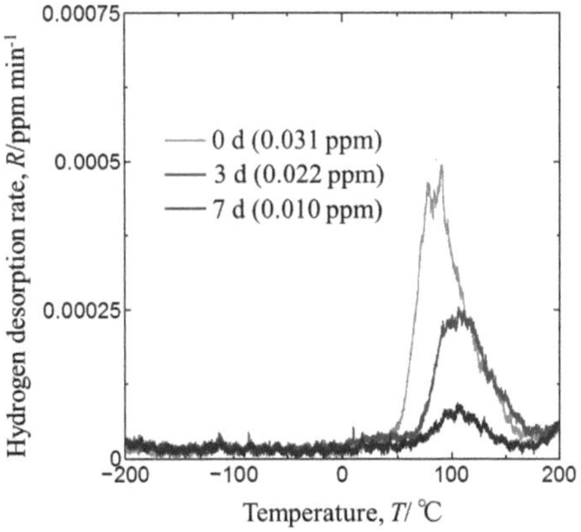

Fig. 5 Tracer hydrogen desorption spectra of specimens aged with hydrogen charging for 0, 3, and 7 days

Figure 4 shows the behavior of hydrogen-enhanced strain-induced vacancies in air at 30 °C. The decreased amount of tracer hydrogen detected and the reduction in the peak height indicate that the hydrogen-enhanced strain-induced vacancies were almost eliminated after 3 d of aging in air. The observed shift in the ranges of tracer hydrogen desorption to higher temperatures indicates that the smaller vacancies were eliminated.

Tracer Hydrogen Desorption Spectra of Pure Iron Aged with Hydrogen Charging

Figure 5 shows tracer hydrogen desorption spectra of specimens aged with hydrogen charging for 3 and 7 d after straining with hydrogen charging. The amount of tracer hydrogen and the height of its peak decreased with increasing aging time. Moreover, the tracer hydrogen desorption of the 3 and 7 d specimens began at approximately 70 °C, whereas the tracer hydrogen desorption of the 0 d specimen began at approximately 50 °C.

Figure 5 shows the behavior of hydrogen-enhanced strain-induced vacancies generated by hydrogen charging. The decrease in hydrogen-enhanced strain-induced vacancies in the specimen aged with hydrogen charging for 3 d was less than that of the specimen aged in air for 3 d. This result suggests that hydrogen decreases the mobility and removal of vacancies.

It is widely recognized that the migration energy of a vacancy can be expressed as follows:

$$E_m = E_m^\circ + E_b$$

where E_m is the migration energy of a vacancy, $E_m{}^\circ$ is the migration energy of a vacancy in the pure metal, and E_b is the binding energy between the trapping site and hydrogen. A previous study reported a value of E_b = 51 kJ/mol in Fe [5]. In the above expression, E_m. increases after hydrogen is absorbed in a trapping site with E_b.. Therefore, it is reasonable that hydrogen decreased the mobility and removal of vacancies in this study.

Conclusions

Iron with a purity of 99.98% containing hydrogen-enhanced strain-induced vacancies generated by straining with hydrogen charging was aged in LN, in air at 30°C, and under hydrogen charging conditions. The tracer hydrogen desorption spectra of specimens were evaluated using low-temperature thermal desorption spectroscopy (L-TDS). The results can be summarized as follows.

(1) Hydrogen-enhanced strain-induced vacancies were not removed by 18 days of aging in LN.
(2) Hydrogen-enhanced strain-induced vacancies were almost removed after 3 days of aging in air at 30 °C. However, compared with the specimens aged in air at 30 °C, hydrogen-enhanced strain-induced vacancies have not been removed for 3 days with hydrogen charging. Therefore, hydrogen probably decreases the mobility and removal of hydrogen-enhanced strain-induced vacancies in iron.

References

1. K. Takai et al., Lattice defects dominating hydrogen degradation of metals. Acta Mater. **56**, 5158–5167 (2008)
2. T. Doshida et al., Enhanced lattice defect formation associated with hydrogen and hydrogen embrittlement under elastic stress of a tempered martensitic steel. ISIJ Int. **52**, 198–207 (2012)

3. K. Takai, R. Watanuki, Hydrogen in trapping states innocuous to environmental degradation of high-strength steels. ISIJ Int. **43**, 520 (2003)
4. H. Suzuki, K. Takai, Summary of round-robin tests for standardizing hydrogen analysis procedures. ISIJ Int. **52**, 174–180 (2012)
5. S.M. Myers et al., Trapping and surface permeation of deuterium in He-implanted Fe. J. Appl. Phys. **53**, 8734 (1982)

Factors Causing Hydrogen Embrittlement of Cold-Drawn Pearlitic Steel Fractured Under Elastic/Plastic Region

R. Konno, T. Manabe, N. Matsui, D. Hirakami and K. Takai

Abstract Factors causing hydrogen embrittlement of cold-drawn pearlitic steel fractured under plastic/elastic region have been investigated from the perspective of lattice defects. Tensile tests were conducted for hydrogen-charged specimens containing 1.5 and 4.0 mass ppm hydrogen to evaluate mechanical properties. The amount of tracer hydrogen, i.e., lattice defects in the specimens unloaded just before tensile fracture strength, was measured using a thermal desorption analysis. Specimens containing 1.5 and 4.0 mass ppm hydrogen fractured under plastic and elastic region, respectively. The specimen fractured under plastic region showed enhanced formation of lattice defects corresponding to vacancies, which directly caused embrittlement. In contrast, the specimen fractured under elastic region showed no enhancement to the formation of lattice defects. These results reveal that one of the factors causing hydrogen embrittlement under plastic region is due to hydrogen-enhanced strain-induced vacancies, whereas the factors causing hydrogen embrittlement under elastic region are due to others.

Keywords Hydrogen embrittlement · Pearlitic steel · Lattice defect · Vacancy · Elastic · Plastic

---

R. Konno (✉)
Graduate School of Science and Technology, Sophia University, 7-1 Kioi-cho, Chiyoda-ku, Tokyo 102-8554, Japan
e-mail: ryosukekonno1010@gmail.com

T. Manabe · D. Hirakami
Nippon Steel & Sumitomo Metal, 20-1 Shintomi, Futtsu, Chiba Prefecture 293-8511, Japan

N. Matsui
Nippon Steel & Sumitomo Metal, 1 Kimitsu, Kimitsu, Chiba Prefecture 299-1141, Japan

K. Takai
Faculty of Science and Technology, Department of Engineering and Applied Science, Sophia University, Tokyo, Japan

The Minerals, Metals & Materials Society, *TMS 2017 146th Annual Meeting & Exhibition Supplemental Proceedings*, The Minerals, Metals & Materials Series, DOI 10.1007/978-3-319-51493-2_55

Introduction

Intergranular and quasi-cleavage fracture modes have been reported as representative hydrogen embrittlement fracture modes in tempered martensitic steel. Hydrogen-enhanced decohesion (HEDE) has been reported as a factor of intergranular fracture, with a crack propagating along a prior austenite grain boundary. Additionally, the accumulation of hydrogen-enhanced lattice defects (HELDs) and hydrogen-enhanced localized plasticity (HELP) have been reported as factors of quasi-cleavage fracture.

Cold-drawn pearlitic steel consists of a ferrite–cementite lamellar structure that elongates in the cold-drawing direction and contains no grain boundaries, as with tempered martensitic steel. The fracture surface of cold-drawn pearlitic steel containing hydrogen is quasi-cleavage, not intergranular. Therefore, HELDs have been reported as one of the causes of hydrogen embrittlement of cold-drawn pearlitic steel [1].

HELDs have been investigated in numerous studies. It has been reported that the formation of lattice defects is enhanced by plastic strain in the presence of hydrogen in pure iron [2], Inconel625 [2], SUS304 [3], and SUS316L [3]. It has also been clarified that HELDs can be formed by applying cyclic prestress [4] with hydrogen in tempered martensitic steel as well as by applying constant elastic tensile stress [5]. These HELDs directly cause embrittlement.

In the present study, factors causing the hydrogen embrittlement of cold-drawn pearlitic steel were investigated. Specimens were fractured in the plastic and elastic regions by changing the hydrogen content and were evaluated from the perspective of HELD formation.

Experimental

Materials

Table 1 gives the chemical composition of the specimens used in this study. Eutectoid steel was drawn to a diameter of 5 mm after patenting treatment. The specimen was aged at 300 °C for 600 s after cold-drawing. The microstructure was pearlite and consisted of a ferrite-cementite lamellar structure that elongates in the cold-drawing direction.

Table 1 Chemical composition of specimen material (mass%)

C	Si	Mn	P	S
0.84	0.25	0.73	0.017	0.005

Hydrogen Analysis and Hydrogen Embrittlement Susceptibility

The surfaces of the specimens were polished using emery paper (800, 1000, and 2000 grit). Two specimens were hydrogen charged using two different methods of electrochemical charging: specimen 1 was prepared in a 0.1 N solution of NaOH maintained at a current density of 150 A m^{-2} and a temperature of 30 °C, and specimen 2 was prepared in a 0.1 N solution of NaOH + 5 g L^{-1} NH_4SCN maintained at a current density of 50 A m^{-2} and a temperature of 30 °C. The charging was performed for 96 h; this was the time required for the hydrogen concentration at the center of the specimens to reach equilibrium. The hydrogen content was analyzed by thermal desorption analysis (TDA) using a gas chromatograph at a heating rate of 100 °C h^{-1}. The hydrogen contents of specimens 1 and 2 were 1.5 and 4.0 mass ppm, respectively.

Tensile tests were conducted on specimens 1 and 2 and a third uncharged specimen (specimen 3) at a crosshead speed of 0.01 mm min^{-1} and a temperature of 30 °C. For specimens 1 and 2, tensile tests were conducted in conjunction with hydrogen charging in the abovementioned solutions to maintain the hydrogen content of the specimens.

Evaluation of the Amount of Lattice Defects

A schematic describing the procedure for evaluating the HELDs after tensile testing with hydrogen is shown in Fig. 1. Specimens 1 and 3 were subjected to a plastic strain to compare the characteristics of fracture in the plastic region with and

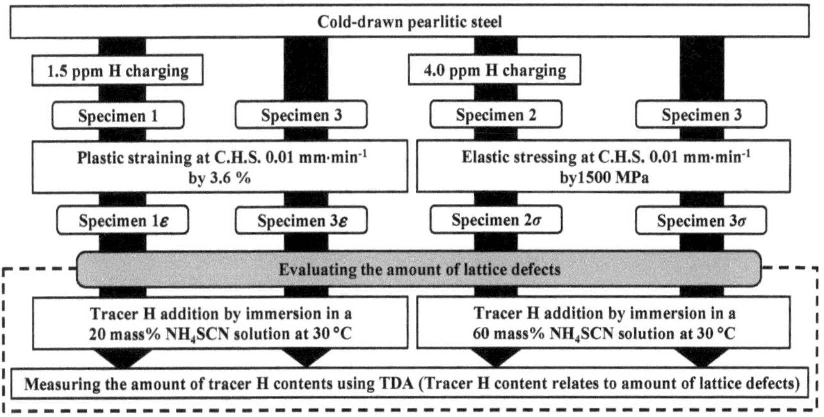

Fig. 1 Procedure for determining the effect of elastic stress and plastic strain on the formation of HELDs

without hydrogen charging. These two specimens, denoted specimens 1ε and 3ε, respectively, were subjected to a plastic strain of 3.6% at a crosshead speed of 0.01 mm min^{-1} and unloaded. As stated above, specimen 1ε was charged with a hydrogen content of 1.5 ppm hydrogen, and specimen 3ε was uncharged. For specimen 1ε, necking began at a strain of 3.6%. Additionally, specimen 1ε was unloaded just before necking began. Both specimens 1ε and 3ε were subjected to a constant plastic strain and underwent uniform elongation.

Specimens 2 and 3 were subjected to elastic stress to compare the characteristics of fracture in the elastic region with and without hydrogen. These two specimens, denoted specimens 2σ and 3σ, were subjected to an elastic stress of 1500 MPa at a crosshead speed of 0.01 mm min^{-1} and unloaded. As stated above, specimen 2σ was charged with a hydrogen content of 4.0 ppm hydrogen, and specimen 3σ was uncharged. Specimen 2σ fractured at a stress of 1580 MPa. It was unloaded just before the stress reached the tensile fracture strength.

To evaluate the amount of lattice defects, hydrogen was used as a tracer [1]. Tracer hydrogen was added to all specimens until they became saturated, and the tracer hydrogen content of each specimen was measured using TDA. The tracer hydrogen content corresponds to the amount of lattice defects. To evaluate the amount of lattice defects in specimens 1ε and 3ε, tracer hydrogen was added by immersion in a 20 mass% NH_4SCN solution at a temperature of 30 °C until saturation. To evaluate the amount of lattice defects in specimens 2σ and 3σ, tracer hydrogen was added by immersion in a 60 mass% NH_4SCN solution at a temperature of 30 °C until saturation.

Evaluation of the Effects of Lattice Defects on Mechanical Properties

Specimens 1ε and 3ε were prepared to evaluate the effects of hydrogen-enhanced strain-induced lattice defects (HESILDs) on the mechanical properties of the specimens. After being subjected to strain in combination with hydrogen charging, specimen 1ε contained hydrogen. It was then aged at a temperature of 50 °C for 3 days to degas the contained hydrogen. Specimen 3ε was aged under the same conditions to prepare a heating history.

Specimens 2σ and 3σ were prepared to evaluate the effects of HELDs in the elastic region. After being subjected to stress in combination with hydrogen charging, specimen 2σ contained hydrogen. It was then aged at a temperature of 50 °C for 4 d to degas the contained hydrogen. Specimen 3σ was aged under the same conditions to prepare a heating history. After aging, tensile tests were conducted at a crosshead speed of 0.1 mm min^{-1} at room temperature for specimens 1ε, 3ε, 2σ, and 3σ.

Results and Discussion

Hydrogen Embrittlement Susceptibility

Figure 2 shows the nominal stress–displacement curves of specimens 1–3 at a crosshead speed of 0.01 mm min^{-1}. The elongation decreased with increasing hydrogen content. Specimens 1 and 2 fractured in the plastic and elastic regions, respectively. For specimen 1, necking began at a strain of 3.6%. Specimen 2 fractured at a stress of 1580 MPa.

Factors Causing Hydrogen Embrittlement Fractures in the Plastic Region

Figure 3 shows the TDA profiles and tracer hydrogen contents of specimens 1ε and 3ε after tracer hydrogen addition by immersion in a 20 mass% NH_4SCN solution at a temperature of 30 °C until saturation. The tracer hydrogen contents of specimens 1ε and 3ε were 3.9 and 3.1 mass ppm, respectively. The tracer hydrogen content of specimen 1ε was larger than that of specimen 3ε even though both specimens were subjected to a constant plastic strain until just before necking began for specimen 1ε.

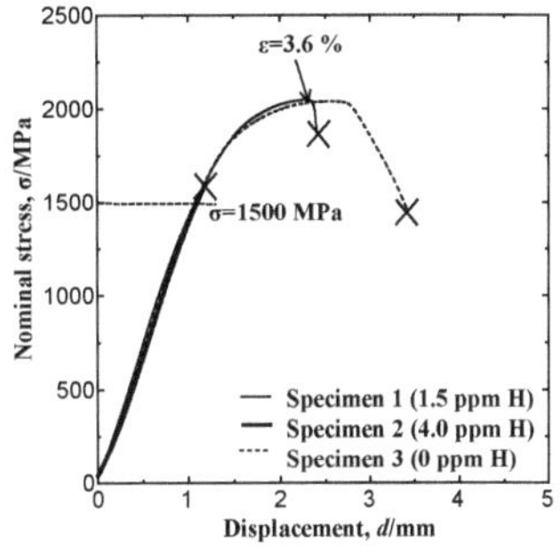

Fig. 2 Nominal stress–displacement curves of specimens 1–3 at a crosshead speed of 0.01 mm min^{-1}

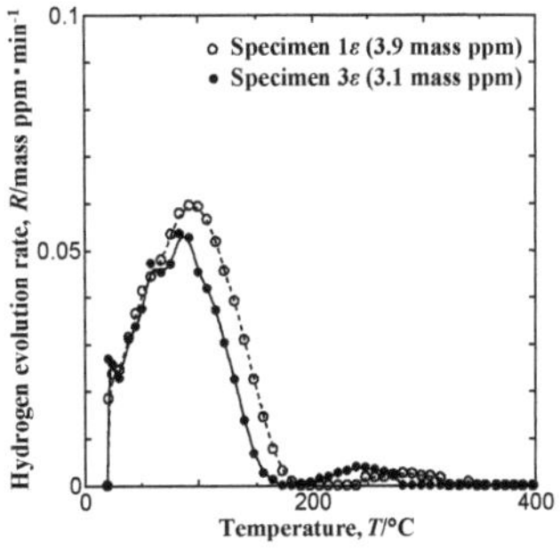

Fig. 3 TDA profiles and tracer hydrogen contents of specimens 1ε and 3ε

This result indicates that plastic strain in the presence of hydrogen increases the amount of lattice defects.

The types of HESILDs were identified using the different disappearance temperatures of the different types of lattice defects. Aging was conducted at temperatures ranging from 50 to 200 °C for 2 h for specimens 1ε and 3ε. After aging, tracer hydrogen was added to the specimens by immersion in a 20 mass% NH_4SCN solution at a temperature of 30 °C until saturation. The tracer hydrogen contents of the two specimens were measured using TDA.

Figure 4 shows ΔC_H, which was calculated by subtracting the tracer hydrogen content of specimen 3ε from that of specimen 1ε, plotted against aging temperature. The tracer hydrogen content of specimen 1ε was larger than that of specimen 3ε before aging and after aging at temperatures of up to 50 °C. After aging at a temperature of 100 °C, ΔC_H was approximately zero. Therefore, HESILDs were nearly eliminated by aging at a temperature of 100 °C. This result indicates that HESILDs are a type of thermally unstable lattice defect, such as vacancies or vacancy clusters.

Figure 5 shows the nominal stress–nominal strain curves of specimens 1ε and 3ε after aging at a temperature of 50 °C for 3 days. Even though hydrogen was absent in both specimens, the elongation of specimen 1ε was smaller than that of specimen 3ε. Specimen 1ε was subjected to plastic strain in the presence of hydrogen,

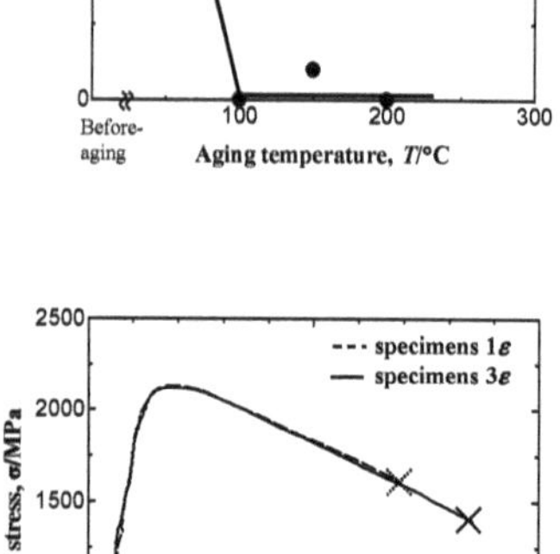

Fig. 4 ΔC_H plotted against aging temperature. ΔC_H was calculated by subtracting the tracer hydrogen content of specimen 3ε from that of specimens 1ε

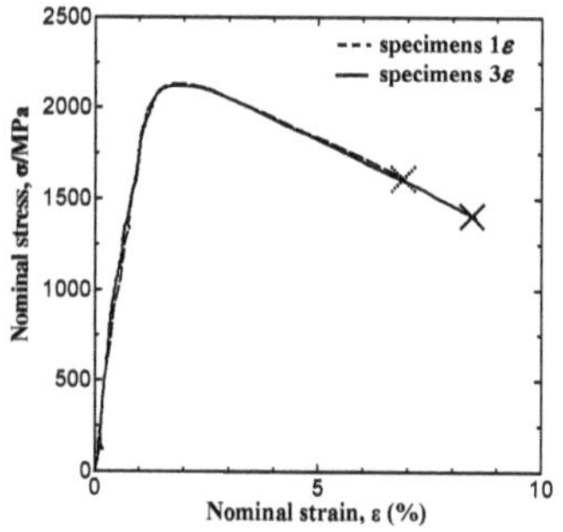

Fig. 5 Nominal stress–nominal strain curves of specimens 1ε and 3ε after aging at 50 °C for 3 days

and HESILDs were formed in this specimen. This result indicates that HESILDs are one of the causes of the hydrogen embrittlement of cold-drawn pearlitic steel that fractures in the plastic region.

Factors Causing Hydrogen Embrittlement Fractures in the Elastic Region

Figure 6 shows the TDA profiles and tracer hydrogen contents of specimens 2σ and 3σ after tracer hydrogen addition by immersion in a 60 mass% NH_4SCN solution at a temperature of 30 °C until saturation. The tracer hydrogen contents of specimens 2σ and 3σ, which were 4.3 and 4.2 mass ppm, respectively, were approximately equal. This indicates that the presence of hydrogen has no effects on the formation of lattice defects under an applied elastic stress.

Figure 7 shows the nominal stress–nominal strain curves of specimens 2σ and 3σ after aging at a temperature of 50 °C for 4 days. Unlike Fig. 5, the elongations of specimens 2σ and 3σ were approximately equal. This is because when specimen 2σ was subjected to elastic stress in the presence of hydrogen, HELDs did not form in the specimen. This result indicates that the presence of hydrogen is not a factor causing hydrogen embrittlement in the elastic region.

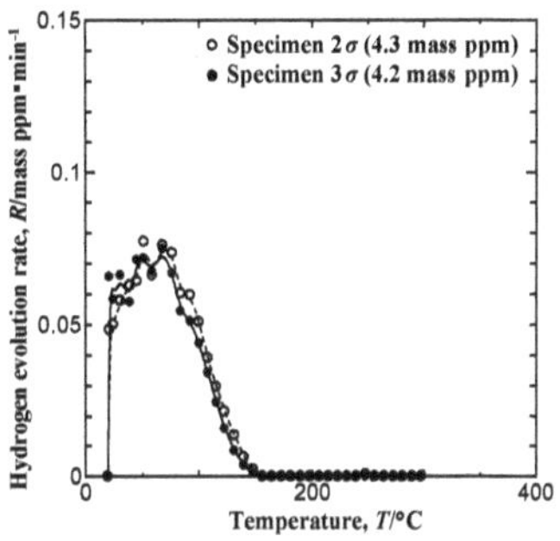

Fig. 6 TDA profiles and tracer hydrogen contents of specimens 2σ and 3σ

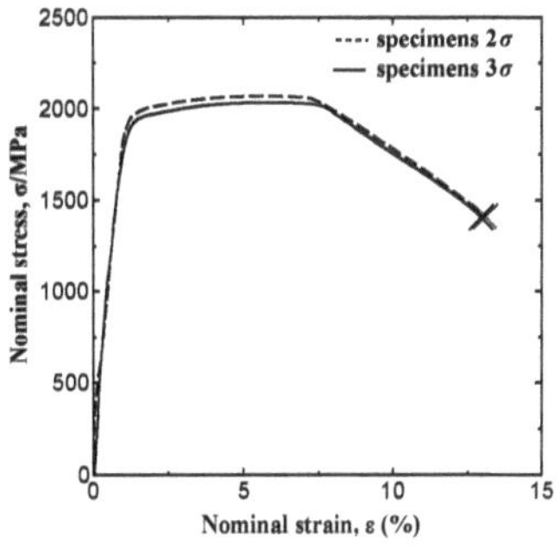

Fig. 7 Nominal stress–nominal strain curves of specimens 2σ and 3σ after aging at 50 °C for 4 days

Conclusions

Factors causing the hydrogen embrittlement of cold-drawn pearlitic steel fractured under elastic/plastic region were investigated from the perspective of lattice defects.

(1) The tensile test results for specimens 1–3, which contained 1.5, 4.0, and 0 ppm hydrogen, respectively, showed that the elongation decreased with increasing hydrogen content. Specimens 1 and 2 fractured in the plastic and elastic regions, respectively.
(2) The factors causing the hydrogen embrittlement of cold-drawn pearlitic steel that fractures in the plastic region were evaluated. The amount of lattice defects in the specimen containing 1.5 ppm hydrogen was enhanced by subjecting it to plastic strain until just before necking. HESILDs were eliminated by aging at a temperature of 100 °C. This result indicates that HESILDs are a type of thermally unstable lattice defect, such as vacancies or vacancy clusters. Moreover, these lattice defects directly decreased the mechanical properties of cold-drawn pearlitic steel. This indicates that HESILDs are a cause of the hydrogen embrittlement of cold-drawn pearlitic steel that fractures in the plastic region.
(3) The factors causing the hydrogen embrittlement of cold-drawn pearlitic steel fracturing in the elastic region were evaluated. The amount of lattice defects of the specimen containing 4.0 ppm hydrogen was not increased by subjecting it to elastic stress until just before tensile fracture. This indicates that lattice defects are not a factor causing the hydrogen embrittlement of cold-drawn pearlitic steel that fractures in the elastic region.

References

1. T. Doshida, K. Takai, Dependence of hydrogen-induced lattice defects and hydrogen embrittlement of cold-drawn pearlitic steels om hydrogen trap state, temperature, strain rate and hydrogen content. Acta Mater. **79**, 91–107 (2014)
2. K. Takai, H. Shoda, H. Suzuki, M. Nagumo, Lattice defects dominating hydrogen degradation of metals. Acta Mater. **56**, 5158–5167 (2008)
3. M. Hatano, M. Fujinami, K. Arai, H. Fujii, M. Nagumo, hydrogen embrittlement of austenitic stainless steels revealed by deformation microstructures and strain-induced creation of vacancies. Acta Mater. **67**, 342–353 (2014)
4. T. Doshida, M. Nagumo, H. Saito, T. Sawada, K. Takai, Hydrogen-enhanced lattice defect formation and hydrogen embrittlement of cyclically prestressed tempered martensitic steel. Acta Mater. **61**, 7755–7766 (2013)
5. T. Doshida, H. Suzuki, K. Takai, N. Oshima, T. Hirade, Enhanced lattice defect formation associated with hydrogen and hydrogen embrittlement under elastic stress of a tempered martensitic steel. ISIJ Int. **52**, 198–207 (2012)

Corrosion of Nickel-Titanium, C110, and Al6061 in Gallium-Based Liquid Metal Alloys

Jacob Mingear and Darren Hartl

Abstract Nickel-titanium is one of the most common shape memory alloys (SMAs). These alloys provide the highest actuation density of all active materials. SMA-based actuators are thermally induced and thus are limited in their actuation response frequency, specifically with regard to their slow speed of cooling. Polyalphaolefin and water have been investigated as fluids for active cooling mechanisms. However, a liquid metal may act as a more effective coolant due to its high thermal conductivity. To date, there has been no research on the liquid metal corrosion effects on nickel-titanium, which is crucial to understanding the viability of using liquid metal as an active coolant for SMA-based actuators. In this work, specimens of nickel-titanium alloys were placed in liquid metal-filled crucibles at 220 °C for 300 h. The responses of these specimens were compared to that of others composed of elemental Ni and Ti, as well as the common aerospace aluminum alloy Al6061. For the purposes of understanding liquid metal electronics challenges, copper alloy C110 was also investigated. Liquid metals investigated included gallium and gallium-indium-tin eutectic. Specimens were characterized using scanning electron microscopy and energy-dispersive spectroscopy to investigate the surface effects, penetration of liquid metal, and reaction layer chemistry associated with prolonged exposure to the liquid metals.

Keywords Liquid metal corrosion · Nickel-titanium · Copper · SMA · Reaction layer

J. Mingear (✉) · D. Hartl
Texas A&M University, Administration Building 400 Bizzell St, College Station, TX 77843, USA
e-mail: mingear@tamu.edu

D. Hartl
e-mail: darren.hartl@tamu.edu

The Minerals, Metals & Materials Society, *TMS 2017 146th Annual Meeting & Exhibition Supplemental Proceedings*, The Minerals, Metals & Materials Series,
DOI 10.1007/978-3-319-51493-2_56

Introduction

Nickel-titanium is one of the most common shape memory alloys (SMAs). Shape memory alloys are a class of active materials that have the ability to recover their shape after significant deformation or a temperature threshold. These alloys provide the highest actuation density of all active materials [1]. The application of SMAs span a wide variety of industrial sectors such as aerospace, automotive, biomedical, and oil exploration [1]. Equiatomic nickel-titanium, usually referred to as Nitinol or NiTi, exhibits fully recoverable strains of around 7%. Increasing nickel content of NiTi from 50at.% Ni to 55% Ni lowers the strain recovery capability to 3% but drastically increases the thermally-induced fatigue life [1, 2]. Thus, both alloys have been heavily studied over recent decades due to their niche capabilities, with the nickel rich variant receiving special attention in the aerospace sector [2, 3].

Currently, many SMA-based actuators are able to be heated and thus actuated rapidly, but a rapid mechanism for cooling these monolithic metallic bodies remains challenging. This limitation has been investigated by Coppola et al. by using polyalphaolefin and water as a fluid for active cooling mechanisms [4]. However, a fluid that has a higher thermal conductivity coefficient could act as an even more effective coolant. Candidate fluids might include low-temperature liquid metals, though these are unfortunately known to be highly corrosive and embrittling to other solid metals [5–11]. Two commonly known liquid metal systems are mercury and gallium-based alloys. However, due to the high toxicity of mercury, only gallium-based alloys arc considered herein due to their extremely low vapor pressure, high boiling point, and generally low toxicity. The specific gallium-based liquid metal alloy under consideration for this cooling mechanism is a widely commercially-available eutectic mixture of gallium, indium, and tin (denoted herein as GaInSn; but also similar in composition to the commercially-available Galinstan).

Corrosion reactions in liquid metal-solid metal systems are governed by several mechanisms including: dissolution, impurity and interstitial reactions, alloying, and compound reduction [12]. The most important mechanisms involved in this current work include dissolution and alloying. Dissolution is the release of metal atoms from the solid metal into solution with the liquid metal. Alloying is the formation of stable products, i.e., intermetallics, on the solid metal surface when the liquid metal and solid metal interact.

Currently, there is no literature on liquid metal-induced corrosion in nickel-titanium alloys, which is crucial to understanding the viability of using liquid metal as an active coolant for SMA-based actuators. This study aims to elucidate the interactions of GaInSn on equiatomic nickel-titanium (NiTi), nickel-rich nickel-titanium (known herein as $Ni_{55}Ti_{45}$), while comparing it to its elemental components: Nickel 200 and Titanium Grade 2. Alloys also investigated include the Copper alloy 110 (C110), and Aluminum Alloy 6061 (Al6061) to assess the feasibility of utilizing potentially corrosive liquid metals in a wider range of applications (e.g. electrical and aerospace). Observations include the surface effects, liquid metal penetration, and reaction layer chemistry that result from prolonged exposure of otherwise untreated surfaces. In studying materials beyond SMA systems, the feasibility of a wide range of potential engineering applications of liquid metal can be assessed and the existing body of data can be expanded.

Experimental

The six alloys that were investigated included equiatomic nickel-titanium (NiTi), nickel-rich nickel-titanium alloy ($Ni_{55}Ti_{45}$), Nickel 200, Titanium Gr. 2, Aluminum alloy Al6061-T6, and Copper alloy C110. Specimen material was obtained from Online Metals.com in sheet form, except NiTi and $Ni_{55}Ti_{45}$, which were procured from extra material remaining from previous studied SMA structures [2, 3]. Specimens were nominally 3.175 mm (1/8 in) thick except for NiTi, which was 2 mm thick. The specimen material was then cut using a diamond rotary saw to produce multiple rectangular prism specimens. Widths ranged from 4 to 5.5 mm, while lengths ranged from 12 to 13 mm. Specimens were ground on all sides using SiC metallurgical paper with grits of 180 and then 320 to remove potential surface defects and preexisting surface oxides.

The gallium used for the liquid gallium-exposed specimens was 99.99% pure and supplied in ingot form by Alfa. The GaInSn was supplied by GalliumSource, LLC, and had a nominal composition of 68.5at.% Ga, 21.5at.% In, and 10at.% Sn.

For the series of liquid gallium tests, fragments of solid gallium were placed in a porcelain crucibles and then heated slightly above 30 °C until molten; one specimen of each metal were then added to its respective liquid gallium-filled crucibles. For the liquid GaInSn tests, an additional set of specimens were each placed first in their respective crucible and then filled with GaInSn. Because of their low densities, Ti and Al6061 floated significantly on the free surface of both liquid metal systems, but still had the appropriate contact with the liquid surface needed to affect any surface effects. Crucibles were complemented with porcelain lids and placed in an oven for 300 h at 220 °C.

After the 300 h corrosion test, all specimens were washed in soapy water above 30 °C and then sonicated in a 1:1 ethanol to acetone mixture to additionally remove the loose gallium-based liquid and solid products from the surface (inspired by Barbier and Blanc [11]). Specimens were sectioned using a diamond rotary saw to expose the cross section and allow assessment of liquid metal penetration. Specimens were mounted in epoxy resin and polished using SiC metallurgical paper of the following grits: 180, 320, 400, 600, and 1200; then, an alumina slurry of 1 μm powder and 0.3 μm powder was used on a soft felt pad to complete the preparation. Specimens were then rinsed with water and dried with compressed air.

The effects of the liquid metal on the specimens were assessed using a Tescan Vega 3 Scanning Electron Microscope (SEM), with both Secondary Electron (SE) imaging and Back-Scatter Electron (BSE) imaging including an Oxford Instruments X-act Energy-Dispersive X-ray Spectroscopy (EDS).

Results

Nickel-Titanium Alloys

NiTi

After the rinse in soapy water and solvent sonication, the nickel-titanium visually appeared to have faint silver colored areas surrounded by a gold-colored oxidation layer on its faces for both the liquid gallium-exposed and GaInSn-exposed specimens. The specimens were then observed in the SEM and both had relatively smooth surfaces at the liquid metal-solid metal interface, though some minor pitting was observed, as seen in Fig. 1. The largest pit measured in the liquid gallium-exposed specimen was about 3 μm deep, while the GaInSn-exposed specimen had about 6 μm deep. EDS was used to identify any penetration of liquid metal into the bulk specimens. Based on this measurement, there were no gallium, indium, or tin products observed near the faces or in the bulk for both the liquid gallium-exposed specimen and the GaInSn-exposed specimen.

$Ni_{55}Ti_{45}$

After the rinse in soapy water and solvent sonication, the $Ni_{55}Ti_{45}$ had a similar visual appearance to NiTi but with larger silver-colored areas surrounded by darker golden-orange colored oxidation layer for the liquid gallium-exposed specimen. The GaInSn-exposed specimen had an even darker colored oxidation layer. Under SEM observation, the liquid metal-solid metal interface surface seemed to be rougher in the GaInSn-exposed specimen as compared to the liquid gallium-exposed specimen, as seen in Fig. 2. It was observed that the GaInSn-exposed specimen had pits of 36 μm deep, and it was noted that the pitting frequency was increased compared to the NiTi specimens. In a manner similar to the NiTi specimens, there were no gallium, indium, or tin products observed near the faces or in the bulk for both the liquid gallium-exposed specimen and the GaInSn-exposed specimen.

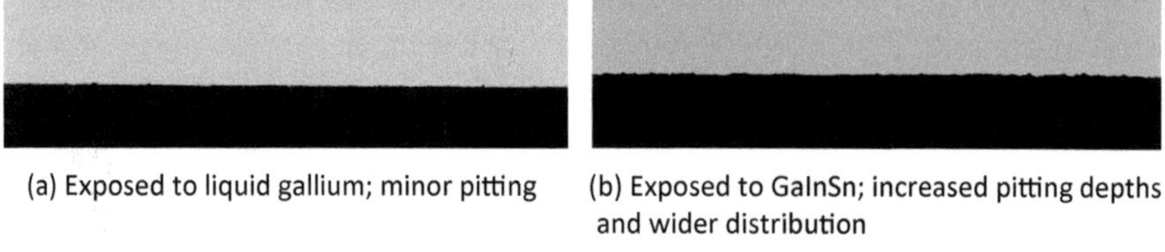

(a) Exposed to liquid gallium; minor pitting (b) Exposed to GaInSn; increased pitting depths and wider distribution

Fig. 1 BSE image of the NiTi surface after exposure to liquid metal systems

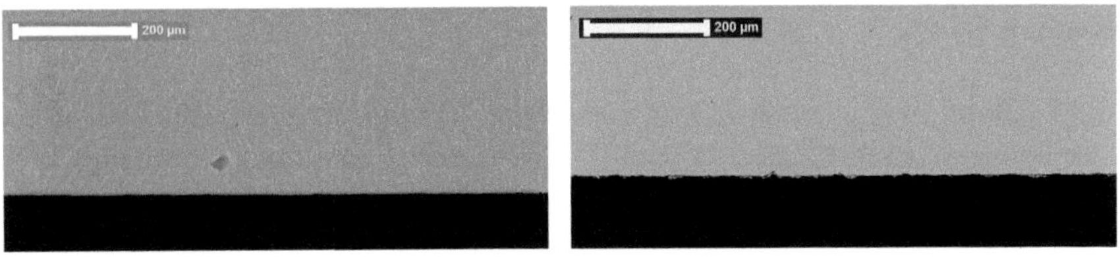

(a) Exposed to liquid gallium; minor pitting (b) Exposed to GaInSn; increased pitting depths and wider distribution

Fig. 2 BSE image of the $Ni_{55}Ti_{45}$ surface after exposure to liquid metal systems

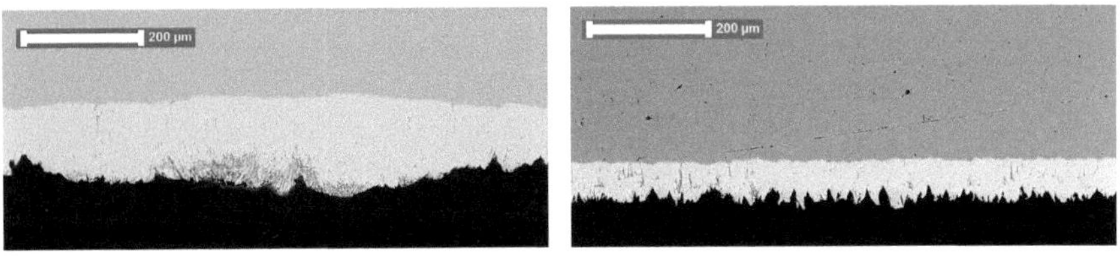

(a) Exposed to liquid gallium; thick reaction layer (b) Exposed to GaInSn; thinner but more uniform reaction layer

Fig. 3 BSE image of the Nickel 200 surface after exposure to liquid metal systems

Elemental Ni and Ti Specimens

To better understand the corrosion of nickel-titanium alloys, Nickel 200 and Titanium Gr. 2 were each independently investigated in the same manner as the SMA specimens. These two materials clearly represent which nickel-titaniums elemental components.

Nickel 200

Nickel 200 reacted significantly with both the liquid gallium and GaInSn. After the cleaning treatment, the Nickel 200 surface was visually dark gray with newly formed products protruding from the faces of both the liquid gallium-exposed specimen and GaInSn-exposed specimen. SEM observation showed that the liquid gallium-exposed specimen formed a reaction layer 140 ± 10 µm thick. The GaInSn-exposed specimen had a reaction layer of 60 ± 20 µm thick. The reaction layer was not a simple monolithic formation but resembled a dense collection of rectangular crystals oriented with their long axis normal to the surface, as seen in Fig. 3. On the corners of both specimens, there seemed to be the initiations of cruciform-type structures resembling those observed by Barbier and Blanc [11]. EDS measurements indicated that this reaction layer was 30at.% Ni and 70at.% Ga for both specimens.

Titanium Grade 2

Both Titanium Gr. 2 specimens, with a density of 4.51 g/cc, floated on the free surface of the liquid metals (6.01 g/cc for liquid gallium and 6.29 g/cc for GaInSn) thus, only the bottom face of the specimens were in contact with the liquid metal alloys.

After the cleaning treatment, visual inspection showed a faint silver-colored area on the face that was in contact with each liquid metal. SEM observation showed some pitting (3–6 μm) in the liquid gallium-exposed specimen while more common 3–6 μm pits were observed in the GaInSn-exposed specimen, as seen in Fig. 4. EDS measurements indicated there were no gallium, indium, or tin products observed near the faces or in the bulk for either the liquid gallium-exposed specimen or the GaInSn-exposed specimen (cf. Section "Nickel-Titanium Alloys").

Engineering Alloys

In addition to investigating the effects of liquid metal corrosion on the elemental components of nickel-titanium alloys, engineering alloys were also considered to assess the feasibility of utilizing potentially corrosive liquid metals in a wider range of applications. For example, actuators are usually coupled with a copper-based electrodes to heat up the SMA via Joule heating. Liquid metal antennas [13], and stretchable electronics [14] have also been considered. Thus, copper alloy C110 was investigated. Additionally, there have been a number of SMA-based actuators considered for aircraft applications. Aluminum alloy 6061 is a common aircraft alloy and interactions between this structural material and common liquid metals should be understood.

Copper Alloy C110

After the cleaning treatment, both C110 specimens were seen to be uniformly covered in a shiny coat of the liquid metal-based products. SEM observation of the

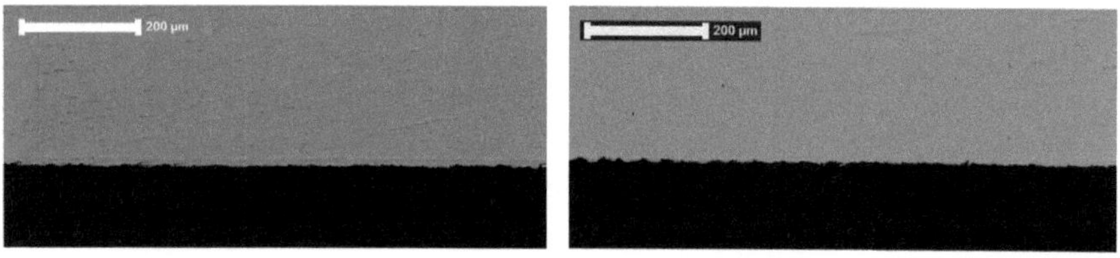

(a) Exposed to liquid gallium; minor pitting (b) Exposed to GaInSn; increased pitting depths and wider distribution

Fig. 4 BSE image of the Titanium Grade 2 surface after exposure to liquid metal systems

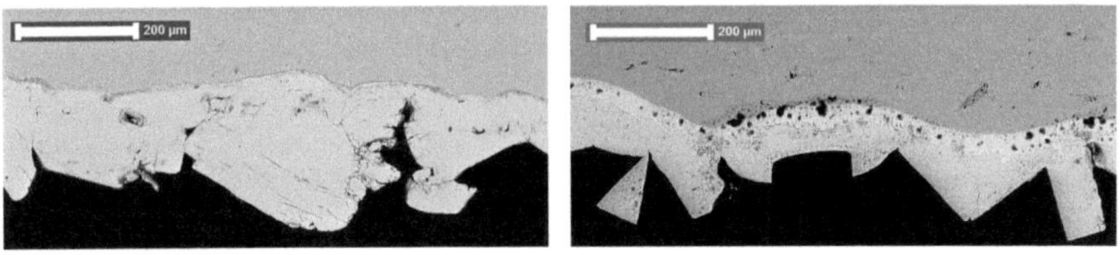

(a) Exposed to liquid gallium; rounded reaction (b) Exposed to GaInSn; sharper reaction layer edges

Fig. 5 BSE image of the C110 surface after exposure to liquid metal systems

liquid gallium-exposed specimen indicated a total reaction layer having a thickness of 150 ± 40 µm, as seen in Fig. 5, with multiple phases comprised of three discrete layers. EDS measurements indicted that the outermost phase consisted of 33at.% Cu and 66at.% Ga; thus, it is most likely the known γ phase $CuGa_2$ intermetallic. The middle layer was 15 ± 5 µm thick had a composition around 60at.% Cu and 40at.% Ga. The inner layer was only 1–2 µm thick with a composition of 66at.% Cu and 33at.%, and most likely the known θ phase Cu_2Ga intermetallic. The bulk maintained a chemistry of 99at.% Cu.

For the GaInSn-exposed specimen, there was a similar three layer reaction but with additional complexity within the second layer. The total thickness of the three layers was shown to be 80 ± 20 µm thick. The outer most layer consisted of a 33at.% Cu and 66at.% Ga; again, most likely the γ phase $CuGa_2$ intermetallic. The middle layer was 30 ± 5 µm thick and had a composition of 60 ± 3at.% Cu, 40 ± 3at.% Ga, and 1 ± 0.2at.% In. However, in that middle layer, there appeared to be dark precipitates in a light matrix observed in the Back-Scatter Electrons (BSE) imaging. EDS measurements indicated the chemical difference was within the margin of error so no significant finding can be claimed at this time regarding the chemical differences between the dark phases and light matrix. The inner layer was around 1–2 µm thick with a composition of 66at.% Cu and 33at.% thus, most likely the Cu_2Ga intermetallic, followed by the bulk 99at.% Cu.

Aluminum Alloy 6061

The Al6061, with a density of 2.7 g/cc, initially floated on the liquid metal free surfaces in a manner similar to the Titanium Gr. 2. However, after completion of the elevated temperature exposure process, the aluminum specimens were found to be no longer floating. For the liquid gallium-exposed specimen, the Al6061 had completely dissolved into the liquid gallium, no solid specimen remained, and no further assessment was possible. In the GaInSn-exposed specimen, the aluminum specimen was so significantly penetrated by the liquid metal that its effective density increased and it was found submerged. There were shiny crack-like features on the GaInSn-exposed Al6061 surface, which is most likely due to the liquid metal penetrating throughout the grains of the specimen. During the attempt to apply the standard cleaning treatment to the GaInSn-exposed specimen, the specimen broke into several pieces upon the moderate force applied by tweezers. These pieces were then placed in warm soapy water, which caused an unexpected reaction to occur, converting the brittle metallic pieces into a fine black powder. It is not clear what exact reaction took place. Regardless, the GaInSn-exposed Al6061 specimen was unable to be further characterized.

Discussion

Nickel-Titanium Alloy Trends

Reviewing the elemental Ni and Ti responses as foundational to the interactions observed in the nickel-titanium alloys, it seems that nickel reacts much more strongly than titanium with gallium-based liquid metal alloys. A significant reaction layer was produced on both Nickel 200 specimens. In contrast, the Titanium Gr. 2 had no reaction layer and only slight pitting was observed. Further, NiTi had similar interactions to the Titanium Gr. 2; only slight pitting and visual discoloration on the surface was observed, thus behaving much like the elemental titanium despite being comprised of 50at.% Ti. In comparison, $Ni_{55}Ti_{45}$ had similar results to the Titanium Gr. 2 but with increased frequency of pitting and more drastic surface discoloration than equiatomic NiTi. Thus, it seems that even 45at.% titanium is enough to prevent reaction layers from forming with the majority nickel content in a nickel-titanium alloy at the elevated temperature considered.

Copper Alloy C110 Trends

C110 showed significant corrosion with multiple reaction layers and complex morphologies when exposed to both liquid metal compositions. The complex reaction layers were closely investigated and it was found that there were three discrete layers of reaction products for both liquid metals. Juxtaposed in Fig. 6, one can see that there is additional complexity within the middle layer of the GaInSn-exposed specimen. Using BSE imaging, dark phases within a light matrix are elucidated. EDS measurements were used find the chemical composition of these the dark phase and light matrix. However, the results were within the margin of error. Uniquely, however, this region was the only region of this specimen to detect indium, as confirmed by EDS measurements. Though, it is not clear at this time how indium or tin played a part in the formation of these complex reaction layers for the GaInSn-exposed specimen as compared to the simpler reaction layers in the liquid gallium-exposed specimen.

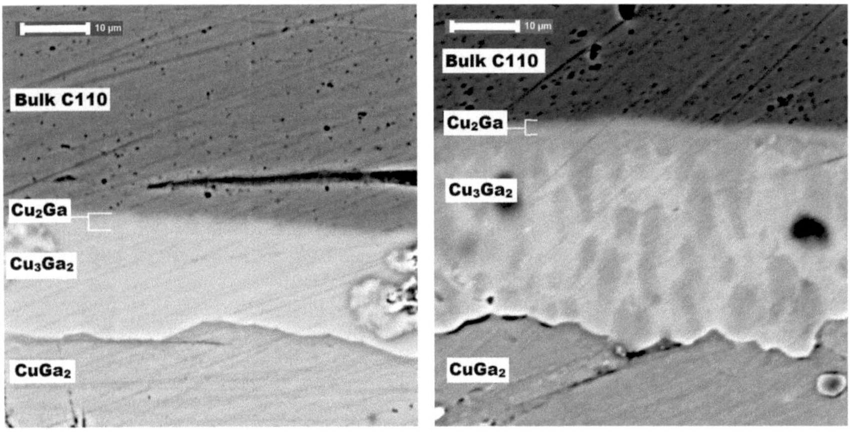

(a) Exposed to liquid gallium; complex reaction layer system

(b) Exposed to GaInSn; additional morphology in the middle layer

Fig. 6 BSE image of the C110 reactions layers. Starting from top and moving down: bulk C110, γ phase Cu_2Ga,γ_3 phase Cu_3Ga_2, θ phase $CuGa_2$

Conclusion

Liquid metal-induced corrosion was observed on the surface of various alloys important for potential aerospace applications, including two SMA compositions, to assess the viability of using liquid metal as a high thermally conductive cooling fluid. Significant reaction layers were formed on the Nickel 200 specimens, while the Ti, NiTi, and $Ni_{55}Ti_{45}$ specimens did not exhibit the formation of such any reaction layers and only had varying degrees of pitting and surface discoloration. It seems that the increased titanium content corresponded to less pitting and surface discoloration. The C110 specimens seemed to form significant reaction layers on their surfaces in response to exposure to each liquid metal. Finally, liquid gallium and GaInSn completely dissolved or greatly embrittled the Al6061 specimens, respectively. Future work would include examining nickel-titanium alloys in liquid metal at higher temperatures and tensile measurements coupled with liquid metal to elucidate the severity of liquid metal-induced embrittlement of nickel-titanium alloys.

Acknowledgements This work is supported in part by the US Air Force Office of Scientific Research under a grant number FA9550-16-1-0087, titled Avian-Inspired Multifunctional Morphing Vehicles monitored by Dr. BL Lee. Additionally, the authors would like to acknowledge Mohammad Naraghi (Department of Aerospace Engineering, Texas A&M University) for providing the oven used.

References

1. D.C. Lagoudas (ed.), *Shape Memory Alloys: Modeling and Engineering Applications*. (Springer, New York, 2008). OCLC: 636747268
2. D.J. Hartl, D.C. Lagoudas, F.T. Calkins, J.H. Mabe, Use of a Ni60ti shape memory alloy for active jet engine chevron application: I. Thermomechanical characterization. Smart Mater. Struct. **19**, 015020 (2010)
3. D.J. Hartl, J.T. Mooney, D.C. Lagoudas, F.T. Calkins, J.H. Mabe, Use of a Ni60ti shape memory alloy for active jet engine chevron application: II. Experimentally validated numerical analysis. Smart Mater. Struct. **19**, 015021 (2010)
4. A.M. Coppola, L. Hu, P.R. Thakre, M. Radovic, I. Karaman, N.R. Sottos, S.R. White, Active cooling of a microvascular shape memory alloy-polymer matrix composite hybrid material: active cooling of a microvascular shape memory. Adv. Eng. Mater. **18**, 1145–1153 (2016)
5. W. Rostoker, J.M. McCaughy, H. Markus, *Embrittlement by Liquid Metals* (Reinhold Publishing Corporation, New York, NY, 1960)
6. K.A. Narh, V.P. Dwivedi, J.M. Grow, A. Stana, W.-Y. Shih, The effect of liquid gallium on the strengths of stainless steel and thermoplastics. J. Mater. Sci. **33**(2), 329–337 (1998)
7. K. Ina, H. Koizumi, Penetration of liquid metals into solid metals and liquid metal embrittlement. Mater. Sci. Eng. A **387–389**, 390–394 (2004)
8. M. Nicholas, C. Old, Liquid metal embrittlement. J. Mater. Sci. **14**(1), 1–18 (1979)
9. M.H. Kamdar, Embrittlement by liquid metals. Prog. Mater Sci. **15**(4), 289–374 (1973)
10. S.P. Yatsenko, N.A. Sabirzyanov, A.S. Yatsenko, Dissolution rates and solubility of some metals in liquid gallium and aluminum. J. Phys. Conf. Ser. **98**, 062032 (2008)
11. F. Barbier, J. Blanc, Corrosion of martensitic and austenitic steels in liquid gallium. J. Mater. Res. **14**(03), 737–744 (1999)
12. P.F. Tortorelli, S.J. Pawel, Corrosion by Liquid Metals, in *Corrosion: Fundamentals, Testing, and Protection*, vol. 13A of *ASM Handbook*, (ASM International, 2003), pp. 129–134
13. D.J. Hartl, G.H. Huff, H. Pan, L. Smith, R.L. Bradford, G.J. Frank, J.W. Baur, Analysis and characterization of structurally embedded vascular antennas using liquid metals, (2016), pp. 980333
14. K. Entesari, A.P. Saghati, Fluidics in Microwave Components. IEEE Microwave Mag. **17**, 50–75 (2016)

Part XIV
General Poster Session

Synthesis and Characterization of Al–B_4C Powders by Mechanical Alloying

Hao Guo, Zhongwu Zhang, Yu Zhao, Songsong Xu, Junpeng Li and Jing Zhang

Abstract Boron carbide reinforced aluminum matrix composites are widely used as neutron absorption materials. Here we report that mechanical alloying has been successfully employed to synthesize metal matrix composite powders with Al as the matrix and B_4C as the reinforcement. The effects of powder blending process, solid sintering and thermo-mechanical processes on the microstructure evolution and mechanical properties are studied. The results show that the density of the composites with 30 wt% boron carbide can reach 98.2% after sintering. Thermomechanical processing can increase the density further to 99.4% and remove the defects, leading to an increase in strength from 362.0 to 388.7 Mpa. The influences of boron carbide and processing routes on the mechanical properties and microstructure evolution, specifically the interface between boron carbide and matrix are discussed.

Keywords Boron carbide · Mechanical alloying · Aluminum matrix composites · Mechanical properties

Introduction

Aluminum matrix composites have attracted interest in structural applications in the aerospace application, military industries and automotive industries [1] by virtue of their good properties. Aluminum matrix composites can be produced by powder metallurgy with various reinforced particulates like oxides, carbides and borides, such as Al_2O_3, SiC, B_4C, TiC, TiB_2 [2–6]. Boron carbide ceramics have excellent

---

H. Guo · Z. Zhang (✉) · Y. Zhao · S. Xu · J. Li · J. Zhang
Key Laboratory of Superlight Materials and Surface Technology, Ministry of Education, College of Materials Science and Chemical Engineering, Harbin Engineering University, Harbin 150001, People's Republic of China
e-mail: zwzhang@hrbeu.edu.cn

H. Guo
e-mail: guohao2016@qq.com

The Minerals, Metals & Materials Society, *TMS 2017 146th Annual Meeting & Exhibition Supplemental Proceedings*, The Minerals, Metals & Materials Series, DOI 10.1007/978-3-319-51493-2_57

properties of high hardness, good wear resistance, low density, excellent chemical resistance and high capability for neutron absorption [7–9]. It has been used in many industrial fields, such as military, engineering and nuclear energy [6].

Mechanical alloying is widely used to produce composite powders [10–14]. Mechanical alloying is a high-energy milling process [10]. In this study, we use mechanical alloying to produce our composite powders. We can get composite powders with uniform distribution and fine microstructure after mechanical alloying.

There are many studies producing B_4C/Al composites have been carried out [15–18]. But there are few studies focus on high B_4C content. Mohammad Sharifi et al. [1] investigated Al–B_4C powders with 5, 10 and 15 wt% amounts of B_4C. Baradeswaran et al. [19] investigated aluminum alloy were reinforced with 10 wt% of B_4C and 5 wt% of graphite. Khakbiz and Akhlaghi [10] investigated 5 and 10 wt% amounts of B_4C particles. Khakbiz and Akhlaghi [10] was just aimed to synthesis and characterization Al–B_4C powders, but there were no composite bulk samples and mechanical property characterization subsequently.

In this study, The effects of powder blending process, solid sintering and thermo-mechanical processes on the microstructure evolution and mechanical properties are studied. The influences of boron carbide and processing routes on the mechanical properties and microstructure evolution, specifically the interface between boron carbide and matrix are discussed. XRD was used analysis the change of phases, optical microscope (OM) were reveal the microstructure, and scanning electron microscope (SEM) showed the size distribution, morphology of powders and the appearance of fracture.

Experimental

High purity (99.9%) Al powders and 92.5% pure boron carbide particles were prepared as starting materials. And the average particle size of Al powders was about 13 μm, the average particle size of boron carbide particles was about 10 μm. Aluminum powders were mixed with boron carbide particles and ball milled to produce Al–B_4C composites with 30 wt% of boron carbide content using a ball milling machine named QM-3SP2 high energy. In this work, we used two zirconia vials containing Al_2O_3 balls. Ball milling parameter: a rotating speed was 300 r/min, a milling time was 8 h and ball to powder weight ratio was 5:1. To minimize cold welding of Al powders, we used 1 wt% of stearic acid as a process control agent.

Then milled powders were put into a steel die for hot pressed and sintered. The die containing milled powders was heated to 640 °C for 2 h and pressed at 20 MPa. The whole process is under high vacuum environment.

The as-sintered samples were then prepared for hot rolling. The samples were heated at 450 °C for 30 min and then hot-rolled with a thickness reduction of

25% and 50%. Between passes samples were returned to heat to 450 °C for 30 min. The tensile tests were tested with a strain rate of 1 × 10^{-3}/s on INSTRON 5565 at room temperature. The gauge length of the tensile specimens is 12.5 mm, width is 4 mm and thickness is 1.5 mm. The yield strength was used the 0.2% off set plastic strain method.

The phase components were investigated by X-ray diffraction (XRD) on Rigaku D/max-TTR-III diffractometer with a Cu Target K alpha ray, 40 kV, 150 mA, scanning angle is 20°–90°, and scanning speed is 5°/min. Optical microscopy (OM) was used to characterize the microstructure and scanning electron microscope (SEM) is for the size distribution, morphology of powders and tensile fracture morphologies.

Results and Discussion

Powder Characterization

Figure 1 shows the SEM micrographs of the mixture of Al-30%B_4C powders after 8 h milling. It can be seen that particles are not spherical after milled for 8 h, particles appear irregular shape. And Al particles seem plate like and fragmentary, Al particles became smaller, but there were limited change happened to B_4C particles.

Figure 2 shows the XRD patterns of the mixture of Al-30%B_4C powders after 8 h milling. It can be seen that the intensities of the Al peaks and B_4C peaks are all clear, and it's obviously to be seen that the intensities of the Al peaks are stronger than B_4C peaks, it means that the content of Al much more than the content of B_4C, in fact, Al content is 70% and the content of B_4C is 30%. On the other hand, we can see that, the peaks are only Al and B_4C, no new phase appeared during ball milling.

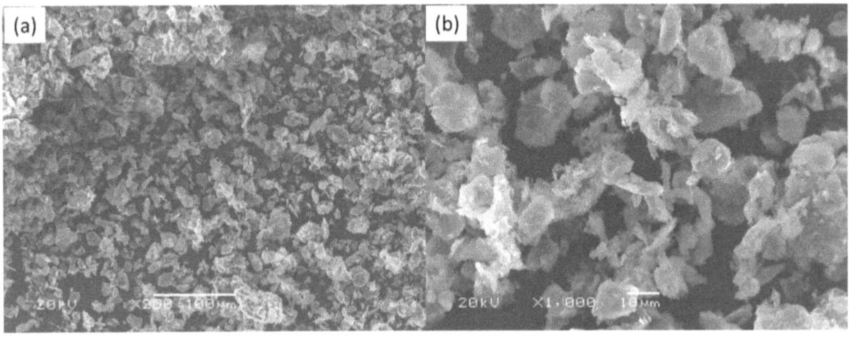

Fig. 1 SEM micrographs of a mixture of Al-30%B_4C powders, milled for 8 h: **a** at low magnifications, **b** at high magnifications

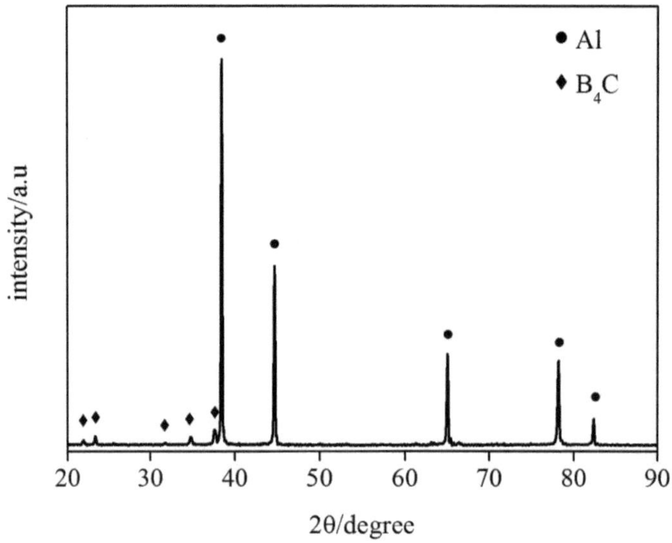

Fig. 2 XRD patterns of Al-30%B_4C powders for 8 h

Microstructure

Figure 3 shows the microstructures of the Al-30%B_4C samples: (a) as sintered at low magnifications, (b) 50% hot-rolled at low magnifications, (c) as sintered at high magnifications and (d) 50% hot-rolled at high magnifications. From the microstructures, it can be seen that very little change was happened between as sintered Al-30%B_4C samples and 50% hot-rolled Al-30%B_4C samples. But the density of the composites with 30 wt% boron carbide can reach 98.2% after sintering. Thermomechanical processing can increase the density further to 99.4% and remove the defects.

Figure 4 shows the XRD patterns of Al-30%B_4C samples after sintered. It can be seen that very little change was happened between peaks of as sintered Al-30%B_4C samples and Al-30%B_4C powders after 8 h milling. The intensities of the Al peaks and B_4C peaks are all clear, and it's obviously to be seen that the intensities of the Al peaks are stronger than B_4C peaks. On the other hand, we can see that, the peaks are only Al and B_4C, no new phase appeared during hot pressing and sintering.

Figure 5 shows the XRD patterns of Al-30%B_4C powders and samples after sintered. It can be seen that peaks offset to the right after hot pressed and sintered. It told us interplanar spacing is widened after hot pressed and sintered. Figure 6 shows the XRD patterns of Al-30%B_4C samples after sintered, 25% hot-rolled and 50% hot-rolled. It can be seen that very little change was happened between peaks of Al-30%B_4C samples after sintered, 25% hot-rolled and 50% hot-rolled.

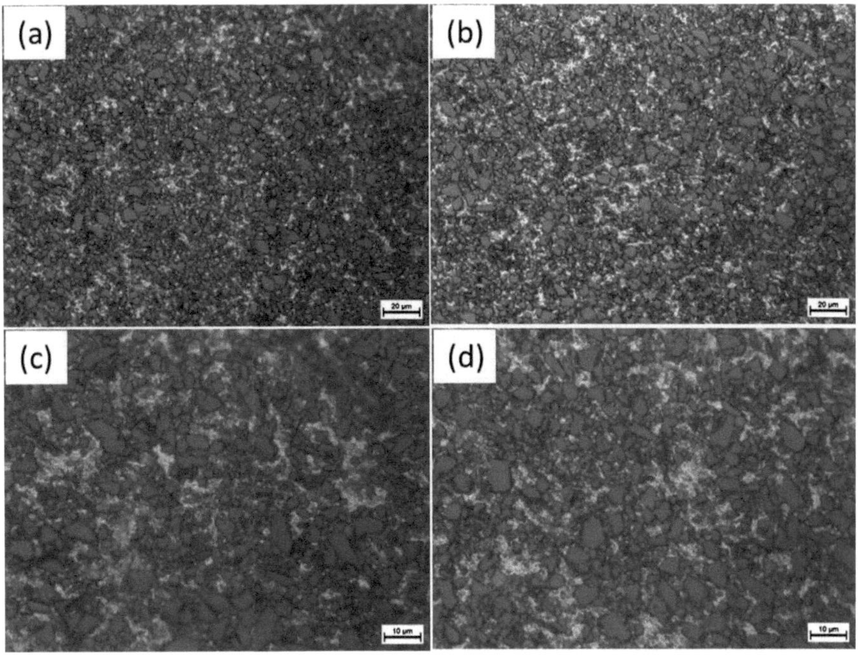

Fig. 3 Microstructures of the samples: **a** as sintered at low magnifications, **b** 50% hot-rolled at low magnifications, **c** as sintered at high magnifications and **d** 50% hot-rolled at high magnifications

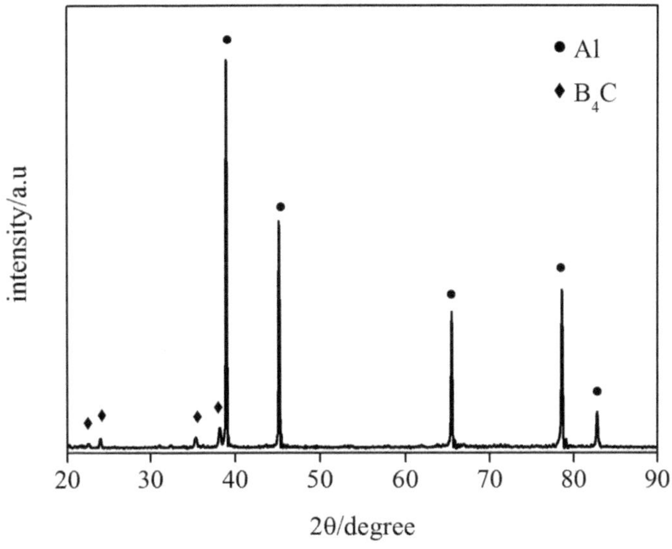

Fig. 4 XRD patterns of Al-30%B_4C samples after sintered

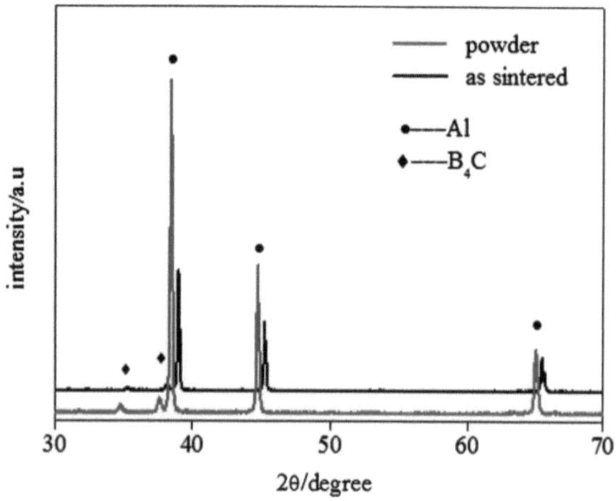

Fig. 5 XRD patterns of Al-30%B_4C powders and samples after sintered

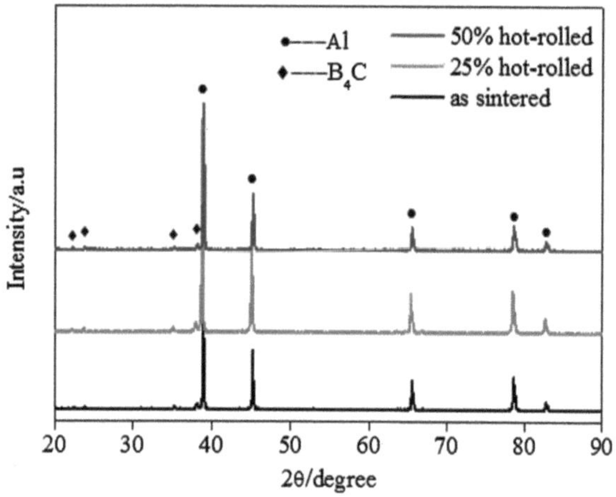

Fig. 6 XRD patterns of Al-30%B_4C samples after sintered, 25% hot-rolled and 50% hot-rolled

Mechanical Properties

The tensile engineering stress-strain curves of the Al-30%B_4C samples after sintered, 25% hot-rolled and 50% hot-rolled measured at ambient temperature are shown in Fig. 7. The ultimate tensile strengths (UTS) of the Al-30%B_4C samples after sintered, 25% hot-rolled and 50% hot-rolled are 362.0 MPa, 379.2 MPa and 388.7 MPa, respectively. As for the ductility, fracture elongations of the Al-30%B_4C samples after sintered, 25% hot-rolled and 50% hot-rolled are 1.33%, 1.48% and 1.28%, respectively. It is obviously that the strength is enhanced significantly meanwhile the ductility changes little by the rolling processes. The results show that an increase in strength from 362.0 to 388.7 Mpa accompanying with the density of the composites with 30 wt% boron carbide increase from 98.2% to 99.4% after Thermomechanical processing. Thermomechanical processing can also remove the defects.

SEM Analysis of Tensile Fracture

Figure 8 shows the fracture surface of the Al-30%B_4C sample. It can be seen by SEM images that the fracture morphology of the Al-30%B_4C sample is including both shallow dimples and brittle fractures. However, the size of dimples was small and shallow. Because it's obviously to be seen the traces of B_4C particles pulled out. Cracks appear at the B_4C particles interface, the interfacial bonding strength between particle and matrix is weak. So the interfacial debonding is the primary cause and the failure mode of Al-30%B_4C sample is the cleavage fracture of particle.

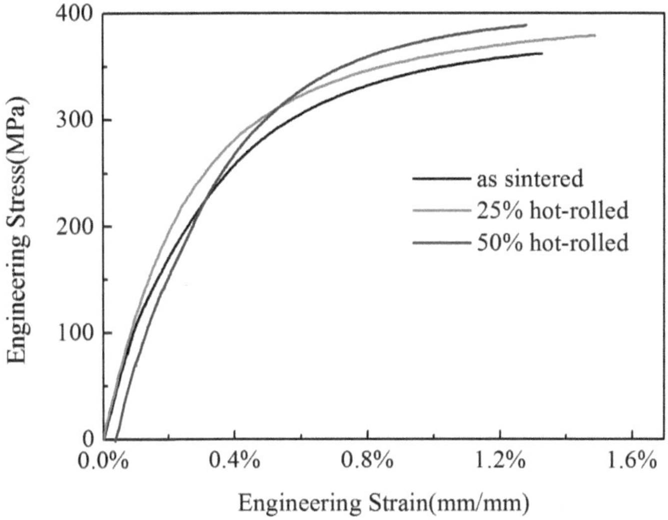

Fig. 7 Tensile properties of the investigated Al-30%B_4C composites

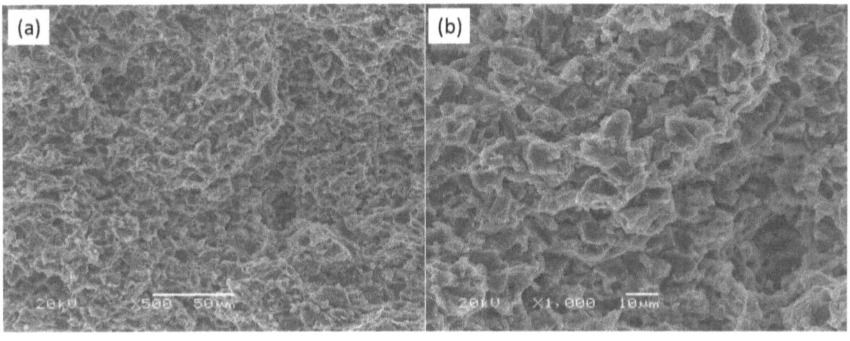

Fig. 8 Tensile fracture morphologies of the samples: **a** at low magnifications, **b** at high magnifications

Conclusions

In this study, synthesis and characterization of Al–B_4C powders and Al-30%B_4C composites by mechanical alloying was investigated. The present work has revealed that:

1. Al particles seem plate like and fragmentary, Al particles became smaller, but there were limited change happened to B_4C particles after milled for 8 h. the XRD peaks are only Al and B_4C, no new phase appeared during ball milling.
2. The density of the composites with 30 wt% boron carbide can reach 98.2% after sintering. Thermomechanical processing can increase the density further to 99.4%. No new phase appeared during hot pressing and sintering. Interplanar spacing is widened after hot pressed and sintered
3. An increase in strength from 362.0 to 388.7 Mpa accompanying with the density of the composites with 30 wt% boron carbide increase from 98.2% to 99.4% after Thermomechanical processing. Thermomechanical processing can also remove the defects.
4. The interfacial debonding is the primary cause and the failure mode of Al-30% B_4C sample is the cleavage fracture of particle.

Acknowledgements This work was supported by the NSFC Funding (51371062 and U1460102), NSFHLJ (ZD201411), the Scientific Research Foundation for the Returned Overseas Chinese Scholars (Heilongjiang Province), the Project for Innovative Talents of Science and Technology of Harbin (2014RFXXJ006) and the Project of High-Tech Ships of Ministry of Industry and Information Technology of the People's Republic of China. Hao Guo is benefited from the International Exchange Program of Harbin Engineering University for Innovation-oriented Talents Cultivation.

References

1. E. Mohammad Sharifi et al., Mater. Des. **32**, 3263–3271 (2011)
2. P. Shen et al., Mater. Sci. Eng. A **454–455**, 300–309 (2007)
3. X.Y. Yue et al., Mater. Sci. Eng. A **559**, 719–724 (2013)
4. W.J. Tian et al., J. Alloy. Compd. **561**, 48–53 (2013)
5. W.-C. Lu et al., J. Alloy. Compd. **682**, 318–325 (2016)
6. X. Yue et al., Mater. Des. **46**, 285–290 (2013)
7. G. Fanchini, V. Gupta, A.B. Mann, M. Chhowalla, J. Am. Ceram. Soc. **91**, 2666–2669 (2008)
8. S. Hayun, D. Rittel, N. Frage, M.P. Dariel, Mater. Sci. Eng. A **487**, 405–409 (2008)
9. L. Jiang et al., Acta Mater. **89**, 327–343 (2015)
10. M. Khakbiz, F. Akhlaghi, J. Alloy. Compd. **479**, 334–341 (2009)
11. H.S. Chen et al., J. Alloy. Compd. **632**, 23–29 (2015)
12. C. Nie et al., J. Alloy. Compd. **454**, 118–122 (2008)
13. R.M. Mohanty et al., Mater. Sci. Eng. A **498**, 42–52 (2008)
14. C. Wu et al., Mater. Sci. Eng. A **675**, 421–430 (2016)
15. H. Yang et al., Mater. Sci. Eng. A **616**, 35–43 (2014)

16. Y.Z. Li et al., Mater. Sci. Eng. A **620**, 445–453 (2015)
17. B. Manjunatha et al., Mater. Sci. Eng. A **632**, 147–155 (2015)
18. D.M. Hulbert et al., Mater. Sci. Eng. A **493**, 251–255 (2008)
19. A. Baradeswaran et al., Mater. Des. **63**, 620–632 (2014)

Direct Conversion of Celestite to $SrCO_3$ by Wet Milling

Raşit Sezer, Ayşegül Bilen, İbrahim Göksel Hizli, Selim Ertürk and Cüneyt Arslan

Abstract Effect of the high energy ball milling on the direct conversion of celestite to strontium carbonate ($SrCO_3$) was studied. Celestite ore was ground by a planetary ball mill with sodium carbonate (Na_2CO_3) solution. Experimental study carried out to determine the effects of milling time (5, 10, 15, 30, 60, and 120 min) and Na_2CO_3 concentration (1, 1.25, 1.5, and 2 M) on the precipitation efficiency. Volume of carbonate solution was calculated for the stoichiometric amount of sulfate in celestite and solid material/ball weight ratio is selected 1/10 as constant. After filtering, washing and drying the precipitated $SrCO_3$ sludge was analyzed by X-Ray Diffractometer (XRD), Scanning Electron Microscope (SEM), and Energy Dispersive Spectroscopy (EDS). The conversion rate of celestite to $SrCO_3$ increased systematically with extending milling time. The strontium carbonate in the product can be concentrated up to 99.8% by 120 min wet milling of celestite in 2 M Na_2CO_3 solution.

Keywords Celestite · Wet milling · Direct conversion · $SrCO_3$

Introduction

Strontium carbonate ($SrCO_3$) is significantly used in the manufacture of X-ray cathode tube glass, ferrite magnets [1], luminescence materials [2], and other strontium chemicals such as strontium nitrate ($Sr(NO_3)_2$), strontium chloride ($SrCl_2$), strontium hydroxide ($Sr(OH)_2$), etc. [3, 4].

R. Sezer (✉)
Karadeniz Technical University, 34000 Trabzon, Turkey
e-mail: rasitsezer@gmail.com; rsezer@ktu.edu.tr

R. Sezer · A. Bilen · İ.G. Hizli · S. Ertürk · C. Arslan
Istanbul Technical University, 34000 Maslak, Turkey

İ.G. Hizli
Istanbul University, 34000 Maslak, Turkey

The Minerals, Metals & Materials Society, *TMS 2017 146th Annual Meeting & Exhibition Supplemental Proceedings*, The Minerals, Metals & Materials Series,
DOI 10.1007/978-3-319-51493-2_58

It is possible to obtain $SrCO_3$ from natural sources but there are only a few feasible strontianite ($SrCO_3$) ore deposits in the world. Thus, celestite ($SrSO_4$) is the main source of strontium ore [5]. There are two different industrial processes for manufacturing $SrCO_3$ from celestite ($SrSO_4$). First one is the pyro-metallurgical technic (black ash); where celestite ore is reduced to sulfide form (SrS) by carbothermic reduction at 900–1100 °C [6]. Then, soluble SrS is dissolved and filtered [7]. $SrCO_3$ can be precipitated either by sodium carbonate (Na_2CO_3), ammonium carbonate ($NH_4(CO_3)_2$), or carbon dioxide gas (CO_2) [8] from pregnant solution. Another technic for manufacturing $SrCO_3$ from celestite is called direct conversion [9]; which is an ion exchange reaction of sulfate and carbonate;

$$SrSO_4 + Na_2CO_{3(aq)} = SrCO_3 + Na_2SO_{4(aq)} \quad (1)$$

This reaction is quite slow under thermodynamic and kinetic aspects which is explained as shrinking core model in the literature [3, 4]. Some researchers suggested more effective methods for direct conversion, such as hydrothermal synthesis [10] and mechano-chemical conversion. Mechanical milling induce chemical reactions owing to mechanical energy and activate mineral by grinding and structural changes [11]. Mechano-chemical direct conversion to $SrCO_3$ tried by both wet milling and dry milling [12] with sodium carbonate (Na_2CO_3) [13] and ammonium carbonate ($NH_4(CO_3)_2$) [14].

In the present study, mechano-chemical conversion of Turkish celestite ore to $SrCO_3$ by wet milling in sodium carbonate (Na_2CO_3) media was studied. Morphologic and crystallographic formation of $SrCO_3$ powders are examined as a function of carbonate concentration and milling time.

Experimental Study

Turkish celestite ore from Barit Maden Türk Ltd. with −2 mm in size, contains 95.5% $SrSO_4$, 3% $CaSO_4$, 0.5% $BaSO_4$, 0.5% SiO_2 and 0.5% Fe_2O_3. Carbonating solutions were prepared from technical grade sodium carbonate (Na_2CO_3) as 1, 1.25, 1.5 and 2 M Na_2CO_3. Wet milling processes run 300 rpm for 5, 10, 15, 30, 60 and 120 min while 5 min milling and 2 min cooling intervals with Fritsch Pulversiette 6 planetary ball mill. Milling media is made out of tungsten carbide (WC) and balls' diameters were 10 mm. For each milling processes 25 g ore as received was milled by 250 g ball. 136 ml carbonate solution was added to milling jar to satisfy the stoichiometry of carbonate for strontium. Shortly after the milling, the sludge was centrifuged to separate solid from the liquid phase. Milled solids were washed four times with 200 ml distilled water, then filtered and dried at 105 °C for 24 h. Characterization of milled solids were done by X-Ray Diffractometer (XRD) with X'pert Pro X-Ray Diffractometer, Scanning Electron Microscope (SEM) with Zeiss Evo LS 100 and Energy Dispersive Spectroscopy (EDS) with Brunker Quantax.

Results and Discussion

Samples taken from wet milled solids were analyzed by XRD and the collected data are shown in Fig. 1. XRD patterns of the $SrSO_4$ and $SrCO_3$ are shown on Table 1.

Depending on the milling time, peak intensities of $SrCO_3$ are increasing while those of $SrSO_4$ are decreasing. Regardless of the carbonate concentration of the solutions, after 5 min milling, XRD patterns start to show $SrCO_3$ peaks such as 25.190°, 36.226° and 36.568° and also the intensities are getting higher.

Peaks at 25.906° belong to $SrSO_4$ give place to peaks 25.839° belong to $SrCO_3$ after 60 min milling in 1 M and 1.25 M carbonate solutions whereas it occurs after 30 min in 1.5 M and 2 M carbonate solutions.

Table 1 XRD patterns of $SrSO_4$ and $SrCO_3$

$SrCO_3$	2θ, degree	*25.190*	*25.839*	*29.650*	*31.543*	*34.563*	*35.137*	*36.226*	*36.568*
	h k l	1 1 1	1 0 2	2 0 0	2 0 1	2 1 0	0 2 0	2 1 1	0 1 3
$SrSO_4$	2θ, degree	**23.580**	**25.906**	**27.014**	**28.042**	**30.029**	**32.739**	**33.472**	**37.777**
	h k l	1 1 1	0 0 2	2 1 0	1 0 2	2 1 1	1 1 2	0 2 0	2 1 2

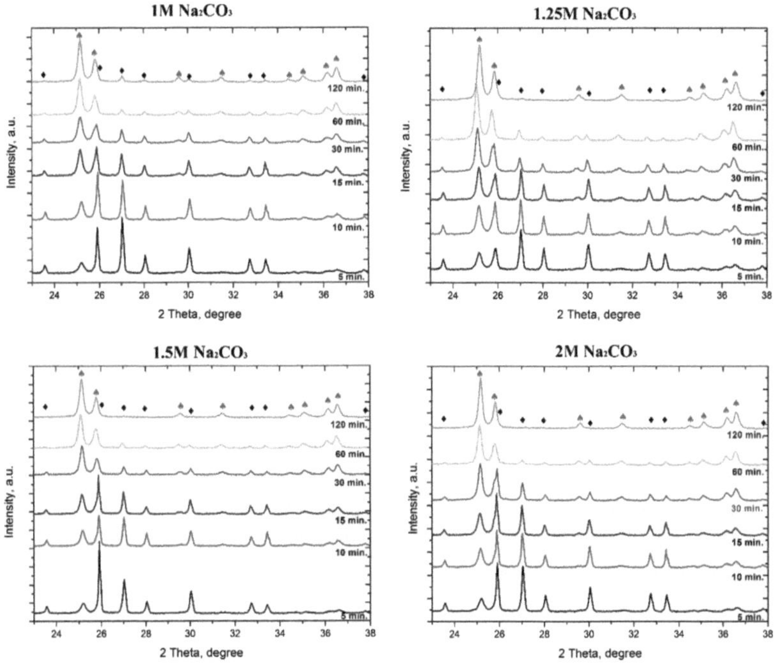

Fig. 1 XRD analyses of wet milled celestite samples; ♦ $SrSO_4$, ♠ $SrCO_3$ (Color figure online)

All the peaks belong to $SrSO_4$ appear slightly after 120 min milling in 1 M carbonate solution despite there is no $SrSO_4$ peak after 120 min milling in 1.25, 1.5 and 2 M carbonate solutions.

Selected samples of dried powders were morphologically analyzed via SEM and chemical composition of the particles with different shapes were analyzed via EDS (Fig. 2).

During the milling process celestite ore particles are ground and small $SrSO_4$ crystals converted to $SrCO_3$. SEM images show small spherical particles and slightly larger angular particles. EDS analyses indicate that spherical particles are precipitated $SrCO_3$ and the angular particles are un-reacted $SrSO_4$ (pointed out by red arrows in Fig. 2).

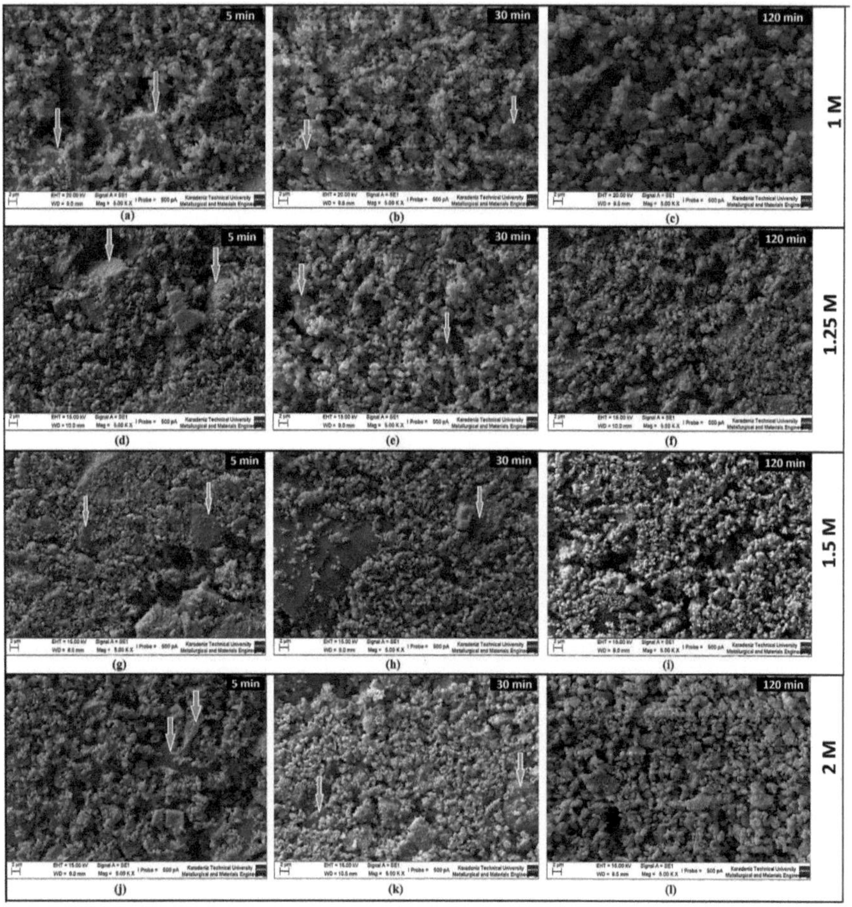

Fig. 2 SEM images of some selected dried sludge; **a–c** 1 M; **d–f** 1.25 M; **g–i** 1.5 M; **j–l** 2 M

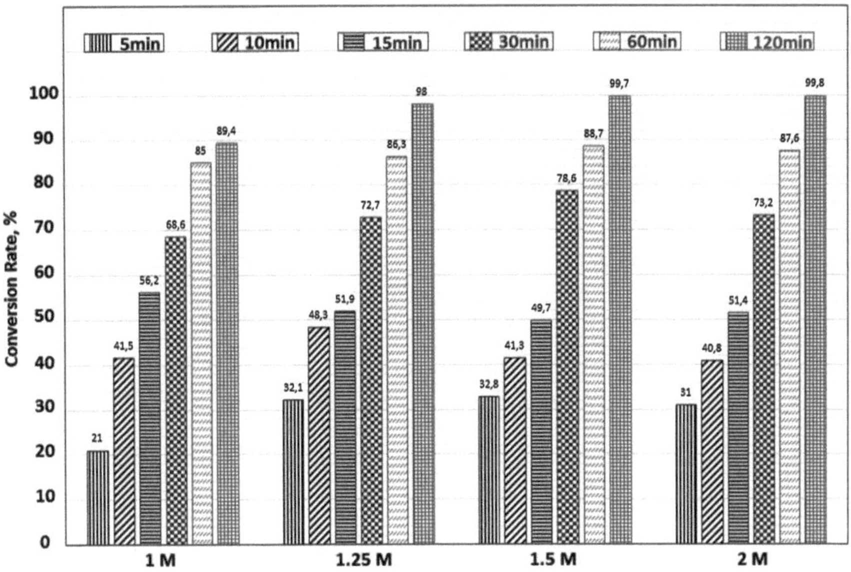

Fig. 3 Conversion rate of $SrCO_3$

By extending the milling time particle size decreases and angular particles diminish, and consequently, the rate of conversion to $SrCO_3$ increases. On the other hand, long milling periods such as 120 min cause agglomeration of the particles.

Conversion rate of the celestite to $SrCO_3$ is calculated via Rietvelt method by X'pert High Score Plus software (Fig. 3). The highest conversion rates were obtained by 120 min wet milling as 89.4%, 98%, 99.7% and 99.8% in 1 M, 1.25 M, 1.5 M, and 2 M carbonate solutions, respectively.

Conclusion

Mechano-chemical conversion to $SrCO_3$ starts at the beginning of wet milling, XRD data shows all experiments have more or less $SrCO_3$ peaks. Total conversion can be obtained in the case of excess carbonate concentration and with 120 min wet milling. Grinding celestite significantly encourage conversion to $SrCO_3$. Extending milling time causes agglomeration of precipitated powders. The highest $SrCO_3$ percentage (99.8%) was obtained by 120 min wet milling in 2 M Na_2CO_3 solution.

Acknowledgements The authors greatly acknowledge the financial support provided by Scientific and Research Council of Turkey (TUBITAK) under the contract no 115M631 and would like to thank to Barit Maden Türk Ltd. for technical and materials support.

References

1. M. Sivakumar et al., Nanophase formation of strontium hexaferrite fine powder by the sonochemical method using $Fe(CO)_5$. J. Magn. Magn. Mater. **268**(1–2), 95–104 (2004)
2. R. Zhang et al., Gel combustion synthesis and luminescence properties of nanoparticles of monoclinic $SrAl_2O_4$:Eu^{2+}, Dy^{3+}. Mater. Chem. Phys. **113**(1), 255–259 (2009)
3. A.H.E. Castillejos, F.P.B. de la Cruz del, A.S. Uribe, The direct conversion of celestite to strontium carbonate in sodium carbonate aqueous media. Hydrometallurgy **40**(1–2), 207–222 (1996)
4. F.R.P. Carrillo, A.S. Uribe, A.H.E. Castillejos, A laboratory study of the leaching of celestite in a Pachuca tank. Miner. Eng. **8**(4–5), 495–509 (1995)
5. H. Dogan, M. Koral, S. Kocakusak, Acid leaching of Turkish celestite concentrate. Hydrometallurgy **71**(3–4), 379–383 (2004)
6. R.S. Sonawane et al., Effect of a catalyst on the kinetics of reduction of celestite ($SrSO_4$) by active charcoal. Metall. Mater. Trans. B **31**(1), 35–41 (2000)
7. M. Erdemoğlu, M. Canbazoğlu, The leaching of SrS with water and the precipitation of $SrCO_3$ from leach solution by different carbonating agents. Hydrometallurgy **49**(1–2), 135–150 (1998)
8. G. Owusu, J.E. Litz, Water leaching of SrS and precipitation of $SrCO_3$ using carbon dioxide as the precipitating agent. Hydrometallurgy **57**(1), 23–29 (2000)
9. H.S. Booth, E.F. Pollard, Conversion of celestile to strontium carbonate. Ind. Eng. Chem. **40** (10), 1986–1988 (1948)
10. R. Suárez-Orduña et al., The conversion of mineral celestite to strontianite under alkaline hydrothermal conditions. J. Phys.: Condens. Matter **16**(14), S1331–S1344 (2004)
11. M. Erdemoğlu, S. Aydoğan, M. Canbazoğlu, A kinetic study on the conversion of celestite ($SrSO_4$) to $SrCO_3$ by mechanochemical processing. Hydrometallurgy **86**(1–2), 1–5 (2007)
12. N. Setoudeh, N.J. Welham, S.M. Azami, Dry mechanochemical conversion of $SrSO_4$ to $SrCO_3$. J. Alloy. Compd. **492**(1–2), 389–391 (2010)
13. A. Obut, P. Baláž, İ. Girgin, Direct mechanochemical conversion of celestite to $SrCO_3$. Miner. Eng. **19**(11), 1185–1190 (2006)
14. D. Bingöl, S. Aydoğan, S.K. Bozbaş, Production of $SrCO_3$ and $(NH_4)_2SO_4$ by the dry mechanochemical processing of celestite. J. Ind. Eng. Chem. **18**(2), 834–838 (2012)

Part XV
High Temperature Electrochemistry III

Electrochemistry in Molten LiF-BeF_2 Salt for Fluoride Salt-Cooled High Temperature Reactor Applications

W.H. Doniger, T. Chrobak, B. Kelleher, K. Dolan, G. Cao, M. Anderson and K. Sridharan

Abstract A dynamic beryllium reference electrode has been developed for in situ electrochemistry studies of molten LiF-BeF_2 (66-34 mol%) (FLiBe) salt which is being considered as the primary coolant for the fluoride salt-cooled high temperature reactor (FHR). Measurement and control of the salt redox potential can mitigate corrosion which is an important challenge in high temperature molten fluoride salts. This three electrode probe has demonstrated good reproducibility in the measurement of the salt's redox potential with an error of ±4 mV at 500 °C. The response of the probe to changes in salt chemistry was investigated by the intentional additions of fluorides of nickel, iron, and chromium, which are corrosion products of common structural materials. The equilibrium potentials of these metal impurity fluorides were measured in solutions containing approximately 250 ppm of each impurity using cyclic voltammetry. This approach is being extended to study electrochemical purification of molten FLiBe.

Keywords FLiBe · Molten fluoride salt · Redox potential · Dynamic reference electrode · Electrochemistry

Introduction

The unique passive safety features, high efficiency, and improved economics compared to current reactors make the Fluoride-salt-cooled High-temperature Reactor (FHR) a likely design for the next generation of civilian nuclear reactors. FHRs combine technologies successfully demonstrated by the Molten Salt Reactor Experiment (MSRE) at Oak Ridge National Laboratory (ORNL) in the 1960s. A reactor coolant composed of molten fluoride salt has a high volumetric heat capacity, atmospheric pressure operation, and good chemical stability under

W.H. Doniger (✉) · T. Chrobak · B. Kelleher · K. Dolan · G. Cao · M. Anderson · K. Sridharan
University of Wisconsin-Madison, Madison, WI, USA
e-mail: doniger@wisc.edu

The Minerals, Metals & Materials Society, *TMS 2017 146th Annual Meeting & Exhibition Supplemental Proceedings*, The Minerals, Metals & Materials Series,
DOI 10.1007/978-3-319-51493-2_59

irradiation. These properties make higher operating temperatures possible while passively preventing dangerous reactor transient. This will improve reactor efficiency and open the door to new industrial applications for nuclear energy [1].

A challenge to the successful development of FHR is the mitigation of corrosion of structural materials in molten FLiBe (66%LiF-34%BeF_2, mole%) which is being considered as the lead candidate primary coolant for the FHR. At the temperatures required for FHR (500–700 °C), the protective oxide layer that is relied upon for high temperature corrosion resistance in most other reactor environments is unstable in molten fluoride salts. Moreover, elements such as chromium in the alloy readily dissolve in molten fluoride salts with accelerated corrosion attack occurring along the grain boundaries [2].

It is well-known that corrosion in molten fluoride salts is driven strongly by the presence of impurities such as moisture, oxygen, and other metallic impurities [3]. The presence of impurities such as moisture for example, can lead to the formation of the highly corrosive hydrogen fluoride. Hydrogen fluoride, an oxidizing agent, can initiate the formation of fluorides of alloying elements in structural materials leading to their dissolution in the molten salt. The stability of various metal fluorides as calculated using HSC Chemistry 9 software is shown in Fig. 1. Fluorides of Ni and Mo are not very stable in FLiBe and consequently these materials are very resistant to corrosion, while Cr which forms a stable fluoride and is therefore prone to corrosion in molten fluoride salts. As shown in Fig. 1, the two salt constituents LiF and BeF_2 are highly stable. Thus, while structural materials are expected to be stable in pure molten FLiBe salt, the inevitable presence of even small amounts of impurities can initiate corrosion. Also, for this reason beryllium additions to the salt can potentially make the salt more reducing and less corrosive by preferentially reacting with the oxidizing agents. Electrochemical methods for

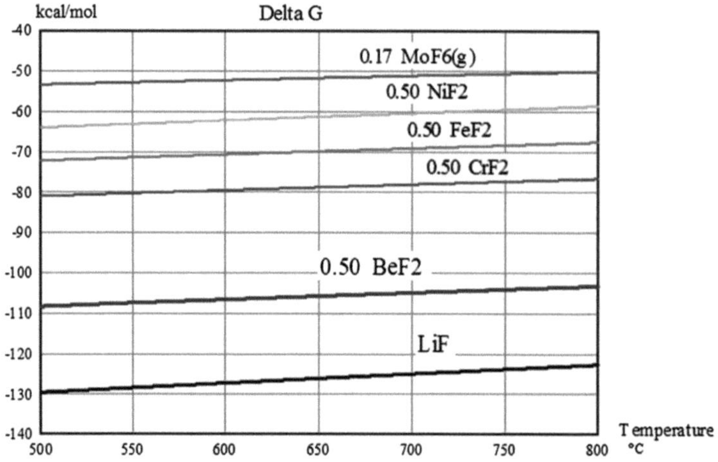

Fig. 1 Gibbs free energy of formation of fluorides of various alloying elements as well as FLiBe constituents LiF and BeF_2. Values are normalized per fluorine atom in each molecule [5]

monitoring and minimizing the presence of impurities in FLiBe are critical to corrosion mitigation and the success of the FHR [4].

The salt redox potential is a laboratory measurable voltage which serves as a metric for the tendency of all constituents of an ionic liquid to undergo an oxidation or reduction reaction. The redox potential, E, of thermodynamically reversible reactions can be related to the relative concentration of all species involved in the reaction through the Nernst Equation

$$E = E^{\circ} - \frac{RT}{nF}\ln(Q) \tag{1}$$

where E° is the standard cell potential, R is the universal gas constant, T is the temperature, n is the number of moles of electrons transferred in the cell reaction, F is the Faraday constant, and Q is the reaction quotient related to the relative concentration of all species in the reaction. The reaction quotient is the ratio of the concentration of products over the concentration of reactants of the two half-cell reactions involved. After sufficient time, the reaction quotient is equal the equilibrium constant when the reactants and products reach equilibrium [4]. The measured redox potential is a "dynamic" quantity since the system is minimally displaced from equilibrium when a potential is applied. This allows for the redox potential to be correlated with the equilibrium constant [6].

Cyclic voltammetry is emerging as a useful tool for detecting the presence of metal impurities in FLiBe. This technique could potentially be used for detecting and quantifying impurities in molten FLiBe salt and could lead to potential approaches for in situ electro-purification of the salt by plating out impurities while imposing a potential.

The goal of this work is to demonstrate the sensitivity of the dynamic reference electrode and cyclic voltammetry to the presence of nickel, iron and chromium fluorides. The ability to perform both beryllium redox potential measurements and cyclic voltammetry with a single probe would be an important step for the development of in situ diagnostics for the molten salt in FHR.

Experimental

Experimental Apparatus

All electrochemical measurements were performed inside an argon atmosphere glovebox which was kept slightly above atmospheric pressure in order to prevent the ingress of ambient atmosphere. Oxygen and moisture levels were maintained below 1 ppm and 2 ppm respectively using a HE-493 Dri-Train gas purification system from Vacuum Atmosphere Company.

Salt was heated to 500 °C by a ceramic radiant full cylindrical heater. A National Instruments (NI) LabVIEW program was used to control the heaters inside the glovebox. Salt was melted inside a GAT 13 glassy carbon crucible produced by HTW Hochtemperatur-Werkstroffe GmbH. The glassy carbon crucible was placed inside a stainless steel crucible which acted as both a grounded Faraday cage and as a back-up for salt containment in the event of a failure of the glassy carbon crucible. Both the glassy carbon and stainless steel were nested inside a larger alumina crucible which provides electrical insulation between the inner crucibles and the heater coils.

Materials

A FLiBe salt purification system was constructed at the University of Wisconsin–Madison to accommodate the large volume of high quality FLiBe salt needed for corrosion, electrochemistry, and future flow loop experiments [4]. This system can synthesize approximately 40 kg of FLiBe from raw materials. Lithium fluoride (99.8% purity) was obtained from Noah Technologies and beryllium fluoride (98.8% purity) was purchased from Materion. A melt of the correct ratio of LiF and BeF2 was then purified at 600 °C. Hydrofluorination was conducted for 72 h. A pure hydrogen sparge was conducted at various rates for approximately 87 h. Salt was transferred into the glovebox using a mobile heater with a nickel container. Chemical analysis of the salt was conducted by the Wisconsin State Laboratory of Hygiene using inductively coupled plasma optical emission spectroscopy (ICP-OES) in 2016. ICP-OES confirmed that the purified FLiBe produced by the University of Wisconsin had a composition of 65.33 mol% LiF–34.53 mol% BeF_2–0.14 mol% impurities.

Electrochemical Probe Design

The UW redox probe is a dynamic reference electrode probe designed to withstand prolonged exposure to the corrosive high temperature environment. Figure 2 shows the UW redox probe along with a cutaway drawing. Based on Afonichkin et al's diaphragm-free three electrode probe, it features a glassy carbon counter electrode (CE) (1), molybdenum working electrode (WE) and a molybdenum quasi-reference electrode (QRE) (2) [6]. Swageloks with Teflon ferrules (3) both allow for the height of the electrodes to be adjusted and provide electrical isolation between electrodes. A KF connection (4) allows the probe to be easily disassembled for maintenance. A Viton O-Ring (5) provides a gas seal for the probe's tube stub. A boron nitride spacer (6) keeps electrodes aligned vertically and electrically isolated [5]. The QRE is in contact with the salt and takes a stationary (quasi-equilibrium) potential close to that of the salt. Potential drift of the electrode due to interaction with the electrolyte is minimized by prohibiting current from passing through the electrode [6].

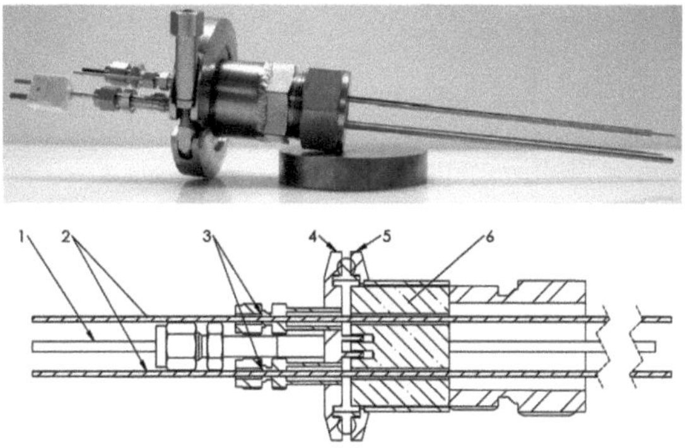

Fig. 2 UW redox probe used for redox potential measurement and cyclic voltammetry [5]

Measurement Procedures

The redox potential measurement and cyclic voltammetry were the two electrochemical techniques used in the present study. Both techniques were performed using the UW redox probe discussed earlier. Electrochemical measurements where conducted with two power supplies. For measurement of the redox potential using the dynamic beryllium reference electrode, an HP 3616 A power supply was used in constant voltage mode. LabVIEW was used to control the HP 3616 A and redox potential data acquisition via a National Instruments cRIO 9074 chassis. Cyclic voltammetry was conducted using a Solartron 1287 A potentiostat featuring: a current range of 2 A to 200 nA, a current resolution of 100 pA, a current accuracy of 0.2% ± 0.1% of the current range, and a bandwidth of 8 Hz.

Redox Potential Measurement

The redox potential measurements consisted of two steps. In step one, a potential is applied between the WE and the CE to make a pulse of current pass between them. The potential of the WE must be negative enough with respect to the potential of the CE to reduce beryllium on the WE. The thin layer formed on the WE is a beryllium deposit which is reversible to Be^{2+} ions in the electrolyte. The thin Be layer forms the dynamic reference electrode. A shift of dynamic reference electrode potential during experiment was not expected because there is no sufficient change of the electrodeposited beryllium activity in the surface layer [6]. In step two, the power supply is cutoff and an open circuit voltage is measured between WE and QRE. The beryllium as well as all other impurity metals reduced on to the WE are allowed to oxidize and re-dissolve back into solution. Beryllium is reduced at a less negative potential than lithium; therefore, it is used as the reference reaction at the dynamic reference electrode. The recombination of these metals with displaced fluorides in solution creates a distinct voltage that can be measured with respect to the QRE using an open circuit voltage. This phenomenon results in a potential plateau that lasts several seconds. Taking the average of this plateau voltages yields the redox potential measurement [5, 6].

Cyclic Voltammetry

Cyclic voltammetry with linear potential sweep was performed on FLiBe salt samples of various purity using the molybdenum QRE. Due to the Solartron potentiostat's relatively low bandwidth, a scan rate of 80 mV/s was chosen because it yielded both reproducible, quasi-reversible waveform behavior and good data resolution.

Results and Discussion

Salt Redox Potential Measurement with the Dynamic Reference Electrode

The salt redox potential measurement with respect to the dynamic reference electrode is designed to serve as one metric by which all FLiBe salts used in corrosion, electrochemistry and flow loop experiments can be compared. The advantages of this method include the durability, reproducibility, and the simplicity of the experimental setup [5].

Electrochemical Validation

In order to facilitate better understanding of corrosion behavior of materials as a function of redox potential, efforts have been made to develop parameters and standard practices for the salt redox potential measurement. The plating time in the initial step was kept constant at 1 s. Care was taken to select a plating voltage that adequately reduces beryllium ions while maintaining the dynamic nature of the experiment. Too small a voltage difference with respect to the QRE can fail to adequately reduce beryllium ions. Too high a voltage difference can reduce lithium onto the WE. In general, plating voltages between 1 and 4 V were used. Several factors can affect the plating voltage required for each experiment. Surface condition of the molybdenum electrodes, for example can have a significant effect. The most significant effect of the plating voltage is the length of time required for dissolution of plated metals back into solution. Often, too large a plating voltage can result in the deposition of large quantities of impurity metals, such as nickel on the electrode which take significant time to re-dissolve [5]. Figure 3 shows typical open circuit voltages measurements between the dynamic reference electrode and QRE in step two of the salt redox potential measurement. The voltage plateau

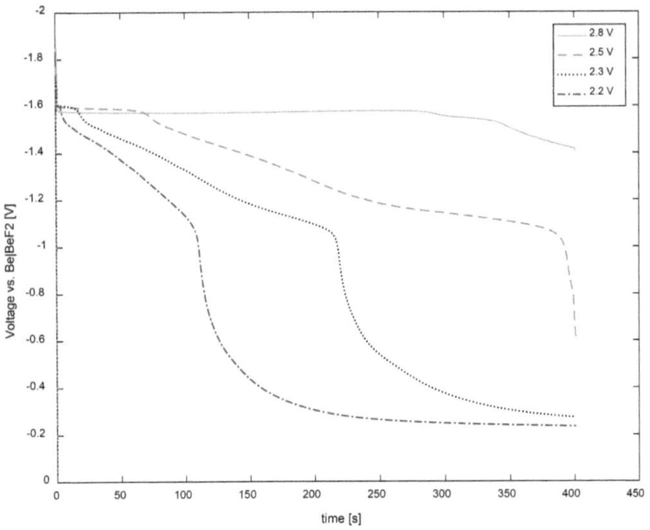

Fig. 3 Typical redox potential measurements using dynamic Be/BeF$_2$ reference electrode of pure FLiBe using varying plating voltages at 500 °C

becomes more distinct with increasing plating voltage. The dynamic reference electrode is capable of reproducible distinct plateau voltages with an error of ±4 mV [5]. This means that the dynamic reference electrode potential does not shift greatly making it a suitable reference voltage with which to compare the quasi-equilibrium potential of the QRE.

Chemical Redox Control

One method of chemically controlling the redox behavior of is by the addition of beryllium metal in the salt, which acts as a reducing agent. When a salt is oxidizing and the concentration of impurities in the salt is high the plateau voltage moves towards −1.7 V. As the salt becomes effectively reduced the redox potential becomes less negative [5].

Metal Fluorides Effects on Redox Potential of FLiBe

The characterization of nickel fluoride's effect on the salt redox potential is particularly important for gauging the performance of nickel based alloys, such as Hastelloy-N, in FLiBe. In two experiments, small additions of nickel (II) fluoride were made and allowed to dissolve overnight at 600 °C. A clear oxidation of both salts was observed when NiF$_2$ was added in both salts with an initial high and low

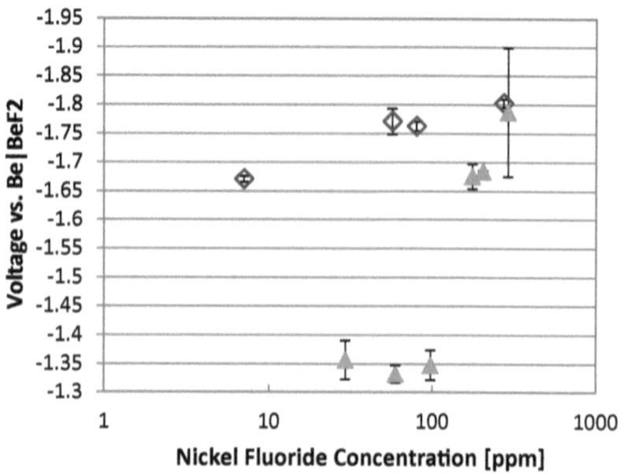

Fig. 4 The redox potential versus Be/BeF_2 of two FLiBe salt samples as a function of added nickel (II) fluoride concentration

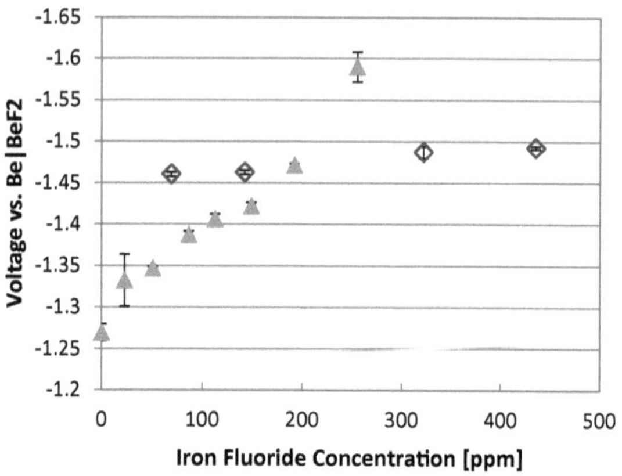

Fig. 5 The beryllium redox potential of two FLiBe salt samples as a function of added iron (II) fluoride concentration

redox potential. Figure 4 shows that the effect of NiF_2 on the redox potential versus Be/BeF_2 reference electrode at 500 °C is linear with respect to the log of added NiF_2 concentration, in good agreement with the Nernst equation.

A remarkable difference in the change in redox potential with increasing NiF_2 concentration is observed between the two salts. The initially oxidizing FLiBe had a much lower (more negative) redox potential than the beryllium reduced FLiBe sample. In the beryllium reduced FLiBe, it was also apparent that the response of the salt redox potential is initially delayed. It is likely that the delay is caused by excess of beryllium metal in the melt which could have a residual reducing effect as NiF_2 was added. The change in slope is possibly a result of differing solubility of NiF_2 in the two experiments. NiF_2 may be less soluble or take longer than 24 h to dissolve at high concentrations in FLiBe.

The salt redox potential behavior of iron fluoride is of particularly important to understanding how iron based alloys, such as stainless steel, perform in FLiBe. Figure 5 shows a plot of the redox potential with respect to added FeF_2 concentration. A clear redox potential change was observed in both salts which initially showed very reducing conditions. Iron's redox behavior with respect to the log of concentration is not linear suggesting that other impurities may be influencing the measurement and preventing it from behaving Nernstian.

Cyclic Voltammetry

Cyclic voltammetry using the identical redox probe is a useful tool for detecting and even quantifying metallic impurities in the molten FLiBe salt. The electrochemical reversibility, or ability of a charge transfer process to behave Nernstian, can be quantified. Analysis of waveforms measured in salts of varying composition showed reproducible, quasi-reversible behavior. This means that the charge transfer process happens on a short enough time scale that the system is always close to equilibrium. The equilibrium potentials of iron and chromium were observed using the salts from the previous redox potentials measurements.

Electrochemical Validation

Linear potential sweep cyclic voltammetry was performed on several batches of FLiBe including a beryllium reduced sample and samples oxidized with chromium and iron. The waveform behavior of several impurity metals satisfied the criteria of quasi-reversibly. They do not fit the description of reversible because the anodic peaks are larger than the cathodic peaks ($i_{pa}/i_{pc} \neq 1$). Figure 6a shows the

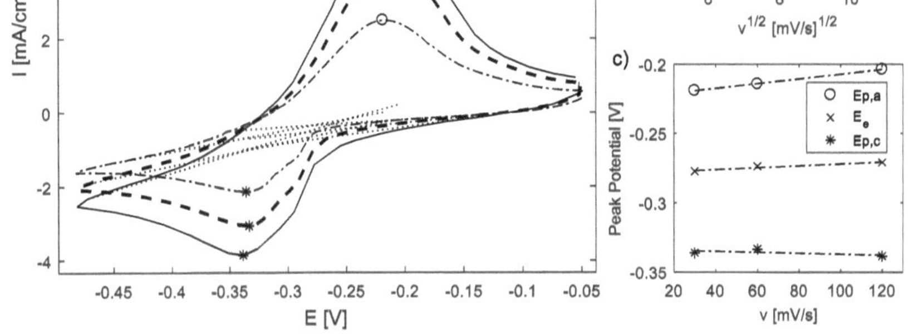

Fig. 6 Cyclic voltammetry of FLiBe salt oxidized with 250 ppm of chromium (II) fluoride at 500 °C. **a** Waveforms and potential-time profile of scans taken at 30, 60 and 120 mV/s. **b** Peak current versus the square root of scan rate, *v*. **c** Peak Potential versus scan rate

reduction of chromium in a salt oxidized with approximately 250 ppm of chromium (II) fluoride. The magnitude of peak anodic current, $I_{p,a}$, and peak cathodic current, $I_{p,c}$, increases linearly with the square root of scan rate, *v*, in Fig. 6b. This is an indication that the reaction follows the Randels-Sevcik equation,

$$i_p = 0.4463nFAC\left(\frac{nFvD}{RT}\right)^{1/2} \tag{2}$$

for the peak current, i_p, of a diffusion rate limited process of an electroactive species. Here *n* is the number of electrons transferred in the redox event, *A* is the electrode surface area, *F* is Faraday's constant, *D* is the diffusion coefficient, *C* is the concentration, *v* is the scan rate, *R* is the gas constant and *T* is temperature. The equilibrium potential, E_e, is the average of the anodic peak potential, $E_{p,a}$, and the cathodic peak potential, $E_{p,c}$. In Fig. 6c, E_e remains relatively constant between scan rates of 30 and 120 mV/s. The standard deviation of E_e is ±2.6 mV. The quasi-reversibility of impurities in FLiBe makes cyclic voltammetry a useful nondestructive tool for studying FLiBe salt chemistry.

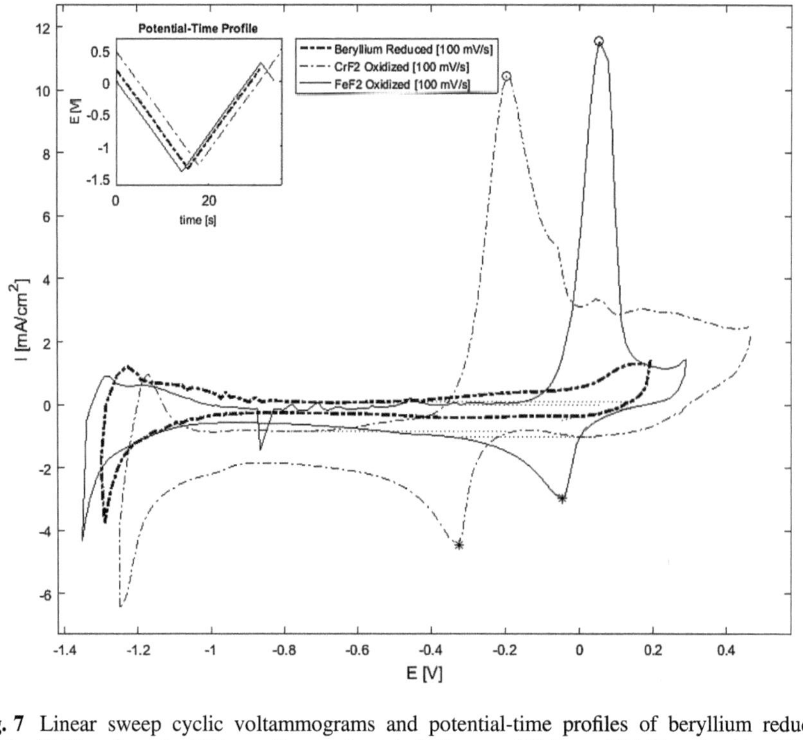

Fig. 7 Linear sweep cyclic voltammograms and potential-time profiles of beryllium reduced FLiBe, FLiBe salt with additions of approximately 250 ppm of CrF_2, and FLiBe salt with addition of approximately 250 ppm of FeF_2. [Scan Rate 100 mV/s]

FLiBe Salt Impurity Equilibrium Potentials

The waveform behavior and equilibrium reduction potentials of chromium and iron impurities have been quantified using cyclic voltammetry. In Fig. 7, FLiBe salts with additions of with 250 ppm of each impurity are compared to a pure FLiBe sample that has been reduced with beryllium. The reduction and oxidation of chromium and iron are clearly distinguishable. The equilibrium potentials, E_e, of chromium and iron are calculated by taking the average of the corresponding anodic and cathodic peak potentials. The E_e of chromium and iron with respect to the QRE are −0.2605 V and 0.0027 V, respectively. Tracking the behavior of these current peaks overtime could lead to a greater understanding of the rate at which different impurities affect the corrosivity of FLiBe salt.

Conclusion

The redox probe discussed in this work is a versatile electrochemical cell for the investigation of the relative corrosivity of molten fluoride salts at temperatures above 500 °C. The salt redox potential measurement and cyclic voltammetry are useful techniques for salt characterization which can be conducted with the probe. The in situ technique allows better understanding of corrosion experiments and determination of when chemical reduction of the salt is necessary. The robustness of the method has been demonstrated by measuring salt redox potential and the oxidation and reduction potentials of the impurity fluorides when Ni, Fe, and Cr fluorides were added to the salt.

Ongoing work is focused on the detection of chromium using the salt redox potential and development of procedures for accurate measurement of the redox current response of impurities using cyclic voltammetry. The current response measurement could be a method for tracking the prevalence of impurities.

References

1. H.G. MacPherson, Development of materials and systems for the molten-salt Reactor concept. React. Technol. **15**(2), 136–155 (1972)
2. G. Zheng, B. Kelleher, G. Cao, M. Anderson, T. Allen, K. Sridharan, Corrosion of 316 stainless steel in high temperature molten Li_2BeF_4 (FLiBe) salt. J. Nucl. Mater. **461**, 143–150 (2015)
3. J.W. Koger, *Effect of FeF_2 addition on mass transfer in a Hastelloy N: LiF–BeF_2–UF_4 thermal convection loop system—ORNL-TM-4188* (Oak Ridge National Lab, Tenn, USA, 1972)
4. B.C. Kelleher, Purification and chemical control of molten Li_2BeF_4 for a fluoride salt cooled reactor. Ph.D., The University of Wisconsin - Madison, United States – Wisconsin (2015)

5. K. Dolan, Redox potential measurement and control for the fluoride-salt-cooled high-temperature reactor. MS thesis, Nuclear Engineering and Engineering Physics, University of Wisconsin - Madison (2015)
6. V.K. Afonichkin, A.L. Bovet, V.V. Ignatiev, A.V. Panov, V.G. Subbotin, A.I. Surenkov, A.D. Toropov, A.L. Zherebtsov, Dynamic reference electrode for investigation of fluoride melts containing beryllium difluoride. J. Fluor. Chem. **130**(1), 83–88 (2009)

Impurity Removal from Titanium Oxycarbide

Farzin Fatollahi-Fard and Petrus Christiaan Pistorius

Abstract A process to remove impurities from titanium oxycarbide–the feedstock used to make titanium via the MER electrochemical process–from impure oxycarbide produced from low-grade feedstocks such as ilmenite is described. Titanium oxycarbide produced by carbothermal reduction from a natural ilmenite concentrate will contain many metallic impurities, such as iron, manganese, and silicon. If these impurities are not removed, then these metallic impurities will contaminate the titanium product. The MER process can be modified into a two-step process to first remove the contaminating metallic impurities before titanium production begins. A simplified system has been used to test this iron removal method, and the titanium oxycarbide produced by this first electrochemical step has been characterized to determine its impurity content.

Keywords MER process · Titanium · Ilmenite

Introduction

In any process which produces titanium, a relatively pure feedstock material is required. In the Kroll process, the feedstock is titanium(IV) chloride, which itself is often made from lower grade ore materials that have been "upgraded" to remove enough of their impurities so the ore can be used for chlorination [5]. In electrochemical processes, such as the FFC Cambridge process, the feedstock needs to be pure TiO_2, otherwise the impurities will end up in the final titanium product [2]. As such, there are currently no processes for making titanium metal which can tolerate a large amount of impurities in its feedstock material.

F. Fatollahi-Fard (✉) · P.C. Pistorius
Carnegie Mellon University, 5000 Forbes Avenue, Pittsburgh, PA 15213, USA
e-mail: ffatolla@andrew.cmu.edu

The Minerals, Metals & Materials Society, *TMS 2017 146th Annual Meeting & Exhibition Supplemental Proceedings*, The Minerals, Metals & Materials Series,
DOI 10.1007/978-3-319-51493-2_60

The MER process is a new electrochemical process for making titanium developed by Materials and Electrochemical Research (MER) Corporation [7]. In the MER process, a source of TiO_2 is reduced with carbon to produce titanium oxycarbide–a solid solution of titanium(II) oxide (TiO) and titanium carbide (TiC), ideally in a 1:1 ratio (Ti_2OC)–which is used as a feedstock in a molten salt electrowinning process. Previous work has shown that titanium oxycarbide can be made from a range of natural titanium materials such as leucoxene and pseudorutile mineral sands [3, 4].

However, the problem with using natural titanium minerals is that they contain a number of impurities. Some of these impurities are inert–such as Al_2O_3, MgO, ZrO_2, and zircon crystals ($ZrSiO_4$)–and will not participate in the electrowinning process as they are both electrically insulating and immiscible in the molten salt used (a eutectic NaCl-KCl mixture). However other impurities such as iron, silicon, manganese, chromium, and vanadium oxides–all present in some level in most natural titanium ores–are reduced under the same carbothermal reduction conditions used to make titanium oxycarbide. In the case of silicon, manganese and chromium, these metals all dissolve in liquid iron, and usually end up in solution with iron (iron is the main impurity element in most titanium ores). Vanadium oxide can reduce to form vanadium oxycarbide, which may end up in solid solution with titanium oxycarbide.

It is important to know whether any of these reduced impurities will plate out during electrowinning. The potentials of these reactions are presented in Table 1. Reactions more negative than the oxycarbide dissolution anode reaction will cause the metal to be put into solution with the salt, while reactions less negative than the titanium plating cathode reaction will allow the metal ions to plate out of the salt.

Table 1 Half-cell potentials for possible reactions in the MER process (1073 K)

Half-cell reaction[a]	Potential Cl_2 ref., V_{Cl_2} (V)	Potential Reference, V_{ref} (V)
$Cl_2 + 2e^- \rightarrow 2Cl^-$	0	+1.06
$2TiCl_2 + CO + 4e^- \rightarrow Ti_2OC + 4Cl^-$	−0.63	+0.43
$2VCl_2 + CO + 4e^- \rightarrow V_2OC + 4Cl^-$	−0.91[b]	+0.15[b]
$AgCl + e^- \rightarrow Ag + Cl^-$	−1.06[c]	0[c]
$FeCl_2 + 2e^- \rightarrow Fe + 2Cl^-$	−1.11	−0.05
$SiCl_4 + 4e^- \rightarrow Si + 4Cl^-$	−1.35	−0.29
$CrCl_2 + 2e^- \rightarrow Cr + 2Cl^-$	−1.36	−0.30
$VCl_2 + 2e^- \rightarrow V + 2Cl^-$	−1.52	−0.46
$MnCl_2 + 2e^- \rightarrow Mn + 2Cl^-$	−1.79	−0.73
$TiCl_2 + 2e^- \rightarrow Ti + 2Cl^-$	−1.79	−0.73
$NaCl + e^- \rightarrow Na + Cl^-$	−3.39	−2.33
$KCl + e^- \rightarrow K + Cl^-$	−3.53	−2.47

[a] Activity of metal/chorides assumed to be 1
[b] Ideal solution assumed for V_2OC
[c] Activity of AgCl = 0.04

The calculations show that all the reducible impurities will be put into solution and plated out during the titanium electrowinning process, contaminating the titanium cathode product.

However, careful control of the anode potential should allow for the dissolution and plating of all these reducible impurities without starting the titanium oxycarbide dissolution (anode) reaction [6]. By this method, the reducible impurities can be removed from the titanium oxycarbide without the need for any sort of chemical leaching process, which often affects the titanium oxycarbide itself [4]. By this method, an oxycarbide made from a low grade ore such as ilmenite can be turned into commercial purity titanium without the need for any additional equipment or chemicals beyond that which is needed for the MER process.

Methods

Titanium oxycarbide was produced by mixing TiO_2 pigment (Alfa, 99.8+% purity, 2 μ m) with carbon black (Fisher Chemical) in a 1:2.2 molar ratio in a high-purity alumina crucible (Coorstek, 99.8%). This mixture was then taken to 1600 °C in a vertical tube furnace under a continuously flowing argon atmosphere (99.995% purity) with a heating rate of 5 °C/min and held at temperature for 2 h.

Iron powder was prepared by using a source of iron powder (Fisher Scientific, 20 mesh) and reducing it under a flowing 10% H_2-90% Ar atmosphere at 800 °C for 1 h, to reduce any oxides that may be present.

The anode was prepared by mixing 900 mg of titanium oxycarbide powder with 100 mg of reduced iron powder and placing it into a graphite anode basket. This graphite anode basket was then suspended from a stainless steel rod using Type-B thermocouple wire (70% Pt-30% Rh). The cathode was a 1.5 cm wide strip of 1016 steel affixed to a stainless steel rod (1 cm of the cathode was immersed into the salt). The reference electrode consisted of a silver wire placed in a fused quartz tube (4 mm ID) with a porous alumina plug on one end. The plugged end was placed into the salt to a depth of 2 cm. Before the start of the experiment, silver was dissolved anodically in the salt to develop a local concentration of 4 mol percent Ag^{+}.

225 grams of electrolyte was prepared using sodium chloride (Fisher Scientific) and potassium chloride (Fisher Scientific). The composition of the salt mixture was eutectic (50.6 molar% NaCl, balance KCl) and put into a 1.75 in. diameter high-purity alumina crucible (Coorstek, 99.8%). 50 mg of ferrous chloride tetrahydrate ($FeCl_2 \cdot 4H_2O$, Fisher Scientific) was also mixed into the salt for the purpose of pre-doping the electrolyte with Fe^{2+} ions. This mixture was put into a stainless steel reactor and sealed. The mixture was heated to 400 °C and put under low vacuum (15–30 kilopascals) for the purpose of dehydrating the salt. The mixture was then heated to 850 °C.

After the salt was fully molten, the anode, cathode, and reference electrodes were inserted into the melt. The reference electrode solution was generated galvanostatically–using a Gamry Reference 3000 potentiostat–by running +50 mA for 220 s with the silver wire as the working electrode and the graphite anode basket as the counter and reference electrode. Potentiodynamic scans were then run on the anode and cathode at a rate of 2 millivolts/second for characterization purposes. Upon completion of characterization, the anode basket was set as the working electrode with the iron cathode as the counter electrode, and a potentiostatic potential of +0.2 volts (relative to reference) applied to the working electrode. This potential is above the iron dissolution potential, but below the oxycarbide dissolution potential. The full-cell potential was also measured using the auxiliary input channel on the Reference 3000.

Upon the completion of the experiment, all electrodes were removed from the salt and the furnace allowed to cool. After the electrodes were fully cooled, the anode basket was removed and placed upside down in a beaker of 80 °C water for 4 h. The contents of the beaker was then filtered through a cellulose thimble. This thimble was then put into a Soxhlet extractor with water as the working fluid for 12–18 h. The sample was then removed and dried at 80 °C. The cathode product was recovered from the salt using a similar procedure.

Different methods of characterization were used at different steps of this procedure. X-ray diffraction (XRD) was performed using a PANalytical X'Pert Pro x-ray diffractometer. Any unknown phases from XRD were identified using QualX2 software [1]. Overall chemical composition was characterized by x-ray fluorescence spectrometry (XRF) using a PANalytical MiniPal 4. Imaging was performed using an FEI/ASPEX Explorer or Phillips XL30 scanning electron microscope (SEM) with energy-dispersive spectroscopy (EDS).

Results

Using the procedures outlined in Sect. Methods, we produced titanium oxycarbide of relatively high purity as shown by the XRD pattern in Fig. 1, which shows the material is mostly titanium oxycarbide, with some amount of titanium(III) oxide (Ti_2O_3) impurity.

At the beginning of the potentiostatic (iron removal and plating) phase, the current was measured to be very high–over 100 mA–as shown in Fig. 2. The current dropped precipitously for the first few hours, until it approached a low of around 20 mA. The current then remained at this level for the rest of the experiment.

The x-ray diffraction pattern of the anode product is also shown in Fig. 1. The pattern here shows the titanium oxycarbide material as unchanged. However, the Ti_2O_3 which was present in the starting material is no longer present, but new unidentified peaks are present. QualX2 shows that these peaks are due to graphite and Ti_6O_{11}.

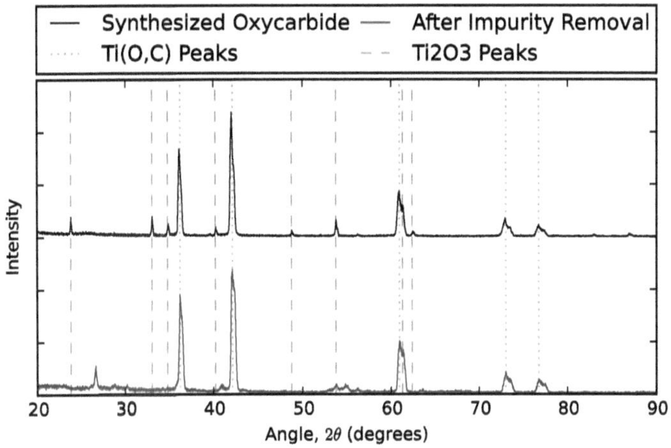

Fig. 1 X-ray diffraction pattern of synthesized titanium oxycarbide compared to pure compounds. Upper pattern is synthesized oxycarbide, and lower pattern is after impurity removal

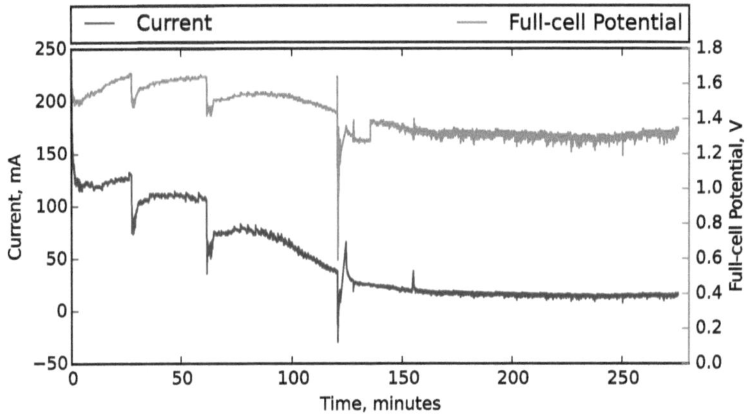

Fig. 2 Potentiostat current and full-cell potential during iron removal (anode potential = +0.2 V_{ref}). Upper curve is potential. Lower curve is current

The anode product after this process is shown in the SEM image in Fig. 3B. Compared to the original oxycarbide powder–shown in Fig. 3A, the particle size and morphology are unchanged. This powder was also subjected to XRF analysis to determine the iron and titanium content. These results are shown in Table 2. The cathode product was also analyzed by SEM and EDS to determine the titanium content. No titanium was detected in the cathode product.

Fig. 3 Secondary SEM images of **A** the titanium oxycarbide starting material, and **B** the remaining titanium oxycarbide material after electrochemical purification

Table 2 Iron and titanium content of the anode product determined by XRF

Element	Before electrolysis[a]	After electrolysis
Titanium	87.4	97.6
Iron	12.6	2.4

Composition in weight percent
[a]Composition calculated by input mass

Discussion

The current passed through the anode as shown in Fig. 2, seems to indicate that the anode current efficiency is very low. Based on the amount of iron present in the anode (100 mg), we can calculate the number of coulombs necessary to put all of the iron into solution with the salt as ferrous chloride ($FeCl_2$). Assuming 100% current efficiency, this should be 344.6 coulombs. Based on the XRF results in Table 2, we know that not all of the iron was put into solution (only 83% was removed). So with 100% current efficiency, the amount of charge used to put the iron into solution was

286 coulombs. Assuming this charge was supplied uniformly over time (constant current), based on the duration of the experiment (6892 s), the actual current that was putting iron into solution was 17.5 mA. This is approximately the same value of the current for the last 2 h of the experiment.

This indicates that there was another reaction (or reactions) taking place in parallel with the iron dissolution reaction. The exact nature of this reaction is not known, though water contamination may be responsible. Despite the vacuum degassing operation before the experiment, condensation was still observed in the exhaust tube, indicating that water was still present in the reactor. Over time, this would be removed, to the point where only the iron reaction is observed.

The full-cell potential over the course of the reaction–seen in Fig. 2–can be used to determine the cathode potential. This is important because we need to be sure that the cathode potential does become negative enough to decompose the electrolyte, as the cathode potential is variable and controlled by the potentiostat automatically. Based on the data, the most negative value of the cathode potential was $-1.46\ V_{ref}$, well above the sodium and potassium plating potentials of -2.33 and $-2.47\ V_{ref}$ respectively. The very negative values of the cathode potential are not surprising, as the cathode reaction is highly diffusion limited, so a large overpotential is required.

It also stands to reason that had the reaction been allowed to continue (for another hour), all of the iron would have been put into solution at the anode, leaving no metallic iron present in the oxycarbide mixture. At this point, we expect the current to have decayed to nearly zero, as there would be no additional reactions possible at the applied anode potential. Had this been allowed to happen, the anode material would only consist of titanium oxycarbide and other inert "anode sludge" compounds. This purified material could be used in making commercially pure titanium powder.

The XRD pattern (Fig. 1) and SEM images (Fig. 3) of the anode product shows the oxycarbide material was unaffected by the electrochemical process. However, in the XRD pattern, the Ti_2O_3 has been oxidized to Ti_6O_{11}. Reoxidation is not likely to occur during the electrochemical process, but is a distinct possibility during the washing and drying process–which take place at 80–100 °C while exposed to air. Therefore, this is not expected to be an issue in actual titanium production, as the oxycarbide will not be washed between the iron removal and titanium electrowinning steps. The presence of graphite is probably due to contamination from the graphite anode basket itself.

While this experiment was done using pure compounds, there is no technical reason why it could not be performed with a titanium oxycarbide material made from a natural ore material. These oxycarbide materials are expected to have reducible impurities of mainly iron, but also containing silicon, manganese and chromium. As seen in Table 1, the dissolution potentials for all of these materials is below the applied anode potential of $+0.2\ V_{ref}$. Additionally, if we wish to not remove the vanadium impurities–to make a titanium-vanadium alloy, the anode potential can be lowered to below the calculated dissolution potential (e.g. $+0.1\ V_{ref}$) of vanadium oxycarbide, leaving it in the anode.

Conclusions

It is possible to remove most–if not all–of the iron from a titanium oxycarbide-iron mixture using the same electrochemical reactor used in the MER process. This is done by careful control of the anode potential during electrolysis. With this method, any of the common metallic impurities in natural titanium oxycarbide can be removed, allowing for the production of commercially pure titanium metal powders from low-grade ores such as ilmenite, without the need for any additional chemicals or processing equipment.

References

1. A. Altomare, N. Corriero, C. Cuocci, A. Falcicchio, A. Moliterni, R. Rizzi, QUALX2.0: a qualitative phase analysis software using the freely available database POW_COD. J. Appl. Crystallogr. **48**(2), 598–603 (2015). doi:10.1107/S1600576715002319. http://scripts.iucr.org/cgi-bin/paper?S1600576715002319
2. L.L. Benson, I. Mellor, M. Jackson, Direct reduction of synthetic rutile using the FFC process to produce low-cost novel titanium alloys. J. Mater. Sci. **51**(9), 4250–4261 (2016). doi:10.1007/s10853-015-9718-1. http://link.springer.com/10.1007/s10853-015-9718-1
3. F. Fatollahi-Fard, P.C. Pistorius, Production of titanium oxycarbide from titania-rich mineral sands. In: EPD Congress (TMS 2015) (2015), pp. 297–304. doi:10.1002/9781119093503.ch34
4. F. Fatollahi-Fard, P.C. Pistorius, Production of titanium oxycarbide from iron-rich titanium ores. In: Proceedings of the 13th World Conference on Titanium. John Wiley & Sons, Inc., Hoboken, NJ, USA (2016), pp. 65–69. doi:10.1002/9781119296126.ch8. http://doi.wiley.com/10.1002/9781119296126.ch8
5. H. Sibum, V. Güther, O. Roidl, F. Habashi, Titanium, titanium alloys, and titanium compounds. Ullmann's Encyclopedia of Industrial Chemistry **37**, 51–82 (2012). doi:10.1002/14356007.a27_095
6. J.C. Withers, Personal Communication (2016)
7. J.C. Withers, R.O. Loutfy, J.P. Laughlin, Electrolytic process to produce titanium from TiO2 feed. Mater. Technol. **22**(2), 66–70 (2007). doi:10.1179/175355507X214078

New Generation Molten Oxide Energy Materials R&D

Valery V. Belousov

Abstract Energy technologies and their supporting materials are crucial in the context of the global challenges facing our present civilization. A great many opportunities remain, particularly for energy vector in relation to the molten oxides. Thanks to their unique electrochemical properties, molten oxide innovations offer significant potential benefits for improving energy efficiency. Advances in the field of solid and liquid state ionics for energy conversion are necessary to address the challenge of attaining cleaner and sustainable sources of energy. In this framework high temperature electrochemical devices based on oxide materials play a crucial role in decarbonising the future energy mix. Current targets of cost and durability necessitate solid oxide fuel cells (SOFCs) to operate in the intermediate temperature range. To achieve these targets, the oxygen ion conductivity of the SOFC ceramic oxide based electrolyte needs to be enhanced. Innovative molten oxide based electrolytes, with highest oxygen ionic conductivity at intermediate temperatures, show some real promise as the alternative materials. We summarize the recent progresses that have been made in the research and development of innovative alternative materials for the fabrication of novel electrochemical devices, such as molten oxide fuel cells (MOFCs) for electric power generation and molten oxide membranes (MOMs) for oxygen separation from air.

Keywords Energy materials · MOM · MOFC · Oxygen transport · Kinetics · Modeling

Introduction

In the first half of the 19th century, Michael Faraday [1] reported: "*Many bodies are decomposed directly by the electric current, their elements being set free: these I propose to call electrolytes*". He confirmed that molten oxides were among those

V.V. Belousov (✉)
A.A. Baikov Institute of Metallurgy and Materials Science,
Russian Academy of Sciences, 49 Leninskii Pr., 119334 Moscow, Russia
e-mail: vbelousov@imet.ac.ru

The Minerals, Metals & Materials Society, *TMS 2017 146th Annual Meeting & Exhibition Supplemental Proceedings*, The Minerals, Metals & Materials Series,
DOI 10.1007/978-3-319-51493-2_61

bodies. Martin and Derge [2] published a review of early 20th century successes with molten oxide electrolytes. This review describes ideas put forth as early as 1906 when Aiken was granted a patent in the United States for the production of iron by molten oxide electrolysis. The second half of the 20th century saw some important developments related to molten oxides in the former USSR where scientists were interested in the ionic nature of oxide melts as well as their potential use for metal production [3, 4]. At present, molten oxides are widely used in metallurgy [5, 6] and crystal growth [7]. They play important role in geological processes [8]. Molten oxide electrolysis is being considered as a promising route for oxygen generation for extra-terrestrial exploration [9, 10]. Furthermore, molten oxides show promise for energy and materials. Recently, the molten oxide fuel cells (MOFC) [11] and molten oxide membranes (MOM) [12] have been developed. The MOFCs are a new class of intermediate temperature fuel cells, which combine the advantages of both solid oxide fuel cells (SOFC) and molten carbonate fuel cells (MCFC) and could be used for electric power generation. The MOMs are a new type of ion transport membranes (ITM) for high purity oxygen separation from air. Renewed interest in these materials as well as processes has become possible because of the recognition of the fact that ITM-based technology modules have the potential to improve the efficiency and performance characteristics of energy generation systems [13, 14]. ITMs can separate highly pure oxygen from air at elevated temperatures and as a result offer the potential to reduce the overall oxygen production cost. Traditionally, the bulk of the commercial oxygen supply is purified by cryogenic distillation and extracted from the air. It has been recognized that the cryogenic oxygen is not cost-affordable for low-cost, next-generation energy production, and other highly oxygen-intensive industries such as steel, glass, non-ferrous metallurgy, refineries, etc. The ITM oxygen technology is a radically different approach to produce low-cost oxygen at temperatures synergistic with power production.

High temperature fuel cells based on oxide materials have a crucial role to play in decarbonising the future energy mix. However, current targets pertaining to both cost and durability necessitate the SOFCs to operate in the intermediate temperature range. To achieve these goals, the oxygen ionic conductivity of SOFC ceramic oxide electrolyte needs to be enhanced [15]. To alleviate this problem, alternative and innovative MOFC electrolytes with highest possible oxygen ionic conductivity at intermediate temperatures have been developed [11, 16]. The conductivities of the innovative MOFC electrolyte materials (as compared to the traditional SOFC and MCFC electrolyte materials [17–19]) are shown in Fig. 1. The superior oxygen transport in these new materials and the capability to identify the governing transport mechanisms in them are the subject matters of the present article.

Molten oxide materials represent composites that consist of either an ion-conducting liquid and electron-conducting solid oxide phases (Mixed Ionic Electronic Conductor, MIEC) or ion-conducting liquid and insulating (or ion-conducting) solid phases (electrolyte). Due to the requirement for mechanical strength of the material, the volume fraction of liquid cannot exceed 30% [12]. A molten oxide material is usually prepared in two steps. In the first step, a ceramic

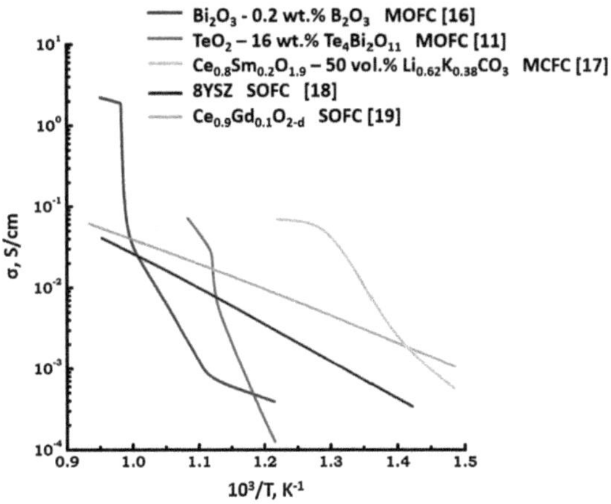

Fig. 1 Temperature dependences of conductivities of innovative (MOFC) and traditional (SOFC and MCFC) electrolytes

composite is sintered. In the second step, the so-called liquid-channel-grain-boundary-structure (LGBS) is formed by the grain boundary (GB) wetting [20]. In order to prepare the ceramic composite, the corresponding powder mixture is pressed and then sintered in air at a temperature below the eutectic melting. In order to obtain the LGBS material, the ceramic composite is then heated above the eutectic melting into two-phase area of corresponding diagram where the solid and liquid phases are in equilibrium. During the ceramic composite heating, the GB wetting by a chemically compatible liquid occurs at the eutectic melting [20]. The result is a LGBS material consisting of solid grains and intergranular liquid channels (or molten oxide material). The intergranular liquid channels provide the material high ionic conductivity, gas tightness, and ductility.

The oxygen transport properties of molten oxide materials are not only of technological interest as noted above, but also of fundamental importance. During the last decade, significant advances have been made in understanding the transport properties of molten oxide materials. Two main factors have contributed to this progress. First, the development of more refined experimental technique that led to a better measurement of the transport properties. The measurements are frequently difficult because of the involvement of high temperatures and the follow up melt-container reactions [21]. A major problem in performing the measurements is the selection of a container material which is inert to the melt. A partly molten oxide method [22] was developed to alleviate this problem. The oxygen permeation fluxes of several extremely reactive oxide melts have been measured by using this method under different oxygen partial pressure gradients and temperatures [23]. Second, the development of powerful models that have explained the oxygen transport mechanism in some oxide melts [24, 25]. This article highlights the theoretical and experimental progresses made in understanding the transport properties of molten oxide materials and the latest progress made in the research and development of innovative as well as alternative materials for both MOFC and MOM.

Transport Properties of MOM and MOFC

Molten Oxide Membranes

Carl Wagner was perhaps the first person to discuss about the interesting properties of mixed-conducting materials [26]. If a mixed-conducting oxide material is placed in between two gases that have different oxygen partial pressures, it can act as an oxygen separation membrane. Later, Elshof et al. [27] suggested that the perovskite-type mixed conductor, Sr(Fe, Co)$O_{3-\delta}$, can be used as a membrane material. Bouwmeester and Burggraaf [28] and Pena and Fierro [29] compiled and reviewed the perovskite-type membrane materials. Mazanec et al. [30, 31] proposed a material design concept for the dual-phase oxygen-permeable composites. A combination of high mixed-oxide ion and electronic conductivity was reported for the composites of solid-oxide-ion electrolyte/noble metal provided a percolating pathway exists for each component. The mixed-conducting cermets, including YSZ/Pd [32], samarium-doped CeO_2/Ag [33] and rare-earth-doped Bi_2O_3/Ag [34, 35] have an appreciable oxygen permeation rate. High oxygen permeation rates show the ceramic composites of undoped solid electrolyte δ-Bi_2O_3/In_2O_3 or NiO semiconductor fabricated by the in situ Bi_2O_3 melt crystallization method [36, 37] $Ce_{0.8}Gd_{0.2}O_{2-\delta}$-$CoFe_2O_4$, [38] and $Ce_{0.8}Sm_{0.2}O_{2-\delta}$-$SrCO_3$-$Co_3O_{43}$ [39]. Sasabe et al. [40–42] showed that a small addition of a transition metal oxide to the molten slags results in considerable increase in the oxygen permeation rate. Recently, high oxygen permeation rate was established for the mixed-conducting molten oxide materials based on Bi_2O_3 and V_2O_5 [23]. These materials show promise for use as MOM for oxygen separation from air [12]. We will divide the MOM materials into two groups containing the Bi_2O_3 and V_2O_5 oxygen ion-conducting melts [43] (i) NiO-Bi_2O_3 [12], In_2O_3-Bi_2O_3 [12], ZnO-Bi_2O_3 [44], Co_3O_4-Bi_2O_3 [45] and CuO-Bi_2O_3 [46] (ii) ZrO_2-V_2O_5 [47], CuO-V_2O_5 [48, 49] and Bi_2O_3-V_2O_5 [22, 50].

The MOM materials are usually prepared in two steps: first, the ceramic composite sintering; second, the LGBS formation [22]. The result is MOM material consisting of electron-conducting solid grains (light constituents in Fig. 2) and ion- or mixed-conducting intergranular liquid channels (dark constituents in Fig. 2). Under an oxygen electrochemical potential gradient, the ambipolar conductivity of oxygen ions and electrons results in an oxygen permeation through the membrane. The overall permeation kinetics can be controlled by both chemical diffusion and surface exchange reactions. Wagner`s equation is usually used to describe the oxygen permeation flux due to the chemical diffusion reaction [12]

$$j_{O_2} = \frac{RT}{16F^2h}\overline{\sigma_{amb}}\ln\frac{P'_{O_2}}{P''_{O_2}} \quad (1)$$

where P'_{O_2} and P''_{O_2} are the oxygen partial pressures at the oxygen-rich and -lean sides of the membrane, j_{O_2} is the oxygen permeation flux through the membrane, h is the membrane thickness, R is the universal gas constant, F is the Faraday

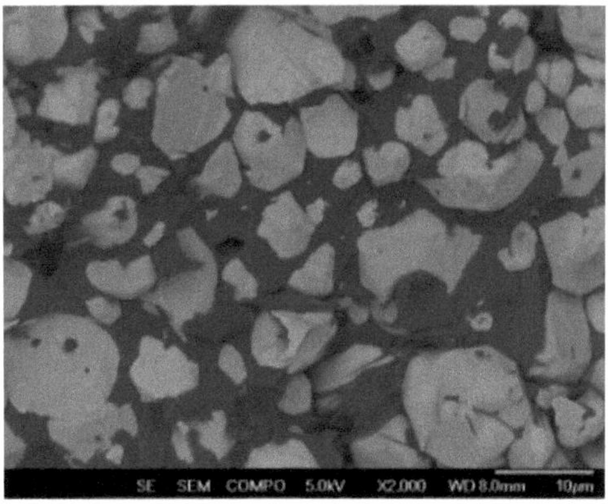

Fig. 2 Scanning electron micrograph of fracture of $BiVO_4$-10 wt% V_2O_5 MOM material after cooling from 640 °C (dark and light constituents are solid V_2O_5 and liquid (molten V_2O_5 and $BiVO_4$), respectively)

constant, T is the temperature (K) and σ_{amb} is the ambipolar conductivity of oxygen ions and electrons:

$$\sigma_{amb} = \frac{\sigma_i \sigma_e}{\sigma_i + \sigma_e} \tag{2}$$

where σ_e and σ_i are the electronic and ionic conductivities respectively. Mixed controlled kinetics dominate for the thin membranes. If the membrane thickness decreases, the limited transfer through the surfaces becomes the rate determining step. In this case, the oxygen permeation flux is described by the Eq. (3) [12]

$$j_{O_2} = \frac{RT}{16F^2(h + 2h_c)} \overline{\sigma_{amb}} \ln \frac{P'_{O_2}}{P''_{O_2}} \tag{3}$$

where h_c is the membrane characteristic thickness (establishes the value at which oxygen permeation is equally governed by the bulk diffusion and kinetics taking place at membrane surface, also known as surface exchange reactions).

The oxygen permeation rate through the MOM depends on three mechanisms. The first is the dissociation of oxygen molecules into oxygen anions at the membrane surface, known as surface exchange kinetics. The second is the migration of oxygen anions through the membrane, known as chemical diffusion and the third is the recombination of oxygen anions back to oxygen molecules on the other side of the membrane. Each mechanism has the potential to become the rate-limiting step. In order to determine whether chemical diffusion or the surface exchange reactions is the rate-limiting step in determining the overall permeation kinetics, the oxygen fluxes were measured for different membrane material thickness [12]. The result of the thickness variation experiments for $BiVO_4$-V_2O_5 MOM material is displayed in Fig. 3. If oxygen permeation through the membrane is totally controlled by a chemical diffusion process as shown in Eq. (1), the oxygen flux would have to be inversely proportional to the membrane thickness for given oxygen partial pressure gradient. The linear relationship of NiO-Bi_2O_3, In_2O_3-Bi_2O_3, Co_3O_4-Bi_2O_3 and ZrV_2O_7-V_2O_5 MOM materials indicates that the overall permeation kinetics is controlled by the

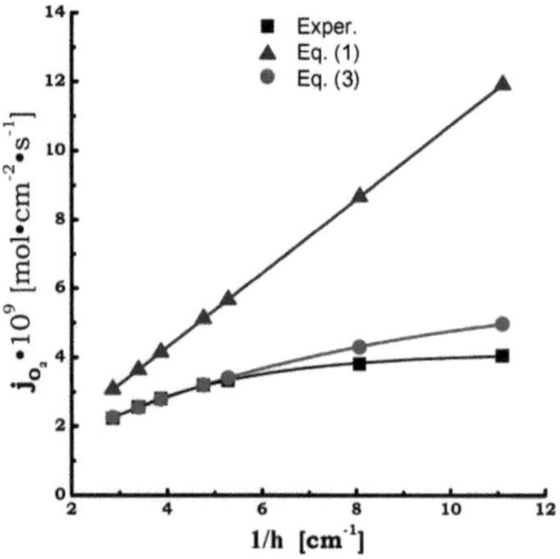

Fig. 3 Experimental and calculated by Eqs. (1) and (3) thickness dependences of oxygen permeation fluxes of $BiVO_4$-10 wt% V_2O_5 MOM material ($\lg(P'_{O_2}/P''_{O_2}) = 2.3, T = 650^\circ C$)

chemical diffusion process [12]. However, the $BiVO_4$-V_2O_5 MOM material is shown to deviate from the linear relationship (Fig. 3). This deviation indicates that the overall permeation kinetics is not only controlled by the chemical diffusion but also by the surface exchange reactions (Eq. 3). Parameters obtained from the fitting (Eqs. 1–3) to the experimental oxygen permeation flux data indicate that at a characteristic thickness (h_c) value of 0.07 cm, the transition from predominantly chemical diffusion control to that of the surface exchange reactions will take place [12]. The calculated value (by Eq. 3) and the experimental curves (Fig. 3) practically coincided in the thickness range of 0.4–0.1 cm (mixed controlled kinetics). With further decrease in the thickness, the limited transfer through the surfaces became the rate determining step. Overall the oxygen permeation fluxes through the NiO-Bi_2O_3, In_2O_3-Bi_2O_3, Co_3O_4-Bi_2O_3 and ZrV_2O_7-V_2O_5 MOM materials are predominantly controlled by a chemical diffusion process in the thickness range 1–5 mm (used in experiments) and at temperatures between 750 and 900 °C. $BiVO_4$-V_2O_5 MOM material was the only exception where the overall permeation kinetics was observed to be controlled by both the chemical diffusion and surface exchange reactions. The mixed controlled kinetics was found to be at thicknesses and temperatures below 0.07 cm and 700 °C respectively.

In order to compare the membrane material performances, the oxygen permeability ($j^*_{O_2}$) was calculated by using the Eq. 4

$$j^*_{O_2} = j_{O_2}\frac{h}{\ln(P'_{O_2}/P''_{O_2})} \tag{4}$$

The values of the MOM oxygen permeability were observed to be comparable to those of the state-of-the-art membranes (Table 1). Therefore, the MOM could be used in electrochemical devices for oxygen separation from the air [54].

Table 1 The oxygen permeability ($j^*_{O_2}$) of ceramic (CER), cermet (CM) and molten (MOM) membranes

Membrane	T (K)	j_{O_2} (mol cm^{-2} s^{-1})	h (mm)	P'_{O_2} (atm)	P''_{O_2} (atm)	$j^*_{O_2}$ (mol cm^{-1} s^{-1})	References
$Pr_{0.6}Sr_{0.4}Co_{0.5}Fe_{0.5}O_{3-\delta}$ (CER)	1173	2.3×10^{-7}	0.6	0.21	0.001	5.7×10^{-9}	[51]
$Ba_{0.5}Sr_{0.5}Co_{0.8}Fe_{0.2}O_{3-\delta}$ (CER)	1173	1.7×10^{-6}	1.3	0.21	0.001	2.9×10^{-8}	[52]
$La_{0.6}Ca_{0.4}Co_{0.8}Fe_{0.2}O_{3-\delta}$ (CER)	1173	1.2×10^{-7}	1.0	0.20	0.0053	3.3×10^{-9}	[53]
In_2O_3-55 wt% Bi_2O_3(CER)	1073	5.3×10^{-8}	2.9	0.21	0.054	1.1×10^{-8}	[36, 37]
$(Bi_2O_3)_{0.75}(Er_2O_3)_{0.25}$-40 vol.% Ag (CM)	953	1.8×10^{-7}	1.3	1.00	0.000002	1.8×10^{-9}	[34, 35]
In_2O_3-48 wt%Bi_2O_3 (MOM)	1123	8.7×10^{-8}	2.6	0.21	0.038	1.3×10^{-8}	[12]
NiO-48 wt%Bi_2O_3 (MOM)	1123	3.8×10^{-8}	2.5	0.21	0.031	5.0×10^{-9}	[12]
Co_3O_4-36 wt% Bi_2O_3 (MOM)	1123	1.9×10^{-7}	2.0	0.21	0.047	2.5×10^{-8}	[45]

Molten Oxide Fuel Cells

An ideal fuel cell is supposed to directly convert the chemical energy into the electrical energy with a relatively higher efficiency and lower emission of pollutants. However, before fuel-cell technology can gain a strong foothold in the domain of the electric power market, important issues need to be addressed. These issues include the development of alternative fuel cell materials in the intermediate temperature range. Here we consider the latest progresses made pertaining to the search and development of innovative potential MOFC materials.

The fuel cells are usually classified by the electrolytes employed in the cell [55]. These can be both low- and high-temperature fuel cells. The low-temperature fuel cells include the Alkaline Fuel Cell (AFC), Polymer Electrolyte Fuel Cell (PEFC), Direct Methanol Fuel Cell (DMFC), and Phosphoric Acid Fuel Cell (PAFC). The high-temperature fuel cells, on the other hand, constitute the Molten Carbonate Fuel Cell (MCFC) and Solid Oxide Fuel Cell (SOFC).

An advantage of the high-temperature fuel cells is the possibility to (i) allow for the internal reforming due to a relatively higher operating temperature (600–1000 °C) and (ii) use the waste heat. High temperature significantly improves the oxygen reduction kinetics thereby eliminating the need for very high loadings of precious-metal catalysts. The high-temperature fuel cell systems can attain electrical efficiencies up to 70% by way of combining the fuel cell with another power generator. Over the years, different concepts for the SOFC have been proposed. However, the conventional high-temperature SOFC (800–1000 °C) has a number of disadvantages, namely insufficient long-term stability of the YSZ electrolyte and high manufacture cost. In order to develop the commercial type intermediate temperature SOFC, numerous research efforts have been primarily focused to explore a variety of ion-conducting materials in the temperature range 600–800 °C as well as various film-preparation techniques (slurry coating, tape casting, screen printing, etc.) to minimize the Ohmic losses by way of reducing the thickness of the electrolyte layer. Recently, more effective salt-ceria-composite electrolytes have been developed for their application in the hybrid molten carbonate/solid oxide fuel cells [17, 56, 57]. These cells require an effective way of mixing of CO_2 and O_2 in the cathode region. To circumvent this problem, a novel molten oxide based fuel cell (MOFC) concept has been suggested [11]. The performance characteristics of the air/chromel/TeO_2-16 wt% $Te_4Bi_2O_{11}$ LGBS/chromel/CO cell have been evaluated by a standard electrochemical method. The MOFC electrolyte consists of TeO_2 solid grains and oxygen ion-conducting intergranular liquid channels (molten TeO_2 and $Te_4Bi_2O_{11}$). The intergranular liquid channels consists of the MOFC electrolyte with a combination of highly useful characteristics (high ionic conductivity, gas tightness and ductility). With air and CO being the cathode and fuel respectively, the cell showed a power level of 11.5 mW/cm^2 when the volume fraction of liquid, electrolyte thickness and the operating temperature were 0.17, 2.5 mm and 640 °C respectively. As the solid grains of TeO_2 do not conduct the oxygen ions, the liquid amount significantly limited the ionic conductivity of the

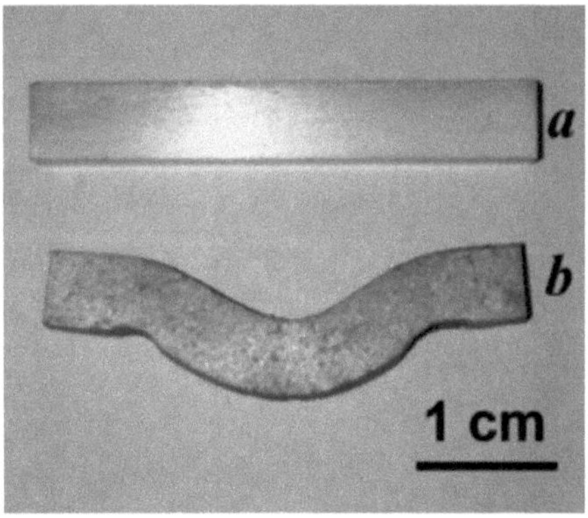

Fig. 4 Photography of Bi_2O_3-0.2 wt% B_2O_3 MOFC electrolyte: **a** initial sample and **b** after its deformation at 750 °C

electrolyte. Lately, a novel δ-Bi_2O_3-0.2 wt% B_2O_3 MOFC electrolyte, where both the δ-Bi_2O_3 solid grains and intergranular liquid channels (molten Bi_2O_3 and B_2O_3) transport the oxygen ions, has been developed [16]. Both solid and molten Bi_2O_3 show the highest ever known oxygen ionic conductivities [43, 58]. However, one major problem of using the solid Bi_2O_3 as an oxygen ion conductor in different electrochemical devices is its cracking due to the large volume change (~11%) associated with the polymorphic transformation of the monoclinic α- into the cubic δ-Bi_2O_3 at 735 °C [36]. Although the δ-phase could be stabilized, it significantly reduced the ionic conductivity [59–61]. To minimize this problem, a gas-tight ductile δ-Bi_2O_3-0.2 wt% B_2O_3 MOFC electrolyte with the highest oxygen ionic conductivity (Fig. 4) has been developed [16]. The temperature dependence of the electrolyte conductivity is presented in Fig. 1. According to the Bi_2O_3-B_2O_3 phase diagram [62], an inflection on the conductivity curve at 620 °C is associated with the eutectic melting and LGBS formation. The conductivity jump at 735 °C induced by α → δ-Bi_2O_3 polymorphic transformation was accompanied by an increase in the oxygen ion conductivity. The high ionic conductivity of δ-Bi_2O_3 is caused by the presence of large number of oxygen vacancies and a high O^{2-} anion mobility. The high concentration of oxygen vacancies (25% of the lattice sites) is a result of obtaining the fluorite structure with a Bi^{3+} cation. The high anion mobility is related to the high polarizability of the Bi^{3+} cation with its lone pair of electrons [63]. The liquid phase, comprising of predominantly molten Bi_2O_3, also conducts oxygen ions. The oxygen ion conductivity in molten Bi_2O_3 was experimentally confirmed [43]. The oxygen ion transport number, determined by the volumetric measurements of the Faradaic efficiency technique, for δ-Bi_2O_3-0.2 wt% B_2O_3 MOFC electrolyte, is 0.98 at 750 °C [16]. Thus, the electrolyte shows real promise for use in the MOFC. Applicability of such molten oxide electrolytes, in view of their insufficient thermodynamic stability in a reducing atmosphere, was discussed elsewhere [11]. Overall development and subsequent implementation of the MOFC concept may result in the generation of new fuel cells that can operate in the intermediate temperature range.

Modeling

A common theme of the molten oxide based materials being reviewed here is the prevailing oxygen transport mechanism. In this Section most of the discussion and results, taken from literature, are based on the existence of a dynamic polymer chain model in these materials [24, 25]. A summary describing the key features of this has been discussed. Detailed discussions pertaining to atomic scale modeling concepts, as applied to molten oxides, have been presented elsewhere [64, 65].

The model adapts Wagner's theory for molten oxides and provides an interesting as well as remarkably simple interpretation of the oxygen ion mobility in the oxide melts. The salient features/assumptions of this model are as follows:

(i) Melt has a polymer structure
(ii) Average length of the polymer chains depends on the operating temperature
(iii) "Connection–disconnection"of polymer chains occurs stochastically in the melt
(iv) A disconnection of the chain leads to the formation of two new chains and oxygen pseudo-vacancy at the end of one of them
(v) Connection of the chains leads to the release of one of the two terminal oxygen ions
(vi) Free oxygen ion is captured by the nearest pseudo-vacancy of oxygen
(vii) Oxygen ions and electrons constitute the most mobile charge carriers in the melt.
(viii) Under an oxygen electrochemical potential gradient, the ambipolar conductivity of the oxygen ions and electrons results in the oxygen permeation through the melt
(ix) Relative migration rates of ions and electrons are balanced in such a way that the net current through the melt is zero.

Figure 5 shows a schematic melt structure for which the model is valid.

Within the framework of this model, the oxygen permeation flux through the MOM can be described by the following equation (Eq. 5)

$$j_{O_2} = \frac{(1-\tau_i)D_i\eta\rho}{480Mh}\ln\frac{P'_{O_2}}{P''_{O_2}} \quad (5)$$

where ρ and M are the density and molar mass of melt, η is the volume fraction of melt, D_i is the oxygen ion diffusion coefficient in melt, and τ_i is the oxygen ion transport number in melt. Process of oxygen transfer through the melt under an oxygen electrochemical potential gradient could be presented as follows.

Oxygen is chemisorbed on the surface of the melt (step 1 in Fig. 5c) and is then captured by the nearest pseudo-vacancy of oxygen (step 2 in Fig. 5c). Then connection of the chains leads to the formation of free oxygen ion (step 3 in Fig. 5c)

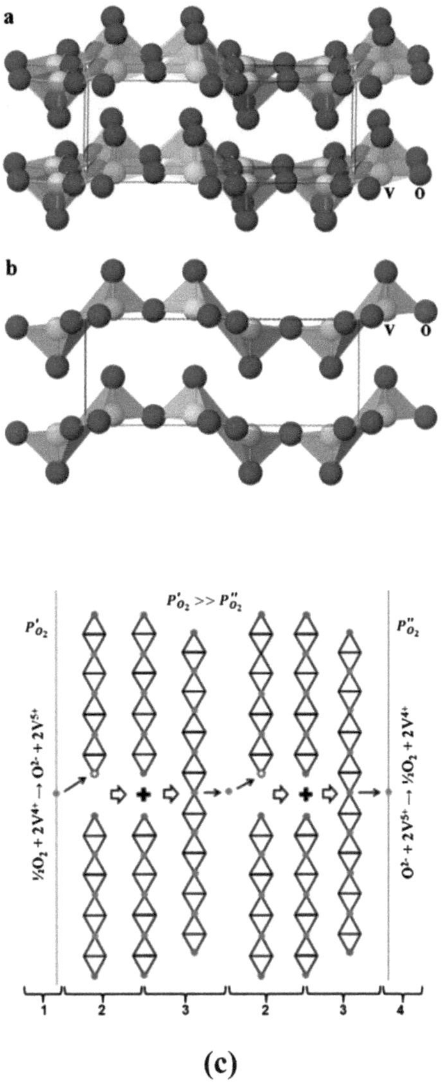

Fig. 5 Structure of **a** solid and **b** molten V_2O_5; **c** schematic of oxygen ion transport in oxide melt based on V_2O_5; ●—mobile oxygen; ●—terminal oxygen; ●—bridging oxygen; ○—pseudo-vacancy of oxygen

which is again captured by the nearest pseudo-vacancy. As a result, the free oxygen ion reaches the opposite surface of the melt (step 4 in Fig. 5c). Electronic conductivity of the molten oxide behaves like a semiconductor [4]. The calculated values of the oxygen permeation fluxes through the MOM, by Eq. 5 are comparable with the experimental data [12]. It should be noted here that this model provides an interpretation of the oxygen mobility in a specific class of molten oxides. Overall the present approach appears to be quite sound and provides the foundation upon which refined models can be developed for a wide-ranging molten oxide systems.

Conclusions and Future Outlook

An improvement in the existing as well as development of new and novel materials, that possess high ionic and mixed ion-electronic conductivities, are being dictated by the demand for rapidly progressive electric power engineering. A possible way to meet this demand is to utilize the molten oxide materials which offer the potential for further improvement in the efficiency of power generation systems. To accomplish this feat in near future, a set of questions must be answered: what is the long-term stability of the molten oxide materials? How the solid/liquid interfaces might affect the transport properties of these materials? It should be emphasized here that the solid/liquid interfaces can significantly affect the transport properties of both MOFC and MOM based materials. In particular, the space charge regions with an enhanced ionic conductivity can form at the interfaces [66–68]. Electrical charge as well as mass transfer rates, along the interfaces, are governed by a change in the interfacial tension which, in turn, can induce Marangoni flow and easy mass transport reactions [69]. However, the structure and transport properties of the solid/liquid interfaces in MOFC and MOM materials have not been extensively studied. The answers to these questions will perhaps make it possible to advance the understanding of both the electrical charge transfer and mass transport processes in molten oxide based materials and their eventual technological deployment.

Acknowledgements I thank the Russian Science Foundation for the generous support through the project 16-19-10608.

References

1. M. Faraday, Philos. Mag. **3**, 163 (1834)
2. A.E. Martin, G. Derge, Trans. Am. Inst. Min. Met. Eng. **154**, 104 (1943)
3. O.A. Esin, P.V. Geld, Physical chemistry of pyrometallurgical processes. Metallurgiya (1966)
4. E.A. Pastukhov, V.I. Musihin, N.A. Vatolin, *Electrical Properties of Non-Stoichiometric Oxide Melts* (Ural Branch of the USSR Academy of Sciences, Sverdlovsk, 1984)
5. A. Allanore, J. Electrochem. Soc. **162**, E13 (2015)
6. R. Boom, S. Riaz, Y. Xiao, K.C. Mills, Steel Res. Int. **84**, 623 (2013)
7. P. Rudolf, *Handbook of Crystal Growth*, vol. 2A. Basic Technologies, Elsevier (2014)
8. D.L. Anderson, *New Theory of the Earth* (Cambridge University Press, Cambridge, 2007)
9. F. Sanderson, Nature. doi:10.1038/news.2009.803
10. A. Allanore, L. Yin, D.R. Sadoway, Nature **497**, 353 (2013)
11. V.V. Belousov, S.V. Fedorov, Fuel Cells **16**, 401 (2016)
12. V.V. Belousov, Ionics **22**, 451 (2016)
13. A.C. Bose, G.J. Stiegel, P.A. Armstrong, B.J. Helper, E.P. Foster, Progress in ion transport membranes for gas separation application, in *Inorganic Membranes for Energy and Environmental Application*, ed. by A.C. Bose (Springer, Berlin, 2009), pp. 3–25
14. S.S. Hashim, A.R. Mohamed, S. Bhatia, Renew. Sustain. Energy Rev. **15**, 1284 (2011)
15. A. Chroneos, B. Yildiz, A. Tarancon, D. Parfitt, J.A. Kilner, Energy Environ. Sci. **4**, 2774 (2011)
16. V.V. Belousov, S.V. Fedorov, Chem. Commun. (2017). doi: 10.1039/c6cc07935k

17. W. Zhu, C. Xia, D. Ding, X. Shi, G. Meng, Mater. Res. Bull. **41**, 2057 (2006)
18. M. Mori, T. Abe, H. Itoh, O. Yamamoto, Y. Takeda, T. Kawahara, Solid State Ionics **74**, 157 (1994)
19. V.V. Kharton, F.M. Figueiredo, L. Navarro et al., J. Mater. Sci. **36**, 1105 (2001)
20. V.V. Belousov, S.V. Fedorov, Russ. Chem. Rev. **81**, 44 (2012)
21. P.N. Quested, B.J. Monaghan, High Temp. Mater. **20**, 219 (2001)
22. S.V. Fedorov, V.V. Belousov, A.V. Vorobiev, J. Electrochem. Soc. **155**, F241 (2008)
23. V.V. Belousov, Metall. Mater. Trans. A **45A**, 4257 (2014)
24. V.V. Belousov, A.A. Klimashin, S.V. Fedorov, Ionics **22**, 369 (2016)
25. A.A. Klimashin, V.V. Belousov, Metall. Mater. Trans. B **47B**, 749 (2016)
26. C. Wagner, Z. Elektrochem. **60**, 4 (1956)
27. J.E. Ten Elshof, H.J.M. Bouwmeester, H. Verweij, Solid State Ionics **81**, 97 (1995)
28. H.J.M. Bouwmeester, A.J. Burggraaf, Dense ceramic membranes for oxygen separation, in *The CRC Handbook of Solid State Electrochemistry*, ed. by P.J. Gellings, H.J.M. Bouwmeester (CRC, New-York, 1997), pp. 481–555
29. M.A. Peña, J.L. Fierro, Chem. Rev. **2001**, 101 (1981)
30. T.J. Mazanec, T.L. Cable, J.G. Frye Jr., Solid State Ionics **53–56**, 111 (1992)
31. T.J. Masanec, Solid State Ionics **70–71**, 11 (1994)
32. K. Huang, M. Schroeder, J.B. Goodenough, Electrochem. Sol. State Lett. **2**, 375 (1999)
33. E. Ruiz-Trejo, P. Boldrin, A. Lubin et al., Chem. Mater. **26**, 3887 (2014)
34. C. Capoen, C. Steil, G. Nowogrocki et al., Solid State Ionics **177**, 483 (2006)
35. C. Capoen, G. Nowogrocki, S.J. Skinner et al., Solid State Ionics **177**, 489 (2006)
36. V.V. Belousov, V.A. Schelkunov, S.V. Fedorov et al., Electrochem. Commun. **20**, 60 (2012)
37. V.V. Belousov, MRS Commun. **3**, 225 (2013)
38. Y. Lin, S. Fang, D. Su, K.S. Brinkman, F. Chen, Nat. Commun. **6**, 6824 (2015)
39. Z. Zhang, W. Zhou, Y. Chen et al., ACS Appl. Mater. Interf. **7**, 22918 (2015)
40. M. Sasabe, K.S. Goto, Metall. Trans. **5**, 2225 (1974)
41. M. Sasabe, J. Masami, Can. Metall. Q. **22**, 29 (1983)
42. Y. Sayadyaghoubi, S. Sun, S. Jahanshahi, Metall. Mater. Trans. B **26B**, 795 (1995)
43. V.R. Yarlagadda, T.V. Nguyen, ECS Trans. **33**, 119 (2011)
44. I.I.V. Kul'bakin, V.V. Belousov, S.V. Fedorov, A.V. Vorobiev, Mater. Lett. **67**, 139 (2012)
45. V.V. Belousov, I.V. Kul'bakin, S.V. Fedorov, P. G. Petrov, ACS Appl. Mater. Interf. **8**, 22324 (2016)
46. V.V. Belousov, J. Mater. Sci. **40**, 2361 (2005)
47. I.V. Kul'bakin, S.V. Fedorov, A.V. Vorobiev, V. V. Belousov, Russ. J. Electrochem. **49**, 878 (2013)
48. V.V. Belousov, A.A. Klimashin, Metall. Mater. Trans. A **43**, 3715 (2012)
49. V.V. Belousov, A.A. Klimashin, Russ. Chem. Rev. **82**, 273 (2013)
50. V.V. Belousov, S.V. Fedorov, A.V. Vorobiev, J. Electrochem. Soc. **158**, B601 (2011)
51. K. Partovi, F. Liang, O. Ravkina, J. Caro, A.C.S. Appl, Mater. Interf. **6**, 10274 (2014)
52. P. Zeng, Z. Chen, W. Zhou, H. Gu, Z. Shao, S. Liu, J. Membr. Sci. **291**, 148 (2007)
53. K. Efimov, T. Klande, N. Juditzki, A. Feldhoff, J. Membr. Sci. **389**, 205 (2012)
54. P. Dyer, R.E. Richards, S.L. Russek, D.M. Taylor, Solid State Ionics **134**, 21 (2000)
55. L. Carrette, K.A. Friedrich, U. Stimming, Chem. Phys. Chem. **1**, 162 (2000)
56. W. Liu,Y. Liu, B. Li, T.D. Sparks, X. Wei, W. Pan, Compos. Sci.Technol **70**, 181 (2010)
57. C. Xia, Y. Li, Y. Tian, Q. Liu, Z. Wang, L. Jia, Y. Zhao, Y. Li, J. Power Sources **195**, 3149 (2010)
58. N.M. Sammes, G.A. Tompsett, H. Näfe, F. Aldinger, J. Eur. Ceram. Soc. **1999**, 19 (1801)
59. N. Jiang, E.D. Wachsman, J. Am. Ceram. Soc. **82**, 3057 (1999)
60. E.D. Wachsman, S. Boyapati, M.J. Kaufman, N. Jiang, J. Am. Ceram. Soc. **2000**, 83 (1964)
61. N. Jiang, R.M. Buchanan, F.E.G. Henn, D.A. Stevenson, E.D. Wachsman, Mater. Res. Bull. **29**, 247 (1994)
62. Yu.F. Kargin, V.Yu. Voevodskii, V.M. Skorikov, Russ. J. Inorg. Chem. **43**, 1946 (1998)
63. P.D. Battle, C.R.A. Gatlow, J.W. Heap, L.M. Moroney, J. Solid State Chem. **63**, 8 (1986)

64. A. Bunde, K. Funke, M.D. Ingram, Solid State Ionics **105**, 1 (1998)
65. E. Thibodeau, I.H. Jung, Metall. Mater. Trans. B **47B**, 355 (2016)
66. V.V. Belousov, J. Eur. Ceram. Soc. **27**, 3459 (2007)
67. J. Maier, Prog. Solid. St. Chem. **23**, 171 (1995)
68. A. Chandra, J. Maier, Solid State Ionics **148**, 153 (2002)
69. F.D. Richardson, Can. Metall. Q. **21**, 111 (1982)

Effects of Oxide Precursor Preparation Parameters on the Electrochemical Reduction of Tantalum Pentoxide in Calcium Chloride Melt

Maureen P. Chorney, Bridger P. Hurley, Prabhat K. Tripathy and Jerome P. Downey

Abstract Successful electrochemical reduction of a metal oxide to its constituent metal depends upon two broad unit operations: (i) preparation of the oxide precursor with ideal morphological features and (ii) electrochemical reduction under optimum process parameters. The critical parameters that affect mechanical strength and open porosity of the precursor are oxide particle size; green pellet preparation; and sintering time, temperature, and atmosphere. A judicious combination of these process parameters will produce an ideal precursor for achieving complete reduction. Uniaxial pressing was utilized to prepare 13 mm diameter cylindrical pellets, which were sintered in either oxidizing (air) or reducing atmospheres. Sintered pellets were characterized by XRD, SEM-EDS, and porosity measurement techniques. Electrochemical reduction experiments were performed with a molten calcium chloride electrolyte in a cell with a three electrode configuration. Transient electrochemical techniques were employed to assess the effects of the precursor parameters on the reducibility of the sintered oxide specimens.

Keywords Electrodeoxidation · Molten salt · Tantalum

Introduction

Tantalum is a rare, highly sought after material due to its high melting point, high electrical and thermal conductivities, and excellent corrosion resistance [1]. The world reserves of tantalum are not abundant, and the United States relies solely on imports from other countries including Australia, Brazil, and Canada [2]. The United States' demand for tantalum has increased by 36% from 2014 to 2015, and it is likely that this demand will continue to increase for the near future [2]. Recycling

M.P. Chorney (✉) · B.P. Hurley · J.P. Downey
Montana Tech of the University of Montana, Butte, USA
e-mail: mchorney@mtech.edu

P.K. Tripathy
Idaho National Laboratory, Idaho Falls, USA

The Minerals, Metals & Materials Society, *TMS 2017 146th Annual Meeting & Exhibition Supplemental Proceedings*, The Minerals, Metals & Materials Series,
DOI 10.1007/978-3-319-51493-2_62

of tantalum within the United States is an important endeavor due to the limited deposits located in the U.S. as well as its continued reliance on other countries for import [2].

Previous studies of the electrochemical reduction of tantalum pentoxide have analyzed the oxide compressed into the form of a pellet [3–6]. These experiments have utilized pellets with varying diameters and thicknesses, as well as sintering time and temperatures. In this study, various pellet preparation and sintering parameters were evaluated to determine the ideal procedure for pellet processing so that the highest reduction efficiencies may be achieved.

Experimental

Oxide Precursor Preparation

The Ta_2O_5 powder was milled for several hours with poly(vinyl butyral-co-vinyl alcohol-co-vinyl acetate) (PVB/PVA) and polyethylene glycol (PEG), all of which was immersed in isopropanol. The initial powder size, prior to milling, was varied between <250 μ, <44 μ, and <20 μ. Utilizing a 13 mm diameter steel die, uniaxial pressing was employed to compress the powder mixture into thin pellets (1.5–3.0 mm) with pressures ranging from 18.6 to 35.9 MPa. The green pellets were then sintered in either air or a reducing atmosphere using an MTI GSL-1100X Series tube furnace. Sintering was generally conducted at 975 °C for 5 h, but sintering time and temperature were sometimes altered to create pellets with varying porosity values for the reduction experiments.

Electrochemical Reduction

The reduction experiments were conducted in a calcium chloride (high purity, anhydrous)—1 wt% calcium oxide (high purity) melt contained in a nickel crucible. A three electrode configuration was utilized—a sintered tantalum pentoxide pellet threaded onto a nickel wire served as the working electrode, and an oxygen evolving electrode and a glassy carbon (GC) rod served as the counter electrode and the reference electrode, respectively. The furnace was heated to 900 °C in an argon atmosphere controlled glove box.

Results

Two sintered oxide pellets were electrochemically reduced in the molten calcium electrolyte. The oxide pellets were prepared by sintering the pellets in air and in a reducing environment (Fig. 1). As expected, the color of the air-sintered pellet was

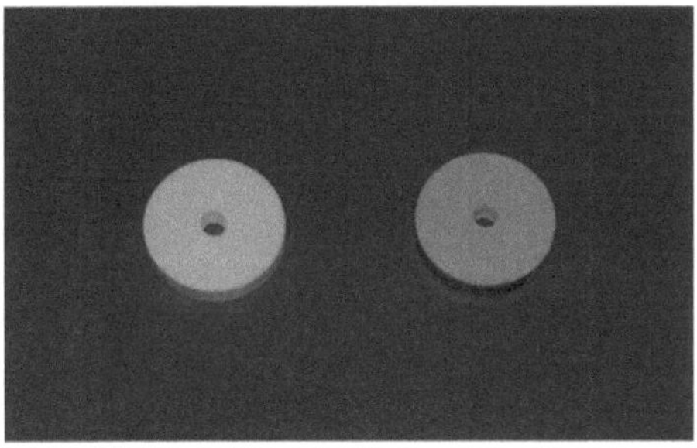

Fig. 1 State of the sintered oxide precursors prior to their electrochemical reduction, *left pellet*: air-sintered (oxide powder size <325 mesh, sintering temperature: 975 °C and duration—5 h; open porosity was 40.8%, Batch-12/17), *right pellet*: sintered pellet, prepared in a reducing atmosphere (oxide powder size <20 μm, sintering temperature: 975 °C and duration—5 h; open porosity was 43.4%, Batch-14/14)

observed to be white. However, the color of the pellet prepared in a reducing atmosphere was observed to be light-grey (Fig. 1), which was ascribed to the formation of a slightly reduced oxide phase (Ta_2O_{5-x}/Ta_xO_y).

The as-prepared sintered oxide pellets were threaded onto a nickel wire and used as the working electrode (cathode) in a pool of high purity and anhydrous (molten) calcium chloride (that contained 1 wt% high purity and ultra-dry CaO). An oxygen evolving electrode and a standard glassy carbon (GC) rod (3 mm dia. and 100 mm long) were used as the counter (anode) and reference electrode respectively.

The electrolytic reduction test runs were carried out in a stainless steel, argon atmosphere controlled, custom-fabricated glove box. A resistance heater externally heated the furnace. The glove box was continuously purged with laboratory-grade argon (Air Liquide), which could maintain very low levels of moisture and oxygen contents (0.1 ppm each). However, during the reduction experiments, the prevailing atmosphere consisted of <0.1 ppm moisture and <10 ppm oxygen. The "O" ring (that acted as the seal between the glove box and the furnace) allowed for the necessary cooling by a glycol-based chiller that was plumbed onto the furnace.

Electrochemical reduction experiments were carried out at a constant potential mode in the range 2.6–2.9 V. The theoretical decomposition potential of Ta_2O_5 to tantalum and oxygen is −1.6 V at 900 °C for the reaction in Eq. 1.

$$Ta_2O_5(s) = 2Ta(s) + 2.5O_2(g) \tag{1}$$

The applied cathode potential, during the reduction experiment was in the range −1.8 to −1.9 V versus GC. The progress of the electrolytic reduction was monitored by recording the current versus time profile (Fig. 2). The current versus time profile showed an initial rise in the current followed by it smooth decline as the reduction reaction progressed with time. The initial, peak and residual current values were 0.29 A, 0.54 A and 0.14 A respectively.

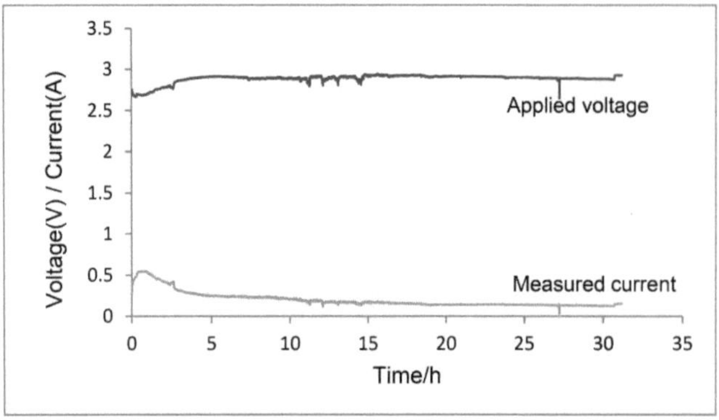

Fig. 2 Current/voltage versus time during the electrochemical reduction of Ta_2O_5: electrolyte temperature—900 °C, duration—31 h, pellet thickness 3 mm, pellet dia. 13 mm, sintering atmosphere—air, peak current—0.54 A, background current (after 31 h)—0.15 A

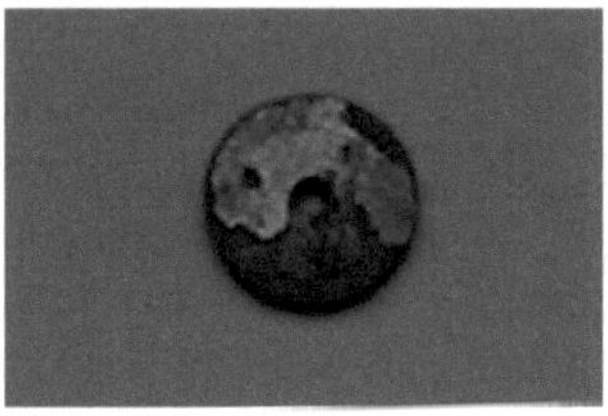

Fig. 3 State of the pellet after the electrochemical reduction of air-sintered Ta_2O_5 (*left-hand side pellet* shown in the Fig. 1), partial *silvery white color* on the surface of the reduced pellet was solidified calcium chloride, electrolysis temperature—900 °C, duration—31 h

The reduced air-sintered pellet is shown in Fig. 3. The pellet was observed to be partially covered with solidified calcium chloride-calcium oxide, a portion of which could be easily scraped from the surface of the reduced pellet.

After soaking in de-ionized water, the reduced pellet crumbled into greyish black powder (Fig. 4). The reduced powder was subsequently treated with ethyl alcohol and acetone. The cleaned and partially dried powder was placed in an oven at 50 °C overnight (~12 h). After drying, the powder was observed to be free-flowing.

In order to estimate the reduction efficiency, the powder was analyzed to determine the residual oxygen content. The reduced powder (from the pellet sintered in air) was determined to contain 1200 ppm of oxygen. This residual oxygen measurement suggested that it could be possible to reduce the oxygen content from an initial value of 18.1% to a final value of 0.12% (99.3% reduction).

Reduction of the sintered pellet, prepared under the reducing atmosphere, also showed a similar current versus time profile, as shown in Fig. 2. Upon analysis, the oxygen content was determined to be 1300 ppm when the reduction was carried out for a total duration of ~15 h. Presence of a higher degree of open porosity, as well as defects/voids, may have accelerated the oxygen removal kinetics.

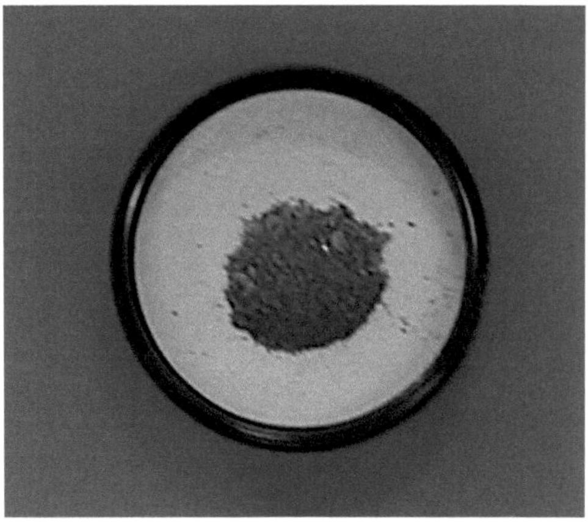

Fig. 4 Cleaned (reduced) tantalum powder

Conclusions

Preliminary results indicate that high tantalum reduction efficiencies can be achieved with the pellets prepared for these experiments. Further experiments will be conducted to provide additional insight into the optimum pellet preparation parameters. Additional analyses will also aid in the determination of the efficacy of the direct oxide reduction (DOR) process to prepare tantalum powder that can be used for a variety of applications.

References

1. R. Callaghan, USGS. Minerals Information: Niobium (Columbium) and Tantalum Statistics and Information. May 2016, http://minerals.usgs.gov/minerals/pubs/commodity/niobium/index.html
2. J.F. Papp US Geological Survey, Mineral Commodity Summary–Tantalum. January 2016, http://minerals.usgs.gov/minerals/pubs/commodity/niobium/mcs-2016-tanta.pdf
3. Q. Song, Q. Xu, X. Kang, J. Du, Z. Xi, Mechanistic insight of electrochemical reduction of Ta_2O_5 to tantalum in a eutectic $CaCl_2$-NaCl molten salt. J. Alloy. Compd. **490**, 241–246 (2010)
4. T. Wu, X. Jin, W. Xiao, X. Hu, D. Wang, G.Z. Chen, Thin pellets: fast electrochemical preparation of capacitor tantalum powders. Chem. Mater. **19**, 153–160 (2007)
5. R.P. Barnett, D.J. Fray, Reaction of tantalum oxide with calcium chloride-calcium oxide melts. J. Mater. Sci. **48**, 2581–2589 (2012)
6. R. Barnett, K.T. Kilby, D.J. Fray, Reduction of tantalum pentoxide using graphite and tin-oxide based anodes via the FFC-Cambridge Process. Metall. Mater. Trans. B **40B**, 150–157 (2009)

The Effect of Temperature on Electrochemical Codeposition of Mg–Ni Hydrogen Storage Alloys from Molten Salt System

Gökçe Hapçı Ağaoğlu and Gökhan Orhan

Abstract A Mg–Ni hydrogen storage alloy was produced as a film by molten salts electrolysis and diffusion process. Mg and Ni was co-deposition on a nickel substrate from 20 wt% $MgCl_2$ + 2 wt% $NiCl_2$ and balans NaCl + KC1 mixture. The electrodeposition was conducted at 700–750 °C and constant current density of 0.5 A cm^{-2} for 60 min. The phase composition and microstructure of Mg–Ni alloy were detected by X-ray diffraction (XRD) and scanning electron microscope (SEM). The electrode performance in 6 M KOH solution were characterized by cyclic charge–discharge experiment. The best initial discharge capacity (375 $mAhg^{-1}$) and acceptable capacity retaining rate (70%) were obtained at the 750 °C alloy.

Keywords Mg–Ni alloys • Hydrogen storage • Molten salt electrolysis

Introduction

For more than a decade, hydrogen as an alternative to traditional energy sources such as oil and natural gas has been the focus of research and development efforts in all technologically advanced countries of the world. It is strongly believed that hydrogen can help to address the growing demand for energy and slow down global climate change. Hydrogen can be stored by several methods [1, 2], i.e. gas compress, liquefaction and metal hydriding, among which, metal hydriding is universally accepted as the promising method in consideration of the hydrogen storage volume density, energy consumption and safety factors. Among Mg-based alloys, Mg–Ni intermetallic compounds have been demonstrated to have excellent hydrogen storage capability [3]. For example, Mg_2Ni intermetallic compound can absorb up to 3.6 wt% hydrogen in addition to having a good hydriding/deydriding

G.H. Ağaoğlu · G. Orhan (✉)
Faculty of Engineering, Department of Metallurgical and Materials Engineering, Istanbul University, Avcılar, 34320 Istanbul, Turkey
e-mail: gorhan@istanbul.edu.tr

The Minerals, Metals & Materials Society, *TMS 2017 146th Annual Meeting & Exhibition Supplemental Proceedings*, The Minerals, Metals & Materials Series, DOI 10.1007/978-3-319-51493-2_63

rate. However due to the high oxidation and reactive nature of Mg, it is difficult to fabricate Mg_2Ni by conventional methods. In this study, we used molten salt electrolysis to produce an MgNi intermetallic compound. The effect of the temperature on the formation of intermetallic Mg–Ni compound formation was also studied.

Experimental Procedure

Electrolyte composition was chosen to contain 20 wt% $MgCl_2$, 2 wt% $NiCl_2$ and equiweight NaCl and KC1. The mixture was first dried under vacuum for more than 48 h at 220 °C to remove excess water, and then melted in an alumina crucible placed in a quartz cell located in an electric furnace. Magnesium and nickel ions were introduced into the bath in the form of dehydrated $MgCl_2$ and $NiCl_2$ powder. Metal ion impurities in the melts were removed by pre-electrolysis at −2.00 V for 2 h. All experiments were performed in an argon atmosphere. An 8 mm diameter graphite was used to the anode. Cathode material was electrolytic Ni (99.8%) which has a surface area of 15 × 15 mm and 0.1 mm thickness. At the beginning of each experiment a new Ni cathode material was prepared, weighed and placed on the cathode. The cathode assembly containing Ni cathode material was placed to the bottom of the alumina crucible. A Good Will, GPR 3510 HD direct current power supply that had a capacity of providing 10 A current at 35 V maximum was used for the experiments. The electrodeposition experiments were performed within galvanostatic regime at 0.5 A cm^{-2} for 60 min and three different temperature was conducted including 700 °C (A700), 725 °C (A725) and 750 °C (A750). Temperature measurements were made with an alumina sheathed alumel–chromel thermocouple placed into the molten salt electrolyte with an accuracy of ±2 °C. All the experiments were carried out under a carefully purified and dehydrated high purity argon atmosphere. After the electrolysis the specimen was cooled to room temperature in the furnace under argon atmosphere for about 8 h.

X-ray diffraction (XRD) analyses were performed using a Rigaku–Miniflex diffractometer with Cu Kα radiation. The morphology and chemical composition of the electrodeposited Mg–Ni were investigated using SEM and EDS analyses. SEM observations were made using a JEOL JSM 5600 microscope operating at 20 kV. In the other hand in order to determine Mg and Ni contents of the samples, each sample was dissolved in aqua regia (HNO_3:HCl:H_2O = 1:3:8, v/v). The solution was diluted and analyzed using an inductively coupled plasma atomic emission spectrometer (ICP-Spectro Ciros Vision).

Electrochemical tests were performed on a Gamry Instrument Model 600 potentiostat/galvanostat/ZRA at room temperature. Working electrodes were prepared by mixing 0.15 g Mg–Ni alloy powder with 0.45 g nickel powder, which acted as the current collector for the electrode, in a weight ratio of 1:3 and then cold pressing into pellets of 10 mm in diameter, under a pressure of 5 ton cm^{-2}. $NiOOH/Ni(OH)_2$ counter electrode and a Hg/HgO reference electrode were used to set up a three-electrode cell in 6 M KOH solution. Charge–discharge cyclic tests were performed using Gamry potentiostat/galvanostat (equipped with PWR800 Electrochemical Energy Software Package) operating in a galvanostatic mode. The working electrodes were charged at a current density of 100 mA g^{-1} and the charging was carried out down to the severe gassing potential. The discharge current density was 25 mA g^{-1} and the discharge cut-off potential was −0.6 $V_{Hg/HgO}$.

Results and Discussion

Considering the majority of the binary hydrogen absorption alloys, consist of metal A, which forms hydride easily, and metal B, which oppositely forms hydride uneasily. When the substrate B is immersed into a molten salt containing A ions and a cathodic current is passed, metal A is electrodeposited on the substrate B. If mutual diffusion and a chemical reaction occur between the electrodeposited A and substrate B an intermetallic compound A–B is formed. In the alloying process between A and B, if the mutual diffusion of A into B or B into A is the rate determining step, the metal phase A must remain on the surface of the specimen. The rate determined step in the reaction determines whether the A-B alloy layer consists of single or multi-layer [4]. When the substrate B is immersed into a molten salt containing A and B ions and a cathodic current is passed, metal A and metal B are also electroco-deposited on the substrate B. Thus, a combined process with molten salts electrolysis and diffusion, i.e. alloying co-deposits with the substrate, has been proposed to produce hydrogen storage alloy films. That is to say, the alloy phase formed on the substrate should be determined by the experimental condition such us temperature, current density, electrolyte composition and electrolysis time.

The electrolysis temperature is one of the factors to affect the co-deposition process and reaction rate of the electroco-deposited Mg and Ni with the Ni substrate. Thus Mg and Ni electrodeposition at several different temperatures was conducted at 0.5 A cm^{-2} for 60 min with fixed salt mixture. These deposits were analyzed by XRD and XRD patterns of the surface were electrodeposits at 700, 725 and 750 °C are shown in Fig. 1. It can be seen that in any electrolysis temperature the intermetallic compounds, $MgNi_2$ and Mg_2Ni, were formed on nickel substrate. However, the mixture phase of Ni and intermetallics was also observed at 700 °C, while the mixture phase of Mg and intermetallics was also observed at 750 °C. Only the $MgNi_2$ and Mg_2Ni mixed intermetallic phases was formed at 725 °C. In the Mg–Ni phase diagram [3] as shown in Fig. 2 no Mg_2Ni exists over 758 °C (melting point of Mg_2Ni 760 °C) and only the solid $MgNi_2$ remained co-existing with the liquid phase. The intermetalic compound of Mg_2Ni was, therefore, formed on the $MgNi_2$ layer during the cooling process.

It is known that for a single layer compound formation in an A–B diffusion couple, if more A diffuses into the B with the marker position fixed, the thickness of the A–B intermetallic layer on the B side would be thicker than that on the A side. It

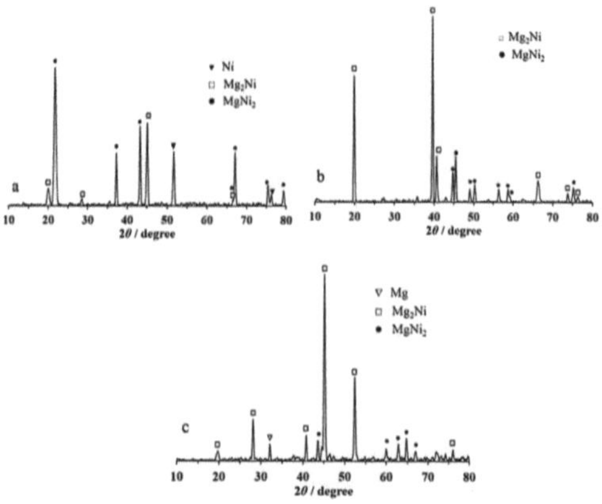

Fig. 1 X-ray diffraction patterns electro co-deposited with 0.5 A cm^{-2} for 60 min under **a** 700 °C, **b** 725 °C and **c** 750 °C

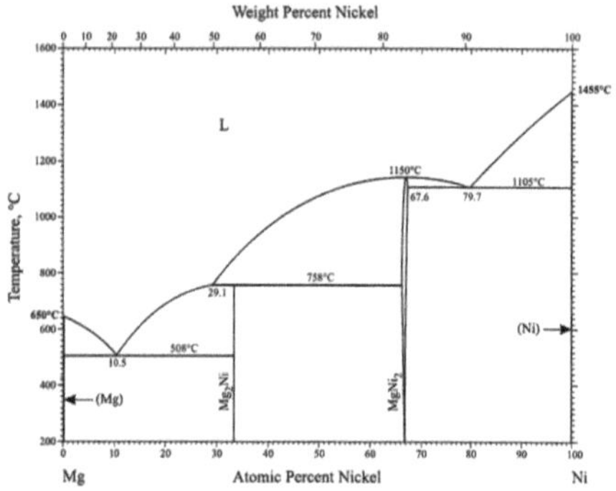

Fig. 2 Mg–Ni phase diagram [3]

is considered that Mg remained on the surface and this followed to form mixed phase of magnesium and Mg_2Ni, since time was enough for magnesium to react with nickel that is because of a higher deposition rate than diffusion and/or chemical reaction rate [4].

As seen from the Ni–Mg binary phase diagram in Fig. 2, two intermetallic compounds (Mg_2Ni and $MgNi_2$) were formed between the Ni-Mg reaction couple. The melting point of Mg_2Ni is about 765 °C, so if the annealing temperature is below 765 °C, only the Mg_2Ni phase would form. The Ni atoms quickly dissolved into the molten Mg to form a eutectic phase (Mg + Mg_2Ni). Consequently, after a short reaction time, the formation of interfacial Mg_2Ni between the molten eutectic Mg–Ni alloy and the Ni substrate actually occurred. On the other hand, both compound phases (Mg_2Ni and $MgNi_2$) formed above the annealing temperature of 765 °C.

Figure 3 showed SEM images of the products after electrolysis at three different molten salt temperature and at 0.5 A cm^{-2}. The alloy, in Fig. 3, deposits exhibited smooth surface. However, it can be say that electrodeposited alloy obtained at 750 °C (Fig. 3c) compared to the ones obtained at other alloy have a different

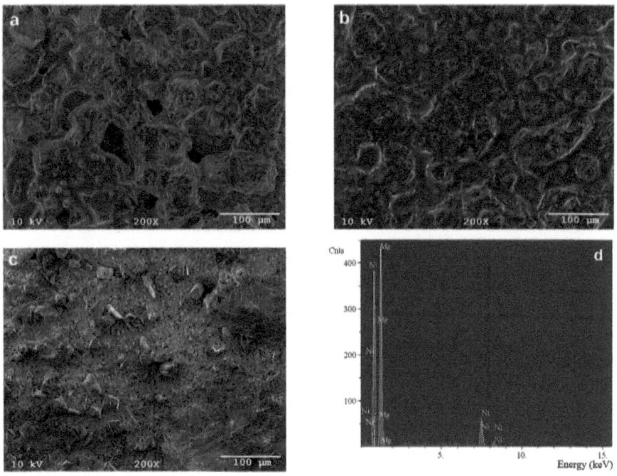

Fig. 3 Surface morphologies of electrodeposited Mg–Ni alloy from different temperature. **a** 700 °C, **b** 725 °C, **c** 750 °C, **d** the EDS analyze of Mg–Ni alloy electrodeposited at 750 °C

surface morphology. It is important to note that in all the electrodeposited samples atomic percentage of the oxygen has been detected, varying between 1 and 3 at.%, while for calculation of the Mg–Ni alloy composition only the Ni and the Mg were taken into account, neglecting oxygen. Ni ratio in the Mg–Ni alloy A700, A725 and A750 wt% 64.7, wt% 45.5 and wt% 49.3 respectively. The EDS analyze of Mg–Ni alloy electrodeposited at 750 °C is in Fig. 3d.

The effect of temperature on the charge (upper curves)–discharge (lower curves) curves is shown in Fig. 4. Only the first cycle charge and discharge curves are given in Fig. 4 for the clear presentation. The change in the discharge capacities of obtained from 750 °C alloy depends on the charge/discharge cycle number are provided in Fig. 5. The cyclic stabilities of the alloys are also evaluated by observing the capacity retaining rates as in Fig. 6. The capacity retaining rate is defined as $C_n/C_{max} \times 100\%$. In this relationship C_{max} is the initial discharge capacity and C_n is the discharge capacity at *n*th charge/discharge cycle step.

A750 alloy has the highest discharge capacity at all the charge/discharge cycles (Fig. 4) and it keeps at least 70% of its initial discharge capacity at the 20th cycle (Fig. 6). The initial discharge capacity of A750 alloy is 375 $mAhg^{-1}$. The capacity decay in Fig. 4 is relatively mild and at twenty cycle MgNi alloy has 261 $mAhg^{-1}$ discharge capacity. It is universally agreed that the rapid degradation in the

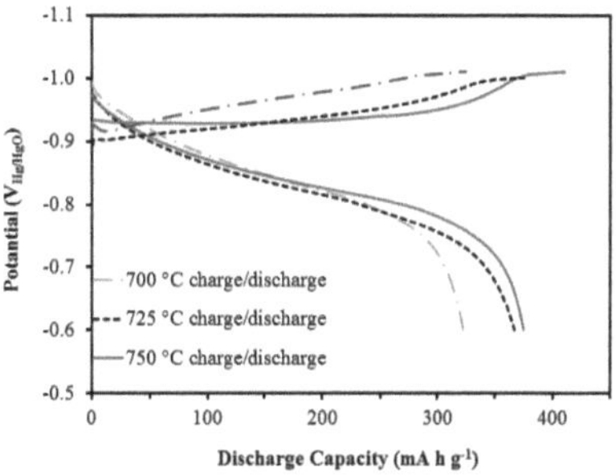

Fig. 4 Effect of electrolysis temperature on charge–discharge characteristic of Mg–Ni alloys

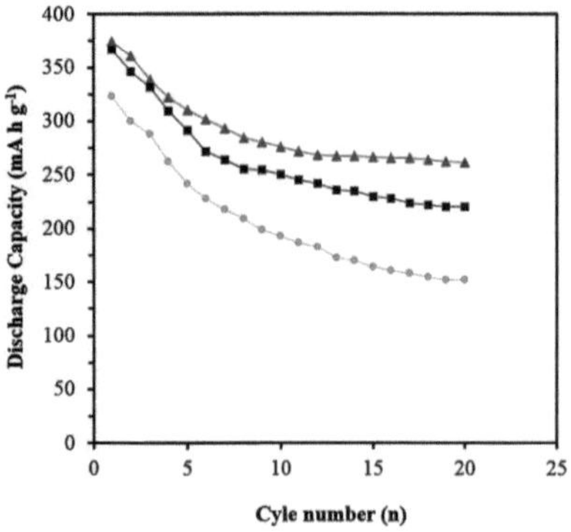

Fig. 5 Discharge capacity as a function of cycle number for electrodeposited Mg–Ni alloy at 750 °C

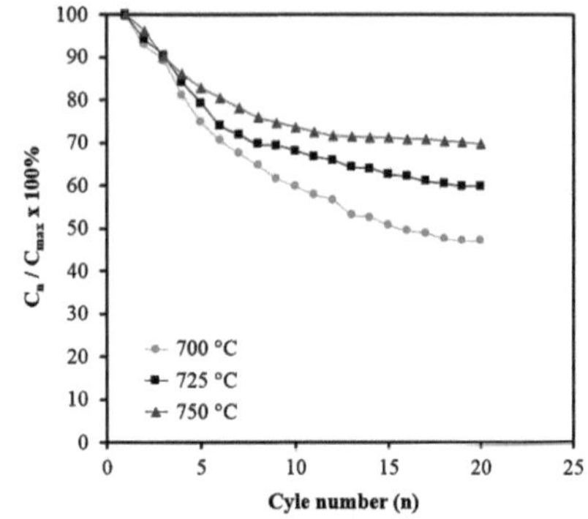

Fig. 6 Cyclic stability of electrodeposited Mg–Ni alloy at 750 °C

discharge capacity during the charge/discharge cycling of Mg based alloy results from the formation and thickening of $Mg(OH)_2$ surface layer, which retards the hydrogen in and out diffusion, in alkaline solutions [5–7].

Conclusion

Electrochemical codeposition behavior of Mg and Ni on Ni electrodes in NaCl–KCl (50:50, wt%) melts containing fixed concentrations of $MgCl_2$ and $NiCl_2$ was investigated at different temperature. According to studies of codeposition conditions, Mg–Ni alloys can be directly prepared via codeposition of Mg and Ni on electrodes from NaCl–KCl–$MgCl_2$–$NiCl_2$ melts. The phases and chemical composition of Mg–Ni alloys were obtained by galvanostatic electrolysis depend on temperature. The best initial discharge capacity (375 $mAhg^{-1}$) and high capacity retaining rate (70%) were obtained for the 750 °C. The chemical composition and phase structure of this alloy coating was ca. 50:50, wt% as Ni and Mg and Mg + Mg_2Ni + $MgNi_2$ mixed phase.

Acknowledgements The authors gratefully acknowledge the financial support of the Scientific Research Projects Coordination Unit of Istanbul University (Project Number 12965).

References

1. J.A. Ritter, A.D. Ebner, J. Wang, R. Zidan, Implementing a hydrogen economy. Mater. Today **6**(9), 18–23 (2003)
2. X. Su Ye, X. Gang Lu, C. He Li, W.Z. Ding, X. Li Zou, Y.H. Gao, Q.D. Zhong, Preparation of Ti-Fe based hydrogen storage alloy by SOM method. Int. J. Hydrogen Energy **1**, 1–7 (2010)
3. C.H. Li, Y.C. Chuang, C.Y. Liu, Fabrication of Mg-based intermetallic compounds by liquid electromigration. J. Electron. Mater. **36**, 1489–1494 (2007)
4. S.L. Cheng, Y.Y. Chen, S.W. Lee, H.F. Hsu, Formation of Mg2Ni alloy layers and kinetic studies in the binary Mg–Ni system. Thin Solid Films **517**, 4745–4748 (2009)
5. N.H. Goo, W.T. Jeong, K.S.J. Lee, Effects of Zr addition on discharge properties of mechanically alloyed Mg_2Ni hydrogen-storage alloy electrode. Power Sources **87**, 118–124 (2000)
6. H.Y. Lee, N.H. Goo, W.T. Jeong, K.S. Lee, The surface state of nanocrystalline and amorphous Mg_2Ni alloys prepared by mechanical alloying. J. Alloy. Compd. **313**, 258–262 (2000)
7. N.H. Goo, K.S. Lee, The electrochemical hydriding properties of Mg–Ni–Zr amorphous alloy. Int. J. Hydrogen Energy **27**, 433–438 (2002)

Part XVI
Materials and Fuels for the Current and Advanced Nuclear Reactors VI

Diffusion Studies in the Development of an FCCI Barrier for High-Burnup Metallic Nuclear Fuel

Daniel Eichel and James Vollmer

Abstract TerraPower, LLC, is developing a high-burnup metallic nuclear fuel design for use in its Traveling Wave Reactor. Many innovative features have been applied to extend the burnup capability of the fuel, including the use of a barrier between the fuel and HT9 steel cladding to mitigate fuel/cladding chemical interaction (FCCI). Different barrier permutations, of various composition and morphology, are being tested in diffusion couple experiments, and the resulting diffusion interfaces are characterized using scanning electron microscopy and energy dispersive spectroscopy. Diffusion couple experiments include control samples in which HT9 steel is placed directly in contact with neodymium, an aggressive, representative surrogate fission product. To varying degrees, FCCI barrier designs offer the potential to greatly reduce the degree to which the cladding is attacked chemically.

Keywords FCCI · FCCI barrier · Nuclear fuel · High-burnup fuel · Metallic fuel · Diffusion couple · HT9 · TerraPower · Traveling wave reactor · Intermetallic · Decarburization

Introduction

The in situ breed-and-burn concept that is being applied at TerraPower in the Traveling Wave Reactor (TWR) design necessitates that the nuclear fuel can reliably achieve a very high degree of burnup, in the range of approximately 30%. This in turn requires innovative design strategies to mitigate the principal factors that typically limit fuel lifetimes. One such factor is fuel/cladding chemical interaction (FCCI), a phenomenon that results in the creation of brittle intermetallic compounds

D. Eichel (✉) · J. Vollmer
TerraPower, LLC, Bellevue, USA
e-mail: deichel@terrapower.com

J. Vollmer
e-mail: jvollmer@terrapower.com

The Minerals, Metals & Materials Society, *TMS 2017 146th Annual Meeting & Exhibition Supplemental Proceedings*, The Minerals, Metals & Materials Series,
DOI 10.1007/978-3-319-51493-2_64

at the inner surface of the cladding of the nuclear fuel pin. This decreases the effective thickness of the cladding, and can ultimately lead to its failure. To address this, TerraPower is currently investigating the potential use of an FCCI barrier, a material placed between the nuclear fuel and cladding that is relatively inert to attack by chemically-active fission product species, notably lanthanides. Out-of-pile diffusion couple tests are being utilized to assess the chemical compatibility and effectiveness of various FCCI barrier systems, and to gain better understanding of the relevant phenomenology. The resulting diffusion interfaces are characterized through the use of scanning electron microscopy (SEM), including compositional analysis using energy dispersive spectroscopy (EDS). At the time of writing, this study is still underway, and complete results from experimental samples are not yet available; however, analysis of the control samples, as discussed here, has yielded insight into some important phenomenology, and preliminary results indicate that barrier layers can indeed be effective in greatly reducing the detrimental effects of FCCI.

Diffusion Couple Experiments

In order to emulate the geometry of an actual nuclear fuel system, a diffusion couple apparatus was conceived in which pieces of neodymium, a representative, chemically-aggressive surrogate fission product, are held in contact with the inside wall of a prototypic cladding tube composed of HT9 steel. Wedges placed inside the tube are forced together in the axial direction by external set screws; this produces an expanding force in the radially-outward direction that forces the neodymium pieces against the inside wall of the cladding tube. This apparatus is depicted below in Fig. 1. Control samples placed the neodymium in direct contact with the HT9 steel, while experimental samples included various permutations of FCCI barrier designs between the two. With the components of the diffusion couple thus held in strong contact together, samples were heat-treated for a period of three months at prototypic temperatures of 590 and 650 °C. Heat treatment was performed in sealed containers to maintain an inert argon atmosphere around the samples, also including an oxygen getter composed of yttrium powder.

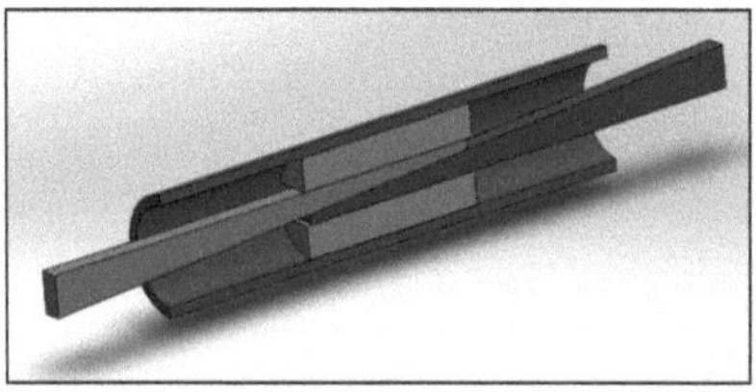

Fig. 1 CAD model showing a cross-section of the diffusion couple apparatus, in which interior wedges are used to hold pieces of neodymium in contact with the inside surface of the cladding tube

Results: Control Samples

Results presented here are largely constrained to those collected from the control samples, in part because the experimental work is still in progress at the time of writing. Nevertheless, the control samples still provide useful insights into the relevant phenomenology that affects a metallic nuclear fuel system, including the observed formation of iron-neodymium intermetallic layers, as well as a decarburization effect that results in rampant grain growth in the HT9 steel. Both of these are detrimental effects that degrade the cladding performance, and are phenomena that should be prevented by any prospective FCCI barrier system.

Formation of Intermetallic Compounds. Inspection of the iron-neodymium phase diagram, provided below in Fig. 2, reveals that diffusion couples operating at the 590 and 650 °C temperatures are expected to form two intermetallic compounds: $Fe_{17}Nd_2$ and $Fe_{17}Nd_5$. Assuming perfect stoichiometry, these compositions approximately correspond to Fe-23 wt%Nd and Fe-43 wt%Nd, respectively.

Consistent with what would therefore be expected, both of these intermetallic compounds were, indeed, observed in the iron-neodymium diffusion interface (although in such a way that is not entirely in agreement with the phase diagram, as will be discussed). Figure 3, provides a representative SEM image showing the diffusion interface in the 590 °C sample, with annotations to identify each of the different compositional regions. The line indicates the approximate location along which an EDS line scan was performed, the results of which are shown in Fig. 4. All SEM analysis in this study was performed using an FEI Quanta FEG 250 instrument.

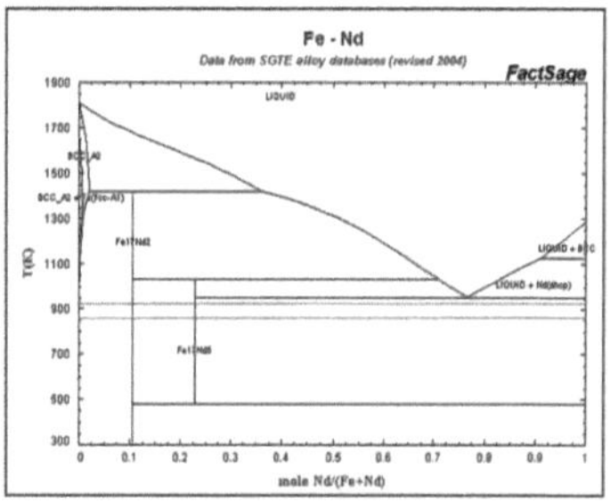

Fig. 2 Phase diagram for the iron-neodymium system, as provided by FactSage [1]. Isotherm lines have been added at the temperatures relevant to this work

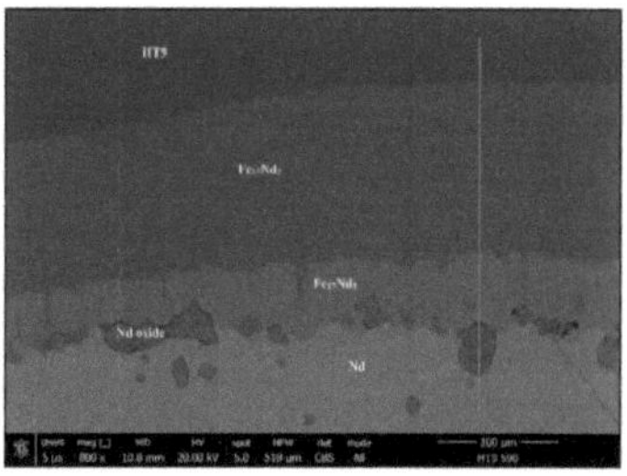

Fig. 3 Representative SEM image showing the diffusion zone in the 590 °C control sample. Total thickness of the interaction zone, including both intermetallic layers, varied around 200 μm

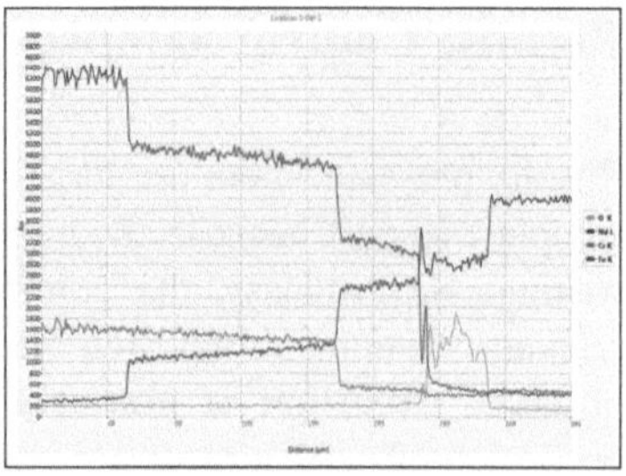

Fig. 4 EDS line scan performed across the interaction zone along the line shown in Fig. 3. *Left-to-right* progression of these traces corresponds to *top-to-bottom* progression along the line in Fig. 3

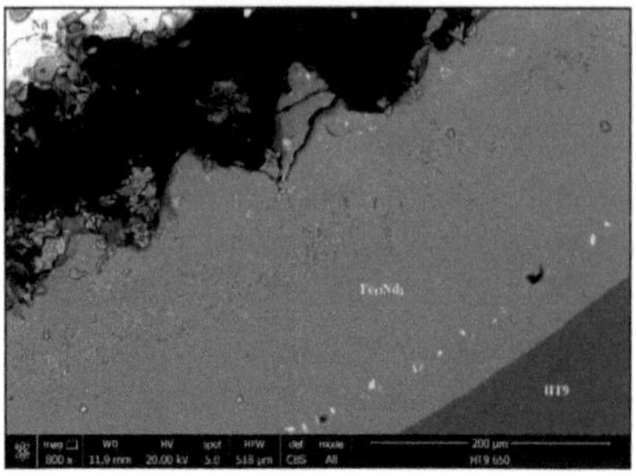

Fig. 5 SEM image showing the diffusion zone in the 650 °C control sample. Only one intermetallic layer has developed; the neodymium-rich intermetallic is conspicuously absent

Composition point analyses of each layer using EDS indicated an average neodymium concentration of 17.98 wt% in the $Fe_{17}Nd_2$ layer and 47.49 wt% in the $Fe_{17}Nd_5$ layer, values that are somewhat but not egregiously disparate from what would be predicted by the chemical formulae. Nevertheless, there can be little doubt that the intermetallic layers observed in the sample are consistent with what would be expected from the phase diagram.

In contrast, however, the diffusion couple treated at 650 °C is not in full agreement with the phase diagram. Rather, only a single intermetallic layer is observed, exhibiting an average neodymium concentration of 19.94 wt%. It is therefore believed that this region is the $Fe_{17}Nd_2$ compound, and that the $Fe_{17}Nd_5$ intermetallic is absent. Although the phase diagram predicts the $Fe_{17}Nd_5$ intermetallic to be stable at a temperature of 650 °C, these experimental observations, as shown in Fig. 5, would seem to indicate otherwise. As an aside, Fig. 5 also depicts what appear to be artifacts of the Kirkendall Effect, although measurement of the depth of these features is compromised by uncertainty as to the location of the original material interface, since inert markers were not included in the diffusion couples. The separation of the neodymium from the diffusion interface is believed to have occurred during cool-down after the heat treatment.

The possibility of perturbing effects caused by other constituents in the system may be one explanation for this disparity between the phase diagram and the experimental data. HT9 steel, of course, is not pure iron, and the significant concentrations of chromium, as well as numerous other alloy constituents (molybdenum, vanadium, tungsten, etc.), could hypothetically have the effect of destabilizing the $Fe_{17}Nd_5$ intermetallic at some temperature threshold between 590 and 650 °C. While not fully understood at this time, this observation is an important finding of this work, and will likely merit further investigation in the future.

Decarburization of the HT9 Steel. Another important phenomenon that was observed in this work involves rampant grain growth in the HT9 steel that is attributed to decarburization of the steel by the neodymium. Initially, the HT9 microstructure features a fine grain structure for resistance to irradiation damage in a reactor environment. However, maintaining this fine grain structure relies in part on carbides in the steel that prevent grain boundary diffusion. Interaction with neodymium depletes the carbides from the steel, allowing the onset of Ostwald ripening and consequent recrystallization and the development of an interaction layer that is characterized by a large grain structure. This layer extends into the steel by 100 μm or more in advance of the intermetallic layer front. Each of these regions can be seen in the annotated high-contrast SEM image shown below in Fig. 6. Note also the frequent occurrence of bright particles along the grain boundaries, identified as Nd-rich particles and believed to be precipitates of $Fe_{17}Nd_2$ that result from grain boundary diffusion of neodymium into the steel.

Estimation of Diffusion Parameters. Using the image analysis software package ImageJ [3], a series of eight measurements was taken at each of two locations to estimate the maximum thickness of each of the intermetallic layers in the diffusion interfaces of the control samples. The resulting findings are presented below in Table 1.

These diffusion distances, together with EDS composition measurements collected at a series of specific locations, allow for the computation of the interdiffusion coefficients by applying Eq. 1. This method, known as the Wagner/

Fig. 6 High-contrast SEM image in which the effects of decarburization in the HT9 can be easily observed: substantial grain growth has occurred in the decarburized region

Table 1 Estimated measurement of the maximum thickness of each intermetallic layer in the control diffusion couples (all values presented in microns)

		$Fe_{17}Nd_2$	$Fe_{17}Nd_5$
590 °C	Peak	147.1	60.7
	Mean	142.7	57.1
	Std. Dev.	3.4	3.1
650 °C	Peak	283.0	–
	Mean	239.1	–
	Std. Dev.	19.3	–

Li-Powell method [5, 6], is applicable in binary single- or multi-phase systems in which the composition and diffusion rates are assumed to be approximately constant across the thickness of a given intermetallic layer.

$$D_i = \frac{(N_i - N_{i-1})(N_{i+1} - N_i)}{(N_i'' - N_i')(N_{i-1}'' - N_{i+1}')} K_{II} \quad K_{II} = \frac{\Delta x^2}{2t} \tag{1}$$

In Eq. 1, D_i is the interdiffusion coefficient in intermetallic layer *i*, K_{II} is termed the *parabolic rate constant*, x is the thickness of the intermetallic layer in meters, and t is the duration of the heat treatment in seconds. All of the composition parameters N are measured concentrations (in atom percent) of a given chemical species at various locations across the diffusion interface, as described by the authors in the original publications.

Application of this analysis method yields the following estimates of the interdiffusion coefficients, presented in Table 2. Since the temperature dependence of the diffusion rate follows an Arrhenius relationship, a plot of *ln(D)* on the ordinate against the inverse of the absolute temperature on the abscissa is expected to result in the data points falling along a straight line; this is shown in Fig. 7. Of course, fitting a straight line with only two temperature points is of limited value, but ongoing diffusion couple experiments at additional temperature points (specifically, 550 and 625 °C) will further inform the expected linearity of the data.

The findings of this study are largely in agreement with those of previous investigations of HT9/Nd interdiffusion phenomena. Helmreich et al. [2] carried out diffusion couple experiments between neodymium and a variety of steels, including HT9, and resulting values for the interdiffusion coefficients are in approximate agreement with the estimates attained in this work. Diffusion coefficient values calculated in this work are consistently slightly greater than those reported by Helmreich et al.; however, the values from both studies are in agreement as to the

Table 2 Estimated interdiffusion coefficients as calculated using the Wagner/Li-Powell method from the control diffusion couple data. Values shown are in (m^2/s)

	$Fe_{17}Nd_2$	$Fe_{17}Nd_5$
590 °C	3.266×10^{-15}	2.159×10^{-15}
650 °C	5.847×10^{-15}	–

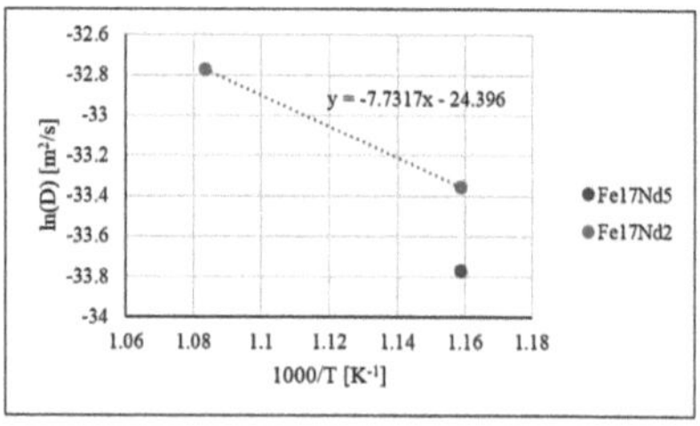

Fig. 7 Arrhenius plot of estimated interdiffusion coefficients. Data to be gathered at additional temperature points in the near future are expected to adhere to the linear trend shown here

Fig. 8 Zirconium coating applied on the inside surface of an HT9 cladding tube

general order of magnitude of this parameter in the temperature ranges that were studied. Interestingly, Helmreich et al. did not observe the $Fe_{17}Nd_5$ phase at all in HT9 diffusion couples, stating that this phase "… was observed only in … SS-316/Nd diffusion couples" [2].

FCCI Barrier and Future Development. Having thus established a baseline for the expected diffusion behavior between neodymium and HT9 steel, under prototypic geometry and temperature conditions, the effects of an FCCI barrier can be isolated for study. Past work [4] has shown that a zirconium "sheath" on the external surface of injection-cast U-Zr fuel irradiated in the Experimental Breeder Reactor II appears to be effective at mitigating FCCI reactions; therefore, zirconium is an obvious candidate material for an engineered FCCI barrier (although perhaps not the only one). Figure 8 shows one example zirconium coating on the interior surface of an HT9 cladding tube. Such experimental barriers, which include various permutations of the design, are currently undergoing diffusion couple testing, with the expectation of subsequent irradiation testing to occur with actual fueled specimens in the future.

Conclusion

In accordance with the need to develop a high-burnup-tolerant nuclear fuel form to enable the realization of the TWR concept, TerraPower is in the process of investigating the prospective use of an FCCI barrier as a means of extending fuel pin lifetimes. Diffusion couple experiments are in progress to attain a more comprehensive understanding of the relevant phenomena in the interaction of HT9 steel and surrogate fission product species, such as neodymium. Data collected to date from control samples has provided insight into the interaction of these species, including observation of the formation of intermetallic compounds as well as decarburization of the HT9. This data was used to estimate values for the interdiffusion coefficient in this system, and findings were largely in agreement with those of other studies. This body of work provides the foundation for ongoing studies into the effect of an FCCI barrier, placed between the exterior surface of the nuclear fuel and the interior surface of the cladding to prevent chemical interaction between the two. Historical data suggests that zirconium may be effective as an FCCI barrier. Ongoing diffusion couple testing will assess the efficacy of the zirconium barrier in the prevention of FCCI, with the expectation that subsequent irradiation testing will be performed in the future. Although significant development work remains, this strategy for mitigating FCCI and achieving long fuel lifetimes holds great promise for ultimate implementation in the TWR.

References

1. FactSage 7.0. (2007), Accessed from http://www.crct.polymtl.ca/fact/phase_diagram.php?file=Fe-Nd.jpg&dir=SGTE2007
2. G. Helmreich, C. Humrickhouse-Helmreich, J. Vollmer, R. Corbin, S. McDeavitt, Diffusion studies on cladding-lanthanide intermetallic formation, in *NuMat: The Nuclear Materials Conference* (2012)
3. ImageJ 1.51f, National Institute of Health (2016)
4. D.D. Keiser, Metal fuel-cladding interaction, In *Comprehensive Nuclear Materials* (Elsevier, 2012), pp. 423–441
5. G. Li, G. Powell, Theory of reaction diffusion in binary systems. Acta Metall. **33**(1), 23–31 (1985)
6. C. Wagner, The evaluation of data obtained with diffusion couples of binary single-phase and multiphase systems. Acta Metall. **17**, 99–107 (1969)

Part XVII
Materials Engineering of Soft Magnets for Power and Energy Applications

Development of Mold Inductor for Power Conversion System

Hyungsuk Kim, Yeon Jun Chung, Seung Nam Yang and Tae Kyung Lee

Abstract As global warming has become an international environmental problem recently, regulations on carbon emissions are being strengthened by each country. Leading automakers are focusing on developing lightweight and highly efficient automotive components in order to provide competitive eco-friendly cars. It is essential for green cars to have a power conversion system to utilize the electric energy from the battery as a power source. An inductor is a component in the power conversion system, which stores and transfers electrical energy and is composed of a magnetic core and electrical wires. While soft magnetic ferrite is a common material for a magnetic core, research on soft magnetic metallic alloy powders are required to make inductors smaller and lighter. In this letter, we present the research of the inductor in a 48/12 V converter for 48 V mild hybrid vehicles. A mold inductor, which embeds electrical wires inside of the inductor filled with metallic alloy magnetic powder combined with epoxy, has been developed and its properties were compared to the soft magnetic ferrite inductor.

Keywords Power conversion · Mold inductor · Soft magnetic ferrite · Soft magnetic powder

H. Kim (✉) · Y.J. Chung
Hyundai Motor Company, 150, Hyundaiyeonguso-ro, Namyang-eup, Hwaseong-si, Gyeonggi-do 18280, Korea
e-mail: khssaga@hanmail.net; hssaga@hyundai.com

Y.J. Chung
e-mail: yjchung@hyundai.com

S.N. Yang · T.K. Lee
Chang Sung Corporation, 11B-9L, Namdong Industrial Area 320, Incheon, Korea
e-mail: snyang@changsung.com

T.K. Lee
e-mail: tklee@changsun.com

The Minerals, Metals & Materials Society, *TMS 2017 146th Annual Meeting & Exhibition Supplemental Proceedings*, The Minerals, Metals & Materials Series,
DOI 10.1007/978-3-319-51493-2_65

Introduction

Eco-friendly cars have a power conversion system to utilize the electric energy from a battery as a power source to run a drive motor as well as electrical systems in the car. An inductor is a component in the power conversion system, which stores and transfers electrical energy and is composed of the magnetic core and electrical wires. While soft magnetic ferrite is a common material for a magnetic core, there has been a recent surge of interest in soft magnetic metallic alloy powders to make inductors smaller and lighter. In this study, we describe the research of the inductor in a converter for green cars. A mold inductor, which embeds electrical wires inside of the inductor case (mold) filled with metallic alloy magnetic powder combined with epoxy, has been developed and its properties were compared to the soft magnetic ferrite inductor. The volume of the mold inductor has been reduced by 10% due to the fact that the effective window area is increased. Core loss of the inductor has also been improved. The total efficiency of the converter has been increased by 0.5% when compared to that of the converter using ferrite inductors.

Material Development

Soft Magnetic Ferrite

In green cars, soft magnetic ferrite is the most frequently chosen magnetic core material for an inductor in a converter system due to its low cost and low loss caused by high resistance. On the other hand, the demerits of ferrite include loss property's heavy dependence on temperature and performance deterioration as a result of an air gap inserted into the inductor to reduce permeability [1–3].

Figure 1 illustrates the ferrite inductor utilized in a converter system. The ferrite inductor is composed of 2 'E'-shaped ferrites with the center detached from each other; in other words, it is often said that an air gap is inserted to the inductor. By virtue of the air gap, inductance is maintained above a useable level without magnetic flux being saturated in the ferrite as depicted in Fig. 2. However, as can be seen in Fig. 1c, the magnetic flux leaked out of the ferrite from the air gap can act as a source for additional loss as a result of heat generation and for electromagnetic noise.

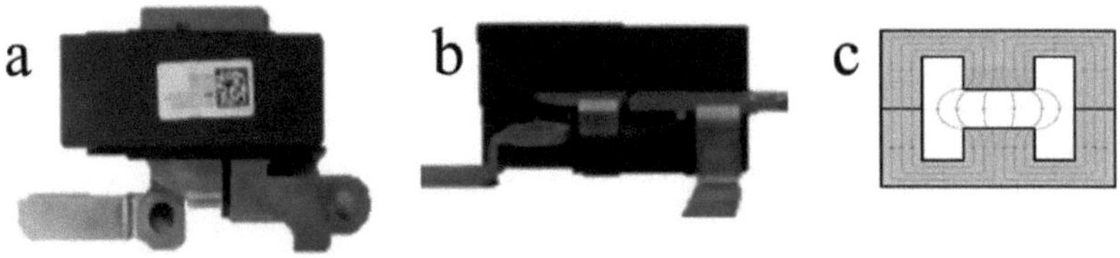

Fig. 1 *Top view* (**a**) and *side view* (**b**) of a ferrite inductor. **c** Schematic picture of magnetic flux in a ferrite inductor

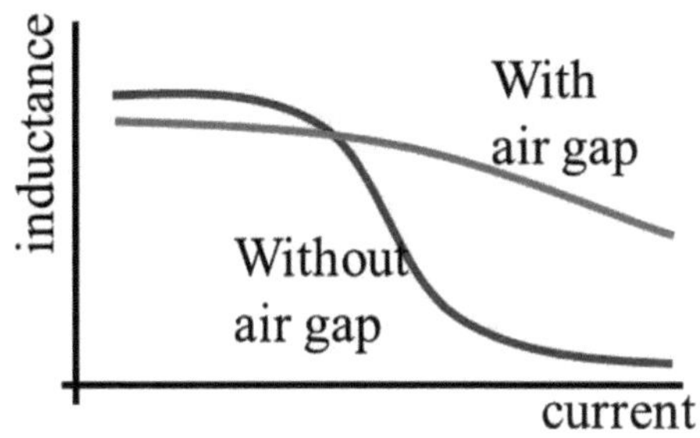

Fig. 2 Inductance of a ferrite inductor as a function of current

Soft Magnetic Iron Alloy Powder

The drawbacks of soft magnetic ferrite mentioned above can be overcome significantly by replacing the material with soft magnetic iron alloy powder. Firstly, the inductance of inductors made from soft magnetic powders remains at much higher levels without deliberately putting in air gaps. Compared with ferrite inductors fabricated through a sintering process, there are spaces between the powders in the magnetic powder inductor and they play the role of the air gaps. Secondly, ferrite inductors have many empty areas due to the fact that coil has to be wound outside of the material, meaning they are not volume-efficient. If the materials were replaced by a magnetic powder so that the powders wrap the coil up without empty spaces, the miniaturization of an inductor could be possible. Additionally, soft magnetic ferrite is brittle as a result of the sintering process. The brittleness of the material can be a crucial problem in the vibration test. The magnetic powder inductors are much less brittle no matter how they are fabricated either by a high pressing process or by a molding process. Lastly, the heavy temperature-dependent loss property of the ferrite inductors can be even more problematic considering the recent trend in electric cars; in order to increase the electric range, batteries are getting bigger and power from the converter systems is getting higher accordingly, which leads to high currents and temperatures. In the temperature range above 120 °C, loss becomes unacceptable for ferrite inductors, making magnetic powder inductors the only available choice for inductors in the converter system.

Optimization of Materials for a Mold Inductor

Not only for the material properties improvement, but also for the miniaturization of inductor, a mold inductor is fabricated. Figure 3 shows the fabrication processes of the ferrite inductor and the mold inductor. As for the mold inductor fabrication process, soft magnetic powders and polymer binder are mixed together and the coil is inserted into the inductor mold. The magnetic mixture is poured into the mold and heat is applied. Compared with the ferrite inductor fabrication process, the

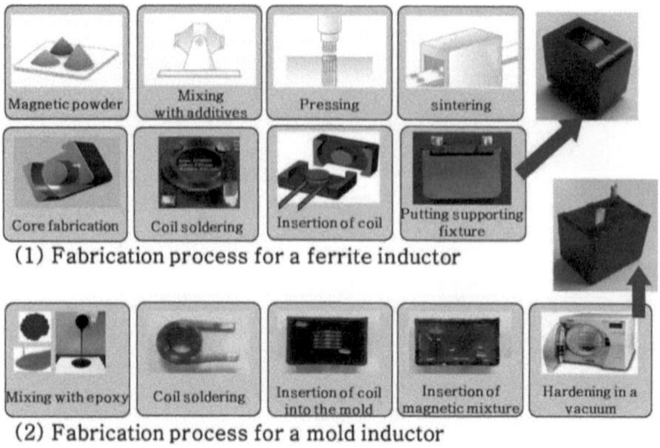

Fig. 3 **1** Ferrite inductor fabrication process. **2** Mold inductor fabrication process

mold inductor does not need a high pressure compacting process and sintering process, which makes it cost-effective.

Sendust which shows low loss property in the high current range and epoxy which is thermosetting polymer were mixed to make magnetic mixture. To achieve better magnetic properties, optimization was performed with regard to putting as little epoxy polymer as possible since more epoxy in a certain volume of magnetic mixture means less magnetic material, which leads to lower magnetic properties.

Table 1 indicates the comparison between early-stage product and final product. As for the final product, the size of the powders were reduced and CIP (Carbonyl Iron Powder) with size less than 2 μm was added to improve density, core loss and permeability. Additionally, degassing in a vacuum while applying heat was conducted to decrease the air gaps in the material as indicated in Fig. 4. In spite of the reduction of air gaps in the material, inductance in the high current range stays relatively unchanged because powders are still distanced from each other by the binder between them.

Table 1 Comparison between early-stage product and final product

	Early-stage product	Final product
Sendust powder size/magnetic powder % by weight (wt%)	100–180 μm/91%	25–75 μm, CIP with size less than 2 μm/97.3%
Core loss/permeability	440 mW/cc/13.28 μ	260 mW/cc/30 μ

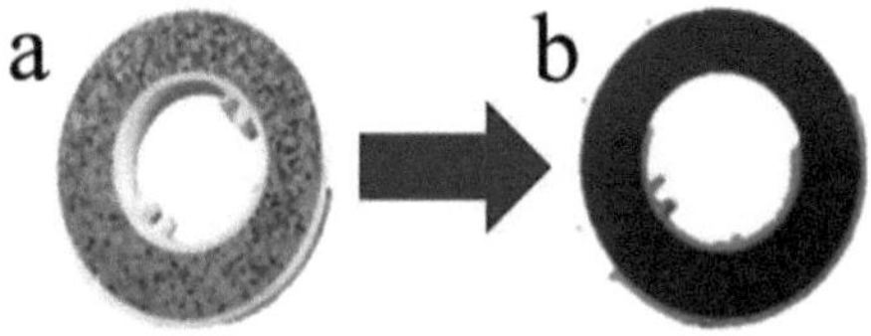

Fig. 4 Hardened material mixture before (**a**) and after (**b**) degassing in a vacuum

Mold Inductor Development

A mold inductor was developed targeting the inductor in the converter system in the eco-friendly cars. The current range was up to approximately 25 A. The inductance of the ferrite inductor stays relatively the same up to 25 A while inductance of the mold inductor decreases. Figure 5 shows the inductance of both inductors as a function of current and in this current range the efficiency of a converter is associated with the inductance. In the early-stage of development, the inductance of the mold inductor and the efficiency of the converter with mold inductors installed, instead of ferrite inductors, were inferior to the ferrite inductor due to the low inductance. The final product was designed to improve the efficiency of the converter by increasing inductance over ferrite inductors'.

Table 2 shows the comparison of inductors in terms of loss property. Even though coil loss was higher for the mold inductor, core loss of the mold inductor was lower due to the absence of the air gaps, leading to lower total loss for the mold inductor. In addition, miniaturization of the mold inductor was accomplished with the volume of the mold inductor being 91% of the ferrite inductor.

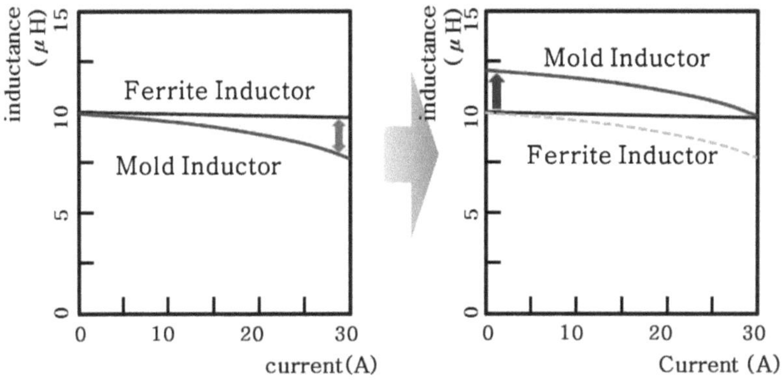

Fig. 5 Inductance as a function of current for early stage (*left*) and final (*right*) product

Table 2 Comparison of inductor properties

		Ferrite inductor	Mold inductor
Inductance (@0 A)		10.4 μH	14.9 μH
Inductance (@max current)		9.9 μH (@26.5 A)	12.1 μH (@24.5 A)
Coil loss (W)	Pdc	1.0	1.43
	Pac	0.53	0.35
	Sum	1.53	1.78
Core loss (W)		2.05	0.59
Total loss (W)		3.58	2.38
Size		10.1 cm^3 (100%)	9.2 cm^3 (91%)

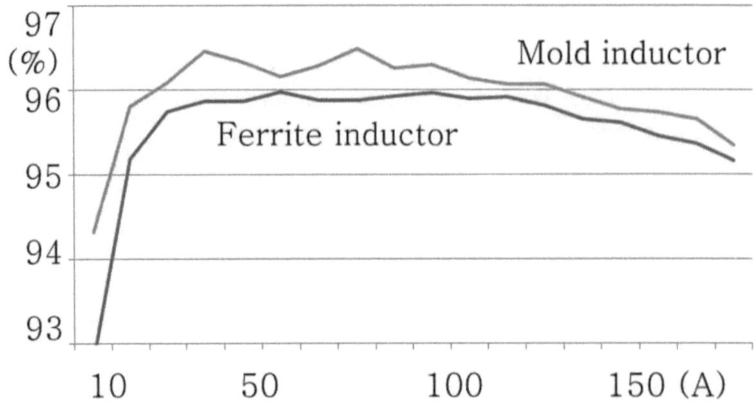

Fig. 6 Efficiency of the converter as a function of current

Converter Efficiency Test

The efficiency of the converter was tested. As can be seen in Fig. 6, not only in the current range of targeted inductor, but also in the current range up to more than 150 A, the mold inductor shows higher efficiency than the ferrite inductor by approximately 0.5%.

Conclusion

(1) Magnetic properties of the mold inductor, such as permeability and loss, were improved by achieving higher density of the magnetic mixture
(2) The air gaps in the magnetic mixture were reduced by applying heat in a vacuum
(3) The converter with the mold inductor installed was more efficient than the converter with the ferrite inductor.

References

1. R.C. O'Handley, *Modern Magnetic Materials Principles and Applications* (Wiley Inter-Science, 2000)
2. B.-G. Yoo, J.-S. Kim, B.-K. Lee, G.-B. Choi, D.-W. Yoo, Optimization of powder core inductors for buck-boost converters for hybrid electric vehicles. J. Electr. Eng. Technol. **6**(4), 527–534 (2011)

Effect of Annealing Time on the Texture of a 2.8% Si Non-Oriented Electrical Steel After Inclined and Skew Rolling

Mehdi Mehdi, Youliang He, Erik J. Hilinski and Afsaneh Edrisy

Abstract Non-oriented electrical steels are widely used in electric motors and generators as core materials to amplify magnetic flux and thus enhancing the conversion of energy. The efficiency of the motors and generators is closely related to the magnetic properties of the lamination core. In order to produce the magnetically favourable <001>//ND texture (θ-fibre) and suppress the unfavourable <111>//ND (γ-fiber) components in non-oriented electrical steels, two *unconventional* cold rolling schemes (*inclined* and *skew* rolling) were employed to process the steel. These rolling schemes have shown great potential in altering the texture of non-oriented electrical steel, especially for the 60° (inclined rolling) and 22.5° (skew rolling) angles. In this paper the effect of annealing time on the texture evolution of a 2.8 wt% Si steel was investigated using EBSD techniques. It was found that, all the unconventional rolling schemes were able to produce a strong θ-fiber texture, but the annealing time to achieve this texture was different.

Keywords Non-oriented electrical steel · Rolling · Texture · Magnetic properties · EBSD · Annealing

M. Mehdi (✉) · Y. He (✉)
CanmetMATERIALS, Natural Resources Canada, Hamilton, ON L8P 0A5, Canada
e-mail: mehdi6@uwindsor.ca

Y. He
e-mail: youliang.he@canada.ca

M. Mehdi · A. Edrisy
Department of Mechanical, Automotive, and Materials Engineering,
University of Windsor, Windsor, ON N9B 3P4, Canada

E.J. Hilinski
Tempel Steel Co., Chicago, IL 60640-1020, USA

The Minerals, Metals & Materials Society, *TMS 2017 146th Annual Meeting & Exhibition Supplemental Proceedings*, The Minerals, Metals & Materials Series,
DOI 10.1007/978-3-319-51493-2_66

Introduction

The final texture of non-oriented electrical steels is determined by the steel chemistry as well as the operational parameters used in the thermomechanical processing routes. Although a lot of researches have been conducted in the past [1–3], limited improvements were made to the final texture using conventional processing methods [2]. During conventional cold rolling and annealing, the favourable hot band texture obtained from hot rolling usually disappears, i.e. the θ-fibre is usually weakened, while the γ fibre is strengthened [3–5]. In recent studies, He et al. [6–8] investigated the influence of initial texture on the final annealing texture of non-oriented electrical steels using two unconventional rolling techniques, i.e. *inclined* rolling and *skew* rolling. It was shown that inclined rolling was able to significantly enhance the cube texture in the final sheets of a 0.9 wt% Si steel when the inclination angle was 60° [6, 7], and significantly different rolling textures were also obtained from the skew rolling technique [8]. This study is intended to characterize the annealing textures of a 2.8 wt% Si steel under such rolling conditions, and investigate the effect of annealing time on the formation of the final texture.

Experimental

The material studied is a high silicon (2.8 wt%) electrical steel, and its chemical composition is listed in Table 1. The steel was melted in a vacuum furnace and cast into a 200 × 200 mm^2 ingot. It was then hot rolled down to a thickness of ~2.5 mm at a temperature of 1311 K (1038 °C). After hot rolling the steel was annealed at 1113 K (840 °C) for 60 h in a dry 100% hydrogen atmosphere. Four small rectangular plates were then cut from the hot band (Fig. 1a): three at 0° and one at 60° to the hot rolling direction (HRD). One 0°-plate and the 60°-plate were conventionally cold rolled, but because of the cutting angle, the 60°-plate was actually rolled with the cold rolling direction (CRD) inclined 60° to the HRD. The other two 0°-plates were skew cold rolled at 22.5° to the rolls: one with *unidirectional* feeding (Fig. 1b), and the other with *bidirectional* feeding (Fig. 1c) [8].

All plates were cold rolled down to a final thickness of ~0.5 mm, and the cold rolled sheets were subsequently annealed at 1023 K (750 °C) for 10, 30, 60, 90 and 120 min in argon protected atmosphere. Cross-sectional samples were cut from the annealed sheets and the microstructure and texture were characterized by electron backscatter diffraction (EBSD). Typical texture components were plotted at

Table 1 Chemical composition of the investigated steel (wt%)

C	Mn	P	S	Si	Al	Fe
0.003	0.30	0.01	0.001	2.77	0.52	Balance

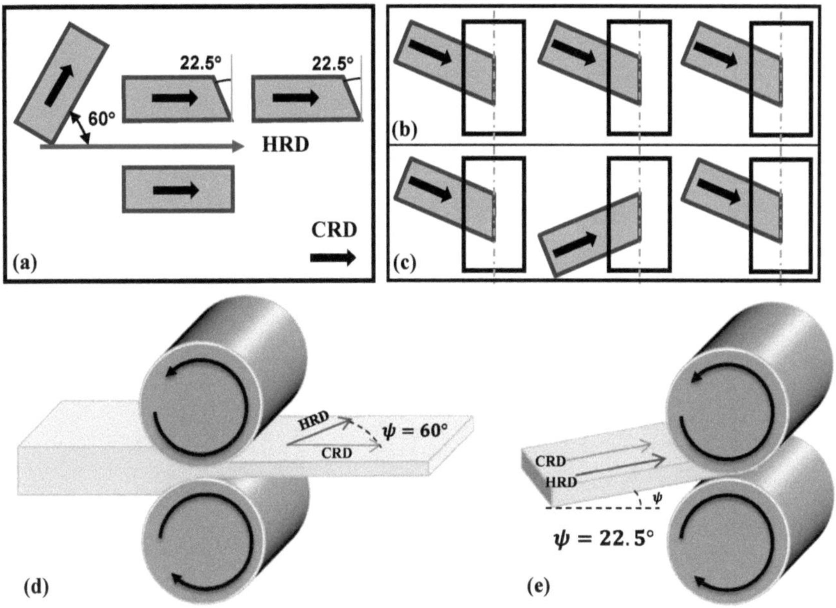

Fig. 1 Schematic illustration of the cold rolling schemes employed in this study: **a** sample cutting from the hot band, **b** unidirectional feeding in skew rolling, **c** bidirectional feeding in skew rolling, **d** inclined rolling, and **e** skew rolling

the $\varphi_2 = 45°$ sections of the Euler space (Bunge notation), and the volume fractions of ideal textures and fibres were calculated from the orientation distribution functions (ODFs) within a 15° tolerance from the exact orientation.

Results

The recrystallization textures (ODFs) obtained after various holding times at 750°C were compared in Fig. 2 for the samples after cold rolling using four different schemes, i.e. conventional rolling, inclined rolling at 60°, and skew rolling at 22.5° with unidirectional and bidirectional feeding. The ODFs were calculated from the EBSD orientation maps obtained from the cross-sections (rolling direction-normal direction). Scanned areas are approximately 0.5 mm (normal direction) × 2.2 mm (rolling direction) for all the samples. Here the textures are all presented in the conventional $\varphi_2 = 45°$ sections with $\varphi_1 = 0°–90°$. However, it should be noted that the skew rolling process does not have an orthorhombic symmetry, thus the texture would not be the same at the four 90° sections in the $\varphi_1 = 0°–360°$ range. Comparisons of the volume fractions of the common texture fibres and components are

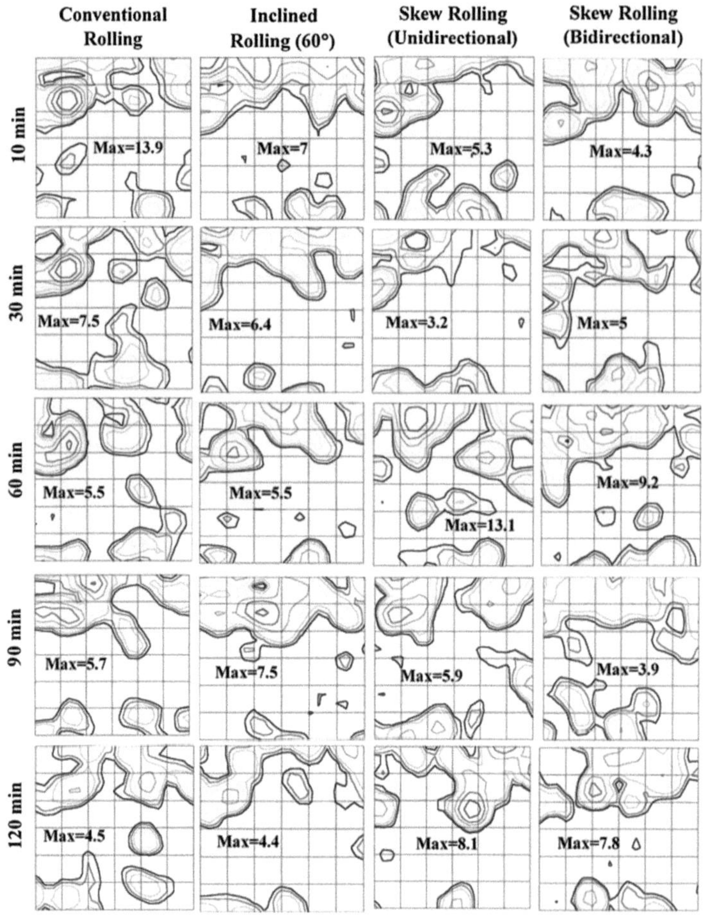

Fig. 2 Annealing textures of the samples after cold rolling with different rolling schemes and annealing times. $\varphi_2 = 45°$ ODF sections of the Euler space (Bunge notation)

illustrated in Figs. 3 and 4, respectively. Surprisingly, the textures after cold rolling with all the different schemes do not show a continuous γ-fibre regardless of the annealing time. Except for conventional rolling, all other rolling schemes are able to produce a strong θ-fibre, although the annealing time to produce it is different.

Conventional Rolling

After 10 min of annealing, a strong {113}<361> ($\varphi_1 \approx 20°$, $\Phi \approx 25°$, $\varphi_2 \approx 45°$) texture presents, which has an intensity of ~14 (Fig. 2). There are also rotated cube, rotated Goss and Brass components. With the increase of the holding time, the

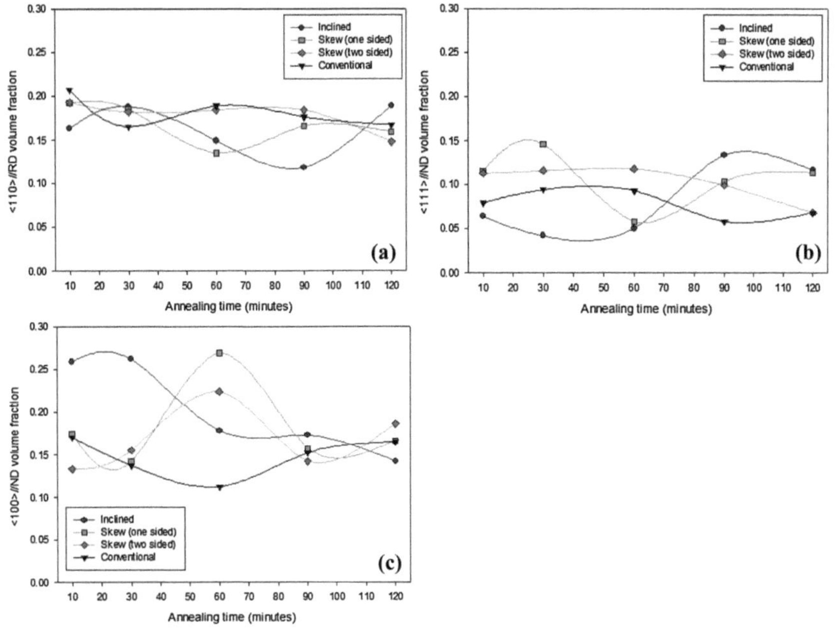

Fig. 3 Variations of the common fibre textures with respect to the annealing time: **a** α-fibre (<110>//RD), **b** γ-fibre (<111>//ND), and **c** θ-fibre (<100>//ND)

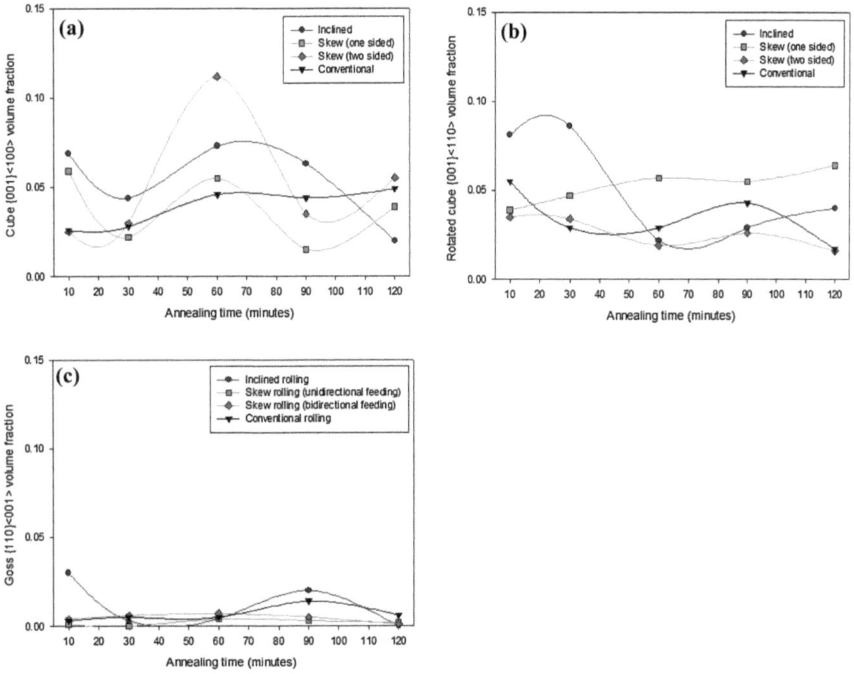

Fig. 4 Variations of the common texture components with respect to the annealing time: **a** cube ({001}<100>), **b** rotated cube({001}<110>), and **c** Goss ({110}<001>)

texture intensity significantly drops, and after 120 min of annealing, the texture is largely randomized, with only a few weak components. However, the {113}<361> component exists in all the times, although the intensity gradually decreases with the increase of the annealing time.

The α-fibre (Fig. 3a) has a high volume fraction of ~21% if the annealing time is 10 min, and it drops to ~16% when the annealing time increases to 30 min. After that, it is relatively stable and the volume fraction only varies slightly with further increase of the annealing time. The γ-fibre texture is also stable (Fig. 3b), with volume fractions varying from ~6 to ~8%. The θ-fibre, however, shows an apparent minimum if the annealing time is 60 min (Fig. 3c). The volume fraction varies from ~11% (60 min) to ~17% (10 and 120 min).

The cube texture is very weak (~2.5%) if the annealing time is 10 min, and it gradually increases to ~4% after the annealing time increases to 60 min or more (Fig. 4a). The rotated cube component has the highest volume fraction (~6%) after 10 min of annealing, and it drops to ~3% when the annealing time increases to 30 or 60 min (Fig. 4b). After 90 min there is a slight increase, and it drops again to ~2% when the annealing time increases to 120 min. The Goss texture (Fig. 4c) is rather weak and the volume fraction is very low, i.e. ~1.5% or less.

Inclined Rolling

The incline-rolled steel shows considerably different textures from the conventionally rolled steel (Fig. 2). After 10 min of annealing, there is a strong θ-fiber (~26% volume fraction, Fig. 3c) and a weak γ-fiber (~6%, Fig. 3b). The highest intensity is at rotated cube (~7.0), while in conventional rolling, it was at {113} <361>. It is also noted that, in this steel, the {113}<361> texture only appears after the annealing time is increased to 60 min or more. Increasing the annealing time to 30 min, the volume fraction of the θ-fibre essentially does not change, but the α-fiber increases considerably from ~16% to ~19%, and the γ-fibre drops slightly from ~7 to ~4% (Fig. 3). Further increasing the annealing time to 60 and 90 min, the θ-fibre drops rapidly to ~17%, and the α-fibre drops to ~12%. In the meantime, the γ-fibre increases significantly to ~13%. If the annealing time increases to 120 min, the θ-fibre is further weakened to ~14%, while the α-fibre is significantly strengthened to ~19%. The γ-fibre, on the other hand, is only slightly weakened.

The cube, rotated cube and Goss components after 10 min of annealing all have the highest volume fractions as compared to samples rolled by other rolling schemes (Fig. 4). The cube component (Fig. 4a) weakens when the annealing time increases to 30 min, and it strengthens again if the time increases to 60 min. After that, the volume fraction of the cube texture decreases with the increase of annealing time. After 120 min, it drops to only ~2% from the initial ~7%. The rotated cube component (Fig. 4b) remains essentially the same if the annealing time is 10 min or 30 min. However, when the annealing time increases to 60 min, the volume fraction rapidly drops to ~2.5% from ~8%. Further increasing the annealing time slightly increases the rotated cube, but the volume fraction is rather low. The Goss component (Fig. 4c) follows a similar trend to the cube, and the volume fractions at 10 and 90 min are considerably higher than those of the samples rolled by other schemes.

Skew Rolling (Unidirectional Feeding)

If the annealing time is 10 min, a strong component ($\varphi_1 \approx 5°$, $\Phi \approx 30°$, $\varphi_2 \approx 45°$) close to {113}<110> develops (Fig. 2). This component weakens gradually with the increase of the annealing time up to 60 min. At 60 min, a strong θ-fibre component ($\varphi_1 \approx 22.5°$, $\Phi \approx 5°$, $\varphi_2 \approx 45°$) develops between the cube and rotated cube, which has a very high intensity of ~13 (Fig. 2). The volume fractions of the θ-fiber after 10 and 30 min annealing (Fig. 3c) are approximately the same as those of the conventionally rolled steel annealed at the same times, but there is a significant increase from ~14 to ~27% when the annealing time increases to 60 min. Further increasing the annealing time to 90 and 120 min decreases the volume fraction again to 15–16%. The γ-fibre essentially follows an opposite trend to the θ-fiber (Fig. 3b), i.e. at 60 min, the volume fraction is the lowest (~5%), while at 30 min, it is the highest (~15%). The variation of the α-fibre with respect to the annealing time is similar to the γ-fibre, i.e. with a minimum (~13%) at 60 min.

The change of the cube component (volume fraction) with respect to the annealing time (Fig. 4a) is very similar to that of the θ-fiber (Fig. 3c), but the cube volume fraction (~5%) at 60 min is much lower than the θ-fibre (~27%) at the same annealing time. The volume fraction of the rotated cube slightly increases from ~4 to ~6% when the annealing time increases from 10 to 120 min. The Goss component (Fig. 4c) is really low in all the cases, and its variation with respect to the annealing time is very small.

Skew Rolling (Bidirectional Feeding)

The variation of the θ-fibre (Fig. 3c) with respect to the annealing time is very similar to that of the steel skew-rolled with unidirectional feeding, i.e. at 60 min, a maximum volume fraction (~22%) is noticed. The volume fraction of the γ-fibre (Fig. 3b) remains approximately the same (~11%) when the annealing time increases from 10 to 60 min, but beyond 60 min the volume fraction starts to decrease, and it drops to a minimum of ~7% at 120 min. The α-fibre volume fraction is also approximately the same at different annealing times (~18%), but it decreases to a minimum of ~15% at 120 min.

The variation of the cube component with respect to the annealing time shows a very similar trend to that of the θ-fibre (and also to that of the steel skew-rolled with unidirectional feeding), i.e. with a maximum volume fraction at 60 min. It should be noted that the cube volume fraction after 60 min is more than 12% in this case, which is considerably higher than that of any other sample after any annealing times (~7% or less). Thus skew rolling with bidirectional feeding can significantly promote the formation of the cube texture if the annealing time is 60 min (at 750 °C). The rotated cube texture slightly decreases with the increase of annealing time, but the variation is very small. The Goss texture is again very weak at all annealing times, with a volume fraction less than 1%.

Discussion

The above analysis of the annealing textures showed that in all the samples cold rolled by the four different schemes, no continuous γ-fibre was observed, and the γ-fibre volume fraction (~5 to ~15%) is relatively low as compared to the θ-fibre (~12 to ~27%) and α-fibre (~13 to ~21%). Inclined rolling and skew rolling with unidirectional feeding significantly promoted the formation of the θ-fibre, with volume fractions larger than 25% after annealing for certain times, while after conventional rolling, the maximum θ-fibre volume fraction is only ~16% regardless of the annealing time. Thus not only do the cold rolling scheme and the resulted cold deformation textures affect the final annealing texture, but the annealing time also plays an important role in determining the final texture.

The mechanisms governing the formation of the annealing textures in deformed metals have been studied for many decades, and the origin of the recrystallization texture has been a subject of dispute for over 50 years [9]. Oriented nucleation theory [9, 10] proposes that there exist preferred crystal orientations for the grain nucleation, while the oriented growth theory [9, 11] suggests that certain crystal orientations grow preferentially over others. Despite the dispute, it is well known that the annealing texture of bcc metals is highly dependent on the processing history (especially the cold deformation texture) and the annealing conditions (temperature, time, heat/cooling rates, etc.). In this study, the inclined rolling essentially changed the initial texture and thus the cold deformation texture. Upon annealing, the recrystallization texture is also altered due to the differences in the nucleation and grain growth associated with the different deformation matrices. Increasing annealing time may induce grain growth, which normally also changes the final texture. Thus the annealing time has an apparent effect on the final texture. Skew rolling also alters the initial texture because of the skew feeding of the material into the rolls. In addition, it was shown [8] to be a three-dimensional deformation process (instead of plane-strain rolling), thus the deformation mode applied to the steel is also different, which results in different deformation textures. The subsequent annealing again will result in different recrystallization texture, and the annealing time also plays an important role in the grain growth process.

Conclusions

Annealing (at 750 °C) of a 2.8 wt% Si non-oriented electrical steel cold rolled with conventional rolling, inclined rolling and skew rolling resulted in the suppression of the γ-fibre regardless of the annealing time.

Inclined cold rolling at 60° to the HRD promoted the formation of a strong θ-fibre and a weak γ-fibre if the annealing time was short (i.e. 10 or 30 min). The θ-fibre rapidly dropped and the γ-fibre increased if the annealing time increased to 60 min or more.

A strong θ-fibre (~27%) formed after 60 min of annealing in the steel sheet skew rolled with unidirectional feeding. At any other annealing time, the θ-fibre was considerably weakened, while the γ-fibre was strengthened.

Skew-rolling with bidirectional feeding also produced a relatively strong θ-fibre after 60 min annealing, but the maximum volume fraction (~22%) was slightly lower. A very strong cube texture (~12%) was produced in this steel if it was annealed for 60 min.

References

1. L. Kestens et al., Orientation selective recrystallization of nonoriented electrical steels. Metall. Mater. Trans. A **27**(8), 2347–2358 (1996)
2. L. Kestens, S. Jacobs, Texture control during the manufacturing of nonoriented electrical steels. Texture Stress Microstruct. **2008**, 1–9 (2008)
3. Jong-Tae Park, Jerzy A. Szpunar, Evolution of recrystallization texture in nonoriented electrical steels. Acta Mater. **51**(11), 3037–3051 (2003)
4. R. Dierk, K. Lücke, Rolling and annealing textures of bcc metals, in *Materials Science Forum*, vol. 157 (Trans Tech Publications, 1994)
5. D. Raabe, K. Lücke, Annealing textures of bcc metals. Scr. Metall. Mater. **27**(11), 1533–1538 (1992)
6. Y. He, E. Hilinski, J. Li, Texture evolution of a non-oriented electrical steel cold rolled at directions different from the hot rolling direction. Metall. Mater. Trans. A **46**(11), 5350–5365 (2015)
7. Y. He, E.J. Hilinski, Texture and magnetic properties of non-oriented electrical steels processed by an unconventional cold rolling scheme. J. Magn. Magn. Mater. **405**, 337–352 (2016)
8. Y. He, E. Hilinski, Skew rolling and its effect on the deformation textures of non-oriented electrical steels. J. Mater. Process. Technol. (2016, in press)
9. R. Anthony et al., *Recrystallization and Related Annealing Phenomena* (Elsevier, 2004)
10. I.L. Dillamore, C.J.E. Smith, T.W. Watson, Oriented nucleation in the formation of annealing textures in iron. Metal Sci. J. **1**(1), 49–54 (1967)
11. P.A. Beck, Annealing of cold worked metals. Adv. Phys. **3**(11), 245–324 (1954)

Multi-Parameter Magnetic Material Characterization for High Power Medium Frequency Converters

Richard Beddingfield and Subhashish Bhattacharya

Abstract With the rapid availability of new magnetic materials and even more options in post-processing techniques such as field and strain annealing, full property maps of materials and final fabricated components are needed for proper system design. This is especially critical in modern wide bandgap power electronics based converters where medium switching frequencies are used to deliver high power. In these systems, the magnetic design has a significant impact in the overall losses. The magnetics constitute a majority of the losses and magnetic properties have a defining influence on the overall systems, e.g. available power flow and soft switching regions. A new lossless structural harness that allows easy testing of cores and the development of property maps is presented. The harness also enables the testing of mechanical variations such as gapping and clamping pressure on the core.

Keywords Magnetic ribbon • Magnetic characterization • Medium frequency • Power electronics

Introduction

There are two significant trends that are enabling high power medium frequency converters. First the rapid commercialization of wide bandgap (WBG) semiconductors has increased the voltage and current rating of devices traditionally used in lower power circuits in high power applications [1]. Second, advancements in magnetic materials have provided the energy storage and coupling capabilities necessary for the realization of these converters. There is ample study and understanding of high power low frequency and low power high frequency magnetic materials such as silicon steel and ferrites respectively. However, there is a need to

R. Beddingfield (✉) · S. Bhattacharya
ECE Department of NSF FREEDM Systems Center, North Carolina State University, 1791 Varsity Drive, Raleigh, NC 27603, USA
e-mail: rbbeddin@ncsu.edu

The Minerals, Metals & Materials Society, *TMS 2017 146th Annual Meeting & Exhibition Supplemental Proceedings*, The Minerals, Metals & Materials Series, DOI 10.1007/978-3-319-51493-2_67

better understand the materials that fall between these two application spaces. Furthermore, the electrical and magnetic waveforms seen by these materials are unique for different WBG converter topologies [2–4]. Therefore traditional characterizations fall short in enabling effective system design.

This paper will describe the construction of a magnetic core harness and test system that allows for material characterization with minimal error with an emphasis on medium frequency, high power converters. First a test bed that enables characterization of the magnetic materials over a variety of operating and environmental parameters will be presented. Next, the sources of characterization error and mitigation techniques will be presented. Finally, an example characterization for a Finemet core with transverse annealing will be presented.

Magnetic Core Test Bed

The test bed consists of three primary entities. First, a harness is needed to enable an appropriate testing domain for the desired material. The harness should allow impact studies of additional environmental factors on the core behavior as well. The second component of the test bed is a DC power supply. This power supply is separate from any necessary control power and should have a sufficient power rating for the excitation energy. Finally, power electronics converter capable of producing the desired excitation waveform is needed. It is important to ensure that this converter has a sufficient power rating necessary to fully energize the core material under test over the extent of the testing domain.

Core Harness

The main purpose of the harness is to provide a platform for controlled and repeatable testing conditions for the magnetic core under test. Figure 1 is a photograph of such a core harness. There is no conductive material in the harness that is in contact or near the core in this harness. The loss errors this can introduce are discussed in the next section. The core is stabilized in both the vertical and horizontal axis. A slotted extruded aluminum superstructure is used to mount all fiberglass stabilizing brackets.

The primary method of holding the core in place is the two horizontal pieces of red GPO3 fiberglass angle. This angle is reinforced with fiberglass bar. GPO3 is preferred for core surface contact because of its superior electrical and temperature ratings of standard fiberglass [5]. With the higher rating, the test domain is expanded. The fiberglass angle is in contact with the core on its corner to maximize holding strength and minimize core surface coverage. This allows more core surface for natural thermal convection.

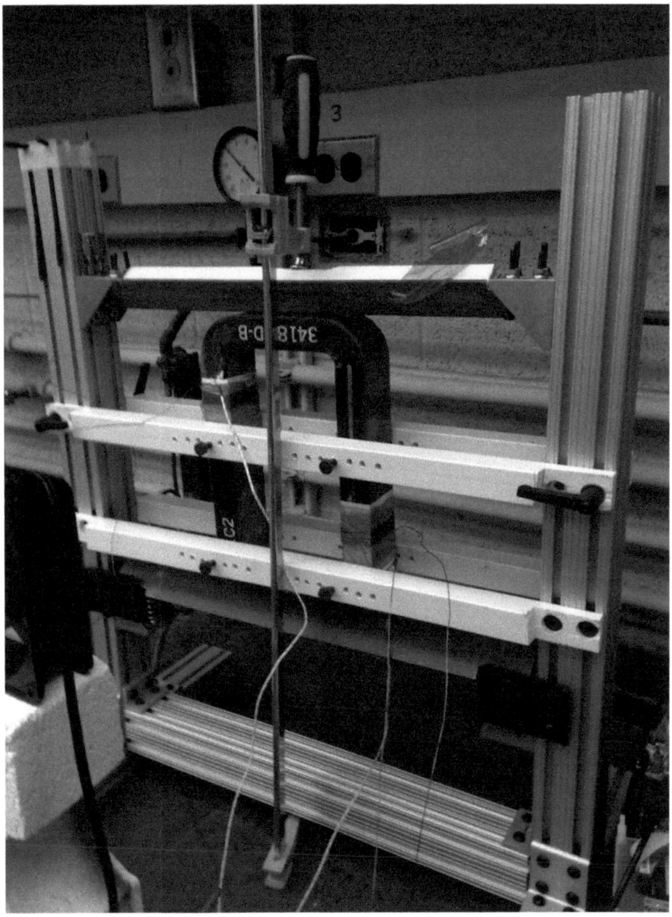

Fig. 1 Magnetic core testbed harness

Eight fiberglass screws are used for horizontal support. These stabilize and ensure proper cut core alignment. The screws are tapped and threaded into the white fiberglass bar seen in the center of Fig. 1 ensuring holding pressure. These screws safeguard from shifting in the horizontal plane. Furthermore, they enable fine-tuned controlled of the cut pieces of magnetic core to ensure proper alignment. Once aligned, the screws prevent the effects that fringing fluxes would have in forcing the core to twist and break out of alignment. The top fiberglass bars have quick release handles enabling easy and rapid test core changes. These screws provide the necessary horizontal position control while minimizing the core surface that is covered. This enables positioning without increased thermal penalties.

Another feature of the harness is the ability to produce consistent and controlled mechanical clamping force on the core under test. There are two calibrated springs that provide the backing for the bottom vertical support angle. A screw clamp is used to apply compressive forces on the core. Then top vertical support is locked in place. The applied compression force is easy measured by determine the compression distance of the springs and using the known spring constant. The compressive force is distributed along the top and bottom of the core. Care must be taken that neither top nor bottom brackets deflect under high compressive forces as this will lead to non-uniform force distribution on the core.

With the variability of support bracket location, the harness provides a flexible platform for a variety of magnetic material tests. There is ample space for various tests such as winding configuration, bobbin design and others. The core surface is exposed as much as possible to ensure good cooling either through forced air or natural convection. Vertical compression enables consistent mechanical studies and provides the flexibility of air gap testing. The presented harness provides the physical framework for producing accurate and repeatable material characterizations while controlling for numerous environmental parameters.

Excitation Power Supply

A voltage controlled DC power supply is needed to provide the core excitation power. For this test bed the Keysight 8937 A 15 kW DC supply was used. This power supply is able to provide the necessary excitation power. The built in current limit functions also provide a level of safety during testing.

One additional consideration for the DC supply is voltage stiffness. While the power supply may have sufficient power rating, and only loss power is being consumed, the voltage may dip during transitions. This is because the converter must supply both loss current and reactive current necessary to magnetize the core. Because this current is cyclic, it will not register in the DC power supplies current reading but it can cause voltage ripples. In order to improve the stiffness and illuminate the spurious ripple, an additional external capacitor bank was added to the output of the supply. This was a capacitance of 3.6 mF. At rated voltage of the converter, 1200 V, this gives an inertia constant (1), of 0.17 s. This is much greater than the 0.0015 s of the supply without the additional capacitance.

$$H = \frac{CV^2}{2P} \tag{1}$$

Excitation Drive

The excitation drive converts the DC power into the desired excitation waveform. In order to focus on the high power medium frequency application domain, Silicon Carbide WBG mosfets are used in an H-bridge topology as seen in Fig. 2.

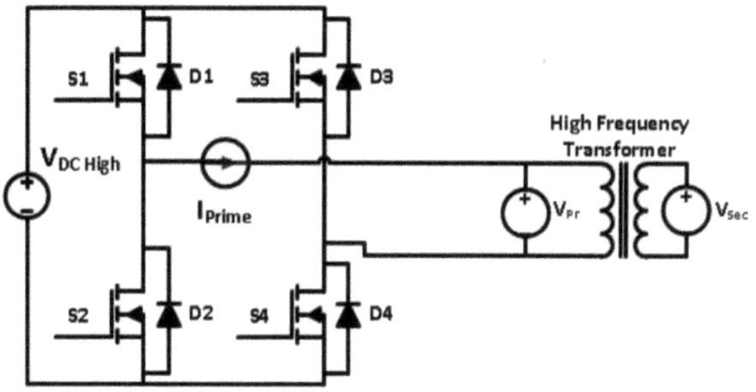

Fig. 2 Circuit topology of magnetic core testbed

C2M0080120D mosfets from Wolfspeed give the converter a 1200 V, 20A rating. The core under test is connected to the two legs of the converter to enable bipolar excitation.

The primary core parameters of core loss, excitation power and relative permeability are easily extracted by measuring the primary current and the open circuit secondary voltage. The magnetizing force and and the flux density are calculated from (2) and (3) respectively. The primary current is scaled by the primary turns, N_p and magnetic path length l_m to determine magnetizing force, H. The induction is calculated by integrating the recorded secondary voltage in post processing and scaling by the secondary turns N_s and core cross sectional area A_c.

$$H(t) = \frac{N_P i_P(t)}{l_m} \tag{2}$$

$$B(t) = \frac{1}{N_s A_c} \int_0^t v_s(\tau) d\tau \tag{3}$$

Using (2) and (3), the relative permeability is calculated as the ratio between the two when the magnetizing force is at its maximum (4).

$$\mu_r = \left.\frac{B}{\mu_0 H}\right|_{H\max} \tag{4}$$

A TI Delfino DSP is used for the control of the mosfet gating pulses. Again, multiple pulse patters can be generated from this controller to achieve the desired excitation waveform. Further protection for the test bed is achieved the gate driver circuit where the mosfet forward conduction voltage is sensed to detect shoot through and over current faults. These faults block all pulses and allow for a safe cool down of the test bed. The control, gatedriver and power circuit are shown from top to bottom respectively below in Fig. 3.

Because the load is predominately inductive, the converter operates in zero voltage switching mode [6]. This allows for effective driving of the circuit in the hundreds of kilohertz where 500 kHz operation has been confirmed for this particular system. With such a high switching frequency, it is easy to generate a variety

Fig. 3 Photograph of magnetic testbed converter stage

of waveforms. A simple LC low pass filter can be used to create sinusoidal excitation while triangular excitation is trivial. This means that the converter can generate the specific waveform that the magnetic core would see in the actual WBG converter.

In many power electronics applications the exciting current is triangular in shape. This is due to the square wave voltages applied to the magnetic components. Figures 4 and 5 are two examples of applied waveforms that the magnetic tesbed is capable of producing. In the two figures, the switching frequency defines the period, T. In cold temperature, baseline tests, a single excitation wave will be used. To study full operation of the magnetic material the exciting waveforms should be applied for a longer duration. An additonal aspect of this longer test is blanking time between excitation waves. Figure 4 shows more traditional excitation waveforms where the

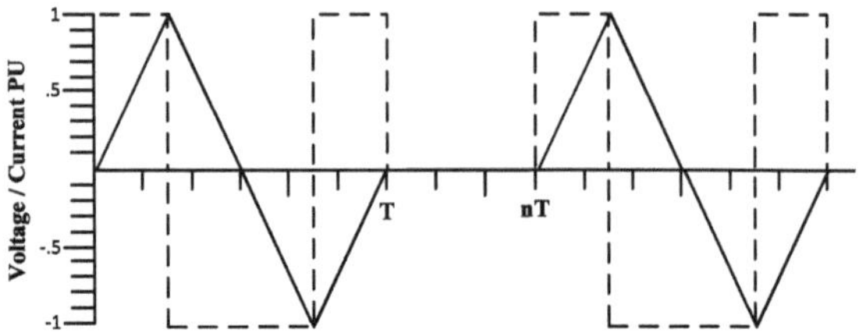

Fig. 4 Continuous excitation waveform template; solid = exciting current, dash = applied voltage

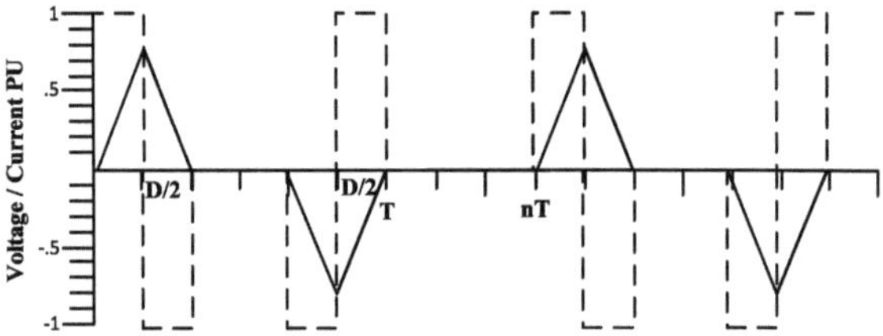

Fig. 5 Discontinuous excitation waveform template. solid = exciting current, dash = applied voltage

flux wave is continuous. Figure 5 can be used to explore modern advanced excitation waveforms specifically used in dual active bridge topologies [6–9].

The magnetic material test bed is a flexible platform that enables the testing of magnetic materials for multiple environmental and application specific parameters. The harness is designed to isolate the core and allow the study of physical stresses that would be present in a practical system. The power converter and supply provide the necessary excitation energy and excitation waveform creation to study the specifics that different high power, medium frequency wide bandgap power converters would use.

Sources of Characterization Error

In order to accurately characterize a material, it is essential to identify and minimize non-core variations. Specifically, measured core losses can be influenced by a variety of sources most notably measurement delays, and harness induced losses. This section will describe best practices for minimizing or mitigating entirely these sources of variation in material characterization.

Current Sensor Introduced Error

First we must understand that the error introduced by the current sensor is due to the delay of the current sensor. For any given signal, this error is shown below in (5). Here, δ is the actual angle of measurement while φ is the angle error introduced by a sensor delay and E_{Tol} is the maximum tolerance of the particular sensor.

$$E_T = 100\frac{\cos(\delta+\phi)-\cos(\phi)}{\cos(\delta)} + E_{Tol} \tag{5}$$

Given the frequency range and ΔB of test, transformer parameters and losses at those particular test points, a profile the current sensor errors can be developed. The values δ can be determined from (6). Here the equivalent core resistance, R_c, and reactance, X_m, can be calculated by (7) and (8) respectively.

$$\delta = \arctan\left(\frac{R_C}{X_m}\right) \tag{6}$$

In (7), N_p is the number of primary turns, μ is the permeability, l_m is mean magnetic path length and A_c is the core area, f is frequency and Δθ is the measured phase difference between voltage and current. In (8), P_{loss} is found using the core loss curves.

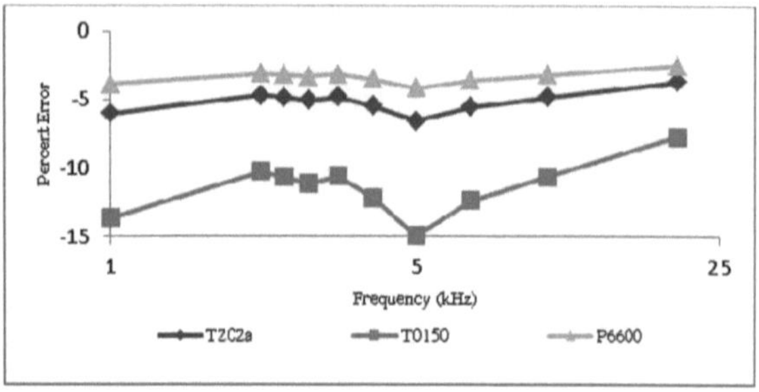

Fig. 6 Maximum error for different current probes delay and tolerance, T is Tektronix and P is Pearson

$$X_m = \frac{2\pi f N_p \mu_r \mu_0 A_c}{l_m} \tag{7}$$

$$R_c = \frac{V_{rms}^2}{P_{loss}} = \frac{\left(2\pi f N_p A_c \frac{\Delta\theta}{2\sqrt{2}}\right)^2}{P_{loss}} \tag{8}$$

It is important to note that the current sensor error is most significant in the low loss situations. A comparison of common laboratory current sensors is shown below in Fig. 6. Only the low frequency, low loss region is shown as the percent error converges to minimal impact at higher losses.

It is clear from Fig. 6 that the Pearson current sensor is the best choice. With the Pearson 6600, the current measurement error is consistently within 5%. Since the manufacturer data is used to determine the potential sensor error, this should only be used as a baseline to ensure that the error is within an acceptable limit.

Stray Eddy Losses in Harness

Another source of error can be introduced by stray eddy currents induced in the harness itself. As mentioned above, great care has been taken to ensure that non-conductive material is the only aspect of the harness in contact with the core. The aluminum super structure is present however and could be such a source of error. Figures 7 and 8, finite element analysis (FEA) verifies that the aluminium frame is sufficiently far enough from stray flux lines.

Figure 7 shows the FEA results for the worst case leakage flux at low excitation levels. This occurs when the core has an air gap and the core is excited at high frequencies. Even in this extreme case no flux lines cross into the aluminum of the core harness so it does not contribute to inflated loss measurements.

The same study is performed for high excitation levels, Fig. 8, where the core is driven near saturation. The same airgap is used and it is also a low frequency excitation. Again, the leakage flux lines do not interact with aluminum core implying that there are no additional eddy currents induced in the core superstructure.

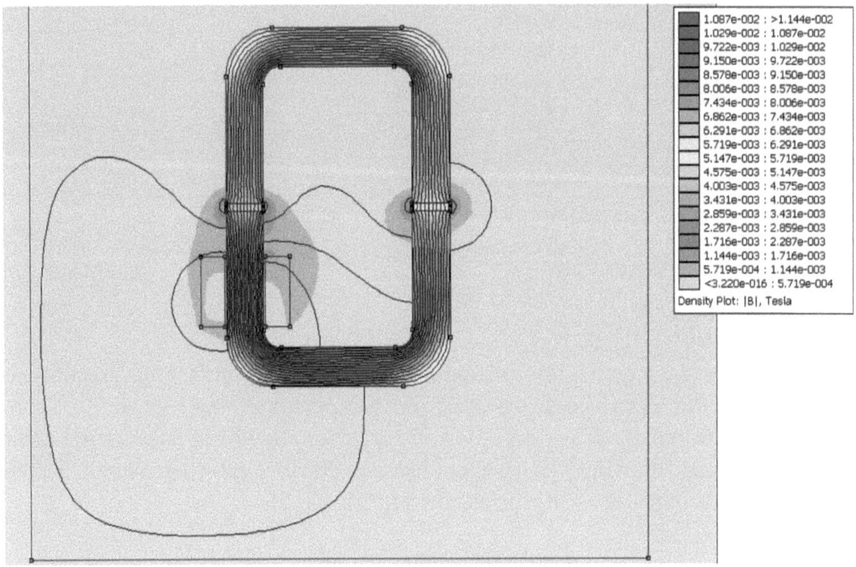

Fig. 7 High frequency, low excitation core harness FEA

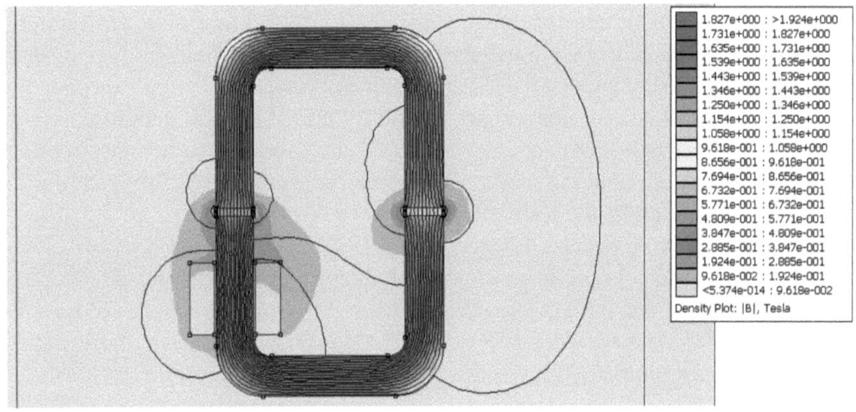

Fig. 8 High frequency, high excitation core harness FEA

Material Characterization

To demonstrate the effectiveness of this magnetic testbed an example characterization is presented below. This results show various parameters extracted from Finemet race track core as seen pictured in Fig. 1. Table 1 lists some of the physical parameters of the core provided by the manufacturer.

Table 1 Physical properties of test core

Property	Value	Units
Magnetic length	28.0614	in
Stacking factor	0.78	
Effective cross sectional area	1.5223	in^2
Weight	11.657	lb
Annealing	Transverse	
Inner corner radius	0.5	in
45° beveling width	0.038	in
Rated core loss	5.0	W/lb at 20 kHz

It is clear from the above information provided by the manufacturer that only an incomplete picture is provided. As such the full parameter map should be determined using the proposed test bed. Baseline parameters will be determined using single cycles and allowing full core cool down. Steady state parameters will be investigated after the core has reached thermal steady state.

Baseline Material Properties

As mentioned above the baseline parameters are determined near room temperature. Three steps were taken to ensure that the core temperature did not vary significantly during testing. First, fans were used to provide forced air cooling. Second, the excitation was applied only for a short duration. Finally, the core temperature was measured using K type thermocouples at the air gap, a corner of the inner bend, at center of the primary winding and at the geometric center of the flux path on the face of the core. During the tests the temperature was held within 5° C and tests were halted if the core became hotter.

Using (2–4) over a broad frequency and induction testing domain the property maps of Figs. 9, 10 and 11 were developed. During testing, the supply voltage was adjusted to reach desired induction using (9) where k is the wave shape factor. Using the pulsing profile of Fig. 4, this leads to k = 1.

$$V_{DC} = 4kfA_cN_pB_{ref} \tag{9}$$

With this loss profile it is easy to curve fit the losses into the Steinmetz form (10) through solving the linear Eq. (11). It is important to note that for this equation is valid only for the triangular excitation. While triangular losses are proportional to sinusoidal losses [10], this method of producing the constants does not allow direct scalability.

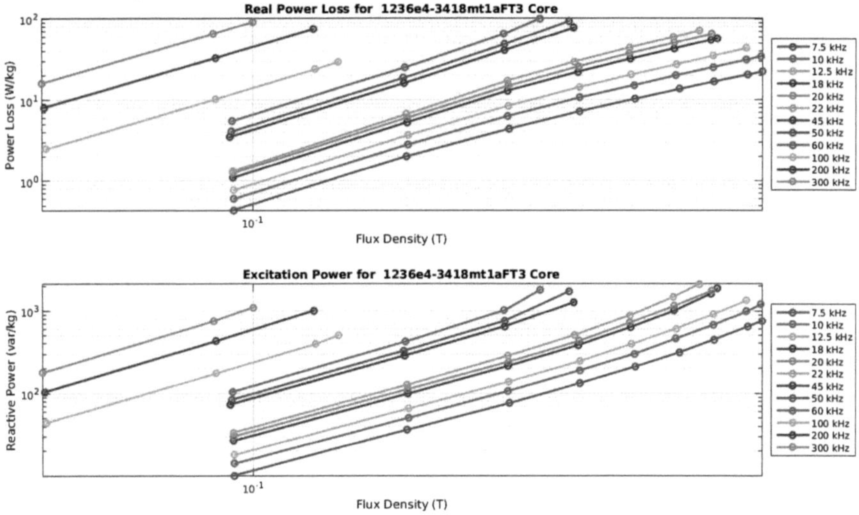

Fig. 9 Core loss and excitation energy at room temperature

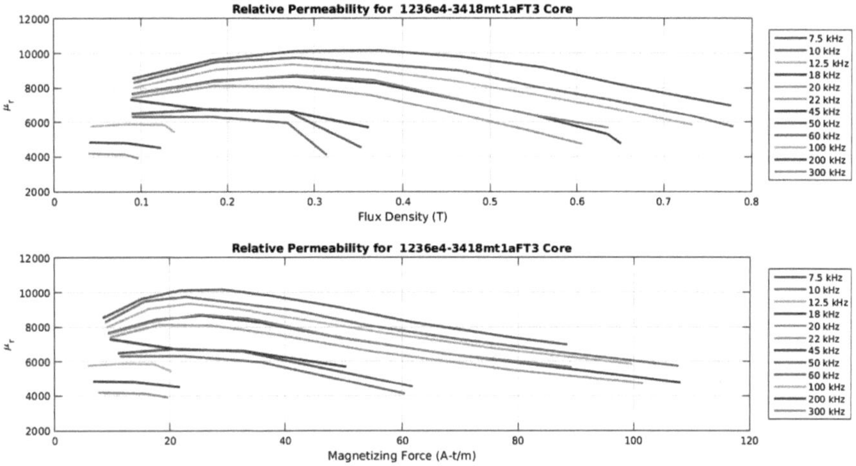

Fig. 10 Relative permeability versus flux density and magnetizing force at room temperature

$$P_{VTriangular} = kf^{\alpha}B^{\beta} \tag{10}$$

$$\begin{bmatrix} P_1 \\ P_2 \\ P_3 \end{bmatrix} = \begin{bmatrix} 1 & \ln(f_1) & \ln(B_1) \\ 1 & \ln(f_2) & \ln(B_2) \\ 1 & \ln(f_3) & \ln(B_3) \end{bmatrix} \begin{bmatrix} \ln(k) \\ \alpha \\ \beta \end{bmatrix} \tag{11}$$

The relative permeability is also calculated according to (4) and plotted against both flux density and excitation force over the test domain as seen below.

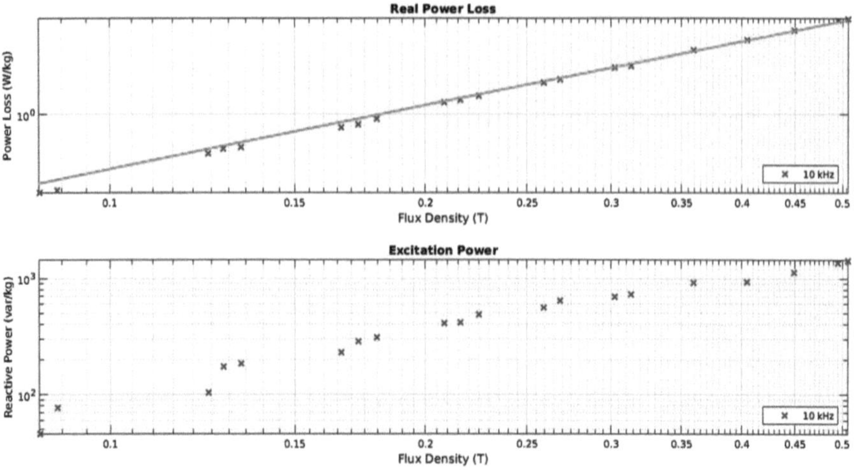

Fig. 11 Core loss and magnetizing power due to gapping for 0.24, 0.48 and 0.72 mm air gaps at room temperature for 10 kHz excitation

Finally, the core was tested at a 10 kHz excitation frequency for various air gap lengths. In Fig. 11 the measured core loss and excitation power are shown. Excitation power increased dramatically as expected due to the reduction in effect magnetizing inductance with the increased air gap. Interestingly, the core loss remained very close to non-gapped predicted values shown by the red trend line. This implies minimal impact from fringing flux or other air gap non-idealities associated with the gap with respect to core loss. In other words, any penalty due to gapping is present once the core is cut and does not change with increasing gap.

Once the base line properties are established it is possible create the first iteration of design. However, a complete design would consider the steady state effects and not rely on parameters established in such an environmentally regulated test.

Practical Material Properties

A practical design will consider the thermal effects on core parameters. The results presented in this test are for a single excitation profile of 10 kHz and 0.6 T flux density. The system was allowed to run for 3 h until thermal steady state was achieved. The excitation was then disable to naturally cool the core. There was no forced air convection for these tests.

The thermal profile is shown in Fig. 12. Snapshots of the primary current and induced voltage were taken as consistent intervals during the test. The face temperature at the geometric center of the magnetic path was used for analysis. The Temperature at the gap was much higher but this was a local effect and therefore not indicative of the magnetic core temperature. A similar argument is made for the

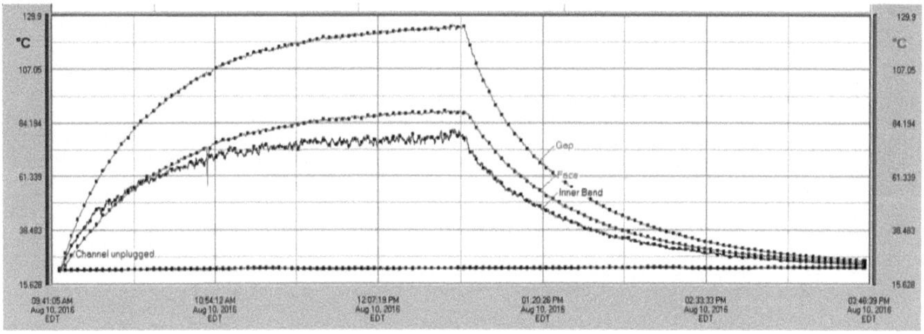

Fig. 12 Temperature profile of long duration test of Finemet magnetic core

inner bend temperature which is cooler due to the increased surface area exposed to cooling.

By repeating the analysis approach used in the baseline tests the thermal impact can also be studied. Figure 13 shows the loss and excitation with respect to temperature. Interestingly, the losses are weakly coupled to temperature variation. The total change in room temperature to the steady state was less than 6%. There is small decrease in losses due increased resistivity reducing eddy currents and

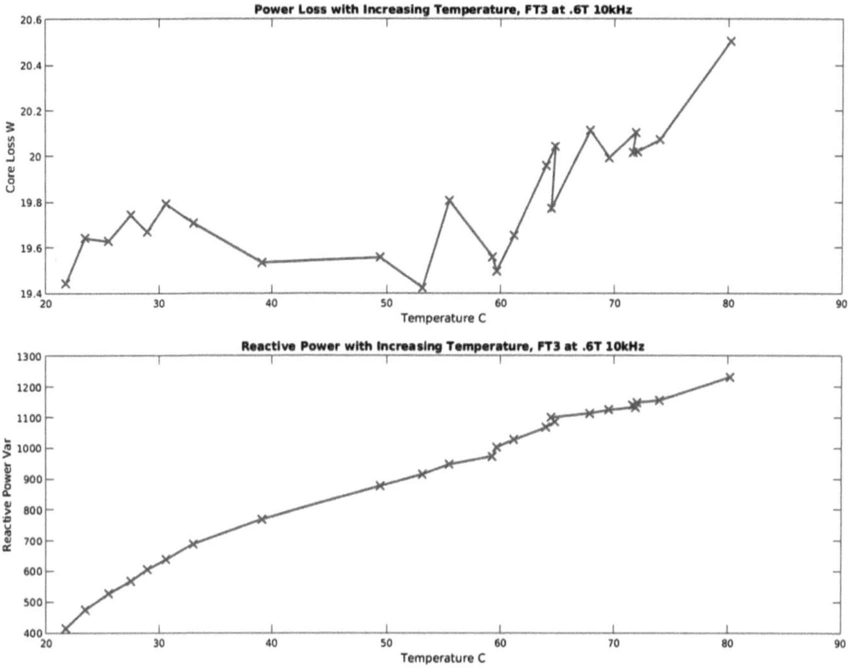

Fig. 13 Core loss and exciting power with respect to face temperature

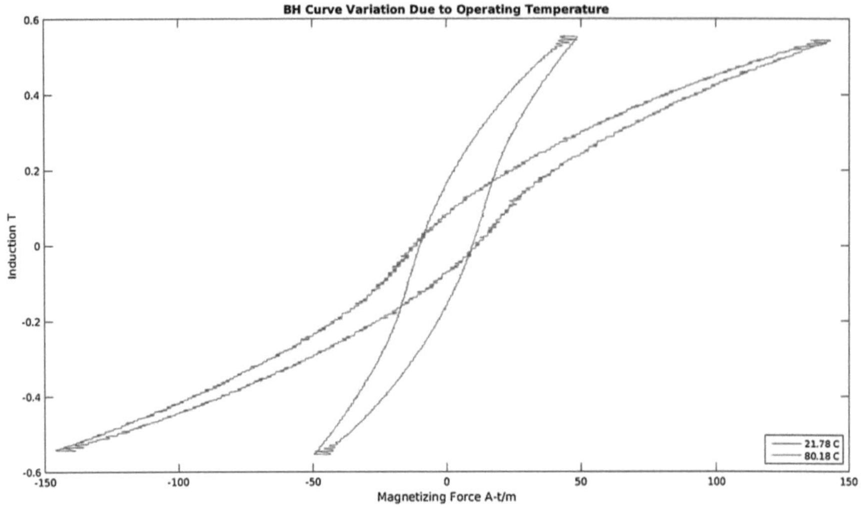

Fig. 14 Hysteresis plots for baseline and thermal steady state of Finemet core

associated losses however it is difficult to extract these effects as the loss variations are near the sensor tolerances.

The exciting power increases dramatically however. This is due to the change in relative permeability caused by increasing temperatures. This is verified by comparing the BH hysteresis plots of the cold temperature and steady state temperature in Fig. 14.

Figure 15 shows the change of relative permeability as calculated over the entire thermal range.

The permeability decay in a practical application presents many design issues. If, as in many designs, the exciting coil covers an extensive region of the core a positive feedback look is created. That is, as the core heats up and permeability decays due to watt loss, the magnetizing inductance will drop. This will cause an

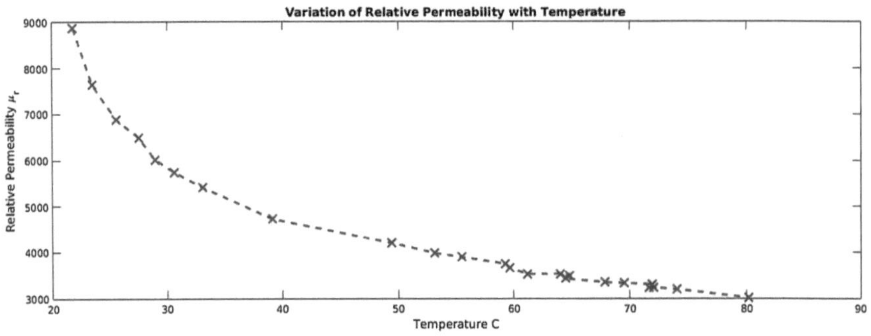

Fig. 15 Decay of relative permeability with increasing temperature for Finemet

increase in the reactive power delivered to the core and consequently increase the current through the coil. This causes an increase in the conduction losses with the square of current. The increased conduction losses can act as a heat source further increasing the core temperature. This results in further permeability decay, and can result in a thermal runaway situation. Careful design must be made to ensure there is enough cooling or thermal stability in the final design.

Future Work

With the completion of the first phase of the test bed there will be continued work in mapping baseline properties to thermal steady state properties. Additionally, all of the current tests have been made with consistent mechanical clamping force. Future work will incorporate a deflection map like that of the thermal change to allow engineers to extrapolate baseline parameters to their specific application.

Another core harness is also in development. This harness will focus on higher temperature applications. In this design ceramic tubing will be used to make contact with the core while remaining nonconductive as in the first design.

Conclusion

The authors presented a magnetic material test bed that enables engineers and designers to evaluate materials over a wide range of operating and environmental parameters. This test bed is focused on the high power medium frequency domain space and utilizes wide bandgap mosfets to achieve practical power converter profiles.

The need for this multi-parameter testing is confirmed by showing the parameter extraction and mapping for a Finemet core. The data presented provides a fuller picture of the material than the information presented by the manufacturer. With this information, more optimal designs are achievable without the need for multiple iterations.

References

1. M. Östling, R. Ghandi, C.M. Zetterling, SiC power devices—present status, applications and future perspective, in *2011 IEEE 23rd International Symposium on Power Semiconductor Devices and ICs*, San Diego, CA, 2011, pp. 10–15
2. A. Rodriguez, A. Vazquez, D.G. Lamar, M.M. Hernando, Three-port power electronic system for energy storage and recovery using a parallel connection of a power factor corrector boost and a dual active bridge, in *2012 15th International Power Electronics and Motion Control Conference (EPE/PEMC)*, Novi Sad, 2012, pp. LS7d.4-1-LS7d.4-8

3. S. Baek, S. Roy, S. Bhattacharya, S. Kim, Power flow analysis for 3-port 3-phase dual active bridge DC/DC converter and design validation using high frequency planar transformer, in *2013 IEEE Energy Conversion Congress and Exposition*, Denver, CO, 2013, pp. 388–395
4. C. Gu, Z. Zheng, Y. Li, A power electronic transformer (PET) with multiport bidirectional resonant DC-DC converters for electric traction applications, in *2015 IEEE Transportation Electrification Conference and Expo (ITEC)*, Dearborn, MI, 2015, pp. 1–6
5. The Gund Company, Inc, NEMA Grade GPO3 Glass Polyester Laminate (The Gund Company Technical Library 2007). http://thegundcompany.com/files/index.cfm?pdfpath=NEMA%20GPO3_IEC%20UPGM%20203.pdf Accessed 20 Aug 2016
6. M.N. Kheraluwala, R.W. Gascoigne, D.M. Divan, E.D. Baumann, Performance characterization of a high-power dual active bridge DC-to-DC converter. IEEE Trans. Ind. Appl. **28**(6), 1294–1301 (1992)
7. S. Kulasekaran, R. Ayyanar, S. Atcitty, Switching frequency optimization of a high-frequency link based energy storage system, in *IECON 2014—40th Annual Conference of the IEEE Industrial Electronics Society*, Dallas, TX, 2014, pp. 1847–1853
8. H. Bai, C. Mi, Eliminate reactive power and increase system efficiency of isolated bidirectional dual-active-bridge DC–DC converters using novel dual-phase-shift control. IEEE Trans. Power Electron. **23**(6), 2905–2914 (2008)
9. G. Xu; D. Sha; J. Zhang; X. Liao, Unified boundary trapezoidal modulation control utilizing fixed duty cycle compensation and magnetizing current design for dual active bridge DC-DC converter, in *IEEE Transactions on Power Electronics*, vol. PP, no. 99, pp. 1–1
10. W. Shen, Design of high-density transformers for high-frequency high-power converters, Ph. D Dissertation, Virginia Polytechnic Institute, July 2006

Part XVIII
Phase Transformations and Microstructural Evolution

An Experimental Assessment of the α + α' Miscibility Gap in Fe-Cr

Alexander Dahlström, Frederic Danoix, Peter Hedström, Joakim Odqvist and Helena Zapolsky

Abstract Phase separation in the Fe-Cr system has been studied experimentally. The upper limit of the miscibility gap and phase separation kinetics during the early stages of decomposition was studied using Vickers micro-hardness tests and atom probe tomography. The results indicate that the upper limit of the miscibility gap in current thermodynamic descriptions is overestimated, but it is necessary to further analyze the transition across the upper limit of the miscibility gap before drawing conclusions. This work demonstrates the efficient combination of micro-hardness measurements and atom probe tomography to investigate the miscibility gap in the Fe-Cr system.

Keywords Phase separation · Fe-Cr alloys · Miscibility gap · Atom probe tomography · Phase diagram

Introduction

The Fe-Cr system is the basis of a frequently used alloy, known as stainless steel. Stainless steels have a great variety of applications in industry and consumer items. Part of this can be explained by their design flexibility with the possibility to tailor material properties for specific applications, and another reason is clearly their innate stainless properties. The mechanical properties of stainless steel depend on the microstructure, commonly Austenitic, Ferritic, Martensitic and Duplex. Among

A. Dahlström (✉) · F. Danoix · H. Zapolsky
CNRS, Groupe de Physique des Matériaux, Normandie University, UNIROUEN, INSA Rouen, 76000 Rouen, France
e-mail: alexander.dahlstrom@etu.univ-rouen.fr

A. Dahlström · P. Hedström · J. Odqvist
Department of Materials Science and Engineering, KTH (Royal Institute of Technology), 100 44 Stockholm, Sweden

The Minerals, Metals & Materials Society, *TMS 2017 146th Annual Meeting & Exhibition Supplemental Proceedings*, The Minerals, Metals & Materials Series,
DOI 10.1007/978-3-319-51493-2_68

some of the industry applications of stainless steels are boilers for chemical and nuclear industries, distillation columns, heat exchangers and pipes. In many of these applications, service temperatures can range from 300 to 600 °C. After long service times during such conditions severe embrittlement can occur in case the stainless steel contains bcc (ferritic or martensitic) phase. The bcc phase is clearly present in ferritic, martensitic and duplex stainless steels but also in austenitic stainless steel welds. The embrittlement will cause a loss of ductility and impact toughness. The phenomenon is often referred to as the "475 °C embrittlement", because the embrittlement occurs at the highest rate at ~475 °C for many of the alloys. The embrittlement in Fe-Cr is a well-known phenomenon that has been studied since the 1950s [1, 2]. Over time the knowledge of this phenomenon has been developed by many authors. Some of the most famous breakthroughs and the theory of phase separation have been developed by Cahn [3], Cook [4] and Langer-Baron-Miller [5].

In spite of the vast literature available discussing the upper limit of the miscibility gap in the Fe-Cr system, it has not been well established. Thus, the purpose of the present investigation is to further study the shape and critical temperatures of the miscibility gap depending on Cr composition. A combination of the experimental techniques atom probe tomography (APT) and Vickers micro-hardness measurements is applied.

Brief Literature Review Spinodal Decomposition Fe–Cr

To date, phase separation in Fe-Cr alloys has been extensively studied using different theoretical and experimental methodologies. A comprehensive evaluation of this prior work have been compiled by Xiong et al. [6]. In terms of assessing the $\alpha + \alpha'$miscibility gap some of the more well-known experimental studies have been made by Miller et al. [7], Kuwano [8] and Williams and Paxton [2]. When it comes to describing phase separation inside the miscibility gap, the most well-known theory was developed by Cahn [3]. According to this theory an analytical solution in one dimension to the difference in free energy between the initial bulk average concentration (C_0) and a decomposed concentration (C) is given by the composition change,

$$C - C_0 = A\,cos\,\beta x. \tag{1}$$

Here A is the amplitude, β is the wave vector and x the direction of that vector. Thus, the total change in free energy (F) per unit volume (V) is given by,

$$\frac{\Delta F}{V} = \frac{A^2}{4}\left(\frac{d^2 f'}{dc^2} + 2x\beta^2\right). \tag{2}$$

Here f' is the Helmholtz free energy of a homogeneous material, the direction vector x originates from the description of the composition gradient energy term. It is a representation of the surface energies between the regions with different compositions. Since the second term in Eq. (2) is positive, the instability of the system with respect to infinitesimal fluctuations of concentration will depend on the sign of the first term. If $d^2f/dc^2 > 0$ then the system is stable for infinitesimal fluctuations of all wavelengths. If $d^2f/dc^2 < 0$ the system becomes unstable with respect to infinitesimal fluctuations of wavelengths greater that $2\pi/\beta_c$ given by,

$$\lambda_c = \frac{2\pi}{\beta_c} = \left[-8\pi^2 x / \left(d^2 f' / dc^2\right)\right]^{1/2}. \quad (3)$$

The concept of a critical wavelength (λ_c) was first developed by Hillert [9]. From Eq. (3) it can be seen that as the spinodal is approached, and the free energy difference decreases towards zero, the critical wavelength goes towards infinity.

The two parameters frequently used to describe the spinodally decomposed structure are: wavelength, which is highly dependent on the driving force of decomposition, and fluctuation amplitude that increases with time. This means that based on the theory of decomposition a higher free energy difference will result in a structure with a smaller characteristic wavelength. The term d^2f/dc^2 in Eq. (3) is highly temperature dependent and it decreases to become more negative inside the miscibility gap.

Experimental Details

The material investigated was provided by OCAS NV. Three 33 kg ingots were prepared in a vacuum induction furnace. A 125 × 125 × 70 mm block was cut, reheated to 1200 °C and then rolled to 15 mm thickness. The chemical composition of the Fe-Cr alloys can be found in Table 1.

Heat treatments were performed on samples with an approximate dimension of 40 × 15 × 0.3 mm. They had been manufactured from cold rolling of a ~1 mm thick sheet. The graph in Fig. 1 shows a general temperature cycle of a sample treated for 25 h.

First the cold rolled samples were solution treated at 1100 °C for two hours, followed by a water quench to room temperature (~25 °C). The solution treatment was evaluated using energy dispersive spectroscopy (EDS), using 3 line scans (900 μm) and 5 spots (6 million counts/spot) respectively. The heat treatments were then

Table 1 Chemical specifications of the alloys

Name	Cr	Ni	Al	C	Fe
Fe-20Cr	19.53	0.0093	0.0456	0.0011	Bal.
Fe-35Cr	34.61	0.0077	0.0407	0.0016	Bal
Fe-50Cr	49.63	–	0.0428	0.0033	Bal.

Realized composition with ICP-OES, combustion and XRF

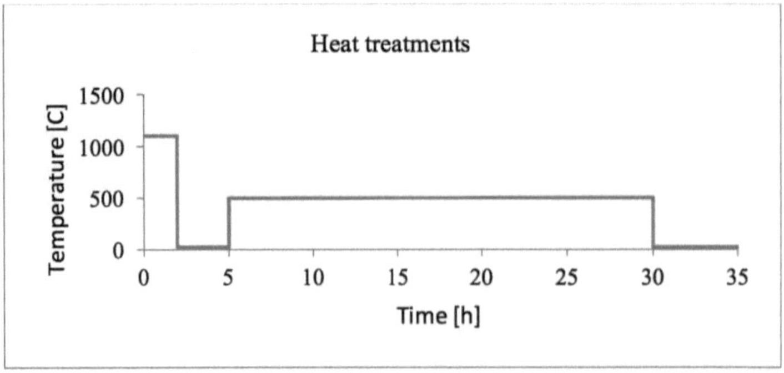

Fig. 1 This graph is a generic representation of the temperature history of the heat treatments, e.g. a 25 h heat treatment cycle

performed in air using a small box furnace. The furnace was calibrated by thermocouples and its accuracy, in the range of the heat treatments, was ±0.9 °C.

Prior to, and after heat treatments the samples were analyzed using 10 Vickers micro-hardness (HV) indents per sample. The indent size was ~50 × 50 μm, sample grain size realized from electron back scatter diffraction (EBSD) was in the range of ~ 200–400 μm. The four extreme values were removed from the dataset and the average was calculated for the middle six data points.

Selected samples were analyzed using APT, prepared by electro polishing. A precision saw was used to cut 0.3 × 0.3 × 12 mm bars. These bars were placed in a nickel holder and electro polished using the classic two step method [10]. First step using perchloric acid/ethanol (25/75 vol.%), the final polishing was conducted using a perchloric acid/ethanol (2/98 vol.%) solution. The polishing was made using a micro-loop device with stepwise decreasing voltage, with the aim of getting the final tip radius to around 50 nm.

A local electrode atom probe CAMECA® LEAP 4000 HR was used for the APT analysis, using the following parameters: detection rate 0.20%, temperature 50 K and a pulse-fraction 20%. The data treatment was performed using the IVAS® software. Concentration profiles were calculated from the top of the tip in a cylinder with a radius of 5 nm and 90 nm down (Z-direction with step size 0.6 nm). The lateral extent was selected to reduce the S/N-ratio for all samples, this means it's not optimized for quantitative determination of the wavelength but to qualitatively follow the wavelength evolution one sample to the other. The Z-step size was selected to average over two α-Fe unit cells. The characteristic wavelength was calculated using autocorrelation function (ACF) from the concentration profiles. The frequency distribution was calculated in a volume 15 × 25 × 75 nm selected to exclude crystallographic poles with a bin size of 50 ions (bin volume ~1.6 nm^3, 1 nm fixed in z/y-directions and variable length in x to minimize bias) and a significance level (σ) of 0.010.

Results and Discussion

The hardness measurements allow following the development of phase separation with temperature since it is known that the hardness can be related to the phase separation and the embrittlement of the material [11]. The Fe and Cr atoms have slightly different atomic size, thus during separation the lattice misfit between Cr- and Fe-rich regions appear and induce internal elastic stress in the system. These elastic stresses eventually cause hardness increase, in such a way that it is possible to detect a difference in mechanical properties, i.e. a relative hardness change. Hence, through the relative hardness change from the re-homogenized state it is possible to detect phase separation and follow the separation kinetics.

The graph in Fig. 2 is an example of how the relative hardness (ΔHV) changes with increasing temperature for the Fe-35Cr sample. Notice the difference in ageing time, the methodology included the development of ΔHV with time. Thus, generally for intermediate ageing ~25 h and 120 h were used to make sure that ΔHV was consistent in time.

This graph shows that at lower temperature inside the miscibility gap where phase separation is expected, the relative hardness has increased. When the top of the miscibility gap is approached, the relative hardness increase becomes less significant and when ageing is performed at a temperature above the top of the miscibility gap no increase is observed. However, it was found that there is a slight decrease of hardness for samples that are expected to be above the top of the miscibility gap. This HV change might be explained by a small recovery of the sample after the first water quenching.

The HV measurements were used to select which samples should be analyzed by APT. The extent of the decomposition has been represented by the wavelength and frequency distributions. The result can be seen in Table 2. In this table the V-parameter is the sum of the normalized integrated area difference between the

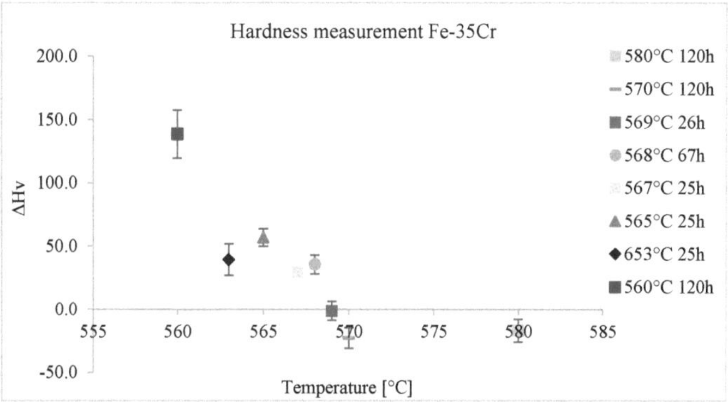

Fig. 2 The graph illustrates the relative change of hardness after a heat treatment, increasing temperature from *left* to *right*, legend markers decreasing from *top* to *bottom*

Table 2 Atom probe analysis of the Fe-35Cr samples

Name	Time (h)	Temperature (°C)	Wavelength (nm)	V-parameter
Fe-35Cr	2	1100	∞	0.021
Fe-35Cr	120	570	>90	0.18
Fe-35Cr	67	568	20	0.49
Fe-35Cr	92	563	12.5	0.43
Fe-35Cr	120	560	7	1.00
Fe-35Cr	120	550	6.5	1.04

observed and the binomial frequency distribution ($V = \sum(| a_{obs} - a_{bin}|)$, $0 \leq V$ 2) [11].

The V-parameter increases as two concentration regions develop. Significant change in microstructure has been observed between 568 and 570 °C. It should be noted that the sample at 570 °C have been aged for a longer time in comparison with the samples aged at 568 °C. However, phase separation is still less developed at 570 °C. There is also a clear V-parameter difference between the 570°C aged and the solution treated sample (1100 °C), this is clearly illustrated by the frequency distribution diagrams in Fig. 3.

In this investigation the Fe-35Cr, 570 °C sample is significantly less decomposed compared to Fe-35Cr, 568 °C. Therefore, it indicates that the $\alpha + \alpha'$ miscibility limit lies in-between these temperatures. This is in agreement with what is indicated by the experimental HV measurements and what was seen by Williams and Paxton [2]. But further kinetic investigations are needed before concluding. It should be noted that the Fe-35Cr, 570 °C is not fully homogeneous as seen in relation to the re-homogenized state, i.e. Fe-35Cr at 1100 °C. This could be related to e.g. short-range order (see e.g. Ref. [12]) and needs further attention.

The trend of wavelength measurements is that deep inside the miscibility gap there is indications of a characteristically shorter wavelength than what has been seen in samples closer to the upper $\alpha + \alpha'$ limit, that was implied by HV measurements. This increase in wavelength follows the predictions of Eq. (3). Thus, it is expected that as the temperature of the Fe-35Cr sample approaches 570 °C the limit of the miscibility gap is also approached. The same methodology has been

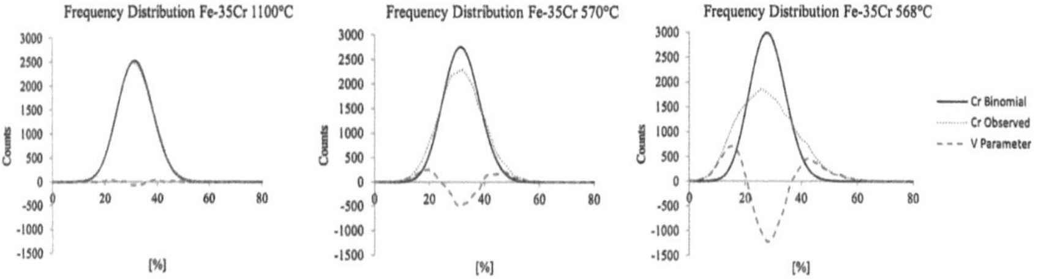

Fig. 3 Frequency distribution diagrams of Fe-35Cr samples inside (568 °C) and outside (570 °C) the miscibility gap, compared with a homogeneous sample (1100 °C)

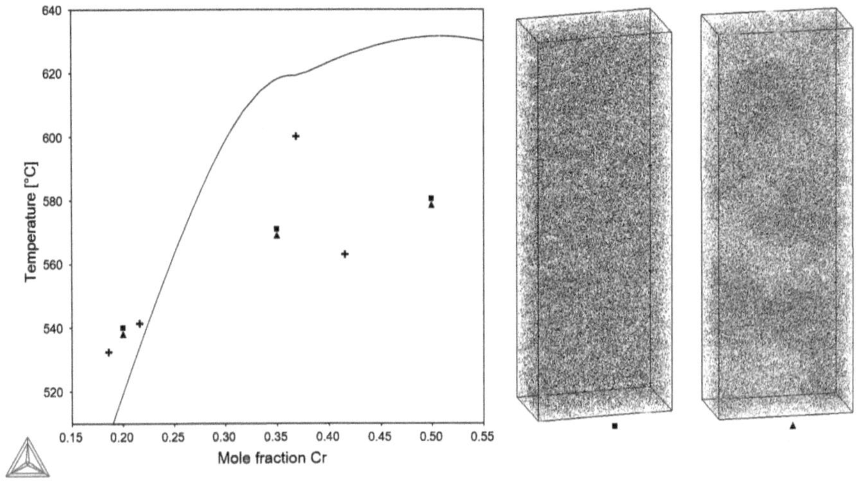

Fig. 4 A comparison between experimental results of samples aged inside (▲) and outside (■) the miscibility gap. FIM literature reference of non-decomposed samples by Miller et al. [7] marked by +, reconstructed volumes of 15 × 25 × 75 nm for 570 °C (*left*) and 568 °C (*right*) Fe-35Cr samples

applied on the Fe-20Cr and Fe-50Cr samples as well. These result have then been used for a comparison in Fig. 4 with the CALPHAD description of the Fe-Cr phase diagram provided by Thermo-Calc AB and previous Field Ion Microscopy (FIM) experiments by Miller et al. [7].

The crosses (+) are from Miller et al. [7] FIM investigation and the markers represent Fe-Cr alloys that were aged above the miscibility gap, i.e. samples where no phase separation were detected. Based on the difference in frequency distributions, HV and what is indicated by the measured wavelengths, two samples have been selected to represent samples that have been aged outside respectively inside the miscibility gap. The temperatures used for aging the Fe-35Cr samples was 568 °C (▲) inside and 570 °C (■) outside the miscibility gap. The corresponding nanostructure is shown in Fig. 4. The Fe-35Cr, 568 °C phase separation is characterized by a wavelength of 20 nm and a significant deviation from the binomial frequency distribution. The Fe-35Cr 570 °C sample did not show an increase in hardness and, moreover, based on the result from atom probe it is more homogeneous with a much lower V-parameter than Fe-35Cr, 568 °C.

Conclusion

Phase separation in the Fe-Cr system has been studied by atom probe tomography and Vickers micro-hardness measurements. It was shown that using the relative hardness of Fe-Cr samples phase separation could be related to the nanostructure evolution. Nonetheless, the frequency distribution obtained from APT experiments is more reliable to detect deviations from the homogeneous concentration distribution. Based on the present results, it is concluded that the upper limit of the miscibility gap should be revisited in thermodynamic descriptions.

Acknowledgements This research has been funded by the Region Normandie and Carl Tryggers Stiftelse.

References

1. H.D. Newell, *Properties and Characteristics of 27% Chromium-Iron*, vol. 49 (1949), pp. 977–1006
2. R.O. Williams, H.W. Paxton, The nature of aging of binary iron chromium alloys around 500 °C. J. Iron Steel Inst. **185**, 358 (1957)
3. J.W. Cahn, On spinodal decomposition. Acta Metall. **9**(9), 795–801 (1961)
4. H.E. Cook, Brownian motion in spinodal decomposition. Acta Metall. **18**(3), 297–306 (1970)
5. J.S. Langer, M. Bar-on, H.D. Miller, New computational method in the theory of spinodal decomposition. Phys. Rev. A **11**(4), 1417–1429 (1975)
6. W. Xiong, M. Selleby, Q. Chen, J. Odqvist, Y. Du, Phase equilibria and thermodynamic properties in the Fe-Cr system. Crit. Rev. Solid State Mater. Sci. **35**(2), 125–152 (2010)
7. M.K. Miller, J.M. Hyde, M.G. Hetherington, A. Cerezo, G.D.W. Smith, C.M. Elliott, Spinodal decomposition in Fe-Cr alloys: experimental study at the atomic level and comparison with computer models—I. introduction and methodology. Acta Metall. Mater. **43** (9), 3385–3401 (1995)
8. H. Kuwano, Mössbauer effect study on the miscibility gap of the iron-chromium binary system. Trans. Jpn. Inst. Metals **26**(7), 473–481 (1985)
9. M. Hillert, A solid-solution model for inhomogeneous systems. Acta. Metallurgica **9**(6), 525–535 (1961)
10. M.K. Miller, R.G. Forbes, *Atom-Probe Tomography* (Springer, US, Boston, MA, 2014)
11. F. Danoix, P. Auger, Atom probe studies of the Fe-Cr system and stainless steels aged at intermediate temperature: a review. Mater. Charact. **44**(1–2), 177–201 (2000)
12. J. Zhou, J. Odqvist, A. Ruban, P. Hedström, Effect of solution treatment on spinodal decomposition during aging of an Fe-46.5 at.% Cr alloy. J. Mater. Sci. (2016)

Phase Transformation, Microstructural Evolution and Property Modification in Rapidly Solidified Grey Cast Iron

Olamilekan Oloyede, Robert F. Cochrane and Andrew M. Mullis

Abstract The phase transformation and microhardness changes of hypoeutectic grey iron subjected to rapid solidification in Helium was studied to showcase the relationship between the evolved microstructure and microhardness of this important engineering material. Droplets samples were prepared, using containerless processing via drop-tube technique. These rapidly solidified samples were collected and sieved into 8 size ranges from 850 to 53 μm diameters with corresponding estimated cooling rate range of ~900 to ~78,000 K s^{-1} respectively. The emerged phases and evolved morphologies were characterized using optical and scanning electron microscopy and X-ray diffraction analysis; while Vickers microhardness tester was used to measure the hardness of the various samples. As the cooling rate increases, the undercooling increased and martensitic or acicular ferrite structure was observed in the very small droplets which confirms the progressive increase in microhardness of the samples from the as-cast to decreasing droplets sizes.

Keywords Phase transformation · Microstructure · Rapid solidification · Grey iron · Microhardness

Introduction

For many years conventionally cooled cast iron materials were limited in their applications because of low tensile strength due to the presence of randomly distributed flake graphite exhibited in the microstructure. However, through the development of rapid solidification processing such as drop-tube technique and other similar experimental methods, researchers over the years have derived means of re-engineering this microstructure by controlling the cooling rate rather than alloying. For instance, commercial grey cast iron remains an important Fe-C-Si alloys which respond basically to composition alteration or thermal treatment or

O. Oloyede (✉) · R.F. Cochrane · A.M. Mullis
Institute for Material Research, University of Leeds, Leeds LS2-9JT, UK
e-mail: pmoro@leeds.ac.uk

The Minerals, Metals & Materials Society, *TMS 2017 146th Annual Meeting & Exhibition Supplemental Proceedings*, The Minerals, Metals & Materials Series,
DOI 10.1007/978-3-319-51493-2_69

both as a result of corresponding phase shift during solidification which affects its morphology, microstructure and microhardness [1]. These basic changes are expected when there is definite phase evolution as a result of transformation through a particular metallurgical processing especially in Fe-C based alloys.

Although, there are still very limited direct resources on rapid solidification of grey cast iron; Kiani-Rashid et al. [2], provided an account of the possible transformation in Nickel-Molybdenum rich grey cast iron alloy by heat treatment and how this influenced its mechanical property. Using a 3 level 'staircase' designed sample of different thickness, they observed that with austenitising and austempering processes of this varied sample thickness, its heat treatment in CO_2 tight underground sand mould between 400 and 900 °C for about 60 min, transformed not only the graphitic microstructure nature of the sample sections but there was definite phase change due to the treatment. Their result show that by increasing the cross sections of the sample, the evolved martensite percentage decreases and the phase proportion of the desired intermediate bainite-ferrite increases which made the alloy reasonably ductile and more malleable. In another similar work using a different method, rapid quenching procedure; Ahmed et al. [3], show the possibility of different phase evolution involving Al-Si rich Ferrous metallic melt at $>10^4$ K s^{-1} which resulted in significant microstructural, constitutional and phase changes in the material. The effect of this transformations was reflected on the microhardness of the rapidly quenched alloy with estimated cooling rate of $\sim 1.2 \times 10^5$ K s^{-1} which caused lowering of Fe content of β-phase and higher solubility of Si in α-Al. Hence, they concluded that the improvement of noticed microhardness was as a result of microstructure refinement and the emerged modified phases. Also, using laser surface processing, Fouquet and Szmatula [4] performed surface melt-hardening on pearlitic grey iron using 1 kW CO_2 continuous wave laser. They revealed that with suitable surface preparation and laser conditions, it is possible to transform the retained dendritic austenite and interdendritic ledeburite into cementite and martensite in an all over lapping cladded regions leading to an increase in the average hardness of the alloy.

In the light of the above different processing that other researchers have employed, the purpose of this study is to showcase possible phase transformation in commercial grey case iron using drop tube technique, a containerless rapid solidification method. Hence, the effect of this on the alloys microhardness was investigated using light optical micrograph, SEM. XRD and Vickers microhardness.

Experimental Procedure

The control sample for this study was obtained from Yorkshire Steel, UK. It is a typical low alloyed commercial grey cast iron with elemental composition as compared to ASM standard shown in Table 1 [5]. It was confirmed using XRF

Table 1 Elemental composition of commercial grey cast iron used analysed by XRF

Element (wt%)	C	Si	Mn	P	S	Fe	C_{eq}
ASTM specification:	2.5–4.0	1.0–3.0	0.2–1.0	0.002–1.0	0.02–0.025	96.28-90-96	Cal.
Grey cast iron used:	2.70	2.83	0.58	0.15	0.05	93.34	3.70

analysis to be hypoeutectic with carbon equivalent, C_{eq}. of roughly 3.70%. Predominate phase in this typical conventionally cooled grey cast iron alloy is ferritic-pearlitic matrix with randomly distributed graphites microstructure. Hence, using 6.5 m high drop tube to produce rapidly solidified droplets under Helium (He) environment, enabled one to compare and contrast noticeable changes in the morphology and mechanical property (microhardness) of the different particle sizes produced as a function of their cooling rate since there was no compositional change. In preparation, the drop tube was evacuated to ~1 Pa and then flushed 3 times with Helium at 50 kPa to ensure appropriate inert environment by means of set of high vacuum pumps. The tube was then finally evacuated to a pressure of 4×10^{-4} Pa before being back filled with 50 kPa of He gas. This procedure took roughly 8–10 h.

To obtain the droplets sample, pieces of the as-received sample weighing ~20 g was used as feedstock for the drop tube experiment. This was loaded into an alumina crucible with 3 laser drilled bottom ejection orifices (roughly 300 μm diameter) and shielded by hollow graphite susceptor for effective induction heating. The entire container was pressure tight and then heated up by RF controlled furnace at the top of the apparatus. The furnace temperature was monitored by R-type thermocouple inserted in the crucible but not touching the alloy and when sufficient superheat was attained through monitoring, the melt was ejected by pressurizing the crucible with 300 kPa of He gas. The pressure sprayed melt cool rapidly into droplets as it fall downward in the tube. Hence, the spherical droplets produced were then collected and sieved into 8 different sizes ranging from the biggest (850 μm) to the smallest (53 μm) sizes. The droplets were thereafter mounted using transoptic resin. These mounted samples were grounded and polished to reveal the particles interior hemisphere using different set/grade of Silicon carbide papers and $6 - 1$ μm diamond paste and after thorough metallography preparation, they were subjected to XRD analysis, SEM characterization and microhardness indentation measurement using a TUKON^{TM} 1202 Wilson Hardness (Vickers) analyser at room temperature with an average of 10 minimum measurements using HV0.05 as indentation load. Full details on droplets production and metallography procedures can be found in Ref. [6]. Figure 1 shows the graphical relationship between the calculated cooling rate and the droplet diameters for each particle sizes produced.

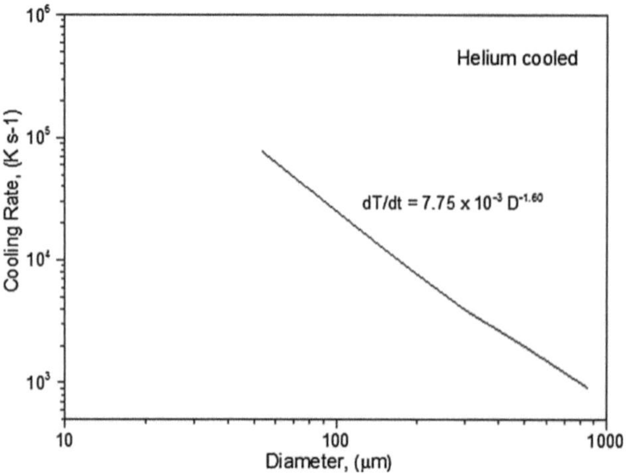

Fig. 1 Estimated cooling rate of He cooled droplets as a function of the particles diameters

Result and Discussion

Figure 2 shows the XRD pattern for the as-cast and the different droplet sizes with identified evolved phases. These helium cooled droplets were mounted on transoptic resin but the interiors were well exposed after thorough metallography preparation. Also, the XRD peak range was chosen from $35^{\circ} \leq 2\theta \geq 95^{\circ}$ to avoid the characteristic "hump" effect of the amorphous resin materials used, which was observed between 20° to 30° of 2θ. In so doing, the graphite peak which occur strongly around $2\theta = 26^{\circ}$ in the as-cast materials but absent in all the droplets have been left out. The main phases identified are predominately ferrite (α-Fe) in the as-cast and varied proportion of all other phases in the different droplet sizes as indicated by the different peak heights in the XRD patterns. However, there is a confirmed trend as the cooling rate increases, (i.e. as the particle sizes reduces) there is a definite transformation from α-Fe to γ-Fe and then to α′-Fe. Although there are more retained austenite (γ-Fe) plus cementite (Fe_3C) in the big or medium particles while martensite or acicular ferrite (α′-Fe) is more dominating in the smaller droplets. These phases evolution and their consequent effect on the property of this alloy is mainly due to the processing route employed; thereby, relating the process —structure—property interconnectivity of this very important commercial alloy.

Figure 3a, d show the light optical micrographs of the unetched as-cast and the biggest droplet at different stages of sample observation. The first morphological changes noticed here is that, the randomly distributed graphite flakes seen in the as-cast sample is absent even in the droplet with the modest cooling rate i.e. the

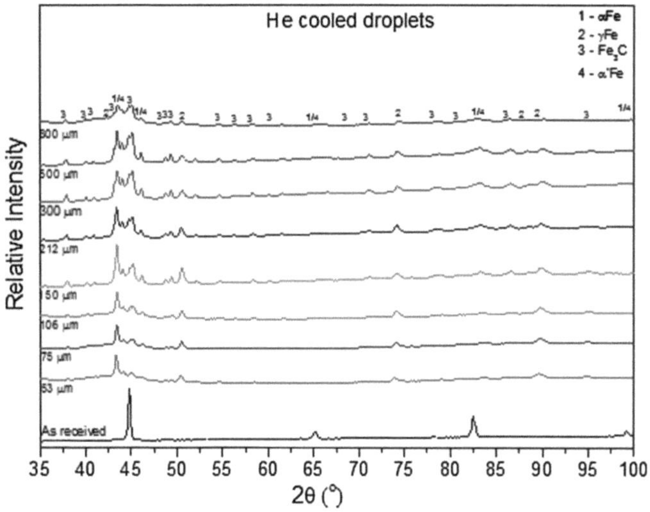

Fig. 2 Shows the XRD peak patterns for the as-received sample with predominately α-Fe identified, the graphite peak was found to be around $2\theta = 26^{\circ}$ but has been omitted as a result of characteristic 'hump' of the amorphous Transoptic resin used; along with profiles of the combined peak patterns of all the droplets with progressive phase transformation from α-Fe through γ-Fe and finally to α′-Fe evolution

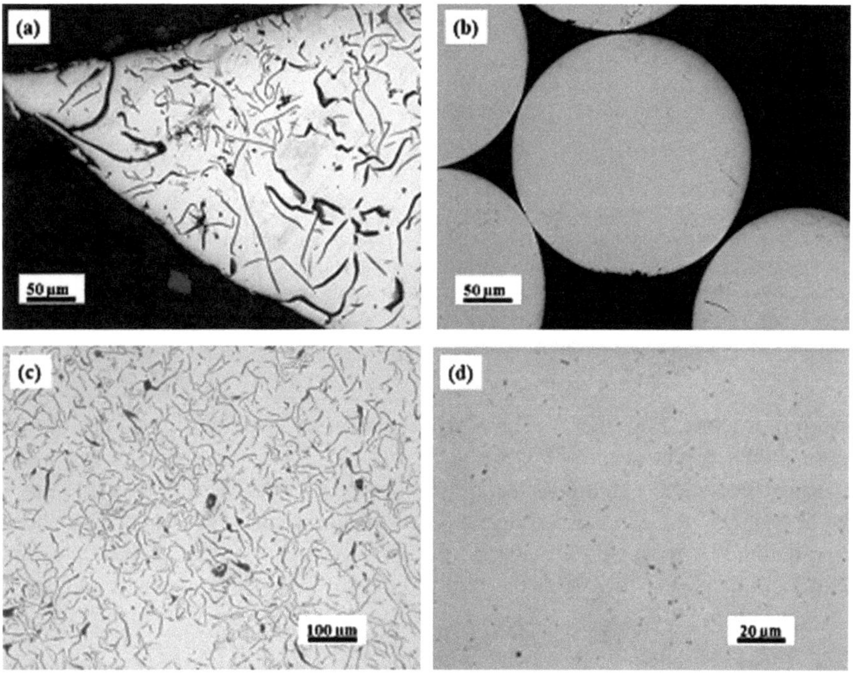

Fig. 3 Unetched light optical micrograph of **a** as-received sample (tip-edge) revealing randomly distributed graphite in predominately ferritic matrix, **b** *middle* section of the as-cast sample with characteristic Type C graphite morphology. **c** A rapidly solidified droplet from drop-tube with no trace of graphite as shown in the enlarged micrograph in **d**

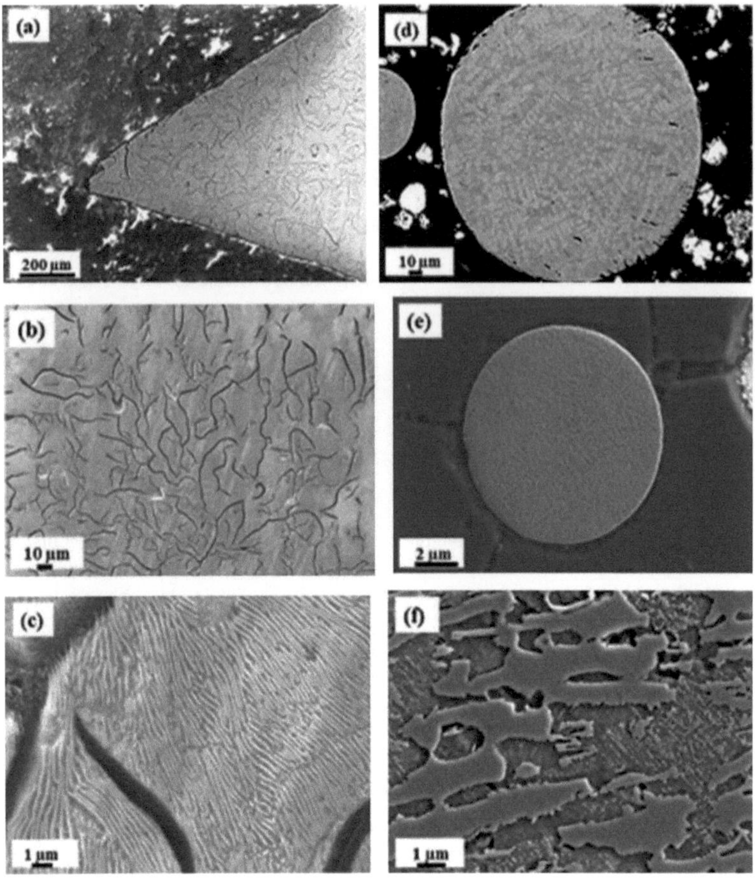

Fig. 4 SEM micrograph of **a** as received tip-edge with **b** showing the graphite flakes and **c** revealing the ferritic-pearlitic lamellar nature of a typical conventionally cooled grey cast iron. On the other hand, **d** shows the morphology of the biggest droplet, with strong primary dendritic morphology; the XRD traces of which confirm no trace of graphite but rather evolved retained γ-Fe along with α-Fe and Fe_3C. Meanwhile, **e** is the smallest droplet with evolving martensitic plates which at higher magnification gives **f**

850 µm droplets, Fig. 3b [7]. The high resolution SEM images of course reveals more morphologies details. For instance, Fig. 4a–f progressively but ultimately revealed the lamellar nature of the graphite rich ferritic–pearlitic microstructure of the as-cast material as shown progressively in Fig. 4a, b and c. Similarly, it shows morphologies whose XRD patterns reveal to be majorly ferrite, austenite and evolving martensitic phases in the big/medium to the smallest droplets as seen in

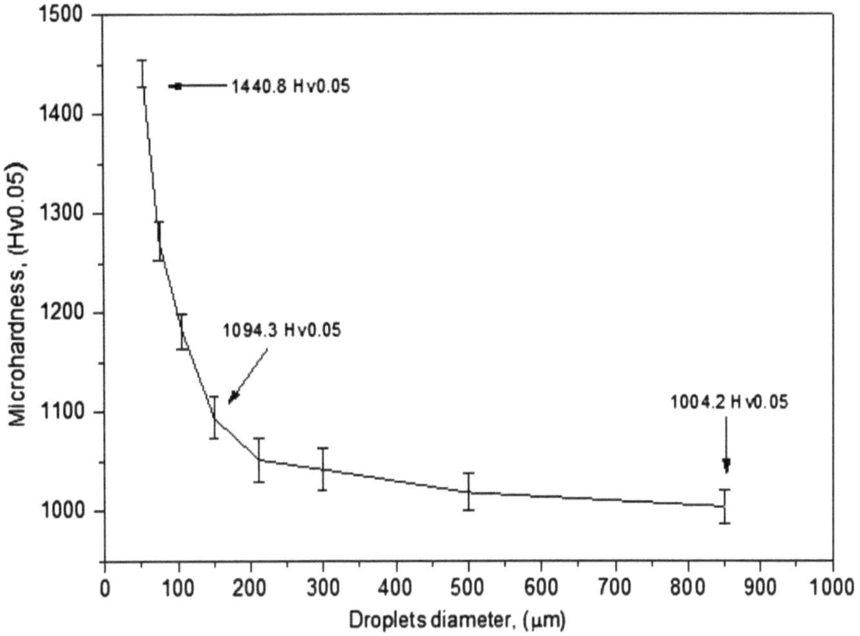

Fig. 5 Microhardness values (Hv0.05) as a function of droplet diameters cooled in He gas

Fig. 4d, e and f respectively. However, since composition is constant; then the observed difference both in phase evolution and microstructure is purely a function of the samples cooling rate which are ~900 and ~78,000 K s^{-1} for the biggest and smallest droplets respectively. Compared to results in our previous publication [4], one can easily confirm that helium gas used here has better thermal conductivity enhancing more undercooling than N_2 gas when used as a cooling medium during drop-tube processing. The result of the Vickers microhardness measured against droplets diameter is as shown in Fig. 5. To obtain true hardness of the as-cast sample, effort was made to avoid the graphites sites as shown in Fig. 6 and the as-cast matrix average microhardness value is ~365 ± 5 Hv0.05. The biggest (850 μm) droplet has lowest hardness (1004 ± 7 Hv0.05); while the smallest droplet size used in this case (53 μm) has the highest (1441 ± 6 Hv0.05). This is in line with the observed progressive XRD evolved phase changes from α-Fe to γ-Fe and finally to α′-Fe with particle size reduction or increasing cooling rate of the droplets [8].

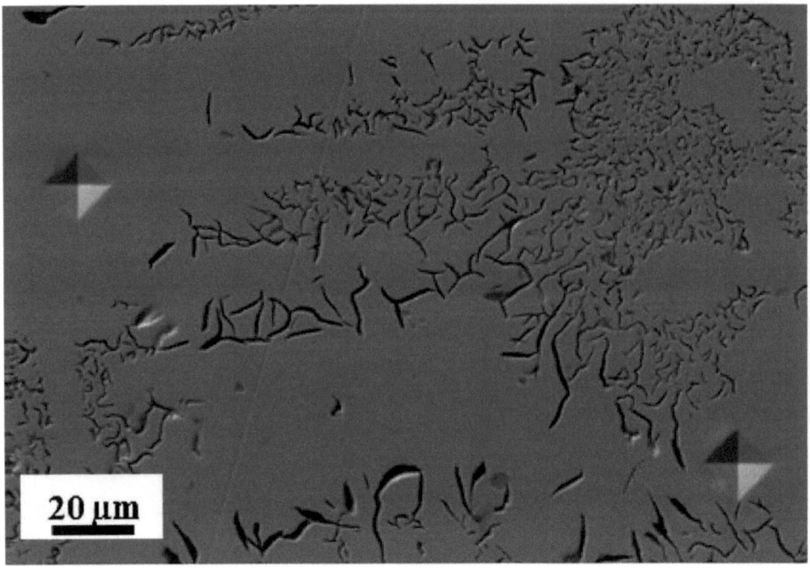

Fig. 6 SEM micrograph showing avoidance of graphite sites while measuring the hardness value in the as-received sample

Conclusion

From the explanation on the solidification morphologies observed and the progressive phase transformation at constant elemental composition of the rapidly cooled droplets, we see that increased cooling rates enhances better undercooling in He gas which resulted in a steady increment in microhardness. The observed phase changes from ferritic matrix to that of retained austenite and eventually to harder martensite in the highly undercooled droplets. Hence, the evidence presented in this study has shown that with rapid solidification processing such as drop-tube and like other previously used methods like laser welding, austenitising/austempering heat treatment, levitation methods and such likes previously used by other researchers on other metallic alloys, it is quite possible to transform not only the microstructure but also the mechanical property of many important commercial alloy such as grey cast iron based on the interdependence of processing—microstructure and mechanical property.

Acknowledgements The authors are grateful to Diane Cochrane for her assistance in sample preparation and metallography. Also, special thanks to Mike, Stuart and John of LEMAS laboratory and other staff in the School of Chemical & Process Engineering, University of Leeds, UK for their support. Oloyede Olamilekan appreciate the Federal Government of Nigeria for sponsorship through PTDF scholarship board.

References

1. J.R. Davis (ed.), *ASM Specialty Handbook: Cast Irons* (ASM international, 1996)
2. A.R. Kiani-Rashid, M. Mostafapour, S.K. Kaboli-Mallak, A.Babakhani, *The Effect of Cooling Rate on Bainite Phase Formation in Austempered Nickel-Molybdenum Gray Cast Iron.* 2011 (ISRN Materials Science, 2011)
3. E.M. Ahmed, M.R. Ebrahim, Microstructure and microhardness evolutions of high Fe containing near-eutectic Al-Si rapidly solidified alloy. J. Metall. (2014)
4. F. Fouquet, E. Szmatula, Laser surface melting of a pearlitic grey cast iron. Mater. Sci. Eng. **98**, 305–308 (1988)
5. O. Oloyede, R.F. Cochrane, A.M. Mullis, Effect of rapid solidification on the microstructure and microhardness of BS1452 grade 250 hypoeutectic grey cast iron. J. Alloy. Compd. (2016)
6. O. Oloyede, T.D. Bigg, R.F. Cochrane, A.M. Mullis, Microstructure evolution and mechanical properties of drop-tube processed, rapidly solidified grey cast iron. Mater. Sci. Eng. A **654**, 143–150 (2016)
7. H.H. Liebermann, *Rapidly Solidified Alloys: Processes, Structures, Properties, Applications* (Marcel Dekker, Inc, 270 Madison Ave, New York, New York 10016, USA, 1993)
8. G.F. Vander Voort (ed.), *ASM Handbook*, vol. 9 (Materials Park, OH: ASM International, 2004)

Morphology of Order-Disorder Structures in Rapidly Solidified L12 Intermetallics

Nafisul Haque, Robert F. Cochrane and Andrew M. Mullis

Abstract Utilization of intermetallics in high temperature applications is limited due to their poor room temperature ductility. One route to overcoming this is disorder trapping (and subsequent anti-phase domain formation) during rapid solidification, motivating the study of disorder trapping in intermetallics. The single-phase, $L1_2$ intermetallic β-Ni_3Ge has been rapidly solidified via drop-tube processing. At low cooling rates (850–500 μm diameter particles, 700–2800 K s^{-1}) the dominant solidification morphology, revealed after etching, is that of isolated spherulites in an otherwise featureless matrix. Selected area diffraction analysis in the TEM reveals the spherulites to be partially disordered β-Ni_3Ge, whilst the featureless matrix is the fully ordered variant of the same compound. Dark-field TEM imaging has confirmed that the spherulites grow as radially emanating fingers of the ordered phase, with disordered material in the space between the fingers.

Keywords Intermetallics · Rapid solidification · Disorder trapping · Spherulites

Introduction

Intermetallic compounds are characterised by strong internal order and mixed covalent/ionic and metallic bonding, which gives rise to mechanical behaviour intermediate between ceramics and metals [1]. They have a range of potential applications as high temperature structural materials due to good chemical stability and high hardness at elevated temperatures. Conversely, poor room temperature

N. Haque (✉) · R.F. Cochrane · A.M. Mullis
Institute for Materials Research, University of Leeds, Leeds LS2 9JT, UK
e-mail: pmnh@leeds.ac.uk

R.F. Cochrane
e-mail: R.F.Cochrane@leeds.ac.uk

A.M. Mullis
e-mail: A.M.Mullis@leeds.ac.uk

The Minerals, Metals & Materials Society, *TMS 2017 146th Annual Meeting & Exhibition Supplemental Proceedings*, The Minerals, Metals & Materials Series,
DOI 10.1007/978-3-319-51493-2_70

ductility limits formability. Low temperature ductility can be increased by rapid solidification processing, wherein a reduction of the degree of order and the formation of antiphase domains increases ductility [2–4]. Moreover, the high temperature properties can be restored by annealing to restore chemical ordering [2].

In this article we present an analysis of the rapid solidification of the single phase, congruently melting, intermetallic β-Ni_3Ge, a compound which at equilibrium orders at all temperatures below the liquidus. The main motivation for this is to study the solidification morphologies associated with disorder trapping at high growth rate without the associated complication of solute trapping. However, as the main solidification morphology observed are spherulites, a relatively uncommon morphology in metals and intermetallics, we first present a review of the spherulite morphology and it previous occurrence in metallic systems.

The crystalline morphology termed spherulitic was first observed by Talbot in 1837 during the crystallisation of borax from phosphoric acid [5]. Spherulites are commonly observed in polymers and small molecule organic crystals, together with some minerals, volcanic rocks, inorganic crystals and a few pure elements (e.g. graphite, sulphur and selenium) [6]. Spherulites are only rarely observed in metals, sometimes as graphite formations in cast irons and otherwise as the result of devitrification of metallic glasses [6].

As mentioned above, spherulites are most common in polymers, particularly those of high molecular weight grown directly from the melt, where topological constraints restricting the reorientation of long chain molecules are encountered. The structure of such spherulites is of multiply branched crystalline arms separated by amorphous regions between the arms [6]. The formation of spherulites is most common in unoriented systems. Where a strong tendency for uniform molecular orientation exists, such as for growth in a strong, externally imposed temperature gradient, spherulite growth appears to be suppressed [7].

Spherulite growth is also observed in a number of metallic systems, particularly during the devitrification of amorphous metals. Chrissafis et al. [8] studied the crystallisation of Fe-Si-B metallic glass in the solid-state, observing the formation of Fe spherulites, the distribution of which was indicative of homogenous nucleation and isolated, diffusion controlled three dimensional growth in the amorphous matrix. Yano et al. [9] observed spherulite formation in $Zr_{50}Cu_{40}Al_{10}$ bulk metallic glass, annealed at 773 K, using TEM and positron annihilation lifetime measurements. Crystallisation proceeded by the formation of spherical agglomerates of radially developing crystalline grains, some 600 nm in diameter. Various other studies [9–12], looking at the crystallisation products of devitrification in binary and multicomponent component metallic glasses, have made similar observations of spherulite structures. The results appear to be insensitive to heating rate, with heating rates as high as 10^3 K s^{-1} being found not to inhibit spherulite formation [13].

Although there is controversy in the literature as to the formation mechanism for spherulitic growth, a number of common requirements for their formation have been identified. One is a tendency towards non-crystallographic, small angle branching [14]. A second is a high viscosity in the medium being crystallised.

The importance of this has been demonstrated unambiguously by Morse et al. [15, 16] who, in a study of the crystallisation behaviour of inorganic salts, identified around 70 salts that would crystallise in spherulitic form if grown in a gel, but not otherwise. It has been suggested that spherulite growth may be favoured in situations where translational diffusion may be significantly easier than rotational diffusion [17] and one can certainly imagine how this situation might arise during the crystallisation of long chain polymers, although it is less obvious how such a situation might arise in metals or other small molecular systems. Despite this, there is some evidence that similar conditions may exist in both organic [18] and metallic [19] undercooled glass forming liquids, with a decoupling of the translational diffusion coefficient from the macroscopic viscosity and a decoupling of the translational and rotational diffusion coefficients. The propensity of metallic glass formers to form spherulites is also suggestive of the requirement for some level of structure in the non-crystalline precursor to be a prerequisite. This also appears to be true of other spherulite formers, such as pure elemental Se, which is reported to have a very peculiar molecular structure, intermediate between truly polymeric and simple molecular liquids [20].

According to the phase diagram of Nash and Nash [21], β-Ni_3Ge is a congruently melting compound with a melting point of 1405 K. It has a homogeneity range of 22.5 to 25 at % Ge and crystallizes to the ordered fcc $L1_2$ structure. The system has previously been studied by Ahmed et al. [22] who, using a flux undercooling technique, observed a maximum undercooling of 362 K, wherein the corresponding growth velocity was measured at 3.55 m s^{-1}. In common with other researchers who have determined the velocity-undercooling curves for intermetallic compounds passing through the order-disorder transformation, Ahmed et al. observed a discontinuous break in the curve at the onset of fully disordered growth. This condition was observed for β-Ni_3Ge at an undercooling of 168 K and at a critical growth velocity, V_C, of 0.22 m s^{-1}. More recently Haque et al. [23] have observed a range of solidification structures in rapidly solidified Ni_3Ge including a dendritic to dendritic seaweed transition.

Experimental Methods

Elemental Ni and Ge were obtained from Alfa Aesar with purity of 99.99 and 99.999%, metals basis, respectively. The alloy, of composition Ni–25 at % Ge, was produced by arc-melting the elemental constituents together under a protective argon atmosphere, with the melting process being repeated 8 times to ensure uniform mixing of the final alloy. Following arc-melting the phase composition of the alloy ingot was checked using a PANalytical Xpert Pro X-ray diffractometer and drop-tube processing was only undertaken on samples that were confirmed as single phase β-Ni_3Ge.

The alloy sample, of approximately 9.5 g mass, was loaded into an alumina crucible which has three 300 μm laser drilled holes in the base. Induction heating of a graphite subsector was used for melting the sample. Temperature determination was by means of an R-type thermocouple which sits inside the melt crucible, just above the level of the melt. When the temperature in the crucible attained 1480 K (75 K superheat) the melt was ejected by pressuring the crucible with ~400 kPa of N_2 gas. The resulting spray of particles solidify in-flight whilst falling the 6.5 m length of the tube, which is filled to 50 kPa with dry, oxygen free N_2 gas, and are collected at the base of the tube. Further details of the Leeds drop-tube and the likely cooling rates that can be thus achieved can be found in [24].

Following retrieval of the sample from the base of the drop-tube the powders were sieved into standard size fractions, with the size fractions, >850 μm and 850–500 μm being selected for analysis here. Using the method given by [24] the mean post-recalescence cooling rate for these two size fractions is estimated at 700 and 2800 K s^{-1} respectively.

The selected size fractions were subject to XRD analysis to ensure they remained single-phase and were then mounted in transopic resin and prepared for microstructural analysis. This involved grinding with SiC grinding media and then polishing with progressively finer grades of diamond paste, finishing with a 1 μm grade. β-Ni_3Ge was found to be highly resistant to most chemical etchants normally used in metallography. The etchant used was the aggressive mix consisting equal parts of undiluted HF, HCl and HNO_3. A Carl Zeiss EVO MA15 scanning electron microscope (SEM) was used to image the microstructure of the droplets revealed by etching while an Oxford Instrument X-Max Energy-Dispersive X-Ray (EDX) detector was used to check the chemical homogeneity of the samples. Selected area diffraction analysis and bright-field imaging was performed using a FEI Tecnai TF20 Transmission Electron Microscope (TEM), with samples, of approximately 10 μm × 7 μm and between 55–70 nm in thickness, being cut using a FEI Nova 200 Nanolab focused ion beam (FIB).

Results & Discussion

Figure 1a shows an SEM micrograph of a polished and etched sample from the 850–500 μm sieve fraction. Numerous spherulite like structures, typically with diameters of the order of 10–20 μm, are evident in an otherwise featureless matrix. Such structures are very rare in metallic systems, although appear to be ubiquitous throughout the drop-tube powders in this size fraction. We note that, except for the spherulites, no other crystalline features are revealed in the sample by etching. XRD analysis (not shown) performed on the powder samples confirms that the material remain single phase β-Ni_3Ge subsequent to drop-tube processing. The contrast between the spherulite and the surrounding matrix material appears not to be the result of compositional differences due to solute partitioning during solidification. The absence of solute partitioning is demonstrated in Fig. 1b which shows an EDX line across the diameter of a spherulite as revealed by etching. From this it is clear that, to within the experimental error associated with the technique, there is little

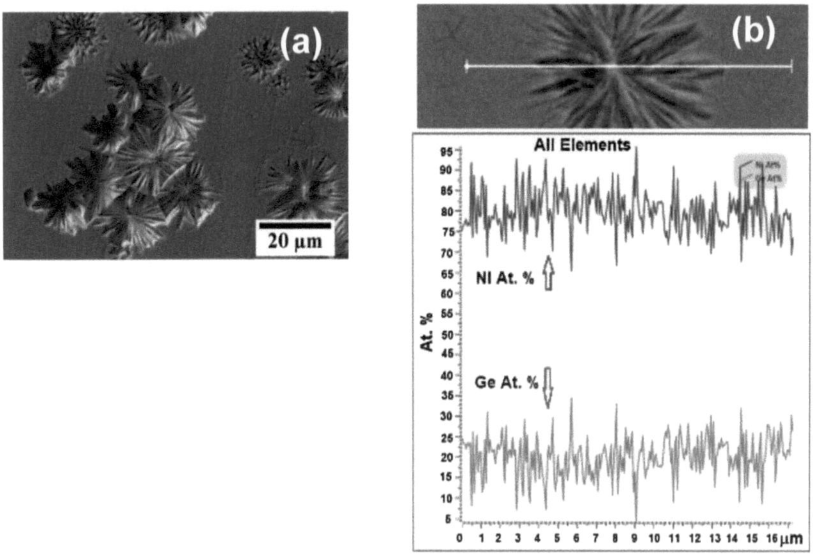

Fig. 1 **a** SEM micrograph of HF etched β-Ni3Ge drop-tube particle from the 850–500 μm size fraction showing spherulites structures in a featureless matrix and **b** EDX line scan across a spherulite trunk showing that the contrast revealed by etching is not the result of solute partitioning

variation in composition between the structures revealed by etching and the surrounding featureless matrix. The contrast revealed by etching does not therefore appear to be related to differences in either phase (XRD) or chemical (EDX) composition.

In order to understand the origins of the spherulite morphology TEM imaging and Selected Area Diffraction (SAD) analysis has been performed. Figure 2a shows typical bright-field TEM image of an isolated spherulite. Figure 2b, c show two selected area diffraction patterns obtained from the microstructure shown in Fig. 2a, taken (i) from the spherulites structure, and (ii) from the initial core of the spherulite. Both are consistent with an fcc cubic structure with the lattice parameters of β-Ni_3Ge. However, (b) also displays superlattice spots consistent with the ordered $L1_2$ crystal structure of β-Ni_3Ge, as expected from the phase diagram, while (c) shows no such superlattice spots and thus appears to be the disordered A1 fcc structure. Beyond the spherulites, similar SAD pattern analysis (not shown) confirms that all of the featureless matrix material has the fully ordered $L1_2$ crystal structure.

Figure 2d shows a TEM dark-field image of a region within a spherulite obtained from one of the superlattice spots within the diffraction pattern. The spherulite appears to consist of fingers of material that are bright and are likely to therefore be of the ordered phase, with a dark, and therefore probably disordered material between the fingers. In common with many polymer spherulites the fingers within a spherulite are generally observed to be of uniform width and in order to

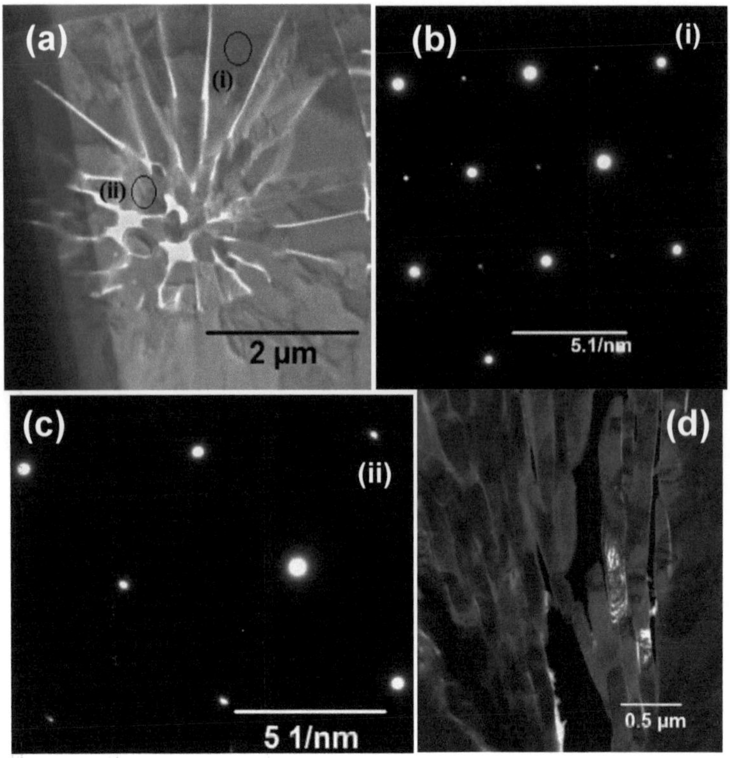

Fig. 2 **a** TEM bright field image of a spherulite, **b** selected area diffraction patterns from regions (i) and **c** (ii) of the spherulite, and **d** TEM dark-field image of a region within a spherulite obtained from one of the superlattice spots within the diffraction pattern

achieve space filling whist maintaining a constant width the fingers undergo non-crystallographic branching [25].

A number of amorphous metallic systems crystalize to a spherulitic morphology upon heating above the glass transition temperature. While in some respects crystallization from a metallic glass precursor is similar to crystallization from the melt, in other respects it is quite different. The main difference is that the viscosity of a glass is much higher than that of a melt and correspondingly, the atomic diffusivity is much lower in a glass, compared to a melt of the equivalent composition. Consequently, crystallization from a glass is often restricted to short range diffusion. This in turn means that growth kinetics dominate over diffusion. An analogous situation may be considered to exist in the growth of β-Ni_3Ge from its parent melt. As the compound is congruently melting there is no requirement for long range solute transport by diffusion, which is consistent with EDX results. Moreover, Ahmed et al. noted the noted the abnormally high viscosity of the Ni_3Ge melt, with undercooled droplets being able to be pulled into teardrop shapes.

This characteristic may be indicative of structure in the liquid, in a fashion similar to that found glass forming melts.

However, the morphological analogies between the spherulite structures observed here and those observed in polymers is significantly greater than previously reported for any form of metallic spherulite. These similarities include the growth of constant width fingers and space-filling by non-crystallographic branching. Moreover, the structure of the Ni_3Ge spherulites, with ordered fingers in a disordered matrix is analogous to the structure of polymer spherulites in which crystalline fingers are imbedded within an amorphous matrix. While the constant width fingers may be explained with recourse to a Mullins and Sekerka type instability, the non-crystallographic branching of an fcc metal is much more difficult to explain. The strong parallels between the spherulites observed here and those in polymers may suggest that some of the features of spherulites which have previously been attributed to the properties of their host polymer may in fact be inherent features of the spherulite morphology.

Acknowledgements Nafisul Haque is thankful to the Higher Education Commission (HEC) Pakistan and NED University of Engineering & Technology for financial support.

References

1. S. Murarka, *Metallization: Theory and Practice for VLSI Application* (1993)
2. R. Cahn, P. Siemers, J. Geiger, P. Bardhan, The order-disorder transformation in Ni_3Al and Ni_3Al-Fe alloys—I. determination of the transition temperatures and their relation to ductility. Acta Metall. **35**, 2737–2751 (1987)
3. A. Inoue, H. Tomioka, T. Masumoto, Microstructure and mechanical properties of rapidly quenched L11 alloys in Ni-Al-X systems. Metall. Trans. A **14**, 1367–1377 (1983)
4. H. Assadi, M. Barth, A. Greer, D.M. Herlach, Microstructural development in undercooled and quenched Ni_3Al droplets, in *Materials Science Forum* (Trans Technical Publication, 1996), pp. 37–44
5. W.H.F. Talbot, Phil. Trans. R. Soc. London (London, 1837)
6. J. Magill, Review spherulites: a personal perspective. J. Mater. Sci. **36**, 3143–3164 (2001)
7. A. Keller, The spherulitic structure of crystalline polymers. Part I. investigations with the polarizing microscope. J. Polym. Sci. **17**, 291–308 (1955)
8. K. Chrissafis, M. Maragakis, K. Efthimiadis, E. Polychroniadis, Detailed study of the crystallization behaviour of the metallic glass Fe75-Si9-B16. J. Alloy. Compd. **386**, 165–173 (2005)
9. T. Yano, Y. Yorikado, Y. Akeno, F. Hori, Y. Yokoyama, A. Iwase, A. Inoue, T.J. Konno, Relaxation and crystallization behavior of the $Zr_{50}Cu_{40}Al_{10}$ metallic glass. Mater. Trans. **46**, 2886–2892 (2005)
10. K. Ziewiec, Kinetics of phase transformations in $Ni_{87}P_{13}$ alloy upon heating. J. Alloy. Compd. **397**, 207–210 (2005)
11. C. Xia, L. Xing, W.-Y. Long, Z.-Y. Li, Y. Li, Calculation of crystallization start line for Zr48-Cu45-Al7 bulk metallic glass at a high heating and cooling rate. J. Alloy. Compd. **484**, 698–701 (2009)
12. J. Liu, H. Zhang, H. Fu, Z.-Q. Hu, X. Yuan, In situ spherical B2 CuZr phase reinforced ZrCuNiAlNb bulk metallic glass matrix composite. J. Mater. Res. **25**, 1159–1163 (2010)

13. H. Sun, K. Flores, Microstructural analysis of a laser-processed Zr-based bulk metallic glass. Metall. Mater. Trans. A **41**, 1752–1757 (2010)
14. H. Keith, F. Padden Jr., A phenomenological theory of spherulitic crystallization. J. Appl. Phys. **34**, 2409–2421 (1963)
15. H.W. Morse, C.H. Warren, J.D.H. Donnay, Artificial spherulites and related aggregates. Am. J. Sci. 421–439 (1932)
16. H. Morse, J. Donnay, Optics and structure of three-dimensional spherulites. Am. Mineral. **21**, 391–427 (1936)
17. L. Gránásy, T. Pusztai, G. Tegze, J.A. Warren, J.F. Douglas, Growth and form of spherulites. Phys. Rev. E **72**, 011605 (2005)
18. S.F. Swallen, P.A. Bonvallet, R.J. McMahon, M. Ediger, Self-diffusion of tris-naphthylbenzene near the glass transition temperature. Phys. Rev. Lett. **90**, 015901 (2003)
19. A. Masuhr, T. Waniuk, R. Busch, W. Johnson, Time scales for viscous flow, atomic transport, and crystallization in the liquid and supercooled liquid states of $Zr_{41.2}Ti_{13.8}Cu_{12.5}Ni_{10.0}Be_{22.5}$. Phys. Rev. Lett. **82**, 2290 (1999)
20. J. Bisault, G. Ryschenkow, G. Faivre, Spherulitic branching in the crystallization of liquid selenium. J. Cryst. Growth **110**, 889–909 (1991)
21. A. Nash, P. Nash, Binary Alloy Phase Diagrams, *US National Bureau of Standards Monograph Series 25* (Elsevier, ASM, Ohio, 1976), p. 35
22. R. Ahmad, R. Cochrane, A. Mullis, Disorder trapping during the solidification of β-Ni_3Ge from its deeply undercooled melt. J. Mater. Sci. **47**, 2411–2420 (2012)
23. N. Haque, R.F. Cochrane, A.M. Mullis, Rapid solidification morphologies in Ni_3Ge: spherulites, dendrites and dense-branched fractal structures. Intermetallics **76**, 70–77 (2016)
24. O. Oloyede, T.D. Bigg, R.F. Cochrane, A.M. Mullis, Microstructure evolution and mechanical properties of drop-tube processed, rapidly solidified grey cast iron. Mater. Sci. Eng. A **654**, 143–150 (2016)
25. B. Crist, J.M. Schultz, Polymer spherulites: a critical review. Prog. Polym. Sci. **56**, 1–63 (2016)

Part XIX
Recent Developments in Biological, Structural and Functional Thin Films and Coatings

Development of Enamel Coatings in Accordance with Recent Regulations of Food Contact Materials

Meltem Ipekci, Kagan Benzesik, Filiz Cinar Sahin and Onuralp Yucel

Abstract The holding ability of enamel on substrate can be provided by different mechanisms such as chemical bonding. Chemical bonding can be achieved by the diffusion of FeO from the substrate to the enamel, and the diffusion of oxides such as NiO and CoO from enamel to the substrate. In this study, we used frit containing NiO and CoO in enamel according to the recent regulations of food contact materials in order to investigate the effects of these metal oxides on the holding ability of enamel. Firing temperature was kept stable at 830 °C during the experiments, and firing times were 2 and 5 min, so we also observed the effect of firing time at stable temperature. After the trials, migration tests and ICP MS analysis were performed to determine the amount of the Ni and Co which can migrate to food from the enamel. Moreover, impact tests were carried out to determine the holding abilities of enamel in terms of firing time.

Keywords Enamel · Food contact materials · Migration

Introduction

Enamel coatings are defined as frit which is matching to metal substrate in terms of thermal expansion coefficient. Frit has a glassy structure, and is produced from inorganic materials by melting in the range of 1000–1500 °C, and after that a fast

M. Ipekci (✉) · K. Benzesik · F.C. Sahin · O. Yucel
Faculty of Chemistry and Metallurgy, Metallurgical and Materials Engineering Department, Istanbul Technical University, 34469 Istanbul, Turkey
e-mail: meltemipekci@gmail.com

K. Benzesik
e-mail: kaganbenzesik@gmail.com

F.C. Sahin
e-mail: cinar@itu.edu.tr

O. Yucel
e-mail: yucel@itu.edu.tr

The Minerals, Metals & Materials Society, *TMS 2017 146th Annual Meeting & Exhibition Supplemental Proceedings*, The Minerals, Metals & Materials Series,
DOI 10.1007/978-3-319-51493-2_71

cooling process. The required properties of enamel such as mechanical strength, hardness, chemically stability, color or easy-cleaning are depending on chemical composition of frit, additive agents in enamel coating suspension, and coating techniques [1, 2].

This type of coatings is used for corrosion and wear protection of metal, besides provides thermal shock endurance, acid and alkali resistances, and decorative looking [3, 4]. In comparison with other types of coatings, enamel can be used in heating and cooking and kitchen devices, food industry, mining, machinery and chemical industries because of its hardness, and thermal and chemical resistances [3].

The quality of the enamel coating is depending on the adhesion ability of the enamel on the substrate material. Adhesion is the most important property of the enamel—substrate system, and adhesion strength can give us the usage life of enamel coated product. Adhesion strength is directly related to the chemical composition, surface tension and roughness of the substrate material [5].

The holding mechanism of the enamel coating is explained by chemical theory, mechanical theory and diffusion theory. According to the chemical theory, adhesion mechanism is provided by saturation of a small area in glassy structure of the enamel with FeO. If there occurs a sufficient diffusion of FeO from the steel as substrate material to the glassy phase of enamel, there forms an oxide interlayer in enamel-substrate interface. In addition to forming of FeO, there are metal oxides such as CoO and NiO in enamel which provides the adhesion. The existences of these metal oxides lead to an intensive roughness on the surface of the substrate material. And this roughness contributes the adhesion between substrate and enamel coating. Also this roughness is based on the electrochemical corrosion between Ni, Co and Fe [5–7].

In coating process, enamel melts and solves this FeO layer during firing because of the higher oxygen affinity of the iron than that of nickel and cobalt at firing temperature. Then CoO and NiO are reduced to elemental Co and Ni, respectively. The reactions at firing temperature and Gibbs free energy values are given below.

$$Fe_3O_4 \rightarrow Fe^{2+} + 2\,Fe^{3+} + 4O^{2-} \quad (1)$$

$$2\,Fe^{3+} + Fe \rightarrow 3\,Fe^{2+} \quad (2)$$

$$Fe + CoO \rightarrow FeO + Co \quad \Delta G = -36893\ J \quad (3)$$

$$Fe + NiO \rightarrow FeO + Ni \quad \Delta G = -52988\ J \quad (4)$$

Since Ni and Co are toxic to organisms, and can migrate from the enamel coated product to the food at different conditions in terms of different temperature and/or different pH, European Directorate for the Quality of Medicines & HealthCare (EDQM) has published international regulations about metals and alloys used in food contact materials in different years. The amount of migrated elements and enamel suitability for use are determined by the tests presented in "1935/2004 EC &

EDQM Metals and Alloys Guidelines & 84/500 EC". The recent regulations were published in 2013, and according to this, we tried to produce enamel coated product according to these regulations. To determine the migration amount of Ni and Co, we simulated some food contact conditions using solutions containing acid such as citric or acetic acid, and high temperature such as 100 °C in terms of cooking under the guidance of 84/500/EEC regulations. After the migration experiment, we performed ICP MS analysis to investigate the amounts of the migrated elements. In addition to chemical analysis, we carried out impact tests to determine the level of the holding ability of enamel.

Experimental Procedure

In this study, the steel plates coded as EN 10209 DC04EK which is defined as drawing enameling quality Al-killed cold rolled steel were used as substrate. Enamel Frits were provided by Gizem Frit A.Ş. Before the trials, both steel plates and frits were characterized by using Bruker-Elemental Q4/VUV optical emission spectrometry (OES) and Perkin Elmer AAnalyst 800 atomic absorption spectroscopy (AAS), respectively. The coated steel plates were in size of 100×100 mm. Before coating, the steel plates were cleaned by detergent and water to get rid of grease and other contaminants. To avoid corrosion and negative side effects in coating, steel plates were dried. Enamel suspension was prepared by using ball mill. Ball mill provided a homogenous mix of frit and water, in addition to this the required particle size of frit can be obtained by ball mill. Enamel suspension was applied on the steel plate as double coating by air aided spray coating technique, and then a single firing process was performed. For the process, spray coating system and Devilbiss Pri-Pro spray gun were used. In this system, enamel suspension was poured in the chamber of the spray gun, and then sprayed on the substrate with an air pressure at 3 bar. After the spray coating, coated steel samples were dried at room temperature for 1.5 h, then in stove at 100 °C for 10 min. So, excessive water in enamel was eliminated before firing. Firing is the last process of enameling. Double layer—single firing was applied to the coated samples at 830 °C for 2 and 5 min in Heraus muffle furnace.

After the enameling process, impact test was applied to the samples according to ASTM B916-01(2007) (Standard Test Method for Adherence of Porcelain Enamel Coatings to Sheet Metal). In this experimental setup, a load of 2 kg is applied from a height of 50 cm. As a result of this, there occurs a deformation. The level of the deformation is determined by the impact scale which can be seen in Fig. 1.

According to the impact test, the holding level of the enamel is expressed by the numbers in the range of 0–5. The percentage of the remains of the enamel coating after the impact test leads to the impact test scale as it can be seen in Table 1.

For the migration test, a setup was prepared by taking into account the surface area and the volume of the solution contacted with the surface according to the regulations as it can be seen in Fig. 2.

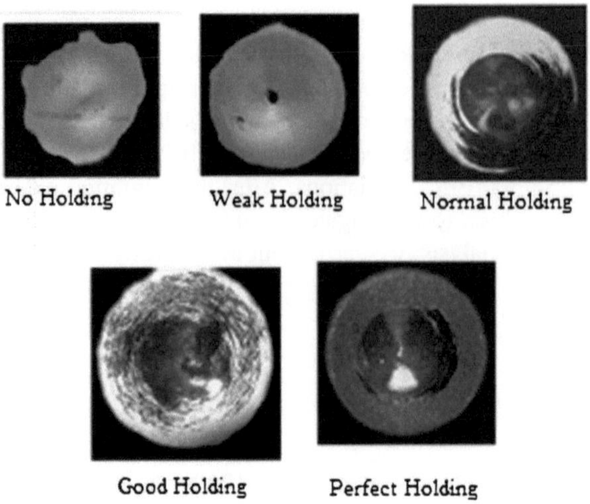

Fig. 1 The impact test scale

Table 1 Forming the impact test scale

Scale	The level of holding	Description
0	No Holding	No remains of the coating
1	Weak holding ability	Remaining 10% enamel on the surface after impact test
2	Normal holding ability	Remaining 25–30% enamel on the surface after impact test
3	Better holding ability	Remaining 50–60% enamel on the surface after impact test
4	Very good holding ability	Remaining 70–80% enamel on the surface after impact test
5	Perfect holding ability	Remaining at least 90% enamel on the surface after impact test

Fig. 2 Experimental setup of migration test

In this setup, there are a cylindrical glass containing acid solution to simulate food, a silicon ring for sealing, and a pressing system for sealing and a complete contacting between surface and the acid solution. Samples were placed between the silicon ring and metal plate of the pressing system for migration test.

According to the regulations, 1 L of acid solution should be used for 6 dm^2 of surface area of enamel. According to the calculations, 100 ml of solution is enough for this setup. For this test, 0.5% solution of citric acid solution was prepared to simulate both acidic and alkali food media. The test was performed at 100 °C for 2 h according to the regulations. Before the test, the coated surface was cleaned by ethanol to eliminate contaminations. During the test, the setup was covered by petri dish to avoid evaporation of the citric acid solution and contamination coming from the stove. Then this setup was located in the stove. After the migration test, ICP MS analysis was performed to determine the amount of the migrated Ni and Co.

Results and Discussion

Characterization of Starting Materials

Before the spray coating process, the starting materials such as frit and substrates were characterized. Frit was characterized by AAS, and its chemical composition can be seen in Table 2.

The steel substrates were characterized by OES, and the chemical composition of this material can be seen in Table 3.

Impact Tests

Impact test shows the holding ability of enamel. The firing time and temperature are important parameters for determination of the holding ability. In this study, firing

Table 2 The chemical composition of Frit

Element	Composition by weight percent
SiO_2	56.12
TiO_2	1.29
Na_2O	22.68
Fe_2O_3	0.13
K_2O	4.30
CaO	1.42
Al_2O_3	7.69
Li_2O	3.39
MnO_2	0.14
NiO	2.17
CoO	0.40

Table 3 The chemical composition of EN 10209 DC04EK coded steel

Element	Composition by weight percent
Fe	99.15
C	0.0058
Mn	0.181
Si	0.030
P	<0.005
S	<0.150
Al	0.051
Ni	0.018
Cu	0.019

temperature was kept stable, and only the effect of the different firing times was investigated in terms of holding.

According to the impact test results, the sample fired at 830 °C for 5 min coded D-18-3 showed perfect holding ability, and the sample fired at 830 °C for 2 min coded D-18-2 showed intermediate holding ability. The results can be observed in Fig. 3.

According to the holding test results, firing time is an important parameter, and the longer firing time leads to the better holding ability of enamel.

Migration Tests

Migration of the toxic elements such as Ni and Co from enamel coated samples to food can be observed by migration tests. As it is mentioned before, migration tests

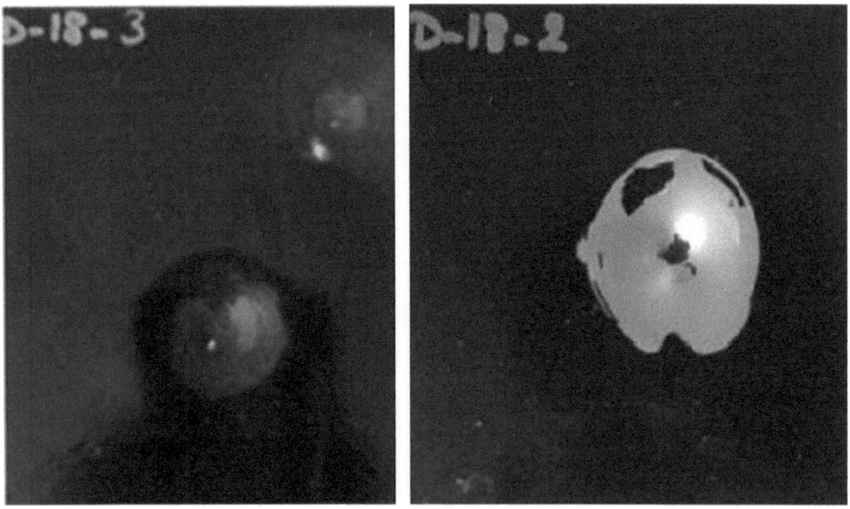

Fig. 3 The impact test results of the samples. D-18-3 was fired for 5 min, D-18-2 was fired for 2 min

Table 4 SRL values of Ni and Co for food contact materials and migration test results of the sample coded D-18-3

Element	SRL (mg element/kg of food)	Migrated value (mg/kg)
Ni	0.14	0.319
Co	0.02	0.215

carried out by simulating food. For this test, acidic solutions and high temperature values in terms of cooking such as 100 °C are used according to the regulations. After the tests, the obtained migration solutions are analyzed by ICP MS. The results of ICP MS analysis show the suitability of the relevant enamel for using as food contact material according to the regulations. Table 4 shows maximum amounts of migrated Ni and Co that must be found in food contact materials. These limits are also called Specific Release Limit (SRL).

As it can see from Table 4, the maximum value of migrated Ni and Co must be very low because of their toxic effect for the living organisms. Since the sample fired at 830 °C for 5 min coded D-18-3 showed perfect holding ability, only the migration of this sample was investigated by ICP MS analysis. The migration test results show that the migration of the toxic elements Ni and Co of the sample coded D-18-3 is greater than the SRL values according to the regulations. That means this enamel coated sample cannot use as food contact material.

Conclusions

In this study, enamel coated steel samples were produced from the very beginning, and characterized in terms of holding ability and elemental migration. The holding ability shows the expected lifetime of the enamel coated products, and the migration tests show if these products are suitable for using as food contact material according to the recent regulations published by European Directorate for the Quality of Medicines & HealthCare (EDQM) in 2013.

According to the impact tests, if the firing temperature is kept stable, the firing time becomes important: the longer firing time, the better holding ability. So, the sample fired at 830 °C for 5 min coded D-18-3 showed a better holding ability than that of the sample fired at 830 °C for 2 min coded D-18-2.

According to the chemical theory of the enamel holding, Co and Ni are important for holding of enamel to the surface of the substrate. However, these elements are toxic for living organisms and they should be kept under control by regulations, and the migrations tests have an important role to determine the maximum values of the toxic migrated elements such as Co and Ni from the coated samples. After the migration tests, ICP MS or any chemical analysis like that must be performed to determine the level of the migrated elements that if they exceed SRL which is determined by EDQM. According to ICP MS analysis of the sample coded D-183, the migrations of Ni and Co are more than the SRL values, and therefore this enamel is not suitable for contacting to food.

For future studies, new compositions of frits and different steel substrates will be investigated, and the most suitable compositions of frits and substrate materials will be determined according to the EDQM regulations and perfect holding abilities of enamels.

Acknowledgements The authors would like to thank the Scientific and Technological Research Council of Turkey (TÜBİTAK-1505-5150062) for its financial support.

References

1. A. Mittal, Steel for enameling and enameled steel user manual, Luxembourg
2. S. Şahin, Emaye okulu ders notları, II, in *International Ceramic, Glass, Porcelain Enamel, Glaze and Pigment Congress*, vol. 11, 2011, pp. 113–116
3. A. Çavuşoğlu, Emaye okulu ders notları, II, in *International Ceramic, Glass, Porcelain Enamel, Glaze and Pigment Congress*, vol. 11, 2011, pp. 13–36
4. W.D. Faust, Enamels, Porcelain or Vitreous, in *Encyclopedia of chemical technology*, 4th edn. vol. 9. ed. by Kirk-Othemer (Wiley, New York, 1994)
5. A.H. Dietzel, Emaillierung: Wissenschaftliche Grundlagen und Grundzüge der Technologie (Springer, 2013)
6. B.W. King, H.P. Tripp, W.H. Duckworth, Nature of adherence of porcelain enamels to metals. J. Am. Ceram. Soc. **42**(11), 504–525 (1959)
7. D.G. Moore, J.W. Pitts, J.C. Richmond, W.N. Harrison, The galvanic corrosion theory for adherence of porcelain enamel ground coats to steel **37**(1), 1–6 (1954)

Transmission Probability of Diffusing Particles—A Case Study

Kinnari Shah and Nuggehalli M. Ravindra

Abstract A diffusing particle is subjected to a variety of collisions that lead to a random or Brownian motion. The aim of this study is to compute the transmission probability and highlight the visualization of Brownian motion under the conditions of varying viscosity, particle size and temperature. As a preliminary study, two simulation results such as transport of diluted species and particle based approach have been compared and the transmission probability is computed by a particle based approach. We validate our results with Stokes-Einstein equation and Fang and Ning's experimental and theoretical work and show that the transmission probability increases with decrease in viscosity and particle size and increase in temperature. The obtained results also describe the Brownian motion for various particle sizes, viscosity and temperature.

Keywords Brownian motion · Diffusion · Particles · Agglomeration

Introduction

Self-assembly is a fundamental mechanism by which various structures are formed from the spatial arrangement of atoms, molecules, macromolecules and colloidal particles. It is known that substances can have vastly different properties depending on the way their building blocks are arranged [1, 2]. Particles self-assemble on surfaces by design or accident. As compared to immediate continuous interfaces such as liquid or gas phase, diffusing particles, with distinguishable unique

K. Shah (✉)
Department of Mechanical and Industrial Engineering, New Jersey Institute of Technology, Newark, New Jersey 07102, USA
e-mail: khs24@njit.edu

N.M. Ravindra
Interdisciplinary Program in Materials Science & Engineering, New Jersey Institute of Technology, Newark, New Jersey 07102, USA
e-mail: nmravindra@gmail.com

The Minerals, Metals & Materials Society, *TMS 2017 146th Annual Meeting & Exhibition Supplemental Proceedings*, The Minerals, Metals & Materials Series, DOI 10.1007/978-3-319-51493-2_72

properties such as dielectric constant and or magnetic susceptibility, acquire a dipole moment in a homogeneous external electric (magnetic) field [3, 4]. The resulting dipolar interactions can lead to assembly/aggregation of the particles depending on the properties of particles and fluids. During the last few decades, building blocks of desired size, shape and functionality have been engineered at the nanometer and micrometer scales. Various new synthesis and fabrication techniques have been a major focus of interest to academia and industry to not only study orientation and self-assembly of hard and soft colloidal size particles on a liquid-liquid and liquid-gas interface but also to make these structures permanent. The evolved new building blocks from the spatial arrangement of atoms, molecules and particles of novel materials, self-assembling into unique structures, have been made possible solely by their design. Interdisciplinary approach has been well documented in the literature based on the efforts made towards the development of a significant variety of nanoscale building blocks that are potentially capable of self assembling into larger and more complex systems.

During the last few decades, the scientific challenge has shifted from making new building blocks to organizing them into one-, two-, and three dimensional (1-D, 2-D, and 3-D) structures by using electrostatic interactions [5–7], external electric fields [8, 9], covalent bonding [10], and capillary forces [11–13]. It is mandatory for colloidal particles to acquire control of particle size and position to create arrays of particles that are periodic for any desired application. Natan et al. [14] first employed the idea of chemical binding to assemble colloidal particles in 2D, and this work was extended by other researchers including Sato et al. [15], He et al. [16], and Zheng et al. [17]. Recently, various researchers have focused on self-assembly of micron and sub-micron sized particles by employing an external field to induce pattern formation in colloid dispersions [18–25]. In order to understand the possible agglomeration/arrangement of such building blocks requires answering the various fundamental questions based on the properties of building blocks, methodology and applications. The answers to these questions lead to the possibility for many interesting discoveries and technologies.

In this connection, we are interested in understanding the diffusion of micron to sub-micron size particles that influence the lateral motion of particles in a fluid. In the literature, efforts have been made to study Brownian motion of nanometer and micrometer scale particles, as well as the important issues for generating and predicting their self-assembly into novel and useful structures. A diffusing particle is subjected to a variety of collisions that lead to random motion or Brownian motion. Brownian motion is the essential process of diffusion and mass transport and is critical to many physical, chemical, and biological processes [26, 27]. It is a physical phenomenon which can be observed, for instance, when a small particle is immersed in a liquid. The particle will move randomly under the influence of random forces in varying direction and magnitude [28]. Movement of particles under random forces can be observed by particle trajectory which has been explained by Wiener process [29–31] and stochastic calculus [32–34]. It has been well studied and observed that the assembly of micron size particles and binary mixture in a bulk and interface can be possible under the influence of external force.

But it is a great challenge to make the assembly of sub-micron size particles. Various researchers have made efforts to experimentally demonstrate self-assembly of sub-micron size particles in fluids under external fields. However, they have faced limitations due to Brownian motion of particles in fluids that makes the self-assembly process difficult and time consuming [35].

In the present approach, a systematic effort has been made to compute the transmission probability from the inner to the outer domain through COMSOL Multiphysics particle tracing modeling study [36]. The study of diffusion modeling and transmission probability is described by either transport of diluted species or a particle based approach. The diffusion equation is used in transport of diluted species approach whereas, in the particle based approach, Brownian and drag forces result in the diffusion of particles from the inner domain to the outer domain. Transmission probability is an important phenomenon to visualize the Brownian motion and to compute particle diffusion. It is very important to compute diffusivity for colloidal sub-micron size particles for varying viscosity, particle size and temperature in order to support the experimental optimization which has not been described till now. The objective of this manuscript is to compute the transmission probability and highlight the visualization of Brownian motion for various viscosity, particle size and temperature.

Theoretical Background

Particle tracing provides a Langrangian description of a problem by solving ordinary differential equation using Newton's laws of motion. Newton's laws of motion require specification of the particle mass, and all the forces acting on the particle. Considering Brownian force and drag force in a particle based approach, the equations of motion are [36]:

$$\frac{d(m_p v_p)}{\mathrm{dt}} = F_D + F_B \tag{1}$$

Drag Force

$$F_D = \frac{(18\eta)}{(\rho_p d_p^2)} m_p (v_f - v_p) \tag{2}$$

where,

$$\frac{(18\eta)}{(\rho_p d_p^2)} = \frac{1}{tp} \tag{3}$$

tp = particle velocity response time in seconds.
Brownian Force

$$F_B = \lambda \frac{6\pi k_b d_p \eta T}{\Delta t} \quad (4)$$

k_b: *Boltzmann Constant*
T: *Absolute fluid temperature*
ρ_p: *Particle density in kg/m³*
η: *Fluid viscosity*
m_p: *Fluid viscosity*
v_p: *Particle mass in kg*
vf: *Fluid velocity in m/s*
Δt: *Time step taken by the solver*
dp: *Particle diameter in m*
λ: *Normally distributed random number*

A different value of λ is created for each particle, at each time step for each component of the Brownian force which leads to spreading of the particles. Transmission Probability from the inner to the outer region is defined as:

$$\text{Transmission Probability} = \frac{\int_o cdS}{\int_I cdS + \int_o cdS} \quad (5)$$

where, c is the concentration of particles. I is the inner domain and O is the outer domain.

Design of Experiment and Simulation Methodology

A detailed experimental setup has been used to carry out the assembly of different types of model particles having varying sizes on air-liquid and liquid-liquid interfaces. The results of these studies have been published elsewhere [37]. It has been observed that the optimization of the experimental conditions for the assembly of sub-micron size particles needs to evaluate in detail the Brownian motion of particles. In the present study, efforts have been made to understand the role of Brownian motion, under the boundary condition validated by the transport of diluted species interface model and the particle based approach [38]. The governing equation for the concentration of particles is given by [36]:

$$\frac{\partial c}{\partial t} + \nabla.(-D\nabla c) = 0 \tag{6}$$

with the initial condition given by a delta function at (0, 0): $c_0 = \delta(0, 0)$. The initial delta function begins to diffuse from (0, 0) radially outwards in all directions. After 100 s, some initial concentration will have diffused from the inner circular domain to the outer domain. The number of particles is assumed to be 5000 which are released at the point (0, 0) with the initial velocity of all components = zero. The radii of outer and inner domain are taken as 500 μ and 250 μ respectively. Particles considered in this diffusion model, with density of 2.5 g/cm^3, are dispersed at room temperature in less viscous liquid.

Results

In this study, efforts have been made to compute the transmission probability through a particle-based approach for 2.25 and 0.480 μ solid glass particles as a model system. This transmission probability represents the diffusivity of the model system. Also, the Brownian motion of the particles in a fluid is the key parameter to investigate and can be visualized under the influence of the properties of particles and fluids. By using the COMSOL Multiphysics module [36], simulations have been performed with particle based approach and the transmission probability has been computed as function of viscosity, particle size and temperature. In this particle based approach, we added Brownian and drag force which are stochastic [32–34] and, therefore, it was required to solve the problem multiple times and a statistical average of the results is obtained.

Particle Tracing of 2.25 μ Silica Particle

The computation of transmission probability can be carried out with two different methods. First is the transport of diluted species at the interface and the second is the particle based approach. The number of particles is assumed to be 5000 which are released at the point (0, 0) with initial velocity of all components = zero. The radii of the outer and inner domain are considered to be 500 μ and 250 μ respectively. The particles, considered in this diffusion model, are 2.25 μ silica particles with density of 2.5 g/cm^3, dispersed at room temperature, in less viscous liquid. The transmission probability from the inner to the outer domain is computed by counting the number of particles in the outer domain and dividing it by the total number of particles. This is the main quantity of interest in a particle tracing model.

Transport of Diluted Species

In the first method, particles begins to diffuse from (0, 0) radially outwards in all directions. After 100 s, some of the initial concentration has diffused from the inner region to the outer region as shown in Fig. 1a. The initial concentration for the diluted species is considered to be 5×10^{-11} mol/m^3. The transmission probability computed by transport of diluted species is 0.2005 as mentioned in Table 1.

Particle Based Approach

In a particle based approach, the combination of drag force and Brownian forces results in the diffusion of particles from regions of high to low number density. In particle tracing, for fluid flow interface, the Brownian and drag force features potentially include random contribution. Because of these features that are included in the model, it is necessary to solve the problem multiple times and take the statistical average of the results. Each time the model is solved, a new set of random numbers is created for each particle, at each time step, which is used to compute the

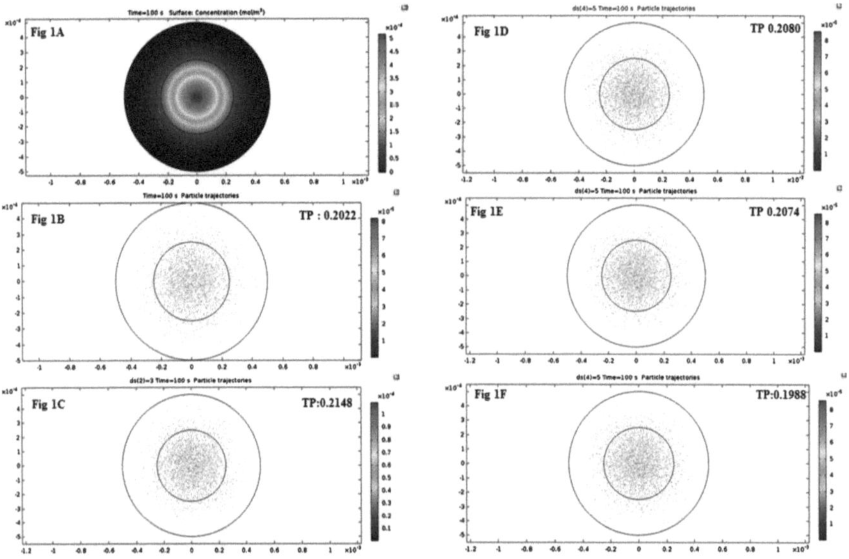

Fig. 1 **b–f**—Plots of location of particles at final solution time for five different runs. Location of particles is shown after 100 s. For each run, different random numbers are generated

Table 1 Transport of diluted species at the interface

Method	Transmission probability
Transport of diluted species at the interface	0.2005

Table 2 Simulated transmission probability by particle based approach

Method	Transmission probability
Particle tracing 1st run	0.2022
Particle tracing 2nd run	0.2148
Particle tracing 3rd run	0.2080
Particle tracing 4th run	0.2074
Particle tracing 5th run	0.1988

force. The Brownian force feature has a parameter called the additional input argument to the random number generator. In this case, the problem is solved 5 times and for each run, the final location of the particles after 100 s is shown in Fig. 1b–f.

The results of the simulation of transmission probability, by particle based approach, are summarized in Table 2. In each case, the transmission probability is slightly different; but in all cases, the results are in accord with those obtained from solving the diffusion by transport of diluted species at the interface (shown in Table 1).

Transmission Probability and Visualization of Brownian Motion at Varying Temperature

Figure 2A shows the transmission probability with temperature for 2.25 μ silica particle. As temperature increases, more and more particles move to the outer domain and hence increase the transmission probability and Brownian motion.

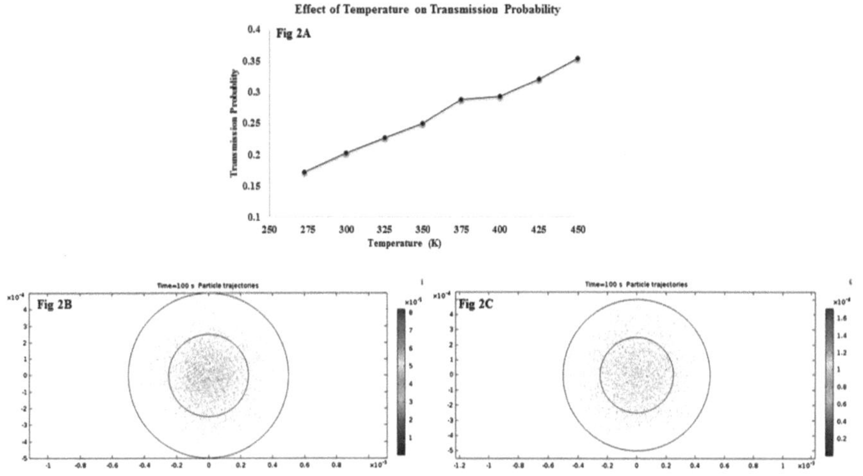

Fig. 2 Effect of temperature on transmission probability—**a** 275–450 K, **b** 300 K, **c** 450 K

From the simulation results, shown in Fig. 2b, c the computed transmission probability, at 300 K and 450 K, is 0.2022 and 0.3534 respectively.

Transmission Probability and Visualization of Brownian Motion at Varying Viscosity

Figure 3a–d show the simulated results of transmission probability and the visualization of Brownian motion for 480 nm silica particles at a constant temperature of 300 K for various viscosities. As viscosity deceases, the particle transmission probability increases. At various viscosities, 2×10^{-5}, 1×10^{-5}, 0.5×10^{-5}, 0.25×10^{-5}, the particle transmission probability is 0.00318, 0.1868, 0.425, and 0.677, respectively. These results indicate that it is easy for the particles to diffuse in lesser viscous fluid. This implies that it is easy to assemble particles in oil than in water.

Transmission Probability and Visualization of Brownian Motion for Varying Particle Size

Figure 4 shows the transmission probability and visualization of Brownian motion for different sizes of silica particles which were used in the self-assembly experiment. Particle sizes such as 0.480, 2.25, 8 and 27 μ are considered in this simulation. Final location of particles, after 100 s, at room temperature in liquid, having the same viscosity, is shown here. With increase in size of the particles, the

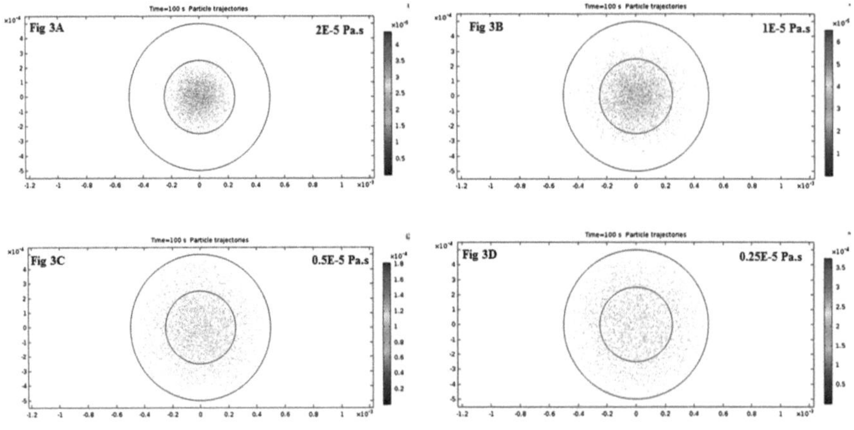

Fig. 3 Effect of viscosity **a** 2×10^{-5} Pa.s, **b** 1×10^{-5} Pa.s, **c** 0.5×10^{-5} Pa.s, **d** 0.25×10^{-5} Pa.s on transmission probability

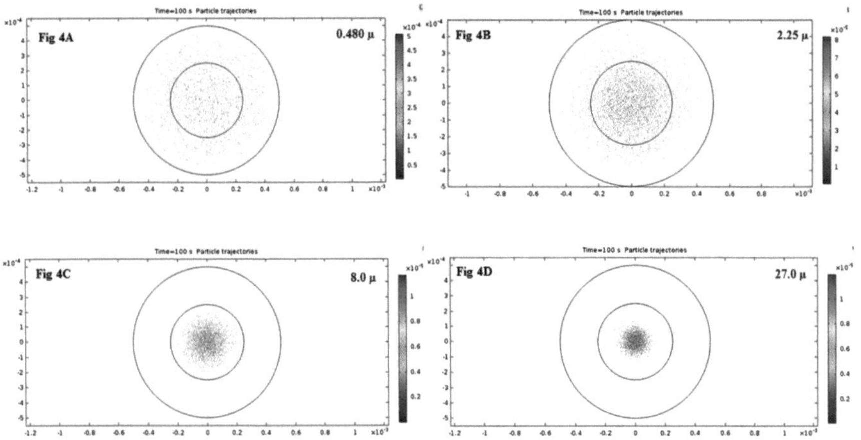

Fig. 4 Effect of particle size **a** 0.480, **b** 2.25, **c** 8 and **d** 27 μ on transmission probability

transmission probability and Brownian motion decreases. This implies that the movement of the particles towards the outer domain may decrease and less number of particles may transfer from the inner region to the outer domain. The results also indicate that the sub-micron sized particles exhibit more Brownian motion than micron size particles. The analysis of transmission probability for particle size of 0.480 μ, 2.25 μ, 8 μ and 27 μ is 0.512, 0.20, 0.00315, and 0 respectively.

Discussion

Transmission probability and Stokes-Einstein Equation

By utilizing COMSOL Multiphysics module [36], simulations have been carried out with particle based approach and the transmission probability, at different viscosity, particle size and temperature, has been computed. Transmission probability increases with decrease in viscosity and particle size and increases with temperature which is in accord with the Stokes-Einstein equation:

$$D = \frac{k_B T}{6\pi\eta r} \tag{7}$$

The above equation relates the diffusion D of a microscopic particle of radius r, undergoing Brownian motion, to the viscosity η of fluid in which it is immersed. This equation is the basis for the theoretical derivation of Fick's laws of diffusion from the consideration of Brownian motion of colloidal particles. Our results also lead to the visualization of Brownian motion for different conditions and are in accord with experimental results from the literature [30–34].

Transmission Probability and Fang and Ning's Results

We also compared our simulation results of transmission probabilities (with viscosity, particle size and temperature) with the experimental results of Fang and Ning [35]. In this work [35], a novel theoretical model of the transport process of micro-particles has been established. Fang and Ning [35] have described and shown that the transport coefficient is a function of the diameter of the micro-particles, density of the fluid, viscosity and temperature. Our results, in Fig. 4, show that the transmission probability is inversely proportional to the particle size. This result is supported by the conclusion of Fang and Ning which indicates that the transport of smaller particles is more extended than the larger ones [35].

Conclusions

The computation of transmission probability from the inner to the outer domain, through COMSOL Multiphysics particle tracing modeling, provides important insight into the design optimization of colloidal size particle assembly/aggregation phenomena and to visualize Brownian motion in two-phase interface under the influence of various parameters such as particle size, temperature and viscosity of the medium. The study of diffusion and transmission probability is described either by transport of diluted species or particle based approach. The results of the computed values of diffusivity for colloidal micron and sub-micron size particles, in terms of varying viscosity, particle size and temperature, show excellent correlation with the experimental results in the literature.

Acknowledgements One of the authors (Kinnari Shah) acknowledges the financial support, in the form of a Teaching Assistantship, by the Department of Mechanical and Industrial Engineering at the New Jersey Institute of Technology.

References

1. S.C. Glotzer, M.J. Solomon, N.A. Kotov, AIChE J. **50**(12), 2978–2985 (2004)
2. J. Park, W. Lu, Appl. Phys. Lett. **93**(24), 243109 (2008)
3. J.N. Israelachvili, *Intermolecular and Surface Forces*, 2nd edn. (Academic Press, Amsterdam, 2003), p. 450
4. A.P. Alivisatos, P.F. Barbara, A.W. Castleman, J. Chang, D.A. Dizon, M.L. Klein, G.L. McLendon, J.S. Miller, M.A. Ratner, P.J. Rossky, S.I. Stupp, M.E. Thompson, Adv. Mater. **10**, 1297 (1998)
5. K.M. Chen, X. Jiang, L.C. Kimerling, P.T. Hammond, Langmuir, **16**(2000), 7825–7834
6. J. Tien, A. Terfort, G.M. Whitesides, Langmuir **13**, 5349–5355 (1997)
7. J. Schmitt, P. Machtle, D. Eck, H. Mohwald, C.A. Helm, Langmuir **15**, 3256 (1999)
8. S. Yeh, M. Seul, B.I. Shraiman, Nature **386**, 57–59 (1997)
9. M. Trau, D.A. Saville, I.A. Aksay, Science **272**, 706–709 (1996)
10. M. Trau, S. Sankaran, D.A. Saville, I.A. Aksay, Nature **374**, 437–439 (1995)

11. R.C. Hayward, D.A. Saville, I.A. Aksay, Nature **404**, 56–59 (2000)
12. H.X. He, H. Zhang, Q.G. Li, T. Zhu, S.F.Y. Li, Z.F. Liu, Langmuir **16**, 3846–3851 (2000)
13. P. Jiang, J.F. Bertone, K.S. Hwang, V.L. Colvin, Chem. Mater. **11**, 2132–2140 (1999)
14. R.G. Freeman, K.C. Grabar, K.J. Allison, R.M. Bright, J.A. Divas, A.P. Guthrie, M.B. Hommer, M.A. Jackson, P.C. Smith, D.G. Walter, M.J. Natan, Science **267**, 1629–1632 (1995)
15. T. Sato, D.G. Hasko, H.J. Ahmed, J. Vac. Sci. Technol., B **15**, 45 (1997)
16. H.X. He, W. Huang, H. Zhang, Q.G. Li, F.Y.S. Li, Z.F. Liu, Langmuir **16**, 517 (2000)
17. J. Zhang, Z. Zhu, H. Chen, Z. Liu, Langmuir **16**, 4409–4412 (2000)
18. N. Aubry, P. Singh, M. Janjua and S. Nudurupati, Proc. Natl. Acad. Sci. U.S.A. 105 (2008), 3711–3714; N. Aubry, and P. Singh, Phys. Rev. E, 77 (2008), 056302
19. D.Y.C. Chan, J.D. Henry Jr., L.R. White, J. Colloid Interface Sci. **79**, 410 (1981)
20. J. Lucassen, Colloids Surf. **65**, 131–137 (1992)
21. M. Janjua, S. Nudurupati, P. Singh, N. Aubry, Electrophoresis **32**, 518–526 (2011)
22. N. Aubry, P. Singh, Phys. Rev. E **77**, 056302 (2008)
23. N. Aubry, P. Singh, M. Janjua, S. Nudurupati, Proc. Natl. Acad. Sci. U.S.A. **105**, 3711–3714 (2008)
24. N. Bowden, A. Terfort, J. Carbeck, G.M. Whitesides, Science **276**(5310), 233–235 (1997)
25. C.D. Dushkin, K. Nagayama, T. Miwa, P.A. Kralchevsky, Langmuir **9**, 3695–3701 (1993)
26. P. Hanggi, F. Marchesoni, Chaos **15**(2), 026101 (2005)
27. E. Frey, K. Kroy, Ann. Phys. **14**, 20–50 (2005)
28. V.I. Tikhonov, M.A. Mironov, *Markov Processes* (Sov. Radio, Moscow, 1977)
29. S.A. Akhmanov, Yu.E. D'yakov, A.S. Chirkin, *Introduction to Statistical Radio Physics and Optics* (Nauka, Moscow, 1981)
30. B.V. Gnedenko, in *Theory of Probability* (Fizmatgiz, Moscow, 1961; Central Books, London, 1970), ed. by V.I. Bogachev, Gaussian Measures (Nauka, Moscow, 1997)
31. I.I. Gikhman, A.V. Skorokhod, *Introduction to the Theory of Random Processes* (Nauka, Moscow, 1965; Saunders, Philadelphia, 1969)
32. R. Elliott, *Stochastic Calculus Applications* (Springer, Berlin, 1983; Mir, Moscow, 1986)
33. A.N. Borodin, P. Salminen, Handbook on Brownian Motion (Birkhäuser Verlag, Basel, 1996; Lan', St.Petersburg, 2000)
34. S.C. Glotzer, M.J. Solomon, N.A. Kotov, AIChE J. **50**, 2985 (2004)
35. H.Y. Fang, M. Ning, J. Hydrodyn. **26**(6), 875–881 (2014)
36. Brownian Motion, Solved COMSOL Multiphys. **5**, 2 (2015)
37. M. Hossain, K. Shah, D. Ju, S. Gurupatham, I.S. Fischer, P. Singh, ASME 2013 Fluids Engineering Summer Meeting (FEDSM 2013), 7–11 July 2013, Nevada, USA
38. M. Kim, A.L. Zydney, J. Colloid Interface Sci. **269**, 425–431 (2004)

Preparation of Porous Titanium Oxide Film by Sol-Gel Method

Guobo Yang, Baoqiang Xu, Xianjun Lei, Hongpin Dai, Xutao Guo, Jian Wu and Bin Yang

Abstract In this paper, porous titanium oxide films were prepared through sol-gel route by using titanium butoxide ($Ti(OC_4H_9)_4$) as a precursor and PEG-2000 (polyethylene glycol) on monocrystalline silicon/stainless steel substrate. Some impact factors for pore characterization were investigated by means of Atomic Force Microscopy (AFM) and Scanning Electron Microscopy (SEM). The results indicate porous films with 20–250 nm spherical pores were prepared. The pore diameter gradually becomes bigger with the dosage of PEG increasing but almost unchangeable through changing the drying temperature. With increases of pulling times and sintering temperature, the pore diameter reduced and enlarged slightly,

B. Xu (✉) · X. Guo · J. Wu · B. Yang
State Key Laboratory of Complex Nonferrous Metal Resources Clear Utilization, Kunming 650093, China
e-mail: kmxbq@126.com

X. Guo
e-mail: 286700647@qq.com

J. Wu
e-mail: 413579358@qq.com

B. Yang
e-mail: kgyb2005@126.com

G. Yang · B. Xu · X. Lei · H. Dai · X. Guo · J. Wu · B. Yang
National Engineering Laboratory for Vacuum Metallurgy, Kunming 650093, China
e-mail: 785265430@qq.com

X. Lei
e-mail: 754824160@qq.com

H. Dai
e-mail: 2606359586@qq.com

G. Yang · B. Xu · X. Lei · H. Dai · X. Guo · J. Wu · B. Yang
Yunnan Provincial Key Laboratory for Nonferrous Vacuum Metallurgy, Kunming 650093, China

G. Yang · B. Xu · X. Lei · H. Dai · X. Guo · J. Wu · B. Yang
Faculty of Metallurgical and Energy Engineering, Kunming University of Science and Technology, Kunming 650093, China

The Minerals, Metals & Materials Society, *TMS 2017 146th Annual Meeting & Exhibition Supplemental Proceedings*, The Minerals, Metals & Materials Series,
DOI 10.1007/978-3-319-51493-2_73

respectively. The different material substrates remarkably affect the surface of porous titanium oxide films.

Keywords Porous titanium oxide film • Pore diameter • PEG • Atomic force microscopy

Introduction

The porous titanium oxide material has widely been investigated because of its remarkable optical and chemical properties, especially in photocatalytic activity and bio-compatibility [1–7]. The research on TiO_2 has become a hotspot in the field of photocatalysis, and the optical properties of titanium oxide thin films were studied early [8]. Basically, preparation process, pore characteristics and film composition would decide physical and chemical properties of film. The titanium oxide with pores can be prepared by many methods, such as the facile PS-templated method, the combining reactive magnetron sputtering, the electrochemical fabrication method and sol-gel method [9–18]. Among them, the sol-gel method is a facile way to prepare metal oxide thin films at low temperatures, with appropriate homogeneity, high purity products, controllable composition and low cost [19]. Although the sol-gel process may appear to be a simple operation, many variables including the dosage of PEG, the drying temperature, the substrate, the pulling times, and sintering temperature can influence the quality of the final film sample [20]. In this study a sol-gel route to prepare porous titanium oxide film is presented and the influence of several preparation parameters on pore sizes and surface was investigated.

Experimental

Preparation Procedures of Porous Titanium Oxide Film

The sol-gel route to prepare porous titanium oxide film is shown in Fig. 1. Sol of titanium oxide was synthesized using titanium butoxide ($Ti(OC_4H_9)_4$), PEG, diethanol amine(DEA), deionized water and ethanol absolute. Eqs. (1) and (2) show the principle of the reaction [21]. sIn ethanol absolute, DEA was added as stabilizing agent to weaken hydrolysis, at the same time the titanium butoxide was added slowly, and then the resulting solution was stirred forcefully at room temperature. After that, the de-ionized water was added to prepare sol with stirred forcefully at room temperature. Finally, the PEG with different dosage (5, 15 and 20 g/L) was added into as-prepared solution until it is dissolved completely. A transparent sol was obtained. The different material substrates were pulled out vertically from sol at 1 cm/min to prepare titanium oxide film. Then, the after-dipping sample was dried

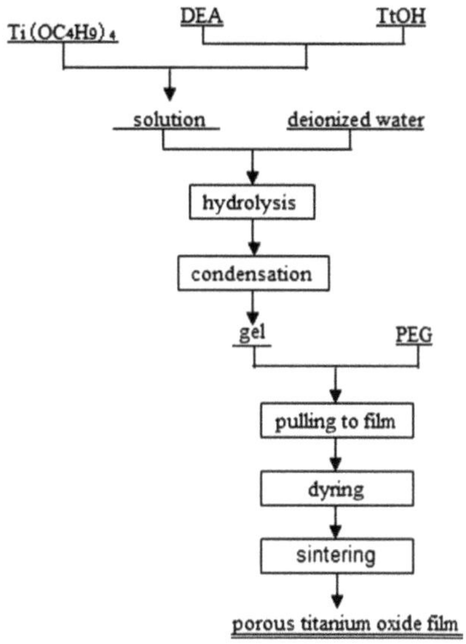

Fig. 1 The PFD of preparing porous titanium oxide films

at different temperatures. The above procedures were repeated for several times. Additionally, parts of dried films samples suffered from a sintering treatment at different temperatures at 2 °C/min ramp rate.

$$\text{Hydrolysis: } M - OR + H_2O \rightarrow M - OH + ROH \tag{1}$$

$$\begin{aligned} \text{Condensation: } & M - OH + RO - M \rightarrow M - O - M + ROH \\ & M - OH + HO - M \rightarrow M - O - M + H_2O \end{aligned} \tag{2}$$

Characteristics of Samples

The pore diameter and surface of films were analyzed by Atomic force microscope (AFM, Agilent-5500 made in Agilent Technologies in USA. AC mode was used in testing process. Silicon probes (NANOSENSORS-NCL-20) with force constant of 21–98 N/m were used to scan.). Additionally, The pore diameter and morphology were observed by SEM (FEI, QUANTA200, USA).

Results and Discussion

Influence of the Dosage of PEG

Figure 2 shows the AFM pictures of the porous titanium oxide films obtained at different dosage of PEG. Although these samples only suffered from drying at

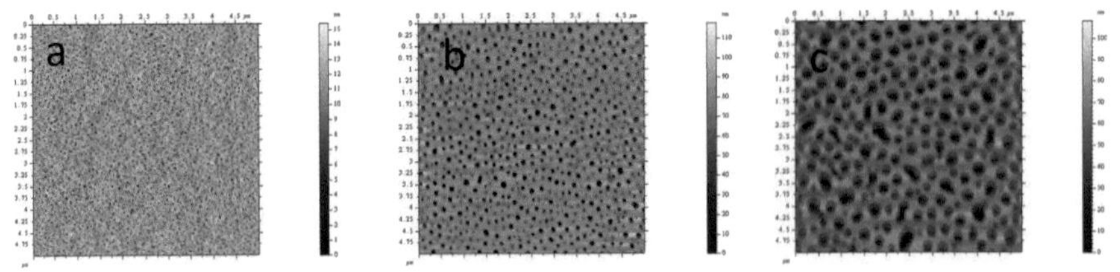

Fig. 2 AFM pictures of the porous titanium oxide films with different dosage of PEG **a** 5 g/L **b** 15 g/L **c** 20 g/L (drying temperature: 80 °C; pulling times: four; monocrystalline silicon substrate; unsintered)

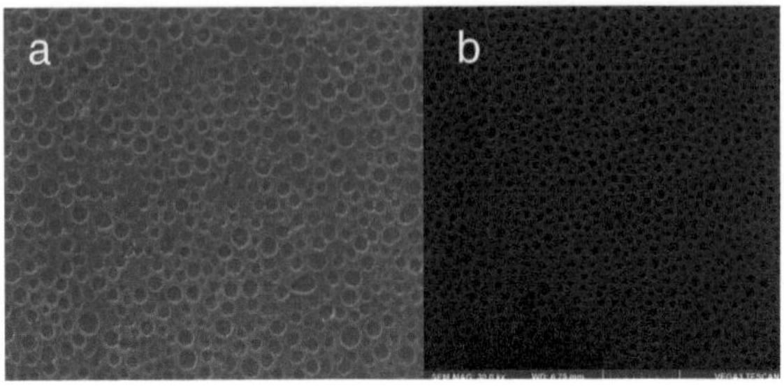

Fig. 3 SEM pictures of porous titanium oxide films with different substrates **a** polished stainless **b** monocrystalline silicon (PEG: 20g/L; pull times: 4; drying temperature: 80°C)

80 °C without sintering, much lower than the boiling temperature of PEG, the definite porous structure has formed. It indicates pore forming for this sol-gel route should owe to volatilization of ethanol absolute not that of PEG. Regardless of dosage of PEG, porous films present perfect round holes. Moreover, the pore diameter of samples remarkably expands from 20 to 250 nm corresponding to PEG concentration increase from 5 to 20 g/L. Basically, more PEG adding into solution obtained higher viscosity sol and impacted on volatilization behavior of ethanol absolute which resulted in different pore diameters. Ethanol absolute was able to easily and freely escape from surface of sol liquid on substrate under 5 g/L dosage of PEG, which resulted in 20 nm diameter pore distribution. While increasing dosage of PEG made higher viscosity sol and restrained the tiny bubbles escaping out from the surface easily, which forced tiny bubbles to aggregate to be bigger bubbles then got out of the surface and resulted in the bigger pore on films. Figure 3 shows the SEM pictures porous titanium oxide films on polished stainless steel and monocrystalline silicon sheet with the dosage of PEG is 20 g/L, it is accordant with the AFM result.

Influence of Drying Temperature

The films were prepared with the dosage of PEG of 20 g/L on monocrystalline silicon substrate. After dip-coating in the Sol, it was dried with three different temperatures (room temperature about 25, 50 and 80 °C). Each process was repeated four times. Figure 4 shows the AFM pictures of the porous titanium oxide films. The pore diameter is within the range of 200–300 nm and some of neighbor pores merge to big elliptical pores with about 500 nm long axis. The unremarkable change of pore size should owe to the easy volatility even at room temperature.

Influence of Substrate

The surface of films prepared on different substrates was investigated. Substrates include common titanium sheet, unpolished stainless steel sheet, polished stainless steel sheet and monocrystalline silicon sheet. Figure 5 shows sample photographs

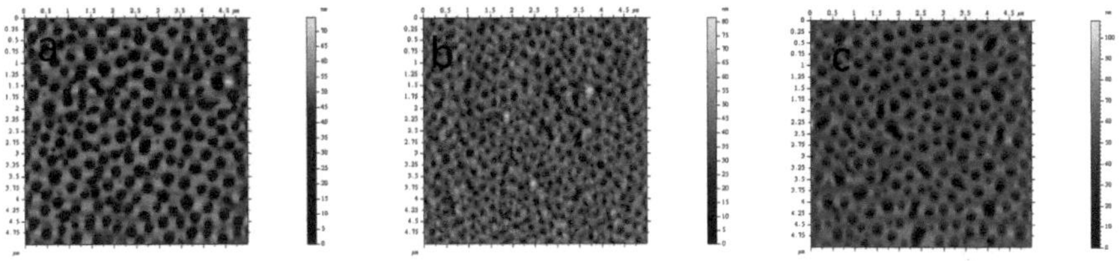

Fig. 4 AFM pictures of the porous titanium oxide films with different drying temperature (PEG: 20 g/L; pull times: 4; drying temperature: 80 °C; monocrystalline silicon substrate. **a** 25 °C, **b** 50 °C **c** 80 °C; unsintered)

Fig. 5 Photographs of films of different substrate **a** common titanium **b** unpolished stainless steel **c** polished stainless **d** monocrystalline silicon (PEG: 20 g/L; pull times: 4; drying temperature: 80 °C; unsintered)

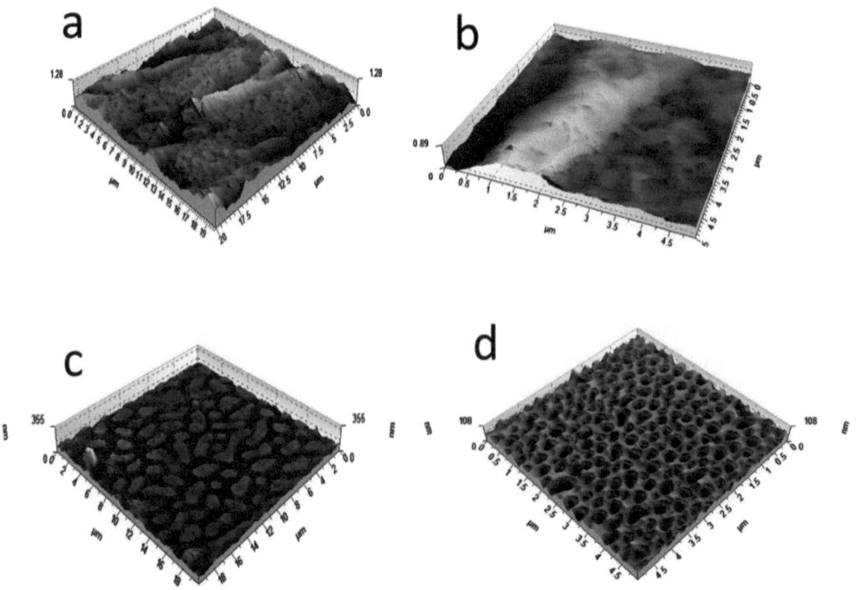

Fig. 6 AFM pictures of films of different substrates **a** common titanium **b** unpolished stainless steel **c** polished stainless **d** monocrystalline silicon (PEG: 20 g/L; pull times: 4; drying temperature: 80 °C; unsintered)

and Fig. 6 shows the AFM pictures of films on different substrates. For the films on common titanium sheet and unpolished stainless steel sheet, their sample surfaces were rugged but others look much more flat according to AFM topography and graphic model. Meanwhile, it had little effect on the surface between polished stainless steel sheet and monocrystalline silicon sheet.

Influence of Pulling Times and Sintering Temperature

The films were prepared with the dosage of PEG of 20 g/L on monocrystalline silicon substrate. After pulling from the Sol, it was dried at 80 °C. This process was repeated different times (four, six and eight). And then the films with pulling four times were sintered at 300 and 600 °C. From Fig. 7a–c, it can be seen the pore diameter decreased slightly with the pulling times increasing. Furtherly, the pore border became much sharper and the skeleton looks shrunken and presents clear texturing as shown in Fig. 8d, e. It was caused by volatilizing of PEG along with pores.

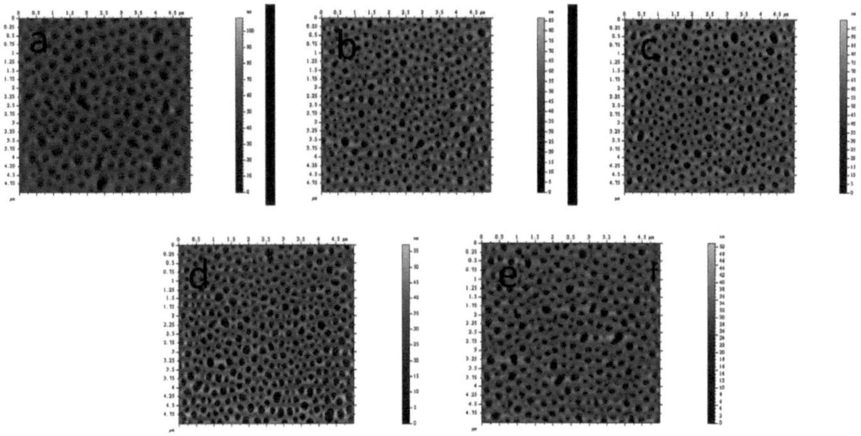

Fig. 7 The AFM pictures of films with different pulling times and sintering temperature (PEG: 20 g/L; drying temperature: 80 °C; monocrystalline silicon substrate. **a** four times; **b** six times; **c** eight times; **a** no sintering; **d** sinter at 300 °C; **e** sinter at 600 °C)

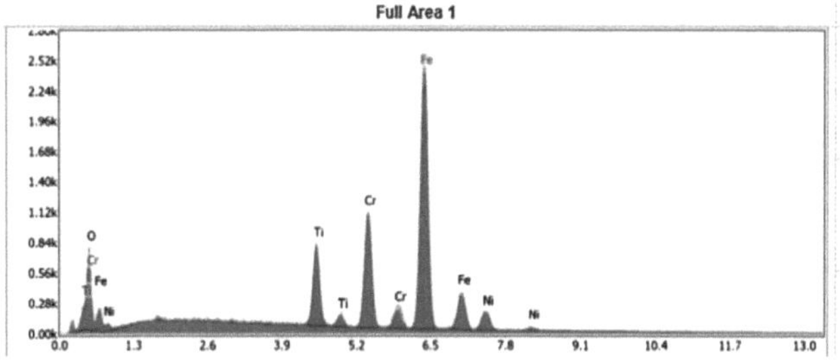

Fig. 8 EDS of porous titanium oxide film on polished stainless substrate (PEG: 20 g/L; pull times: 4; drying temperature: 80 °C; unsintered)

EDS Analysis of the Film

EDS pattern was employed to test the composition of the porous film on polished stainless steel substrate as presented in Fig. 8 and the element content was presented in Table 1. Fe, Cr and Ni in stainless steel substrate were detected. The atomic ratio of Ti and O is 1:3.5, lower than the stoichiometric value of TiO_2 molecule should be caused by the surface of substrate oxidized when it suffered from sintering.

Table 1 Element content of porous titanium oxide film on polished stainless substrate (PEG: 20 g/L; pull times: 4; drying temperature: 80 °C; unsintered)

Element	Wight (%)	Atom (%)	Intensity	Error (%)	K ration	Z	R	A	F
O K	4.57	14.19	126.63	9.27	0.03	1.25	0.87	0.45	1
Ti K	0.96	1.00	45.48	13.27	0.01	0.98	0.98	0.97	1.2
Cr K	16.10	15.40	629.46	2.44	0.21	0.99	1	0.99	1.32

Conclusions

Preparation of porous titanium oxide film by Sol-Gel method was investigated. The pore formation on titanium oxide films for this sol-gel route resulted from volatilization of ethanol absolute not that of PEG. The pore diameter remarkably expands from 20 to 250 nm corresponding to PEG concentration increase from 5 to 20 g/L. It was found that the factor of drying temperature rarely worked in pore diameter because the easy volatility of ethanol absolute. The substrates of different material greatly influence the surface of porous titanium oxide films. Polished stainless steel sheet and monocrystalline silicon sheet are helpful to prepare the more flat films. The pore diameter decreased slightly with the pulling times increase, and the pore border became much sharper and the skeleton looks shrunken and presents clear texturing after sintering.

Acknowledgements This work was supported by National Science Foundation of China (No. 51464023), National Science Foundation of China (No. 51674129), Science and technological leading talents project of Yunnan province, China (No. 2014HA003) and the Program for Innovative Research Team in Ministry of Science and Technology of China (No. 2014RA4018).

References

1. P. Novotna, J. Krysa, J. Maixner et al., Photo-catalytic activity of sol–gel TiO_2 thin films deposited on soda lime glass and soda lime glass pre-coated with a SiO_2 layer. Surf. Coat. Technol. **204**, 2570–2575 (2010)
2. P. Li, I. Kangasniemi, K. Groot, T. Kokubo, Bonelike Hydroxyapatite Induction by a Gel-Der ived Titania on a Titanium Substrate. J. Am. Ceram. Soc. **77**(5), 1307–1312 (1994)
3. T. Kokubo, H.M. Kim, M. Kawashita, T. Nakamura, Bioactive metals: preparation and properties. J. Mater. Sci. Mater. Med. **15**(2), 99–107 (2004)
4. B. Yang, M. Uchida, H.-M. Kim, X. Zhang, T. Kokubo, Preparation of bioactive titanium metal via anodic oxidation treatment. Biomaterials **25**(6), 1003–1010 (2004)
5. M. Uchida, H.M. Kim, T. Kokubo, S. Fujibayashi, T. Nakamura, Structural dependence of apatite formation on titania gels in a simulated body fluid. J. Biomed. Mater. Res., Part A **64** (1), 164–170 (2003)
6. M. Wei, M. Uchida, H.-M. Kim, T. Kokubo, T. Nakamura, Apatite-forming ability of CaO-containing titania. Biomaterials **23**(1), 167–172 (2002)
7. T. Kasuga, H. Kondo, M. Nogami, Apaptite Formation on TiO_2 in Simulated Body Fluid. J. Cryst. Growth, **235**(1), 235–240 (2002)

8. D. Mardare, G.I. Rusu, The influence of heat treatment on the optical properties of titanium oxide thin films. Mater Lett. **56**(3), 210–214 (2002)
9. X. Zheng, Z. Wei, H. Chen et al., In-situ fabrication of dual porous titanium dioxide films as anode for carbon cathode based perovskite solar cell. J. Energy Chem. **24**(6), 736–743 (2015)
10. J. Dervaux, P.A. Cormier, S. Konstantinidis et al., Deposition of porous titanium oxide thin films as anode material for dye sensitized solar cells. Vacuum **114**, 213–220 (2015)
11. Y.S. Joung, B.M. Figliuzzi, C.R. Buie, Design of capillary flows with functionally graded porous titanium oxide films fabricated by anodization instability. J. Colloid Interface Sci. **423** (3), 143–150 (2014)
12. Z.N. Kayani, Z. Saleem, S. Riaz, S. Naseem, F. Saleemi, Deposition of porous titanium oxide thin films as anti-fogging and anti-reflecting medium. Optik **127**(12), 5124–5127 (2016)
13. M.G. Choi, Y.M. Sung, Preparation of double layer porous films of Titanium/Titanium oxide for photoelectrochemical cells application. Opt. Mater. **36**(8), 1430–1435 (2014)
14. P.S. Liu, H.Y. Du, F.J. Xia, Fabrication and characterization of a porous titanium dioxide film with sub-micropores. Mater. Des. **51**(5), 193–198 (2013)
15. A.G. Macedo, L.L. Mattos, E.R. Spada et al., Preparation of porous titanium oxide films onto indium tin oxide for application in organic photovoltaic devices. Appl. Surf. Sci. **258**, 5375–5379 (2012)
16. N. Agoudjil, T. Benkacem, Synthesis of porous titanium dioxide membranes. Desalination **206**(1), 531–537 (2007)
17. I. Piwoński, Preparation method and some tribological properties of porous titanium dioxide layers. Thin Solid Films **515**(7–8), 3499–3506 (2007)
18. T. Tsuru, T. Kan-no, T. Yoshioka, M. Asaeda, A photocatalytic membrane reactor for VOC decomposition using Pt-modified titanium oxide porous membranes. J. Membr. Sci. **280**(1), 156–162 (2006)
19. M. Alzamani, A. Shokuhfar, E. Eghdam, S. Mastali, Influence of catalyst on structural and morphological properties of TiO_2 nano-structured films prepared by sol–gel on glass. Prog. Nat. Sci. **23**(1), 77–84 (2013)
20. C. Sanchez, J. Livage, M. Henry, F. Babonneau, Chemical modification of alkoxide precursors. J. Non-Cryst. Solids **100**(1–3), 65–76 (1988)
21. J. Livage, D. Ganguli, Sol-gel electrochromic coating and devices: a review. Sol. Energy Mater. Sol. Cells **68**(3-4), 365–381 (2001)

Part XX
Solid State Precipitation

Hydride Precipitates in Zirconium Alloys: Evolution of Dissolution and Precipitation Temperatures During Thermal Cycling Correlated to Microstructure Features

Egle Conforto, Stephane Cohendoz, Patrick Girault, Cyril Berziou and Xavier Feaugas

Abstract The fast and spontaneous hydrogen diffusion in zirconium alloys used in the nuclear industry leads to the hydride precipitation which is often pointed as causing embrittlement and rupture. Our studies using X-ray Diffraction (XDR), Transmission Electron Microscopy (TEM) and Scanning Electron Microscopy coupled to Back-Scatter Diffraction (SEM-EBSD) have been demonstrating that the nature of the hydride phase precipitate depends on the hydrogen content, and can show crystallographic orientation relationships (ORs) with the substrate. Differential Scanning Calorimetry (DSC) has been used to identify the dissolution and precipitation energies at global scale. The difference of both can be associated to the misfit dislocations contribution to the precipitation. Local In-situ TEM dissolution observations confirm the dissolution temperature identified at a global scale and show the depinning of some misfit dislocations during dissolution process. The consequence of this mechanism is that dissolution and precipitation temperatures shift during thermal cyclic loading. This situation will be correlated to the nature of crystallographic hydride phases and their ORs.

Keywords Zirconium alloys • Hydrides • Dissolution • Precipitation • Misfit dislocations • Orientation relationships

Introduction

Hydrogen diffuses very rapidly into materials with close-package hexagonal (CPH) structure as titanium, zirconium and their alloys. The hydrogen content in Ti and Zr can easily exceed the solubility threshold, leading to the spontaneous nucleation and growth of hydride precipitates. This phase transformation has a technological importance, because it generates a volume expansion (around 15%)

E. Conforto (✉) · S. Cohendoz · P. Girault · C. Berziou · X. Feaugas
LaSIE, University of La Rochelle, UMR CNRS 7356 Avenue Michel Crépeau, 17042 La Rochelle CEDEX, France
e-mail: egle.conforto@univ-lr.fr

The Minerals, Metals & Materials Society, *TMS 2017 146th Annual Meeting & Exhibition Supplemental Proceedings*, The Minerals, Metals & Materials Series,
DOI 10.1007/978-3-319-51493-2_74

which induces high internal stresses [1, 2] at the origin of a decreased toughness [3]. On the crystallographic point of view, the coherence between the hexagonal substrate and the hydride precipitates is fairly high. In previous works, TEM Bright-Field (BF) images and Selected Area Electron Diffraction (SAED) patterns allowed the identification of four particular epitaxial orientation relationships (OR) [4–8] between the CPH matrix and face-centered tetragonal (FCT) and face-centered cubic (FCC) hydrides. We list below the parallel planes and zone axes between the two phases for each OR as well as those at the interface plane:

OR1: (0001) // (001), $[1\bar{2}10]$ // [110], interface plane $(10\bar{1}0)$ // $(2\bar{2}0)$
OR2: (0001) // $(1\bar{1}1)$ (angle of 4°), $[1\bar{2}10]$ // [110], interface plane $(10\bar{1}3)$ // $(2\bar{2}0)$
OR3: $(10\bar{1}0\ )$ // $(\bar{1}11)$, $[1\bar{2}10]$ // [110], interface plane (0001) // $(2\bar{2}4)$
OR4: $(\bar{1}011)$ // (001), $[1\bar{2}10]$ // [110], interface plane $(10\bar{1}1)$ // $(\bar{1}11)$

The OR2 which has been observed in both Ti and Zr alloys does not show exactly (0001) // (111). Indeed, the two families of planes are not exactly parallel but make an angle of 4°. This point is surprising because the orientation (0001) // (111) which corresponds to the best fit between the dense planes of the two phases is a priori the most natural one. This has however been accounted for by the edge-to-edge matching theory, detailed in reference [7]. According to this theory, the two phases are oriented in such a way that the Δg vectors of the reciprocal space connecting the g vectors of the two phases are parallel to each other and perpendicular to the interface plane.

The morphology and the crystalline structure of the hydride phases depend on the available hydrogen content in the metal. Table 1 summarizes some characteristics of these phases.

Thermodynamic parameters play a crucial role in the dissolution/precipitation mechanism of hydride phases in zirconium alloys. Many authors have been measuring the activation energy, the entropy and enthalpy for these processes in different zirconium alloys [10–14]. These values correspond to measurements at a global approach (over a large density of hydrides). However, only few data are available at a local approach (over a single hydride precipitate) to explain the energies values and the mechanism acting close to the hydride/substrate interface during hydride dissolution and reprecipitation. The goal of this work is to compare

Table 1 Characteristics of various hydride phases formed in α-Ti and in α-Zr [2, 9]

Hydride	Composition	Orientation relationship	Volume expansion (%)
γ (fct)	TiH	OR1, OR2	12,5–14
	ZrH	OR1, OR2	12,3–15,7
δ (fcc)	TiH_x, $1.5 < x < 1.99$	OR1, OR3, OR4	21–29
	ZrH_x, $1.5 < x < 1.9$	OR1, OR2, OR3	16,3–17,2
ε (fct)	TiH_x, $1.8 < x \leq 2$	OR3, OR1	3–14,1
	ZrH_x, $1.7 < x \leq 2$		

energies values experimentally obtained at global and local approaches correlated to misfit dislocations and microstructure evolutions.

Experimental

Samples examined in this work were small tubes with external diameter of 9 mm constituted of two types of zirconium alloys, both containing oxygen, iron, sulfur, tin and chromium in very low contents. Alloy #1 is zircalloy-4, here after labeled "Zy-4". Alloy #2 contains niobium. They were charged by gaseous and electrochemical routes with different hydrogen contents, and recrystallized.

The hydrogen content was measured by fusion using a Thermal Desorption Spectroscopy (TDS) EMGA-621W equipment on polished samples after cutting in small tiles. For Zy-4, two hydrogen contents were measured: 112 wppm and 356 wppm. For alloy 2, contents fall in the 3.3–152 wppm range.

Hydride phases were initially identified in 5 × 5 mm^2 tiles by XRD. Patterns were obtained with a Brucker AXS D8-Advanced equipment using a Cu-Kα radiation and the theta-2theta method between 29 and 72°, with a step of 0.005° each 6 s. Peaks were identified using JCPDS cards: 01-078-2921, 00-034-0649, 00-036-1340 and 00-036-1339. The Zr hydride phases identified are: *γ-ZrH* (metastable), with a FCT crystallographic structure (a = 0.459 nm, c = 0.497 nm); and *δ*–FCC structure (a = 0.478 nm, stable), with chemical composition in the $ZrH_{1.5}$–$ZrH_{1.66}$ range.

Hydride/Zr substrate orientation relationships were examined by Transmission Electron Microscopy using a JEOL JEM 2011 apparatus equipped with an Energy Dispersive Spectroscopy (EDS) OXFORD-INCA Analyzer. Observations were performed at 200 kV in Bright and Dark Field modes. Crystallographic correlation between phases was obtained directly from SAED patterns and through stereographic projections. In-situ hydride dissolution was performed using a GATAN double-tilt heating holder with a tantalum furnace in the 300–930 K temperature range. The SEM used in this work was a FEI/Philips Quanta 200 ESEM/FEG equipped with an EDAX-EDS and an EDAX/TSL-EBSD.

Dissolution and re-precipitation of zirconium hydrides were studied a DSC Q100 V9.9 from TA Instruments. All cycles were performed at a speed of 20 °C/min between 40 °C and 550 °C under pure nitrogen flow of 25 mL/min.

Results and Discussion

Statistical Approach of ORs (Alloy-2)

X-rays measurements (not shown) were performed as an initial, general investigation on the hydride phases present in all Zy-4 and "alloy 2" samples.

In Fig. 1a an intragranular, δ-hydride precipitate is observed by TEM in OR1 in a pre-hydrogenated "alloy 2" sample. In the upper part at left, a Bright Field (BF) image shows the precipitate inside a grain of the zirconium alloy. At the center the SAED pattern of both structures shows the parallelisms indicated in the stereographic projections below it: [020] // [0001] and (10$\bar{1}$) // ($\bar{1}1\bar{2}0$), which are typical for the *OR1*. At right, a Dark-Field (DF) image of the δ-hydride precipitate was obtained using the diffraction of the (200) planes. In Fig. 1b a transgranular, δ-hydride grain is shown in [$\bar{1}$10] // [$\bar{1}\bar{1}$20] substrate zone axis. Moreover, (11$\bar{1}$) // (0001) of the substrate, which characterizes the *OR2*.

Using EBSD technique, Zr hydrides of an "alloy-2" sample were identified at a scale comparable to that of TEM, as shown in Fig. 2. The hydrides in the mapped zone are of "δ" type (FCC). By comparing the phase map (a) with the crystallographic orientation map (b), the hydride morphology observed is in good agreement

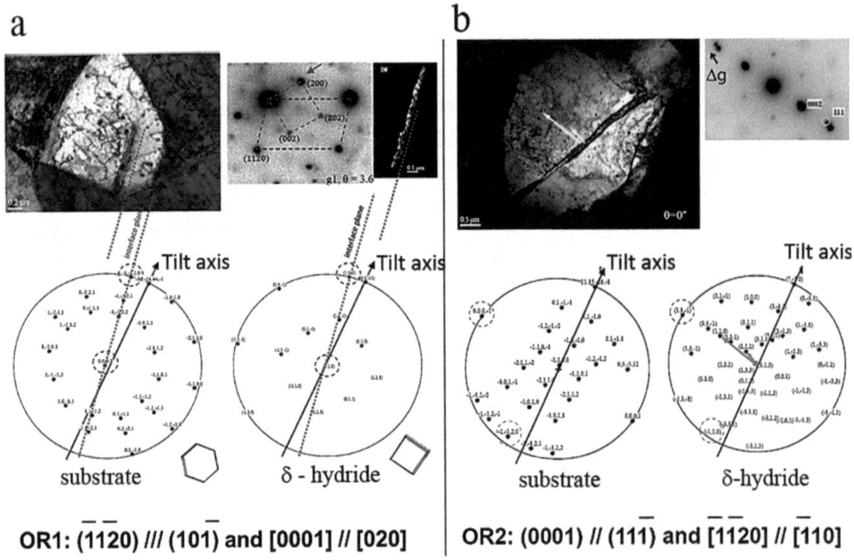

Fig. 1 **a** At *top*: at left, TEM-BF image showing an intragranular hydride precipitate whose direction is indicated by a *dashed line*; at the center, a SAED pattern showing both structures in zone axis; at *right*, a DF image of the hydride precipitate. At *bottom*: stereographic projections allowing to identify the OR1. **b** At *top*: TEM-BF image showing a transgranular hydride precipitate and its SAED pattern. *Arrows correspond to the deviation of 4° between (0001) and* (1$\bar{1}$1). At *bottom*: stereographic projections allowing to identify the *OR2*

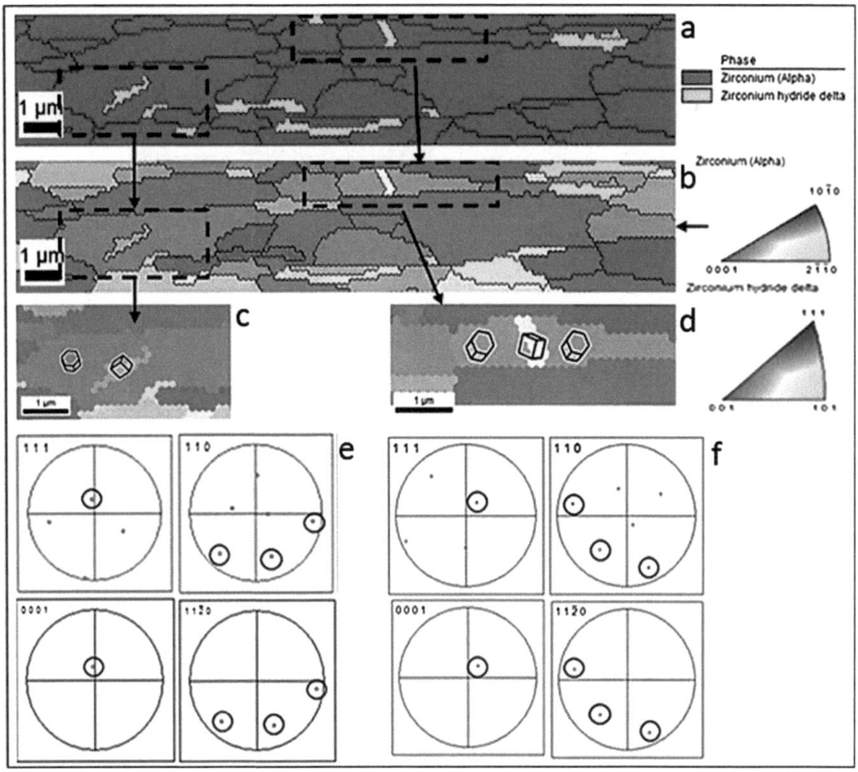

Fig. 2 EBSD maps of a δ-hydride. **a** Phase map; **b** crystallographic orientation map; **c** and **d** details of two regions in (**b**) containing hydride precipitates; **e** and **f** stereographic projections showing the orientation relationship between hydrides and the substrate: *OR2* ((0001) // (111) and (11$\bar{2}$0) // (110))

with that observed by TEM. Details of two regions containing hydrides are shown in Fig. 2c and d. Stereographic projections corresponding to each hydride precipitate and its substrate allowed retrieving the orientation relationship between them, which is *OR2* ((0001) // (111) and (11$\bar{2}$0) // (110)) in both cases (Fig. 3e, f). Consequently, EBSD technique seems to be appropriate to conduce statistical analyze for polycrystalline structure.

Dissolution and Precipitation of Hydrides at a Global Scale

Thermodynamic investigations at a global scale were performed by thermal-cycling samples using DSC over a large number of cycles aiming to determine the hydride dissolution and re-precipitation temperatures. The goal was, in particular,

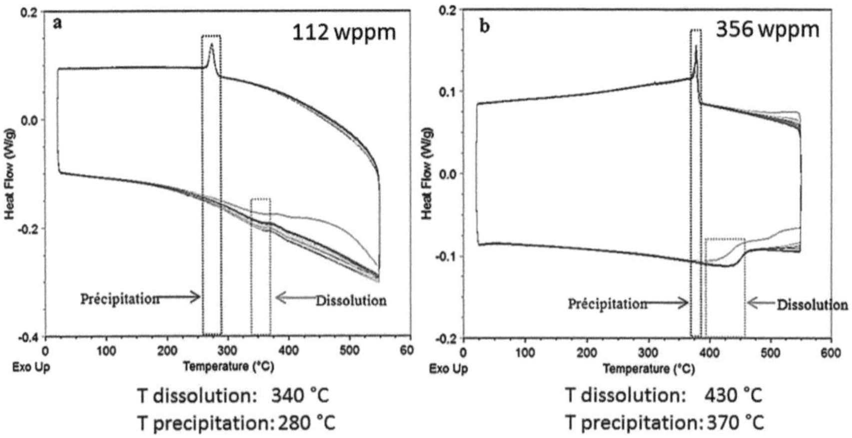

Fig. 3 Thermal cycling of Zy-4 samples containing 112 and 356 wppm in H by DSC. Curves show 20 cycles between 40 and 550 °C where the first is different from the 19 followings due to dislocation relaxation

to determine the activation energies of both processes and the stability of hydrides submitted to a thermal cycling. Figure 3 shows curves superimposing 20 thermal cycles on Zy-4 samples containing 112 and 356 wppm of hydrogen. A large endothermic peak for increasing temperatures is observed, followed by a narrower exothermic peak during cooling. According to previous works, the first peak (T_d) is attributed to the hydrides dissolution and second (T_p) to their re-precipitation [11–15]. Both temperatures are lower for the poorer-hydrogen sample.

Both dissolution and re-precipitation temperatures increase with the number of cycles for the two samples. Figure 4 shows the evolution of T_d (Fig. 4a) and T_p (Fig. 4b) as a function of the number of cycles for Zy-4 containing 112 wppm and 356 wppm of H. From Fig. 4a, $dT_d/dN = 0.038$ (112 wppm) and $dT_d/dN = 0.2$ (356 wppm). From Fig. 4b, dT_p/dN (112 wppm) $= 0.033$ and dT_p/dN (356 wppm) $= 0.073$. If extrapolated to 100 cycles, the variation in the dissolution temperature would be respectively $\Delta T = 4$ °C and 20 °C and that of precipitation, $\Delta T = 3$ °C and 7 °C.

Figure 5 shows the evolution of the critical temperatures T_p and T_d associated with the precipitation and dissolution of hydrides as a function of the hydrogen content C_H for the first cycle. The temperature values correspond to the peaks measured in several alloy #2 samples (indicated by circles) containing only δ-hydride precipitates. The values for the two different H contents in Zy-4 (squares) were superimposed to those of alloy #2 (circles) for comparison. We observe that the dissolution temperature is higher than that of precipitation independently of the hydrogen content in the range investigated.

The difference in temperature between these two curves can be discussed in terms of energy. In fact, according to Fukai [10] the equilibrium between the δ-hydride phase and hydrogen atoms in solid solution in zirconium leads to the

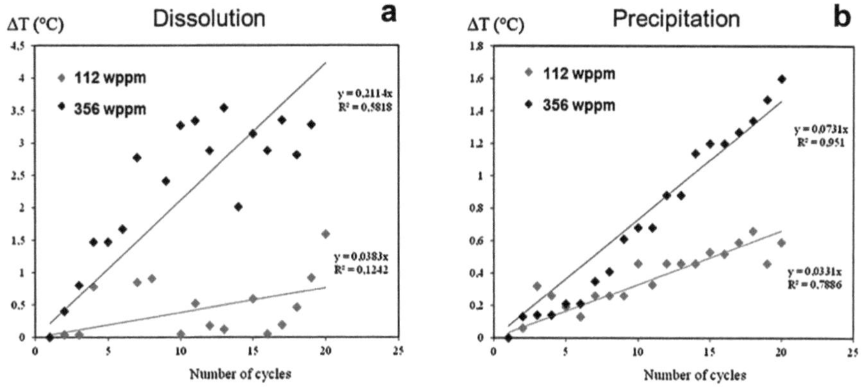

Fig. 4 Evolution of T_d and T_p as a function of the number of cycles for Zy-4 samples containing 112 wppm and 356 wppm of H

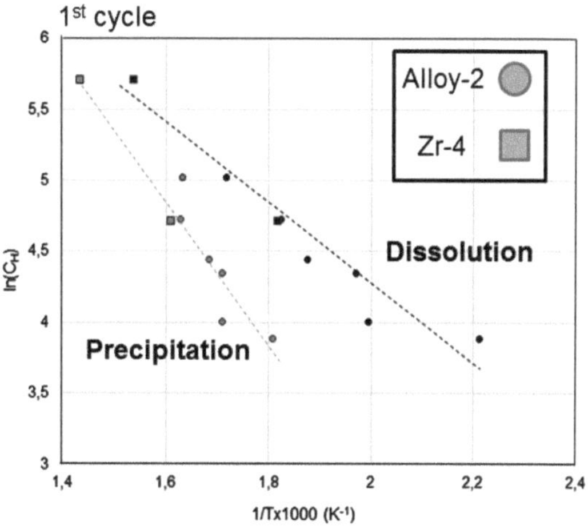

Fig. 5 Evolution of the critical temperatures T_p and T_d associated to the precipitation and dissolution of hydrides as a function of the hydrogen content C_H

equality of chemical potentials described as: ln $(C_H) = \Delta S/R + \Delta H/RT$ where R is the constant of perfect gazes, ΔS and ΔH are respectively the entropy and the enthalpy of the dissolution or the hydride precipitation. The dissolution enthalpy obtained for alloy #2 (112 wppm) is 48 kJ/mol, which is higher than values reported in the literature for zirconium alloys [11–14]. The precipitation enthalpy for the same sample is 20 kJ/mol, which is lower than those reported by the previous authors. Then, these values seem to be correlated to the chemical composition of alloys as well as to their metallurgical state. The precipitation enthalpy is a function of the hydride solution enthalpy ΔH_{hyd} and the elastic (W_{el}) and plastic (W_{pl}) energies associated to the phase transformation: $\Delta H_p = \Delta H_{hyd} - (W_{el} + W_{pl})$. However, the dissolution enthalpy is only a function of ΔH_{hyd} and the storage elastic energy W_{el}, which favors the dissolution: $\Delta H_d = \Delta H_{hyd} + W_{el}$. The difference in energy between them is thus equal to $2W_{el} + W_{pl}$, i.e., 28 kJ/mol and can be correlated to misfit dislocations.

Dissolution of Hydrides at a Local Scale

The hydride dissolution was performed by In-situ TEM on Zy-4 samples with 112 and 356 wppm in H. The aim of this experience is to promote and to observe the equivalent effect, at a local scale, of the fist DSC cycle. In Fig. 6a the steps of the γ-hydride dissolution in a Zy-4 sample with 112 wppm in H are shown, from 300 K up to the complete process at 700 K. The dissolution becomes observable from 430 K. In fact, the SAED pattern at 300 K shows the presence of both structures with the FCT one in the [112] zone axis, meanwhile at 700 K only the diffraction of the CPH is observed. In Fig. 6b the hydride is observed though another tilt angle at the initial and final temperatures. Before heating misfit dislocations are visible around γ-FCT hydride precipitates. The depinning of a misfit dislocation is observed at 700 K, which means that part of hydrogen atoms from the dissolved precipitate is driven away from the precipitation region and is no more available to reprecipitate. In Fig. 6c, the heating rate as a function of the temperature at a global scale by DSC shows the temperature range (600–730 K) for dissolution. In fact in TEM images we observed that this is the temperature range in which the main volume of the hydride dissolves. This indicate a good agreement between global measurements (DSC) and In-situ TEM observations at a local scale.

The same experience was performed in Zy-4 samples with 356 wppm in H. Prior to the hydride dissolution the crystallographic correlation between the FCC δ-hydride precipitate and the hexagonal substrate was investigated (Fig. 7a). We observe that $(\overline{1}\overline{1}20)$ ≈// (010) and (1010) ≈// $(\overline{1}\overline{1}1)$. In Fig. 7b the steps of the δ-hydride dissolution are shown from 300 K up to 930 K, when the process is complete. At 480 K the precipitate length seems to increase and to approach the secondary phase close to it. In fact, some dislocation movement and the beginning of the dissolution process were observed at a local scale. The main dissolution process however is observed between 680 and 780 K, which is confirmed by the curve of the heating rate as a function of the temperature at a global scale (DSC measurement) in Fig. 7d. Back to 300 K after hydride dissolution and cooling, the diffraction observed is that of the hexagonal Zr alloy (Fig. 7c). However, in the bright-field image we still observe some traces and dislocations remaining in the substrate after the precipitate dissolution (Fig. 7b).

The decreasing in concentration of hydrogen atoms available to precipitate can explain the difference in energy between dissolution and precipitation after a thermal cycling at a global approach. Moreover, the accumulated effect of multiples misfit dislocation depinning after a large number of cycles can lead to the precipitation of hydrogen-poorer hydride phases. In fact, the simultaneous presence of the

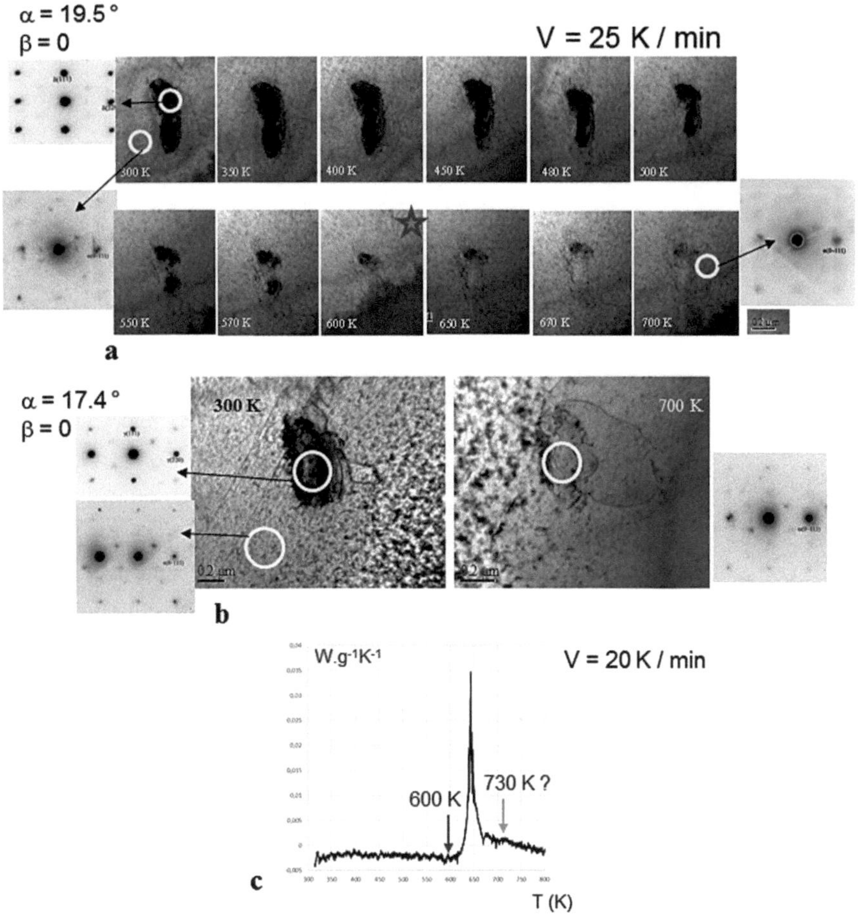

Fig. 6 **a** In-situ TEM images of FCT γ-hydride dissolution (300 K < T < 700 K) in Zy-4 sample with 112 wppm in H showing the beginning of the process around 430 K, observable at local scale. The main dissolution is observed between 600 and 730 K, in good agreement with the heating rate as a function of the temperature at a global scale by DSC in (**c**). **b** Depinning of a misfit dislocation allowing hydrogen atoms to diffuse far from the reprecipitation region

initial hydrogen-rich phase (the δ-FCC) and a hydrogen-poorer phase (the γ-FCT) was observed at the hydride-substrate interface after cycling, as shown in Fig. 8. The reprecipitation of the δ-hydride after 20 thermal cycles is still highly coherent with respect to the substrate (orientation close to an OR3) and the (220) planes of the γ-FCT phase are parallel to those of the δ-FCC (Fig. 8c, d).

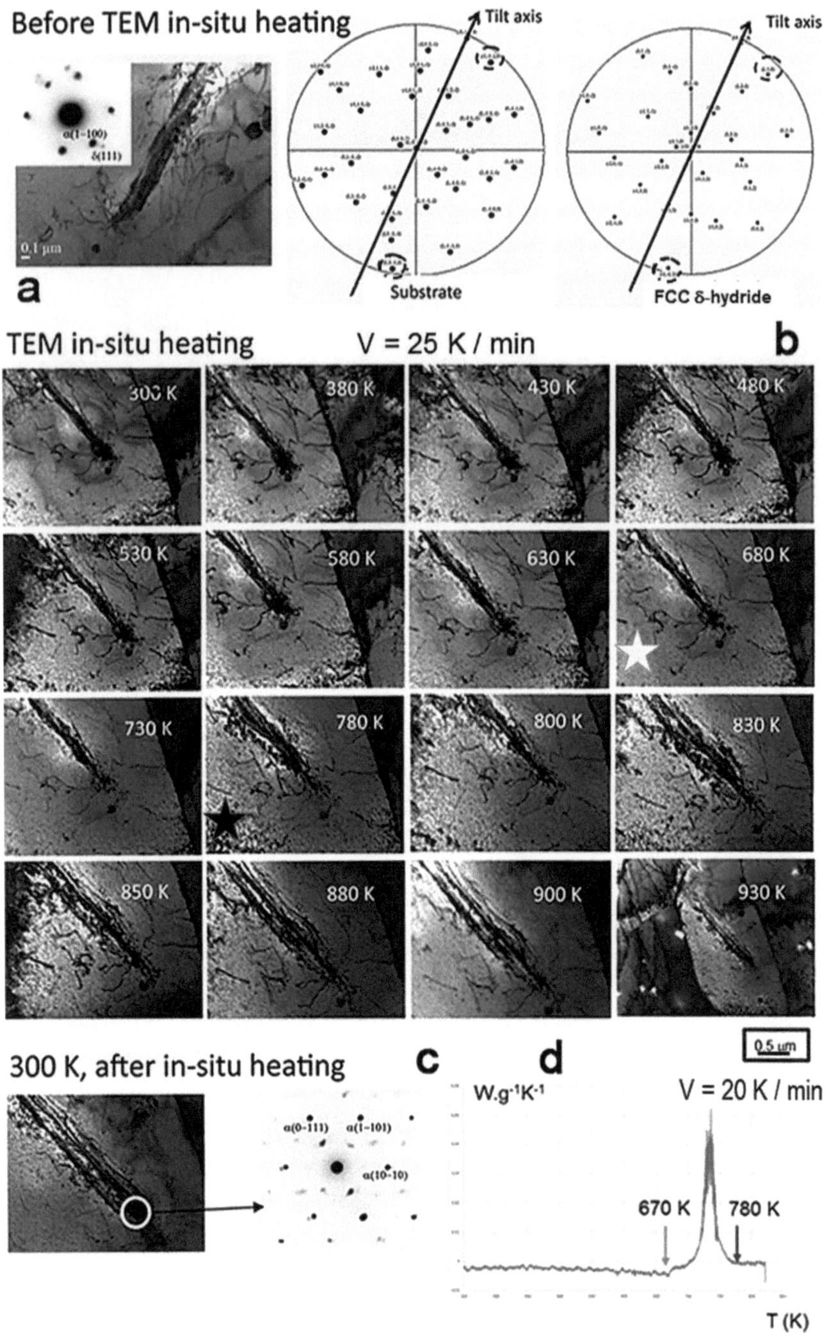

Fig. 7 **a** δ-$ZrH_{1.66}$ / α-Zr orientation before heating. **b** TEM in-situ dissolution (300K < T < 930K) of the precipitate. The main process is observed in the 680–780 K range, which is confirmed by DSC at a global scale in (**d**). The local scale dissolution starts at temperatures slightly higher than those observed for the γ-hydride in Fig. 6, but both mechanisms seem to be similar

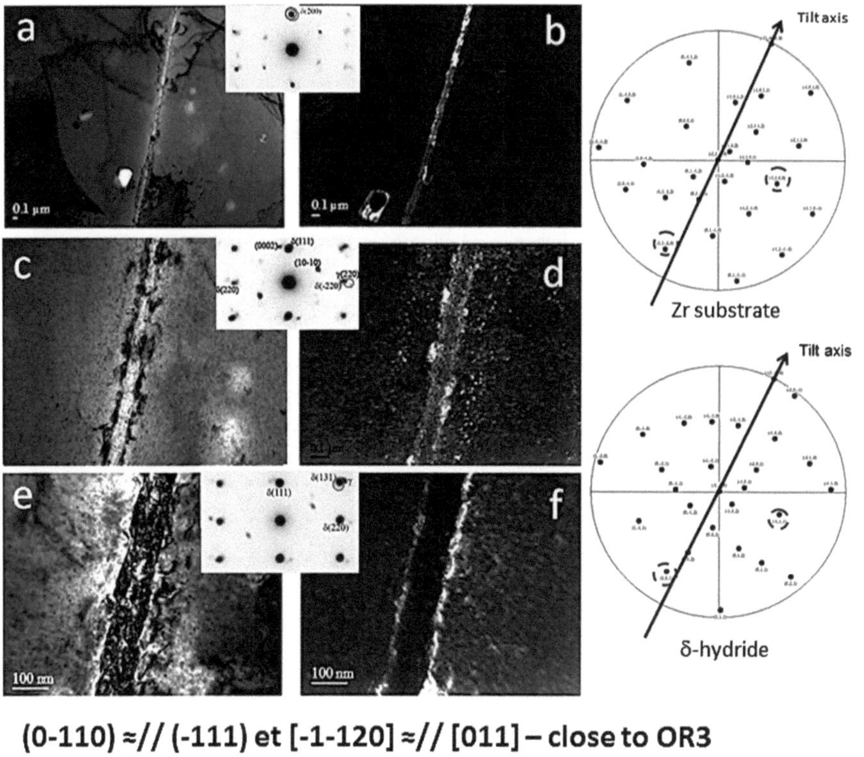

Fig. 8 TEM images and SAED patterns showing the simultaneous presence of a δ- and a γ-hydrides in a Zy-4 sample with 356 wppm in H after 20 thermal cycles. Before cycling only the presence of δ-hydride precipitates was detected in this sample

Conclusion and Perspectives

In conclusion, the evolution of the dissolution and precipitation energies is correlated to the concentration of hydrogen atoms available to reprecipitate, which is submitted to a diffusion controlled by the misfit dislocation migration. This demonstrates the high importance of the hydrogen diffusion in the dissolution-reprecipitation process. In the next step of this project the nature of misfit dislocation will be studied as well as the impact of ORs in dissolution-precipitation energies.

Acknowledgements This work was funded by the French Agency (ANR) under Contract ANR-SIZHYP.

References

1. I. Guillot, X. Feaugas, M. Clavel, Dislocation-hydride interactions at low plastic strain in titanium. Scr. Mater. **44**, 1011–1017 (2001)
2. X. Feaugas, E. Conforto, Influence of hydrogen on the malformation mechanisms and damage to alloys of titanium and zirconium, in *Conference on Mechanisms and Mechanics of Plasticity-Environment Interactions PlastOx 2007*, May 2007 Argeles sur Mer, France (EDP Sciences, 2009), pp. 161–178
3. J. Huez, X. Feaugas, A.L. Helbert, I. Guillot, M. Clavel, Damage process in commercially pure α-titanium alloy without (Ti40) and with (Ti40-H) hydrides. Metall. Mater. Trans. A: Phys. Metall. Mater. Sci. **29**(6), 1615–1628 (1998)
4. H. Numakura, M. Koiwa, Hydride precipitation in Titanium. Acta Metall. **32**, 1799–1807 (1984)
5. O.T. Woo, G.C. Weatherly, C.E. Coleman, R.W. Gilbert, The precipitation of γ-deuterides (hydrides) in Titanium. Acta Metall. **33**, 1897–1906 (1985)
6. A. Bourret, A. Lasalmonie, S. Naka, In-situ high-resolution observation of hydride precipitation in Titanium. Scr. Metall. **20**, 861–866 (1986)
7. E. Conforto, D. Caillard, A fast method for determining favourable orientation relationships and interface planes: application to Titanium-Titanium hydrides transformations. Acta Mater. **55**, 785–798 (2007)
8. E. Conforto, D. Caillard, Edge-to-edge matching at Ti-TiH interfaces: kinetics of hydride growth and clustering of precipitates with different orientation relationships. Solid State Phenom. **172–174**, 242–247 (2011)
9. E. Conforto, S. Cohendoz, C. Berziou, P. Girault, X. Feaugas, Formation and dissolution of hydride precipitates in zirconium alloys: crystallographic orientation relationships and stability after temperature cycling. Mat. Sci. Forum **879**, 2330–2335 (2017)
10. Y. Fukai, *The Metal-Hydrogen System: Basic Bulk Properties* (Springer, New York, 2005)
11. K. Une, S. Ishimoto, Dissolution and precipitation behavior of hydrides in Zircaloy-2 and high Fe Zircaloy. J. Nucl. Mater. **322**, 66–72 (2003)
12. R. Tang, X. Yang, Dissolution and precipitation behaviors of hydrides in N18, Zry-4 and M5 alloys. Int. J. Hydrogen Energy **34**, 7269–7274 (2009)
13. P. Vizcaino, R.O. Rios, A.D. Banchik, Hydrogen determinations in a zirconium based alloy with a DSC. Thermochim. Acta **429**, 7–11 (2005)
14. J.P. Giroldi, P. Vizcaino, A.V. Flores, A.D. Banchik, Hydrogen terminal solid solubility determinations in Zr–2.5Nb pressure tube microstructure in an extended concentration range. J. Alloys Compd. **474**, 140–146 (2009)
15. D. Khatamian, DSC "peak temperature" versus "maximum slope temperature" in determining TSSD temperature. J. Nucl. Mater. **405**, 171–176 (2010)

Analysis of β′ Cu_4Ti Precipitation in Cu-Ti Alloys by Conventional and Diffusion-Couple Methods

Felipe Hernandez-Santiago, Victor M. Lopez-Hirata, Pamela E. Hernandez-Duran, Maribel L. Saucedo-Muñoz and Erika O. Avila-Davila

Abstract An analysis of the Cu_4Ti precipitation process was carried out in binary Cu-Ti alloys using the conventional method and subsequently employing a diffusion-couple based method, known as the Macroscopic-Gradient Method. The phase separation took place via the mechanism of spinodal decomposition at early stages of aging. The precipitation of the β′ (Cu_4Ti) phase preceded to that of the β (Cu_3Ti) phase, which is formed by cellular precipitation. The highest hardness and fastest precipitation kinetics occurred at aging of 400 and 600 °C, respectively. In the case of the macroscopic gradient method, a diffusion couple of Cu-4 wt. Ti alloys and pure copper was prepared by a diffusion annealing. This analysis enabled to determine the equilibrium and coherent lines in this alloy system. The Gibbs-Thomson equation was verified for this type of precipitation.

Keywords Cu_4ti β′ precipitation · Cu-Ti · Alloys · Diffusion-couple · Age hardening

Introduction

Cu-Ti alloys are susceptible to age-hardening. Besides, these alloys were considered to be a possible substitute for other age-hardenable Cu base alloys. Besides, Cu-Ti alloy have been considered to be ultra-high strength conductive materials for applications such as conductive springs, interconnections, etc. [1]. The precipitation process and strengthening of these alloys have been a subject of different studies [1–5]. The phase decomposition in the early stage of aging has been shown to occur by the mechanism of spinodal decomposition [2]. Additionally, it has been reported [3–7] that an ordering reaction takes place to form metastable β′(Cu_4Ti) precipi-

F. Hernandez-Santiago · V.M. Lopez-Hirata (✉) · P.E. Hernandez-Duran
M.L. Saucedo-Muñoz · E.O. Avila_Davila
Instituto Politécnico Nacional (ESIQIE-ESIME), UPALM Edif. 7,
07300 Mexico, D.F, Mexico
e-mail: vlopezhi@prodigy.net.mx

The Minerals, Metals & Materials Society, *TMS 2017 146th Annual Meeting & Exhibition Supplemental Proceedings*, The Minerals, Metals & Materials Series,
DOI 10.1007/978-3-319-51493-2_75

tates, which contribute significantly to the maximum strengthening of these alloys. Either prolonged aging or aging at high temperatures promotes the cellular or discontinuous precipitation with the formation of equilibrium β (Cu_3Ti) precipitates [1, 2].

Miyazaki et al. proposed [8, 9] a new experimental method was proposed to study the precipitation process in binary alloys, called macroscopic composition gradient method. This method permits to determine solubility limits and phase equilibria and it is based on the microstructural observation of different composition alloys formed by a continuous composition gradient. The macroscopic composition gradient can be created in a specimen by diffusion coupling, imperfect arc melting of sandwiched metals, imperfect homogenization of coarse precipitates, etc.

Thus, the purpose of this work is to analyze the precipitation process in Cu-Ti alloys by the aging of a Cu-4at. %Ti alloy at different temperatures, and with the Gibbs-Thomson relation using the macroscopic composition gradient method using a diffusion couple between pure Cu and a Cu-4at. %Ti alloy, after aging at 600 °C.

Experimental Procedure

A Cu-4 wt%Ti was produced by melting of 99.99% purity titanium and 99.999% purity copper in an alumina crucible under an argon gas atmosphere. The ingot was homogenized at 950 °C for 10 days and then cold rolled to obtain specimens of 30 × 10 × 5 mm. Alloy specimens were solution treated at 950 °C for 1 h, and subsequently aged at 400, 500 and 600 °C for several times in an electrical resistance furnace, controlled within ±2 °C. The diffusion couple consisting of two specimens, Cu-4 wt%Ti and pure Cu, with polished flat surfaces was placed in an austenitic stainless steel holder with screws. Specimens were pressed by the screw and diffusion welded by heating at 900 °C for 1 day under an argon atmosphere. The diffusion couple was sealed in an evacuated quartz tube and annealed at 900 °C for 10 days and subsequently quenched in ice-water in order to obtain a solution treated specimen. After this heat treating procedure, specimens were aged at 600 °C for times from 600 to 3600 s. The aged specimens were also observed with an SEM at 15 kV. The specimens for transmission electron microscopy TEM were prepared by the twin-jet electropolishing method at–60 ° C with an electrolyte composed of 25% nitric acid and 75% methanol in volume. The TEM observation of specimens was pursued at 200 kV. The bulk heat treated diffusion-couples were cut and ground to cylinders with diameters of about 3 mm so that the interface could be included in the middle along the cylinder axis. Discs with about 0.15 mm were obtained by electroerosion of cylinders. These discs were thinned using the same procedure described above. TEM specimens were mounted on a Be-specimen holder and also observed with a TEM at 200 kV, equipped with EDXS analysis. The Vickers hardness of heat-treated samples was determined using a load of 100 g.

Results and Discussion

Microstructural and Mechanical Characterization of Cu-4 Wt%Ti Alloy

Bright Field BF-TEM micrographs of the precipitation evolution of specimens solution treated and then aged at 400 °C for different times are shown in Fig. 1a–d. The TEM micrograph of solution treated sample shows no presence of precipitation, Fig. 1a. The corresponding electron diffraction pattern indicates only the spots characteristic of an fcc crystalline structure. This suggests that the quenching rate was fast enough to suppress the phase decomposition during the quench. The microstructure of specimen, aged for 5 h, consisted of a modulated structure with sidebands flanking the 200 reflections, Fig. 1b. This fact confirms that the phase decomposition takes place via spinodal decomposition [1, 2].

After prolonged aging, the specimen microstructure was composed of a periodic array of coherent cuboid precipitates aligned on the elastically soft < 100 > matrix crystallographic directions, Fig. 1d. Superlattice reflections were also observed in the corresponding electron diffraction pattern, Fig. 4f. These have been identified [2]

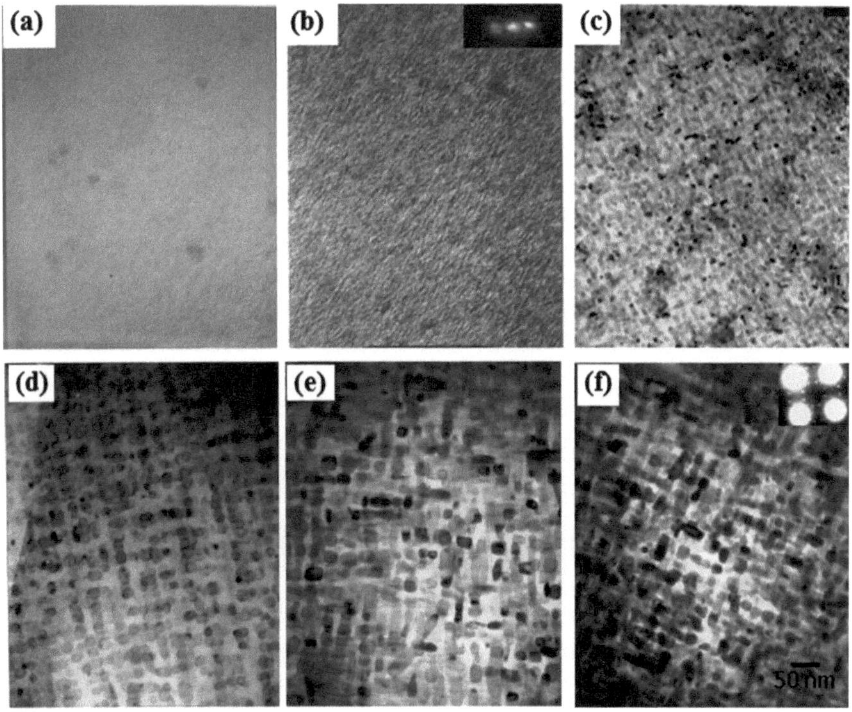

Fig. 1 BF-TEM micrographs of Cu-4 wt%Ti alloy aged at 400 °C for **a** 0 h, **b** 5 h, **c** 25 h, **d** 75 h, **e** 100 h and **f** 150 h

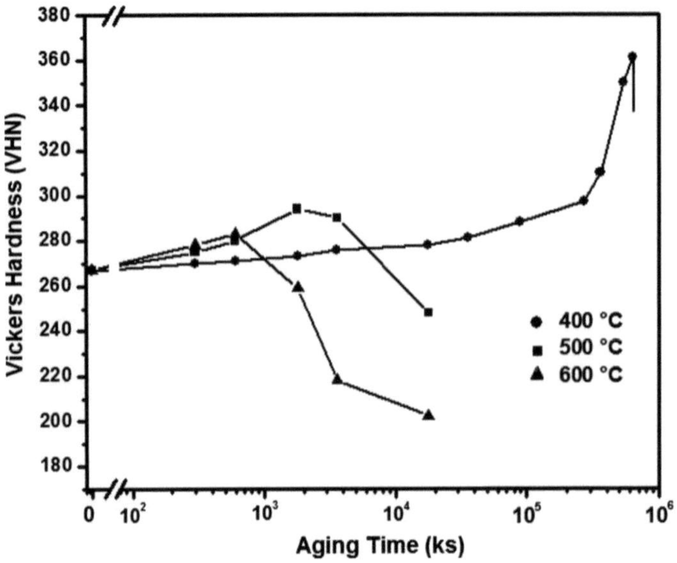

Fig. 2 Aging curves for Cu-4 wt%Ti alloy for 400, 500 and 600 °C

to match with a Ni_4Mo-type structure. This suggests that the cuboid particles correspond to the metastable β′ (Cu_4Ti) phase [5]. The β′ phase has been reported [2] to be formed from a spinodal decomposition mechanism involving clustering and ordering. The stable β phase was present in the lamellar structure, nucleated on grain boundaries due to the cellular precipitation, $\alpha \rightarrow \alpha + \beta$ [4, 5]. The Ostwald ripening of β′ precipitates is also observed in these micrographs. The morphology of β′ precipitates was observed to change from cuboids to block-like rods elongated on < 100 > matrix directions during coarsening. The aging curves for the Cu-4 wt% Ti alloy aged at 400, 500 and 600 °C are shown in Fig. 2. The highest hardness of about 360 VHN was obtained for specimen aged at 400 °C. This increase in hardness seems to be associated with the formation of β′ phase. The lowest peak of hardness was detected for specimens aged at 600 °C. The increase in hardness was the lowest one and the overaging occurred rapidly for this aging temperature. The overaging process can be associated with both the coarsening of β′ phase and discontinuous precipitation. The drastic drop in hardness for prolonged aging at 400 °C seems to be related to the coarsening process of precipitates after a long aging, about 270 h.

Analysis of Precipitation by Diffusion Couples

Figure 3 presents the concentration profile determined on the surface of diffusion-couple after aging at 600 °C for 10 min. Several BF-TEM micrographs

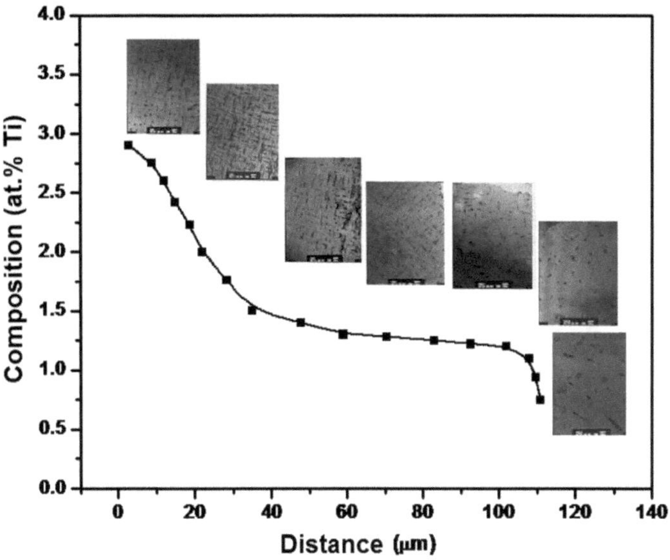

Fig. 3 Concentration profile of the diffusion-couple, after aging at 600 °C for 10 min and BF-TEM micrographs corresponding to some positions of diffusion couple

corresponding to different compositions are also shown in this figure. These micrographs indicate the precipitation evolution for different titanium compositions at the same aging temperature. Three different zones can be noticed in the concentration profile. Precipitates were mainly cuboids in the case of compositions higher than 1.5 at. % Ti. These cuboids were aligned in the elastically-softest < 100 > direction of the Cu-rich matrix. The analysis of the corresponding electron diffraction pattern indicates that the precipitates correspond to the metastable β′ (Cu_4Ti) phase. The dark and bright contrast of precipitates is due to the coherency-strains and this suggests that the β′ precipitates are coherent with the matrix. The β′ precipitates show an elongated plate shape in the < 100 > direction of the matrix for the compositions about between 1.5 and 1.0 at. % Ti. The interface between matrix and precipitates seems to be semicoherent because it can be observed the coherency-strains in the longest side of plates.

Figure 4 shows the plot of the equilibrium matrix composition $C_e(r)$ as a function of the reciprocal value of the mean radius r of precipitates for the semi-coherent precipitates in Cu-Ti alloys. There is a linear relationship, with an adjusted contribution ratio R^2 of about 0.96116, between these two parameters. The empirical $C_e(r)$ of the matrix phase seems to be inversely proportional to the particle radius r.

In the limiting case of binary two-phase alloy with almost pure A, $C_\alpha \rightarrow 0$, and pure B, $C_\beta \rightarrow 1$, the Gibbs-Thomson equation predicts a variation of $C_e(r)$ with r as follows [10, 11]:

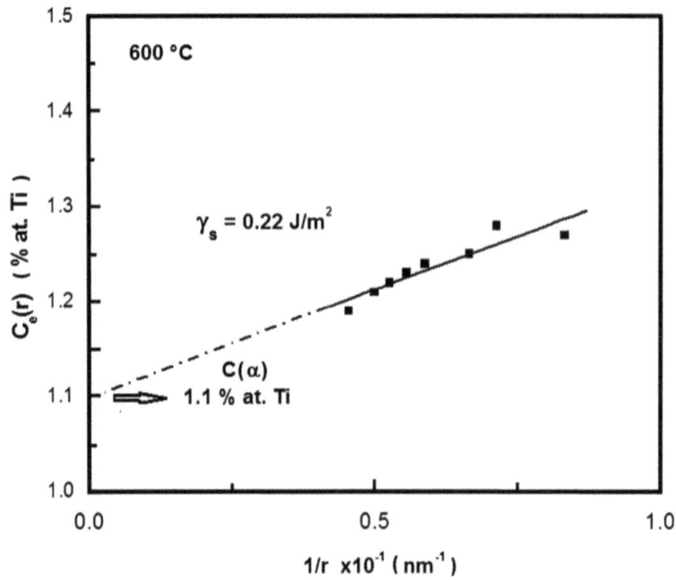

Fig. 4 Gibbs-Thomson relation for the semicoherent β′ precipitates in Cu-Ti alloys at 600 °C

$$C_e(r) = C(\infty).\exp[1 + (2\gamma V_m/rRT)] \quad (1)$$

Where γ is the interfacial energy between precipitates and matrix, V_m the molar volume of precipitate, R the gas constant, T the temperature and C(∞) the equilibrium composition at the precipitate with the infinite size. If the radius is larger than a few nanometers, Eq. (1) reduces to the more commonly quoted form of the Gibbs-Thomson equation:

$$C_e(r) = C(\infty)[1 + (2\gamma V_m/rRT)] \quad (2)$$

Considering a V_m value of about 7.11×10^{-6} m^3 mol^{-1} for the β′ precipitates, the interfacial energy γ can be determined using the slope value of the straight line shown in Fig. 3. The interfacial energy γ was determined to be about 0.31 ± 0.02 J m^{-2}, which is in good agreement with the reported values for semicoherent interfaces, $\gamma > 0.25$ J m^{-2} [12]. The equilibrium composition at the precipitate with the infinite size, C(∞), is about 1.054 ± 0.02 at. % Ti, which is close to the equilibrium value of about 1 at. % Ti at 600 °C in the equilibrium phase diagram [13].

Conclusions

The macroscopic composition gradient method permitted to analyze the precipitation process in different Cu-Ti alloy compositions. Besides, the precipitation process followed the Gibbs-Thomson relationship for the precipitation of the metastable and semicoherent (Cu_4Ti) β′phase. The interfacial energy between the Cu-rich matrix and β′precipitates was determined to be about 0.31 ± 0.02 J m^{-2}. The conventional method permits to analyze the morphology evolution of precipitates, as well as its effect on the age hardening.

Acknowledgements The authors wish to thank the financial support from SIP-COFAA-IPN and CONACYT.

References

1. W.A. Soffa, D.E. Laughlin, High-Strength age hardening Copper-Titanium Alloys: redivivus. Prog. Mater Sci. **49**, 347–366 (2004)
2. D.E. Laughlin, J.W. Cahn, Spinodal decomposition in age hardening Copper-Titanium Alloys. Acta Metall. **23**, 329–339 (1975)
3. A. Datta, W.A. Soffa, The structure and properties of age hardened Cu-Ti Alloys. Acta Metall. **24**, 987–1001(1976)
4. J. Dutkiewwicz, Electron microscopy study of the effect of deformation on precipitation and recrystallization in Copper-Titanium Alloys. Metall. Trans. **8A**, 751–761 (1977)
5. S. Nagarjuna, K. Balasubramanian, D.S. Sarma, Effect of prior cold work on mechanical properties and structure of an Aged-Hardened Cu-1.5wt% Ti Allo. J. Mater. Sci. **32**, 3375–3385 (1997)
6. S. Nagarjuna, K. Balasubramanian, D.S. Sarma, Effect of cold work on precipitation hardening of Cu-4.5 mass % Ti Alloy. Mater. Trans. JIM **36**, 1058–1066 (1995)
7. A.A. Hameda, L. Blaz, Microstructure of hot-deformed Cu-3.45 wt. % Ti Alloy. Mater. Sci. Eng. **254A**, 83–89 (1998)
8. T. Miyazaki, T. Koyama, S. Kobayashi, A new characterization method of the microstructure using the macroscopic composition gradient in Alloys. Metall. Mater. Trans. **27A**, 945–949 (1996)
9. T. Miyazaki, S. Kobayashi, T. Koyama, Determination of the critical nucleus size of precipitates using the macroscopic composition gradient. Metall. Mater. Trans. **30A**, 2783–2789 (1999)
10. J.W. Martin, R.D. Doherty, B. Cantor, *Stability of Microstructures in Metallic Systems* (Cambridge University Press, New York USA, 1997), p. 243
11. J.A. Ardell, Observations of the effect of volume fraction on the coarsening of γ′ precipitates in binary Ni-Al Alloys. Scr. Metall. et Mater. **24**, 343–346 (1990)
12. D.A. Porter, K.E. Easterling, *Phase Transformations in Metals and Alloys* (Chapman & Hall, London UK, 1997), p. 145
13. H. Baker (ed.), *Alloy Phase Diagrams* (ASM International; Materials Park (OH) USA, 1993), p. 173

Carbide Precipitation in a Low Alloy Ferritic Steel

M.L. Saucedo-Muñoz, R. Gómez-Martínez, A. Ortiz-Mariscal, V.M. Lopez-Hirata, J.D. Villegas-Cardenas and J.L. Gonzalez-Velazquez

Abstract the precipitation evolution was studied during aging of 2.25Cr-1Mo steel at 550 °C. The as-received steel was aged in two different ways: the first one was the a continuous isothermal treatment at 550 °C for times up to 1000 h and the second aging was performed using isothermal aging cycles which consisted of aging at 550 °C for 60 min and then water quenched at room temperature. $M_{23}C_6$ and M_6C precipitation occurred intragranular and intergranular in both types of aging. The coarsening of the carbides was observed to occur with the increase in aging time. Nevertheless, the growth kinetics coarsening was faster in the case of isothermal aging cycles. The hardness decreased with aging time in both cases; however, it occurred in a shorter aging time for the cyclic aging. Nanoindentation tests indicated the increase of ductility as the aging time increased.

Keywords Carbide precipitation · Isothermal aging · Cycle aging · Growth kinetics · Hardness

Introduction

Ferritic low alloy steels are predominant materials for manufacturing different industrial components such as, heat exchanger and direct fired heaters tubes since they are economically accessible and possess good mechanical properties. This type of steels are usually used for not severe corrosion conditions at temperatures from 30 to 700 °C [1–4] in energy power and petrochemical plants. This material is commonly used at temperatures higher that 475 °C for very prolonged times and thus the deterioration in mechanical properties such as toughness and creep strength are associated to the long exposure of heating which originates microstructural changes such as grain boundary segregation and coarsening of precipitates [5, 6].

M.L. Saucedo-Muñoz · R. Gómez-Martínez · A. Ortiz-Mariscal · V.M. Lopez-Hirata (✉) · J.D. Villegas-Cardenas · J.L. Gonzalez-Velazquez
Instituto Politécnico Nacional (ESIQIE), UPALM Edif. 7, Mexico 07300, USA
e-mail: vlopezhi@prodigy.net.mx

The Minerals, Metals & Materials Society, *TMS 2017 146th Annual Meeting & Exhibition Supplemental Proceedings*, The Minerals, Metals & Materials Series,
DOI 10.1007/978-3-319-51493-2_76

Additionally, the operation of these components may be interrupted because of either failures or maintenance service which requires to cool and heat those components for several times during its operation life. Thus, it is important to analyze the microstructural changes not only during isothermal heating, but also during cycles heating. Therefore, the purpose of this paper is to know the effect of microstructural changes on mechanical properties for ferritic 2.25Cr-1Mo steel during isothermal and cycle aging at 550 °C for times up to 1000 h.

Experimental Procedure

The chemical composition of 2.25Cr-1Mo steel is shown in Table 1. The steel plate has a thickness of about 12.7 mm and it was austenitized austeniteat 950 °C for 8 min then air cooled, normalized, and subsequently tempered at 750 °C for 16 min and then air cooled. This plate was cut to obtain specimens of about 10 × 10 10 mm. Two aging treatments were conducted: continuous aging at 550 °C for times up to 2000 h and cycle aging at the same temperature. The former treatment was carried out as follows: specimens were heated at 550 °C for 1 h and then air cooled to room temperature and this procedure was repeated for 300 times. Both aged specimens were prepared metallographically using emery papers and alumina solution, and then etched with Nital etchant to be observed wit optical microscope (OM), and conventional and high-resolution scanning electron microscopes (SEM) equipped with EDS analysis. Vickers hardness od specimens was determined wit 50 g and 12 s.

Results and Discussion

Microstructural Evolution

Figure 1 shows the conventional SEM micrographs of the microstructure evolutions for 2.25Cr-1Mo steel aged continuously at 550 °C for different times up to 1000 h. SEM micrograph of as-received steel, 0 h, indicates the presence of equiaxed ferrite and bainite which was formed during the normalizing process. Besides, inter—and intragranular fine precipitation seems to be present. As the aging progresses, two events are evident: first the disappearance of bainite and the coarsening of precipitates. Additionally, a slight increase in ferritic grain size was observed to occur after aging for 1000 h.

Table 1 Chemical composition of steel

C	Cr	Mo	Mn	S	Si	Cu	Ni	P	Nb	V
0.133	2.408	0.911	0.522	0.009	0.07	0.041	0.053	0.005	0.002	0.005

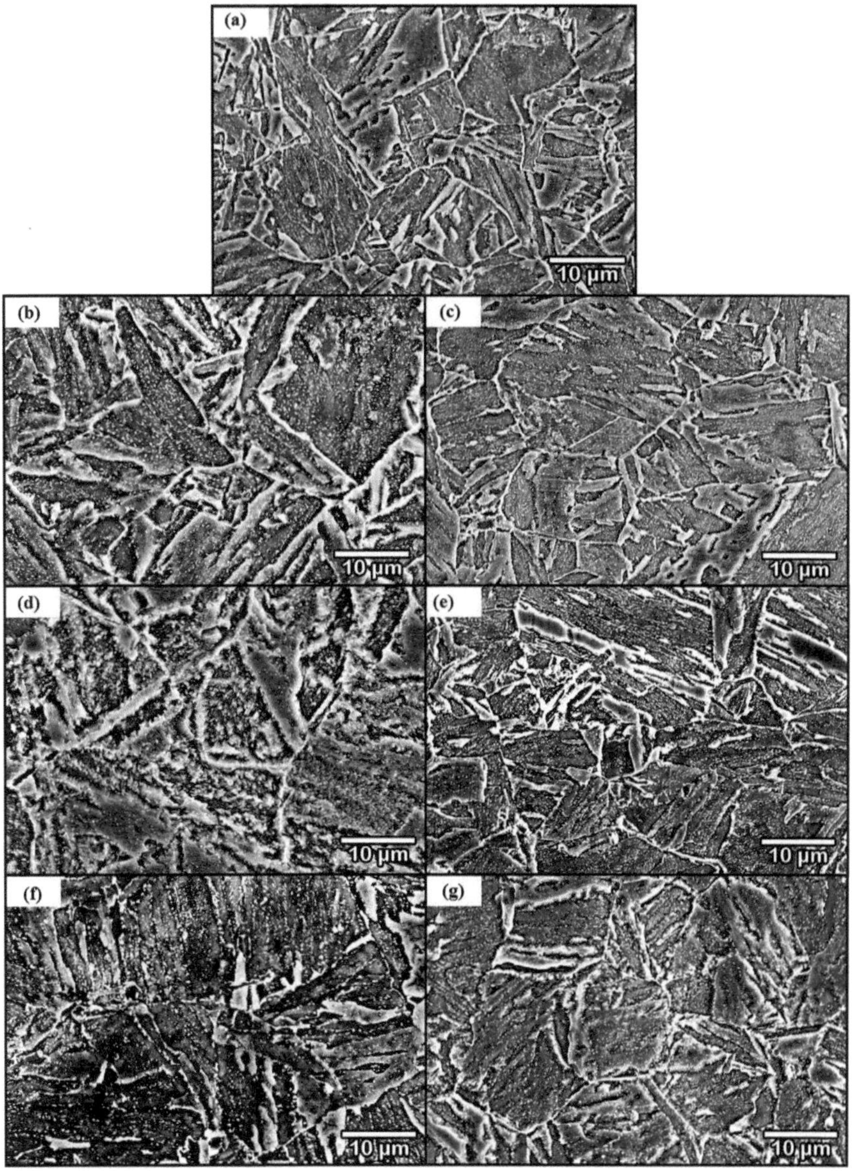

Fig. 1 Conventional SEM micrographs of 2.25Cr-1Mo steel aged continuously at 550 °C for **a** 0 h, **b** 50 h, **c** 100 h, **d** 300 h, **e** 500 h, **f** 700 h and **g** 1000 h

On the other hand, the conventional SEM micrographs corresponding to the microstructure evolution of the cycle aged specimens is shown in Fig. 2 for times up to 300 h. A similar microstructure evolution can be observed in these micrographs; however, the growth kinetics is faster than that of the continuous aging. For

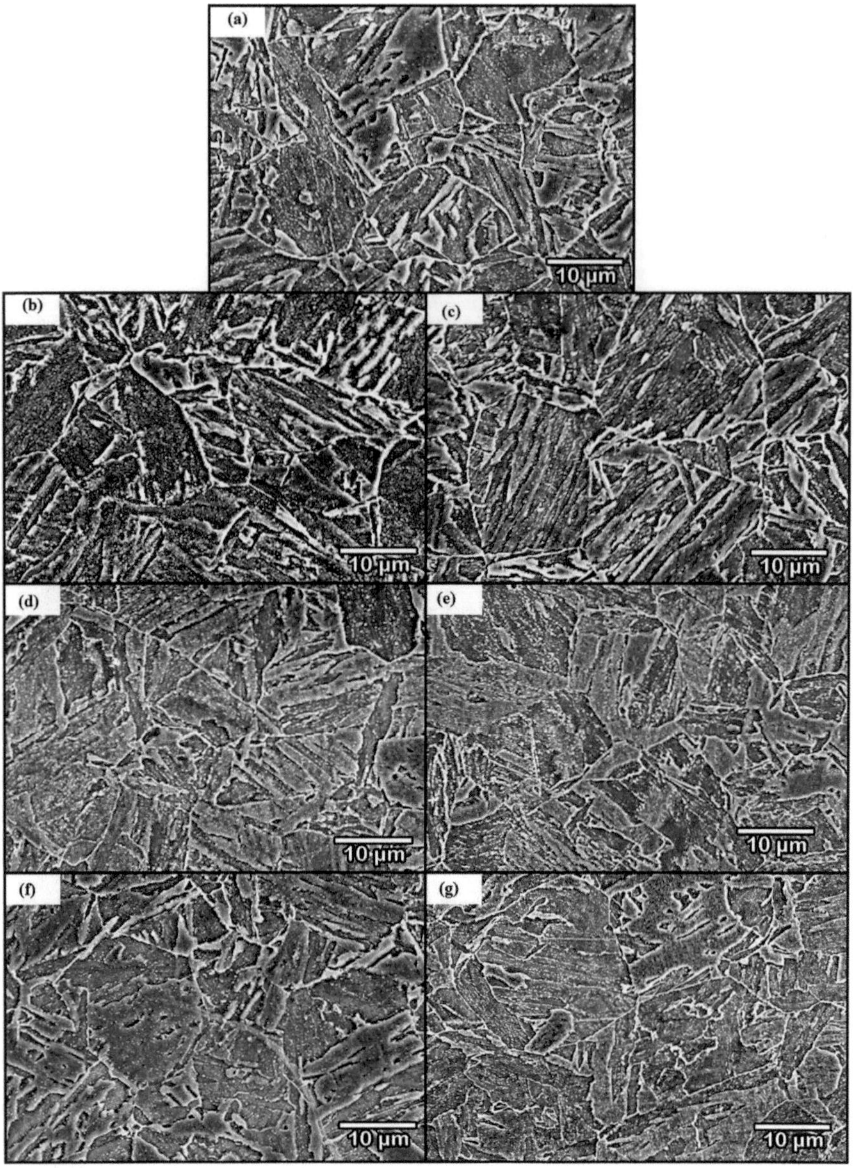

Fig. 2 Conventional SEM micrographs of 2.25Cr-1Mo steel aged cyclically **a** 0 h, **b** 5 h, **c** 10 h, **d** 50 h, **e** 100 h, **f** 200 h and **g** 300 h

instance, Figs. 3 and 4 present HR-SEM micrographs corresponding to the specimens aged continuously and cyclically, respectively. The comparison of these two figures makes evident the difference of precipitate size. That is, the precipitate size is larger in the cyclically aged steel. Additionally, two types of morphology can be

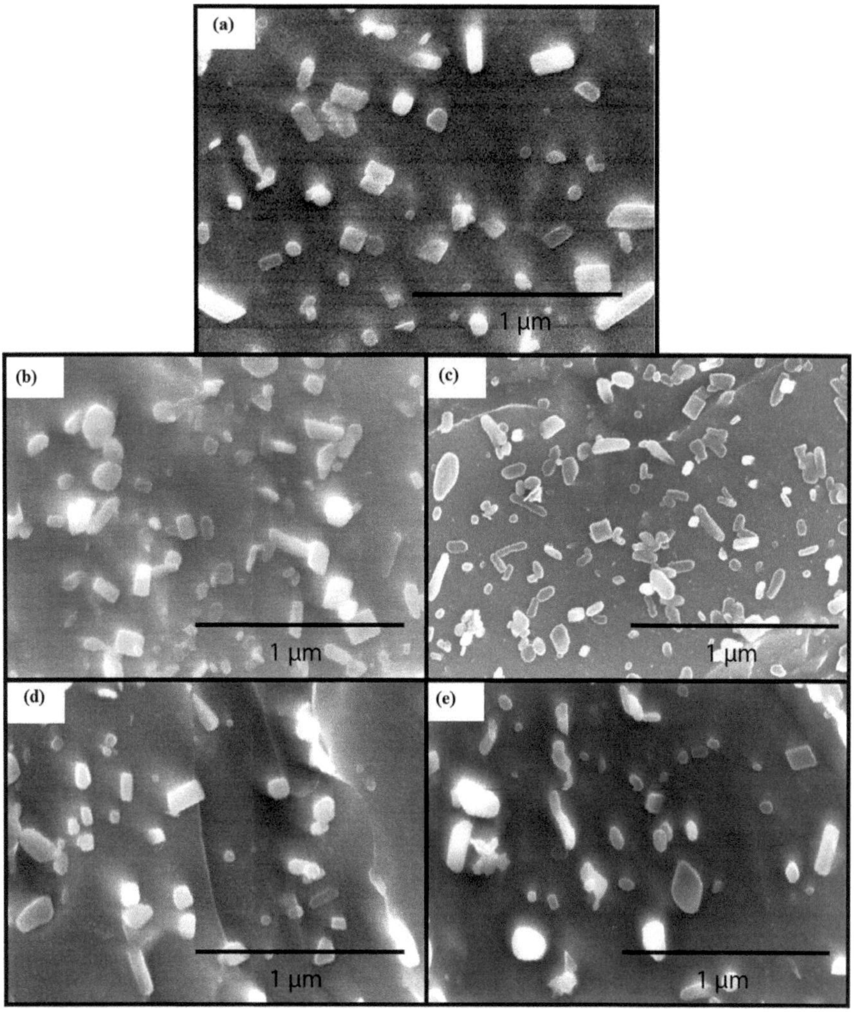

Fig. 3 High-Resolution SEM micrographs of 2.25Cr-1Mo steel aged continuously at 550 °C for **a** 0 h, **b** 100 h, **c** 300 h, **d** 500 h and **e** 1000 h

observed in both aged specimens, spheroids or ellipsoids and plates or cubes. The first morphology has been associated [6] to the presence of M_6C carbides with a hexagonal structure, while the second morphology is related to $M_{23}C_6$ carbides with a cubic structure.

The EDS-SEM microanalysis of some precipitates indicates that the $M_{23}C_6$ carbides are richer in chromium as shown in the EDS spectrum, Fig. 5b: however, Fe and Mo are also present. In contrast, M_6C carbides also has Fe and Cr, but the presence of Mo is more notorious, Fig. 6b.

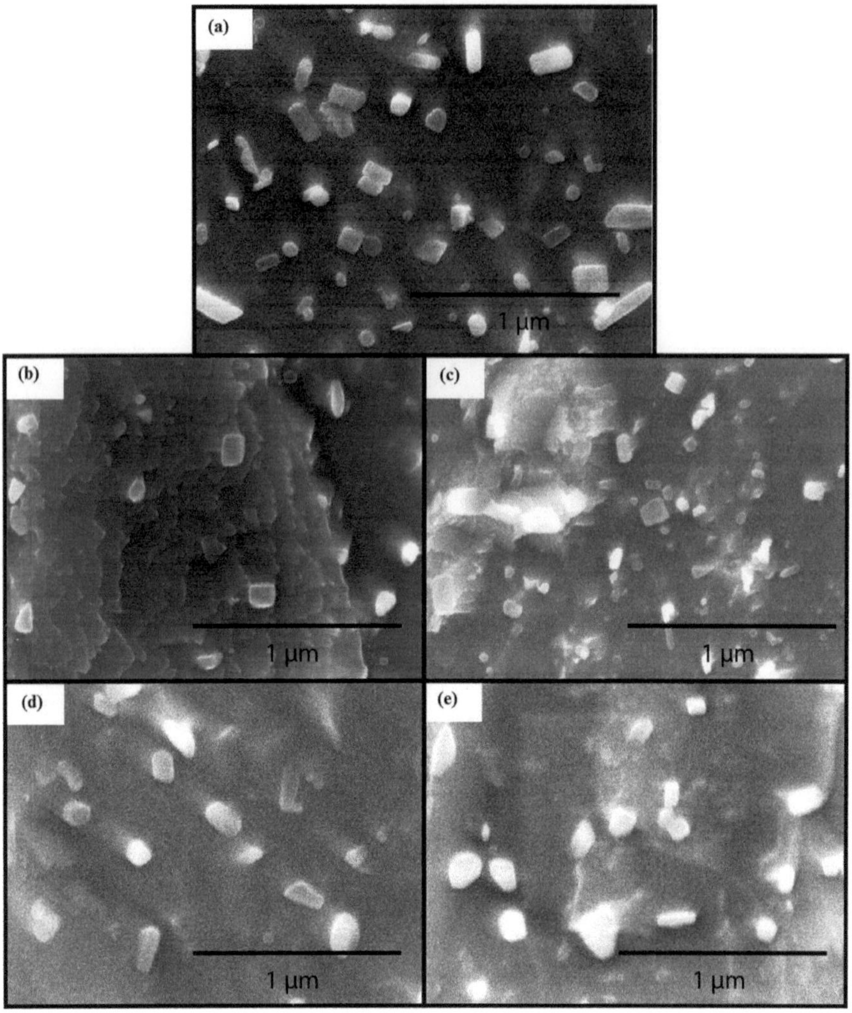

Fig. 4 High-Resolution SEM micrographs of 2.25Cr-1Mo steel aged cyclically at 550 °C for **a** 0 h, **b** 50 h, **c** 100 h, **d** 200 h and **e** 300 h

Coarsening Kinetics

Figure 7a, b present the growth kinetics of coarsening for both types of carbides in the continuously and cyclically aged steels, respectively. These plots the variation of the cube value the equivalent radius of precipitates, *r*, as a function of time, t. It can be noticed that there is a linear relationship between them, which suggests that the coarsening process is controlled by the atomics diffusion as predicted in the Lifshitz-Slyozov-Wagner theory [7]. Additionally, it is evident the faster growth

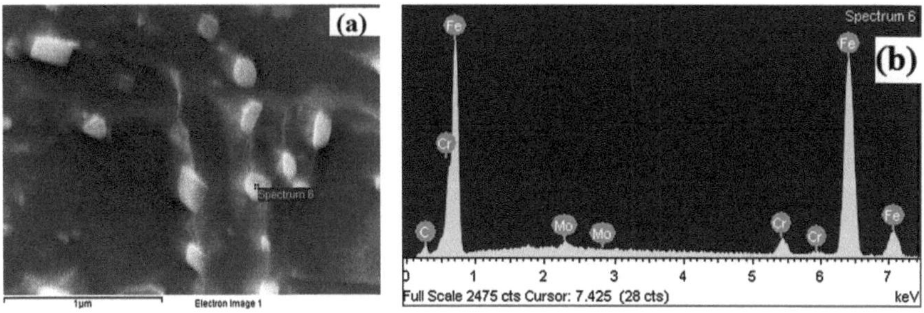

Fig. 5 **a** HR-SEM micrograph of 2.25Cr-1Mo steel aged continuously at 550 °C for 1000 h and **b** EDS spectrum of a spheroid $M_{23}C_6$ carbide

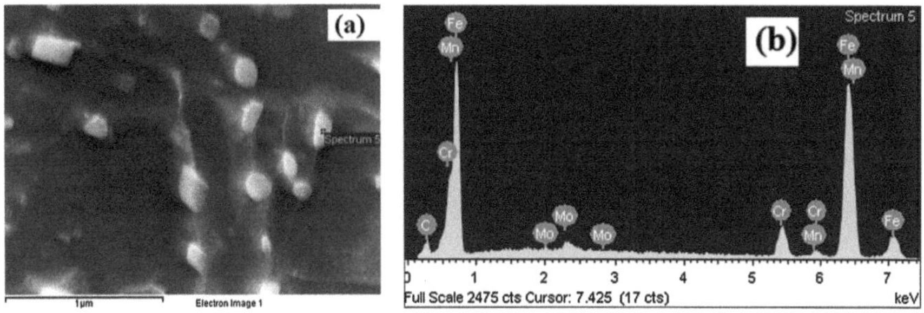

Fig. 6 **a** HR-SEM micrograph of 2.25Cr-1Mo steel aged continuously at 550 °C for 1000 h and **b** EDS spectrum of a cuboid M_6C carbide

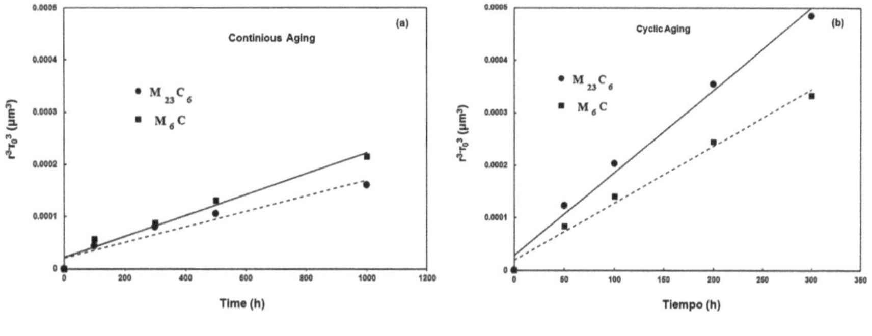

Fig. 7 Plot of r^3-r_0^3 versus aging time for 2.25Cr-1Mo steel aged **a** continuously and **b** cyclically at 550 °C

kinetics of coarsening for the cyclically aged steel in comparison to the continuously aged specimens. We believe that this behavior seems to be related to the refresh of solute supersaturation during the cooling down and heating up process, but it is not confirmed yet.

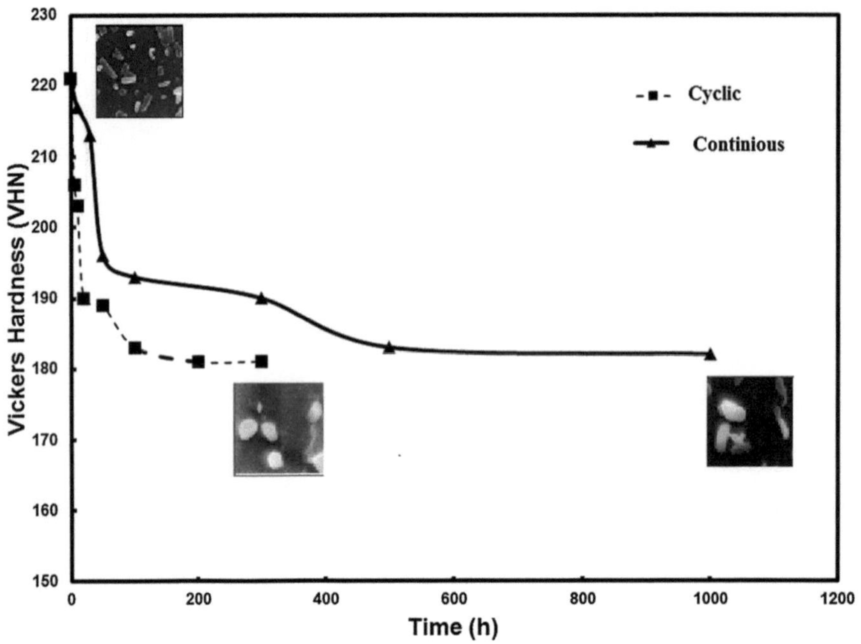

Fig. 8 Plot of Vickers hardness versus aging time for 2.25Cr-1Mo steel aged continuous and cyclically at 550 °C

The aging curve for both the continuous and cyclic aging is shown in Fig. 8. The hardness of both aged specimens decreased rapidly for short aging times since bainite is transformed into ferrite. The steel hardness continues decreasing with aging time and this fact can be attributed to the coarsening process of carbides, as indicated by the SEM micrographs include in this figure.

Conclusions

This study shows clearly that the cyclic aging causes faster growth kinetics of precipitate coarsening which is responsible for a decrease in hardness during aging at 550 °C in comparison to the continuous aging at the same temperature. The presence of $M_{23}C_6$ and M_6C precipitates was detected for both continuously and cyclically aged specimens.

Acknowledgements The authors wish to thank the financial support from SIP-COFAA-IPN and CONACYT.

References

1. F. Abe, T.U. Kern, R. Viswanathan, *Creep Resistant Steels* (Cambridge, England, Woodhead 2008), pp. 18–64
2. D.G. Mitchell, C.J. Ball, A quantitative x-ray diffraction and analytical electron microscopy study of service-exposed 2.25Cr–1Mo steels. Mater. Charact. **47**, 17–26 (2001)
3. B. Fujio, Creep resistance steel. Nature. **424**, 294–296 (2003)
4. J. Tan, Y.J. Chao, Effect of service exposure on the mechanical properties of 2.25Cr–1mo pressure vessel steel in a hot-wall hydrofining reactor. Mater. Sci. Eng., A **405**, 214–220 (2005)
5. M. Tamura, Larson–miller constant of heat-resistant steel. Met. Mater. Trans. **44A**, 2645–2661 (2013)
6. R. Viswanathan, *Damage Mechanisms and Life Assessment of High Temperature Components* (ASM, OH USA, 1989), pp. 85–110
7. G. Kostorz, *Phase Tranformation in Material* (Wiley-VCH, Germany, 2001), pp. 309–408

Part XXI
Student-Run Symposium: Building Bridges —Connecting Academic and Industry Research

Fundamental Principles for a Successful Collaboration Between University and Metalworking Industries

Silvia Lombardo, Federico Gobber and Mario Rosso

Abstract The gap in research between universities and industries is commonly considered to be fairly wide. Especially in metalworking industries, where the production volumes are huge, it seems very difficult for universities that operate with small-scale laboratories, to have a successful collaboration. On one hand, it is impossible to have the same equipment and installations of the industry inside universities, and to carry out the exact simulation of a certain industrial condition becomes challenging. On the other hand several advantages emerge and are directly related to laboratory investigations. Thanks to the experience that our research group gained over the years with the continuous collaboration with companies, it has been possible to underline and develop the fundamentals of a successful collaboration with an industrial reality. In this work two real cases of collaboration will be presented: an aluminum foundry and a mechanical industry. Company and university are members of the same team, they have to be able to communicate and joint efforts to reach the same goal.

Keywords University-industry collaboration · Metalworking industries · Connections to companies · Academic entrepreneurship

Introduction

Universities and industries in the past always seemed to be two different worlds, very far from each other. These two seemingly different protagonists actually spent their energy for the development of new technologies, always more efficient and environmentally compatible.

S. Lombardo (✉) · F. Gobber
Department of Applied Science and Technology, Politecnico di Torino sede Alessandria, 5 V.le Teresa Michel, 15121 Alessandria, Italy
e-mail: silvia.lombardo@polito.it

M. Rosso
Department of Applied Science and Technology, Politecnico di Torino, 24 C.so Duca degli Abruzzi, 10129 Turin, Italy

The Minerals, Metals & Materials Society, *TMS 2017 146th Annual Meeting & Exhibition Supplemental Proceedings*, The Minerals, Metals & Materials Series, DOI 10.1007/978-3-319-51493-2_77

The academic reality can appear very far from the real production world, too focused on theories and methods unsuitable for practical use and too complex to be applied to a larger scale; on the other hand industries goal is to achieve the best production in terms of quality and costs. The technological researches can be the means towards these final points. However, in the last years, industries and universities started to create new partnerships [1], that now are becoming increasingly numerous [2]. Companies are interested on this kind of relations in order to develop and increment researches on specific aspects of their sector. Universities obtain from these collaborations investments in their projects, while the academic researchers that work in close cooperation with the industry reality benefit from a fuller knowledge of the production process. The entering of universities in industries is, therefore, imperative, also when the companies are very distant from the idea of academic laboratory [3].

Our research group, belonging to the applied science and technology department, for several years now, is involved in the field of steels, light alloys, tools, and had the chance to work during the years with many different industrial realities in the metallurgical field. The metalworking industries can appear very divergent from the "small and clean" world of industrial laboratories and can be difficult for academics to enter and be accepted in a company reality.

Analyzing our experiences (and reporting two real cases) a review of the project management has been made, in order to extrapolate the main points for a successful collaboration and to determine what should be improved to be more effective.

Case Study I: Aluminum Foundry

The department of material science during the years had collaborated with different foundries of light alloys. For simplicity just one case study has been reported, a medium size aluminum foundry, that has been subsequently analyzed and discussed.

The first approach between the university and the company started with a student internship, where the student worked on its project master's thesis.

This first connection permitted to the company to discover first of all the potentialities and resources of the university; on the other hand the student got in contact with a more practical world and learned about the production process of the industry. In addition, the university member got confident with the foundry environment and with the people working there.

After this first working experience, both university and foundry, where interested to develop a productive ad continuous collaboration: the same student that was on a previous internship program was hired as part of the research group of the university, operating as a bridge between these two realities. The academic partnership formed implies:

1. the development of a new project where were present also other partners (European project)

2. the optimization of the actual production process
3. the formation of workers, problem that often afflicts the in metal foundry industry where technically skilled people are missing [4].

Initially the academic was seen as an outside figure in the foundry: first of all because of the time spent inside the foundry (not full time); secondly for personal characteristic depending on age, sex, and level of education. The partial time spent working in the company was perceived as if the project was a partial work, being of less importance than other topics.

From the point of view of personal characteristics, in some situations a young age can be perceived as a lack of experience and knowledge (very probable the first case, not certain the second one), generating on the first approach a lack of confidence in the researcher. On the other hand, in different conditions, it is the level of education that can create barriers between the workers and the university members.

After an initial moment of adaptation to this new figure inside the company, university and industry members worked together to implement different goals in a quite good way. It has been very important the high communication between the two worlds and the mutual respect developed during the experience. Both sizes had to adapt sometimes to the other's technical vocabulary and spent time understanding each other, in order to solve a problem or to develop a new concept in the fastest and effective way possible. Respect and confidence on the other partners work forms the basis of a professional and effective collaboration.

One of the problems encountered during this project was linked to the difference between "production time" and "laboratory time". In fact, even if the company facilities in terms of materials, machinery and equipment availability would permit the production of a new component in a very short time, the laboratory times and the development times have a different weight (typically higher) if compared to production time and requests. On the other hand, thanks to university resources it has been possible to solve problems and doubts easily and rapidly (example of fractography analysis performed using electron scanning microscope, instrument whose potential would not be fully exploited in an aluminum foundry). Because of these reasons a review of the Gantt chart for the key operational activities was necessary at a certain point, taking in particular account that some operation would have been slower in the academic laboratory or in the industry.

The industry tends to consider the university as an instrument to solve its immediate problems, like a technical instrument. University is instead an organ that can potentiate a company in terms of innovation and long-term quality. One of its main roles is to transmit this message and try to meet the needs, in this case, of the aluminum foundry.

The company received suggestions from the university during the analysis of the production process, ideas and hints of technological innovativeness for the development of a more successful system. However, ideas are obviously not enough to persuade to make investments and it is necessary to show clearly advantages and disadvantages of a determined change. At the same time the university member can better understand which are the needs and the main problematic in a foundry, important clues for future projects and researches.

The formation in the foundry was specifically designed to address the workers of the company. The lectures were structured with the aim of transmitting concept of process optimization more focused on the specific company reality: examples of defects, treatment problems and process critical situation that may quite commonly arise in the foundry. This technical formation was successful and appreciated especially by less skilled operators.

Case Study II: Mechanical Industry

In this second case study, the collaboration between the university and another industrial partner is presented: a mechanical industry producing bolting for special applications (oil and gas, energy production). The link between university and industry, in this reality, was established and toughened through several project: trust and reciprocal knowledge were already developed and a common result—oriented way to work was acknowledged on both sides.

In this project a young researcher from the university was employed under a PhD program and acted as a direct link between the university and the industries with the main task of characterizing the materials on a laboratory scale, operate a choice based on engineering parameters and then scale the result to the level of industrial production. Two research groups that are into different fields of research participate to this project with the aim of synergically unite different experts to reach an optimal solution and transfer multidisciplinary knowledge to the industrial partners. One research group belongs to the Dept. of Applied Science and Technology (DISAT) and works in the field of metallurgy and surface engineering while the other group is from the Dept. of Production Engineering (DIGEP) and works in process optimization and numerical finite element simulation.

The idea to set up a research project originates from the implementation of new cut-off techniques in the production cycle of bolting. A new generation highly automated cut-off station using tungsten carbide tipped circular saw blades (TCT saw) was installed as first step of the production cycle in order to guarantee high production rates (more than 1400 pieces/h), constant surface finishing and tolerances as required by customers. This mechanical industry has decided to set up a research project at an industrial level involving the university, a local toolmaking company and a company that deposits protective hard coatings with PVD technology with the goal of designing a specific tool for their cut-off stations.

The role of the university during the preliminary part of the collaboration was to write the research program assembling the contributions derived from the know—how of each partner. The research proposal was checked and approved by the partners and it was then broken up following the work breakdown structure (WBS) approach dividing it into tasks assigned to each partner.

Small-medium enterprises (SMEs) usually have a deep knowledge in their field of work but seldom they have an office devoted to project management for example. Skills can be present in different people that are not devoted to a single job such as technicians or managers. The presence of university is necessary in order to have one or more persons focused on the project. One of the goals of the academic is to unite the company experience and knowledge background to make it available and exploitable during the development of the project.

During the operative part of the project the young researcher's goal is to operate a constant definition of improvements for cutting tools via the continuous interpretation of the results obtained both at the lab level and at the industrial level. In this project the researchers at university have spent their skills and have gained some new ones from both working with industrial partners and academic partners from another department.

New fields of research for the academics were explored and the project gave the opportunity to design and built a new experimental machine for testing fatigue resistance of PVD coatings. The direct contact with partners, the regular production and presentation of the research results by means of oral presentations and written reports, and moreover proactiveness in managing unexpected drawbacks were demonstrated to be key features for a successful collaboration between university and industries.

University inside the project also operates in promoting the transfer of technological and scientific knowledge to the industry. This knowledge, based on researches and results of projects published and spread in the scientific community, can be applied to this specific case study, yielding added value in a number of ways. For example the researcher has proposed new materials (coating architecture), different cutting geometries for machining specific materials (cutting force reduction) but also has suggested and developed instruments that could facilitate the defect control in the production process.

These features directly affect the industry that gains new and fresh knowledge and implements its know—how going deeper into subjects that despite not being its core business are functional to it.

Location

A very important aspect of the collaboration experiences was the presence of the university inside the company. The academic organ does not have to be seen as an external element but part of the industrial system. Being was fundamental in order to know the productive process, its limits and weaknesses. Without this knowledge it would have been impossible to create a correct formation project and to establish a picture of the overall production situation.

In most of the cases it is impossible to create a pole of the university inside the company, especially if we are talking about SME; however an important presence of academics has to be conceived. The size of industries itself has both advantages and drawbacks: dealing with a SME is usually easier in terms of communication, personal contact and reduced bureaucracy while the distribution of expertise can be sometimes confuse or can lack totally in certain areas; while in large companies entire departments are deputed to carry out a specific job and more people with different skills are employed there.

Although for the second case study the distances do not permit the daily presence of the researcher, it has been possible a good collaboration exploiting different methods of communication and optimizing the amount of time spent in the industry depending on the experimental activities.

People

A continuous and proficient contact with people is one of the keys for a successful collaboration: knowing each other helps to strengthen the team. The development of a good relation, where everybody knows the role that every component plays in the process, leads to a superior mutual respect and trust.

The involvement of all the people working inside the industry (the machine technician, the metalworker, the welder, the toolmaking tech, …) makes them more sensitive and active participants to the development of a new project.

However, this is possible in a small reality. In the case of bigger companies, this idea has to be applied to a determined organ of the industry directly involved in the project. It is important to get in contact with workers in order to know the dynamics of the production line and detect omissions or inaccuracies that can be problematic for the experimentation, and an eventual margin for improvements.

Technical formation can be helpful in involving the workers and to outline those who could aspire to more qualified positions due to high proactivity and interest.

Communication

One of highest barriers that came out from the collaboration experiences was created by the wrong communication between university and industry. This point is strictly correlated with the previous one: it is important to know industrial team and to learn how to dialog with these group members for the development of a successful collaboration.

The communication has to be simple and straightforward, in order to be as efficient as possible. The way of expression has to be understandable for both, using a language without obscure technicalities from both sizes.

Furthermore the means of communication are basic: most of the time a face to face meeting allows to resolve problems faster and to individualize new action options.

In the second case study, the number of partners is higher. The relation between team members at the beginning can be difficult, because a young research is asking to other team members to perform some deliveries. In this case should be the company flexible and open to the discussion in order to facilitate the work of the member that works as an intermediate. In this project another part of collaboration is another department of the university. In this case the communication was easier due to same backgrounds and knowledge of the academic system from the experimental point of view (analysis results, interpretation of data) and time schedules. The collaboration was very useful in order to implement lacks that our formative background had in some topics more concerning with mechanics.

All the components, from company or university, should be on the same footing during the discussion; listening and participation have to be encouraged and those ideas considered successful in perspective have to be developed regardless of who conceived it.

Goal

Have the same goal: all the energy has to be focused to solve a problem or to improve a process. Of course, from the development of the matter, the two systems will obtain different results. Industries can continue the production, potentiate the process, be more competitive in the market, and be more environmental friendly. The interpretation and the analysis of the results, especially in the second case study, were mainly done by the university. The conclusions are usually reported to the partners of the research during periodical meetings that usually consist of an oral presentation followed by an open discussion.

Universities increase their knowledge, publish scientific studies on new topics and technological developments throughout the scientific community.

Time

The timeframes of the two realities have to be clear to all the components of the project. A great part of a project is made if time is well managed from the beginning, in this situation all the partners can focus on their scope monitoring their time properly. The Gantt chart is the most diffuse tool used to visualize the results of the WBS and it has to be done with the collaboration of all the partners. The project manager needs to have direct feedback about time from who has to develop single tasks. On the contrary there is a high probability that he will underestimate or overestimate time.

The university has to remember that industries usually work on short-term. Although a project can take several years, it is important to set realistic and concrete targets that have to be reach in shorter times, in order to make the project more successful and supported by the company.

When many partners have an active role in a project it can be possible also for industrial partners to have delays. One cause encountered is specifically linked to the experimentation: if a new solution has to be experimented it can mean for an industry to spend extra—time from that related to production in manufacturing a new tailored tool with unique specifications and quality or setting up a precise sequence of actions to prepare the following experimentation. It is fundamental to be clear and sincere with other partners because all the team has the same goal and long delays can just have negative effects on the whole project.

Conclusions

The main goal moving the production of this work was to carry out an assessment of previous works and collaborations between our research group and metalworking industries, in order to analyze successes and the weaknesses met during the experience.

The main elements that have been underlined from our review are: the importance of been present as university inside the company, developing a good relation with the people working there in order to facilitate the communication inside the team. The group has to concentrate on the same goal, in order to finish all the duties in time as organized since the beginning of a project or activity.

What can be developed with a good collaboration between industries and companies can be synthetized as:

1. Universities help the development of SME
2. Industries become centre of excellence
3. Industries support university-industry project.

References

1. E. Mansfield, Academic research and industrial innovation 1–12 (1991). Elsevier Science Publishers B.V
2. J.J. Watson-Capps, T.R. Cech, Companies on campus. Nature **514**, 297–298 (2014)
3. M. Perkmann et al., Academic engagement and commercialization: a review of the literature on university-industry relations. Res. Policy **42**, 423–442 (2013)
4. M.F. Vaachia, H. Nel, R. Bakker, University-industry collaboration in the metal casting industry in south africa, in *EDULEARN12 Proceedings* (2012), pp. 697–704

Retraction Note to: Surface Reaction and Transport in Oxides Formed on FeCrAl Alloys in High Temperature Nitridation Environments

Christine Geers, Vedad Babic, Lars-Gunnar Johansson and Itai Panas

Retraction Note to: Chapter 36 in: The Minerals, Metals & Materials Society (ed.), *TMS 2017 146th Annual Meeting & Exhibition Supplemental Proceedings*, The Minerals, Metals & Materials Series, DOI 10.1007/978-3-319-51493-2_36

The Minerals, Metals & Materials Society, editor of the publication *TMS 2017 146th Annual Meeting & Exhibition Supplemental Proceedings*, is retracting the book chapter, "Surface Reaction and Transport in Oxides Formed on FeCrAl Alloys in High Temperature Nitridation Environments" (DOI 10.1007/978-3-319-51493-2_36). The corresponding author requested this retraction because an image was used for which the authors do not hold copyright. The authors could not obtain permission to use the image.

The retracted online version of this chapter can be found at http://dx.doi.org/10.1007/978-3-319-51493-2_36

The Minerals, Metals & Materials Society, *TMS 2017 146th Annual Meeting & Exhibition Supplemental Proceedings*, The Minerals, Metals & Materials Series,
DOI 10.1007/978-3-319-51493-2_78

Author Index

The Minerals, Metals & Materials Society, *TMS 2017 146th Annual Meeting & Exhibition Supplemental Proceedings*, The Minerals, Metals & Materials Series,
DOI 10.1007/978-3-319-51493-2

Subject Index

The Minerals, Metals & Materials Society, *TMS 2017 146th Annual Meeting & Exhibition Supplemental Proceedings*, The Minerals, Metals & Materials Series, DOI 10.1007/978-3-319-51493-2

MIX
Papier aus verantwortungsvollen Quellen
Paper from responsible sources
FSC® C105338

If you have any concerns about our products,
you can contact us on
ProductSafety@springernature.com

In case Publisher is established outside the EU,
the EU authorized representative is:
Springer Nature Customer Service Center GmbH
Europaplatz 3, 69115 Heidelberg, Germany

Printed by Libri Plureos GmbH
in Hamburg, Germany